Fluid Mechanics, Hydraulics, Hydrology and Water Resources for Civil Engineers

Fluid Mechanics, Hydraulics, Hydrology and Water Resources for Civil Engineers

Amithirigala Widhanelage Jayawardena

CRC Press
Taylor & Francis Group
Boca Raton London New York

CRC Press is an imprint of the
Taylor & Francis Group, an **informa** business

First edition published 2021
by CRC Press
2 Park Square, Milton Park, Abingdon, Oxon, OX14 4RN

and by CRC Press
6000 Broken Sound Parkway NW, Suite 300, Boca Raton, FL 33487-2742

CRC Press is an imprint of Informa UK Limited

British Library Cataloguing-in-Publication Data
A catalogue record for this book is available from the British Library

ISBN: 9781138390805 (hbk)
ISBN: 9781138390812 (pbk)
ISBN: 9780429423116 (ebk)

Typeset in Sabon
by codeMantra

Contents

6   Ideal fluid flow   73

18 Gradually varied flow in open channels

Preface

Water is a precious resource essential for all forms of life. Although abundant in nature, fresh water is not always available when and where it is needed. It is also a fact that the per capita share of fresh water is decreasing with time resulting in competition from different sectors. Optimal and equitable management of this resource without causing negative effects on the physical and biotic environment plays an important role in maintaining and improving the quality of life of all human beings. Water professionals, including engineers, civil engineers in particular, have the duty and responsibility of ensuring this task is carried out for the benefit of all concerned. The necessary background and training for civil engineers to carry out this task begin in universities and other institutions of higher learning where the relevant courses are taught.

Fluid mechanics, hydraulics, hydrology and water resources are core courses taught in most, if not all, civil engineering curricula in most universities. They cover the basic scientific, mathematical and empirical knowledge necessary for civil engineering professionals to carry out their functions. Although there are several books that treat these individual subjects but to the best of the author's knowledge, there is no single book that covers the entire spectrum of the four core areas. The motivation to compile this book came out of the need to fill this gap and to put on record the teaching material that the author has assimilated during his academic career in the Department of Civil Engineering of the University of Hong Kong, the International Centre for Water Hazard and Risk Management (ICHARM) under the auspices of UNESCO, Public Works Research Institute, Japan, the Department of Civil and Environmental Engineering of the Hong Kong University of Science and Technology, and the Department of Civil Engineering of Chu Hai College of Higher Education, Hong Kong. Related research work carried out by the author and his co-workers and graduate students are also embedded in places where they fit in.

This book is targeted at undergraduate and graduate students in civil engineering. It can also be a source of reference for practising civil engineers to refresh and update their skills in the three core areas highlighted in this book. Several worked examples are also presented to help students to consolidate the theoretical and empirical contents.

Acknowledgments

The author would like to record his appreciation to the Japan Society for the Promotion of Science (JSPS) for awarding an Invitational Fellowship to be attached to the Institute of Industrial Science, University of Tokyo and to Professor Taikan Oki who initiated the application process and hosted me in his laboratory during July to August 2018. My appreciation is also extended to the members of his laboratory. During the tenure of this Fellowship, discussions were also held with Professors Katsumi Musiake of Tokyo University (Emeritus), Kuniyoshi Takeuchi of Yamanashi University, Kaoru Takara of Kyoto University, So Kazama of Tohoku University, Tomohito Yamada of Hokkaido University and Junichi Yoshitani of Shinshu University on topics related to this book and other research topics of mutual interest. Acknowledgements are also due specially to Luo Shuxin and Zhang Junwei of the Department of Civil Engineering of the University of Hong Kong who helped me with several figures in this book as well as to Zhang Kun and Yang Yang, also in the Department of Civil Engineering of the University of Hong Kong who occasionally helped me out with some of the figures in this book.

The author also benefitted from the published work of others. Appropriate acknowledgements have been made in citing such work in relevant parts of the text. Last but not least, the author expresses his gratitude to Tony Moore, Senior Editor of the publisher Taylor & Francis, who initially proposed the idea and gave continuous encouragement. His patience is gratefully appreciated. Thanks are also due to Gabby Williams, who looked after the project initially, and Frazer Merritt, who took over from Gabby towards the latter part.

This book contains more than 1850 equations, some simple and some not so simple. Despite the care taken to ensure the correctness of these equations and other material presented, it is still possible that there may be typographical errors and/or omissions due to oversight. The author would be grateful if the readers would kindly bring to his attention if they find any such errors and/or omissions. After all, to err is human and to forgive is divine.

Amithirigala Widhanelage Jayawardena
May 24, 2020

Author

Amithirigala Widhanelage Jayawardena, Ph.D., is an adjunct professor in the Department of Civil Engineering, University of Hong Kong, and a visiting professor in the Department of Civil Engineering, Chu Hai College of Higher Education in Hong Kong. Previously he was a technical advisor to the Research and Development Centre, Nippon Koei Co. Ltd, Japan, the research and training advisor to International Centre for Water Hazard and Risk Management (ICHARM) and concurrently a professor in the National Graduate Institute for Policy Studies (GRIPS), also in Japan, and, an honorary professor in the Department of Statistics and Actuarial Sciences of the University of Hong Kong. He is the author of *Environmental and Hydrological Systems Modelling* (CRC Press, 2014).

Chapter 1

Introduction

Water is a precious resource essential for all forms of life. It is abundant in nature but has significant temporal and spatial variability. As a result, water is not always available when and where it is needed. Oceans hold approximately 97.5% of planet earth's water, leaving with only about 2.5% in land areas. Of this 2.5%, approximately 1.925% is locked in ice caps and glaciers and not easily accessible for human needs. Some countries are water-rich, whereas some others are water-poor. With the unabated increase in population, the per capita share of water availability is decreasing with time, resulting in water stresses, water shortages and water scarcity in some countries and regions and during certain times of the year. Lack of safe drinking water is a major problem for over a billion inhabitants of the earth. Too much water also brings about misery, agony and destruction to many people, places and infrastructure. The former may be attributed to the physical lack of water, pollution or unaffordability, and the latter is attributed mainly to urbanization and livelihood issues. A major challenge facing humankind and the environment is how to share this precious freshwater resource in an equitable and optimal manner. This requires contributions from many disciplines, and civil engineering is one of them.

Civil engineering is a broad discipline encompassing key areas such as structural engineering, geotechnical engineering, water resources engineering, transportation engineering, environmental engineering, etc. These are also core subject areas taught in any civil engineering curriculum in any university. The three components of water resources engineering are fluid mechanics, hydraulics and hydrology. The objective of this book is to combine all these three core areas into a single source of reference targeted towards civil engineering students and practicing civil engineers.

The contents of this book are organized into the four sub-themes: fluid mechanics, hydraulics, hydrology and water resources. This chapter begins with a brief introduction to fluid mechanics.

1.1 FLUID MECHANICS

1.1.1 Definition of a fluid

Most materials are designated as solids, liquids or gases. Some materials have a dual designation, e.g. jellies, paints and polymer solutions (solid and fluid). All materials are deformable. Deformations of solids are small even for large external shear forces and do not continue to deform, whereas in the case of fluids, they are large even for small external shear forces and have no fixed shape.

Also, in solids the strong molecular forces between molecules tend to restore the deformed body to its original state (shape) when the external force is removed. There are limits to this, i.e. within the elastic limit (Figure 1.1).

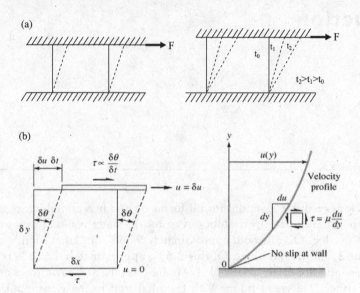

Figure 1.1 (a) Deformation of a solid and a fluid; (b) deformation of a fluid and velocity profile.

A fluid continuously deforms when subjected to a shear stress. It cannot sustain a shear stress when at rest, implying that shear stresses exist only when the fluid is in motion.

In a **real fluid** (or **viscous fluid**), the shear stresses exist when the fluid is in motion. On the other hand, an **ideal fluid** (or **inviscid fluid**) has no shear stress when in motion.

Gases and liquids are both fluids. The most significant properties that distinguish gases from liquids are the bulk modulus of elasticity and density. Gases are compressible and are much lighter under ordinary conditions.

Figure 1.1 shows the deformation of a solid and a fluid under the action of a constant shear force F. The liquid continues to deform as long as the force is applied. Note also that at the fixed boundary, the velocity of the fluid is the same as the velocity of this boundary, implying that there is '**no slip**' at the boundary. This is for a real fluid. But, for an ideal fluid, the fluid 'slips' at the boundary without wetting (Figure 1.2). This is **unreal**, but is a powerful theory used for solving fluid flow problems when the effect of viscosity is small.

Fluid mechanics can be broadly divided into Fluid Statics and Fluid Dynamics. The former deals with fluids at rest, whereas the latter deals with fluids in motion. Fluid dynamics can be further sub-divided into Fluid Kinematics, which deals with the geometry of motion considering displacements, velocities and accelerations, and Fluid Kinetics, which deals with displacements, velocities, accelerations and forces. Kinetics has been the terminology used for dynamics in the early days.

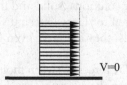

Figure 1.2 Velocity profile for an ideal fluid.

1.1.2 Illustrations of fluid mechanics in everyday life

1.1.2.1 Blood flow

Blood is a complex viscous fluid consisting of formed elements suspended in plasma. Plasma is an aqueous saline solution containing proteins that behaves like a Newtonian fluid. The plasma contains 91.5% water, 7% proteins and 1.5% other solutes. The formed elements are platelets, white blood cells and red blood cells. The difference between an ideal Newtonian fluid and blood is due to the presence of the formed elements and their interaction with plasma molecules.

Blood flow in the cardiovascular system of mammals is one of the main illustrations of fluid mechanics in everyday life. It is a complex process due to the complex geometry of the blood vessels and the pulsating character of the flow.

The cardiovascular system consists of the heart which acts as a pump, the arterial system which carries oxygenated blood from the heart to other organs of the body through relatively large blood vessels (of the order of 20 mm diameter), the venous system which carries the oxygen-depleted blood through blood vessels slightly larger than arterial vessels back to the heart and the capillary system with very small blood vessels (of the order of 5,000 μm). The function of the cardiovascular system is to transport blood between different organs of the body of mammals. Through this convective transport process, diffusive transport of oxygen, carbon dioxide, nutrients and other solutes takes place at cellular level in the tissues.

The fluid mechanics of the heart is strongly determined by inertial forces and the pressure forces which balance with each other. The arterial system transforms the pulsating flow produced by the heart to near-steady flow in smaller arteries. The arterial system also maintains a relatively high pressure. The venous system, which carries blood back to the heart, has a larger volume than the arterial system but with velocities and pressures much lower than those of the arterial system. The thickness of the blood vessels in the venous system is smaller than the corresponding thickness of arteries of the same diameter.

From the fluid mechanics point of view, fully developed flow is rarely attained because of the short lengths of blood vessels. The circulation of blood starts from the left side of heart to the systemic arteries, then to the capillaries during the forward cycle and from systemic veins to the right side of heart to the pulmonary system (lungs) and finally to the left side of heart during the reverse cycle.

The heart normally pumps about 5 litres of blood per minute which can go up to about 25 litres per minute during exercise.

1.1.2.2 Water flow

Examples of water flow include

    a. Flow of water in pipes
    b. Pumping of water
    c. Flood damage – energy of flow
    d. Seepage, infiltration
    e. Waves – energy
    f. Forces on seawalls, etc.
    g. Coastal erosion due to waves
    h. Hydraulic structures – energy dissipaters
    i. Spillways
    j. Irrigation canals, waterways

k. Hydro-electric machinery
l. Hydraulic machinery

I.I.2.3 Air flow

Example of air flow include

a. Airplanes, jets
b. Lift and drag
c. Rockets, re-entry
d. Automobiles – streamlining shape

I.I.2.4 Wind flow

a. Cyclones/anti-cyclones
b. Tornadoes
c. Forces on buildings
d. Energy

I.I.2.5 Environmental flow

a. Smoke plumes
b. Pollutant dispersion

I.I.3 Continuum approach

A fluid can be considered as an infinitely divisible substance, and the behaviour of individual molecules is of no concern for civil engineers. The macroscopic effect in which the fluid is considered as a continuum is of interest for most practical problems. Thus, each fluid property has a definite value at each point in space. Therefore, the fluid properties are considered as functions of space and time. Consider, for example, the density of a fluid at a point.

The mean density of fluid is $\frac{m}{V}$, which in general is not equal to the density at a point $C(x_0, y_0, z_0)$. To determine the density at C, a small volume δV surrounding C (Figure 1.3) is chosen. Then, the density is $\frac{\delta m}{\delta V}$. The density will fluctuate as the size of δV gets smaller and smaller, but approach a mean value as shown in Figure 1.4. However, if $\delta V \to 0$, it would be difficult to define δm, as the molecules will cross the boundaries of δV in and out resulting in violent fluctuations. Therefore, the density is defined as

$$\rho = \underset{\delta v \to \delta v'}{\text{Limit}} \frac{\delta m}{\delta V} \tag{1.1}$$

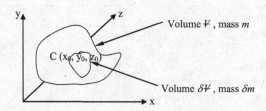

Figure I.3 Continuum.

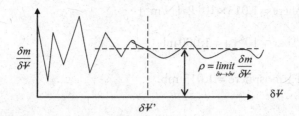

Figure 1.4 Definition of density at a point.

where $\delta V'$ is the lower limit of the volume that is used for defining the density.

The continuum approach, however, may fail in the study of gases at very low pressures when the average molecular spacing can be comparable with distances considered in the problem.

1.1.4 Properties of fluids

1.1.4.1 Density

$$\rho = \frac{\text{Mass}}{\text{Volume}} \quad (\text{ML}^{-3})$$

$$\rho_{\text{water}} = 1,000 \text{ kg/m}^3$$

$$\rho_{\text{air}} = 1.24 \text{ kg/m}^3 \text{ at atmospheric pressure}$$

1.1.4.2 Specific volume

$$\text{Specific volume} = \frac{1}{\text{Density}} = \frac{1}{\rho}$$

1.1.4.3 Specific weight

$$\gamma = \rho g$$

$$\gamma_{\text{water}} = \rho_{\text{water}} \times 9.81$$

$$= 1,000 \times 9.81$$

$$= 9,810 \text{ N/m}^3$$

Specific weight is not a fluid property because it depends on g, which varies from place to place.

$$\text{Specific gravity} = \frac{\rho}{\rho_{\text{water}}}$$

1.1.4.4 Pressure

$$\text{Pressure} = \text{Normal force per unit area}$$

$$= \text{Shear stress} - \text{Shear force (Tangential force) per unit area}$$

$$\text{Standard atmosphere} = 1.013 \times 10^5 \, \text{Pa} \left(\text{N/m}^2 \right)$$

$$1 \, \text{bar} = 10^5 \, \text{Pa} = 1,000 \, \text{mb}$$

Therefore, standard atmosphere = 1,013 mb.

1.1.4.5 Vapour pressure

At the interface of a liquid and a gas, there is continuous exchange of molecules from the liquid to the gaseous phase and vice versa. When conditions are right, an equilibrium state is attained between the rate at which liquid molecules escape from the liquid phase and the rate at which molecules enter the liquid phase from the gaseous phase. The molecules above the interface that return to the liquid phase exert a pressure known as the partial pressure of the vapour. If the space above the interface contains different gases, the sum total of the individual partial pressures constitutes the total pressure. The molecules that escape from the liquid phase give rise to the vapour pressure, which depends upon the rate at which molecules escape from the liquid phase. When the space above the interface is saturated with the vapour, further escape from the liquid phase is not possible. The magnitude of the vapour pressure at this point is known as the saturation vapor pressure which varies with temperature. The higher the temperature, the greater the saturation vapour pressure. At 100°C, the saturation vapour pressure of water is about 100 kPa, which is also the atmospheric pressure at sea level. This means that water will boil at this temperature and pressure. If the pressure is lower, water will boil at a lower temperature.

Different liquids have different vapor pressures. Mercury, which has a high density, has a very low vapour pressure (about 0.16 Pa at 20°C) which is the main reason for using it as a fluid in barometers.

1.1.4.6 Compressibility and elasticity

All fluids can be compressed, but liquids compress to a very much lesser degree. Therefore, liquids are generally classified as incompressible fluids.

The bulk modulus of elasticity, κ, depends on pressure for gases and pressure and temperature (though slightly) for liquids. The limiting value of the above parameter, when the ratio volume to pressure change becomes infinitesimal, is the true bulk modulus of elasticity.

$$\kappa = -\frac{dp}{d\forall / \forall} = \frac{dp}{d\rho/\rho} \tag{1.2a}$$

$$\left(d(\rho\forall) = 0 \Rightarrow \rho d\forall + \forall d\rho = 0 \Rightarrow \frac{d\forall}{\forall} = -\frac{d\rho}{\rho} \right)$$

The value of κ for water at 20°C is about 2.18×10^9 Pa at atmospheric pressure and varies almost linearly to about 2.86×10^9 Pa at a pressure of about 1,000 atmospheres. If κ is constant, the density is a function of pressure only. Such a fluid is called a **barotropic fluid**. The value of κ depends upon the compression process:

- If isothermal (constant temperature)

$$p\forall = RT \Rightarrow p \Rightarrow \rho RT$$

Differentiation gives

$$\frac{dp}{p} = \frac{d\rho}{\rho}\left[p = \rho RT \Rightarrow dp = RTd\rho \Rightarrow dp = \left(\frac{p}{\rho}\right)d\rho \Rightarrow \frac{dp}{p} = \frac{d\rho}{\rho}\right]$$

Therefore,

$$\kappa_{\text{isothermal}} = \frac{dp}{d\rho\Big/\rho} = p \tag{1.2b}$$

- If adiabatic (isentropic or reversible adiabatic) ($p V^{-\gamma}$ = Constant or $\frac{p}{\rho^\gamma}$ = Constant; $\gamma = \frac{c_p}{c_v}$)

Differentiating $\frac{p}{\rho^\gamma}$ = Constant gives

$$\frac{dp}{p} = \gamma \frac{d\rho}{\rho}$$

Therefore,

$$\kappa_{\text{adiabatic}} = \frac{dp}{d\rho\Big/\rho} = \gamma p \tag{1.2c}$$

The bulk modulus is important in fluid mechanics as well as in acoustics:

$$c = \sqrt{\frac{\kappa}{\rho}} = \text{Velocity of sound in any medium} \tag{1.2d}$$

1.1.4.7 Surface tension and capillarity

In the absence of external forces, a drop of liquid takes the shape of a sphere since a sphere has the least surface area for a given volume. It is also the shape that gives the minimum potential energy, a condition for stable equilibrium. Small drops of liquids in a gas and small gas bubbles in a liquid in the absence of external forces have surface tensions. Examples include spoonfuls, glassfuls, rise in capillary tubes, etc.

Surface tension σ is defined as the energy per unit area, which is equal to the force per unit length.

For a spherical drop or a spherical gas bubble of radius r which expands to $r + dr$ as a result of additions of mass, the increase in energy σdA can be expressed as

$$\sigma dA = \sigma d\left(4\pi r^2\right) = \sigma\left(8\pi r dr\right)$$

When there is no external force involved (or temperature, phase change, etc.), this increase in energy is equal to the work done by the pressure forces arising from the differences in pressure inside and outside:

$$\Delta p \left(4\pi r^2 \right) dr = 8\sigma \pi r dr$$

$$\Delta p = \frac{2\sigma}{r} \tag{1.3}$$

For a general curved surface of principal radii r_1 and r_2,

$$\Delta p = \sigma \left(\frac{1}{r_1} + \frac{1}{r_2} \right) \tag{1.4}$$

For a sphere, $r_1 = r_2$.
   Therefore,

$$\Delta p = \frac{2\sigma}{r}$$

For a cylinder, $r_1 = r$; $r_2 = \infty$.
   Therefore,

$$\Delta p = \frac{\sigma}{r} \tag{1.5}$$

Gas–liquid–solid interface

$\sigma_{gs} = \sigma_{sl} + \sigma_{gl} \cos \theta$ (scalar condition!), from which the contact angle θ may be determined (Figure 1.5).

If $\theta < \frac{\pi}{2}$, the liquid wets the surface and the degree of wetting increases as θ decreases to zero; e.g. water ($\theta \approx 0°$) (Figure 1.6).

If $\theta > \frac{\pi}{2}$, the liquid is non-wetting, e.g. Hg ($\theta \approx 130–150°$).

Some surfaces treated with wax (some fabrics too!) have $\theta > \frac{\pi}{2}$.

Water, with a water–air interface, has a surface tension of about 0.073 N/m, while Mercury has a value of about 0.48 N/m.

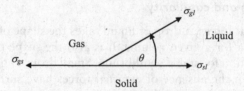

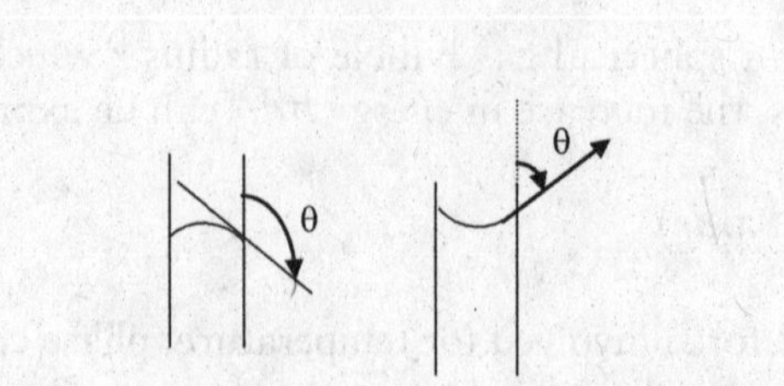

Figure 1.5 Gas–liquid–solid interface.

Figure 1.6 Surface tension effect.

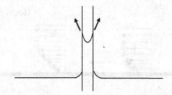

Figure 1.7 Capillary action.

Capillary action: It is the rise of liquid in a column due to surface tension. By equating forces (Figure 1.7),

$$\rho g h \pi r^2 = \left(\Delta p \pi r^2 \right)$$

$$= \sigma 2 \pi r \cos \theta \qquad (1.6)$$

$$\Rightarrow h = \frac{2\sigma}{\rho g r} \cos \theta$$

If $\theta < \frac{\pi}{2}$, then $h > 0$ implying capillary rise. If $\theta > \frac{\pi}{2}$, then $h < 0$, implying a depression.

Surface tension forces are small in relation to gravity, viscosity and pressure forces.

For water/air interface, $\sigma = 0.0731$ N/m, which may be reduced to about one-half by the addition of wetting agents.

For Hg/air interface, $\sigma = 0.435$ N/m.

For Kerosene/air, $\sigma = 0.023 - 0.032$ N/m.

For CCl$_4$/air interface, $\sigma = 0.027$ N/m.

Surface tension decreases with temperature.

1.1.4.8 Viscosity

Viscosity is the fluid property that offers resistance to flow. It distinguishes real fluids from ideal fluids. The viscosity of a gas increases with temperature because of the greater molecular activity. For liquids, molecular spacings are much smaller than for gases and the molecular cohesion is strong. Increased temperature decreases the molecular cohesion and results in a decrease of viscosity.

Newton's law of viscosity: According to Newton's law of viscosity,

Shear stress $\propto$ velocity gradient

$$\tau \propto \frac{du}{dy} \Rightarrow \tau = \mu \frac{du}{dy} \qquad (1.7)$$

where the constant of proportionality μ is the dynamic viscosity (M/LT), which has units of kg/m s.

$$\text{Kinematic viscosity}, v = \frac{\text{Dynamic viscosity}}{\text{Density}} = \frac{\mu}{\rho} = \frac{M/LT}{M/L^3} = L^2 T^{-1} \left(m^2/s \right)$$

Velocity profile: In the case of the flow past a flat plate kept in a flow field, the velocity at the boundary is zero because of the 'no-slip' condition. At point A' (Figure 1.8), the velocity is u_0, and the velocity changes from zero to u_0 in a smooth and monotonic manner. At point

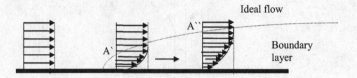

Figure 1.8 Boundary layer development.

A", the velocity does not, in general, attain u_0 at the same distance from the boundary. This is because of the development of the boundary layer.

Ideal and real flow regimes: Ideal and real flow regimes are separated by the boundary layer.

Examples

- Flow between parallel plates when one of the plates is stationary, and the other is moving with a constant velocity V (Figure 1.9).
- Axial motion – Shaft and sleeve of constant radii r_1 and r_2 (Figure 1.10).

Shearing force over the shaft = Shearing force over the sleeve

i.e. $\mu \left(\dfrac{dv}{dr} \right)_1 2\pi r_1 L = \mu \left(\dfrac{dv}{dr} \right)_2 2\pi r_2 L$

$\Rightarrow \left(\dfrac{dv}{dr} \right)_1 \Big/ \left(\dfrac{dv}{dr} \right)_2 = \dfrac{r_2}{r_1}$

i.e. velocity gradient at the shaft (r_1) > velocity gradient at sleeve (r_2)

- Rotation – Bearing and shaft

Torque on shaft = Torque on bearing

Therefore,

$\mu \left(\dfrac{dv}{dr} \right)_1 (2\pi r_1) r_1 L = \mu \left(\dfrac{dv}{dr} \right)_2 (2\pi r_2) r_2 L \, (L = \text{Length of bearing})$

$\Rightarrow \left(\dfrac{dv}{dr} \right)_1 \Big/ \left(\dfrac{dv}{dr} \right)_2 = \dfrac{r_2^2}{r_1^2}$

$\Rightarrow \left(\dfrac{dv}{dr} \right)_1 > \left(\dfrac{dv}{dr} \right)_2$

Newtonian and non-Newtonian fluids: In Newtonian fluids, the viscosity μ depends on temperature but independent of shear rate (velocity gradient).

i.e. τ vs. $\dfrac{du}{dy}$ is a straight line (e.g. water, air and gasoline).

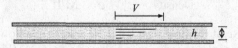

Figure 1.9 Flow between parallel plates.

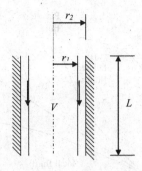

Figure 1.10 Axial motion.

Any other fluid is a non-Newtonian fluid (e.g. toothpaste, paint). Fluids can also be classified as

- **Purely viscous fluids**

    Both Newtonian and Non-Newtonian for which τ depends only on $\dfrac{du}{dy}$ but not on time, e.g. air, water.

    A number of equations are used to describe Non-Newtonian fluids:

$$\tau = \mu \left(\frac{du}{dy} \right)^{n} \tag{1.8}$$

   $n = 1$ – Newtonian
   $n < 1$ – Pseudoplastic (e.g. slurries, polymer solution)
   $n > 1$ – Dilatant (e.g. suspensions with high solid concentrations)

- **Bingham fluids**

$$\tau = \tau_1 + \mu_B \frac{du}{dy} \, (\text{e.g. slurries, drilling, mud, oil paints}) \tag{1.9}$$

- **Time dependent fluids**

    Viscosity decreases with time under constant shear rate (velocity gradient). For example, gypsum pastes, slurries, etc.
- **Viscoelastic fluids** – e.g. flour dough.

1.1.4.9 Stress-rate of strain relationships

Different fluids have different stress – rate of strain relationships as shown in Figure 1.11.

1.2 HYDRAULICS

Hydraulics is a topic in applied science and engineering dealing with the application of the principles of fluid mechanics to practical engineering problems. Topics in hydraulics include, but are not limited to, the generation, control and transmission of power by the use of

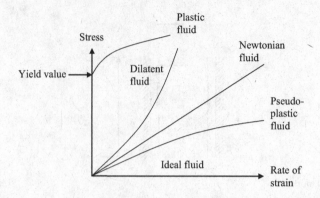

Figure 1.11 Stress–strain relationships for different types of fluids.

pressurized liquids, pipe flow, dam design, pumps and turbines, hydropower, flow measurement, river channel behavior, and sedimentation and erosion.

Historically, notable applications of hydraulics for the utilization of water for human needs include the diversion of waters of River Nile in Egypt and Rivers Tigris and Euphrates in Mesopotamia for irrigation, canals constructed in India, also for irrigation, invention of 'Egyptian bucket wheel' (Noria) in Egypt, construction of the Grand Canal in China for the transportation of grains via barges, qanats (underground canals) in Mesopotamia, irrigation canals and control structures in Anuradhapura and Polonnaruwa areas of ancient Sri Lanka and Roman aqueducts to convey water from areas of abundance to areas of shortage. These engineering marvels of the time date back to time periods ranging from about 6000 BC to about 300 BC. Pre-historic developments of irrigation canals also have taken place in Mexico. Such engineering feats have been achieved without much theoretical knowledge of fluid mechanics or hydraulics. They were motivated and driven by necessity and natural intelligence.

Major hydraulic structures in recent times include the Suez Canal in Egypt that connects the Mediterranean Sea and the Red Sea and the Panama Canal managed by the Panama Canal Authority that connects the Atlantic Ocean and the Pacific Ocean. These two waterways cut the respective sea voyages by about 7,000 and 12,875 km.

Theoretical and experimental developments in hydraulics started with the postulation of the well-known Archimedes (285–212 BC) principle of buoyancy, which has been applied to determine the gold content of the crown of King Hiero of Greece, and the invention of the Archimedes screw pump for lifting fluids. Since then, there have been significant theoretical and empirical developments in both fluid mechanics and hydraulics resulting from contributions made by Newton, Euler, Bernoulli, Rayleigh, Reynolds, Navier, Stokes, Prandtl, von Karman, Boussinesq, Taylor, Richardson and Kolmogorov, etc.

In the period from the 17th to the 20th centuries, contributions on measurement, instrumentation and hydraulic machines came from Torricelli who invented the barometer, Pitot who invented the Pitot tube, Venturi who invented the Venturi meter, Bramah who invented the hydraulic press, Bourdon who invented the Bourdon pressure gauge, Pelton who invented the Pelton wheel, Francis who invented the Francis turbine, and Kaplan who invented the Kaplan turbine.

Hydraulics is not a branch of pure science. Although fluid mechanics form the backbone of hydraulics, it also has a great deal of contributions based on empirical studies. Such contributions came from Weisbach who studied friction in pipe flow, Darcy who studied flow through porous media, Chezy and Manning for their respective equations for uniform steady flow, Moody for Moody diagram for friction factors, and Nikuradse also for a clearer understanding of friction in pipe flows.

These and other significant developments in fluid mechanics and hydraulics are summarized in Chapter 2. Modern-day major hydraulic structures in the world are briefly described in Chapter 24.

1.3 HYDROLOGY

Hydrology can be defined as the science that treats the occurrence, circulation and distribution of earth's water resources and their interaction with the environment, including living things. It is a multi-disciplinary subject with contributions from civil engineering, soil physics, agriculture, forestry, geomorphology, geography, etc. with emphasis on measurement, recording and publication of basic data, analysis of such data to develop and expand fundamental theories, and application of the theories and data to practical problems. Although the definition covers the entire spectrum of processes in the hydrological cycle, the treatment in this book is confined to hydro-meteorology, surface water and groundwater. Snow hydrology, which is important in cold regions, is not covered in this book.

The modern-day understanding of hydrology follows the concept of the hydrological cycle conceived by Leonardo da Vinci in the 16th century. Since then, there has not been much significant developments in the understanding of the hydrological cycle until about 1674. Perrault in 1674 and Mariotte in 1686 independently carried out experiments on runoff and concluded that rainfall is sufficient to produce streamflow. Chapter 25 gives an introduction to the hydrological cycle as well as significant developments that have taken place over the years.

1.4 WATER RESOURCES

Water resources is a very general topic which has technical inputs from Fluid mechanics, hydraulics and hydrology but also non-technical inputs from social sciences, economics, environment, population etc. What is described in this book is only a brief review of the present state of the resource including water demands and future challenges in coping with water problems.

1.5 SCOPE AND LAYOUT

The contents of this book are arranged in 41 chapters. Chapter 2 gives a review of the historical development of the science of water, including a list of pioneers in the development of quantitative water science. Chapters 3–16 cover topics in fluid mechanics. In particular, fluid statics in Chapter 3, fluid kinematics in Chapter 4, governing equations of fluid motion in Chapter 5, ideal fluid flow in Chapter 6, viscous fluid flow and boundary layer in Chapter 7, dimensional analysis in Chapter 8, fluid flow measurements in Chapter 9, pipe flow in Chapter 10, pipe networks in Chapter 11, fluid machinery in Chapter 12, applications of the governing equations of fluid flow equations in Chapter 13, turbulence in Chapter 14, turbulence modelling in Chapter 15 and computational fluid dynamics in Chapter 16.

Chapters 17–24 cover topics in hydraulics. In particular, open channel flow in Chapter 17, gradually varied flow in open channels in Chapter 18, rapidly varying flow in Chapter 19, hydraulics of alluvial channels in Chapter 20, channel stability analysis in Chapter 21, sediment transport and deposition in Chapter 22, environmental hydraulics in Chapter 23 and major hydraulic structures in the world in Chapter 24.

Chapters 25–38 cover topics in hydrology. In particular, hydrological cycle and its principal processes in Chapter 25, hydro-meteorology in Chapter 26, precipitation in Chapter 27, evaporation and evapo-transpiration in Chapter 28, infiltration in Chapter 29, runoff in Chapter 30, analysis and presentation of rainfall data in Chapter 31, unit hydrograph methods in Chapter 32, rainfall-runoff modelling in Chapter 33, flood routing in Chapter 34, flow through saturated porous media in Chapter 35, statistical methods in hydrology in Chapter 36, systems theory approach to hydrological modelling in Chapter 37, and time series analysis and forecasting in Chapter 38.

Chapters 39–41 cover general topics in water resources. In particular, water state of the resource in Chapter 39, sources and demand for water in Chapter 40, and challenges in coping with water problems in Chapter 41, the last chapter.

In addition to the descriptive parts of the various topics, worked examples for most of the chapters as well as lists of references are given at the end of each chapter.

Chapter 2

Historical development of the science of water

2.1 INTRODUCTION

Water is essential for all forms of life, and food comes next. Management of the available water resources for direct consumption and for the production of food, therefore, goes as far back as one can trace the beginning of civilization. In ancient times, this has been achieved through conventional wisdom, or natural intelligence, without much technological inputs. Community spirit, cultural values and in some cases religious and political guidance have helped to harness the available water resources for the benefit of not only human beings but for all living things. Examples of attempts to harness the waters on earth for the well-being of all living things through natural intelligence include cuts made across embankments of the River Nile to divert water to irrigable areas circa 6000 BC in Egypt, utilization of the waters of Tigris and Euphrates rivers in Mesopotamia (present Iraq and Iran) circa 4000 BC, irrigation canals constructed in India circa 2600 BC, 'Egyptian bucket wheel' invented circa 750 BC, Grand Canal of China circa 600 BC, qanats in Mesopotamia circa 550 BC, irrigation canals in Mexico circa 500 BC, irrigation canals and cascades of reservoirs in Sri Lanka circa 300 BC, Roman aqueducts circa 300 BC, and, irrigation canals constructed in China circa 256 BC. One can only attribute such wonders to natural intelligence proving the saying that 'necessity is the mother of invention'.

In the ancient world, the demand for water has been mainly for direct consumption and for growing food. In the modern world, there are additional demands such as power generation, prevention of flooding, recreation, environmental conservation and sometimes for transportation. Some of these demands at times can be conflicting and to manage the available resources which fluctuate with time and space requires a more sophisticated and scientific approach. The three main components in water resources management are fluid mechanics, hydraulics and hydrology.

Fluid mechanics provides the basic scientific backbone of water science, followed by hydraulics which provides the applications of fluid mechanics to practical problems. Hydrology deals with the occurrence and movement of water in the earth system via the different processes of the hydrological cycle. Developments in each of these three areas have taken place since ancient times through contributions made by various philosophers, mathematicians, scientists and engineers from various regions. A partial list of the pioneers in these fields is given at the end of this chapter.

The development of fluid mechanics started with the contributions made by Archimedes (285–212 BC), who postulated the well-known Archimedes principle of buoyancy. This principle was applied to determine the gold content of the crown of King Hiero I of Greece. Archimedes is also credited for the development of the Archimedes Screw, which has been used to lift water from a low level to a higher level. At about the same time, the Romans built an extensive network, better known as Roman aqueducts, to transport fresh water from snowmelt Alps to cities in the valleys below. These include the first aqueduct, the 'Aqua Appia', built in 312 BC and the last, 'Aqua Alexandrina' built in 226 AD. Over a period of 500 years, 11 aqueducts, some below ground surface and some elevated have been built. Roman aqueducts consisted of infiltration galleries, steep chutes or drop shafts, settling tanks, tunnels, covered trenches, bridges to support the aqueduct, siphons and a distribution system at the destination. Pre-historic developments of irrigation canals also have taken place in Mexico circa 500 BC. Other significant waterworks of the time include the use of water wheels to power mills in Greece, the invention of the Egyptian water wheel known as Noria, the installation of a weather vane in Acropolis in Greece, the invention of conical valve and the introduction of early automatic controls in fluid mechanics by Banu Mūsā (circa 800–860) in Iran, the application of experimental scientific methods to fluid statics such as determining the specific weights by Abu Rayhan Biruni (circa 973–1048) in Iran, and the contributions by Al-Khazini (circa 1115–1130), also in Iran.

Since then, there has not been much significant developments until Leonardo da Vinci (1459–1519) carried out several experiments in fluid mechanics as well as introducing the concept of the hydrological cycle as it is understood today. The basic laws of fluid motion were introduced by Isaac Newton (1649–1727) together with the linear law of viscosity, which identifies a fluid as either Newtonian or non-Newtonian. Newton's laws paved the way to describe fluid motion in differential equations for inviscid fluids. These include the Euler equation (Leonhard Euler, 1707–1783) and the Bernoulli equation (Daniel Bernoulli, 1700–1782) for incompressible fluids, which have now become household names in fluid mechanics. Dimensional analysis, which is a powerful technique for model testing, was developed by Lord Rayleigh (1849–1919). The familiar dimensionless number in fluid mechanics, the Reynolds Number, was introduced by Osborn Reynolds (1849–1912) based on his extensive experimental studies with pipe flow. The general equations of fluid flow that includes fluid friction were developed independently by Claude Louis Navier (1785–1836) and George Gabriel Stokes (1819–1903), although at that time the equations had no simple solutions. At present, these equations form the basis of almost all fluid flow problems. Modern-day Computational Fluid Dynamics (CFD) software solve these equations numerically using various types of numerical techniques. The next important contribution came from Ludwig Prandtl (1875–1953) through the introduction of the boundary layer theory that enabled a fluid flow to be divided into two regions: a layer of flow near the wall known as the boundary layer where fluid friction is taken into account and a layer outside the boundary layer where the fluid friction is negligible. Outside the boundary layer, the Euler equation and the Bernoulli equation are applicable.

In the 20th century, notable contributions to the development of fluid mechanics have been made by Theodore von Karman (1875–1963) and Sir Geoffrey Ingram Taylor (1886–1975). An area of fluid mechanics still not well understood is turbulence where important contributions have been made by Joseph Valentin Boussinesq (1877), who hypothesized that turbulent stresses are linearly proportional to the large scale mean strain rates, Ludwig Prandtl (1925) who introduced the mixing length theory and the logarithmic velocity profile near a solid wall, G. I. Taylor (1921) who introduced the idea of presenting turbulence in statistical terms as well as the concept of mixing length and the statistical theory of turbulence, Lewis Fry Richardson (1922) who introduced the concept of energy cascade which was followed by Andrey Kolmogorov (1941) who postulated that the statistics of small scales are isotropic and uniquely determined by the length scale, l, the kinematic viscosity, ν, and an average rate of kinetic energy dissipation per unit mass, ε, and, Edward Norton Lorenz (1963) who proposed a link between chaos and turbulence.

The developments in fluid mechanics have taken place along two fronts, one made by mathematicians and physicists for fluid flow under ideal conditions, and the other made by engineers to solve practical problems. The latter type of developments can be considered as belonging to hydraulics where a great deal of empirical knowledge has been added based on observations. Experimentalists of the time include Chézy, Pitot, Borda, Weber, Francis, Hagen, Poiseuille, Darcy, Manning, Bazin, and Weisbach, who have contributed empirical data on open channels, ship resistance, pipe flows, waves, and turbines, etc. These two fronts of development complement each other.

In hydrology, the concept of the hydrological cycle has been understood in crude form as early as 3000 BC when the Egyptians learned to harness the waters of the Nile and measure the rise and fall of the river. Descriptive hydrology was conceived by Plato and Aristotle circa 400 BC. Romans constructed aqueducts without complete knowledge of quantitative hydrology. A reasonably clear idea of the hydrological cycle was put forward by Vitruvius circa 27 to 17 BC, but the concept of the hydrological cycle in its present form is only after Leonardo da Vinci (1452–1519). Since then, there has not been much significant developments in the understanding of the hydrological cycle until Perrault in 1674, and Mariotte in 1686 independently carried out experiments on runoff and concluded that rainfall is sufficient to produce streamflow. Other significant developments in chronological order include the introduction of Darcy's law for flow through porous media in 1856, unit hydrograph theory by Sherman in 1932, infiltration theory by Horton in 1933, extreme value theory by Gumbel in 1941, tank model for rainfall-runoff modelling by Sugawara in 1950, infiltration theory by Philip in 1954 and the kinematic wave theory by Lighthill and Witham in 1955. Institutional developments that promoted the development of hydrology include the establishment of the International Association of Hydrological Sciences (IAHS) in 1922, International Hydrological Decade (IHD) from 1965 to 1974 and the International Hydrological Programme (IHP) from 1975 onwards.

2.2 PIONEERS IN THE DEVELOPMENT OF QUANTITATIVE WATER SCIENCE

- **Plato (390 B.C.):** Plato, a Greek philosopher, described the origins of rivers as water escaping through holes in the ground.
- **Archimedes (285 to 212 B.C.):** Contributions by Archimedes, also from Greece, include the principle or law that states the upward buoyancy force of a floating or submerged body is equal to the weight of the fluid displaced by the body and acts at the centre of mass of the displaced fluid. He formulated the laws of buoyancy and applied them to floating and submerged bodies, and derived a form of differential calculus as part of the analysis. He is also known for the invention of Archimedes screw which is used to lift liquids from a lower level to a higher level by a screw.
- **Banū Mūsā (circa 800–860):** Musa brothers of Persian origin in their *Book of Ingenious Devices* described a number of early automatic controls in fluid mechanics. They are also credited for the invention of the conical valve.
- **Abu Rayhan Biruni (973–1048):** Biruni, regarded as one of the greatest scholars of the medieval Islamic era from Iran, was the first to apply experimental scientific methods to fluid statics such as determining specific weights. He discovered a correlation between the specific gravity of an object and the volume of water it displaces.
- **Al-Khazini (flourished during 1115–1130):** Al-Khazini, an Iranian scholar, followed Biruni's work, and also is credited for his book entitled *The Book of the Balance*

of Wisdom which is an encyclopaedia of medieval mechanics and hydrostatics, in particular hydrostatic balance.

- **Leonardo da Vinci (1452–1519):** da Vinci, the Italian scientist and painter, well known for his painting of 'Mona Lisa' has many scientific contributions. Among them are his accurate descriptions of waves, jets, hydraulic jumps, eddy formation and drag through experimental analysis. He derived the equation of conservation of mass in one-dimensional steady state flow, developed the first turbine water wheel and the parachute and investigated the capillary movement of liquids in small tubes. He is also credited with the introduction of the concept of the hydrological cycle.
- **Benedetto Castelli (1577–1643):** Castelli, an Italian mathematician, is credited for establishing the continuity principle as a basis of fluid kinematics.
- **Evangelista Torricelli (1608–1647):** Torricelli, an Italian scientist and a disciple of Galilio, invented the mercury barometer. The term Torr, a unit of pressure commonly used in vacuum measurements, is named after him.
- **Pierre Perrault (1608–1680):** Perrault, a French hydrologist, in 1674, and Mariotte in 1686 independently carried out experiments on runoff and concluded that rainfall is sufficient to produce streamflow.
- **Edme Mariotte (1620–1684):** Mariotte, a French hydrologist in 1686, independently carried out experiments on runoff and concluded that rainfall is sufficient to produce streamflow. Perrault, also a French hydrologist concluded the same in 1674. Mariotte is also known to have built the first wind tunnel and tested models in it and carried out a great variety of well-conducted experiments on the motion of fluids at Versailles and Chantilly.
- **Blaise Pascal (1623–1662):** Pascal, a French scientist, made important contributions to the study of fluids. His inventions include the hydraulic press and the syringe. He demonstrated the laws of the equilibrium of liquids in the simplest manner and amply confirmed by experiments. The fact that pressure is always acting normal to the surface of contact is known as Pascal's law.
- **Isaac Newton (1642–1727):** the British scientist, postulated his laws of motion and the law of viscosity of linear fluids, now called Newtonian. The theory first yields the frictionless assumption which led to several beautiful mathematical solutions. He threw much light upon several branches of hydromechanics, including the effects of friction and viscosity in reducing the velocity of running water, which can be noticed in the *Principia* of Sir Isaac Newton. He also showed that the velocity of any stratum of the vortex is an arithmetical mean between the velocities of the strata which enclose it. From this, it evidently follows that the velocity of a filament of water moving in a pipe is an arithmetic mean between the velocities of the filaments which surround it.
- **Daniel Gabriel Fahrenheit (1686–1736):** Fahrenheit, a Dutch-German-Polish physicist, inventor, and scientific instrument maker, was one of the notable figures in the Golden Age of Dutch science and technology. The Fahrenheit scale of measuring temperature is after him who was a pioneer of exact thermometry.
- **Henri Pitot (1695–1771):** Pitot, Italian born French hydraulic engineer showed that the retardations arising from friction are inversely proportional to the diameters of the pipes in which the fluid moves. He invented the Pitot tube in which the height of the fluid column is proportional to the square of the velocity of the fluid at the depth of the inlet to the **Pitot** tube.
- **Daniel Bernoulli (1700–1782):** Bernoulli, a Swiss scientist, is well known in fluid mechanics for his Bernoulli equation. He and Euler studied the velocity and pressure variation of the flow of blood. His principle is used even nowadays to measure the

speed of air passing an aircraft. In 1738, Daniel Bernoulli published 'Hydrodynamica' which discusses the pressure and velocity of fluids and describes the Bernoulli equation.

- **Anders Celsius (1701–1744):** Celsius, a Swedish astronomer, physicist and mathematician, invented the Celsius scale of temperature measurement.
- **Leonhard Euler (1707–1783):** Euler, a Swiss scientist, developed both the differential equations of motion and their integral form, now called Bernoulli equation. This calculus was first applied to the motion of water by d'Alembert. It enabled both him and Euler to represent the theory of fluids in formulas restricted by no particular hypothesis.
- **Jean le Rond d'Alembert (1717–1783):** D'Alembert, a French mathematician, philosopher and writer, used the equation developed by Euler to show his famous paradox that a body immersed in a frictionless fluid has zero drag.
- **Antoine Chézy (1718–1798):** Chezy, a French hydraulic engineer is well known for his Chezy equation which expresses the mean flow velocity in terms of channel roughness, hydraulic radius, and bed slope in open-channel flows.
- **Jean-Charles de Borda (1733–1799) and Lazare Carnot (1753–1823):** Borda, a surveyor, mathematician, political scientist and physicist, and Carnot, an engineer, mathematician and legislator, both of French origin are known for their Borda-Carnot equation that describes empirically the mechanical energy loss of a fluid due to a sudden flow expansion which is in contrast to Bernoulli equation where the total head is constant along a streamline.
- **Joseph-Louis Lagrange (1736–1813):** Lagrange, a French mathematician and astronomer, re-formulated classical Newtonian mechanics to simplify formulas and ease calculations. These mechanics are called Lagrangian mechanics. He introduced a system of co-ordinates now known as Lagrangian co-ordinates to describe fluids in motion where the observer follows an individual fluid parcel as it moves through space and time. Plotting the position of an individual parcel through time gives the pathline of the parcel. The description of flow by following a fluid particle is known as the Lagrangian approach, as opposed to the Eulerian approach in which the fluid passing through a fixed position in space (control volume) is considered.
- **Giovanni Battista Venturi (1746–1822):** Venturi, an Italian physicist, is the inventor of the Venturi effect, which is used in the Venturimeters for measuring the flow rates.
- **Joseph Bramah (1748–1814):** Bramah, an English inventor and locksmith, developed the hydraulic press. He is also recognized for improving and patenting the flushing toilet.
- **Pierre-Simon, marquis de Laplace (1749–1827):** Laplace, a French scholar in engineering, mathematics, statistics, physics and astronomy is well known for his 'Laplace transform' in solving differential equations and Laplace equation that describes many physical phenomena including groundwater flow and potential flow.
- **Francis Joseph von Gerstner (1756–1832):** Gerstner, a German-Bohemian physicist and engineer is known for providing a solution for the phase speed of deep ocean waves with finite amplitude in which the fluid particles have a circular orbit.
- **Louis Marie Henri Navier (1785–1836) and George Gabriel Stokes (1819–1903):** Navier, a French scientist and Stokes, a British scientist together developed the well-known Navier-Stokes equations by adding Newtonian viscous terms to the governing equations of motion. The Navier–Stokes equations describe the motion of fluids by applying Newton's second law to fluid motion together with the assumption that the fluid stress is the sum of a diffusing viscous term (proportional to the velocity gradient) and a pressure term. The Navier-Stokes equations now are the basis of all fluid flows.

- **Jean Léonard Marie Poiseuille (1797–1869):** Poiseuille, a French physicist and physiologist experimentally derived, formulated and published Poiseuille's law which states that the velocity of a liquid flowing through a capillary is directly proportional to the pressure of the liquid and the fourth power of the radius of the capillary and is inversely proportional to the viscosity of the liquid and the length of the capillary.
- **Gotthilf Heinrich Ludwig Hagen (1797–1884):** Hagen, a German Civil Engineer, describes laminar flow properties. The Hagen–Poiseuille equation which gives the pressure drop in a fluid flowing through a long circular pipe is named after Hagen and Poiseuille.
- **Adhémar Jean Claude Barré de Saint-Venant (1797–1886):** St. Venant, a French theoretician, formulated the equations of unsteady flow in open channels, now known as the St. Venant's equations.
- **Henry Philibert Gaspard Darcy (1803–1858):** Darcy, a French engineer is known for several contributions in hydraulics and hydrology of which the Darcy equation for flow through porous media is one which is widely used.
- **Julius Weisbach (1806–1871):** Weisbach, a German mathematician and engineer is well known for the Darcy-Weisbach friction factor used to calculate frictional losses in pipe flow.
- **Eugene Bourdon (1808–1884):** Bourdon, a French engineer, is known for the invention of the Bourdon pressure gauge.
- **William Froude (1810–1879):** Froude, a British hydrodynamics engineer and naval architect and his son developed laws of model testing. Froude Number is named after him.
- **James Bicheno Francis (1815–1892):** Francis, a British-American Civil Engineer developed the radial flow turbine, now known as Francis turbine used in hydro-electric schemes. Francis turbines usually operate on low head high flow conditions.
- **Robert Manning (1816–1897):** Manning, an Irish hydraulic engineer, is well known for his equation for flow in open channels referred to as the Manning's equation in which the roughness is termed as Manning's 'n'.
- **Hermann Ludwig Ferdinand von Helmholtz (1821–1894):** Helmholtz, a German physician and physicist, has made significant contributions in several scientific fields. In fluid mechanics, he established his three 'laws of vortex motion' which lead to the significance of vorticity in fluid mechanics and science in general.
- **Lester Allan Pelton (1829–1908):** Pelton, an American inventor invented the unique water wheel that can convert the kinetic energy of the water to mechanical energy and then to electrical energy. The Pelton wheel, named after him is widely used to generate hydro-electricity in many countries where there are high heads.
- **Henri-Émile Bazin (1829–1917):** Bazin, a French engineer followed Darcy's work on frictional losses and completed it after Darcy's death. He later studied the problem of wave propagation and the contraction of fluid flowing through an orifice.
- **Ernst Waldfried Josef Wenzel Mach (1838–1916):** Mach, an Austrian physicist and philosopher studied shock waves. The well-known Mach number in aerodynamics which is the ratio of the speed to that of sound is named after him.
- **Osborne Reynolds (1842–1912):** Reynolds, a British scientist, published the classic pipe experiment and showed the importance of the dimensionless Reynolds number, named after him. Reynolds number is used to distinguish laminar flow from turbulent flow. He also described turbulent shear stresses.
- **Lord Rayleigh (1842–1919):** Rayleigh, a British engineer, proposed dimensional analysis. This was done while he was trying to understand why the sky is blue.

- **Joseph Valentin Boussinesq (1842–1929):** Boussinesq, a French mathematician and physicist, made significant contributions to the theory of hydrodynamics, vibration, light and heat. In fluid dynamics, the Boussinesq approximation for water waves is named after him. The Boussinesq approximation for water waves takes into account the vertical structure of the horizontal and vertical flow velocity leading to non-linear partial differential equations called Boussinesq-type equations. He also made significant contributions to the understanding of turbulence. Boussinesq introduced eddy viscosities and hypothesized that the Reynolds stress could be linked to the mean rate of deformation, which is widely used in turbulence modelling.
- **Vincenc Strouhal (1850–1922):** Strouhal was a Czech physicist who studied experimental physics. The Strouhal number (St) in dimensional analysis, which describes oscillating flow mechanisms is named after him. It is an integral part of the fundamentals of fluid mechanics.
- **Edgar Buckingham (1867–1940):** Buckingham, an American physicist, is known for his contributions to dimensional analysis and in particular the Buckingham's π theorem.
- **L. K. Sherman (1869–1954):** Sherman is known for introducing the concept of Unit Hydrograph in 1932 which has become the backbone of hydrology.
- **Moritz Weber (1871–1951):** The Weber number which is the ratio of the inertial force to the surface tension force which indicates whether the kinetic or the surface tension energy is dominant is named after Weber.
- **Robert Elmer Horton (1875–1945):** Horton, an American civil engineer and soil scientist, is known for introducing the infiltration equation widely known as the 'Horton equation' for infiltration.
- **Ludwig Prandtl (1875–1953):** Prandtl, a German engineer and aerodynamicist, pointed out that fluid flows with small viscosity, such as water flows and airflows, can be divided into a thin viscous layer (or boundary layer) near solid surfaces and interfaces, patched onto a nearly inviscid outer layer, where the Euler and Bernoulli equations apply. He obtained a solution to Navier Stokes equation for boundary layer flow. The boundary layer concept was one of the most important developments in fluid mechanics.
- **Viktor Kaplan (1876–1934):** Kaplan, an Austrian engineer developed the axial flow turbine named after him as Kaplan turbine used in hydro-electric projects.
- **Boris Alexandrovich Bakhmeteff (1880–1951):** Bakhmeteff, an engineer and businessman of Russian origin who later was a professor of Civil Engineering at Columbia University is known for introducing the concept of specific energy in the context of open channel flows.
- **Lewis Ferry Moody (1880–1953):** Moody, a US engineer, is well known for the Moody diagram which relates the Darcy-Weisbach friction factor to the relative roughness height and the Reynolds number.
- **Lewis Fry Richardson (1881–1953):** Richardson, an English mathematician, physicist, meteorologist, psychologist and pacifist pioneered modern mathematical techniques of weather forecasting, and introduced the concept of energy cascade in turbulent flows where the process of passing down the energy from the largest scale to the smallest scale is considered to take place sequentially.
- **Theodore von Karman (1881–1963):** von Karman, a Hungarian American, was a physicist, mathematician and aerospace engineer with various contributions in aerodynamics and fluid mechanics. He studied under aerodynamicist Ludwig Prandtl at the University of Göttingen for 6 years, during which time he first explained what has become known as the 'Kármán vortex street'. The Jet Propulsion Laboratory of California Institute of Technology was established under his guidance.

- **Paul Richard Heinrich Blasius (1883–1970):** Blasius, a German fluid dynamics physicist, described the steady two-dimensional laminar boundary layer that forms on a semi-infinite plate, which is held parallel to a constant unidirectional flow. He was Prandtl's doctoral student.
- **Geoffrey Ingram Taylor (1886–1975):** Taylor, a British physicist and mathematician, was a pioneer in fluid dynamics and wave theory. He also contributed to the understanding of turbulent flows. He introduced the idea of presenting turbulence in statistical terms such as correlations, Fourier transforms and power spectra as well as the concept of mixing length, and subsequently introduced the statistical theory of turbulence.
- **Emil Julius Gumbel (1891–1966):** Gumbel, a German mathematician is known for the extreme value distribution named after him which is widely used in statistical hydrology.
- **Johann Nikuradse (1894–1979):** Nikuradse, a German engineer and physicist who later was a Ph. D. student of Prandtl, measured the frictional resistance in turbulent flow in rough pipes using sand grains with varying roughnesses and found that the rougher the surface, the greater the friction, and hence the greater pressure loss. He found a relationship linking pipe flow, friction factor and Reynolds number.
- **Andrey Kolmogorov (1903–1987):** Kolmogorov, a Soviet mathematician postulated that the statistics of small scales are isotropic and uniquely determined by the length scale, l, the kinematic viscosity, ν, and an average rate of kinetic energy dissipation per unit mass, ε. He subsequently presented a statistical approach of describing turbulence.
- **Hans Albert Einstein (1904–1973):** Einstein, a Swiss-American hydraulic engineer and educator is known for his bed load function in sediment transport that uses probability concepts.
- **Albert Frank Shields (1908–1974):** Shields, a US experimentalist who studied in Germany, is known for his pioneering studies in sediment transport research. In particular, he is well-known for his Shields diagram (1936) that gives a relationship between the dimensionless critical shear stress and Reynolds number at the point of incipient motion of sediments.
- **Ray Keyes Linsley (1917–1990):** Linsley, an American hydrologist, water resources engineer and educator, has a multitude of contributions to hydrology and water resources. He and his co-worker N. H. Crawford at Stanford University developed perhaps the first mathematical model using digital computers in hydrology known as the 'Stanford Watershed Model'.
- **Edward Norton Lorenz (1917–2008):** Lorenz, an American mathematician and meteorologist was a pioneer in chaos theory and presented the well-known 'butterfly effect'. He later proposed a link between chaos and turbulence.
- **Masami Sugawara (1919–2011):** Sugawara, of Japan, is well known for the development of the 'tank model' as a lumped mathematical model for hydrological prediction.
- **Ren Jun Zhao (1924–1993):** Zhao, a Chinese hydrologist with his team at Hohai University developed the first indigenous conceptual hydrological model widely used in China since 1980, known as the Xinanjiang model. Xinanjing is a name of a river in China.
- **Michael James Lighthill (1924–1998):** Lighthill, a British mathematician is known for introducing the kinematic wave theory to flood flow problems. He compared traffic flow to flood flow.
- **Gerald B. Witham (1927–2014):** Witham, a British born American applied mathematician, known for his contribution to the kinematic wave theory for flood flows together with his doctoral advisor M. J. Lighthill.

Chapter 3

Fluid statics

3.1 INTRODUCTION

Fluid statics refers to the mechanics of a fluid at rest. The main concerns in fluid statics are the pressure variation in a fluid at rest and the forces on various objects immersed in a fluid. Forces include body forces and surface forces. Body forces are forces developed without physical contact and distributed over the volume of the fluid. Examples include gravitational and electromagnetic forces. Surface forces are forces acting on the boundaries of a fluid through direct contact such as shear forces and normal forces.

In a fluid at rest, there are no shear forces. The only surface force is, therefore the pressure force. Pressure is a scalar quantity and in general varies from point to point as a function of the position co-ordinates as

$$p = p(x, y, z) \tag{3.1}$$

The pressure at any two (or more) points at the same horizontal elevation of the same fluid is the same (Pascal's law). The pressure variation in the vertical direction is given by

$$\frac{\partial p}{\partial z} = -\rho g; \quad \text{or, } p = -\rho g z \tag{3.2}$$

The negative sign indicates that pressure is decreasing with increasing z (z measured positive upwards). In Fluid Mechanics, it is more common to express pressure as $\dfrac{p}{\rho g}$ which is a height (depth, head). Net pressure force resulting from pressure variations can be obtained by summation of forces acting on the fluid element. When pressure acts upon an area, it gives rise to a pressure force.

Area is a vector quantity, and the area vector points in a direction normal to the area and its magnitude is equal to the magnitude of the area. Pressure force is a vector and is considered positive when acting inwards, producing a compressive stress.

3.2 BASIC EQUATIONS OF FLUID STATICS

Considering a fluid element at rest as shown in Figure 3.1, the only forces to be considered are the pressure forces and body forces. Resolving forces in the three directions, x, y and z, they take the form

$$p\,dydz - \left(p + \frac{\partial p}{\partial x}\,dx\right)dydz + \rho\,dxdydz\,B_x = 0 \tag{3.3a}$$

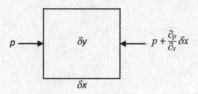

Figure 3.1 Definition sketch for pressures and forces acting on a fluid element.

$$pdzdx - \left(p + \frac{\partial p}{\partial y} dy\right)dzdx + \rho dxdydzB_y = 0 \qquad (3.3b)$$

$$pdxdy - \left(p + \frac{\partial p}{\partial z} dz\right)dxdy + \rho dxdydzB_z = 0 \qquad (3.3c)$$

where B_x, B_y and B_z are the body forces per unit mass is the three directions. Simplifying,

$$-\frac{\partial p}{\partial x} + \rho B_x = 0 \qquad (3.4a)$$

$$-\frac{\partial p}{\partial y} + \rho B_y = 0 \qquad (3.4b)$$

$$-\frac{\partial p}{\partial z} + \rho B_z = 0. \qquad (3.4c)$$

In other words,

Pressure forces per unit volume at a point + body forces per unit volume at a point = 0

In a fluid system at rest, $B_x = 0$, $B_y = 0$ and, $B_z = -g$ (negative because z is measured positive upwards). Therefore,

$$\frac{\partial p}{\partial x} = 0 \qquad (3.5a)$$

$$\frac{\partial p}{\partial y} = 0 \qquad (3.5b)$$

$$\frac{\partial p}{\partial z} = -\rho g \qquad (3.5c)$$

The pressure p is a function of z only ($p = p(z)$), and x and y have no effect. The partial derivative can therefore be replaced by the total derivative as

$$\frac{dp}{dz} = -\rho g \qquad (3.6)$$

and Eq. 3.1, as

$$p = p(z) \tag{3.7}$$

3.3 PRESSURE

3.3.1 Pressure variation in a fluid at rest

3.3.1.1 Incompressible fluid ($\rho = constant = \rho_0$)

For constant gravity,

$$\frac{dp}{dz} = -\rho_0 g = \text{Constant}$$

$$\int_{p_0}^{p} dp = -\int_{z_0}^{z} \rho_0 g dz \tag{3.8}$$

$$p - p_0 = -\rho_0 g (z - z_0)$$

In Eq. 3.8, p_0 is the pressure at the reference height z_0. For liquids, the free surface is taken as the reference level, and distances are measured positive downwards from the surface. Then, if h is the depth below the reference level (i.e. $z_0 - z = h$),

$$p = p_0 + \rho_0 g h \tag{3.9}$$

This is a basic equation used for pressure variation calculations. The following rules are useful:

- Any two points at the same elevation in a continuous length of the **same** liquid will be at the same pressure, and,
- Pressure increases along the depth downwards.

3.3.1.2 Compressible fluid

For compressible fluids, to carry out the integration of Eq. 3.6, the density ρ must be expressed as a function of pressure p or elevation z. Several conditions can be applied to represent this variation.

Isothermal (constant temperature): Under isothermal conditions, the perfect gas law can be applied:

$$pv = RT \quad \text{or} \quad p = \rho RT \Rightarrow \frac{p}{\rho} = \frac{p_0}{\rho_0} \tag{3.10}$$

where the subscript refers to the reference level. Substituting Eq. 3.10 in Eq. 3.6,

$$\frac{dp}{dz} = -\frac{p}{p_0} \rho_0 g \tag{3.11}$$

which when integrated gives

$$\ln\left[\frac{p}{p_0}\right] = -\frac{\rho_0 g}{p_0}(z - z_0) \implies p = p_0 \exp\left(-\frac{\rho_0 g}{p_0}(z - z_0)\right) \tag{3.12}$$

Isentropic (adiabatic reversible): The process equation

$$pv^\gamma = \text{Constant} \tag{3.13}$$

is applicable. Eq. 3.13 can be written as

$$\frac{p}{\rho^\gamma} = \text{Constant} = \frac{p_0}{\rho_0^\gamma} \implies \left[\frac{\rho}{\rho_0}\right]^\gamma = \frac{p}{p_0} \tag{3.14}$$

Substituting in Eq. 3.6,

$$\frac{dp}{dz} = -\rho g = -\rho_0 g \left[\frac{p}{p_0}\right]^{1/\gamma} \implies \frac{dp}{p^{1/\gamma}} = -\frac{\rho_0}{p_0^{1/\gamma}} g\, dz \tag{3.15}$$

Integration gives

$$p^{(\gamma-1)/\gamma} = p_0^{(\gamma-1)/\gamma} - \frac{(\gamma-1)}{\gamma}\frac{\rho_0}{p_0^{1/\gamma}} g(z - z_0) \tag{3.16}$$

In a barotropic process, the density is a function of pressure only. In a polytropic process the relationship $pv^n = \text{Constant}$ can also be isentropic or isothermal depending on the value of the index n.

Temperature decreasing linearly with elevation: Here, the basic equations are

$$\frac{dp}{dz} = -\rho g \quad \text{and} \quad p = \rho RT \tag{3.17}$$

which lead to

$$\frac{dp}{dz} = -\frac{p}{RT}g \tag{3.18}$$

Since the lapse rate α is given by $\dfrac{dT}{dz} = -\alpha, \implies T = T_0 - \alpha(z - z_0)$, and therefore,

$$\frac{dp}{p} = -\frac{g\, dz}{R[T_0 - \alpha(z - z_0)]}.$$

Integrating

$$\ell n\left(\frac{p}{p_0}\right) = \frac{g}{\alpha R} \ell n \frac{[T_0 - \alpha(z - z_0)]}{T_0}$$

$$= \frac{g}{\alpha R} \ell n\left(\frac{T}{T_0}\right) \tag{3.19}$$

Therefore,

$$\frac{p}{p_0} = \left(\frac{T}{T_0}\right)^{\frac{g}{\alpha R}} \tag{3.20}$$

For example, if the temperature at the surface is 20°C, and the difference is 10°C,

$$\frac{T}{T_0} = \frac{273+10}{273+20} = \frac{283}{293} = 0.966$$

$g = 9.81$; $R = 287$ Nm/Kg°K K, $\alpha = 6.5$°C/km and, $\dfrac{g}{\alpha R} = \dfrac{9.81 \times 1,000}{6.5 \times 287} = 5.25$

Therefore,

$$\frac{p}{p_0} = (0.966)^{5.25} = 0.833$$

3.3.2 Standard atmosphere

There are several definitions of standard atmosphere. There is also an international standard atmosphere (ISA) as defined in document 7488/2 of the International Civil Aviation Organization (ICAO) which gives the variations of temperature, pressure and density as functions of altitude above mean sea level. For practical purposes, the following values are taken to represent the standard atmosphere:

- Temperature: 288°K (15°C)
- Pressure: 101.3 kPa (abs) or 1,013 mb or 760 mm of Hg at sea level
- Density: 1.225 kg/m^3
- Viscosity μ: 1.781×10^{-5} kg/m.s
- Lapse rate: 6.5°C/km up to 11 km altitude

3.3.3 Gauge and absolute pressure

Absolute pressure is the pressure above a vacuum, whereas gauge pressure is the pressure above a reference pressure, usually taken to be atmospheric pressure. Absolute pressures must be used in ideal gas or other equations of state.

3.3.4 Measurement of pressure

Absolute pressure is measured by means of a **barometer**. A simple barometer consists of a tube with its air evacuated placed on a reservoir of a fluid upside down (Figure 3.2). Since the air in the tube is evacuated, the only pressure above the meniscus of the fluid is that due to the vapour pressure of the fluid. Therefore,

$$p_{\text{atmosphere}} = \rho g h + p_{\text{vapour}} \tag{3.21}$$

Mercury, a fluid that has a very low vapour pressure, is often used as the manometric fluid.

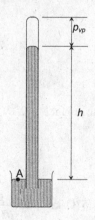

Figure 3.2 Mercury barometer.

3.4 MANOMETRY

Manometer is a device to measure differences in pressure. There are many types of manometers.

3.4.1 Simple (open) manometers

The simplest type is usually called a **piezometer** which measures the pressure in a liquid when it is **above atmospheric**. The pressure at a point A is given as the head of liquid h above the point A or as $h\rho g$ in absolute units (Figure 3.3a). This type is not suitable for measuring large pressures, as it would require a very long tube.

For small positive or negative pressures (above or below atmospheric), a manometer as shown in Figure 3.3b can be used. Depending on whether the pressure is above or below

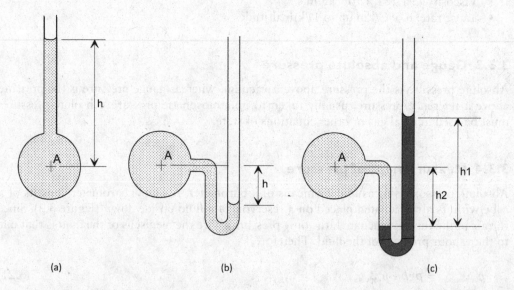

(a)   (b)   (c)

Figure 3.3 Simple manometers: (a) positive pressure at A; (b) negative pressure at A; (c) manometer with two fluids.

atmospheric, the meniscus will rest above or below the level of A. Then, the pressure is given by the column of liquid h, as $h\rho g$ (above or below atmospheric).

For larger positive or negative pressures, a second liquid of higher density can be used (Figure 3.3c). The two fluids used must be immiscible. The common fluids used are water and mercury, although gases can also be used. The pressure at A can then be expressed as

$$p_A + h_2\rho_1 g - h_1\rho_2 g = 0 \tag{3.22}$$

In working out pressures in manometric systems, the following procedure should be followed:

- Start at one end and write down the pressure there in any unit
- Add to this the change in pressure, in the same unit, from one meniscus to the next (+ if the next meniscus is lower and – if higher)
- Continue until the other end is reached and equate the expression to the pressure at that point.

Sometimes, inclined manometers are used to get a better resolution of manometric readings.

3.4.2 Differential manometers

This type is used to measure the difference in pressures at two points when the actual pressure at any point in the system cannot be determined. Referring to Figure 3.4a,

$$p_A - h_1\rho_1 g - h_2\rho_2 g + h_3\rho_3 g = p_B \Rightarrow p_A - p_B = h_1\rho_1 g + h_2\rho_2 g - h_3\rho_3 g \tag{3.23}$$

Similarly, for Figure 3.4b,

$$p_A + h_1\rho_1 g - h_2\rho_2 g - h_3\rho_3 g = p_B \Rightarrow p_A - p_B = -h_1\rho_1 g + h_2\rho_2 g + h_3\rho_3 g \tag{3.24}$$

The common fluids used in manometers are mercury, water and benzene. In the choice of a manometric fluid, the vapour pressure is important. Mercury has a very low vapour pressure compared to water or benzene. There is a significant difference between the heights of a barometer with mercury, water and benzene as shown in the table below.

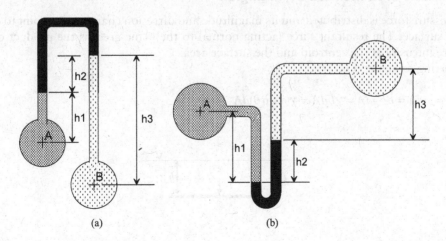

(a)   (b)

Figure 3.4 Differential manometers: (a) Eq. 2.23; (b) Eq. 3.24.

	Vapour Pressure Pa (abs)	γ_b (ρg) (N/m³)	$\gamma_b h_b$ ($\rho g h$) (kPa)	h_b (m)
Mercury	0.18	133,200	101.3	0.76
Water	2,340	9,810	99.0	10.09
Benzene	9,990	8,670	91.3	10.53

With the three types of manometric fluids, the atmospheric pressures would be

- $101.3 + 0.00018 = 101.3$ kPa with Hg

- $99.0 + 2.34 = 101.34$ kPa with water

- $91.3 + 9.99 = 101.29$ kPa with Benzene.

The rise in a barometer is not directly proportional to the specific weight of the liquid.

3.5 FORCES ACTING ON SUBMERGED PLANE AREAS

Since there are no shear forces in a static fluid, the hydrostatic force is normal to the surface (Pascal's law).

3.5.1 Flat plane area

A flat surface has the same pressure intensity at every point (Figure 3.5).
  The total (resultant) force is given by

$$F = \rho g h A = \gamma h A \qquad (3.25)$$

and is equal and opposite in direction because the pressure is the same at the top and bottom surfaces provided the thickness of the plate is small.

3.5.2 Plane area inclined at an angle θ

The pressure force is distributed and its magnitude and direction change from point to point on the surface. The resultant force (acting normal to the plane area) is the product of the pressure intensity at the centroid and the surface area A
  Referring to Figure 3.6,

$$dF = p\,dA = h\rho g\,dA = \gamma h\,dA = \gamma(y\sin\theta)\,dA$$

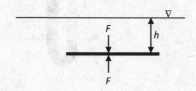

Figure 3.5 Force on a flat plane area.

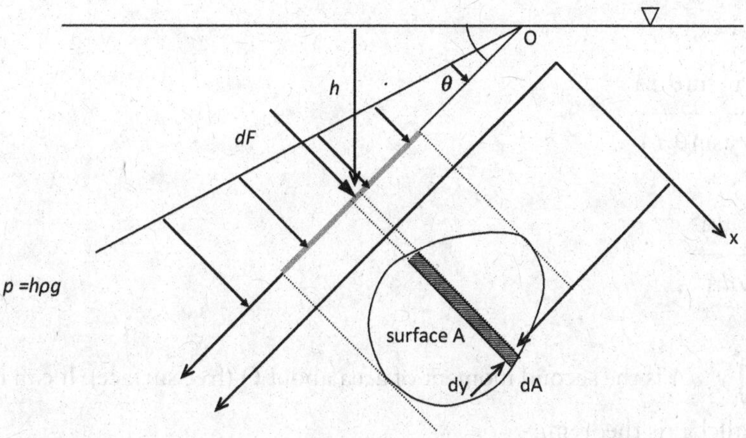

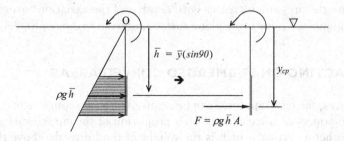

Figure 3.6 Force on an inclined plane area.

Therefore,

$$F = \int_A \gamma y \sin\theta \, dA = \gamma \sin\theta \int_A y \, dA \qquad (3.26)$$

But $\int_A y \, dA$ = First moment of area about $O(x\text{-axis}) = \bar{y}A$ $\qquad$ (3.27)

where $\bar{y}$ – distance from O (free surface) to the centroid of surface. Therefore,

$$F = \gamma \sin\theta \bar{y}A = \gamma \bar{h}A = \bar{p}A \qquad (3.28)$$

where $\bar{h} = \bar{y}\sin\theta$ and $\bar{p}$ is the pressure at the centroid.

Line of action of the resultant force: The force F is **not** applied at the centroid but at a point below the centre of gravity (CG). By taking moments, the distance y_{cp} is given (centre of pressure) as (Figure 3.7):

$$Fy_{cp} = \int \gamma y^2 \sin\theta dA$$
$$\left(\text{or } Fy_{cp} = \int_A y \, dF \right) \qquad (3.29)$$

Figure 3.7 Line of action of pressure force.

Therefore,

$$y_{cp} = \frac{\int \gamma y^2 \sin\theta \, dA}{\int \gamma y \sin\theta \, dA}$$

$$= \frac{\int y^2 \, dA}{\int y \, dA} \tag{3.30}$$

In Eq. 3.30, $\int y^2 \, dA$ is the second moment of area about O (free surface). It can be expressed as (using parallel axis theorem):

$$\int y^2 \, dA = I_{CG} + \bar{y}^2 A \tag{3.31}$$

where I_{CG} is the second moment of area about the CG. Therefore,

$$y_{cp} = \frac{I_{CG} + \bar{y}^2 A}{\int y \, dA}$$

$$= \frac{I_{CG}}{\int y \, dA} + \frac{\bar{y}^2 A}{\int y \, dA}$$

$$= \frac{I_{CG}}{\bar{y} A} + \frac{\bar{y}^2 A}{\bar{y} A}$$

$$= \frac{I_{CG}}{\bar{y} A} + \bar{y} \tag{3.32}$$

Similarly, the x-co-ordinate of the centre of pressure can be found by taking moments about the y-axis:

$$x_{cp} = \frac{1}{\bar{y} A} \int_A xy \, dA = \frac{I_{xy}}{\bar{y} A} = \bar{x} + \frac{I_{xy,c}}{\bar{y} A} \tag{3.33}$$

As the depth increases, the distance between the centre of pressure and the centroid decreases because the pressure increases with depth and the variations over the submerged area become negligible. I_{CG} for some plane surfaces is given in Appendix 3.1A.

3.6 FORCES ACTING ON SUBMERGED CURVED AREAS

On curved surfaces, the resultant force can be resolved into two components, one horizontal which is the component of force exerted on a projection of the curved surface in a vertical plane, and the other a vertical which is the weight of fluid directly above the surface and applied at the centroid of the fluid.

The main difference between plane and curved surfaces is that in a curved surface the directions of the normal forces differ from point to point because of the curvature.

Considering a curved surface and an element of area dA on it, the force in the x-direction is $F_x = F_H$ (both in magnitude and line of action). In the z-direction, it is $F_z = F_1 + W$ where W is the weight of liquid inside OPQR (refer to Figure 3.8 (right)) and $F_1 = \rho g h_1 \times$ (area of OPR) $= \rho g \times$ (volume above OPR), making F_z, the weight of liquid above the curved surface PQR and below the free surface (Figure 3.8).

In general,

$$dF = pdA \Rightarrow F = \int p\,dA \tag{3.34}$$

Because the curved surface is three-dimensional, there will be three components as given below:

$$F_x = \int p\,dA_x \tag{3.35a}$$

$$F_y = \int p\,dA_y \tag{3.35b}$$

$$F_z = \int p\,dA_z \tag{3.35c}$$

where dA_x is the projection of dA on yz plane, dA_y is the projection of dA on zx plane, dA_z is the projection of dA on xy plane, and

$$dA_x = dA\cos\theta_x \tag{3.36a}$$

$$dA_y = dA\cos\theta_y \tag{3.36b}$$

$$dA_z = dA\cos\theta_z \tag{3.36c}$$

where θ_x, θ_y, and θ_z are the angles between dA and x-, y- and z-axes, respectively.

These equations indicate that the three components must be evaluated separately by integration. For the vertical component,

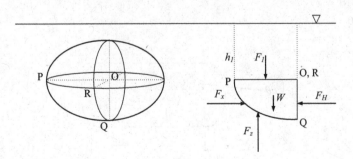

Figure 3.8 Forces on curved surfaces.

$$dF_z = p\, dA_z; \quad p = \int_{z_1}^{z_2} \rho g\, dz \tag{3.37}$$

Then,

$$dF_z = \int_{z_s}^{z_0} \rho g\, dz\, dA_z \tag{3.38a}$$

and

$$F_z = \int_{A_z} \int_{z_s}^{z_0} \rho g\, dz\, dA_z \tag{3.38b}$$

where z_s is the z-co-ordinate of the surface (of the body) and z_0 is the z-co-ordinate of free surface. The magnitude of this component represents the weight of the liquid directly above the surface of the object. In summary,

- The horizontal component in a given direction on a curved surface is equal to the force on the projection of the surface on a vertical plane perpendicular to the given direction. The line of action is the same as that of the force on the vertical projection.
- The vertical component is equal to the weight of the fluid extending above the surface of the object to the free surface. It acts through the centre of gravity (CG) of that volume of fluid.

    (**Note:** It may not be possible to define a resultant force where the three force components meet at a point in space.)

For cylindrical surfaces (i.e. surfaces with constant R), with width w (Figure 3.9)

$$dA = wR\, d\theta \tag{3.39}$$

$$F_z = \int p\, dA \cos\theta = \int_{\theta_1}^{\theta_2} p \cos\theta w R\, d\theta \tag{3.40}$$

The lines of action of the resultant force can be obtained by taking the moments of the resultant force and the corresponding distributed components as follows:

$$r_x F_x = \int r\, dF_x \tag{3.41a}$$

$$r_y F_y = \int r\, dF_y \tag{3.41b}$$

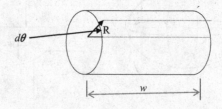

Figure 3.9 Force acting on a cylindrical surface.

$$r_z F_z = \int r \, dF_z \tag{3.41c}$$

3.6.1 Two-dimensional case

$$F_y = \int dF_y = \int p \, dA_y = p \, dA \cos \theta_y \tag{3.42a}$$

$$F_z = \int dF_z = \int p \, dA_z = p \, dA \sin \theta_z \tag{3.42b}$$

Lines of action of forces:

$$z' F_y = \int z \, dF_y = \int z p \, dA_y \tag{3.43a}$$

$$y' F_z = \int y \, dF_z = \int y p \, dA_z \tag{3.43b}$$

Hence (y, z') can be obtained. The lines of action of the components may not coincide; they may give rise to a force and a couple.

3.7 BUOYANCY AND STABILITY

Buoyancy is the force (only vertical) acting on a body floating on a liquid surface due to liquid pressure. There is no horizontal thrust since pressures are equal at the same horizontal level. It is due to the pressure difference between the upper and lower surfaces of the body. It acts upwards in the vertical direction because the pressure at the lower surface is always higher than that at the upper surface. For a static fluid,

$$p = h\rho g \Rightarrow \frac{dp}{dh} = \rho g \Rightarrow p = p_0 + \rho g h \tag{3.44}$$

The net vertical force on the element is

$$dF = (p_0 + \rho g h_2) dA - (p_0 + \rho g h_1) dA = \rho g (h_2 - h_1) dA = \rho g \, dV \tag{3.45}$$

where dV in the volume of the element. Therefore,

$$F = \int dF = \int_V \rho g \, dV = \rho g V \tag{3.46}$$

Where V is the volume of the object. Thus the buoyancy force is equal to the weight of the liquid displaced by the object and acts at the centroid of the displaced fluid. This is 'Archimedes' principle' (220 BC). The location of the line of action of the buoyancy force determines its stability. The Centre of Buoyancy is the centroid of the displaced fluid.

3.7.1 Metacentre

When the floating body is in equilibrium condition (Figures 3.10 and 3.11), the CG and the centre of buoyancy (B) are on the same vertical line. The centre of buoyancy changes when

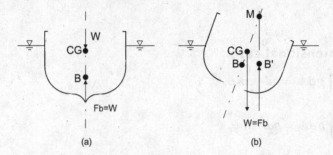

Figure 3.10 Center of buoyancy and metacentre. (a) Equilibrium condition and (b) disturbed condition.

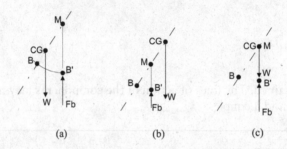

Figure 3.11 Conditions of equilibrium. (a) Stable, (b) unstable, and (c) neutral.

the body is displaced. In the new position, it is at B' (Figure 3.10 and 3.11). Where a vertical line through B' intersects the line through B and CG is called the Metacentre, M. When M is above the CG, the body is stable. When M is below CG, the body is unstable. When they coincide, the body is in neutral equilibrium.

The distance between CG and M is called the **Metacentric Height, GM**. It should be positive for stability and negative for instability. The relative position of M and CG determines the stability of the body. For stability, the CG should be as low as possible. The greater the distance between CG and M, the greater will be the stability.

3.7.1.1 Calculation of metacentric height GM

Referring to Figure 3.12, considering an elemental area dA at a distance x from the centreline of the floating object in the equilibrium position, the elemental volume displaced is given by

$$d\forall = zdA$$

This when integrated should be equal to

$$\forall \bar{x}_0 = \int (zdA)x \qquad (3.47)$$

where $\bar{x}_0$ is the x coordinate of the centre of buoyancy and $\forall$ is the immersed volume. If the submerged part is symmetrical about the yz plane, $\bar{x}_0 = 0$. After a small tilt θ, the new $\bar{x}$ is given by

$$\forall \bar{x} = \int \left[(z + x\tan\theta)dA \right]x \qquad (3.48)$$

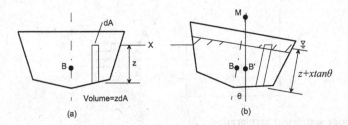

Figure 3.12 Calculation of metacentric height: (a) Equilibrium position; (b) Displaced position.

Therefore,

$$\Psi(\bar{x} - \bar{x}_0) = \int x^2 \tan\theta \, dA$$

$$= \tan\theta \int x^2 \, dA$$

$$= I \tan\theta \qquad (3.49)$$

where $\int x^2 dA$ is the second moment of area of plane of floatation about the centroidal axis normal to the plane of rotation, and I is the moment of inertia of the floating body in the plane of the displaced surface (longitudinal axis).

If $\bar{x}_0 = 0$, then

$$\Psi \bar{x} = I \tan\theta \Rightarrow \frac{\Psi \bar{x}}{\tan\theta} = I$$

But $\dfrac{\bar{x}}{\tan\theta} = \mathrm{BM}$ and $\mathrm{GM} = \mathrm{BM} - \mathrm{BG}$ and therefore $\mathrm{GM} = \dfrac{I}{\Psi} - \mathrm{BG}$, which for stability should be positive and negative for instability. If it is zero, then the body is in neutral equilibrium. The distance BM is the metacentric radius.

3.8 FLUIDS IN RIGID BODY MOTION

Rigid body motion does not induce shear stresses. It retains its shape because the fluid does not deform. For rigid body motion, Newton's second law of motion gives the resulting pressure field. Examples include a liquid inside a container undergoing linear acceleration and a liquid in a rotating container.

3.8.1 Linear acceleration of a fluid in a container

Consider a liquid inside a container that moves with a constant acceleration a_x in the x direction and a_z in the z-direction. All liquid particles are undergoing the same acceleration. Considering a fluid element and applying Newton's second law ($F = ma$), the resulting equation takes the form (Figure 3.13)

$$p(\delta y \delta z) - \left(p + \frac{\partial p}{\partial x}\delta x\right)(\delta y \delta z) = \rho(\delta x \delta y \delta z)a_x \Rightarrow \frac{\partial p}{\partial x} = -\rho a_x \qquad (3.50)$$

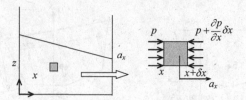

Figure 3.13 Rigid body with linear acceleration.

Also $\dfrac{\partial p}{\partial z} = -\rho(g + a_z)$, and $\dfrac{\partial p}{\partial y} = 0$

The total differential of $p = p(x, y, z)$ is

$$dp = \frac{\partial p}{\partial x}dx + \frac{\partial p}{\partial y}dy + \frac{\partial p}{\partial z}dz = -\rho a_x dx - \rho(g + a_z)dz \tag{3.51}$$

Along the lines of constant pressure, the differential is zero: $dp = 0$ and thus

$$\frac{dz}{dx} = -\frac{a_x}{a_z + g} \tag{3.52}$$

The free surface is a line of constant pressure ($p = p_{atm}$), and it is sloping backwards at an angle $\theta = \tan^{-1}\left(\dfrac{a_x}{a_z + g}\right)$. The unequal pressure in the x-direction provides the force required to accelerate the liquid (Figure 3.14).

3.8.2 Rigid-body rotation

Considering a liquid inside a container of the shape of a circular cylinder rotating at an angular velocity ω, the force balance equation can be written as (here it is more convenient to use the cylindrical co-ordinate system (z, r, θ)).

$$prd\theta dz - \left(p + \frac{\partial p}{\partial r}dr\right)(r + dr)d\theta dz = \rho(rd\theta)drdz(-\omega^2 r) \tag{3.53}$$

$$\Rightarrow \frac{\partial p}{\partial r} = \rho\omega^2 r$$

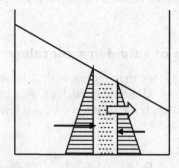

Figure 3.14 Force imbalance due to unequal pressure.

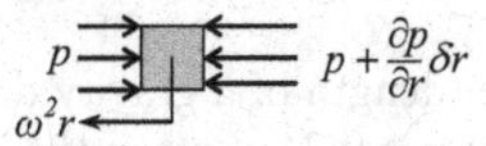

$$p \rightrightarrows \quad \leftleftarrows p + \frac{\partial p}{\partial r}\delta r$$
$$\omega^2 r \leftarrow$$

Figure 3.15 Forces acting on a fluid element.

All liquid particles are under a centripetal acceleration $\omega^2 r$. There will be a pressure gradient along the r direction (Figure 3.15). The pressure gradients in the other directions are

$$\frac{\partial p}{\partial z} = -\rho g; \quad \frac{\partial p}{\partial \theta} = 0 \tag{3.54}$$

The total pressure differential is given by

$$dp = \frac{\partial p}{\partial r}dr + \frac{\partial p}{\partial z}dz + \frac{\partial p}{\partial \theta}d\theta \Rightarrow dp = \rho\omega^2 r dr - \rho g dz \tag{3.55}$$

Lines of constant pressure lie on

$$dp = \rho\omega^2 r dr - \rho g dz = 0 \tag{3.56a}$$

or

$$\frac{dz}{dr} = \frac{\omega^2 r}{g} \tag{3.56b}$$

Integrating and applying the boundary condition $z(r = 0) = z_0$

$$z = z(r) = \frac{\omega^2 r^2}{2g} + z_0 \tag{3.57}$$

The elevation of the free surface varies with r in the shape of a paraboloid (Figure 3.16). Rigid body rotation of a fluid by an external force is equivalent to a forced vortex.

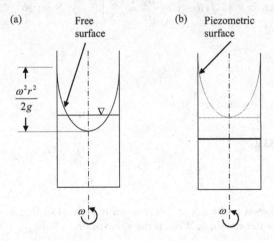

Figure 3.16 Surface profiles for rigid body motion ((a) open water surface; (b) closed water surface).

Example 3.1

Calculate the force acting per unit length of a gravity dam with its upstream face in the shape of a parabola described by $y = 0.25x^2$ when the depth of water on the upstream face is 10 m.

$$F_x = \int p\, dA_x = \int p\, dy \quad \text{(per unit length of dam)}$$

$$F_y = \int p\, dA_y = \int p\, dx \quad \text{(per unit length of dam)}$$

$p = (h - y)\rho g$ (h is the depth of water; y is measured from the bottom of the dam vertically upwards)

$$y = 0.25x^2 \Rightarrow x = 2\sqrt{y} \Rightarrow dx = \frac{dy}{\sqrt{y}}$$

$$F_x = \int_0^h \rho g(h - y)\, dy = \frac{1}{2}\rho g h^2 = 0.5 \times 1{,}000 \times 9.81 \times 100 = 490.5 \text{ kN}$$

$$F_y = \int_0^h \rho g(h - y)\, dx = \int_0^h \frac{\rho g(h - y)}{\sqrt{y}}\, dy = \frac{4}{3}\rho g h^{3/2} = \frac{4}{3} \times 1{,}000 \times 9.81 \times 10^{3/2} = 413.6 \text{ kN}$$

Direction

$$\tan\alpha = \frac{F_y}{F_x} = \frac{4}{3}\frac{1}{\sqrt{0.25}}h^{-1/2} = 0.843 \Rightarrow \alpha = 40.14°.$$

Position of the line of action

$$x'F_y = \int x\, dF_y = \int xp\, dx = \int_0^h 2\sqrt{y}(h - y)\rho g \frac{dy}{\sqrt{y}} = \frac{\rho g}{0.5}\int_0^h (h - y)\, dy = \rho g h^2$$

$$x' = \frac{\rho g h^2}{413.6} = 2.37 \text{ m}$$

$$y'F_x = \int y\, dF_x = \int y(h - y)\rho g\, dy = \frac{1}{6}\rho g h^3$$

$$y' = \frac{\rho g h^3}{6 \times 490.5} = 3.33 \text{ m}$$

Example 3.2

The square gate shown in the figure is eccentrically pivoted so that it will automatically open at a certain water depth, h. What is the value of h?

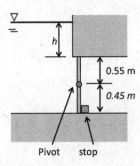

Gate will open when centre of pressure is at the pivot

$$y' = y_c + \frac{I_{cc}}{Ay_c}$$

$$h + 0.55\,\text{m} = h + 0.5\,\text{m} + \frac{\frac{1}{12} \times b \times (1\,\text{m})^3}{b \times (1\,\text{m}) \times (h + 0.5\,\text{m})}$$

$$0.05\,\text{m} \times (1\,\text{m}) \times (h + 0.5\,\text{m}) = \tfrac{1}{12} \times (1\,\text{m})^3$$

$$\underline{h = 1.167\,\text{m}}$$

3.1A APPENDIX

I_{CG} for some plane surfaces.

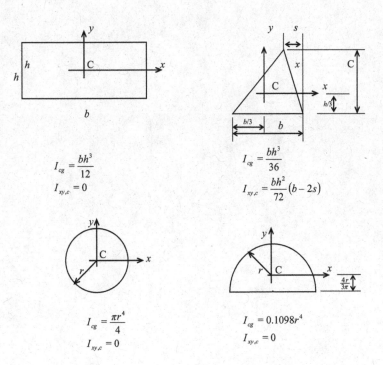

$$I_{cg} = \frac{bh^3}{12}$$
$$I_{xy,c} = 0$$

$$I_{cg} = \frac{bh^3}{36}$$
$$I_{xy,c} = \frac{bh^2}{72}(b - 2s)$$

$$I_{cg} = \frac{\pi r^4}{4}$$
$$I_{xy,c} = 0$$

$$I_{cg} = 0.1098r^4$$
$$I_{xy,c} = 0$$

Chapter 4

Fluid kinematics

4.1 INTRODUCTION

Fluid Kinematics refers to the description of fluid flow without considering forces. In fluid flow there are five basic variables, namely the three velocity components and two thermo-dynamic properties. The two thermodynamic properties considered are usually the pressure and density. With these five properties, a fluid flow field is completely defined. These five variables are usually functions of space and time:

$$u = u(x,y,z,t) \tag{4.1a}$$

$$v = v(x,y,z,t) \tag{4.1b}$$

$$w = w(x,y,z,t) \tag{4.1c}$$

$$p = p(x,y,z,t) \tag{4.1d}$$

$$\rho = \rho(x,y,z,t) \tag{4.1e}$$

4.2 DESCRIPTION OF FLOW

4.2.1 Lagrangian approach

The motion of a fluid particle is followed (Figure 4.1) during its motion through space and time. In this approach, the same mass of fluid is followed all the time, and therefore the basic laws of physics can be applied directly. However, because of the problem of identifying the mass at different times, this approach is not very popular.

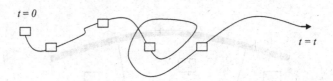

Figure 4.1 Lagrangian description of flow.

In the Lagrangian approach, the position of a particle is specified with reference to an initial position x_0, y_0, z_0, at an initial time t_0. At a later time t,

$$x = F_1(x_0, y_0, z_0, t_0) \tag{4.2a}$$

$$y = F_2(x_0, y_0, z_0, t_0) \tag{4.2b}$$

$$z = F_3(x_0, y_0, z_0, t_0) \tag{4.2c}$$

The velocities and accelerations are given as

$$u = \left(\frac{dx}{dt}\right)_\xi = \left(\frac{\partial x}{\partial t}\right)_\xi ; \quad a_x = \left(\frac{\partial^2 x}{\partial t^2}\right)_\xi \tag{4.3a}$$

$$v = \left(\frac{dy}{dt}\right)_\xi = \left(\frac{\partial y}{\partial t}\right)_\xi ; \quad a_y = \left(\frac{\partial^2 y}{\partial t^2}\right)_\xi \tag{4.3b}$$

$$w = \left(\frac{dz}{dt}\right)_\xi = \left(\frac{\partial z}{\partial t}\right)_\xi ; \quad a_z = \left(\frac{\partial^2 z}{\partial t^2}\right)_\xi \tag{4.3c}$$

where $\xi = x_0 i + y_0 j + z_0 k$ is the initial position.

In this approach, the history of the fluid particle is known and since the same particle is considered, the mass conservation is satisfied, which is an advantage. On the other hand, the equations of motion are non-linear, and also the use of steady state flow does not simplify the equations appreciably, which is a disadvantage. Most practical applications of the Lagrangian approach are limited to one-dimensional flows only.

4.2.2 Eulerian approach

In this approach, the fluid passing through a given fixed position in space is chosen. The fixed space is called a **control volume** (Figure 4.2), and the fluid inside the control volume is continually changing with time. The laws of physics defined for a fixed mass can be modified to suit this approach. This approach is universally used for fluid flow description. The governing equations may be written either in the differential form or in the integral form. In the integral form, the gross effects are considered while in the differential form, point-to-point details are considered.

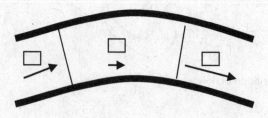

Figure 4.2 Control volume.

4.3 TYPES OF FLOW

i. **Uniform**: Flow is the same at all corresponding points in space

$$\frac{\partial u}{\partial x} = \frac{\partial v}{\partial y} = \frac{\partial w}{\partial z} = \frac{\partial p}{\partial x} = \frac{\partial \rho}{\partial x} = 0 \qquad (4.4a)$$

However, in fluid mechanics, uniform flow also implies

$$\frac{\partial u}{\partial x} = \frac{\partial v}{\partial y} = \frac{\partial w}{\partial z} = 0 = \frac{\partial u}{\partial y} = \frac{\partial u}{\partial z} \qquad (4.4b)$$

Also, it is sometimes possible to have uniform flow across a cross section, but non-uniform flow longitudinally.

ii. **Non-uniform**: Flow varies in space

iii. **Steady**: Fluid variables are time invariant

$$\frac{\partial}{\partial t}\left(\text{all flow variables}\right) = 0$$

iv. **Unsteady**: Fluid variables are time variant

All four combinations of flow are possible.

Uniform steady – e.g., flow in a long pipe at a constant rate.
Uniform unsteady – e.g., flow in a pipe at a variable rate.
Non-uniform steady – e.g., flow in a contracting/expanding pipe at a constant rate.
Non-uniform unsteady – e.g., flow in a contracting/expanding pipe at a variable rate.

Flows can also be classified as laminar or turbulent, viscous or inviscid (ideal), rotational (vortex flow) or irrotational.

4.4 ACCELERATION IN FLUID FLOW

Since u, v and w are functions of space and time co-ordinates as shown in Eq. 4.1,

$$\delta u = \frac{\partial u}{\partial x}\delta x + \frac{\partial u}{\partial y}\delta y + \frac{\partial u}{\partial z}\delta z + \frac{\partial u}{\partial t}\delta t \qquad (4.5)$$

$$\frac{Du}{Dt} = \frac{\delta u}{\delta t} = u\frac{\partial u}{\partial x} + v\frac{\partial u}{\partial y} + w\frac{\partial u}{\partial z} + \frac{\partial u}{\partial t} \qquad (4.6a)$$

Similarly, other components of acceleration are

$$\frac{Dv}{Dt} = \frac{\delta v}{\delta t} = u\frac{\partial v}{\partial x} + v\frac{\partial v}{\partial y} + w\frac{\partial v}{\partial z} + \frac{\partial v}{\partial t} \qquad (4.6b)$$

$$\frac{Dw}{Dt} = \frac{\delta w}{\delta t} = u\frac{\partial w}{\partial x} + v\frac{\partial w}{\partial y} + w\frac{\partial w}{\partial z} + \frac{\partial w}{\partial t} \qquad (4.6c)$$

The total acceleration is, therefore, the sum of convective acceleration and local acceleration. The local acceleration refers to the net rate of change of velocity within the control

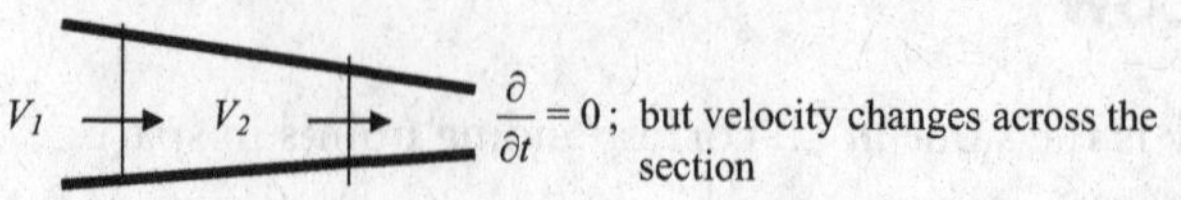

$$\frac{\partial}{\partial t} = 0 \text{ ; but velocity changes across the section}$$

Figure 4.3 Steady flow with convective acceleration.

volume, whereas the convective acceleration refers to the net rate of efflux across the control surfaces. Zero local acceleration implies a steady flow (Figure 4.3). Zero convective acceleration implies uniform flow.

In cylindrical co-ordinates (r, θ, z) with velocity components v_r, v_θ and v_z, the acceleration components are (a_r, a_θ, a_z) as shown below. Velocity components have dimension LT^{-1}, whereas acceleration components have dimension LT^{-2}.

$$a_r = v_r \frac{\partial v_r}{\partial r} + \frac{v_\theta}{r} \frac{\partial v_r}{\partial \theta} + v_z \frac{\partial v_r}{\partial z} - \frac{v_\theta^2}{r} + \frac{\partial v_r}{\partial t} \tag{4.7a}$$

$$a_\theta = v_r \frac{\partial v_\theta}{\partial r} + \frac{v_\theta}{r} \frac{\partial v_\theta}{\partial \theta} + v_z \frac{\partial v_\theta}{\partial z} + \frac{v_r v_\theta}{r} + \frac{\partial v_\theta}{\partial t} \tag{4.7b}$$

$$a_z = v_r \frac{\partial v_z}{\partial r} + \frac{v_\theta}{r} \frac{\partial v_z}{\partial \theta} + v_z \frac{\partial v_z}{\partial z} + \frac{\partial v_z}{\partial t} \tag{4.7c}$$

4.5 STREAMLINES, PATHLINES AND STREAKLINES (GEOMETRICAL CONCEPTS)

4.5.1 Streamlines

Streamline is a line (or curve), to which, at any instant, the velocity vector is tangential (Figure 4.4a). There is no flow across a streamline, and that streamlines cannot intersect themselves or with other streamlines. In steady flows, the same streamline pattern holds at all times. Therefore the path of a particle is a streamline. In unsteady flow, the streamline patterns would change with time.

Differential equation of a streamline can be obtained by equating the cross product of **V** and **ds**. Since the particles move in the direction of the streamline at any point in time, the displacement **ds** has the same direction as the velocity vector **V**.

$$V \times ds = 0 \text{ (Cross product of two parallel vectors is zero).} \tag{4.8a}$$

Since, $V = ui + vj + wk$, and

$$ds = dxi + dyj + dzk,$$

$$V \times ds = \begin{vmatrix} i & j & k \\ u & v & w \\ \delta x & \delta y & \delta z \end{vmatrix} = 0 \tag{4.8b}$$

or

$$(v\delta z - w\delta y)i - (u\delta z - w\delta x)j + (u\delta y - v\delta x)k = 0$$

$$\Rightarrow v\delta z - w\delta y = 0; \; u\delta z - w\delta x = 0; \text{ and } u\delta y - v\delta x = 0 \tag{4.8c}$$

or

$$\frac{\delta x}{u} = \frac{\delta y}{v} = \frac{\delta z}{w} \tag{4.8d}$$

This equation can also be obtained as follows:

$$u = V\cos\theta; \quad v = V\sin\theta$$

$$\frac{v}{u} = \tan\theta = \frac{dy}{dx} \qquad \text{(velocity and displacement vectors are similar)}$$

Equation 4.8d is a general equation for 3D fluid flow steady or unsteady, uniform or non-uniform, viscous or ideal, compressible or incompressible.

A **stream surface** (Figure 4.4c) is generated by a large number of closely spaced streamlines that pass through an arbitrary curve AB as shown in the figure. A closed arbitrary curve forms a **stream tube** (Figure 4.4d).

4.5.2 Pathlines

Pathline (Figure 4.4e) is the trajectory of a given fluid particle as it moves from an initial state to a final state. They are history lines and may intersect themselves (Refer to Lagrangian description of flow). In steady flow, streamlines and pathlines coincide because the velocity vector does not change. The equation of a pathline may be obtained as

$$x = x(t); \quad y = y(t); \quad z = z(t) \tag{4.9}$$

from a knowledge of u, v and w using

$$u = \frac{dx}{dt} \rightarrow x = \int u\,dt \tag{4.10a}$$

$$v = \frac{dy}{dt} \rightarrow y = \int v\,dt \tag{4.10b}$$

$$w = \frac{dz}{dt} \rightarrow z = \int w\,dt \tag{4.10c}$$

with the initial conditions given at x_0, y_0, z_0, t_0 (e.g. $x_0 = x(t_0)$ etc.).

4.5.3 Streaklines

Streakline (Figures 4.4f and 4.5) is the locus of locations of all the fluid particles that have passed through a fixed point in a flow field, at any instant of time. For steady flow, pathlines, streamlines and streaklines coincide. If a dye is injected into a liquid at a fixed point in the flow field, then at a later time t, the dye will indicate the end points of pathlines of particles that have passed through the point of injection. A photograph of a dye introduced into a flow field will be a streakline.

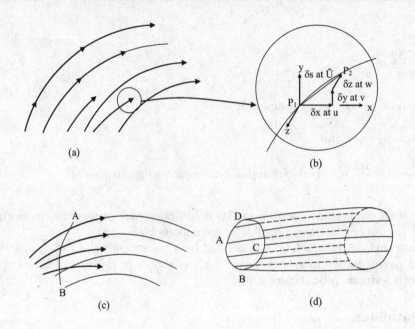

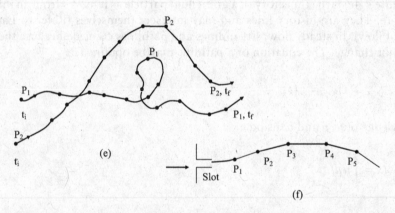

Figure 4.4 Streamlines, pathlines and streaklines. (a,b) Streamlines (at $t = t_1$), (c) steamsurface (at $t = t_1$), (d) steamtube (at $t = t_1$), (e) pathlines of P_1 & P_2 (from $t = t_i$ to $t = t_f$), and (f) steakline (at $t = t_1$).

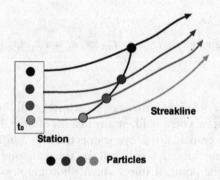

Figure 4.5 Streakline.

4.6 CONTINUITY EQUATION: EXISTENCE OF FLOW

A fluid consisting of matter must obey the law of conservation of matter (mass). This determines whether a given velocity field constitutes a possible flow. Considering a control volume, the continuity equation can be obtained by applying the law of conservatives of mass (see also Chapter 5):

$$\frac{\partial}{\partial x}(\rho u) + \frac{\partial}{\partial y}(\rho v) + \frac{\partial}{\partial z}(\rho w) + \frac{\partial \rho}{\partial t} = 0 \tag{4.11}$$

This is valid for 3D, compressible or incompressible, viscous or inviscid, steady or unsteady, uniform or non-uniform flows.

In cylindrical polar co-ordinates, the continuity equation takes the form: (v_r, v_θ, v_z)

$$\frac{\partial}{\partial r}(\rho v_r) + \frac{1}{r}\frac{\partial(\rho v_\theta)}{\partial \theta} + \frac{\partial(\rho v_z)}{\partial z} + \rho\frac{V_r}{r} + \frac{\partial \rho}{\partial t} = 0 \tag{4.12}$$

which for 2D flow in the $r - \theta$ plane becomes

$$\frac{\partial}{\partial r}(\rho v_r) + \frac{1}{r}\frac{\partial}{\partial \theta}(\rho v_\theta) + \frac{\rho v_r}{r} + \frac{\partial \rho}{\partial t} = 0 \tag{4.13}$$

For incompressible 2D flow, (r, θ)

$$\frac{\partial v_r}{\partial r} + \frac{1}{r}\frac{\partial v_\theta}{\partial \theta} + \frac{v_r}{r} = 0 \tag{4.14}$$

For axisymmetric incompressible flow, (r, z)

$$\frac{\partial v_r}{\partial r} + \frac{v_r}{r} + \frac{\partial v_z}{\partial z} = 0 \tag{4.15}$$

The continuity equation can be written in many other forms depending upon its use.

4.7 STREAM FUNCTION, ψ

The concept of stream function was suggested by the French mathematician Joseph Louis Lagrange in 1781. It is obviously related to streamlines. Streamlines have constant values of stream function ψ. The difference between the magnitudes of stream functions between two adjacent streamlines gives the flux rate per unit length normal to the plane of the stream-lines. For instance, referring to Figure 4.6, the flow across $A_1A_2 \equiv$ flow across $B_1B_2 \equiv$ flow across A_1B_2, or, flow across $12 = u\delta y - v\delta x =$ flow across 23 + flow across 31.

Stream function is defined as

$$d\psi = V ds \tag{4.16a}$$

In general, since $\psi = \psi(x, y)$ (stream functions are defined for 2D flows only)

$$d\psi = \frac{\partial \psi}{\partial x}dx + \frac{\partial \psi}{\partial y}dy \tag{4.16b}$$

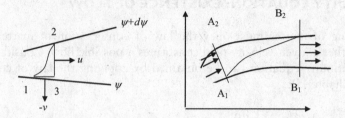

Figure 4.6 Stream function.

where $d\psi$ is the flux across two streamlines separated by stream function values ψ and $\psi + d\psi$, and ds is the length of a line normal to the streamlines; V is the velocity across the line ds. The sign convention can be arbitrary. For example, if V is from left to right (clockwise), it can be taken as positive and vice versa. Referring to Figure 4.6 for 2D incompressible flow,

$$dQ = udy - vdx = d\psi \tag{4.16c}$$

$$Q = \int dQ = \int d\psi = \psi_2 - \psi_1 \tag{4.16d}$$

In polar co-ordinates, $\psi = \psi(r, \theta)$, and therefore

$$d\psi = \frac{\partial \psi}{\partial r} dr + \frac{\partial \psi}{\partial \theta} d\theta \tag{4.17}$$

If the streamline is horizontal, $V = u$, and $ds = dy$; therefore, $d\psi = udy$, giving

$$u = \frac{d\psi}{dy} \tag{4.18a}$$

If the streamline is vertical, $V = v$, and $ds = dx$; therefore, $d\psi = -vdx$, giving

$$v = -\frac{d\psi}{dx} \tag{4.18b}$$

where u and v are the velocity components in the x and y directions, respectively.

In polar co-ordinates (r, θ), the directions are radial and tangential. If the flow is radial, $\theta = 0$, and $V = v_r$, and $ds = rd\theta$, giving

$$v_r = \frac{1}{r} \frac{d\psi}{d\theta} \tag{4.19a}$$

If the flow is tangential, $\theta = 90°$, and $V = v_\theta$; $ds = dr$, giving

$$v_\theta = -\frac{d\psi}{dr} \tag{4.19b}$$

where v_r and v_θ are the velocity components in the radial and tangential directions. The sign convention, as before, can be positive when the flow is from left to right.

Stream function is defined at a point with respect to a reference point. It represents the volume rate of flow across any line (or curve), joining the point to the reference point. The stream function at the reference point is arbitrarily assumed to be zero. The stream function has a constant value along a streamline. It is assumed to be positive when the flow is from left to right about the reference point.

Since $\psi = \psi(x, y)$, as shown in Eq. 4.16b, $d\psi = 0$ if ψ is a constant. Therefore,

$$\frac{\partial \psi}{\partial x} dx + \frac{\partial \psi}{\partial y} dy = 0 \Rightarrow -v dx + u dy = 0 \Rightarrow \frac{dy}{dx} = \frac{v}{u} \tag{4.20}$$

which is the equation of a streamline.

If $d\psi = 0$, $\psi = $ constant implying streamlines.

For 2D incompressible flow, the continuity equation is

$$\frac{\partial u}{\partial x} + \frac{\partial v}{\partial y} = 0 \tag{4.21}$$

Substituting for u, v in terms of ψ, the continuity equation is satisfied. Therefore, the existence of a stream function also implies the existence of flow.

Note: For compressible 2D flow also, stream functions can be defined by relating them to the mass rate of flow and dividing by the relevant density. However, for 3D flow stream function is **not** defined because the stream surfaces are arbitrary.

4.8 DEFORMATION OF A FLUID ELEMENT

In general, an element of fluid in motion can have four types of deformations due to spatial variations of velocity:

   i. Translation
  ii. Linear deformation
 iii. Angular deformation
  iv. Rotation

These may occur in isolation or in any combination.

Referring to Figure 4.7,

  i. Translation in the x, y and z directions are udt, vdt and wdt.
 ii. Linear deformations can be considered as the change in length in each direction (change in velocity in each direction × time)

$$x\text{-direction} : \frac{\partial u}{\partial x} \delta x dt$$

$$y\text{-direction} : \frac{\partial v}{\partial y} \delta y dt$$

$$z\text{-direction} : \frac{\partial w}{\partial z} \delta z dt$$

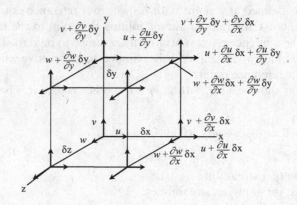

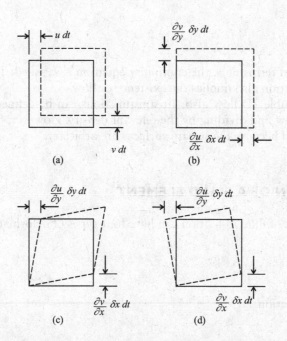

Figure 4.7 Deformation of a fluid element: (a) Translation; (b) Linear deformation; (c) Angular deformation; (d) Rotation.

$$\text{Rate of change of volume} = \frac{\text{Change in volume}}{(\text{Original volume}) (\text{time})}$$

$$= \frac{\left(\delta x + \frac{\partial u}{\partial x} \delta x dt\right)\left(\delta y + \frac{\partial v}{\partial y} \delta y dt\right)\left(\delta z + \frac{\partial w}{\partial z} \delta z dt\right) - \delta x \delta y \delta z}{\delta x \delta y \delta z dt}$$

$$= \left(\frac{\partial u}{\partial x} + \frac{\partial v}{\partial y} + \frac{\partial w}{\partial z}\right)$$

$$= div\, V \tag{4.22}$$

For an incompressible fluid div $\mathbf{V} = 0$ implies no rate of change of volume. It is also the continuity equation.

iii. Angular deformation can be considered positive anti-clockwise and negative clockwise. The deformation is the difference between the angular deformation of the individual sides δx and δy, δy and δz and δz and δx.

$$\delta x \text{ and } \delta y := \left(\frac{\frac{\partial v}{\partial x}\delta x dt}{\delta x}\right) - \left(\frac{-\frac{\partial u}{\partial y}\delta y dt}{\delta y}\right)$$

$$= \left(\frac{\partial v}{\partial x} + \frac{\partial u}{\partial y}\right) dt \tag{4.23a}$$

$$\delta y \text{ and } \delta z := \left(\frac{\partial w}{\partial y} + \frac{\partial v}{\partial z}\right) dt \tag{4.23b}$$

$$\delta z \text{ and } \delta x := \left(\frac{\partial u}{\partial z} + \frac{\partial w}{\partial x}\right) dt \tag{4.23c}$$

In a fluid, there is no resistance to deformation, but there is resistance to time rate of deformation. Therefore, the rates of distortion (rates of strain)[1] are used in place of strain in solid mechanics.

The rate of distortion is defined as half the rate of angular deformation. Therefore,

$$\dot{\varepsilon}_{xy} = \dot{\varepsilon}_{yx} = \frac{1}{2}\left(\frac{\partial v}{\partial x} + \frac{\partial u}{\partial y}\right) \tag{4.24a}$$

$$\dot{\varepsilon}_{yz} = \dot{\varepsilon}_{zy} = \frac{1}{2}\left(\frac{\partial w}{\partial y} + \frac{\partial v}{\partial z}\right) \tag{4.24b}$$

$$\dot{\varepsilon}_{zx} = \dot{\varepsilon}_{xz} = \frac{1}{2}\left(\frac{\partial u}{\partial z} + \frac{\partial w}{\partial x}\right) \tag{4.24c}$$

These can also be called the rates of shearing strains. The normal rates of strains are

$$\dot{\varepsilon}_{xx} = \frac{\delta x + \frac{\partial u}{\partial x}\delta x dt - \delta x}{\delta x dt} = \frac{\partial u}{\partial x} \tag{4.25a}$$

$$\dot{\varepsilon}_{yy} = \frac{\delta y + \frac{\partial v}{\partial y}\delta y dt - \delta y}{\delta y dt} = \frac{\partial v}{\partial y} \tag{4.25b}$$

$$\dot{\varepsilon}_{zz} = \frac{\delta z + \frac{\partial w}{\partial z}\delta z dt - \delta z}{\delta z dt} = \frac{\partial w}{\partial z} \tag{4.25c}$$

---

[1] These are related to stress by the relationship $\sigma = \mu$ (rate of strain) giving $[\mu] = \text{Ns/m}^2$.

iv. Rotation is the average of the rotation of the sides δx and δy (clockwise considered negative), δy and δz, and δz and δx. (The two angles are positive, but the tangent for y is negative.) The rotations in the three planes are, therefore,

$$xy \text{ plane} : \frac{1}{2}\left(\frac{\partial v}{\partial x} - \frac{\partial u}{\partial y}\right)dt \tag{4.26a}$$

$$yz \text{ plane} : \frac{1}{2}\left(\frac{\partial w}{\partial y} - \frac{\partial v}{\partial z}\right)dt \tag{4.26b}$$

$$zx \text{ plane} : \frac{1}{2}\left(\frac{\partial u}{\partial z} - \frac{\partial w}{\partial x}\right)dt \tag{4.26c}$$

The rate of rotation is the angular velocity about an axis normal to the plane, which can be expressed as

$$w_z = \frac{1}{2}\left(\frac{\partial v}{\partial x} - \frac{\partial u}{\partial y}\right) : xy \text{ plane} \tag{4.27a}$$

$$w_x = \frac{1}{2}\left(\frac{\partial w}{\partial y} - \frac{\partial v}{\partial z}\right) yz \text{ plane} \tag{4.27b}$$

$$w_y = \frac{1}{2}\left(\frac{\partial u}{\partial z} - \frac{\partial w}{\partial x}\right) zx \text{ plane} \tag{4.27c}$$

with the angular velocity vector ω expressed as

$$\omega = i\omega_x + j\omega_y + k\omega_z$$

or

$$\omega = \frac{1}{2}\text{curl } V = \frac{1}{2}\begin{vmatrix} i & j & k \\ \dfrac{\partial}{\partial x} & \dfrac{\partial}{\partial y} & \dfrac{\partial}{\partial z} \\ u & v & w \end{vmatrix} \tag{4.28}$$

Translation and rotation do not induce any deformation in the body because they are rigid body displacements. However, linear and angular deformations involve changes in shape of the fluid element. Through these linear and angular deformations, energy is dissipated as a result of viscous action of a fluid. These relationships that link the stresses and rates of strain lead to the Navier–Stokes equations.

4.9 CIRCULATION AND VORTICITY

Circulation and vorticity are the two primary measures of rotation in a fluid. Circulation is a macroscopic measure of rotation for a finite area of the fluid, whereas vorticity is a microscopic measure of rotation at any point. Vorticity gives a measure of the local rotation. Circulation is a scalar integral quantity, whereas vorticity is a vector field. It can be considered as the amount of force that pushes along a closed boundary or path, such as a circle. Vorticity can be considered as the tendency for a fluid element to spin.

Circulation Γ is defined as the line integral of the tangential component of velocity along a closed path

$$\Gamma = \oint V\, ds = \oint (u\, dx + v\, dy) = \iint \left(\frac{dv}{dx} - \frac{du}{dy} \right) dx\, dy \tag{4.29}$$

where V denotes the tangential component of velocity along an element of the fluid. It is the Curl of the velocity field. The rate of rotation about the z-axis is the average of the two angles of rotation. It is also seen that the double integral in x-y plane can be reduced to a line integral along a closed curve. It is equal to the total strength of all vortex filaments that pass a closed curve. Circulation has the dimension $L^2 T^{-1}$.

Circulation is also given as $\Gamma = 2\pi K = \oint V\, ds = \oint d\phi$ where $K = r v_\theta$.

Considering a 2D small element of fluid of sizes δx and δy with velocities u and v which vary across the element, circulation Γ_z is given as

$$\Gamma_z = u\delta x + \left(v + \frac{\partial v}{\partial x} \delta x \right) \delta y - v\delta y - \left(u + \frac{\partial u}{\partial y} \delta y \right) \delta x = \frac{\partial v}{\partial x} \delta x \delta y - \frac{\partial u}{\partial y} \delta y \delta x$$

$$\Rightarrow \frac{\Gamma_z}{\delta x \delta y} = \frac{\partial v}{\partial x} - \frac{\partial u}{\partial y} = 2\omega_z \tag{4.30}$$

$$\text{Vorticity } \Omega = \frac{\text{Circulation}}{\text{Area}} = \frac{\partial v}{\partial x} - \frac{\partial u}{\partial y} \tag{4.31}$$

For irrotational flow, both circulation and vorticity are zero.

From the definition of vorticity which for irrotational flow is zero, and using the stream function to define velocities,

$$\omega_z = \frac{\partial v}{\partial x} - \frac{\partial u}{\partial y} = 0 \Rightarrow \left(\frac{\partial^2 \psi}{\partial x^2} + \frac{\partial^2 \psi}{\partial y^2} \right) = 0 \Rightarrow \nabla^2 \psi = 0 \tag{4.32}$$

which is the Laplace equation.

The irrotationality condition is satisfied when u, v are expressed in terms of ϕ:

$$\frac{\partial^2 \phi}{\partial x \partial y} - \frac{\partial^2 \phi}{\partial x \partial y} = 0 \tag{4.33}$$

Vorticity vector Ω is defined as $\Omega = 2\omega$ where ω is the angular velocity vector. The angular velocity vector ω (which is the same as rotation excepting a factor of 2) is given by

$$\omega = i\omega_x + j\omega_y + k\omega_z$$

where ω_x, ω_y and ω_z are as given by Eq. 4.27. Therefore,

$$\text{Vorticity } \Omega = 2\omega = \text{Curl } V$$

If a line is drawn in the fluid so that it is tangential to the vorticity vector at that point, the line is called a vortex line (corresponding to a streamline in linear flow). A vortex tube is bounded by vortex lines (corresponding to a stream tube). A vortex filament is a vortex tube of infinitesimal cross section. Vorticity has dimension T^{-1}.

The equations of vortex lines can be obtained by making the cross product of the vorticity vector and an element of length ds in the direction of vorticity vector equal to zero (compare with the equation of a stream line, $V \times ds = 0$):

$$\Omega \times ds = 0 \tag{4.34}$$

where

$$\Omega = i\Omega_x + j\Omega_y + k\Omega_z$$

$$ds = i\delta x + j\delta y + k\delta z$$

Therefore,

$$\Omega \times ds = \begin{vmatrix} i & j & k \\ \Omega_x & \Omega_y & \Omega_z \\ \delta x & \delta y & \delta z \end{vmatrix} \tag{4.35}$$

which gives

$$\frac{\delta x}{\Omega_x} = \frac{\delta y}{\Omega_y} = \frac{\delta z}{\Omega_z} \tag{4.36}$$

Compare Eqs. 4.8b and 4.8d for streamlines with Eqs. 4.35 and 4.36 for vortex lines.

4.9.1 Vorticity in polar co-ordinates

In 2D polar co-ordinates (r, θ), the angular velocity of segment δr is given by (Figure 4.8)

$$\frac{v_\theta + \dfrac{\partial v_\theta}{\partial r}\delta r - v_\theta}{\delta r} = \frac{\partial v_\theta}{\partial r} \tag{4.37a}$$

Angular velocity of segment $r\delta\theta$ is given by

$$-\left[\frac{v_r + \dfrac{\partial v_r}{\partial \theta}\delta\theta - v_r}{r\delta\theta}\right] = -\frac{1}{r}\frac{\partial v_r}{\partial \theta} \tag{4.37b}$$

Angular velocity about the centre O (in addition to the above) is $\dfrac{v_\theta}{r}$ (from the relation $v_\theta = \omega r$). Therefore, vorticity is given by

$$\Omega_z = 2\omega_z = \frac{\partial v_\theta}{\partial r} - \frac{1}{r}\frac{\partial v_r}{\partial \theta} + \frac{v_\theta}{r} \tag{4.38a}$$

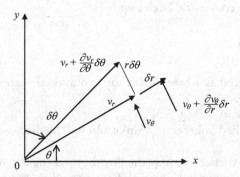

Figure 4.8 Polar co-ordinates.

Similarly, in the other two directions,

$$\Omega_r = 2\omega_r = \frac{1}{r}\frac{\partial v_z}{\partial \theta} - \frac{\partial v_\theta}{\partial z} \tag{4.38b}$$

$$\Omega_\theta = 2\omega_\theta = \frac{\partial v_r}{\partial z} - \frac{\partial v_z}{\partial r} \tag{4.38c}$$

If the vorticity is zero, the flow is called **irrotational**. It means that the fluid particles do not rotate about their own axes, but rotates about a fixed axis. For example, in 2D flows in the horizontal plane, the fluid particles move in concentric circles but without each particle rotating about its own axis. Then, Curl $V = 0$, and

$$\frac{\partial v}{\partial x} = \frac{\partial u}{\partial y}; \frac{\partial w}{\partial y} = \frac{\partial v}{\partial z} \text{ and } \frac{\partial u}{\partial z} = \frac{\partial w}{\partial x} \tag{4.39}$$

(Obtained by setting i, j and k each equal to zero.)

In polar coordination (r, θ)

$$\omega_z = 0 \Rightarrow \frac{\partial v_\theta}{\partial r} - \frac{1}{r}\frac{\partial v_r}{\partial \theta} + \frac{v_\theta}{r} = 0 \Rightarrow \frac{1}{r}\left\{\frac{\partial}{\partial r}(rv_\theta) - \frac{\partial v_r}{\partial \theta}\right\} = 0 \tag{4.40}$$

which, if $r \neq 0$, gives

$$\frac{\partial}{\partial r}(rv_\theta) - \frac{\partial v_r}{\partial \theta} = 0 \tag{4.41}$$

Circulation, which is equal to vorticity times the area, is analogous to the volumetric flow rate which is velocity times the area. According to Kelvin's theorem on circulation in a system where the external forces are conservative, the circulation along a closed contour remains constant with time for an incompressible inviscid fluid.

4.9.1.1 Special cases

a. **Pure circulatory motion:** $v_r = 0$; $v_z = 0$. Therefore, from Eq. 4.40,

$$\frac{\partial}{\partial r}(rv_\theta) = 0 \Rightarrow rv_\theta = \text{Constant}$$

The streamlines are concentric circles and

$$v_\theta \propto \frac{1}{r} \text{ except at } r = 0$$

This can be considered as a **Free Vortex** (or Irrotational vortex). The names are somewhat contradicting!!

b. **Pure radial flow:** i.e. $v_\theta = 0$ $(v_z = 0)$

These are also called **Sources** and **Sinks** and will be discussed in Chapter 6.

Examples of irrotational vortices include the flow beyond the core of a tornado, which is a combination of a irrotational vortex and sink flow, as well as, the case of a cylinder rotated at a constant speed in a viscous liquid of large extent. A rotational vortex is produced when a fluid rotates about an axis as a solid body. In this case, $v_\theta = \omega r$ where ω is the angular velocity of rotation.

Example 4.1

Find the vorticity of the fluid motion for the following velocity components:

a. $u = A(x + y)$; $v = -A(x + y)$
b. $u = 2Axz$; $v = A(C^2 + x^2 - z^2)$
c. $u = Ay^2 + By + C$; $v = 0$

Since the flow is 2D, the angular velocity is about the z-direction. Therefore,

$$\omega_z = \frac{1}{2}\left(\frac{\partial v}{\partial x} - \frac{\partial u}{\partial y}\right)$$

and

$$\Omega = 2\omega = \frac{\partial v}{\partial x} - \frac{\partial u}{\partial y}$$

a. $= -A - A = -2A$
b. $= 2Ax - 0 = 2Ax$
c. $= 0 - (2Ay + B) = -2Ay + B$

Example 4.2

Let x and y components of velocity in steady 2D, incompressible flow be linear functions of x and y such that

$$V = (ax + by)i + (cx + dy)j$$

where a, b, c, d are constants.

a. For what conditions is continuity satisfied?
b. What is the vorticity?
c. For what conditions is the flow irrotational?

a. $u = ax + by;\ v = cx + dy$

   2D continuity equation is

$$\frac{\partial u}{\partial x} + \frac{\partial v}{\partial y} = 0$$

   Substituting, $a + d = 0$

b. Vorticity $\Omega_z = 2\omega_z$

$$= 2 \times \frac{1}{2}\left(\frac{\partial v}{\partial x} - \frac{\partial u}{\partial y}\right) = c - b$$

c. For irrotational flow $\Omega = 0$, therefore $b = c$.

Chapter 5

Governing equations of fluid motion

5.1 INTEGRAL FORMS OF GOVERNING EQUATIONS

In fluid mechanics, the basic laws contain integral quantities of interest. For example,

- Volume rate is the integral of velocity over the area
- Force is the integral of stress over area
- Mass is the integral of density over volume, etc.

5.2 BASIC LAWS

5.2.1 Conservation of mass

The mass of a system remains constant. In integral equation form, it is

$$\frac{D}{Dt} \int_{\text{system}} \rho d\forall = 0 \tag{5.1}$$

where $\forall$ is the volume.

5.2.2 First law of thermodynamics

The rate of heat transfer to a system minus the rate at which the system does work is equal to the rate of change of energy of the system. In equation form,

$$\dot{Q} - \dot{W} = \frac{D}{Dt} \int_{\text{system}} e\rho d\forall \tag{5.2}$$

where $\dot{Q}$ is the rate of heat transfer, $\dot{W}$ is the rate of work done and e is the specific energy, which is the sum of kinetic, potential and internal energies per unit mass.

5.2.3 Newton's second law

The resultant force acting on a system is equal to the rate of change of momentum. In equation form,

$$\sum F = \frac{D}{Dt} \int_{\text{system}} V\rho d\forall \tag{5.3}$$

where V is the velocity vector and F is the force vector. If ρ and V are constant, this reduces to $\sum F = ma$, where m is the mass and a is the acceleration.

5.2.4 Moment of momentum equation (resulting from Newton's second law)

The resultant moment acting on a system is equal to the rate of change of angular momentum. In equation form,

$$\sum M = \frac{D}{Dt} \int_{\text{system}} rV \rho d\cancel{V} \tag{5.4}$$

where M is the moment and r is the lever arm. In general, the left hand side (LHS) of Eq. 5.1 and the right hand side (RHS) of Eqs. 5.2–5.4 can be written as $\frac{D}{Dt} N_{\text{system}}$ where N_{system} represents the integral quantity, which is an extensive[1] property (either scalar or vector) such as mass, momentum and energy.

5.2.5 Second law of thermodynamics

Heat cannot be transferred from a body of lower temperature to one of higher temperature without other simultaneous changes occurring in the two systems of their environments.

Heat from a single source cannot be transformed into mechanical work without other simultaneous changes in the system or their environment. The transfer of mechanical work into heat by friction is irreversible. In an isolated system, the entropy cannot decrease.

$$dS - \frac{dQ}{T} \geq 0 \tag{5.5a}$$

where dS is the change in entropy (J/kg°K) of the system, dQ is heat transferred to the system and T is the temperature (absolute).

Entropy is defined as the quantitative measure of disorder or randomness in a system. The concept comes out of thermodynamics, which deals with the transfer of heat energy within a system. Instead of talking about some form of 'absolute entropy', physicists generally talk about the change in entropy that takes place in a specific thermodynamic process.

There are several types of entropies defined in the literature. In the context of thermodynamics, entropy refers to the amount of 'disorder' in the system – the higher the entropy, the higher the amount of disorder. For a closed thermodynamic system, it is the measure of the amount of thermal energy not available to do mechanical work.

In statistical mechanics, it refers to the amount of uncertainty in the system. In information theory, it is a measure of the uncertainty associated with a random variable. Shannon entropy (Shannon, 1948) refers to a measure of the average information content that is missing by not knowing the value of the random variable. In statistical thermodynamics,

---

[1] **Intensive property** does not depend on the size of the system. Examples include temperature, pressure, density, etc. **Extensive property** depends upon the size of the system. Examples include mass, volume, momentum, energy, entropy, etc. Ratio of two extensive properties of the same object is an intensive property. Examples include the ratios of mass/volume which is density, momentum/mass which is velocity, etc.

it measures the degree to which the probability of the system is spread out over all possible sub-states.

The equality applies to a reversible (frictionless) process, whereas the inequality applies to an irreversible process. If the fluid flow is steady and adiabatic, then, the above relationship becomes

$$dS \geq 0 \qquad (5.5b)$$

when $dS = 0$, the process is called **isentropic**.

Note: Entropy is a derived property and not a basic property. It has the units of energy/°K but is not a form of energy.

In the control volume form,

$$\frac{\partial}{\partial t} \int_{CV} s\rho \, dV + \int_{CS} s\rho V \cdot dA - \int_{CS} \frac{\bar{q}}{T} \, dA \geq 0 \qquad (5.6)$$

where s is the entropy per unit mass and $\bar{q}$ is the heat flux vector (rate of heat transfer per unit area).

5.2.6 Third law of thermodynamics

At absolute zero, the entropy of a pure substance in some perfect crystalline form becomes zero. This is called **Nernst's theorem**, i.e. all possible states of a thermodynamic system at absolute zero have the same entropy that can be taken to be any constant whatever, and in particular zero.

5.3 SYSTEM TO CONTROL VOLUME TRANSFORMATION

Figure 5.1 illustrates the difference between a system and a control volume. A system contains the same mass, whereas in a control volume the mass contained is changing with time. A system can change its boundaries, but a control volume is fixed in space. Referring to Figure 5.2,

$$\text{Flux across } dA = \eta \rho \hat{n} \cdot V dA \qquad (5.7)$$

where $\hat{n}$ is the outward unit vector normal to the area dA, and η is the intensive property of the system. Positive $\hat{n} \cdot V$ implies outward flux, negative implies inward flux and zero implies no flux. The velocity vector V can be at any angle, and the dot product (·) accounts for the appropriate component. ($\hat{n} \cdot V = V \cos\alpha$, where α is the angle between the velocity vector and the outward normal to the surface dA.)

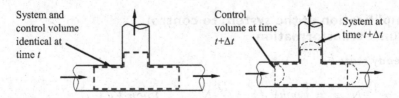

Figure 5.1 System and control volume.

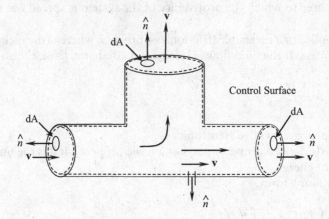

Figure 5.2 Control volume.

It can be shown that (Potter et al., 1997; pp 122–124)

$$\frac{D}{Dt} N_{\text{system}} = \frac{d}{dt} \int_{CV} \eta\rho \, d\mathcal{V} + \int_{CS} \eta\rho\hat{n} \cdot V dA \tag{5.8a}$$

This is also called **Reynolds transport theorem**. Different intensive properties of the system are as follows:

$\eta = 1$ for mass per unit mass
$\eta = V$ for momentum per unit mass
$\eta = r \times V$ for angular momentum

The intensive property is obtained by dividing the **extensive property** by the mass of the system. Temperature and pressure are intensive properties, whereas mass, momentum and energy are extensive properties. Extensive properties depend on the mass of the system.

By transforming the surface integral to a volume integral using the Gauss divergence theorem, it is possible to simplify the above integral equation to (Streeter et al., 1998; pp 195–196):

$$\frac{dn}{dt} = \frac{\partial}{\partial t}(\rho\eta) + \nabla \cdot (\eta\rho V) \tag{5.8b}$$

where η is the intensive quantity per unit volume. This is the Reynolds differential transport theorem. In the above equations, N is the volume integral of the intensive quantity n, i.e.

$$N = \int_{CV} n \, d\mathcal{V} \tag{5.8c}$$

5.3.1 Simplification of the system to control volume transformation

5.3.1.1 Steady state

$$\frac{d}{dt} = 0 \Rightarrow \frac{D}{Dt} N_{\text{system}} = \int_{CS} \eta\rho\hat{n}V \, dA \Rightarrow \frac{D}{Dt} N_{\text{system}} = \int_{CS} \eta\rho\hat{n} \cdot V dA \tag{5.9}$$

5.3.1.2 Steady state, single entrance and single exit, both normal to the flow direction

At entrance, $\hat{n} \cdot V_1 = -V_1$ over area A_1

At outlet, $\hat{n} \cdot V_2 = V_2$ over area A_2

Therefore,

$$\frac{D}{Dt} N_{system} = \int_{A_2} \eta_2 \rho_2 V_2 A_2 - \int_{A_1} \eta_1 \rho_1 V_1 A_1 \tag{5.10}$$

5.3.1.3 Uniform properties over each plane area

$$\frac{D}{Dt} N_{system} = \eta_2 \rho_2 V_2 A_2 - \eta_1 \rho_1 V_1 A_1 \tag{5.11}$$

5.3.1.4 Continuity equation

$$\frac{D}{Dt} N_{system} = \frac{D}{Dt} \int_{system} \rho d\cancel{V} = 0 \, (\text{for mass}, \eta = 1) \tag{5.12a}$$

$$\Rightarrow 0 = \frac{d}{dt} \int_{CV} \rho d\cancel{V} + \int_{CS} \rho \hat{n} \cdot V dA$$

If the flow is steady

$$\Rightarrow \int_{CS} \rho \hat{n} \cdot V \, dA = 0 \tag{5.12b}$$

For uniform flow, $\rho_2 V_2 A_2 = \rho_1 V_1 A_1$, where $\hat{n} \cdot V_1 = -V_1$; $\hat{n} \cdot V_2 = V_2$.

If the density is constant, $V_2 A_2 = V_1 A_1$.

5.4 MOMENTUM EQUATION (DIFFERENTIAL FORM)

In a fluid (just like in a solid) there are two types of stresses, the **normal stresses** resulting from pressure and the **shear stresses** resulting from friction.

These lead to two types of forces acting on a body, namely, surface forces and body forces. The surface forces result from the internal stresses on the body and act on the surface, whereas the body forces result from causes external to the body. Pressure and shear forces belong to the former type, whereas gravity, Coriolis and electromagnetic forces belong to the latter type.

In the three-dimensional stress system for a fluid element shown in Figure 5.3, the normal stresses are denoted by σ_{xx}, σ_{yy} and σ_{zz}, and the shear stresses by τ with two subscripts. The first subscript corresponds to the axis to which the shear stress is normal, and the second subscript corresponds to the axis to which the shear stress is parallel. For example, τ_{xy} is perpendicular to the x-axis and parallel to y-axis.

These stresses in general will vary between parallel faces. For example, σ_{xx} will have a stress on the opposite face with magnitude

$$\sigma_{xx} + \frac{\partial}{\partial x} (\sigma_{xx}) \delta x$$

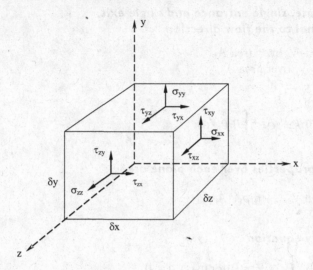

Figure 5.3 Stresses on a fluid element.

Summing up the contributions from all such stresses, the net surface forces in the x, y and z directions are

$$x : \frac{\partial}{\partial x}(\sigma_{xx})\delta x \delta y \delta z + \frac{\partial}{\partial y}(\tau_{yx})\delta x \delta y \delta z + \frac{\partial}{\partial z}(\tau_{zx})\delta x \delta y \delta z$$

$$y : \frac{\partial}{\partial x}(\tau_{xy})\delta x \delta y \delta z + \frac{\partial}{\partial y}(\sigma_{yy})\delta x \delta y \delta z + \frac{\partial}{\partial z}(\tau_{zy})\delta x \delta y \delta z$$

$$z : \frac{\partial}{\partial x}(\tau_{xz})\delta x \delta y \delta z + \frac{\partial}{\partial y}(\tau_{yz})\delta x \delta y \delta z + \frac{\partial}{\partial z}(\sigma_{zz})\delta x \delta y \delta z$$

Using Newton's second law of motion, the force-acceleration equation can be written as follows:

$$x : \frac{\partial}{\partial x}(\sigma_{xx})\delta x \delta y \delta z + \frac{\partial}{\partial y}(\tau_{yx})\delta x \delta y \delta z + \frac{\partial}{\partial z}(\tau_{zx})\delta x \delta y \delta z + B_x \delta x \delta y \delta z$$

$$= \frac{\partial}{\partial t}(\rho u)\delta x \delta y \delta z + \frac{\partial}{\partial x}(\rho u u)\delta x \delta y \delta z + \frac{\partial}{\partial y}(\rho u v)\delta x \delta y \delta z + \frac{\partial}{\partial z}(\rho u w)\delta x \delta y \delta z$$

$$y : \frac{\partial}{\partial x}(\tau_{xy})\delta x \delta y \delta z + \frac{\partial}{\partial y}(\sigma_{yy})\delta x \delta y \delta z + \frac{\partial}{\partial z}(\tau_{zy})\delta x \delta y \delta z + B_y \delta x \delta y \delta z$$

$$= \frac{\partial}{\partial t}(\rho v)\delta x \delta y \delta z + \frac{\partial}{\partial x}(\rho v u)\delta x \delta y \delta z + \frac{\partial}{\partial y}(\rho v v)\delta x \delta y \delta z + \frac{\partial}{\partial z}(\rho v w)\delta x \delta y \delta z$$

$$z : \frac{\partial}{\partial x}(\tau_{xz})\delta x \delta y \delta z + \frac{\partial}{\partial y}(\tau_{yz})\delta x \delta y \delta z + \frac{\partial}{\partial z}(\sigma_{zz})\delta x \delta y \delta z + B_z \delta x \delta y \delta z$$

$$= \frac{\partial}{\partial t}(\rho w)\delta x \delta y \delta z + \frac{\partial}{\partial x}(\rho w u)\delta x \delta y \delta z + \frac{\partial}{\partial y}(\rho w v)\delta x \delta y \delta z + \frac{\partial}{\partial z}(\rho w w)\delta x \delta y \delta z$$

where B_x, B_y and B_z represent the body force components per unit volume in the x-, y- and z-directions. The last three terms on the right hand side of the above equations correspond to the rate of momentum efflux across the control surfaces. In the x-direction, they take the form[2]

$$x : \left[(\rho uu)\delta y\delta z + \frac{\partial}{\partial x}(\rho uu)\delta x\delta y\delta z \right] - (\rho uu)\delta y\delta z$$

$$+ \left[(\rho uv)\delta z\delta x + \frac{\partial}{\partial y}(\rho uv)\delta x\delta y\delta z \right] - (\rho uv)\delta z\delta x$$

$$+ \left[(\rho uw)\delta x\delta y + \frac{\partial}{\partial z}(\rho uw)\delta x\delta y\delta z \right] - (\rho uw)\delta x\delta y$$

which when simplified is

$$x : \frac{\partial}{\partial x}(\rho uu)\delta x\delta y\delta z + \frac{\partial}{\partial y}(\rho uv)\delta x\delta y\delta z + \frac{\partial}{\partial z}(\rho uw)\delta x\delta y\delta z$$

For the y and z directions, there will be similar terms.

The right hand side of these equations after dividing by $\delta x\delta y\delta z$ will be

$$x : \frac{\partial}{\partial t}(\rho u) + \frac{\partial}{\partial x}(\rho uu) + \frac{\partial}{\partial y}(\rho uv) + \frac{\partial}{\partial z}(\rho uw)$$

$$y : \frac{\partial}{\partial t}(\rho v) + \frac{\partial}{\partial x}(\rho vu) + \frac{\partial}{\partial y}(\rho vv) + \frac{\partial}{\partial z}(\rho vw)$$

$$z : \frac{\partial}{\partial t}(\rho w) + \frac{\partial}{\partial x}(\rho wu) + \frac{\partial}{\partial y}(\rho wv) + \frac{\partial}{\partial z}(\rho ww)$$

When expanded, they take the form

$$x : \rho \frac{\partial u}{\partial t} + u \frac{\partial \rho}{\partial t} + u \frac{\partial(\rho u)}{\partial x} + \rho u \frac{\partial u}{\partial x} + u \frac{\partial(\rho v)}{\partial y} + \rho v \frac{\partial(u)}{\partial y} + u \frac{\partial(\rho w)}{\partial z} + \rho w \frac{\partial u}{\partial z}$$

$$y : \rho \frac{\partial v}{\partial t} + v \frac{\partial \rho}{\partial t} + v \frac{\partial(\rho u)}{\partial x} + \rho u \frac{\partial v}{\partial x} + v \frac{\partial(\rho v)}{\partial y} + \rho v \frac{\partial(v)}{\partial y} + v \frac{\partial(\rho w)}{\partial z} + \rho w \frac{\partial v}{\partial z}$$

$$z : \rho \frac{\partial w}{\partial t} + w \frac{\partial \rho}{\partial t} + w \frac{\partial(\rho u)}{\partial x} + \rho u \frac{\partial w}{\partial x} + w \frac{\partial(\rho v)}{\partial y} + \rho v \frac{\partial(w)}{\partial y} + w \frac{\partial(\rho w)}{\partial z} + \rho w \frac{\partial w}{\partial z}$$

which can be written as

---

[2] $V \cdot dA = u\delta y\delta z + v\delta z\delta x + w\delta x\delta y$;　　　$V(V \cdot dA) = u\,(V \cdot dA) + v\,(V \cdot dA) + w\,(V \cdot dA)$

$$x : u\left[\frac{\partial \rho}{\partial t}+\frac{\partial (\rho u)}{\partial x}+\frac{\partial (\rho v)}{\partial y}+\frac{\partial (\rho w)}{\partial z}\right]+\rho\left[\frac{\partial u}{\partial t}+u\frac{\partial u}{\partial x}+v\frac{\partial u}{\partial y}+w\frac{\partial u}{\partial z}\right]$$

$$y : v\left[\frac{\partial \rho}{\partial t}+\frac{\partial (\rho u)}{\partial x}+\frac{\partial (\rho v)}{\partial y}+\frac{\partial (\rho w)}{\partial z}\right]+\rho\left[\frac{\partial v}{\partial t}+u\frac{\partial v}{\partial x}+v\frac{\partial v}{\partial y}+w\frac{\partial v}{\partial z}\right]$$

$$z : w\left[\frac{\partial \rho}{\partial t}+\frac{\partial (\rho u)}{\partial x}+\frac{\partial (\rho v)}{\partial y}+\frac{\partial (\rho w)}{\partial z}\right]+\rho\left[\frac{\partial w}{\partial t}+u\frac{\partial w}{\partial x}+v\frac{\partial w}{\partial y}+w\frac{\partial w}{\partial z}\right]$$

The terms within the first square brackets above represent the continuity equation. Therefore,

$$\frac{\partial}{\partial x}\left(\sigma_{xx}\right)+\frac{\partial}{\partial y}\left(\tau_{yx}\right)+\frac{\partial}{\partial z}\left(\tau_{zx}\right)+B_x = \rho\left[\frac{\partial u}{\partial t}+u\frac{\partial u}{\partial x}+v\frac{\partial u}{\partial y}+w\frac{\partial u}{\partial z}\right]=\rho\frac{Du}{Dt} \tag{5.13a}$$

$$\frac{\partial}{\partial x}\left(\tau_{xy}\right)+\frac{\partial}{\partial y}\left(\sigma_{yy}\right)+\frac{\partial}{\partial z}\left(\tau_{zy}\right)+B_y = \rho\left[\frac{\partial v}{\partial t}+u\frac{\partial v}{\partial x}+v\frac{\partial v}{\partial y}+w\frac{\partial v}{\partial z}\right]=\rho\frac{Dv}{Dt} \tag{5.13b}$$

$$\frac{\partial}{\partial x}\left(\tau_{xz}\right)+\frac{\partial}{\partial y}\left(\tau_{yz}\right)+\frac{\partial}{\partial z}\left(\sigma_{zz}\right)+B_z = \rho\left[\frac{\partial w}{\partial t}+u\frac{\partial w}{\partial x}+v\frac{\partial w}{\partial y}+w\frac{\partial w}{\partial z}\right]=\rho\frac{Dw}{Dt} \tag{5.13c}$$

This is the general momentum equation for any fluid in motion.

5.5 CONSTITUTIVE RELATIONSHIPS (STRESS RATES OF STRAIN RELATIONSHIPS)

In fluids, there is no resistance to deformation, but there is resistance to the time rate of deformation. Therefore, the rates of deformations (rates of strain) are used in place of strain in solid mechanics. These rates are then linked to stresses by some constitutive relationships. The relationships depend on the classification of fluids into two broad classes, Newtonian fluids and non-Newtonian fluids (see Figure 1.11 also).

5.5.1 Newtonian fluids

Fluids that conform to Newton's law of viscosity are called Newtonian fluids. The law, which states that the stress is proportional to the rate of strain, leads to the relationship

$$\tau = \mu\frac{\partial u}{\partial x};[\mu] = ML^{-1}T^{-1}(\text{In SI units, it is N/m}^2\text{ s}) \tag{5.14}$$

in which the coefficient of viscosity, μ, depends on temperature but independent of shear rate, i.e. velocity gradient.

5.5.2 Non-Newtonian fluids

A number of equations exist for describing non-Newtonian fluids. Examples include

(a) $\tau = \mu\left\{\frac{\partial u}{\partial y}\right\}^{n}$ $\tag{5.15}$

with $n = 1$ for Newtonian fluids (water, air, gasoline etc.)
$n < 1$ for pseudoplastic fluid (slurries, polymer solutions, clay, milk etc.)
$n > 1$ for dilatant fluids (suspensions with high solid concentrations)

(b) $\tau = \tau_1 + \mu \dfrac{\partial u}{\partial y}$ (Bingham fluids, e.g. slurries, drilling mud, oil paints) $\hfill (5.16)$

(c) Time-dependent fluids: Viscosity decreases with time under constant shear rate (velocity gradient): e.g. gypsum pastes.

(d) Visco-elastic fluids: e.g., flour dough.

5.5.3 Stokes' hypothesis

Stokes, based on Newton's law of viscosity, hypothesized that

$$\sigma_{xx} = -p + 2\mu \frac{\partial u}{\partial x} - \frac{2}{3}\mu \left[\frac{\partial u}{\partial x} + \frac{\partial v}{\partial y} + \frac{\partial w}{\partial z} \right] \tag{5.17a}$$

$$\sigma_{yy} = -p + 2\mu \frac{\partial v}{\partial x} - \frac{2}{3}\mu \left[\frac{\partial u}{\partial x} + \frac{\partial v}{\partial y} + \frac{\partial w}{\partial z} \right] \tag{5.17b}$$

$$\sigma_{zz} = -p + 2\mu \frac{\partial w}{\partial x} - \frac{2}{3}\mu \left[\frac{\partial u}{\partial x} + \frac{\partial v}{\partial y} + \frac{\partial w}{\partial z} \right] \tag{5.17c}$$

$$\tau_{xy} = \mu \left(\frac{\partial u}{\partial y} + \frac{\partial v}{\partial x} \right) \tag{5.18a}$$

$$\tau_{yz} = \mu \left(\frac{\partial v}{\partial z} + \frac{\partial w}{\partial y} \right) \tag{5.18b}$$

$$\tau_{zx} = \mu \left(\frac{\partial u}{\partial z} + \frac{\partial w}{\partial x} \right) \tag{5.18c}$$

5.6 NAVIER-STOKES EQUATIONS

C.L.M.H. Navier (1785–1836) and G.B. Stokes (1819–1903) independently derived the generalized momentum equation for a fluid element, which is now referred to as the Navier-Stokes equations. It is the basis of all fluids in motion.

Substituting Stokes' hypothesis into the momentum equation

$$\rho \frac{Du}{Dt} = -\frac{\partial p}{\partial x} + \rho g_x + \mu \left(\frac{\partial^2 u}{\partial x^2} + \frac{\partial^2 u}{\partial y^2} + \frac{\partial^2 u}{\partial z^2} \right) + \frac{\mu}{3}\frac{\partial}{\partial x} \left(\frac{\partial u}{\partial x} + \frac{\partial v}{\partial y} + \frac{\partial w}{\partial z} \right) \tag{5.19a}$$

$$\rho \frac{Dv}{Dt} = -\frac{\partial p}{\partial y} + \rho g_y + \mu \left(\frac{\partial^2 v}{\partial x^2} + \frac{\partial^2 v}{\partial y^2} + \frac{\partial^2 v}{\partial z^2} \right) + \frac{\mu}{3}\frac{\partial}{\partial y} \left(\frac{\partial u}{\partial x} + \frac{\partial v}{\partial y} + \frac{\partial w}{\partial z} \right) \tag{5.19b}$$

$$\rho \frac{Dw}{Dt} = -\frac{\partial p}{\partial z} + \rho g_z + \mu \left(\frac{\partial^2 w}{\partial x^2} + \frac{\partial^2 w}{\partial y^2} + \frac{\partial^2 w}{\partial z^2} \right) + \frac{\mu}{3}\frac{\partial}{\partial z} \left(\frac{\partial u}{\partial x} + \frac{\partial v}{\partial y} + \frac{\partial w}{\partial z} \right). \tag{5.19c}$$

These equations (momentum) in cylindrical co-ordinates (r, θ, z) take the form

$$\rho\left(\frac{\partial v_r}{\partial t} + v_r\frac{\partial v_r}{\partial r} + \frac{v_\theta}{r}\frac{\partial v_r}{\partial \theta} + v_z\frac{\partial v_r}{\partial z} - \frac{v_\theta^2}{r}\right)$$

$$= -\frac{\partial p}{\partial r} + \mu\left\{\frac{\partial}{\partial r}\left[\frac{1}{r}\frac{\partial(rv_r)}{\partial r}\right] + \frac{1}{r^2}\frac{\partial^2 v_r}{\partial \theta^2} + \frac{\partial^2 v_r}{\partial z^2} - \frac{2}{r^2}\frac{\partial v_\theta}{\partial \theta}\right\} + \rho g_r$$

(5.20a)

$$\rho\left(\frac{\partial v_\theta}{\partial t} + v_r\frac{\partial v_\theta}{\partial r} + \frac{v_\theta}{r}\frac{\partial v_\theta}{\partial \theta} + v_z\frac{\partial v_\theta}{\partial z} + \frac{v_r v_\theta}{r}\right)$$

$$= -\frac{\partial p}{\partial \theta} + \mu\left\{\frac{\partial}{\partial r}\left[\frac{1}{r}\frac{\partial(rv_\theta)}{\partial r}\right] + \frac{1}{r^2}\frac{\partial^2 v_\theta}{\partial \theta^2} + \frac{2}{r^2}\frac{\partial v_r}{\partial \theta} + \frac{\partial^2 v_\theta}{\partial z^2}\right\} + \rho g_\theta$$

(5.20b)

$$\rho\left(\frac{\partial v_z}{\partial t} + v_r\frac{\partial v_z}{\partial r} + \frac{v_\theta}{r}\frac{\partial v_z}{\partial \theta} + v_z\frac{\partial v_z}{\partial z}\right)$$

$$= -\frac{\partial p}{\partial z} + \mu\left\{\frac{1}{r}\frac{\partial}{\partial r}\left[r\frac{\partial v_z}{\partial r}\right] + \frac{1}{r^2}\frac{\partial^2 v_z}{\partial \theta^2} + \frac{\partial^2 v_z}{\partial z^2}\right\} + \rho g_z$$

(5.20c)

5.7 SPECIAL CASES OF THE NAVIER-STOKES EQUATION

5.7.1 Incompressible fluids (density is constant)

The quantity within the brackets in the last term, being the continuity equation for incompressible fluids, is zero. Therefore,

$$\rho\frac{Du}{Dt} = -\frac{\partial p}{\partial x} + \rho g_x + \mu\left(\frac{\partial^2 u}{\partial x^2} + \frac{\partial^2 u}{\partial y^2} + \frac{\partial^2 u}{\partial z^2}\right)$$

(5.21a)

$$\rho\frac{Dv}{Dt} = -\frac{\partial p}{\partial y} + \rho g_y + \mu\left(\frac{\partial^2 v}{\partial x^2} + \frac{\partial^2 v}{\partial y^2} + \frac{\partial^2 v}{\partial z^2}\right)$$

(5.21b)

$$\rho\frac{Dw}{Dt} = -\frac{\partial p}{\partial z} + \rho g_z + \mu\left(\frac{\partial^2 w}{\partial x^2} + \frac{\partial^2 w}{\partial y^2} + \frac{\partial^2 w}{\partial z^2}\right)$$

(5.21c)

5.7.2 Euler (named after Leonhard Euler (1707–1783)) equation (frictionless fluids (ideal or inviscid))

$$\rho\frac{Du}{Dt} = -\frac{\partial p}{\partial x} + \rho g_x$$

(5.22a)

$$\rho\frac{Dv}{Dt} = -\frac{\partial p}{\partial y} + \rho g_y$$

(5.22b)

$$\rho\frac{Dw}{Dt} = -\frac{\partial p}{\partial z} - \rho g_z$$

(5.22c)

Since z is measured positive vertically upwards and that gravity is vertically downwards, $g_z = -g$. Also, g_x and g_y are both zero.

5.7.3 Bernoulli (named after Daniel Bernoulli (1700–1782)) equation

If the Euler equation is integrated along a streamline for a constant density fluid, the resulting equation is the familiar Bernoulli equation. Assuming the direction of the streamline is s, then, the Euler equation in the s-direction is

$$\rho \frac{DV}{Dt} = -\frac{\partial p}{\partial s} - \rho g \frac{\partial z}{\partial s} \tag{5.23}$$

where V is the velocity in the streamline direction, and the second term on the RHS is the component $\rho g \sin \alpha$ of the gravitational force in the direction of the streamline. Substituting for the acceleration term, the equation reduces to

$$\rho \frac{\partial V}{\partial t} + \rho V \frac{\partial V}{\partial s} = -\frac{\partial p}{\partial s} - \rho g \frac{\partial z}{\partial s} \tag{5.24}$$

For steady state conditions, the time derivative is zero. Then, the partial derivatives can be replaced by their total derivatives because s is the only independent variable. The resulting equation is

$$\rho V \frac{dV}{ds} + \frac{dp}{ds} + \rho g \frac{dz}{ds} = 0 \tag{5.25a}$$

$$\frac{d}{ds}\left(\rho \frac{V^2}{2} + p + \rho g z \right) = 0 \tag{5.25b}$$

which when integrated along a streamline gives the Bernoulli equation

$$\rho \frac{V^2}{2} + p + \rho g z = \text{Constant} \tag{5.26a}$$

This equation is sometimes written as

$$\frac{p}{\rho g} + \frac{V^2}{2g} + z = \text{Constant} \tag{5.26b}$$

5.7.4 Hydrostatic equation

For flow in a horizontal plane, the vertical velocity component w is zero. Therefore, the Euler equation

$$\rho \frac{Dw}{Dt} = -\frac{\partial p}{\partial z} - \rho g$$

simplifies to

$$\frac{\partial p}{\partial z} = -\rho g \tag{5.27}$$

This is the **hydrostatic equation**. It gives the pressure gradient in the vertical direction.

5.8 CONTINUITY EQUATION

The continuity equation can be derived from the Reynolds differential transport theorem or by consideration of mass entering, leaving and change in a control volume. If the former is used, then the total time rate of change of mass per unit volume must be zero and, η, the mass per unit mass is 1. Then,

$$\frac{\partial \rho}{\partial t} + \Delta \cdot (\rho V) = 0 \tag{5.28}$$

which is

$$\frac{\partial \rho}{\partial t} + \frac{\partial(\rho u)}{\partial x} + \frac{\partial(\rho v)}{\partial y} + \frac{\partial(\rho w)}{\partial z} = 0 \tag{5.29}$$

This is the continuity equation for all flows. For incompressible fluids, it simplifies to

$$\frac{\partial u}{\partial x} + \frac{\partial v}{\partial y} + \frac{\partial w}{\partial z} = 0 \tag{5.30}$$

In cylindrical co-ordinates (r, θ, z), the continuity equation takes the form

$$\frac{\partial \rho}{\partial t} + \frac{1}{r}\frac{\partial(\rho r v_r)}{\partial r} + \frac{1}{r}\frac{\partial(\rho v_\theta)}{\partial \theta} + \frac{\partial(\rho v_z)}{\partial z} = 0 \tag{5.31}$$

REFERENCES

Potter, M. C., Wiggert, D. C. and Hondzo, M.C. (1997): *Mechanics of Fluids*, (2nd Edition), Prentice Hall, Upper Saddle River, NJ, 689 pp.

Shannon, C. E. (1948): *A Mathematical Theory of Communication*, Bell Telephone System Technical Publications (http://cm.bell-labs.com/cm/ms/what/shannday/paper.html).

Streeter, V. L., Wylie, E. B. and Bedford, K. W. (1998): *Fluid Mechanics*, (9th Edition), McGraw Hill, New York, 740 pp.

Chapter 6

Ideal fluid flow

6.1 INTRODUCTION

Many researchers in the past (e.g., Archimedes, Leonardo da Vinci, Bernoulli, Euler etc. among others (see Chapter 2 for a partial list of pioneers in fluid mechanics)) have contributed to the development of fluid mechanics. The mathematical representations of fluid flow can be considered as belonging to 'classical hydrodynamics'. Developments also have taken place in experimental fluid flow analysis. Mathematical representations are based on two assumptions; namely, the fluid is inviscid and incompressible. In reality, however, such a fluid does not exist. All fluids have some viscosity and are compressible if sufficient pressure is applied. These assumptions are sometimes relaxed when the mathematical theories have to be applied to practical problems involving real fluids. For practical purposes, the concept of ideal fluid offers a way to distinguish the differences between real fluids and fluids that appear to behave like ideal fluids. For example, the viscous effects in a flowing flow field are significant in a narrow region close to the boundary of flow known as the boundary layer (Figure 6.1). Outside this region, the fluid can be assumed as ideal, and all the laws and equations derived for ideal fluids can be applied in the flow region outside the boundary layer. The separation takes place at the boundary layer where the velocity profile changes from zero (no-slip boundary) to the free stream velocity at a depth known as the boundary layer thickness δ. Within the boundary layer, the fluid is considered as real, and the viscous effects are taken into consideration. Therefore, any real fluid flow can be divided into a region where the ideal fluid flow laws and equations can be applied and to a narrow region where such laws and equations cannot be applied. The concept of ideal fluids also helps in describing fluid flow mathematically. In this chapter, the theoretical basis for ideal fluid flow is presented.

Ideal fluid flow is sometimes referred to as potential flow. Potential flows are irrotational flows whose velocity components may be derived from velocity potential functions. They apply to incompressible fluids, and the Bernoulli equation may be used throughout the flows.

Fluid flows can also be considered as rotational flows and irrotational flows. Rotational flows are flows where the fluid particles rotate about their own axes while flowing along streamlines. The solar system is rotational. Irrotational flows are flows where the fluid particles do not rotate about their own axes while flowing along streamlines. It is a simplified assumption since no real fluid is really irrotational. A fluid inside a rotating tank is irrotational once it has reached steady-state conditions since the flow is in rigid body motion. However, at the centre of rotation, the angular velocity is so high resulting in the flow to be rotational.

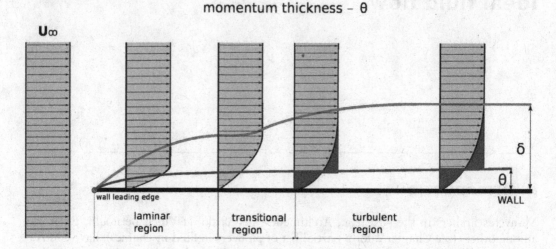

Figure 6.1 Boundary layer as the boundary between real fluids and ideal fluids.

6.2 STREAM FUNCTION AND VELOCITY POTENTIAL

6.2.1 Stream function

Stream function and velocity potential are two complementary functions from which the velocity components of a flow field can be determined. Details of stream function are given in Section 4.7.

6.2.2 Velocity potential

For two-dimensional flow (potential flow exists for three-dimensional compressible fluids also), a velocity potential function ϕ exists, and is defined as the integral of Vds along the length of the line as

$$\phi = \int V \, ds \tag{6.1}$$

If the flow is horizontal, $V = u$, and $ds = dx$, giving $u = \dfrac{d\phi}{dx}$ (6.2a)

If the flow is vertical, $V = v$, and $ds = dy$, giving $v = \dfrac{d\phi}{dy}$ (6.2b)

If the flow is radial, $V = v_r$, and $ds = dr$, giving $v_r = \dfrac{d\phi}{dr}$ (6.3a)

If the flow is tangential, $V = v_\theta$, and $ds = rd\theta$, giving $v_\theta = \dfrac{1}{r}\dfrac{d\phi}{d\theta}$ (6.3b)

The velocity V is zero normal to a streamline. The lines of constant velocity potential are orthogonal to streamlines. The set of streamlines and velocity potential lines is called a flow net.

Combining the equations for stream functions and velocity potentials, the velocity components can be written as

$$u = \frac{d\psi}{dy} = \frac{d\phi}{dx}; \quad v = -\frac{d\psi}{dx} = \frac{d\phi}{dy} \tag{6.4a}$$

$$v_r = \frac{1}{r}\frac{d\psi}{d\theta} = \frac{d\phi}{dr}; \quad v_\theta = -\frac{d\psi}{dr} = \frac{1}{r}\frac{d\phi}{d\theta} \tag{6.4b}$$

and

$$V = \nabla\phi = \text{grad }\phi = \frac{\partial\phi}{\partial x}i + \frac{\partial\phi}{\partial y}j + k\frac{\partial\phi}{\partial z} \tag{6.5}$$

Note: Sometimes, a negative sign is used in the definition of u, v and w. When using the negative sign, the potential decreases in the direction of flow. Velocity potential is not a physical quantity that can be measured. It is only a mathematical entity. Velocity potential is applicable for irrotational flows only, but stream function is applicable to both rotational and irrotational flows.

It is also important to note that in the definition of ϕ and ψ the **Cauchy-Riemann**[1] condition must be satisfied. This means

$$\frac{\partial\phi}{\partial x} = \frac{\partial\psi}{\partial y}, \quad \frac{\partial\phi}{\partial y} = -\frac{\partial\psi}{\partial x} \tag{6.6}$$

Substituting for the velocity u, v, w in the continuity equation

$$\frac{\partial u}{\partial x} + \frac{\partial v}{\partial y} + \frac{\partial w}{\partial z} = 0 \tag{6.7}$$

gives

$$\frac{\partial^2\phi}{\partial x^2} + \frac{\partial^2\phi}{\partial y^2} + \frac{\partial^2\phi}{\partial z^2} = 0 \tag{6.8a}$$

or

$$\nabla^2\phi = 0.$$

which is **Laplace equation**. The corresponding Laplace equation in terms of stream function is

$$\frac{\partial^2\psi}{\partial x^2} + \frac{\partial^2\psi}{\partial y^2} = 0 \tag{6.8b}$$

It is to be noted that $\phi = \phi(x, y, z)$ and $\psi = \psi(x, y)$.

---

[1] This comes from complex variables. Since ϕ and ψ satisfy Laplace equation, they are harmonic functions and forms an analytic function $\phi + i\psi$ called the complex velocity potential.

Equipotential lines are lines of constant ϕ, and streamlines are lines of constant ψ. They always intersect at right angles $\left(\text{because } u = \dfrac{\partial \phi}{\partial x} = \dfrac{\partial \psi}{\partial y}; v = \dfrac{\partial \phi}{\partial y} = -\dfrac{\partial \psi}{\partial x}\right)$, and the two families of lines constitute a **flow net**.

The equation of a streamline is

$$\frac{dx}{u} = \frac{dy}{v} = \frac{dz}{w} \qquad (6.9a)$$

which gives

$$u\,dy - v\,dx = 0 \qquad (6.9b)$$

$$\text{or, } \left(\frac{dy}{dx}\right)_{\psi} = \frac{v}{u}, \qquad (6.9c)$$

Along an equipotential line,

$$\phi = \int u\,dx = \int v\,dy \Rightarrow u\,dx = v\,dy \qquad (6.10a)$$

$$u\,dx - v\,dy = 0 \qquad (6.10b)$$

$$\text{or, } \left(\frac{dy}{dx}\right)_{\varphi} = \frac{u}{v} \qquad (6.10c)$$

Therefore,

$$\left(\frac{dy}{dx}\right)_{\phi}\left(\frac{dy}{dx}\right)_{\psi} = 1 \qquad (6.11)$$

Sometimes, a negative sign is used in the above equation.

The Laplace equation that describes the velocity potential ϕ and the stream function ψ is linear. Therefore, superposition of ϕ's and ψ's is possible.

The stream function is based on the principle of continuity and therefore is applicable to rotational as well as irrotational flows. However, the velocity potential is applicable to irrotational flow only.

In polar co-ordinates (Eqs. 6.3a and 6.3b),

$$v_r = \frac{\partial \phi}{\partial r}; v_\theta = \frac{1}{r}\frac{\partial \phi}{\partial \theta}$$

In terms of the stream function, v_r and v_θ are (Eqs. 4.19a and 4.19b)

$$v_r = \frac{1}{r}\frac{\partial \psi}{\partial \theta}; \quad v_\theta = -\frac{\partial \psi}{\partial r}$$

The Laplace equations in polar co-ordinates can be obtained by substituting the velocity components in the continuity equation in polar co-ordinates.

$$\frac{\partial(rv_r)}{\partial r} + \frac{\partial v_\theta}{\partial \theta} = 0 \Rightarrow \frac{\partial v_r}{\partial r} + \frac{1}{r}\frac{\partial v_\theta}{\partial \theta} + \frac{v_r}{r} = 0$$

$$\frac{\partial}{\partial r}\left(\frac{\partial \phi}{\partial r}\right) + \frac{1}{r}\left(\frac{1}{r}\frac{\partial^2 \phi}{\partial \theta^2}\right) + \frac{1}{r}\frac{\partial \phi}{\partial r} = 0$$

$$\frac{\partial^2 \phi}{\partial r^2} + \frac{1}{r}\frac{\partial \phi}{\partial r} + \frac{1}{r^2}\frac{\partial^2 \phi}{\partial \theta^2} = 0 \quad \text{in } (r, \theta) \text{ plane.} \tag{6.12}$$

In three dimensions (r, θ, z), the equation takes the form

$$\frac{\partial^2 \phi}{\partial r^2} + \frac{1}{r}\frac{\partial \phi}{\partial r} + \frac{1}{r^2}\frac{\partial^2 \phi}{\partial \theta^2} + \frac{\partial^2 \phi}{\partial z^2} = 0 \tag{6.13}$$

where the first two terms correspond to $\dfrac{1}{r}\dfrac{\partial}{\partial r}\left(r\dfrac{\partial \phi}{\partial r}\right)$

In terms of the stream function, Laplace equation in (r, θ) plane is

$$\frac{\partial^2 \psi}{\partial r^2} + \frac{1}{r}\frac{\partial \psi}{\partial r} + \frac{1}{r^2}\frac{\partial^2 \psi}{\partial \theta^2} = 0 \tag{6.14}$$

The latter is obtained by the condition of irrotationality.

The Laplace equation has many applications in physics. For example, in heat conduction, equipotential lines are lines of constant temperature, and streamlines are lines indicating the direction of heat flow; in electricity flows, equipotential lines are lines of constant voltage and streamlines are lines indicating the direction of current flow. In fluid flow, equipotential lines are **not** lines of constant pressure, ϕ being a fictitious quantity. Laplace equation is elliptic and the solution does not pose a discontinuity (hyperbolic equations such as the wave equation have discontinuities).

Note: Existence of ϕ implies irrotationality and existence of ψ implies continuity. The two functions ϕ and ψ respectively satisfy these two conditions. Laplace equations are obtained from the condition of continuity for ϕ and the condition of irrotationality for ψ.

The two dimensional velocity component definitions using stream function and velocity potential in different co-ordinate systems can be summarised as follows:

Cartesian co-ordinates (x, y):(u, v)

$$u = \frac{\partial \psi}{\partial y}; \quad v = -\frac{\partial \psi}{\partial x} \tag{Eqs. 4.18a and 4.18b}$$

$$u = \frac{\partial \phi}{\partial x}; \quad v = \frac{\partial \phi}{\partial y} \tag{Eqs. 6.2a and 6.2b}$$

Polar cylindrical co-ordinates (r, θ): (v_r, v_θ)

$$v_r = \frac{1}{r}\frac{\partial \psi}{\partial \theta}; \quad v_\theta = -\frac{\partial \psi}{\partial r} \tag{Eqs. 4.19a and 4.19b}$$

$$v_r = \frac{\partial \phi}{\partial r}; \quad v_\theta = \frac{1}{r}\frac{\partial \phi}{\partial \theta} \tag{Eq. 6.4b}$$

Axisymmetric cylindrical co-ordinates (r, θ, z) : (v_r, v_θ, v_z)

$$v_r = -\frac{1}{r}\frac{\partial \psi}{\partial z}; \quad v_z = \frac{1}{r}\frac{\partial \psi}{\partial r}$$

$$v_r = \frac{\partial \phi}{\partial r}; \quad v_\theta = \frac{1}{r}\frac{\partial \phi}{\partial \theta}; \quad v_z = \frac{\partial \phi}{\partial z}$$

6.3 CIRCULATION AND VORTICITY

Circulation and vorticity are the two primary measures of rotation in a fluid. Details of these two topics are given in Section 4.9.

6.4 FLOW NETS (2D)

A line (or a curve) of constant ϕ is called an equipotential line (or curve). Similarly, a line (or curve) of constant ψ is called a streamline. A line of constant ψ is always tangential to the velocity vector and orthogonal to the equipotential lines. The velocity vector is always normal to the equipotential lines. A set of streamlines and equipotential lines with the spacing between streamlines the same as the spacing between equipotential lines is called a Flow Net. There are several methods of obtaining flow nets, such as

- Freehand drawing
- Plotting the equations of ψ and ϕ
- Conformal mapping in the complex plane
- Numerical solution of the Laplace equation.

Flow through a sluice gate will have a flow net as shown in Figure 6.2a in which ϕ and ψ (constants), respectively, represent equipotential lines and streamlines. Figure 6.2b shows the flow net for flow over a weir.

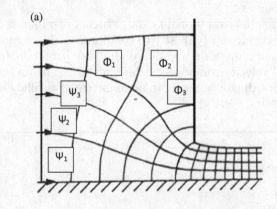

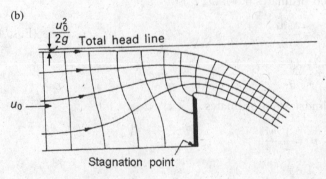

Figure 6.2 (a) Flow net for flow through an opening in a tank. (b) Flow net for flow over a weir.

The existence of a flow net implies

- Possible flow
- Irrotational flow
- 1D, 2D and axisymmetric
- Instantaneous conditions

If the flow is rotational, ϕ is not defined, but ψ is defined. Hence streamlines only can be drawn. Similarly, if the flow is 3D ψ is not defined but ϕ is defined, and hence the velocity potential lines only can be drawn.

The flow net tends to a square net as the spacing between equipotential lines and streamlines tends to zero. The velocity is obtained as the ratio of discharge to the flow area. The flow rate across two stream functions ψ and $\psi + \Delta\psi$ is $\Delta\psi$, and if the spacing between equipotential lines is Δs, then the velocity is given as (Figure 6.3)

$$V \approx \frac{\Delta\varphi}{\Delta s} \approx \frac{\Delta\psi}{\Delta n} \tag{6.15}$$

which means that the velocity increases with decreasing spacing of lines (pressure decreases from Bernoulli's equation). In these equations, s is the direction tangential to the streamline and n is the direction normal to the direction of s and towards the centre of curvature of the streamline.

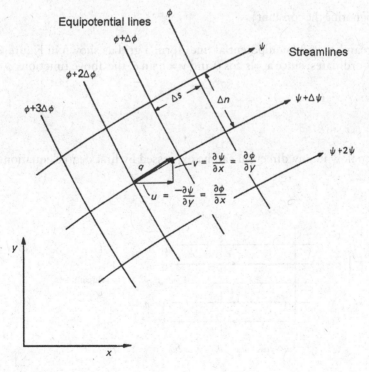

Figure 6.3 Elements of a flow net.

6.5 EXAMPLES OF FLOW NETS

6.5.1 Uniform flow

6.5.1.1 Uniform rectilinear flow in the x-direction

Defined by $u = u_0$, $v = 0$, then

$$\frac{\partial \phi}{\partial x} = u_0 \Rightarrow \phi = u_0 x + (f(y)) + \text{constant}$$

$$\frac{\partial \phi}{\partial y} = 0 \Rightarrow \phi = f(x) + \text{constant (not a function of } y)$$

Therefore, $\phi = u_0 x$ (ignoring the constant)   (6.16)
   Similarly, ψ can be obtained as,

$$u = \frac{\partial \psi}{\partial y} = u_0; \quad v = -\frac{\partial \psi}{\partial x} = 0$$

$$\frac{\partial \psi}{\partial y} = u_0 \Rightarrow \psi = u_0 y + (f(x)) + \text{constant}$$

$$\frac{\partial \psi}{\partial x} = 0 \Rightarrow \psi = \text{constant} + f(y)$$

Therefore,

$$\psi = u_0 y \,(\text{ignoring the constant}).$$ (6.17)

Hence the streamlines and equipotential lines form a grid as shown in Figure 6.4.
   In polar co-ordinates, since $x = r \cos \theta$ and $y = r \sin \theta$, the above functions ϕ and ψ will be

$$\phi = u_0 x = u_0 r \cos \theta$$ (6.18)

$$\psi = u_0 y = u_0 r \sin \theta$$ (6.19)

Note: Uniform flow in any direction can be expressed by first degree equations in x and y.

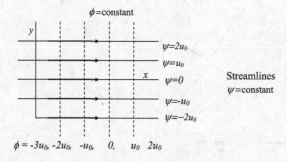

Figure 6.4 Flow net for uniform rectilinear flow (in the x-direction).

6.5.1.2 Uniform rectilinear flow in the y-direction

The velocities are

$$u = 0, \quad v = v_0$$

$$\Rightarrow \phi = v_0 y, \text{ and } \psi = -v_0 x$$

(6.20)

The net will look the same, but the streamlines will be in the vertical direction.

6.5.1.3 Uniform flow at an angle θ to the horizontal

$$\phi = u_0(x\cos\theta + y\sin\theta)$$

$$\psi = u_0(y\cos\theta - x\sin\theta)$$

$$u = u_0\cos\theta$$

$$v = u_0\sin\theta$$

The streamlines will be straight lines inclined at an angle θ.

6.5.2 Source and sink – radial flow

A source is a flow field in which fluid flows from a point radially outwards in equal amounts in all directions (Figure 6.5). The flow lines are straight radial lines. As the flow expands outwards, the velocity decreases because the area of flow increases. The velocity at a given radius is the same. Near the centre, all radial lines get close to each other; the flow area becomes smaller and smaller, resulting in an infinitely high velocity at the centre. The centre $(r = 0)$ is therefore considered as a singular point. The velocity field is given by

$$v_r = v(r); \quad v_\theta = 0$$

(6.21a)

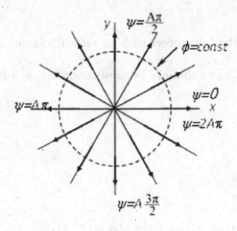

Figure 6.5 Flow net for a source.

From the equation of continuity for radial flow, which is

$$\frac{\partial v_r}{\partial r} + \frac{v_r}{r} = 0,$$

(6.21b)

from which

$$\frac{\partial v_r}{v_r} = -\frac{\partial r}{r}$$

(6.21c).

Integrating

$$\ln(v_r) = -\ln(r) + \text{constant}$$

$$\Rightarrow \ln(v_r) + \ln(r) = \text{constant}$$

$$\Rightarrow \ln(rv_r) = \text{constant}$$

(6.21d)

$$\Rightarrow rv_r = \text{constant} = K$$

The constant K is obtained by equating the rate of flow Q (volume rate per unit length) through the curve bounding the source. Then,

$$Q = \int_0^{2\pi} v_r(r\,d\theta)$$

(6.21e)

which gives

$$Q = 2\pi r v_r \text{ or } v_r = \frac{Q}{2\pi r}, \text{ for } r \neq 0.$$

(6.21f)

Substituting back

$$K = \frac{Q}{2\pi}$$

(6.21g)

and

$$v_r = \frac{Q}{2\pi r} = \frac{K}{r}; v_\theta = 0$$

(6.21h)

Q is defined as the **strength of the source** (volume rate of flow/unit depth). A sink is a negative source.

The stream function, ψ can be obtained by substitution in Eqs. 4.19a and 4.19b as follows:

$$v_r = \frac{1}{r}\frac{\partial \psi}{\partial \theta}; v_\theta = -\frac{\partial \psi}{\partial r}.$$

Therefore,

$$\frac{Q}{2\pi r} = \frac{1}{r}\frac{\partial \psi}{\partial \theta} \text{ and } -\frac{\partial \psi}{\partial r} = 0$$

$$\Rightarrow \psi = \frac{Q}{2\pi}\theta + \left[f(r) \text{ or, } a \text{ constant}\right] \text{ and } \psi = f(\theta) + \text{constant}$$

Comparing the two expressions for ψ, it can be seen that $f(r)$ is a constant which can be made equal to zero ($\psi = 0$ when $\theta = 0$). Then,

$$\psi = \frac{Q}{2\pi}\theta \tag{6.22}$$

Whether or not a velocity potential exists can be checked by substituting in the vorticity equation,

$$\omega_z = \frac{1}{2}\left[\frac{\partial v_\theta}{\partial r} - \frac{1}{r}\frac{\partial v_r}{\partial \theta} + \frac{v_\theta}{r}\right]$$

$$= 0 - \frac{1}{r}0 + 0 = 0 \tag{6.23}$$

Therefore, ϕ exists for sources and sinks, and it is given by Eqs. 6.3a and 6.3b:

$$v_r = \frac{\partial \phi}{\partial r} \quad \text{and} \quad v_\theta = \frac{1}{r}\frac{\partial \phi}{\partial \theta}$$

which when integrated as before, gives

$$\phi = \frac{Q}{2\pi}\ln(r) + \text{constant} \tag{6.24}$$

For a source, $Q > 0$, and for a sink, $Q < 0$.

In the above formulations, it was implied that the source/sink is located at the origin. If the source/sink is moved to a different location (a, b), the resulting stream function and the velocity potential function take the form

$$\psi = \frac{Q}{2\pi}\theta' = \frac{Q}{2\pi}\tan^{-1}\left(\frac{y-b}{x-a}\right) \tag{6.25}$$

$$\phi = \frac{Q}{2\pi}\ln(r') = \frac{Q}{2\pi}\ln\left(\sqrt{(x-a)^2 + (y-b)^2}\right) \tag{6.26}$$

where r' and θ', respectively, are the distance and the inclination from the horizontal to the line joining the new location to the point where the stream function and velocity potential are sought. These equations simplify to

$$\psi = \frac{Q}{2\pi}\theta \quad \text{and} \quad \phi = \frac{Q}{2\pi}\ln(r) \quad \text{when } a = 0 \text{ and } b = 0. \tag{6.27}$$

6.5.3 Vortices

In fluid dynamics, a vortex is a region where the flow rotates around an axis, which may be straight or curved. Vortices exist when there is a circulation of fluid. They can be seen in smoke rings, whirlpools, cyclones, tornados etc. The streamlines are concentric circles. They are described by the velocity distribution, vorticity, curl and circulation. Vortex flow is an important component of turbulent flow. Vortices can move, stretch, twist and interact in complex ways and carry angular momentum, linear momentum, energy and mass. In most vortices, the fluid velocity is greatest near the axis of rotation and decreases with increasing distance from the axis of rotation. In the context of vortices, a vortex line is analogous to a streamline, and a vortex tube to a stream tube. In a streamline, the local velocity vector is tangential to the streamline. In a vortex line, the local vorticity is tangential to the vortex line.

Vorticity is a vector that describes the local rotation of a point in the fluid, as seen by an observer moving along with it (recall Lagrangian approach of flow description). Mathematically, the vorticity is defined as the curl (or rotation) of the velocity field of the fluid, usually denoted by ω which should not be confused with the angular velocity vector of that portion of the fluid with respect to the external environment or to any fixed axis. Vortex line is related to the vorticity in the same way as streamline is related to the velocity as shown by their respective equations:

$$\frac{dx}{u} = \frac{dy}{v} = \frac{dz}{w} \quad (\text{streamline}) \tag{6.28}$$

$$\frac{dx}{\omega_x} = \frac{dy}{\omega_y} = \frac{dz}{\omega_z} \quad (\text{vortex line}) \tag{6.29}$$

Vortex lines passing through a closed loop is called a vortex tube. The strength of a vortex tube is the circulation around a closed loop on the surface of the tube and equal to the product of the mean vorticity and the cross sectional area. Vorticity is given as

$$\Omega_\zeta = \frac{\partial v}{\partial x} - \frac{\partial u}{\partial y} \tag{6.30}$$

Basically, there are two types of vortices: irrotational (free) and rotational (forced).

6.5.3.1 Irrotational (free) vortex

In a free vortex, there is no energy added or taken away. Therefore, the total energy (head) remains constant at all radii. The fluid particles rotate about their axis, but they do not spin around the axes of individual particles. The streamlines are still circles although they are called irrotational vortices. The angular velocity varies with radius whereas for a forced vortex it is the same for all radii.

For an irrotational[2] or free vortex, the streamlines are concentric circles, and the total head across streamlines does not vary (Figure 6.6). This means

$$v_r = 0; \quad v_\theta = v_\theta(r) \tag{6.31}$$

Since it is irrotational, it must satisfy the irrotationality condition, $\omega_z = 0$. Here r represents the radius of curvature of the streamlines.

$$\omega_z = \frac{\partial v_\theta}{\partial r} - \frac{1}{r}\frac{\partial v_r}{\partial \theta} + \frac{v_\theta}{r} = 0 \quad (\text{in polar co-ordinates}) \tag{6.32}$$

$$\Rightarrow \frac{\partial v_\theta}{\partial r} + \frac{v_\theta}{r} = 0$$

---

[2] The term "irrotational vortex" is slightly misleading.

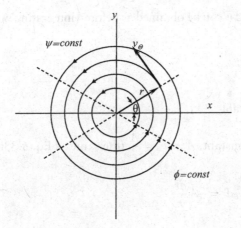

Figure 6.6 Flow net for an irrotational vortex.

which, when integrated, gives

$$\ln(rv_\theta) = \text{constant} \tag{6.33a}$$

or

$$v_\theta = \frac{A}{r} \tag{6.33b}$$

where A = constant which is determined from the definition of circulation (flow along a closed curve), Γ.[3] (The constant A here is the same as K in some other places.)

$$\Gamma = \int_0^{2\pi} v_\theta r\, d\theta = 2\pi r v_\theta = 2\pi A \tag{6.34}$$

In the above equation, $rd\theta$ is the area normal to v_θ.
Therefore,

$$A = \frac{\Gamma}{2\pi} \tag{6.35}$$

Hence,

$$v_r = 0 \quad v_\theta = \frac{\Gamma}{2\pi r} \text{ (except at } r = 0) \tag{6.36}$$

At $r = 0$, v_θ tends to infinity, but this is not possible. Therefore, $r = 0$ for an irrotational vortex is a singular point, and the fluid rotates like a solid body. The equation rv_θ = constant is not valid at the centre.

---

[3] Γ is also the strength of the vortex. Circulation is a vector, whereas strength is a scalar.

The two functions ϕ and ψ can be obtained as before $\left(\text{integrating } v_\theta = \dfrac{1}{r}\dfrac{\partial \phi}{\partial \theta} \text{ and } v_\theta = -\dfrac{\partial \psi}{\partial r}\right)$. They are

$$\phi = \frac{\Gamma}{2\pi}\theta + \text{constant} \tag{6.37}$$

$$\psi = -\frac{\Gamma}{2\pi}\ln(r) + \text{constant} \tag{6.38}$$

For a free vortex $rv_\theta = \text{Constant}$, $A \Rightarrow v_\theta = \dfrac{A}{r}$ (the same as Eq. 6.33b)

$$v_\theta = -\frac{d\psi}{dr} \Rightarrow d\psi = -v_\theta dr = -A\frac{dr}{r}$$

Stream function $\psi = -\displaystyle\int_{r_1}^{r_2} A\frac{dr}{r} = -A\ln\frac{r_2}{r_1}$ \hfill (6.39)

For a free vortex, the angular velocity of the streamlines varies with radius ($\omega = v_\theta r$). For a forced vortex it is the same. The surface profile of a free vortex can be obtained by considering a small annular element between two streamlines separated by dr, having a depth difference of dh. The centrifugal force acting on the element gives rise to a pressure difference between the inner and outer streamlines (p and $p + dp$).

Centrifugal force $= m\dfrac{v_\theta^2}{r} = \left(2\pi r dh dr\right)\rho\dfrac{v_\theta^2}{r}$

Pressure force $= \text{Pressure} \times \text{area} = dp(2\pi r dh)$

Equating the two forces

$$dp(2\pi r dh) = (2\pi r dh dr)\rho\frac{v_\theta^2}{r} \Rightarrow \frac{dp}{dr} = \rho\frac{v_\theta^2}{r}$$

Since $dp = \rho g dh$, $\dfrac{dp}{dr} = \rho g\dfrac{dh}{dr}$

Therefore, $\dfrac{dh}{dr} = \dfrac{1}{\rho g}\dfrac{dp}{dr} = \dfrac{1}{\rho g}\rho\dfrac{v_\theta^2}{r} \Rightarrow \dfrac{dh}{dr} = \dfrac{v_\theta^2}{gr} = \left(\dfrac{A}{r}\right)^2\dfrac{1}{gr}$

Integrating, $dh = \dfrac{A^2}{g}\dfrac{1}{r^3}dr$, the surface profile of a free vortex is

$$h_2 - h_1 = \frac{A^2}{2g}\left(\frac{1}{r_2^2} - \frac{1}{r_1^2}\right) \tag{6.40}$$

Note 1: Here too the vortices are assumed to be placed at the origin. If they are moved to a new location (a, b), the corresponding equations for the stream function and the velocity potential are

$$\psi = \frac{\Gamma}{2\pi}\ln(r') = \frac{\Gamma}{2\pi}\ln\left(\sqrt{(x-a)^2 + (y-b)^2}\right) \qquad (6.41)$$

$$\phi = -\frac{\Gamma}{2\pi}\theta' = \frac{\Gamma}{2\pi}\tan^{-1}\left(\frac{y-b}{x-a}\right) \qquad (6.42)$$

which simplify to the original equations when $a = 0$ and $b = 0$. (Compare Eqs. 6.41 and 6.42 with Eqs. 6.25 and 6.26.)

Note 2: The flow nets for a vortex and source (or sink) are the same except that the streamlines and equipotential lines are interchanged.

Note 3: The condition that the total head across streamlines is a constant can be obtained by substituting the irrotationality condition in the Navier–Stokes equations (together with the continuity equation). The proof is given in pp. 121–122 of Fluid Dynamics by J.W. Daily and D.R.F. Harleman (1966).

The fact that vorticity = 0, also **normally** implies circulation $(\Gamma) = 0$, because

$$\Gamma = \text{Vorticity} \times \text{Area}$$

Circulation depends upon the path chosen. It is zero except for circles and at $r = 0$.

But in the case of the irrotational vortex, this condition is not satisfied at $r = 0$. i.e., the circulation for a contour enclosing the origin $(r = 0)$ is **not** zero. This means that either the fluid does not occupy the space at $r = 0$, or that the fluid at $r = 0$ is rotational. In an inviscid fluid, an irrotational vortex, once formed, cannot be stopped. Theoretically, it cannot also be generated in practice. Examples of irrotational flow include tornados and water spouts.

The total head, H, throughout the fluid is given by

$$H = \frac{p}{\rho g} + h + \frac{v_\theta^2}{2g} = \text{Constant} \qquad (6.43)$$

Using the boundary conditions,

$$\left. \begin{array}{l} h = h_0 \\ p = p_0 \\ v_\theta \to 0 \end{array} \right\} \text{ as } r \to \infty$$

leads to

$$p - p_0 = \rho g(h_0 - h) - \frac{1}{2}\rho v_\theta^2$$

If the surface is exposed to the atmosphere, $p - p_0 = 0$. Therefore,

$$h_0 - h = \frac{v_\theta^2}{2g} \qquad (6.44)$$

which is the drawdown of the water surface (Figure 6.7).

6.5.3.2 Rotational (forced) vortex

A forced vortex occurs when the fluid is rotating about an axis as a rigid body. This can be demonstrated by rotating a cylindrical vessel containing a fluid about a vertical axis with

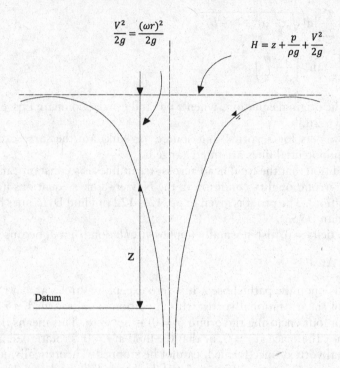

$$\frac{V^2}{2g} = \frac{(\omega r)^2}{2g}$$

$$H = z + \frac{p}{\rho g} + \frac{V^2}{2g}$$

z

Datum

Figure 6.7 Water surface profiles for an irrotational vortex.

constant angular velocity, or a paddle rotated in a fluid. An external force is required to start a forced vortex. In a forced vortex, all particles of the fluid have the same angular velocity about some fixed axis. The fluid rotates like a solid about that axis.

In such a situation, the fluid will displace and assume the free surface position shown in Figure 3.16a, if the vessel is open. If it is closed, the pressure of the fluid will rise, and the piezometric head will have a profile similar to the free surface (Figure 3.16b). The piezometric pressure increases with radius. This is the principle of the centrifugal pump where the fluid is supplied at the centre and discharged at the periphery at a much higher pressure.

In an open container, the pressure at the free surface is atmospheric. The free surface rises according to

$$z - z_0 = \frac{\omega^2 r^2}{2g} \tag{6.45}$$

In a forced vortex, there is no deformation of fluid. (There is no shear strain because there is no relative motion.) The vortex formation is purely due to the centrifugal action $\left(m v_\theta^2 / r\right)$.

The equation of a forced vortex is obtained from the consideration of the forces acting on the fluid element shown (Figure 6.8) as

$$p\left(r\delta\theta \times 1\right) - \left(p + \frac{\partial p}{\partial r}\delta r\right)(r + \delta r)\delta\theta \cdot 1 = -(r\delta\theta)\delta r \rho \cdot \frac{v_\theta^2}{r} \tag{6.46}$$

$$\Rightarrow \frac{\partial p}{\partial r} = \frac{\rho v_\theta^2}{r} = \rho \frac{(\omega r)^2}{r} = \rho \omega^2 r$$

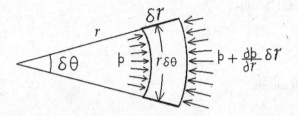

Figure 6.8 Force diagram for a radial fluid element.

This equation gives the pressure variation radially.

If an element at the same radius but with elevation varying is considered, then

$$p\delta A - \left(p + \frac{\partial p}{\partial z}\delta z\right)\delta A - \rho g\delta A\delta z = 0$$

$$\Rightarrow \frac{\partial p}{\partial z} = -\rho g$$

Since $p = p(z, r)$,

$$dp = \frac{\partial p}{\partial r}dr + \frac{\partial p}{\partial z}dz$$

$$= \rho\omega^2 rdr - \rho gdz$$

which, upon integration, gives

$$p = \rho\omega^2\frac{r^2}{2} - \rho gz + \text{constant} \tag{6.47a}$$

6.5.3.2.1 Boundary conditions

When $r = 0$; $v_\theta = 0$
When $r = R$; $v_\theta = \omega R$
When $r = 0$; $z = z_0$, $p = p_0$ (datum for elevation and pressure)

Substituting the boundary conditions, the constant can be determined. Then,

$$p = \rho\omega^2\frac{r^2}{2} - \rho g(z - z_0) + p_0 \tag{6.47b}$$

The locus of the free surface is obtained by substituting $p = p_0$. Then,

$$(z - z_0) = \frac{\omega^2 r^2}{2g} \tag{6.47c}$$

which is a paraboloid of revolution. This is the height of the free surface from the datum level where the pressure is atmospheric (drawdown). In an open container, the free surface will rise according to the above equation.

In this case, the total head is

$$H = \frac{p}{\rho g} + z + \frac{v_\theta^2}{2g},\qquad (6.48)$$

which increases with radius.

6.6 SUPERPOSITION OF FLOWS

Stream functions and velocity potentials satisfy the Laplace equation which is linear. Therefore, superposition of stream functions and velocity potentials is possible. Simple flow systems can be combined to give complex flow systems using this property. Some examples are highlighted below:

6.6.1 Source + sink

When a source and sink of equal strength Q are placed at a distance $2a$ apart (Figure 6.9a and b) the equations for ϕ and ψ are obtained by superposition.

$$\phi = \frac{Q}{2\pi}\left(\ln(r) - \ln(r')\right) + \text{constant}$$

$$= -\frac{Q}{2\pi}\ln\frac{r'}{r} + \text{constant} \qquad (6.49)$$

(a)

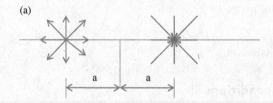

(b)

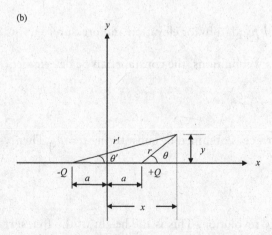

Figure 6.9 (a) Superposition of source and sink of equal strength Q. (b) Definition sketch for the superposition of source and sink of equal strength Q.

which can also be written as

$$\phi = \frac{Q}{2\pi} \ln \sqrt{\frac{(x+a)^2 + y^2}{(x-a)^2 + y^2}}$$

where $r^2 = (x+a)^2 + y^2$; $r'^2 = (x-a)^2 + y^2$

$$\psi = \frac{Q}{2\pi}(\theta - \theta') \tag{6.50}$$

For any streamline ψ is a constant. Therefore $(\theta - \theta')$ is a constant. The streamlines are therefore, circles passing through the origins of the source and the sink.

The streamlines can be constructed as follows:

$$\tan\theta = \frac{y}{x-a}; \quad \tan\theta' = \frac{y}{x+a}$$

and, from trigonometry

$$\tan(\theta - \theta') = \frac{\tan\theta - \tan\theta'}{1 + \tan\theta \tan\theta'}$$

Substituting for $\tan\theta$ and $\tan\theta'$,

$$\tan(\theta - \theta') = \frac{2ay}{x^2 + y^2 - a^2}$$

Substituting into the expression for ψ,

$$\tan\frac{2\pi\psi}{Q} = \frac{2ay}{x^2 + y^2 - a^2}$$

or

$$\psi = \frac{Q}{2\pi} \arctan\left(\frac{2ay}{x^2 + y^2 - a^2}\right) \tag{6.51a}$$

or

$$x^2 + (y - ka)^2 = \left(1 + k^2\right)a^2 \tag{6.51b}$$

where $k = \cot\dfrac{2\pi\psi}{Q}$ that can be considered as an arbitrary constant.

These are equations of circles with centres on the y-axis at distances 'ak' from the origin and with radii equal to $a\sqrt{1 + k^2}$. Hence the streamlines are as shown in Figure 6.10.

Similarly, the equipotential lines can be obtained by making ϕ = constant. Then,

$$y^2 + \left(x + \frac{c+1}{c-1}a\right)^2 = \left(\frac{a\, 2\sqrt{c}}{c-1}\right)^2 \tag{6.52}$$

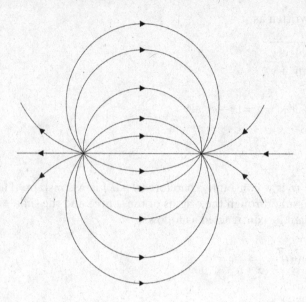

Figure 6.10 Streamlines for a source and a sink.

where $c = e^{\frac{4\pi\phi}{Q}}$ is an arbitrary constant.

These are also circles with the centres on the x-axis at distances $\dfrac{a(c+1)}{(c-1)}$ from the origin

and with radii $\dfrac{2a\sqrt{c}}{(c-1)}$. They will be orthogonal to the set of streamlines.

6.6.2 Doublet

This is a special case when the source and sink are allowed to approach each other, keeping the product of their strength and the distance apart constant. It has infinite strength and an infinitesimal distance of separation. In this case, it is easier to use

$$r' = r + dr; \quad \theta' = \theta - d\theta; \quad \text{and } 2a = ds$$

Then, by substituting in Eq. 6.49,

$$\phi = -\frac{Q}{2\pi} \ln\left(1 + \frac{dr}{r}\right) + \text{constant}$$

$$= -\frac{Q}{2\pi}\left[\frac{dr}{r} - \frac{1}{2}\left(\frac{dr}{r}\right)^2 + \cdots\right] + \text{constant}$$

$$\approx -\frac{Q}{2\pi}\frac{dr}{r} + \text{higher order terms} + \text{constant} \tag{6.53}$$

Substituting the trigonometric relationship,

$$r^2 = \left(ds\right)^2 + \left(r + dr\right)^2 - 2ds\left(r + dr\right)\cos\left(\theta - d\theta\right)$$

which is approximately equal to

$$\frac{dr}{r} = \frac{ds}{r}\cos\theta$$

into the equation for ϕ (Eq. 6.53), gives

$$\phi = \text{limit}\left\{-\frac{Q}{2\pi}\frac{ds}{r}\cos\theta\right\} + \text{constant}$$

$$ds \to 0$$

$$Q \to \infty$$

$$= -\frac{\mu}{2\pi r}\cos\theta + \text{constant} \tag{6.54}$$

where $\mu = Q\Delta s$ is the strength of the doublet, with the doublet oriented in the negative x-direction.

The corresponding velocity components are

$$v_r = \frac{\partial\phi}{\partial r} = \frac{\mu\cos\theta}{2\pi r^2} \tag{6.55}$$

$$v_\theta = \frac{1}{r}\frac{\partial\phi}{\partial\theta} = \frac{\mu\sin\theta}{2\pi r^2} \tag{6.56}$$

Similarly, the stream function is given as

$$\psi = \frac{\mu\sin\theta}{2\pi r} + \text{constant} \tag{6.57}$$

The constants in both the above equations (Eqs. 6.54 and 6.57) can be made equal to zero by taking $\psi = 0$ at $\theta = 0$ (or $\phi = 0$ at $\theta = \pi/2$).

For any arbitrary value of ψ, if Eq. 6.57 is written as

$$r_0 = \frac{\mu}{2\pi\psi} \quad \text{(for } \psi = 0, \pm 1, \pm 2, \ldots) \tag{6.58}$$

then,
$r = r_0 \sin\theta$ for streamlines,
and
$r = R_0 \cos\theta$ for equipotential lines
if

$$R_0 = \frac{\mu}{2\pi\phi} \quad \text{(for } \phi = 0, \pm 1, \pm 2, \ldots) \tag{6.59}$$

These two equations represent circles passing through the origin with centres on the x-axis (for ϕ) and y-axis (for ψ) with diameters r_0 (for ψ) and R_0 (for ϕ). The set of streamlines and equipotential lines is shown in Figure 6.11.

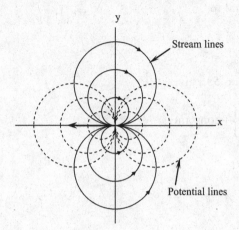

Figure 6.11 Streamlines and equipotential lines for doublet.

6.6.3 Vortex pair – dipole

This is analogous to a doublet, and is defined by a pair of vortices of equal strength but of opposite rotation separated by a distance Δs. If the distance $\Delta s \to 0$, and the strength tends to infinity keeping the product (*strength* $x\Delta s$) a constant, the result is a **dipole** ($\Gamma\Delta s =$ constant $= v$, as $\Gamma \to \infty$ and $\Delta s \to 0$). In some text books, a doublet is also referred to as a dipole. To avoid confusion, the term vortex pair is more appropriate.

The velocity potential and stream functions are obtained by superposition of vortex flows:

$$\phi = -\frac{\Gamma\theta}{2\pi} + \frac{\Gamma}{2\pi}(\theta + d\theta) + \text{constant}$$

$$= \frac{\Gamma}{2\pi}d\theta + \text{constant} \tag{6.60}$$

In a manner similar to a Doublet, it can be shown that the velocity potential of a dipole is identical to that of a doublet if doublet strength, μ is replaced by the dipole strength, v.

$$\phi = \frac{v\cos\theta}{2\pi r} + \text{constant} \tag{6.61}$$

$$\psi = -\frac{v\sin\theta}{2\pi r} + \text{constant} \tag{6.62}$$

6.6.4 Uniform flow + source (Rankine half body)

In this combination, the rectilinear uniform flow will have straight streamlines from left to right on the far upstream side of the source. When the uniform flow meets the source, the streamlines will be deflected but will become straight far downstream of the source. The flux from the source divides the flow pattern to the upper part and a lower part with the x-axis passing through the origin of the source as the centreline. A close example of a half body is the flow past a bridge pier.

For uniform flow,

$$\phi = u_0 x = u_0 r \cos\theta$$

$$\psi = u_0 y = u_0 r \sin\theta$$

For a source,

$$\phi = \frac{Q}{2\pi} \ln(r)$$

$$\psi = \frac{Q}{2\pi} \theta$$

Therefore, the velocity potential for the combined flow is

$$\phi = \frac{Q}{2\pi} \ln(r) + u_0 r \cos\theta$$

and the stream function for the combined flow is

$$\psi = \frac{Q}{2\pi} \theta + u_0 r \sin\theta \tag{6.63}$$

Method of plotting the flow net

a. Draw streamlines for the source and number them in accordance with the convention that flow is from left to right across the direction of positive grad ψ. The direction of streamlines can be obtained by calculating the velocities $u = \dfrac{\partial\psi}{\partial y}$; $\quad v = -\dfrac{\partial\psi}{\partial x}$.

b. Same for rectilinear flow
c. Add ψ values where they intersect
d. Join points of common new ψ values to give the combined flow streamlines.

The streamlines are symmetrical about the x-axis.

The velocity components (v_r, v_θ) from the stream function are

$$v_r = \frac{1}{r}\frac{\partial\psi}{\partial\theta}; \quad v_\theta = -\frac{\partial\psi}{\partial r}$$

$$= \frac{1}{r}\left[\frac{Q}{2\pi} + u_0 r \cos\theta\right] = \frac{Q}{2\pi r} + u_0 \cos\theta; \quad v_\theta = -u_0 \sin\theta \tag{6.64}$$

and

$$V^2 = v_r^2 + v_\theta^2$$

At the stagnation point, the velocity components vanish. Therefore,

$$\frac{Q}{2\pi r} + u_0 \cos\theta = 0$$

$$\Rightarrow r \cos\theta = -\frac{Q}{2\pi u_0} \Rightarrow x = -\frac{Q}{2\pi u_0} \tag{6.65}$$

and

$$-u_0\sin\theta = 0 \Rightarrow \sin\theta = 0 \Rightarrow \theta = 0, \pi, 2\pi \text{ etc.}$$

$$y = r\sin\theta = 0$$

Therefore, the stagnation point is $\left(-\dfrac{Q}{2\pi u_0}, 0\right)$.

Considering the streamline $\psi = \psi_0$ through the stagnation point ($\theta = \pi$),

$$\psi_0 = \left(\frac{Q}{2\pi}\pi + u_0 r\sin\pi\right) = \frac{Q}{2} \tag{6.66}$$

For other values of θ along this streamline,

$$y = \frac{Q}{2u_0}\left(1 - \frac{\theta}{\pi}\right) \tag{6.67}$$

At $\theta = 0$, $y = \dfrac{Q}{2u_0}$ – maximum ordinate

$\theta = \pi/2$, $y = \dfrac{Q}{4u_0}$ – upper ordinate at the origin

$\theta = \pi$, $y = 0$ – leading point ($x = -\dfrac{Q}{2\pi u_0}$)

$\theta = 3\pi/2$, $y = -\dfrac{Q}{4u_0}$ – lower ordinate at the origin

$\theta = 2\pi$, $y = -\dfrac{Q}{2u_0}$ – maximum ordinate (negative)

The above equation for y (Eq. 6.67) represents the streamline passing through the stagnation point. This is called the **zero streamline** (for convenience), and it divides the flow into two regions. This dividing streamline can be replaced by a solid wall because there is no flow across this zero streamline. The source is enclosed in the solid wall. The external uniform stream in the presence of this wall is given by the equation derived above. The flow net is shown in Figure 6.12.

Note 1: The height of the wall $\dfrac{Q}{2u_0}$ is attained asymptotically (i.e., $\theta = 0$ and $x \to \infty$). Physically, this combination can be considered as the flow past one end of a long bluff body in a stream. It is also called a **Rankine half body**.

Note 2: The stagnation point is the safest point in a wind flow past a bluff body.

6.6.5 Uniform flow + doublet (flow past a circular cylinder)

Flow past a circular cylinder is equivalent to the superposition of uniform flow and a doublet. The velocity potential, the stream function and the respective velocity components are

$$\phi = u_0 r\cos\theta + \frac{\mu}{2\pi r}\cos\theta \tag{6.68a}$$

$$\psi = u_0 r\sin\theta - \frac{\mu}{2\pi r}\sin\theta \tag{6.68b}$$

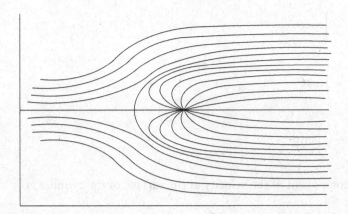

Figure 6.12 Uniform flow + source (Rankine half body).

$$v_r = \left(u_0 - \frac{\mu}{2\pi r^2} \right) \cos\theta \tag{6.68c}$$

$$v_\theta = -\left(u_0 + \frac{\mu}{2\pi r^2} \right) \sin\theta \tag{6.68d}$$

where $-\mu$ is the doublet strength with the source placed upstream. Note that the equations derived for a doublet have the source at the downstream end, and therefore the strength is considered as $+\mu$.

For $\psi = 0$ (or constant if the constant in the ψ function is considered), $\sin\theta = 0$ giving $\theta = 0, \pi$; or,

$$r^2 = \frac{\mu}{2\pi u_0} = \text{constant} = a^2 \quad (a \text{ is the radius of the cylinder})$$

Therefore, $\psi_0 = 0$ represents a streamline of a circle of radius a and passing through the x-axis.

At the surface of the cylinder where $r = a$, the radius of the cylinder, the radial component of velocity is zero which leads to the relationship

$$\theta = \frac{\pi}{2}, \frac{3\pi}{2}$$

or

$$u_0 = \frac{\mu}{2\pi a^2}$$

When this is substituted into the equations for velocity potential, stream function and the velocity components, they simplify to

$$\psi = u_0 r \left(1 - \frac{a^2}{r^2} \right) \sin\theta \tag{6.69a}$$

$$\phi = u_0 r \left(1 + \frac{a^2}{r^2}\right) \cos\theta \tag{6.69b}$$

$$v_r = u_0 \left(1 - \frac{a^2}{r^2}\right) \cos\theta \tag{6.69c}$$

$$v_\theta = -u_0 \left(1 + \frac{a^2}{r^2}\right) \sin\theta \tag{6.69d}$$

The tangential component of the velocity at the surface of the cylinder is

$$v_\theta = -u_0 \left(1 + \frac{a^2}{a^2}\right) \sin\theta = -2u_0 \sin\theta \tag{6.70}$$

which attains its maximum value of $2u_0$ for $\theta = \dfrac{\pi}{2}, \dfrac{3\pi}{2}$. The stagnation points correspond to $\theta = 0, \pi$ when $v_\theta = 0$.

The combined velocity potential and stream function become

$$\phi = u_0 r \cos\theta + \frac{\Gamma}{2\pi r} \cos\theta \quad \left(\phi = u_0 r \cos\theta - \frac{\mu}{2\pi r} \cos\theta\right) \tag{6.71a}$$

$$\psi = u_0 r \sin\theta - \frac{\Gamma}{2\pi r} \sin\theta \tag{6.71b}$$

The velocity components are then

$$v_r = \cos\left(u_0 - \frac{\mu}{2\pi r^2}\right) \tag{6.71c}$$

$$v_\theta = -\sin\left(u_0 + \frac{\mu}{2\pi r^2}\right) \tag{6.71d}$$

The velocity components at the surface where $r = a$ are

$$v_r = u_0 \left(1 - \frac{a^2}{a^2}\right) \cos\theta = 0 \tag{6.72a}$$

$$v_\theta = -u_0 \left(1 + \frac{a^2}{a^2}\right) \sin\theta = -2u_0 \sin\theta \tag{6.72b}$$

$$v_r = 0 \quad \text{when} \quad u_0 = \frac{\mu}{2\pi r^2}$$

$$\phi = u_0 x + \frac{\mu \cos\theta}{2\pi r}$$

$$= u_0 \left(r + \frac{a^2}{r^2}\right) \cos\theta \tag{6.72c}$$

Similarly, the stream function for the combined flow is

$$\psi = u_0 y - \frac{\mu \sin \theta}{2\pi r} = u_0 r \sin \theta - \frac{\mu \sin \theta}{2\pi r} = \left(u_0 r - \frac{\mu}{2\pi r} \right) \sin \theta \qquad (6.72d)$$

For $\psi = 0$ (or constant if the constant in the ψ function is considered), $sin\ \theta = 0$ giving $\theta = 0$, π; or,

$$r^2 = \frac{\mu}{2\pi u_0} = \text{constant} = a^2 \quad (a \text{ is the radius of the cylinder})$$

Therefore, $\psi_0 = 0$ represents a streamline of a circle of radius R and passing through the x-axis. Hence

$$\psi = u_0 \left(r - \frac{a^2}{r} \right) \sin \theta \text{ (substituting for } \mu) \qquad (6.73)$$

The cylinder of radius a may be taken as the solid boundary.

The velocity components are

$$v_r = \frac{1}{r} \frac{\partial \psi}{\partial \theta} = u_0 \left(1 - \frac{a^2}{r^2} \right) \cos \theta \qquad (6.74a)$$

$$v_\theta = -\frac{\partial \psi}{\partial r} = -u_0 \left(1 + \frac{a^2}{r^2} \right) \sin \theta \qquad (6.74b)$$

On the surface of the cylinder $r = a$, and therefore $v_r = 0$, $v_\theta = -2u_0 \sin \theta$.

The stagnation points in the flow correspond to $v_r = 0$, $v_\theta = 0$ (i.e. when $r = a$, $\theta = 0$, π). v_θ varies from zero at $\theta = \pi$ to $2u_0$ at $\theta = \pi/2$ and $\theta = -\pi/2$ (i.e., at the top and bottom).

Both ψ and ϕ satisfy Laplace equation. The boundary conditions which need to be satisfied are

i. $v_r = \dfrac{\partial \phi}{\partial r} = 0$ at $r = a$

ii. $\left. \begin{array}{l} u = u_0 \\ v = 0 \end{array} \right\}$ as $r \to \infty$

The tangential velocity components on the surface of the cylinder are

$$v_\theta = -2u_0 \sin \theta \text{ which has a maximum value of } |2u_0| \qquad (6.75)$$

The surface pressure is calculated using Bernoulli's equation:

$$\frac{p_0}{\rho g} + \frac{u_0^2}{2g} = \frac{p_s}{\rho g} + \frac{u_s^2}{2g}$$

where the subscript s denotes the surface of the cylinder. Since $u_s = -2u_0 \sin \theta$, the surface pressure can be calculated as

$$p_s = p_0 + \frac{1}{2} \rho u_0^2 \left(1 - 4 \sin^2 \theta \right) \qquad (6.76)$$

The drag and lift forces are given by

$$D = -\int_0^{2\pi} p_s a \cos\theta \, d\theta = -\int_0^{2\pi} p_0 a \cos\theta \, d\theta - \frac{\rho u_0^2}{2}\int_0^{2\pi}\left(1 - 4\sin^2\theta\right)\cos\theta \, d\theta \tag{6.77a}$$

$$L = -\int_0^{2\pi} p_s a \sin\theta \, d\theta = -\int_0^{2\pi} p_0 a \sin\theta \, d\theta - \frac{\rho u_0^2}{2}\int_0^{2\pi}\left(1 - 4\sin^2\theta\right)\sin\theta \, d\theta \tag{6.77b}$$

From the symmetry of the pressure distribution, it can be shown that there is no drag or lift on the cylinder. The integration of Eq. 6.77 yields $D = 0$ and $L = 0$. However, physical evidence shows that drag exists. This is called the **D'Alembert's paradox** (after Jean le Rond D'Alembert (1717–1783)). The non-zero value for the drag is caused by the viscosity of fluid, which has been assumed to be an ideal fluid. The actual flow is however, very nearly the same as the potential flow solution up to the point of separation (see Figures 6.13 and 6.14).

6.6.6 Uniform flow + source + sink (Rankine oval)

When a source and a sink of equal strength and placed equidistant (*a*) from the origin together with a uniform flow, the resultant leads to an oval-shaped set of streamlines known as Rankine Oval (Figure 6.15). The velocity potential is given as

$$\phi = u_0 r \cos\theta + \frac{Q}{2\pi}\ln\left\{\frac{\sqrt{r^2 + a^2 + 2ar\cos\theta}}{\sqrt{r^2 + a^2 - 2ar\cos\theta}}\right\} \tag{6.78}$$

and the stream function as

$$\psi = u_0 r \sin\theta - \frac{Q}{2\pi}\tan^{-1}\left(\frac{2ar\sin\theta}{r^2 - a^2}\right) = u_0 y - \frac{Q}{2\pi}\tan^{-1}\left(\frac{2ay}{x^2 + y^2 - a^2}\right) \tag{6.79}$$

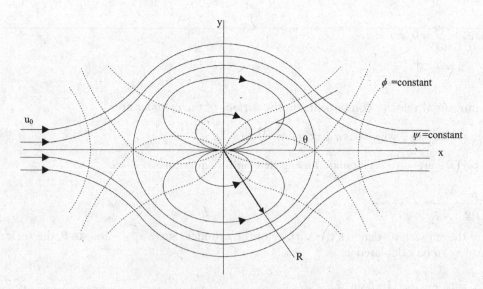

Figure 6.13 Flow net for combined uniform flow and doublet (potential flow solution).

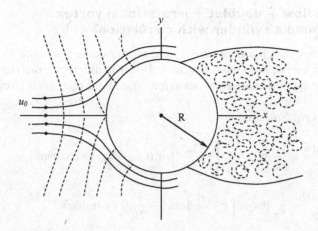

Figure 6.14 Actual flow net for the combined uniform flow and a doublet.

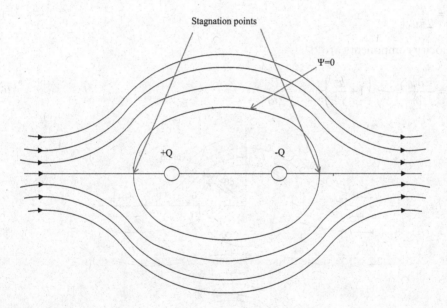

Figure 6.15 Rankine oval.

The velocity components are given as

$$u = \frac{Q}{2\pi} \left\{ \frac{r\cos\theta + a}{r^2 + a^2 + 2ar\cos\theta} - \frac{r\cos\theta - a}{r^2 + a^2 - 2ar\cos\theta} \right\} \tag{6.80a}$$

$$v = \frac{Q}{2\pi} \left\{ \frac{r\sin\theta}{r^2 + a^2 + 2ar\cos\theta} - \frac{r\sin\theta}{r^2 + a^2 - 2ar\cos\theta} \right\} \tag{6.80b}$$

The stagnation points are at $x = \pm l$ where l, the distance from the origin is given by

$$l = \sqrt{\frac{Qa}{\pi u_0} + a^2} \tag{6.81}$$

6.6.7 Uniform flow + doublet + irrotational vortex (flow around a cylinder with circulation)

This refers to the flow when a doublet is placed in a uniform flow with circulation (uniform flow + irrotational vortex + doublet) (Figure 6.16). It is different from the examples considered so far because the flow pattern is not symmetric about the x-axis (vortex assumed to be clockwise).

Adding the relevant functions

$$\psi = u_0 y - \frac{\mu \sin \theta}{2\pi r} + \frac{\Gamma}{2\pi} \ln(r) = u_0 \left(r - \frac{a^2}{r} \right) \sin \theta + \frac{\Gamma}{2\pi} \ln(r) + \text{constant} \qquad (6.82)$$

$$\phi = u_0 x + \frac{\mu \cos \theta}{2\pi r} - \frac{\Gamma}{2\pi} \theta = u_0 \left(r + \frac{a^2}{r^2} \right) \cos \theta - \frac{\Gamma}{2\pi} \theta + \text{constant} \qquad (6.83)$$

(The negative sign accounts for the clockwise circulation.)

$$a^2 = \frac{\mu}{2\pi u_0} \qquad (6.84)$$

The velocity components are

$$v_\theta = -\frac{\partial \psi}{\partial r} = -u_0 \left(1 + \frac{a^2}{r^2} \right) \sin \theta - \frac{\Gamma}{2\pi r} \qquad (6.85a)$$

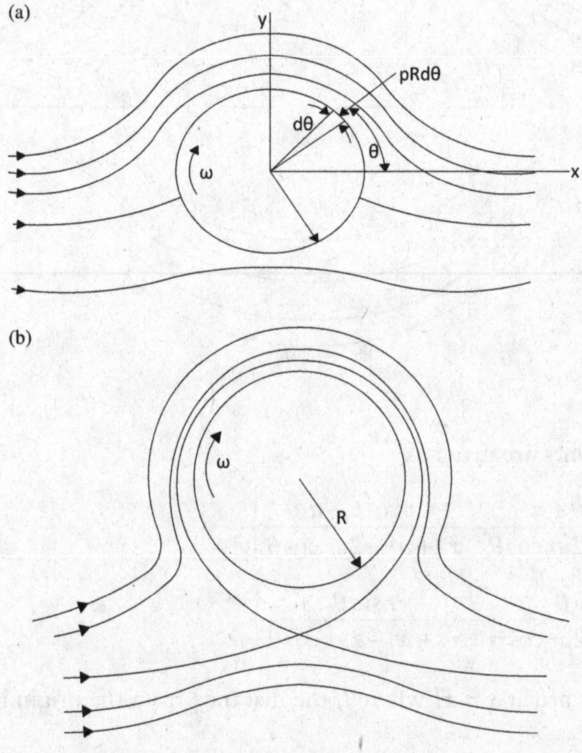

(a)

(b)

Figure 6.16 Flow around a rotating circular cylinder. (a) Stagnation points on the surface of the cylinder; (b) stagnation point outside the surface of the cylinder.

$$v_r = \frac{1}{r}\frac{\partial \psi}{\partial \theta} = u_0\left(1 - \frac{a^2}{r^2}\right)\cos\theta \tag{6.85b}$$

It can be seen that v_r is the same as for the case with no circulation. v_θ also is unaffected as $r \to \infty$ $\left(\text{because } \frac{\Gamma}{2\pi r} \to 0\right)$. Both ϕ and ψ satisfy the Laplace equation.

At the stagnation point, both v_r, v_θ must vanish. Therefore,

$$v_r = 0 \text{ implies } r = a \text{ or}\left(\theta = \frac{\pi}{2}, \frac{3\pi}{2},\dots\text{etc}\right).$$

$$v_\theta = 0 \text{ implies} -u_0\left(1 + \frac{a^2}{r^2}\right)\sin\theta = \frac{\Gamma}{2\pi r}$$

Because $r = a$ at the surface of the cylinder,

$$-u_0\left(1 + \frac{a^2}{a^2}\right)\sin\theta = \frac{\Gamma}{2\pi a}$$

$$\Rightarrow \sin\theta = -\frac{\Gamma}{4\pi au_0} \tag{6.86}$$

The tangential velocity on the surface of the cylinder is

$$v_\theta = -2u_0\sin\theta - \frac{\Gamma}{2\pi a} \tag{6.87a}$$

and

$$v_r = 0 \tag{6.87b}$$

At the stagnation point $v_\theta = 0$ gives

$$\sin\theta = -\frac{\Gamma}{4\pi au_0} \Rightarrow \sin(-\theta) = \frac{\Gamma}{4\pi au_0} = \left[\sin(\pi + \theta)\right]$$

(a) $\Gamma < 4\pi au_0,\ \sin(-\theta) < 1$
   There are two stagnation points diametrically opposite to each other.
(b) $\Gamma = 4\pi au_0,\ \sin(-\theta) = 1$

$$\theta = -\frac{\pi}{2}$$

   implying the two stagnation points coincide
(c) $\Gamma > 4\pi au_0,\ \sin(-\theta) > 1,\ \theta$ is imaginary
   implying the stagnation point is outside the surface.

For case (c), since $r > a$, θ must be $\frac{\pi}{2}$ or $\frac{3\pi}{2}$ (from the condition $v_r = 0$)

For the condition $v_\theta = 0 \left(\sin\theta = -\dfrac{\Gamma}{4\pi u_0 a} \right)$ to be satisfied, θ can only be $\dfrac{3\pi}{2}$ because Γ is not negative. Therefore, r is given by $\left(\text{at } \theta = \dfrac{3\pi}{2} \right)$

$$u_0\left(1 + \frac{a^2}{r^2}\right) = \frac{\Gamma}{2\pi r}$$

leading to

$$r = \frac{\Gamma}{4\pi u_0} + \frac{1}{2}\sqrt{\left(\frac{\Gamma}{2\pi u_0}\right)^2 - 4a^2} \qquad (6.88)$$

The pressure at any point in the field of flow (surface of cylinder) is obtained by Bernoulli's equation:

$$\frac{p - p_0}{\rho g} = \frac{u_0^2}{2g} - \frac{v_\theta^2}{2g} \quad (v_r = 0)$$

$$= \frac{u_0^2}{2g} - \frac{1}{2g}\left(2u_0\sin\theta + \frac{\Gamma}{2\pi a}\right)^2$$

The drag and lift are given by

$$\text{Drag} = -\int_0^{2\pi}(p - p_0)\cos\theta(a\,d\theta) = 0 \qquad (6.89a)$$

and

$$\text{Lift} = \rho u_0 \Gamma \quad \text{per unit length of cylinder.} \qquad (6.89b)$$

This relationship (Eq. 6.89b) is also known as the Kutta–Joukowski theorem. This effect is called Magnus effect in honour of Heinrich Magnus (1802–1870). It can be seen when spinning balls such as golf, baseball and tennis when hit in a horizontal direction follows a curved trajectory.

Note: Zero radial velocity is expected around a circular impervious contour.

6.6.8 Bathtub vortex

Superposition of a sink and a vortex gives what is known as a 'bathtub vortex'. An example of a bathtub vortex is when water drains from a rotating cylindrical container through a bottom hole. The respective velocity potential and stream function are

$$\phi = \frac{Q}{2\pi}\ln(r) + \frac{\Gamma}{2\pi}\theta \qquad (6.90a)$$

$$\psi = \frac{Q}{2\pi}\theta - \frac{\Gamma}{2\pi}\ln(r) \qquad (6.90b)$$

There is some confusion about the direction of swirl of the vortex in the drain of the bathtub or the kitchen sink. Some claim that the direction of swirl is anti-clockwise in the northern hemisphere and clockwise in the southern hemisphere. However, this claim is based on the effect of Coriolis acceleration that applies to large scale motions. The draining of a bathtub or a kitchen sink is a small-scale motion that has no influence of the Coriolis effect.

Example 6.1

A flow field is given by $\psi = x^3 y$. Determine the velocity potential ϕ for the flow.

Existence of ψ implies the existence of flow. However, for ϕ to exist, the flow must be irrotational. If the flow is irrotational, the Laplace equation should be satisfied.

Therefore,

$$\frac{\partial^2 \psi}{\partial x^2} + \frac{\partial^2 \psi}{\partial y^2} = 0$$

$$6xy + 0 = 0 \,(\text{not satisfied})$$

Hence, ϕ is not defined for this flow.

Example 6.2

What type of flow does the following flow field constitute?

$$V = (a + by - cz)i + (d - bx + ez)j + (f + cx - ey)k$$

where a, b, c, d, e, and f are constants.

   i. First, check the continuity equation

$$\frac{\partial u}{\partial x} = 0; \;\; \frac{\partial v}{\partial y} = 0 \text{ and } \frac{\partial w}{\partial z} = 0$$

      Hence 3D incompressible flow is possible.

  ii. Check whether the flow is irrotational:

      The vorticity components are

$$\omega_z = \frac{1}{2}\left(\frac{\partial v}{\partial x} - \frac{\partial u}{\partial y}\right) = -b$$

$$\omega_x = \frac{1}{2}\left(\frac{\partial w}{\partial y} - \frac{\partial v}{\partial z}\right) = -e$$

$$\omega_y = \frac{1}{2}\left(\frac{\partial u}{\partial z} - \frac{\partial w}{\partial x}\right) = -c$$

      Ω exists and the flow is therefore rotational.

 iii. Check whether there are rates of strain (shearing and normal) components.

$$\dot{\varepsilon}_{xy} = \frac{1}{2}\left(\frac{\partial v}{\partial x} + \frac{\partial u}{\partial y}\right) = 0$$

$$\dot{\varepsilon}_{yz} = \frac{1}{2}\left(\frac{\partial w}{\partial x} + \frac{\partial u}{\partial y}\right) = 0$$

$$\dot{\varepsilon}_{zx} = \frac{1}{2}\left(\frac{\partial v}{\partial x} + \frac{\partial u}{\partial y}\right) = 0$$

and

$$\dot{\varepsilon}_{xx} = \frac{\partial u}{\partial x} = 0$$

$$\dot{\epsilon}_{yy} = \frac{\partial v}{\partial y} = 0$$

$$\dot{\epsilon}_{zz} = \frac{\partial w}{\partial z} = 0$$

This means that the fluid has no rates of strains, but has rotation, which suggests a rigid body rotation at constant angular velocity ($\Omega = (\omega_x, \omega_y, \omega_z)$ is constant). Since u, v, w exist, the flow will be rotation plus translation.

REFERENCES

Daily, J. W. and Harleman, D. R. F. (1966): *Fluid Dynamics*, Addison Wesley, pp. 454

Lagrange, J.-L. (1868): "Mémoire sur la théorie du mouvement des fluides (in: Nouveaux Mémoires de l'Académie Royale des Sciences et Belles-Lettres de Berlin, année 1781)", Oevres de Lagrange, Tome IV, pp. 695–748.

Chapter 7

Viscous fluid flow – boundary layer

7.1 INTRODUCTION

Fluids, in general, can be classified as ideal fluids and viscous or real fluids. Ideal fluids are assumed to have no viscosity although all fluids, in reality, have some viscosity. The viscous effect in fluid flow is confined to a thin layer near the boundary of flow, which is known as the boundary layer. Outside the region of the boundary layer, the fluid is assumed to be ideal, and the flow is sometimes referred to as potential flow. The general concept of boundary layer was introduced by Prandtl in 1904. It is an important and significant concept in Fluid Mechanics and provides a useful link between ideal fluids and real fluids. For fluids having relatively small viscosity, the effect of internal friction in a fluid is appreciable only in a narrow region surrounding the fluid boundaries.

In the case of a flat plate placed in a flow field, the velocity (relative to the boundary) of the fluid at the boundary is zero. This is called the no-slip boundary condition. The flow velocity across the flow, therefore, varies from zero to its full value with a steep velocity gradient. Because of the viscosity of the fluid, this velocity gradient gives rise to shear stresses, which retard the flow flowing near the boundary. The region affected is called the **boundary layer**, or friction layer.

The velocity in the boundary-layer approaches the velocity in the main flow asymptotically. The boundary layer is very thin at the upstream end of a streamlined body at rest in an otherwise uniform flow. As the layer moves downstream, the continual action of shear stress tends to slow down additional fluid particles causing the thickness of the boundary layer to increase in the downstream direction.

For smooth upstream boundaries, the boundary layer starts as **laminar**. As the thickness increases, the flow becomes turbulent and transforms into **turbulent boundary layer**. The point at which the transition from laminar to turbulent depends upon the roughness of the surface, the turbulence in the mainstream, the pressure gradient in the mainstream just outside the boundary layer and the local Reynolds number. Even within the turbulent boundary layer, there exists a very thin layer close to the boundary where the viscous effects are significant. This is called the **laminar sub-layer,** where the velocity profile is assumed to be linear. Turbulent boundary layer has a steeper velocity profile than that of the laminar boundary layer.

The concept of boundary layer enables the simplification of fluid dynamics to a great extent. The motion within the boundary layer is considered as fluid flow where internal friction is taken into account (real fluids) whereas the motion outside the boundary layer is considered as ideal (or potential) flow in which internal friction is ignored.

Hence, the boundary-layer concept helps to link ideal fluid flow with real fluid flow. It also helps to determine the important frictional drag forces on moving bodies such as ships, aircraft wings, automobiles etc.

7.2 DEFINITIONS OF BOUNDARY-LAYER THICKNESS

There are four distances normal to the solid surface that can be used as measures of the thickness of the boundary layer (see also Figure 6.1).

7.2.1 Boundary-layer thickness, δ

Boundary-layer thickness is defined as the distance at which the velocity is 99% (or a fixed percentage) of the free stream velocity. It is not easy to measure this because the velocity approaches that of the free stream asymptotically.

7.2.2 Displacement thickness $\delta*$

Displacement thickness $\delta*$ is defined by

$$\rho u_0 \delta^* = \int_0^\infty \rho \left(u_0 - u \right) dy \tag{7.1}$$

where u_0 is the free stream velocity, u is the velocity at a distance y from the boundary and normal to the direction of flow. The right hand side (RHS) of this equation represents the difference between the mass rate of flow in the boundary layer and the mass rate of flow of an inviscid fluid in the same space.

It is, therefore, the deficiency of mass flow rate caused by the boundary layer. Therefore, the displacement thickness is seen to be the thickness of a layer of the free stream whose mass rate of flow is equal to the deficiency. It represents the distance by which external streamlines are shifted owing to the formation of the boundary layer. In other words, it amounts to shifting of the flow bed by an amount equal to the displacement thickness.

For the flow past two fixed parallel plates, it is approximately equal to one-third of the boundary-layer thickness. The proof is as follows:

Considering a fluid element of depth Δy at a depth of y measured from the centerline between the two fixed parallel plates, the forces acting are

$$p\Delta y - \left(p + \frac{\partial p}{\partial x} \Delta x \right) \Delta y - \tau \Delta x + \left(\tau + \frac{\partial \tau}{\partial y} \Delta y \right) \Delta x = 0$$

$$\frac{\partial \tau}{\partial y} \Delta y \Delta x = \frac{\partial p}{\partial x} \Delta x \Delta y \Rightarrow \frac{\partial \tau}{\partial y} = \frac{\partial p}{\partial x}$$

The shear stress for a Newtonian fluid is given by

$$\tau = \mu \frac{\partial u}{\partial y} \Rightarrow \frac{\partial p}{\partial x} = \mu \frac{\partial^2 u}{\partial y^2}$$

Rearranging

$$\frac{\partial^2 u}{\partial y^2} = \frac{1}{\mu} \frac{\partial p}{\partial x}$$

Integrating

$$\frac{\partial u}{\partial y} = \frac{1}{\mu}\frac{\partial p}{\partial x}y + C_1$$

Integrating again

$$u = \frac{1}{\mu}\frac{\partial p}{\partial x}\frac{y^2}{2} + C_1 y + C_2$$

Using the boundary conditions $u = 0$ when $y = \pm h$, where h is the distance from the centerline of the two parallel plates to the upper and lower plates, $C_1 = 0$, and, $C_2 = -\frac{1}{2\mu}\frac{\partial p}{\partial x}h^2$.

Substituting the boundary conditions,

$$u = \frac{1}{\mu}\frac{\partial p}{\partial x}\frac{y^2}{2} - \frac{1}{\mu}\frac{\partial p}{\partial x}\frac{h^2}{2} = \frac{1}{2\mu}\frac{\partial p}{\partial x}(y^2 - h^2)$$

The velocity u becomes a maximum when $y = 0$.

$$u_{max} = u_0 = -\frac{1}{2\mu}\frac{dp}{dx}h^2$$

Therefore, $\dfrac{u}{u_0} = -\left(\dfrac{y^2}{h^2} - 1\right)$

The boundary-layer thickness δ can be approximated to h. Therefore,

$$\frac{u}{u_0} \approx -\left(\frac{y^2}{\delta^2} - 1\right).$$

$$\delta^* = \int_0^\delta \frac{u_0 - u}{u_0}dy = \int_0^\delta dy - \int_0^\delta \frac{u}{u_0}dy = \delta + \int_0^\delta \left(\frac{y^2}{\delta^2} - 1\right)dy = \delta + \frac{\delta^3}{3\delta^2} - \delta = \frac{\delta}{3}$$

7.2.3 Momentum thickness δ_M

Momentum thickness is defined by

$$\rho u_0^2 \delta_M = \int_0^\infty \rho u(u_0 - u)dy \qquad (7.2a)$$

$$\delta_M = \int_0^\infty \frac{u}{u_0}\left(1 - \frac{u}{u_0}\right)dy \qquad (7.2b)$$

The RHS of Eq. 7.2 is the deficiency in momentum flux caused by the boundary layer. The momentum thickness is, therefore, the thickness of a layer of the free stream whose momentum flux equals this deficiency. It is to be noted that the momentum of a mass in the boundary layer is equal to $(\rho u dy)\, u$ per unit width, whereas the momentum of the same mass outside the boundary layer is equal to $(\rho u dy)\, u_0$ per unit width.

7.2.4 Energy thickness, δ_E

The energy thickness is defined by

$$\rho u_0^3 \delta_E = \int_0^\infty \rho u \left(u_0^2 - u^2\right) dy \tag{7.3a}$$

$$\delta_E = \int_0^\infty \frac{u}{u_0}\left(1 - \left(\frac{u}{u_0}\right)^2\right) dy \tag{7.3b}$$

In Eqs. 7.1–7.3 and in most other equations where a function of $(u_0 - u)$ occurs as the integrand,

$$\int_0^\infty [\,]\, dy = \int_0^\delta [\,]\, dy$$

because for $y \geq \delta,\, u = u_0$

$$\text{and} \quad \int_\delta^\infty [\,] dy = 0$$

7.2.5 Typical values of the order of magnitude of boundary-layer thickness

Table 7.1 shows some typical values of the order of magnitude of boundary layer thickness in air and water.

7.3 EXTERNAL FLOWS – FLOW OVER FLAT PLATES

Flow over a flat plate is typical of external flows. At the leading edge, the boundary-layer thickness is zero, and it increases along the length of the plate. At the upstream end of the plate, the boundary layer is laminar but may develop into a turbulent boundary-layer downstream. The laminar region is usually steady and two-dimensional, but the transitional and the fully turbulent regions are unsteady and three-dimensional.

The exact solutions to the boundary-layer equations are difficult, and solutions to simple cases only can be obtained. The solution to the flow past a flat plate has been given by Blasius (1908) in the form of an infinite series (thus approximate).

Table 7.1 Order of Magnitudes of Boundary-Layer Thickness for Some Objects

Object	Fluid	Flow Velocity (m/s)	Order of δ
Supersonic fighter aircraft wing	Air	500	Few mm
Glider wing 1 m wide	Air	20	Few cm
Ship 200 m long	Water	10	1 m
Smooth sea	Air	10	30 m
Land	Air	10	100 m

7.3.1 Prandtl boundary-layer equations

Prandtl (1904) simplified the **Navier–Stokes** equations by carrying out an order-of-magnitude analysis. He assumed that the boundary-layer thickness δ is small compared to any other dimension of the boundary. The distances and velocities were assumed to be of order 1 in the x-direction and order δ in the y-direction. From dimensional analysis, it can also be shown that $\delta \propto \sqrt{\nu}$. The steady-state two-dimensional Navier–Stokes equations for incompressible Newtonian fluids ignoring the gravity terms are

$$u\frac{\partial u}{\partial x}+v\frac{\partial u}{\partial y}=-\frac{1}{\rho}\frac{\partial p}{\partial x}+\nu\left(\frac{\partial^2 u}{\partial x^2}+\frac{\partial^2 u}{\partial y^2}\right) \quad \text{in the } x\text{-direction} \tag{7.4}$$

$$u\frac{\partial v}{\partial x}+v\frac{\partial v}{\partial y}=-\frac{1}{\rho}\frac{\partial p}{\partial y}+\nu\left(\frac{\partial^2 v}{\partial x^2}+\frac{\partial^2 v}{\partial y^2}\right) \quad \text{in the } y\text{-direction} \tag{7.5}$$

and, the continuity equation under the same conditions is

$$\frac{\partial u}{\partial x}+\frac{\partial v}{\partial y}=0 \tag{7.6}$$

In the continuity equation (Eq. 7.6), u is of order 1; v is of order δ. Therefore, $\dfrac{\partial u}{\partial x}$ is of order $\dfrac{1}{1}$ and $\dfrac{\partial v}{\partial y}$ is of order $\dfrac{\delta}{\delta}$, which is also of order 1. The Bernoulli equation for ideal fluids (applicable to flow outside the boundary layer) is of the form

$$\frac{p}{\rho g}+\frac{u_0^2}{2g}+z = \text{Constant} \tag{7.7a}$$

which, at the same elevation becomes

$$\frac{p}{\rho g}+\frac{u_0^2}{2g} = \text{Constant}. \tag{7.7b}$$

When Eq. 7.7b is differentiated

$$\frac{1}{\rho}\frac{\partial p}{\partial x}+u_0\frac{\partial u_0}{\partial x}=0 \Rightarrow \frac{\partial p}{\partial x}=-\rho u_0\frac{\partial u_0}{\partial x} \tag{7.7c}$$

The second term in Eq. 7.7c is of order $1\times(1/1)$ if ρ is assumed to be of order 1, then $\dfrac{\partial p}{\partial x}$ has to be of order 1. In the Navier–Stokes equation in the x-direction (Eq. 7.4), the orders of magnitude of each term are

$$u\frac{\partial u}{\partial x}+v\frac{\partial u}{\partial y}=-\frac{1}{\rho}\frac{\partial p}{\partial x}+\nu\left(\frac{\partial^2 u}{\partial x^2}+\frac{\partial^2 u}{\partial y^2}\right) \tag{7.8}$$

$$1\times\frac{1}{1} \quad \delta\times\frac{1}{\delta} \quad 1 \quad \nu\left(\frac{1}{1^2} \quad \frac{1}{\delta^2}\right)$$

If ν is of order δ^2, then all terms in the Navier–Stokes equation except $\nu \dfrac{\partial^2 u}{\partial x^2}$ are of order 1. Hence, the Navier–Stokes equation can be approximated as

$$u\frac{\partial u}{\partial x} + v\frac{\partial u}{\partial y} = -\frac{1}{\rho}\frac{\partial p}{\partial x} + v\left(\frac{\partial^2 u}{\partial y^2}\right) \tag{7.9}$$

Similarly, the Navier–Stokes equation in the y-direction is

$$u\frac{\partial v}{\partial x} + v\frac{\partial v}{\partial y} = -\frac{1}{\rho}\frac{\partial p}{\partial y} + v\left(\frac{\partial^2 v}{\partial x^2} + \frac{\partial^2 v}{\partial y^2}\right) \tag{7.10}$$

$$1\times\frac{\delta}{1} \quad \delta\times\frac{\delta}{\delta} \quad ? \quad \delta^2\left(\frac{\delta}{1} \quad \frac{\delta}{\delta^2}\right)$$

Since all terms except $\dfrac{\partial p}{\partial y}$ are of order δ, $\dfrac{\partial p}{\partial y}$ must be also of order δ. Therefore, the y-direction Navier–Stokes equation (Eq. 7.10) can be ignored. The simplified Navier–Stokes equation then becomes

$$u\frac{\partial u}{\partial x} + v\frac{\partial u}{\partial y} = -\frac{1}{\rho}\frac{\partial p}{\partial x} + v\frac{\partial^2 u}{\partial y^2} \tag{7.11}$$

The pressure gradient term in the y-direction is zero.

$$\frac{\partial p}{\partial y} = 0 \tag{7.12}$$

Equations 7.11 and 7.12 together with the continuity equation (Eq. 7.6) are known as the Prandtl boundary-layer equations with the boundary conditions $u = v = 0$ when $y = 0$ and $u = u_0$ when y tends to infinity (in reality when y tends to δ). They are applicable to 2-D, steady incompressible flows. When Eq. 7.7c is substituted into Eq. 7.11, the simplified equation is

$$u\frac{\partial u}{\partial x} + v\frac{\partial u}{\partial y} = u_0\frac{\partial u_0}{\partial x} + v\frac{\partial^2 u}{\partial y^2} \tag{7.13}$$

Equations 7.6 and 7.13 have two unknown u and v compared to the three unknowns u, v and p in the original equations. The integration of the simplified equation can be further simplified by introducing the stream function ψ, thereby reducing the number of variables. Substituting for u and v in terms of ψ,

$$u = \frac{\partial \psi}{\partial y}; \quad v = -\frac{\partial \psi}{\partial x} \tag{7.14}$$

Equation 7.13 becomes

$$\frac{\partial \psi}{\partial y}\frac{\partial^2 \psi}{\partial x\partial y} - \frac{\partial \psi}{\partial x}\frac{\partial^2 \psi}{\partial y^2} = v\frac{\partial^2 \psi}{\partial y^2} + u_0\frac{\partial u_0}{\partial x} \tag{7.15}$$

with the boundary conditions $\frac{\partial \psi}{\partial x} = \frac{\partial \psi}{\partial y} = 0$ when $y=0$; and, $\frac{\partial \psi}{\partial y} = u_0(x)$ when y tends to infinity (in reality when y tends to δ). The continuity equation is automatically satisfied by the stream function.

7.3.2 Blasius solution

Blasius (1908) obtained a solution to these equations (continuity and momentum equations, Eqs. 7.6 and 7.13) for two-dimensional, steady, incompressible flows with a constant free stream velocity u_0. He converted the two partial differential equations to a single differential equation using the stream function ψ and a dimensionless form the stream function $f(\eta)$ where η is a dimensionless length parameter defined as

$$\eta = f\left(\frac{y}{\delta}\right) = \frac{y}{\sqrt{\frac{vx}{u_0}}} = y\sqrt{\frac{u_0}{vx}} \text{ since } \delta \propto \sqrt{\frac{vx}{u_0}}. \tag{7.16}$$

The procedure for flow past a thin flat plate of infinite length is as follows (Yuan, 1967; pp 309–317):

$$\psi = \int u \, dy = \frac{u_0}{\sqrt{u_0/vx}} \int F(\eta) \, d\eta = \sqrt{vu_0 x} f(\eta) \tag{7.17}$$

Then,

$$u = \frac{\partial \psi}{\partial y} = \frac{\partial \psi}{\partial \eta} \frac{\partial \eta}{\partial y} = u_0 f'(\eta) \text{ where } f'(\eta) = \frac{\partial f}{\partial \eta}. \tag{7.18}$$

$$v = -\frac{\partial \psi}{\partial x} = -\frac{1}{2}\sqrt{\frac{vu_0}{x}}f(\eta) - \sqrt{u_0 vx}\left(-\frac{\eta}{2x}\right)f'(\eta) \tag{7.19a}$$

$$\left(\frac{\partial \psi}{\partial x} = f(\eta)\frac{\partial}{\partial x}\left(\sqrt{vxu_0}\right) + \sqrt{vxu_0}\frac{\partial f}{\partial x}; \text{ and } \frac{\partial f}{\partial x} = \frac{\partial f}{\partial \eta}\frac{\partial \eta}{\partial x}\right)$$

$$= \frac{1}{2}\sqrt{\frac{vu_0}{x}}\left[\eta f'(\eta) - f(\eta)\right] \tag{7.19b}$$

For constant u_0, Eq. 7.13 simplifies to

$$u\frac{\partial u}{\partial x} + v\frac{\partial u}{\partial y} = v\frac{\partial^2 u}{\partial y^2} \tag{7.19c}$$

Substituting in the simplified **Navier–Stokes** equation for constant u_0 (Eq. 7.19c) in the x-direction gives

$$-\frac{u_0^2}{2}\frac{\eta}{x}ff'' + \frac{u_0^2}{2x}[\eta f' - f]f'' = \frac{u_0^2}{x}f''' \tag{7.20}$$

which upon simplification becomes

$$ff'' = 2f''' = 0 \tag{7.21}$$

where $f' = \dfrac{\partial f}{\partial \eta}$; $f'' = \dfrac{\partial^2 f}{\partial \eta^2}$; $f''' = \dfrac{\partial^3 f}{\partial \eta^3}$

It is also helpful to note the expressions

$$\frac{\partial u}{\partial x} = \frac{\partial^2 \psi}{\partial x \partial y} = -\frac{u_0}{2}\frac{\eta}{x} f''(\eta)$$

$$\frac{\partial u}{\partial y} = u_0 \sqrt{\frac{u_0}{vx}} f''(\eta)$$

$$\frac{\partial^2 u}{\partial y^2} = \frac{u_0^2}{vx} f'''(\eta)$$

The boundary conditions are $f=0$, $f'=0$, when $\eta=0$; and, $f'=1$ when $\eta=\infty$.

Equation 7.21 is a third-order non-linear differential equation that has no analytical solution. Blasius solved it by expanding $f(\eta)$ as a power series. The results agreed well with the experimental observations of **Nikuradse**. The values of f, f', f'' obtained by Howarth (1938) using the series solution for $\eta=0$ are $f=0$, $f'=0$ and $f'' = 0.332$.

7.3.2.1 Shearing stress

$$\tau_0 = \mu \left(\frac{\partial u}{\partial y} \right)_{y=0}$$

$$= \mu \left\{ u_0 \sqrt{\frac{u_0}{vx}} f''(\eta) \right\}_{\eta=0} \tag{7.22}$$

Substituting the numerical value for $f''(0)=0.332$, which Howarth (1938) obtained

$$\tau_0 = \mu \left\{ \rho u_0^2 \sqrt{\frac{u_0}{\mu \rho^2 u_0^2 x}} 0.332 \right\}$$

$$= \frac{0.332 \rho u_0^2}{\sqrt{Re_x}} \tag{7.23}$$

7.3.2.2 Drag and drag coefficient

Local skin friction coefficient C_f is given by

$$C_f = \frac{\tau_0}{\dfrac{1}{2}\rho u_0^2} = \frac{0.664}{\sqrt{Re_x}} \tag{7.24}$$

Drag force $D = \int_0^\ell \tau_0 \, dx$

$$= 0.664 \rho u_0^2 \sqrt{\frac{v\ell}{u_0}} = \frac{0.664 \rho u_0^2 l}{\sqrt{\mathrm{Re}_l}}$$

(7.25)

Drag coefficient $C_D = \dfrac{D}{\dfrac{1}{2}\rho u_0^2 \ell} = \dfrac{1.328}{\sqrt{\mathrm{Re}_\ell}}$

(7.26)

7.3.3 von Karman momentum integral equation

In 1921, **von Karman** proposed an approximate, but quite a reasonable method of analyzing the boundary layer using the momentum integral equation. The accuracy of the method depends upon the velocity profile chosen for the momentum equation. It is used for determining the boundary shear stress for laminar flow.

Considering the flow over a flat plate as illustrated in Figure 7.1, and a control volume ABCD as marked, and assuming

- unit width normal to flow,
- pressure variations normal to flow (i.e. $\perp^r$ to the plate) negligible,
- steady-state flow,
- fluid is incompressible,

the forces acting in the horizontal direction are

$$F_x = p\delta - \left(p + \frac{\partial p}{\partial x}dx\right)(\delta + d\delta) - \tau_0 dx$$

(7.27)

where p is the pressure, δ is the boundary-layer thickness and τ_0 is the boundary shear stress. Neglecting second-order terms (δ is the first order and $d\,\delta$ is the second order)

$$F_x = -\delta \frac{\partial p}{\partial x}dx - \tau_0 dx$$

(7.28)

The momentum integral equation for steady-state conditions is

$$F_x = \int_{cs} \rho V(V \cdot dA)$$

(7.29)

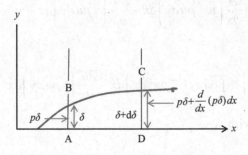

Figure 7.1 Boundary layer over a flat plate.

In the control volume, the boundary layer and the streamlines do not necessarily coincide. In general, the streamlines cross the boundary layer. Therefore, in the control volume ABCD, mass flux takes place across three faces, AB, BC and CD. Mass flux across face AB is $\int_0^\delta \rho u \, dy$ and the mass flux across DC is $\int_0^\delta \rho u \, dy + \dfrac{d}{dx}\left(\int_0^\delta \rho u \, dy\right) dx$. Therefore, the mass flux $\dot{m}$ over the top face is given by (the difference between incoming and outgoing in the x-direction).

$$\dot{m} = \int_0^\delta \rho u \, dy - \left[\int_0^\delta \rho u \, dy + \frac{d}{dx}\left(\int_0^\delta \rho u \, dy\right) dx\right]$$

$$= -\frac{\partial}{\partial x}\left(\int_0^\delta \rho u \, dy\right) dx \tag{7.30}$$

The momentum flux across the top face in the horizontal direction (the velocity of this mass flux in the horizontal direction is u_0, the free stream velocity) is given by

$$\dot{M}_x = \dot{m} u_0$$

$$= -u_0 \frac{\partial}{\partial x}\left(\int_0^\delta \rho u \, dy\right) dx \tag{7.31}$$

Taking the total momentum flux and substituting in the momentum integral equation,

$$F_x = \int_{cs} \rho V(V \cdot dA) = \int_0^\delta \rho u^2 \, dy + \frac{\partial}{\partial x}\left(\int_0^\delta \rho u^2 \, dy\right) dx - u_0 \frac{\partial}{\partial x}\left(\int_0^\delta \rho u \, dy\right) dx - \int_0^\delta \rho u^2 \, dy \tag{7.32}$$

which is equal to F_x given by Eq. 7.28.
   Therefore,

$$\tau_0 dx + \delta \frac{\partial p}{\partial x} dx = u_0 \frac{\partial}{\partial x}\left(\int_0^\delta \rho u \, dy\right) dx - \frac{\partial}{\partial x}\left(\int_0^\delta \rho u^2 \, dy\right) dx \tag{7.33}$$

The first term on the right hand side of this equation can be integrated by parts as

$$u_0 \frac{\partial}{\partial x}\left(\int_0^\delta \rho u \, dy\right) dx = \frac{\partial}{\partial x}\left(u_0 \int_0^\delta \rho u \, dy\right) dx - \frac{\partial u_0}{\partial x}\left(\int_0^\delta \rho u \, dy\right) dx \tag{7.34}$$

Substituting back

$$\tau_0 dx + \delta \frac{\partial p}{\partial x} dx = \frac{\partial}{\partial x}\left(u_0 \int_0^\delta \rho u \, dy - \int_0^\delta \rho u^2 \, dy\right) dx - \frac{\partial u_0}{\partial x}\left(\int_0^\delta \rho u \, dy\right) dx \tag{7.35}$$

Since u_0 does not vary with y,

$$u_0 \int \rho u \, dy = \int \rho u u_0 \, dy \tag{7.36}$$

Then, simplifying by dividing by dx,

$$\tau_0 + \delta \frac{\partial p}{\partial x} = \frac{\partial}{\partial x}\left[\int_0^\delta \rho u(u_0 - u)dy\right] - \frac{\partial u_0}{\partial x}\int_0^\delta \rho u\, dy \tag{7.37}$$

Since the boundary-layer thickness is very small compared with x, the vertical variation of pressure can be neglected. Therefore, outside the boundary layer where the flow is assumed to be ideal (potential), Bernoulli equation can be applied.

Assuming the pressure within the boundary layer to be the same as that outside, the differentiated Bernoulli equation (Eq. 7.7c) can be substituted in the momentum equation. Then,

$$\tau_0 - \delta \rho u_0 \frac{\partial u_0}{\partial x} = \frac{\partial}{\partial x}\left[\int_0^\delta \rho u(u_0 - u)dy\right] - \frac{\partial u_0}{\partial x}\int_0^\delta \rho u\, dy \tag{7.38}$$

Since $\delta = \int_0^\delta dy$, the second term on the left hand side can be replaced by the integral. Then, upon simplification,

$$\tau_0 = \frac{\partial}{\partial x}\left[\int_0^\delta \rho u(u_0 - u)dy\right] + \frac{\partial u_0}{\partial x}\int_0^\delta \rho(u_0 - u)dy \tag{7.39}$$

Recalling the definitions of displacement thickness (Eq. 7.1) and momentum thickness (Eq. 7.2), for incompressible flows

$$\left[\begin{array}{l} \delta_* = \dfrac{1}{u_0}\displaystyle\int_0^\infty (u_0 - u)dy \\[3mm] \delta_M = \dfrac{1}{u_0^2}\displaystyle\int_0^\infty u(u_0 - u)dy \end{array}\right]$$

Equation 7.39 can be written as

$$\tau_0 = \frac{d}{dx}\left(\rho u_0^2 \delta_M\right) + \frac{du_0}{dx}\left(\rho u_0 \delta_*\right) \tag{7.40}$$

This is the momentum integral equation and is applicable to laminar, turbulent and transitional boundary layers. For flow past a flat plate with no pressure gradient $\left(\dfrac{\partial p}{\partial x} = 0,\text{ and}\right.$ hence u_0 is constant),

$$\tau_0 = \rho u_0^2 \frac{d}{dx}\delta_M \tag{7.41}$$

7.3.3.1 Laminar boundary layer

The momentum integral equation alone does not permit an evaluation of δ without further assumptions. For example, the velocity distribution $[u = u(y)]$ must be known. A profile of the form

$$\frac{u}{u_0} = f\left(\frac{y}{\delta}\right) = f(\eta) \tag{7.42}$$

which has been experimentally verified is usually assumed. The choice of the function f should satisfy certain boundary conditions. They are

- $\dfrac{u}{u_0} = 0$ when $\eta = 0$

- $\dfrac{u}{u_0} = 1$ when $\eta = 1$

- $f(\eta)$ should be monotonically increasing with η, and,
- velocity distribution has the same form at every value of x (this is not a boundary condition).

The functions are of the form

$$u = a + by + cy^2 + dy^3, \text{ and,}$$

the coefficients are obtained from the above boundary conditions.
 The following functions have been found to satisfy the above boundary conditions:

- $\dfrac{u}{u_0} = \eta \begin{cases} \dfrac{u}{u_0} = 0 & \text{when } \eta = 0 \\[2mm] \dfrac{u}{u_0} = 1 & \text{when } \eta = 1 \end{cases}$ (7.43a)

- $\dfrac{u}{u_0} = 2\eta - \eta^2 \begin{cases} \dfrac{u}{u_0} = 0 & \text{when } \eta = 0 \\[2mm] \dfrac{u}{u_0} = 1 & \text{when } \eta = 1 \\[2mm] \dfrac{du}{dy} = 0 & \text{at } \eta = 1 \end{cases}$ (7.43b)

- $\dfrac{u}{u_0} = \dfrac{3}{2}\eta - \dfrac{1}{2}\eta^3 \begin{cases} \dfrac{u}{u_0} = 0 & \text{when } \eta = 0 \\[2mm] \dfrac{u}{u_0} = 1 & \text{when } \eta = 1 \\[2mm] \dfrac{du}{dy} = 0 & \text{at } \eta = 1 \\[2mm] \dfrac{d^2u}{dy^2} = 0 & \text{at } \eta = 0 \end{cases}$ (7.43c)

δ^* for this is 0.375δ.

The condition $\dfrac{d^2u}{dy^2} = 0$ at $\eta=0$ is obtained from **Prandtl**'s simplification of the **Navier–**

Stokes equation. This implies $\dfrac{du}{dy} =$ Constant at $\eta=0$ (i.e., linear velocity variation at $\eta=0$).
At a boundary,

$$\tau_0 = \mu\left(\frac{\partial u}{\partial y}\right)_{y=0} = \mu\left(\frac{du}{dy}\right)_{y=0} \quad \text{(for steady state)} \, u = u(y) \tag{7.44}$$

Since $u = u_0 f(\eta)$ and for the third type of function (Eq. 7.43c),

$$\frac{du}{dy} = \frac{du}{df}\frac{df}{dy} = u_0\frac{df}{d\eta}\frac{1}{\delta} \tag{7.45}$$

Therefore,

$$\tau_0 = \left(\mu\frac{u_0}{\delta}\frac{df}{d\eta}\right)_{\eta=0}$$

$$= \frac{\mu u_0}{\delta}\left(\frac{3}{2}-\frac{3}{2}\eta^2\right)_{\eta=0}$$

$$= \frac{3}{2}\frac{\mu u_0}{\delta} \tag{7.46}$$

Substituting for u [in terms of $f(\eta)$] in the integral equation (Eq. 7.39 for constant u_0),

$$\tau_0 = \frac{\partial}{\partial x}\left[\int_0^\delta \rho u(u_0 - u)\,dy\right]$$

$$= \rho\frac{\partial}{\partial x}\left[\int_0^1 \left\{u_0^2 f(\eta) - [u_0 f(\eta)]^2\right\}\delta\,d\eta\right]$$

$$= \rho u_0^2\frac{\partial\delta}{\partial x}\left[\int_0^1 \left\{f(\eta) - [f(\eta)]^2\right\}d\eta\right]$$

$$= \rho u_0^2\frac{\partial\delta}{\partial x}\left[\int_0^1 \left\{\left(\frac{3\eta}{2}-\frac{\eta^3}{2}\right) - \left[\frac{3\eta}{2}-\frac{\eta^3}{2}\right]^2\right\}d\eta\right]$$

$$= \rho u_0^2\frac{\partial\delta}{\partial x}\int_0^1 \left(\frac{3\eta}{2}-\frac{\eta^3}{2}\right) - \left[\frac{9\eta^2}{4}-\frac{3\eta^4}{2}+\frac{\eta^6}{4}\right]d\eta$$

$$= \rho u_0^2\frac{\partial\delta}{\partial x}\left[\frac{3}{4}-\frac{1}{8}-\frac{3}{4}-\frac{1}{28}+\frac{3}{10}\right]$$

$$= \rho u_0^2\frac{\partial\delta}{\partial x}\frac{39}{280}$$

$$= 0.139\rho u_0^2\frac{\partial\delta}{\partial x} \tag{7.47}$$

Equating the expressions for τ_0 (Eqs. 7.46 and 7.47),

$$\frac{3}{2}\frac{\mu u_0}{\delta} = \frac{39}{280}\rho u_0^2 \frac{\partial \delta}{\partial x} \tag{7.48}$$

Integrating (after separation of variables) gives

$$\delta d\delta = \frac{140}{13}\frac{\mu}{\rho u_0}dx \tag{7.49}$$

(Since δ is a function of x only),

$$\frac{\delta^2}{2} = \frac{140}{13}\frac{\mu}{\rho u_0}x + \text{constant} \tag{7.50}$$

Since $\delta = 0$ at $x = 0$, constant$=0$.

$$\left(\frac{\delta}{x}\right)^2 = \frac{280}{13}\frac{\mu}{\rho u_0 x} \tag{7.51}$$

$$\frac{\delta}{x} = \frac{4.64}{\sqrt{\rho u_0 x/\mu}} = \frac{4.64}{\sqrt{\text{Re}_x}} \tag{7.52}$$

where $\text{Re}_x =$ Reynolds number with respect to x.
   Substituting back in Eq. 7.46,

$$\tau_0 = \frac{3}{2}\frac{\mu u_0}{\dfrac{4.64x}{\sqrt{\rho u_0 x/\mu}}}$$

$$= \frac{3}{2}\frac{1}{4.64}\frac{\mu u_0}{x}\sqrt{\frac{\rho u_0 x}{\mu}}$$

$$= 0.323\sqrt{\frac{\mu \rho u_0^3}{x}} \text{ or } \frac{0.323\rho u_0^2}{\sqrt{\text{Re}_x}}. \tag{7.53}$$

7.3.3.2 Drag and drag coefficients

Drag is the component of the resultant force acting on a body parallel to the undisturbed initial velocity. The component normal to the drag is known as **lift**.
   The local drag coefficient (skin-friction coefficient) is defined as

$$C_f = \frac{\tau_0}{\dfrac{1}{2}\rho u_0^2} = \frac{\text{Local wall shear stress}}{\text{Dynamic pressure of the free stream}} \tag{7.54a}$$

Therefore,

$$C_f = \frac{2 \times 0.323}{\sqrt{\text{Re}_x}} = \frac{0.646}{\sqrt{\text{Re}_x}} = f(\text{distance}) \tag{7.54b}$$

The average drag coefficient over the length of the plate is given by

$$C_D = \frac{\text{Drag force}}{\text{Dynamic pressure} \times \text{area}}$$ (7.55)

The drag force on one side of the plate of unit width is given by

$$D = \int_0^l \tau_0 \, dx \times 1$$ (7.56a)

Substituting for τ_0 from Eq. 7.53,

$$D = 0.323 \int_0^l \left(\mu \rho u_0^3 \right)^{1/2} x^{-1/2} dx$$

$$= 0.323 \sqrt{\mu \rho u_0^3} \int_0^l \frac{dx}{\sqrt{x}}$$

$$= 0.323 \sqrt{\mu \rho u_0^3} \left(2\sqrt{l} \right)$$

$$= 0.646 \sqrt{\mu \rho u_0^3 l} = \frac{0.646 \rho u_0^2 l}{\sqrt{\text{Re}_l}}$$ (7.56b)

Therefore,

$$C_D = \frac{0.646 \sqrt{\mu \rho l u_0^3}}{\frac{1}{2} \rho u_0^2 l} l = \frac{1.292}{\sqrt{\text{Re}_l}}$$ (7.57)

and, the average Drag Coefficient=2 (local drag coefficient) at $x = l$, the length of the plate.

It is to be noted that the C_D values given in Eq. 7.57 are slightly different from those given in Eq. 7.26. This is due to the different velocity profiles used in deriving the two equations. Similar differences can be seen in the boundary shear stress and the boundary-layer thickness as well as other parameters of the boundary layer.

7.3.4 Turbulent boundary layer

When the Reynolds number for the plates reaches a value of about 500,000, the boundary layer becomes turbulent. The turbulent boundary layer is usually thicker than the laminar boundary layer, and the velocity gradient is steeper at the boundary.

The momentum integral equations derived earlier can be used to obtain the turbulent boundary-layer thickness as well. A velocity profile, different to that used for the laminar case is used. The boundary shear stress τ_0 is expressed in a form similar to that obtained from measurements on pipe flow under turbulent conditions.

The experimental work in this field is mainly due to **Prandtl**, who studied the turbulent boundary layer on flat plates and compared with that in pipe flow. He concluded that the two cases are not very different.

Although the transition from laminar to turbulent boundary layer is gradual, it is assumed, for the purpose of analysis that the turbulent boundary layer begins at the leading edge of the plate. This is a contradiction, but for long plates, the length of the laminar bound-ary layer is small compared with the length of the turbulent boundary layer. In practice, the turbulent boundary layer can be generated at the leading edge by having an artificial device that will create the turbulence. When the length of the flat plate is small, the laminar boundary-layer length may be a significant proportion of the total length. The drag is then calculated by considering the turbulent boundary layer to start at the leading edge, subtract-ing the drag in the turbulent boundary layer up to x_0 and adding the drag in the laminar boundary layer as follows:

$$D = \left(\frac{1.328}{\mathrm{Re}_{x_0}^{1/2}} \cdot x_0 + \frac{0.074l}{\mathrm{Re}_l^{1/5}} - \frac{0.074x_0}{\mathrm{Re}_{x_0}^{1/5}} \right) \frac{B\rho u_0^2}{2} \tag{7.58}$$

The first term on the RHS of the above equation corresponds to the laminar case for the length x_0, the second term corresponds to the turbulent boundary layer for the entire length and the last term corresponds to the turbulent boundary layer upto length x_0.

   Prandtl proposed an analogy between pipe flow and flow over flat plates. In a pipe flow, the fully developed boundary-layer thickness is the radius and the maximum velocity (centre velocity) corresponds to the free stream velocity.

7.3.4.1 Power law

For turbulent boundary layers, the velocity profile is normally assumed to follow the power law. It is of the form

$$\frac{u}{u_0} = \left(\frac{y}{\delta} \right)^{1/7} \tag{7.59}$$

This is sometimes called **Prandtl's one-seventh power law** and is based on the same profile for turbulent flow in pipes, which is

$$\frac{V}{V_{max}} = \left(\frac{y}{R} \right)^{1/7}, \tag{7.60}$$

where R is the radius of pipe. This relationship is true for Reynolds numbers from about 5×10^5 to 10^7 for flat plates. (For pipes, based on the diameter of pipe, $Re < 10^5$.)

   However, this equation does not hold for $y = 0$, i.e., at the boundary, because

$$\frac{\partial u}{\partial y} = \frac{1}{7} \frac{u_0}{\delta} \left(\frac{y}{\delta} \right)^{-6/7} \rightarrow \infty \ \text{at} \ y = 0$$

and therefore τ_0 cannot be estimated using the relationship

$$\tau_0 = \mu \left(\frac{\partial u}{\partial y} \right)_{y=0}$$

It is also not true for the laminar sub-layer, where a linear velocity profile is normally assumed.

Blasius obtained the following empirical formula for pipe flow:

$$C_f = \frac{0.079}{\left(\mathrm{Re}_D\right)^{1/4}} \tag{7.61}$$

where $\mathrm{Re}_D = \dfrac{VD}{v}$, and V is the average velocity.

By definition,

$$C_f = \frac{\tau_0}{\frac{1}{2}\rho V^2}$$

Therefore,

$$\tau_0 = \frac{1}{2}C_f \rho V^2$$

For a pipe, $V = 0.817\, V_{\max}$.
   Therefore,

$$\tau_0 = \frac{1}{2}C_f \rho \left(0.817 V_{\max}\right)^2$$

$$= 0.334 \rho C_f V_{\max}^2 \tag{7.62}$$

Substituting the **Blasius** empirical formula for C_f, (Eq. 7.61), the turbulent shear stress for pipes can be written as

$$\tau_0 = 0.334 \rho \frac{(0.079)V_{\max}^2}{\left(0.817 V_{\max} 2R/v\right)^{1/4}}$$

$$= 0.334 \times 0.06987 \frac{\rho V_{\max}^2}{\left(\dfrac{V_{\max}R}{v}\right)^{1/4}}$$

$$= 0.0233 \frac{\rho V_{\max}^2}{\left(\mathrm{Re}_R\right)^{1/4}} \tag{7.63}$$

where $\mathrm{Re}_R = \dfrac{V_{\max}R}{v}$ is the Reynolds number based on maximum velocity and radius R. By comparing the turbulent pipe flow with turbulent flow over a flat plate, it can be seen that $V_{\max}$ corresponds to u_0 and R corresponds to δ. Therefore, for a flat plate, by analogy,

$$\tau_0 = \frac{0.0233 \rho u_0^2}{\left(\dfrac{u_0 \delta}{v}\right)^{1/4}} \tag{7.64}$$

From the momentum integral equation with $\frac{\partial p}{\partial x} = 0$, and the one-seventh power law for velocity distribution,

$$\tau_0 = \rho \frac{d}{dx} \int_0^{\delta} u(u_0 - u)\,dy$$

$$= \rho \frac{d}{dx} \int_0^1 u_0 \eta^{1/7} \left(u_0 - u_0 \eta^{1/7}\right) d\eta \delta$$

$$= \rho u_0^2 \frac{d\delta}{dx} \int_0^1 \eta^{1/7} \left(1 - \eta^{1/7}\right) d\eta$$

$$= \frac{7}{72} \rho u_0^2 \frac{d\delta}{dx} \qquad\qquad\qquad (7.65)$$

Equating the two expressions for τ_0,

$$0.0233 \rho u_0^2 \left(\frac{\nu}{u_0 \delta}\right)^{1/4} = \frac{7}{72} \rho u_0^2 \frac{d\delta}{dx}$$

$$0.2397 \left(\frac{\nu}{u_0}\right)^{1/4} dx = \delta^{1/4} d\delta$$

Upon integration,

$$\frac{4}{5} \delta^{\frac{5}{4}} = 0.2397 \left(\frac{\nu}{u_0}\right)^{1/4} x + \text{constant}$$

$$\delta^{\frac{5}{4}} = 0.2996 \left(\frac{\nu}{u_0}\right)^{1/4} x + \text{constant}$$

Assuming the boundary condition that $\delta = 0$ at $x = 0$ (this contradicts the earlier concept of having a laminar boundary layer followed by turbulent boundary layer), the constant $= 0$. Therefore,

$$\delta^{\frac{5}{4}} = 0.2996 \left(\frac{\nu}{u_0}\right)^{\frac{1}{4}} x$$

$$\delta = (0.2996)^{\frac{4}{5}} \left(\frac{\nu}{u_0}\right)^{\frac{1}{5}} x^{\frac{4}{5}}$$

$$\delta = 0.381 \left(\frac{\nu}{u_0}\right)^{\frac{1}{5}} x^{\frac{4}{5}}$$

or

$$\frac{\delta}{x} = 0.381 \left(\frac{v}{u_0}\right)^{\frac{1}{5}} \frac{x^{\frac{4}{5}}}{x}$$

$$= 0.381 \left(\frac{v}{u_0 x}\right)^{\frac{1}{5}}$$

$$\delta = \frac{0.381x}{(\mathrm{Re}_x)^{\frac{1}{5}}} \tag{7.66}$$

7.3.4.2 Boundary shear stress τ_0

By back substitution, in Eq. 7.64,

$$\tau_0 = \frac{0.0233 \rho u_0^2}{\left[\dfrac{u_0}{v} \dfrac{0.381x}{(u_0 x/v)^{\frac{1}{5}}}\right]^{\frac{1}{4}}}$$

$$= \frac{0.02966 \rho u_0^2}{\left(\dfrac{u_0 x}{v}\right)^{\frac{1}{5}}}$$

$$= \frac{0.02966 \rho u_0^2}{(\mathrm{Re}_x)^{\frac{1}{5}}} \tag{7.67}$$

7.3.4.3 Drag and drag coefficient

$$\mathrm{Drag} = \int_0^\ell \tau_0 dx$$

$$= 0.02966 \rho u_0^2 \left(\frac{v}{u_0}\right)^{\frac{1}{5}} \int_0^\ell \frac{1}{x^{\frac{1}{5}}} dx$$

$$= 0.02966 x \frac{5}{4} \rho u_0^2 \left(\frac{v}{u_0}\right)^{\frac{1}{5}} \cdot \ell^{\frac{4}{5}}$$

$$= 0.037 \rho u_0^2 \left(\frac{v}{u_0 \ell}\right)^{\frac{1}{5}} \ell^{\frac{4}{5}} \ell^{\frac{1}{5}}$$

$$= \frac{0.037 \rho u_0^2 \ell}{(\mathrm{Re}_\ell)^{\frac{1}{5}}} \tag{7.68}$$

Drag coefficient C_D is given by

$$C_D = \frac{0.037 \rho u_0^2 \ell / \left(\dfrac{u_0 \ell}{v}\right)^{\frac{1}{5}}}{\dfrac{1}{2} \rho \mu_0^2 \ell}$$

$$= \frac{0.074}{(\mathrm{Re}_\ell)^{\frac{1}{5}}} \tag{7.69}$$

These equations are valid only for the ranges where **Blasius** resistance equation is valid, which is about 5×10^5 to 10^7 in terms of Reynolds numbers. Below $Re = 5 \times 10^5$, the boundary layer is normally laminar. For $Re = 4 \times 10^5 - 10^6$, the exponent in the power law changes to 1/8.

For larger Re values ($>10^7$), the logarithmic law is used instead of the power law. For $10^7 < Re < 10^9$), an equation that uses the velocity profile given in Eq. 7.71 takes the form

$$C_D = \frac{0.455}{\left(\log Re_\ell\right)^{2.58}} \tag{7.70}$$

There are other relationships given in the literature for other ranges of Reynolds numbers.

7.3.4.4 Law of the wall

There are three regions of the boundary layer: the viscous (laminar) sublayer, the transition zone and the turbulent boundary layer. These profiles can be plotted on logarithmic scales with dimensionless axes:

Ordinate $\dfrac{u}{u_*}$ where $u_* = \sqrt{\tau_0 / \rho}$ is the shear velocity

Abscissa yu_* / v

The relationship between these two dimensionless variables is known as the 'law of the wall'. It is a universal plot for smooth surfaces.

7.3.4.5 Universal velocity profile near a wall

The universal velocity profile, obtained by integrating the **Navier–Stokes** equation (Yuan, 1967; pp 372–374) is given by

$$\frac{u}{u_*} = \frac{1}{k}\left(\ell n \frac{u_* y}{v} - \ell n \beta \right) \tag{7.71}$$

where k and β are experimentally determined.

With **Nikuradse**'s experimental results, it takes the form

$$\frac{u}{u_*} = 5.75 \log\left(\frac{u_* y}{v} \right) + 5.5 \tag{7.72}$$

In the laminar sublayer, the linear velocity profile is valid:

$$\frac{u}{u_*} = \frac{u_* y}{v} \tag{7.73}$$

In the transition zone, **von Karman** suggested

$$\frac{u}{u_*} = 11.5 \log\left(\frac{u_* y}{\nu}\right) - 3.05 \tag{7.74}$$

7.4 VISCOUS FLOW BETWEEN PARALLEL PLATES

When a viscous fluid flows between two fixed parallel plates, the resulting flow is known as Poiseuille flow. This is discussed in Section 7.2.2. The velocity profile is

$$u = \frac{1}{\mu}\frac{\partial p}{\partial x}\frac{y^2}{2} - \frac{1}{\mu}\frac{\partial p}{\partial x}\frac{h^2}{2} = \frac{1}{2\mu}\frac{\partial p}{\partial x}\left(y^2 - h^2\right) \tag{7.75}$$

and the maximum velocity is

$$u_{\max} = u_0 = -\frac{1}{2\mu}\frac{dp}{dx}h^2 \tag{7.76}$$

The discharge $Q = \int\limits_{-h}^{h} bu\,dy = \int\limits_{-h}^{h} b\frac{1}{2\mu}\frac{dp}{dx}(y^2 - h^2)dy = \frac{2h^3 b}{3\mu}\left(-\frac{dp}{dx}\right) \tag{7.77}$

where b is the width of the plate and the average velocity V is given as

$$V = \frac{Q}{A} = \frac{h^2}{3\mu}\left(-\frac{dp}{dx}\right) \tag{7.78}$$

Shear stress $\tau_{xy} = \mu\frac{du}{dy} = \mu\frac{h^2}{2\mu}\left(-\frac{dp}{dx}\right)\left(-\frac{2y}{h^2}\right) \tag{7.79}$

Wall shear stress $(\tau_0)_{y=h} = \mu h\left(-\frac{dp}{dx}\right) \tag{7.80a}$

$$(\tau_0)_{y=-h} = -\mu h\left(-\frac{dp}{dx}\right) \tag{7.80b}$$

When one plate is fixed and the other one is moving at a constant velocity, the resulting flow is known as Couette flow. The governing equations in this case can be written as follows:

$$u = u(y); \quad v = 0$$

The momentum equations as before in the x and y directions, respectively, are

$$0 = -\frac{dp}{dx} + \mu\frac{d^2 u}{dy^2}$$

$$0 = -\frac{dp}{dy}$$

When the momentum equation is integrated subject to the boundary conditions $u = u_0$ when $y = h$, and $u = 0$ when $y = 0$, where h is the distance of separation of the two plates, the result is

$$\frac{u}{u_0} = \frac{y}{h} + \left(-\frac{dp}{dx}\right)\left(\frac{h^2}{2\mu u_0}\right)\left(\frac{y}{h}\right)\left(1 - \frac{y}{h}\right) \tag{7.81}$$

Using a dimensionless pressure gradient C and a dimensionless distance between the plates defined as

$$C = \left(-\frac{dp}{dx}\right)\left(\frac{h^2}{2\mu u_0}\right) \text{ and } Y = \frac{y}{h} \tag{7.82}$$

Equation 7.81 can be written as

$$\frac{u}{u_0} = \frac{y}{h} + C\left(\frac{y}{h} - \left(\frac{y}{h}\right)^2\right) = Y + C\left(Y - Y^2\right), \text{ or } \frac{u}{u_0} = (1+C)Y - CY^2 \tag{7.83}$$

Equation 7.83 can be considered under different dimensionless pressure gradients. When $\frac{dp}{dx} = 0$ or $C = 0$, the velocity becomes $\frac{u}{u_0} = \frac{y}{h}$ (linear velocity profile). This condition is also known as simple Couette flow. When $\frac{dp}{dx} > 0$ or $C < 0$, using a nominal value of -1 for C, the velocity becomes $\frac{u}{u_0} = Y^2$, which is a quadratic velocity profile. In this case, the pressure increases with increasing distance resulting in a decrease in velocity. Flow separation can take place when the pressure gradient is positive. When $\frac{dp}{dx} < 0$ or $C > 0$, using a nominal value of $+1$ for C, the velocity becomes $\frac{u}{u_0} = 2Y - Y^2$. In this case, the pressure decreases with distance resulting in an increase in velocity.

The average velocity is given as

$$V = \int\left[u_0\frac{y}{h} + Cu_0\left(\frac{y}{h} - \frac{y^2}{h^2}\right)d\left(\frac{y}{h}\right)\right] = \left(\frac{1}{2} + \frac{C}{6}\right)u_0 \tag{7.84}$$

For simple Couette flow, Eq. 7.84 simplifies to $V = \frac{u_0}{2}$. When $C = -3$, $V = 0$.

$$\text{Flow rate } Q = hV = \left(\frac{1}{2} + \frac{C}{6}\right)u_0 h \tag{7.85}$$

$$\text{Maximum velocity } u_{max} = \frac{u_0}{4}\left(\frac{(1+C)^2}{C}\right) \quad \text{for} \quad C \geq 1 \tag{7.86}$$

$$\text{Minimum velocity } u_{min} = \frac{u_0}{4}\left(\frac{(1+C)^2}{C}\right) \quad \text{for} \quad C \leq -1 \tag{7.87}$$

$$\text{Shear stress } \tau = \mu \frac{du}{dy} = \mu \frac{u_0}{h} + \frac{\mu u_0 C}{h}\left(1 - 2\frac{y}{h}\right) \tag{7.88}$$

7.5 BOUNDARY-LAYER SEPARATION

In the discussion so far, the pressure gradient has been assumed to be zero i.e. $\left(\frac{\partial p}{\partial x} = 0\right)$.

But there are instances where the pressure gradient is not zero, such as for example, when a fluid is flowing over a curved surface where the radius of curvature is much greater than the boundary-layer thickness. The flow in converging sections gives negative pressure gradients while diverging flow gives positive pressure gradients.

A positive pressure gradient is called an adverse pressure gradient and occurs in the flow around curved boundaries and towards a stagnation point. Increase in pressure is of course accompanied by a reduction of velocity, which at some point along the length may decrease to zero. When this happens, the flow separates from the boundary at a point of inflexion of the velocity profile and is called the **boundary-layer separation**. The separation is followed by a region called the **wake** where eddies dissipate energy.

When the pressure gradient is negative, it is called a favourable pressure gradient. When it is so, there is a resultant force in the direction of flow due to pressure. This pressure acts against the boundary shear, thereby reducing the retarding action of the shear. Therefore, the boundary-layer thickness increases more slowly compared with the case when the pressure gradient is zero.

On the other hand, if the pressure gradient is positive, the net pressure force will act in a direction opposing the flow, thereby complementing the boundary shear. In such situations, the boundary-layer thickness increases at a faster rate.

The effect of the combined action of the pressure force and the shear force is to bring the fluid near the boundary to rest because the momentum of the fluid as it enters the region quickly runs out. Thus, the adverse pressure gradient has the effect of reversing the flow to maintain the momentum conservation:

$$\text{Rate of changes of momentum} = F = \left(\frac{\partial p}{\partial x}\delta x + \tau_0 \delta x\right)1$$

Separation can take place only when there is an adverse pressure gradient. When separation occurs, the fluid near the boundary is stationary. Therefore,

$$\left(\frac{\partial u}{\partial y}\right)_{y=0} = 0$$

Separation can take place in both laminar and turbulent boundary layers. Laminar boundary layers are more prone to separation than turbulent boundary layers.

When the boundary layer separates from the surface of a cylinder placed with its axis normal to the free stream, the points of separation move downstream as the Reynolds number increases. In the laminar range, eddies are symmetrical, and the streamlines join together on the downstream side. When the Reynolds number is around 70, the eddies detach themselves and get carried downstream, forming what is known as Karman vortex street. As the Reynolds number is further increased to the turbulent range, the wake becomes narrower.

Example 7.1

Calculate δ^*, δ_M & δ_E for the simple case of a linear velocity distribution shown below

$$\left[u = u_0 \left(\frac{y}{\delta} \right) \right]$$

$$\delta^* = \int_0^\delta \frac{u_0 - u}{u_0} \, dy$$

$$= \int_0^\delta \left(1 - \frac{u}{u_0} \right) dy$$

$$= \int_0^\delta \left(1 - \frac{y}{\delta} \right) dy = \frac{\delta}{2}$$

$$\delta_M = \int_0^\delta \frac{u}{u_0} \left(1 - \frac{u}{u_0} \right) dy$$

$$= \int_0^\delta \left\{ \frac{y}{\delta} - \left(\frac{y}{\delta} \right)^2 \right\} dy = \frac{\delta}{6}$$

$$\delta_E = \int_0^\delta \frac{u}{u_0} \left\{ 1 - \left(\frac{u}{u_0} \right)^2 \right\} dy$$

$$= \int_0^\delta \frac{y}{\delta} \left\{ 1 - \left(\frac{y}{\delta} \right)^2 \right\} dy = \frac{\delta}{4}$$

Example 7.2

Design an experiment that will give a boundary-layer thickness large enough to be visible to the naked eye.

For the laminar boundary layer,

$$\frac{\delta}{x} = \frac{4.64}{\sqrt{R_x}}$$

$$R_x = \frac{u_0 x}{v}$$

For water $v = 1.004 \times 10^{-6}$ m^2/s (at 20°C)

Therefore, $R_x = u_0 \, x \times 10^6$.

Then,

$$\delta = \frac{4.64 x \times 10^{-3}}{\sqrt{u_0 x}} = 4.64 \sqrt{\frac{x}{u_0}} \times 10^{-3}$$

Assuming $\delta = 1$ cm ($= 10^{-2}$ m),

$$10^{-2} = 4.64 \sqrt{\frac{x}{u_0}} \times 10^{-3}$$

$$\sqrt{\frac{u_0}{x}} = 0.464 \Rightarrow \frac{u_0}{x} = 0.215 \Rightarrow u_0 = 0.215x$$

u_0 (m/s)	0.1	0.2	0.3	1.0
x(m)	0.465	0.93	1.395	4.65

Therefore, a smooth flat plate 0.465 m long plated in a water flow of 0.1 m/s will give a boundary-layer thickness of 1 cm. (This is a laminar boundary layer; $R_L = 46,500$.)

REFERENCES

Blasius, H. (1908): Grenzschichten in Flussigkeiten mit kleiner Reibung, *Zeitschrift fur Mathematik und Physik*, vol. 56, no. 1, pp. 1–37. (NACA Technical Memorandum No. 1256).

Howarth, L. (1938): On the solution of the laminar boundary layer equations, *Proceedings of the Royal Society of London*, London, A164, 547.

Prandtl, L. (1904): Uber Flussigkeitsbewegungen bei sehr kleiner Reibung, *Verhandlg. III. Intern. Math. Kongr.*, Heidelberg, Teubner, pp. 484–491.

von Karman, T. (1921): Uber laminare und turbulente Reibung, *ZAMM. Z. Angew. Math. Mech. Bd.*, vol. 1, pp. 233–252. (English translation: On laminar and turbulent friction, NACA Technical Memorandum 1092).

Yuan, S. W. (1967): *Fundamentals of Fluid Mechanics*, Prentice Hall, Englewood, NJ, 608 pp.

Chapter 8

Dimensional analysis

8.1 INTRODUCTION

Very often, real problems in fluid mechanics do not have analytical solutions. For example, there are no analytical means of determining the drag force acting on a ship or an airplane. Other examples include

- Energy dissipation downstream of a dam
- Discharge characteristics of a spillway
- Effect of reclamation on the water movement in a coastal area
- Wind forces on a tall building
- Wave forces on structures.

Solutions to such problems are obtained by experimental analysis. But experiments can only be done by building scale models for obvious reasons. However, a model need not always be smaller than the prototype. Examples of smaller size models include testing of aircrafts in wind tunnels, ships towed in a water tank, a dam constructed in a laboratory, etc. Examples of large sized models include testing of micro-pumps and other hydraulic machines, flow of viscous fluids in blood vessels, etc. The basis for model building is dimensional analysis.

For simplicity, consider the force acting on a stationary smooth sphere by a moving fluid (Figure 8.1). It will depend on the

a. size of sphere characterized by its diameter D
b. velocity of fluid, V,
c. viscosity of fluid, μ, and,
d. density of fluid ρ.

There may be other parameters that affect the drag force, such as the surface texture of the sphere. But one can consider only those factors that are more important and those that

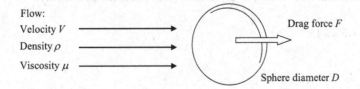

Figure 8.1 Factors affecting the drag force on a stationary sphere.

are **controllable** and **measurable**. Clearly the above four parameters satisfy these criteria. Therefore, it is possible to write

$$F = f(D, V, \mu, \rho) \tag{8.1}$$

If an experiment is carried out to establish the relationship of F with each of the parameters $D, V, \rho,$ and, μ. 10^4 experiments would be needed assuming each relationship requires ten trials. For example,

F vs. D: keeping V, ρ, μ constant require ten trials
F vs. V: keeping D, ρ, μ constant require ten trials
F vs. μ: keeping D, V, ρ constant require ten trials
F vs. ρ: keeping D, V, μ constant require ten trials

If each experiment takes 1 hour, the whole exercise would take 10,000 hours. Assuming 8 hours working day, this is equivalent to 1,250 days or 50 months assuming 25 working days per month or 4 years.

This is both impractical in terms of time and the handling of the enormous amount of data. Meaningful results can be obtained with less effort by the use of dimensional analysis by expressing the relevant parameters in dimensionless form:

$$\frac{F}{\rho V^2 D^2} = f\left(\frac{\rho V D}{\mu}\right) \tag{8.2}$$

The function "f" still needs to be determined experimentally. However, we need only ten experiments rather than 10,000 experiments. It is no longer necessary to select ten different sizes or ten different fluids. Instead, only ten different combinations of the basic parameters would be needed.

8.2 BUCKINGHAM'S PI (π) THEOREM

Suppose the dependent parameter of a physical problem is expressed as a function of $n - 1$ independent parameters as

$$q_1 = f(q_2, q_3, ..., q_n) \tag{8.3}$$

which may be written as

$$g(q_1, q_2, ..., q_n) = 0 \tag{8.4}$$

where g is a function different from f

Then, Buckingham's π theorem states that the n parameters may be grouped into $(n - m)$ independent dimensionless ratios or π parameters, expressed in the form

$$G(\pi_1, \pi_2, ..., \pi_{n-m}) = 0 \tag{8.5a}$$

or

$$\pi_1 = G_1(\pi_2, \pi_3, ..., \pi_{n-m}) \tag{8.5b}$$

where m is usually, but not necessarily equal to the minimum number of independent dimensions required to specify all the parameters $q_1, q_2, ..., q_n$. The functions G or G_1 must still be determined experimentally. A π parameter is not independent if it can be obtained from the product, power or quotient of other π parameters. The number of π functions is always less than the number of physical quantities, i.e. $n - m < n$.

8.2.1 Procedure for the use of the π-theorem

Selection of parameters is based on physical reasoning. Start by including all parameters. If any is insignificant, it will only produce an extra π parameter, which may be eliminated from experimental consideration.

Procedure for determining the π groups

- List all the parameters (say n).
- Select a set of fundamental dimensions, e.g. M, L and T. In heat transfer, the temperature is also a fundamental quantity.
- List the dimensions of all parameters in terms of the fundamental quantities.
- Select the number of repeating parameters. Guidelines for selecting repeating variables:
  - Repeating variables must include among them all of the 'm' fundamental dimensions.
  - Dependent variables should not be used as repeating variables
  - For a fluid flow system, the repeating variables should include
    - a geometrical characteristic
    - a fluid characteristic and
    - a flow characteristic.
- Set up dimensionless equations combining the repeating variables with each of the other variables.

The functions G and G_1 must be experimentally determined. The $(n - m)$ π groups are independent but not unique. If a different set of repeating parameters is chosen, different π's will be obtained. Common usage dictates which groups are preferred.

8.3 PHYSICAL SIGNIFICANCE OF DIMENSIONLESS PARAMETERS

All dimensionless parameters can be obtained by taking ratios of significant forces that exist in the flow. The significant forces are

- Inertia force = (mass) × (acceleration) = $\left(\rho L^3\right)\left(\dfrac{L}{T^2}\right) = [\rho]L^2V^2$

    It can also be written as (dynamic pressure) × (area) = $\rho V^2 L^2$

- Viscous force $\tau A = \mu\left(\dfrac{du}{dy}\right)A = [\mu]\dfrac{V}{L}L^2 = \mu VL$

- Gravity force = $mg = \rho L^3 g$
- Pressure force = (Δp) Area = $\Delta p L^2$
- Surface Tension force = σL
- Compressibility force = κL^2

8.3.1 Force ratios

The common force ratios are

Reynolds number: After Osborne Reynolds (1842–1912), a U.K. engineer who realized the difference between laminar and turbulent flows through experimentation (Reynolds, 1883) and introduced the dimensionless Reynolds number:

$$Re = \frac{\text{Inertia force}}{\text{Viscous force}} = \frac{\rho L^2 V^2}{\mu V L} = \frac{\rho V L}{\mu} = \frac{VL}{\nu} \tag{8.6}$$

V, L are characteristic velocity and length. It is used as a criterion to identify the type of flow.

Froude number: after William Froude (1810–1879), a British naval architect

$$Fr = \frac{\text{Inertia force}}{\text{Gravity force}} = \frac{\rho L^2 V^2}{\rho L^3 g} = \frac{V^2}{Lg} \tag{8.7a}$$

or

$$Fr = \frac{V}{\sqrt{gL}} \tag{8.7b}$$

This is important for free surface flows.

Mach number (1880s): after the Austrian physicist Ernest Mach (1838–1916)

$$M = \frac{\text{Inertia force}}{\text{Compressibility force}} = \frac{\rho L^2 V^2}{\kappa L^2} = \frac{V^2}{\kappa / \rho} \tag{8.8a}$$

or

$$\frac{V}{\sqrt{\kappa / \rho}} \left(= \frac{V}{c} \right) \tag{8.8b}$$

where c is the local speed of sound. This is applicable for compressible flows at high velocities.

Euler number (Pressure coefficient): after French mathematician Leonhard Euler (1707–1783)

$$E_u \left(\text{or } C_p \right) \frac{\text{Pressure force}}{\text{Inertia force}} = \frac{(\Delta p) L^2}{\rho V^2 L^2} = \frac{\Delta p}{\rho V^2} \tag{8.9a}$$

or

$$\frac{1}{2} \rho V^2 \tag{8.9b}$$

Δp is the piezometric head difference between two points in the flow. The quantity $\frac{1}{2} V^2$ represents the dynamic pressure.

Weber number: after German naval architect Moritz Weber (1871–1951)

$$We = \frac{\text{Inertia force}}{\text{Surface tension force}} = \frac{\rho L^2 V^2}{\sigma L} = \frac{V^2}{\sigma / \rho L} \text{ or } \frac{V}{\sqrt{\sigma / \rho L}} \tag{8.10}$$

Cauchy number: after the French mathematician A.L. Cauchy (1789–1857)

$$C_n = \frac{\text{Inertia force}}{\text{Elastic force}} = \frac{\rho L^2 V^2}{\kappa L^2} = \frac{\rho V^2}{\kappa} \tag{8.11}$$

where κ is the bulk modulus of elasticity.

Drag coefficient

$$C_D = \frac{\text{Drag force}}{\frac{1}{2}(\text{Dynamic pressure})\text{Area}} = \frac{F_D}{\frac{1}{2}\rho V^2 A} \tag{8.12}$$

Lift coefficient

$$C_L = \frac{\text{Lift force}}{\frac{1}{2}(\text{Dynamic pressure})\text{Area}} = \frac{F_L}{\frac{1}{2}\rho V^2 A} \tag{8.13}$$

Cavitation number

$$\sigma = \frac{p - p_{\text{ref}}}{\frac{1}{2}\rho V^2} \tag{8.14}$$

where p_{ref} is taken as the vapour pressure.

8.4 SIMILARITY AND MODEL TESTS

When the performance of a large system has to be determined, a scaled-down version of the original problem is studied by model tests. The results of the model tests are then scaled up to obtain the performance of the prototype. The measurements may include velocities, pressures, forces, moments, etc. There are three types of similarities that must be satisfied:.

8.4.1 Geometrical similarity

The model and prototype must be of the same shape. All linear dimensions of the model must be a constant scale factor of the corresponding dimensions of the prototype, i.e.

$$\frac{L_m}{L_p} = \text{Constant} \tag{8.15a}$$

8.4.2 Kinematic similarity

The model and prototype flows must be similar, which means that the corresponding velocity ratios must be a constant, i.e.

$$\frac{V_m}{V_p} = \text{Constant} \tag{8.15b}$$

8.4.3 Dynamic similarity

The model and prototype forces must be similar, i.e.

$$\frac{F_m}{F_p} = \text{Constant} \tag{8.15c}$$

Geometrical and kinematic similarities are pre-requisites for dynamic similarity.

For complete dynamic similarity, all dimensionless numbers must be the same. However, it is not always possible in model tests to have this condition satisfied. Inflow conditions where friction has to be considered, it is difficult to have an equivalent friction in the model as that of the prototype. The same problem arises in modelling sediment transport.

When inertia and viscous effects are dominant, the Reynolds number is kept constant. For small values of Re, the flows are laminar and viscous. When gravity terms are dominant, the Froude number should be the same. When surface tension effects are dominant, the Weber number must be the same. For compressible flows, the Mach number must be the same.

8.5 MODELLING RATIOS

8.5.1 Velocity ratio, V_r

8.5.1.1 Using Fr criterion (flows where gravity is dominant)

$$\frac{V_p}{\sqrt{gL_p}} = \frac{V_m}{\sqrt{gL_m}} \Rightarrow \frac{V_p}{V_m} = \sqrt{\frac{L_p}{L_m}} = \sqrt{L_r} \tag{8.16a}$$

i.e. $V_r = \sqrt{L_r}$ \hfill (8.16b)

8.5.1.2 Using Re criterion (for viscous flows)

$$\frac{V_p D_p}{v_p} = \frac{V_m D_m}{v_m} \Rightarrow \frac{V_p}{V_m} = \frac{v_p}{v_m} \frac{D_m}{D_p} \tag{8.17a}$$

i.e. $V_r = \dfrac{v_r}{L_r}$ \hfill (8.17b)

8.5.1.3 Using We criterion (surface tension)

$$\frac{V_p}{\sqrt{\sigma_p/\rho_p L_p}} = \frac{V_m}{\sqrt{\sigma_m/\rho_m L_m}} \Rightarrow \frac{V_p}{V_m} = \sqrt{\frac{\sigma_p}{\sigma_m} \frac{\rho_m}{\rho_p} \frac{L_m}{L_p}} \tag{8.18a}$$

i.e. $V_r = \sqrt{\sigma_r/\rho_r L_r}$ \hfill (8.18b)

8.5.1.4 Using Mach number criterion

$$\frac{V_p}{\sqrt{\kappa_p/\rho_p}} = \frac{V_m}{\sqrt{\kappa_m/\rho_m}} \Rightarrow \frac{V_p}{V_m} = \sqrt{\frac{\kappa_p}{\kappa_m} \frac{\rho_m}{\rho_p}} \Rightarrow V_r = \sqrt{\frac{\kappa_r}{\rho_r}} \tag{8.19}$$

8.5.2 Angular velocity ratio, ω_r

Angular velocity ω is given by $V = \omega r$. Therefore

$$\omega = \frac{V}{r} = \frac{\text{Velocity}}{\text{Length}}$$

and

$$\frac{\omega_p}{\omega_m} = \frac{V_p}{L_p}\frac{L_m}{V_m} = \frac{V_p}{V_m}\frac{L_m}{L_p} \Rightarrow \omega_r = \frac{V_r}{L_r} = L_r^{-\frac{1}{2}} \tag{8.20}$$

8.5.3 Discharge ratio, Q_r

Discharge $Q = (\text{Area})\times(\text{velocity}) \propto L^2 V$

$$\frac{Q_p}{Q_m} = \frac{L_p^2}{L_m^2}\frac{V_p}{V_m} \Rightarrow Q_r = L_r^2 V_r = L_r^{5/2} \tag{8.21}$$

8.5.4 Time ratio, t_r

$$\text{Time} = \frac{\text{Distance}}{\text{Velocity}}$$

$$\frac{t_p}{t_m} = \frac{L_p}{V_p}\frac{V_m}{L_m} = \frac{V_m}{V_p}\frac{L_p}{L_m} \Rightarrow t_r = \frac{L_r}{V_r} = L_r^{1/2} \tag{8.22}$$

8.5.5 Force ratio, Fr

8.5.5.1 Re modelling

$$\text{Viscous force} = \tau A = \mu\left(\frac{du}{dy}\right)A = \mu VL$$

$$Fr = \mu_r V_r L_r = \mu_r \frac{V_r}{L_r}L_r = \mu_r v_r = \frac{\mu_r^2}{\rho_r} \tag{8.23}$$

8.5.5.2 Froude modelling

Inertia force $=$ Mass $\times$ Acceleration $= \rho L^3 L T^{-2} = \rho L^2 V^2$

$$Fr = \rho_r L_r^2 V_r^2 = \rho_r L_r^2 L_r = \rho_r L_r^3 \tag{8.24}$$

$(= L_r^3$ if the same fluid or a fluid with the same density is used.)

8.6 DISTORTED MODELS

If the depth of flow becomes too small that surface tension and viscosity effects affect the flow, then distorted modelling is preferred. If the prototype is not influenced by surface tension and viscosity effects, then the model also should not be affected. In a distorted model, velocity is determined by gravity forces in the vertical direction.

$$\frac{V_p}{V_m} = \left(\frac{L_p}{L_m}\right)^{\frac{1}{2}}_{\text{Vertical}}$$

(8.25)

Discharge $= (\text{Velocity}) \times (\text{area})$ in a vertical plane

$\qquad = (\text{Velocity}) \times (\text{vertical length} \times \text{horizontal length})$

Therefore,

$$\frac{Q_p}{Q_m} = \frac{V_p A_p}{V_m A_m} = \left(\frac{L_p}{L_m}\right)^{\frac{1}{2}}_V \left(\frac{L_p}{L_m}\right)_H \left(\frac{L_p}{L_m}\right)_V = \left(\frac{L_p}{L_m}\right)_H \left(\frac{L_p}{L_m}\right)^{3/2}_V$$

(8.26)

$$\text{Time} = \frac{\text{Horizontal distance}}{\text{Velocity}}$$

(8.27)

$$\left(\frac{t_p}{t_m}\right) = \left(\frac{L_p}{L_m}\right)_H \left(\frac{V_m}{V_p}\right) = \left(\frac{L_p}{L_m}\right)_H \left(\frac{L_m}{L_p}\right)^{1/2}_V$$

8.7 INCOMPLETE SIMILARITY

For many flow systems, there is more than one force ratio involved. For these, it is not possible to attain complete dynamic similarity except for **Full**-scale modelling. Then the result is incomplete similarity.

For example, the drag force on a ship is due to viscous shear (Re) and due to gravity waves (Fr). For Froude similarity, $V_r = \sqrt{L_r}$. For Reynolds similarity, $V_r = \frac{v_r}{L_r}$. Equating the velocity ratios (for Re and Fr velocity ratios),

$$v_r = L_r^{\frac{3}{2}}$$

Therefore, if $L_r = 10$, $v_r = 31.7$.

This means that the prototype fluid must have a viscosity 31.7 times that of the model. It is not possible to find such a liquid. If the same liquid at the same temperature is used, $v_r = 1$, and,

$$L_r^{\frac{3}{2}} = 1 \Rightarrow L_r = 1$$

This means that the model and the prototype are of the same size. In such situations, only Froude similarity law is tested, and estimation of viscous forces is made as a percentage of the gravity forces. This is not very accurate! In general, Reynolds modelling is used for pipe flow, drag and lift of aerofoils, drag on any shape in incompressible flow, Froude modelling for wave resistances of ships, tidal models of harbours, wave phenomena, beach erosion, river modelling, hydraulic structures and Mach modelling for gas flows at Mach numbers > 0.3.

8.8 ROTATING SYSTEMS

Pumps, turbines, compressors, propellers, etc. are rotating systems, and the discharge through such systems can be expressed as

$$Q = f(gH, P, N, D, \rho, \mu, g) \tag{8.28a}$$

where H is the head, P is the power, N is the rotational speed (usually in revolutions per minute, or, rpm), D is the diameter of the rotating element (impeller, rotor, fan, etc.), ρ is the density of fluid and μ is the viscosity of fluid. The effect of g is negligible and therefore g and H are lumped as gH. This also eliminates the difference between D and H as a length parameter. Then,

$$F(Q, gH, P, N, D, \rho, \mu) = 0. \tag{8.28b}$$

The number of independent variables, $n = 7$ and $m = 3$, and therefore there are four π functions. Choosing the diameter D as a length, which is a geometric variable, the speed N as a flow variable and the density ρ as a fluid property as repeating variables, the π functions will be of the form

$$\pi_1 = D^{a_1} N^{b_1} \rho^{c_1} Q \tag{8.29a}$$

$$\pi_2 = D^{a_2} N^{b_2} \rho^{c_2} gH \tag{8.29b}$$

$$\pi_3 = D^{a_3} N^{b_3} \rho^{c_3} P \tag{8.29c}$$

$$\pi_4 = D^{a_4} N^{b_4} \rho^{c_4} \mu \tag{8.29d}$$

Repeating variable must include among them all the fundamental quantities, and that dependent variables should not be used as repeating variables.

$$\pi_1 : L^{a_1} T^{-b_1} \left(ML^{-3} \right)^{c_1} L^3 T^{-1}$$

$$\pi_2 : L^{a_2} T^{-b_2} \left(ML^{-3} \right)^{c_2} L^2 T^{-2}$$

$$\pi_3 : L^{a_3} T^{-b_3} \left(ML^{-3} \right)^{c_3} ML^2 T^{-3}$$

$$\pi_4 : L^{a_4} T^{-b_4} (ML^{-3})^{c_4} ML^{-1} T^{-1}$$

giving

$$\pi_1 : \quad c_1 = 0; \; b_1 = -1; \; a_1 = -3; \Rightarrow \pi_1 = \frac{Q}{ND^3} \tag{8.30a}$$

$$\pi_2 : \quad c_2 = 0; \; b_2 = -2; \; a_2 = -2; \Rightarrow \pi_2 = \frac{gH}{N^2 D^2} \tag{8.30b}$$

$$\pi_3: \quad c_3 = -1; \, b_3 = -1; \, a_3 = -2; \Rightarrow \pi_3 = \frac{\mu}{N\rho D^2} \tag{8.30c}$$

$$\pi_4: \quad c_4 = -1; \, b_4 = -3; \, a_4 = -5; \Rightarrow \pi_4 = \frac{P}{N^3 \rho D^5} \tag{8.30d}$$

and

$$F\left(\frac{Q}{ND^3}, \frac{gH}{N^2 D^2}, \frac{\mu}{N\rho D^2}, \frac{P}{D^5 N^3 \rho}\right) = 0 \tag{8.31}$$

8.8.1 Significance of parameters

All lengths are proportional to D, and all areas are proportional to D^2. The average velocity $\frac{Q}{A}$ is proportional to $\frac{Q}{D^2}$. The peripheral velocity of impeller (rotating element) is proportional to ND. Therefore

$$\pi_1 = \frac{Q/D^2}{ND} \, \alpha \, \frac{\text{Fluid velocity}}{\text{Blade velocity}} = \text{Discharge number} \tag{8.32}$$

If ND is regarded as a typical velocity,

$$\pi_3 = \frac{\mu}{\rho ND^2} = \frac{\mu}{\rho DV} = \frac{1}{Re} \tag{8.33}$$

$$\frac{\pi_4}{\pi_1 \pi_2} = \frac{P}{D^5 N^3 \rho} \frac{ND^3}{Q} \frac{N^2 D^2}{gH} = \frac{P}{\rho g HQ} \tag{8.34}$$

(*proportional to* some function of efficiency, η)

$$\frac{\pi_1^{1/2}}{\pi_2^{3/4}} = \frac{Q^{1/2}}{N^{1/2} D^{3/2}} \frac{N^{3/2} D^{3/2}}{(gH)^{3/4}} = \frac{NQ^{1/2}}{(gH)^{3/4}} = \text{Specific speed parameter}, \tag{8.35}$$

an important parameter in roto-dynamic machines such as centrifugal pumps.

8.9 SHIP RESISTANCE

Any solid body experiences drag when moving through a fluid. Drag consists of two components: skin friction and form drag. The latter, in nautical terminology, is referred to as eddy-making resistance. A ship, however, is partially immersed in the fluid and gives rise to waves on the surface which require energy derived from the motion. As a result, a ship experiences an increased resistance when in motion.

Waves in a liquid can come from capillary action due to surface tension as well as from gravitational effects. The former type is negligible in ship motion. The latter type can arise at the bow and at the stern as well as minor waves normal to the direction of motion of the ship.

In a wave, some particles of water are above the mean level and some below the mean level. When particles are raised, work must be done against gravity, so gravity plays an important part in ship resistance. The relevant variables for the description of the resistance forces include gravity, viscosity, density, velocity and some characteristic length to represent the length of the ship. The most general form from dimensional analysis can be of the form

$$F = \rho u^2 L^2 \phi \left(\frac{\rho u L}{\mu}, \frac{u^2}{gL} \right) \Rightarrow F = \rho u^2 L^2 \phi \left(Re, Fr^2 \right)$$

For complete dynamic similarity, both Re and Fr must be the same for model and prototype. However, it is impractical to achieve both these conditions simultaneously.

A way out of this situation has been suggested by Froude using the assumption that the total resistance is composed of a wave-making resistance, skin friction and eddy-making resistance, and, further assuming that wave-making resistance is independent of Re (unaffected by viscosity) that skin friction depends only on Re and that eddy-making resistance is small and varies little with Re and therefore lumped with wave-making resistance.

This leads to representing the equation as

$$F = \rho u^2 L^2 \left(\phi_1 \left(Re \right) + \phi_2 \left(Fr \right) \right) \tag{8.36}$$

The skin friction may be obtained by assuming that it is the same as that for a thin flat plate having the same length and wetted surface area and moving through water at the same velocity (Massey, 2006, p. 184). The resistance due to wave-making and eddy-making should, therefore, be the difference between the total resistance and the skin-friction obtained under the above assumption. Model tests are then carried out on the basis of Froude similarity since Reynolds similarity has already been taken care of in the skin friction estimation.

Example 8.1

Force on a smooth sphere placed in a flow field:

$$F = f(D, V, \rho, \mu)$$

$$n = 5 \, (F, D, V, \rho, \mu)$$

$$m = 3 (\text{MLT})$$

$$F = \text{MLT}^{-2}; \, D = \text{L}; \, V = \text{LT}^{-1}$$

$$\rho = \text{ML}^{-3}; \, \mu = \text{ML}^{-1}\text{T}^{-1}$$

Repeating variables:

    i. Geometric characteristic D
   ii. Fluid characteristic ρ
  iii. Flow characteristic V

Therefore, the number of π functions is $n - m = 2$.

$$\pi_1 = D^{a_1} V^{b_1} \rho^{c_1} \mu$$

$$\pi_2 = D^{a_2} V^{b_2} \rho^{c_2} F$$

For the π's to be dimensionless, any one quantity in each π may appear to the first power and the others will appear to some other powers (e.g. a, b, c, etc.).

$$\pi_1 = L^{a_1} (LT^{-1})^{b_1} (ML^{-3})^{c_1} ML^{-1}T^{-1}$$

$$\pi_2 = L^{a_2} (LT^{-1})^{b_2} (ML^{-3})^{c_2} MLT^{-2}$$

Equating the exponents of MLT to be zero (for π's to be dimensionless)

$$\pi_1 : \begin{cases} c_1 + 1 = 0 \quad (M) \\ a_1 + b_1 - 3c_1 - 1 = 0 \quad (L) \\ -b_1 - 1 = 0 \quad (T) \end{cases}$$

which gives

$$c_1 = -1$$

$$b_1 = -1$$

$$a_1 = -1$$

Therefore,

$$\pi_1 = \frac{\mu}{DV\rho}$$

Similarly, for π_2,

$$c_2 + 1 = 0$$

$$a_2 + b_2 - 3c_2 + 1 = 0$$

$$-b_2 - 2 = 0$$

which gives

$$c_2 = -1$$

$$b_2 = -2$$

$$a_2 = -2$$

and therefore,

$$\pi_2 = \frac{F}{\rho D^2 V^2}$$

The functional relationship then is

$$G(\pi_1, \pi_2) = 0 \Rightarrow G\left(\frac{\mu}{\rho DV}, \frac{F}{\rho D^2 V^2}\right) = 0$$

or

$$\pi_1 = G_1(\pi_2) \Rightarrow \frac{\mu}{\rho DV} = G_1\left(\frac{F}{\rho D^2 V^2}\right)$$

The functions G and G_1 must be experimentally determined.

Example 8.2

Pressure drop, Δp, for steady incompressible viscous flow through a straight horizontal pipe:

$$\Delta p = f(\ell, V, \mu, D, \rho, e)$$

where
  ℓ – length of pipe
  e – average roughness height
  V – average velocity

$$g(\Delta p, \ell, V, \mu, D, \rho, e) = 0 \Rightarrow n = 7; m = 3(M, L, T)$$

Dimensions of parameters:

$$\Delta p = ML^{-1}T^{-2} \text{ (Force/Area)}$$

$$\ell = L$$

$$V = LT^{-1}$$

$$\mu = ML^{-1}T^{-1}$$

$$D = L$$

$$\rho = ML^{-3}$$

$$e = L$$

Repeating parameters:
  Geometric: D
  Fluid property: ρ
  Flow: V
  No. of π-groups = $n - m = 7 - 3 = 4$. Therefore,

$$\pi_1 = D^{a_1} \rho^{b_1} V^{c_1} \Delta p$$

$$\pi_2 = D^{a_2} \rho^{b_2} V^{c_2} l$$

$$\pi_3 = D^{a_3} \rho^{b_3} V^{c_3} \mu$$

$$\pi_4 = D^{a_4}\rho^{b_4}V^{c_4}e$$

i.e. $\pi_1 = L^{a_1}\left(ML^{-3}\right)^{b_1}(LT^{-1})^{c_1}ML^{-1}T^{-2}$

$$\pi_2 = L^{a_2}\left(ML^{-3}\right)^{b_2}(LT^{-1})^{c_2}L$$

$$\pi_3 = L^{a_3}(ML^{-3})^{b_3}(LT^{-1})^{c_3}ML^{-3}T^{-1}$$

$$\pi_4 = L^{a_4}(ML^{-3})^{b_4}(LT^{-1})^{c_4}L$$

For π's to be dimensionless,

$\pi_1 : b_1 + 1 = 0; \Rightarrow a_1 = 0$

$a_1 - 3b_1 + c_1 - 1 = 0 \Rightarrow b_1 = -1$

$-c_1 - 2 = 0; \Rightarrow c_1 = -2$

giving

$$\pi_1 = \frac{\Delta p}{\rho V^2}$$

$\pi_2 : b_2 = 0; \Rightarrow a_2 = -1$

$a_2 - 3b_2 + c_2 + 1 = 0 \Rightarrow b_2 = 0$

$-c_2 = 0; \Rightarrow c_2 = 0$

giving

$$\pi_2 = \frac{l}{D}$$

$\pi_3 : b_3 + 1 = 0; \Rightarrow a_3 = -1$

$a_3 - 3b_3 + c_3 - 1 = 0 \Rightarrow b_3 = -1$

$-c_3 - 1 = 0; \Rightarrow c_3 = -1$

giving

$$\pi_3 = \frac{\mu}{\rho DV}$$

$\pi_4 : b_4 = 0; \Rightarrow a_4 = -1$

$a_4 - 3b_4 + c_4 + 1 = 0; \Rightarrow b_4 = 0$

$-c_4 = 0; \Rightarrow c_4 = 0$

giving

$$\pi_4 = \frac{e}{D}$$

Therefore,

$$G\left(\frac{\Delta p}{\rho V^2}, \frac{\ell}{D}, \frac{\mu}{\rho D V}, \frac{e}{D}\right) = 0$$

or

$$\frac{\Delta p}{\rho V^2} = G_1\left(\frac{\ell}{D}, \frac{\mu}{\rho D V}, \frac{e}{D}\right)$$

REFERENCE

Massey, B. S., (2006), *Mechanics of Fluids*, Eighth edition, Taylor and Francis Group, London.

Chapter 9

Fluid flow measurements

9.1 INTRODUCTION

The common variables of fluid flow include the properties of the fluid such as the density and viscosity, and flow variables such as the depth of flow, velocity of flow and flow rate. In gases, the pressure and temperature also need to be considered. To understand the state of a fluid in motion the magnitudes (and sometimes directions) of these variables need to be known and the best way is to measure them. The instruments and measuring techniques have evolved with time and their accuracies and repeatabilities have also improved with time. This chapter describes some of the common methods of fluid flow measurement.

9.2 PROPERTIES OF FLUIDS

The main properties of fluids in the context of Fluid Mechanics are the density, viscosity and surface tension. The two thermodynamic properties pressure and temperature are also relevant in compressible fluids. For incompressible fluids, the density is assumed to be constant and the temperature has no significant role to play except that the viscosity is dependent on the temperature. In the case of water, the density is known to be $1{,}000\,\text{kg/m}^3$, and the surface tension forces are small in comparison to other forces influencing fluids at rest as well as in motion. Thus the only property that needs measurements is viscosity.

9.2.1 Viscosity

Viscosity is a measure of the internal resistance of a real fluid. Different layers of fluid in general flow at different velocities within the boundary layer. Outside the boundary layer, the fluid is assumed to be inviscid and the flow is considered as ideal. There are several methods of measuring the viscosity of a fluid.

9.2.1.1 Falling sphere method

When a solid body is allowed to fall under gravity through a viscous medium, a period of initial acceleration is followed by motion at a uniform terminal velocity. Measurement of this terminal velocity affords a means of determining the viscosity of the medium.

Considering a sphere of diameter D falling through a homogeneous fluid of infinite extent and assuming that the motion is slow for the inertia terms to be negligible, the resistance to motion can be written as (Stokes' law)

$$F = 3\pi D\mu V \tag{9.1}$$

where μ is the viscosity and V is the velocity.

The driving force due to the difference in density between the sphere and the fluid is

$$\frac{\pi D^3 (\sigma - \rho) g}{6} \tag{9.2}$$

where σ is the density of the sphere and ρ is the density of the fluid.

Equating the two forces,

$$\mu = \frac{g D^2 (\sigma - \rho)}{18V} \tag{9.3}$$

Introduction of finite boundaries involves the use of a correction. A correction introduced by Ladenburg (1907) is

$$\mu_{\text{true}} = \frac{\mu_{\text{measured}}}{\left(1 + 2.4 \dfrac{r}{R}\right)\left(1 + 3.3 \dfrac{r}{h}\right)} \tag{9.4}$$

where R is the radius of the cylinder containing the fluid, r is the radius of the sphere and h is the height of the cylinder. The first term in the denominator is a correction for the effect of the wall, and the second term is for the finite depth. He has tested the formula for $\dfrac{r}{R} < 0.09$ and $\dfrac{r}{h} < 0.008$. Other suggested corrections for the wall effect include that proposed by Francis (1933), which is of the form

$$\mu_{\text{true}} = \mu_{\text{measured}} \left(1 - \frac{r}{R}\right)^{2.25} \tag{9.5}$$

and by Faxen (1922), which is of the form

$$\mu_{\text{true}} = \mu_{\text{measured}} \left[1 - 2.104 \frac{r}{R} + 2.09 \left(\frac{r}{R}\right)^3 - 0.95 \left(\frac{r}{R}\right)^5\right] \tag{9.6}$$

In general, the end effects can be neglected if the middle third of the tube is used. British Standard (BS) 188 states that measurements shall be taken over a length of fall of 150 mm between two reference marks that are not less than 55 mm from top and bottom.

For the wall correction, the BS states that $R > 10r$, and that the measured viscosity shall be multiplied by a correction factor $(1 - 2.1 \dfrac{r}{R})$, which is equivalent to the first two terms of Faxen's correction.

9.2.1.2 Rotating cylinder method

In this method, a cylinder is rotated inside another concentric cylinder in which the narrow gap between them is filled with the fluid of interest (Figure 9.1). The resistance to rotation is measured by the torque on the stationary cylinder. The basis of the method is Newton's law of viscosity, which is stated as

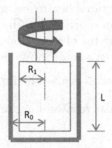

Figure 9.1 Rotating viscometer.

$$\tau = \mu \frac{du}{dy} \tag{9.7}$$

where τ is the shear stress, μ is the coefficient of viscosity and $\frac{du}{dy}$ is the velocity gradient which can be determined if the speed of rotation is known. By measurement of the torque T_c on the stationary cylinder, the shear stress can be determined. The velocity gradient is given as

$$\frac{du}{dy} = \frac{2\pi R_0 N}{60b} \tag{9.8}$$

where N is the speed of rotation in revolutions per minute, L is the height of the cylinder and b is the clearance between the two cylinders ($b = R_0 - R_1$). This equation is based on $b \lll R_0$. The torque on the inner cylinder is measured by a torsion wire from which it is suspended.

If the torque due to the fluid at the bottom of the cylinder is neglected,

$$\tau = \frac{T_c}{2\pi R_1^2 L} \tag{9.9}$$

By substitution, μ can be determined as

$$\mu = \frac{15 T_c b}{\pi^2 R_0 R_1^2 L N}. \tag{9.10}$$

The unit of viscosity is Pascal seconds (Pa s). In this method, the resistance due to the fluid below the bottom of the rotating cylinder is neglected. Corrections can be made in a manner similar to the falling sphere method.

9.2.1.3 Anomalous fluids

The theory described above is applicable only to Newtonian fluids. Sometimes, it is necessary to examine the behaviour of non-Newtonian fluids. The falling sphere method can still be used to obtain an apparent viscosity, but it is not adequate.

For a Newtonian fluid, the relationship between the velocity V and $(\sigma - \rho)$ is linear for a sphere of any radius, and the straight lines pass through the origin. In the case of a non-Newtonian fluid having a yield value, straight lines will have an intercept Δ on the x-axis. This is referred to as the yield value. Until the yield value, Δ for the density difference is reached the spheres will not move.

Minimum density difference is required before any movement of the sphere takes place. This driving force may be taken as proportional to the yield value. The difficulty arises in the calculation of the yield value from the observed value of Δ.

The following analysis is taken from a study by Williams and Fulmer (1938):

For a Newtonian fluid, the shear stress is given by Eq. 9.7 as $\tau = \mu \dfrac{du}{dy}$.

For a non-Newtonian fluid, a similar relationship can be written as follows:

$$\tau - \tau_\Delta = \mu' \frac{du}{dy} \tag{9.11}$$

where τ_Δ is the yield value of shear stress and μ' is the analogous viscosity of the non-Newtonian fluid. At the yield value,

$$\tau = \tau'. \tag{9.12}$$

It has been shown by Lamb (1906) that the shear stress at the equator of the sphere moving through an infinite medium is

$$\tau = \frac{3\mu V}{2r} \tag{9.13}$$

Using Stokes' equation, τ may be written as

$$\frac{3\mu V}{2r} = \frac{1}{3}(\sigma - \rho)gr \tag{9.14a}$$

giving

$$\mu = \frac{2gr^2(\sigma - \rho)}{9V} \tag{9.14b}$$

Substituting for μ from Eq. 9.14 in Eq. 9.13,

$$\tau = \frac{gr(\sigma - \rho)}{3} \tag{9.15}$$

For a Bingham material, when $(\sigma - \rho) = \Delta$, $\tau = \tau'$, and therefore,

$$\tau' = \frac{gr\Delta}{3} \tag{9.16}$$

Pros and cons of using this approach as well as the validity of using Eq. 9.13 have been discussed by Williams and Fulmer (1938). Fulmer and Williams (1936) give an equation for wall effect correction. Some mathematical details related to viscosity can be found in the book by Lamb (1906).

9.3 MEASUREMENTS OF STAGE (SURFACE ELEVATION)

Stage (surface elevation) of a fluid enables the determination of the depth of fluid above a given datum. It can be measured by a point gauge in which the point of contact with the liquid surface is from above or by a hook gauge in which the point of contact with the fluid surface is from below (Figure 9.2).

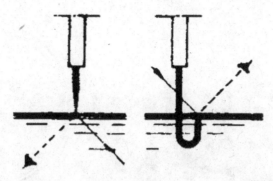

Figure 9.2 Point gauge and hook gauge.

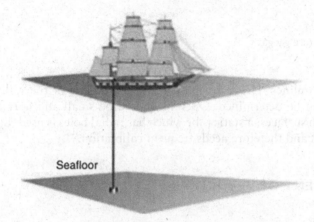

Seafloor

Figure 9.3 SONAR system.

9.3.1 Sea bed depth measurement

For measuring the depth to the sea bed or a reservoir bed, an acoustic method based on the speed of sound is used. Sending a sound signal from a stationary floating body and measuring the time it takes for the reflected signal to reach the emitting point and knowing the speed of sound in the medium, the depth to the reflecting surface (bed) can be calculated (Figure 9.3). The technique is known as SONAR (SOund Navigation And Ranging).

9.3.2 Stage recorders

Stage recorders enable a continuous measurement of stage. The recording device can be a chart or a data logger. They are usually unattended except for periodical downloading of the recordings.

9.4 MEASUREMENT OF PRESSURE

Fluids exert pressures that may be static when the fluid is at rest or undisturbed and dynamic when the fluid is in motion. Different types of pressure gauges (Figure 9.4) are available for their measurement. When the flow is parallel, the pressure variation is hydrostatic and

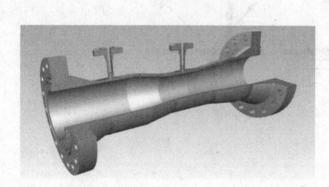

Figure 9.4 Typical pressure gauge.

normal to the streamlines. Hence by measuring the pressure at the wall, the pressure at any other point can be determined. Opening should be small and normal to the surface (smooth). For rough surfaces, a static tube which has radial holes is used. Errors could occur due to disturbances and therefore needs frequent calibration.

9.5 MEASUREMENT OF VELOCITY

9.5.1 Pitot tube

The pitot tube is an example of the application of Bernoulli's equation and is used to measure air (or liquid) speed in many instants, e.g. in aircrafts. A non-rotating obstacle placed in the stream produces a stagnation point where the velocity is zero.

For example, in the case of a flow past a cylinder (Figure 9.5), the streamlines closest to the boundary must divide into two halves, one going along the upper surface and the other along the lower surface. This can only happen if the flow is momentarily brought to rest at a point called the stagnation point before dividing into two halves. Therefore, the pressure of a fluid particle of velocity v is increased at the stagnation point to $p + \frac{1}{2}\rho v^2$. Considering two points at the same elevation and applying Bernoulli equation,

$$\frac{p}{\rho g} + \frac{v^2}{2g} = \frac{p_0}{\rho g} + 0 \Rightarrow p_0 = p + \frac{1}{2}\rho v^2 \qquad (9.17)$$

The pressure, p_0 is known as stagnation pressure, and the term $\frac{1}{2}\rho u^2$ is called the dynamic pressure. Therefore, if the stagnation pressure as well as the static pressure p can be measured, then the velocity can be determined. This is the principle of the **Pitot tube**, which is named after Henry Pitot (1695–1771). It can be used to measure the velocities of liquids and gases (Figure 9.6).

$$p_1 = \rho g y_1$$

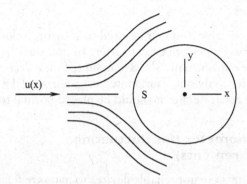

Figure 9.5 Stagnation point in a flow past a cylinder.

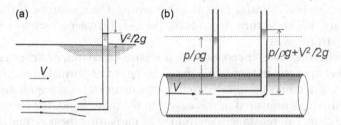

Figure 9.6 (a) Pitot tube in free flowing fluid. (b) Pitot tube in a confined flow.

$$p_0 = \rho g (y_1 + h)$$

$$\frac{p_1}{\rho g} + \frac{v_1^2}{2g} = \frac{p_0}{\rho g} + 0 \Rightarrow v_1 = \sqrt{2gh} \tag{9.18}$$

For a horizontal pipe connected to a differential manometer in which the differential pressure is shown by the difference in the manometric fluid h_m of density ρ_m,

$$p_1 + y_1 \rho g + h_m \rho_m g = p_0 + (y_1 + h_m) \rho g$$

$$\Rightarrow p_0 - p_1 = h_m (\rho_m - \rho) g \tag{9.19}$$

$$\Rightarrow v = \sqrt{\frac{2g(p_0 - p_1)}{\rho g}} = \sqrt{\frac{2(p_0 - p_1)}{\rho}} \tag{9.20}$$

The pressure difference is measured by a manometer as h, and density corrections are required at different levels.

9.5.1.1 *Pitot static tube*

Pitot static tube is a combination of the above two into one unit. The static tube surrounds a total head tube.

9.5.2 Current meters

Current meters are more reliable and can be used to measure velocity at any point in the cross section. There are two types of current meters in use: the cup type and the propeller type. In the cup type, the cups (usually six) are mounted on a vertical axis, whereas in the propeller type the propeller (only one) is mounted on a horizontal axis. The propeller type has less drag and is more accurate. But it should be placed normal to the flow direction.

9.5.3 Hot-wire anemometer (for fluctuating velocity measurements)

Pilot tube and current meters are not reliable devices to measure fluctuating velocities. Hot-wire anemometers are used instead for such measurements. The principle of the hot-wire anemometer is that the temperature attained by a heated body in an airstream depends on the velocity of the stream so that the body can be calibrated to indicate the velocity in terms of the temperature or heat transfer rate. In other words, the resistance of the wire depends on the temperature, which in turn depends on the heat transfer which is a function of the velocity of fluid past the wire.

A typical hot-wire anemometer consists of a fine wire (platinum, nickel or tungsten) heated electrically held between the ends of two pointed prongs to be normal to the direction of flow (Figure 9.7). The current may be adjusted to keep the temperature constant and, therefore, its resistance. The current is measured in this case. On the other hand, the current may be kept constant, and the change in resistance determined by measuring the potential difference using a Wheatstone bridge circuit so that it can operate at constant current or at constant resistance (constant temperature). The anemometer has to be calibrated frequently because it is sensitive to dust in air and slime in water. The wire is about 6 mm long and less than 0.15 mm in diameter. Hot-wire anemometers are mainly used for measuring velocities of gases.

Hot-film anemometer is very similar except that the hot wire is replaced by a more robust hot film (platinum) enclosed in a glass support.

9.5.4 Laser Doppler velocity measurement

Laser Doppler anemometer (LDA), sometimes referred to as laser Doppler velocitometer (LDV) is a non-invasive method of measuring velocities in flowing liquids and gases, especially in a laboratory environment. The technique has been well established about three

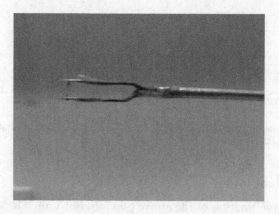

Figure 9.7 Hot-wire anemometer.

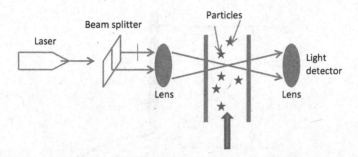

Figure 9.8 Laser anemometry system.

decades ago and has the advantages of being able to give spatial and temporal high resolution and does not need calibration. The technique also has the capability to measure reversing flows. The components of an LDA system consist of a continuous wave laser, transmitting optics and a signal processor. The velocity information comes from light scattering from particles carried in the fluid. Fluids normally have particles naturally, but gases need to be seeded with particles. The particles, which can be solids (powder) or liquids (droplets), typically have sizes ranging from 1 to 10 μm.

The scattered light contains a Doppler shift (difference between the incident and scattered light frequencies), which is proportional to the velocity component perpendicular to the bisector of the two orthogonal laser beams. The fringe spacing provides information about the distance travelled by the particle, and the Doppler frequency provides the information about the time. The velocity can therefore be obtained from the distance and the time (or frequency).

The basic configuration is shown in Figure 9.8. The scattered light after passing through a lens is focused onto a photodetector which converts the light signals to electrical signals. The signal processor filters and amplifies the Doppler bursts and determines the Doppler frequency for each particle using some frequency analysis techniques such as Fast Fourier Transform.

LDA can measure velocities ranging from zero to supersonic in three orthogonal directions, both instantaneous and time-averaged.

9.5.5 Acoustic devices

Since the speed of sound is precisely known, the time to travel a sound wave between two stations can be used to measure the velocity of a flowing fluid. If the fluid is also moving with a velocity V, then the time to travel between the two stations separated by a distance L is given by

$$t = \frac{L}{C_s + V} \tag{9.21}$$

where C_s is the speed of sound in still fluid.

9.6 MEASUREMENT OF DISCHARGE

9.6.1 Orifices (sharp edged)

Sharp-edged orifices (Figure 9.9) provide minimum contact area, thereby reducing friction. If the edges are not sharp, the flow depends upon the thickness of orifices and other boundary

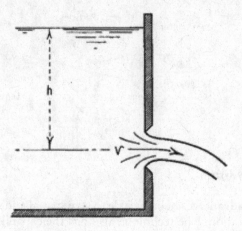

Figure 9.9 Flow through a sharp edged orifice.

conditions. Other dimensions in consideration are very much larger than the dimensions of the orifices. It is also assumed that the velocity at the free surface is negligible.

At steady state, neglecting frictional effects, Bernoulli equation may be applied between a point on the upstream water surface (1) and at the Vena Contracta (2) as follows:

$$\frac{p_1}{\rho g} + \frac{v_1^2}{2g} + h = \frac{p_0}{\rho g} + \frac{v_2^2}{2g} + 0 \tag{9.22}$$

If point (1) is sufficiently far from the orifice, v_1 can be assumed to be zero. This means that the point (1) can be at any point. Therefore it is chosen to be at the surface of reservoir where $p_1 = p_0$ (atmospheric). Then,

$$\frac{p_0}{\rho g} + 0 + h = \frac{p_0}{\rho g} + \frac{v_2^2}{2g} \Rightarrow h = \frac{v_2^2}{2g}$$

or

$$v = \sqrt{2gh} \tag{9.23}$$

This is the theoretical velocity, and the equation is known as **Torricelli's** equation after Evangelista Torricelli (1608–1647), a pupil of Galileo, who demonstrated it experimentally in 1643.

This equation gives the velocity at the **Vena Contracta** where the cross section is a minimum and where uniform flow with parallel streamlines would exist. In the plane of the orifice, the velocity is less than the above and is not known.

9.6.2 Some definitions

$$\text{Coefficient of contraction}, C_c = \frac{\text{Area of Vena Contracta}}{\text{Area of Orifice}} = \frac{A_c}{A_0} \tag{9.24a}$$

$$\text{Coefficient of velocity}, C_v = \frac{\text{Actual velocity}}{\text{Ideal velocity}} \tag{9.24b}$$

Ideal velocity (friction ignored) $= \sqrt{2gh}$ \hfill (9.24c)

Actual velocity (friction considered) $= C_v\sqrt{2gh}$ \hfill (9.24d)

Ideal discharge $= \sqrt{2gh}\,(\text{Area of vena contracta}) = \sqrt{2gh}\,(A_0\,C_c)$ \hfill (9.24e)

Actual discharge $= C_v\sqrt{2gh}$ (Area of vena contracta)

$$= C_v\sqrt{2gh}\,(A_0 C_c)$$

$$= (C_c C_v)\,A_0\sqrt{2gh}$$ \hfill (9.24f)

$$\text{or} = C_d A_0 \sqrt{2gh}$$

where the

Coefficient of discharge, $C_d = C_c C_v$ \hfill (9.24g)

Values of the various coefficients are determined experimentally. For example, for sharp-edged circular orifices, $C_v \approx 0.97 \sim 0.99$; $C_c \approx 0.61 \sim 0.66$; $C_d \approx 0.6 \sim 0.65$.

Note: For large orifices, the velocity in the plane of the vena contracta varies with the depth below free surface level. The total discharge then should be calculated by integration. For a rectangular orifice of contracta width b_c and height d_c,

$$Q = \int_{H_c - \frac{d_c}{2}}^{H_c + \frac{d_c}{2}} C_v \sqrt{2gh}\, b_c dh \hfill (9.25)$$

where H_c is the depth to the centre of vena contracta.

9.6.2.1 Submerged orifices

The outflow from submerged orifices is not exposed to the atmosphere. The downstream pressure p_2 should be taken into consideration when using Bernoulli equation. Referring to Figure 9.10 consisting of a point on the upstream water surface and a point at the centre of the orifice,

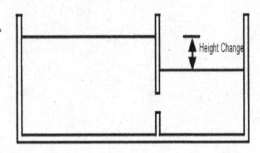

Figure 9.10 Submerged orifice.

$$\frac{p_1}{\rho g} + \frac{v_1^2}{2g} + z_1 = \frac{p_2}{\rho g} + \frac{v_2^2}{2g} + z_2$$

which simplifies to

$$h_1 + 0 = h_2 + \frac{v_2^2}{2g} \Rightarrow v_2 = \sqrt{2g(h_1 - h_2)} \tag{9.26}$$

where h_1 and h_2 are the heights of water surfaces on the upstream and downstream of the orifice.

9.6.2.2 Losses in submerged orifices

Since some of the energy of the jet is destroyed due to turbulence, there will be a drop in the energy level on the downstream side as shown in Figure 9.11.

9.6.3 Orifice meter

An orifice meter (Figure 9.12) is a device where a constriction is made in the form of an orifice installed inside a pipe. The flow converges to the orifice resulting in a drop in the pressure and a corresponding increase in velocity. On the downstream side of the orifice, the velocity decreases, and the pressure increases as the flow returns to normal conditions gradually. Using Bernoulli equation between a point on the upstream side and at the vena contracta,

$$\frac{v_1^2}{2g} + z_1 + \frac{p_1}{\rho g} = \frac{v_2^2}{2g} + z_2 + \frac{p_2}{\rho g}$$

$$\Rightarrow \frac{v_2^2}{2g} = \frac{v_1^2}{2g} + \Delta h^* \tag{9.27a}$$

where

$$\Delta h^* = \left(\frac{p_1}{\rho g} + z_1\right) - \left(\frac{p_2}{\rho g} + z_2\right) \tag{9.27b}$$

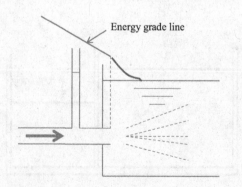

Figure 9.11 Losses in a submerged orifice.

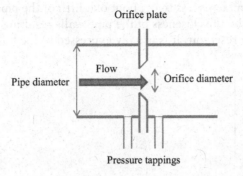

Figure 9.12 Orifice meter.

is the piezometric head difference. The term $\left(\dfrac{p}{\rho g} + z\right)$ is called the **piezometric head,** and not just the manometric head differential. Therefore,

$$v_2 = C_v \left\{ 2g\left(\Delta h^* + \frac{v_1^2}{2g}\right) \right\}^{\frac{1}{2}}$$

From continuity,

$$v_1 = v_2 \frac{A_2}{A_1} = v_2 C_c \frac{A_0}{A_1}$$

where C_c – coefficient of contraction
A_0 – area of the orifice
C_v – coefficient of velocity

$$v_2 = C_v \left\{ 2g\left(\Delta h^* + C_c^2 \frac{A_0^2}{A_1^2} \frac{v_2^2}{2g}\right) \right\}^{\frac{1}{2}}$$

$$v_2^2 \left\{ 1 - C_v^2 C_c^2 \frac{A_0^2}{A_1^2} \right\} = C_v^2 2g\Delta h^* \tag{9.28}$$

$$v_2 = C_v \left\{ \frac{2g\Delta h^*}{1 - C_v^2 C_c^2 \dfrac{A_0^2}{A_1^2}} \right\}^{\frac{1}{2}}$$

Also $Q = A_2 v_2 = C_c A_0 v_2$

$$Q = C_d A_0 \left\{ \frac{2g\Delta h^*}{1 - C_d^2 \cdot \dfrac{A_0^2}{A_1^2}} \right\}^{\frac{1}{2}} \tag{9.29}$$

Usually, if the orifice diameter is less than about one-fifth of the pipe diameter, the contraction of the jet is affected by the closeness of the pipe walls resulting in a higher value for C_d compared with that for a reservoir. It is usually expressed in the following form:

$$C = C_d \left\{ \frac{1 - \left(\dfrac{A_0}{A_1}\right)^2}{1 - C_d^2 \left(\dfrac{A_0}{A_1}\right)^2} \right\}^{\frac{1}{2}} \tag{9.30}$$

Then,

$$Q = \frac{CA_0 \left(2g\Delta h^*\right)^{\frac{1}{2}}}{\left\{1 - \left(\dfrac{A_0}{A_1}\right)^2\right\}^{\frac{1}{2}}} \tag{9.31}$$

Equations 9.29 and 9.31 are identical.

The downstream manometer connection should strictly be made to the section where the vena contracta occurs (this is practically difficult).

9.6.4 Venturi meter (after Italian, Giovanni Battista Venturi (1746–1822))

The Venturi meter (Figure 9.13) is used to measure flow rates. It has a throat section where the velocity is greater because of the smaller cross section resulting in a drop in pressure between the inlet and the outlet. The measurement of difference in pressure enables an evaluation of flow rate. Using Bernoulli equation between sections 1 at upstream and 2 at the throat, and ignoring frictional losses,

$$\frac{p_1}{\rho g} + \frac{v_1^2}{2g} + z_1 = \frac{p_2}{\rho g} + \frac{v_2^2}{2g} + z_2$$

From continuity, assuming steady-state conditions,

$$A_1 v_1 = A_2 v_2 \Rightarrow v_2 = v_1 \frac{A_1}{A_2}$$

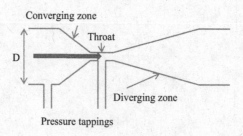

Figure 9.13 Venturi meter.

Substituting,

$$\frac{v_1^2}{2g} + \frac{p_1}{\rho g} + z_1 = \frac{p_2}{\rho g} + z_2 + \frac{v_1^2}{2g}\left(\frac{A_1}{A_2}\right)^2$$

$$\frac{v_1^2}{2g}\left(\left(\frac{A_1}{A_2}\right)^2 - 1\right) = \left(\frac{p_1}{\rho g} + z_1\right) - \left(\frac{p_2}{\rho g} + z_2\right) = \Delta h^*$$

$$v_1 = \left\{\frac{2g\Delta h^*}{\left[\left(\frac{A_1}{A_2}\right)^2 - 1\right]}\right\}^{\frac{1}{2}} \tag{9.32a}$$

$$Q_{ideal} = A_1 v_1 = A_1\left\{\frac{2g\Delta h^*}{\left[\left(\frac{A_1}{A_2}\right)^2 - 1\right]}\right\}^{\frac{1}{2}} \tag{9.32b}$$

$$Q_{actual} < Q_{ideal}$$

To account for the difference, a coefficient of Venturi discharge ($C_d \approx 0.98$) is introduced. Then,

$$Q_{actual} = C_d Q_{ideal} \tag{9.33}$$

The function of the divergence is to restore the pressure and velocity to their original values as nearly as possible. The divergence should not be too large because the flow tends to separate and form eddies resulting in a loss of energy. The greater the angle of divergence, the greater the energy loss. If it is small, the length of divergence is largely resulting in frictional losses. An angle of about 6° has been found to be optimal.

9.6.5 Flow nozzle (nozzle meter)

The nozzle meter or flow nozzle (Figure 9.14) is very much similar to the Venturi-meter without the diverging part. The equations are the same.

9.6.6 Notches and weirs

A Notch is a sharp-edged obstruction over which a fluid flows and where the depth of flow above the base is related to the quantity of flow. Therefore it is a useful device to measure the flow rates. Usually, they are either triangular- or rectangular-shaped. A rectangular one is often called a weir, whereas a triangular one is called a V-notch.

The flow over a notch is quite complex. The pressure is not uniform at any section of flow over a notch, unlike in an orifice. Therefore, the variation of velocity cannot be measured accurately. In addition, turbulence and frictional effects also enter the scene making it more complicated.

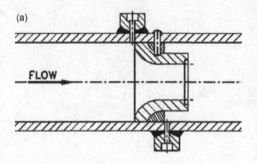
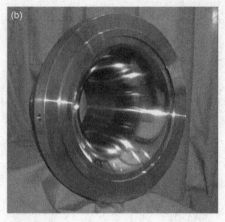

Figure 9.14 Flow nozzle: (a) Cross sectional sketch (b) Actual view.

Therefore an analytical relationship between the rates of flow and the depth of flow is sought by idealizing the problem. However, such equations need modifications by comparing with experiments.

Assumptions:

- The velocities are uniform and parallel on the upstream side of the notch. Therefore the pressure varies according to the hydrostatic equation (Installing baffles may produce uniform velocity conditions).
- The free surface remains horizontal as far as the plane of the notch and all particles passing through the notch move horizontally.
- The pressure at the entire nappe is atmospheric (over the weir section)
- The effects of viscosity and surface tension are negligible.

9.6.6.1 Rectangular sharp-crested weirs

Applying Bernoulli equation for any streamline in the flow between sections 1 and 2,

$$H + z + \frac{v_1^2}{2g} = (H + z - h) + \frac{v_2^2}{2g}$$

(9.34)

$$\Rightarrow v_2 = \sqrt{2gh + v_1^2}$$

The discharge over the weir is given by

$$dQ = v_2 (dA_2)$$

$$Q_{ideal} = \int_0^H v_2 b dh$$

$$= b \int_0^H \left(\sqrt{2gh + v_1^2} \right) dh = b \left[\frac{2}{3} \frac{1}{2g} \left(2gh + v_1^2 \right)^{\frac{3}{2}} \right]_0^H$$

$$= \frac{2}{3} \frac{b}{2g} \left[\left(2gH + v_1^2 \right)^{\frac{3}{2}} - v_1^3 \right]$$

(9.35)

In this equation, both Q and v_1 are unknown. Therefore, a trial and error approach is necessary. However, if the upstream body of water is large in comparison with the weir opening, it is possible to assume that $v_1 \approx 0$. Then,

$$Q_{\text{ideal}} = \frac{2}{3} b \sqrt{2g} H^{3/2} \tag{9.36}$$

$$Q_{\text{actual}} = C_d \frac{2}{3} b \sqrt{2g} H^{3/2} \tag{9.37}$$

Surface tension and viscosity effects are less important for large values of H. The coefficient of discharge C_d depends upon H and $\dfrac{H}{z}$ according to

$$C_d = 0.611 + 0.075 \frac{H}{z} \text{(Given by von Misses, 1917)} \tag{9.38}$$

for values of $\dfrac{H}{z}$ up to about 5. When $\dfrac{H}{z}$ is between 5 and 10, C_d does not obey the above equation. When $\dfrac{H}{z} = 10$, C_d attains a value of 1.135. In Eq. 9.38, z is the weir height.

If $z = 0$, $C_d \to \infty$, which is the case corresponding to a free overfall. In this situation, $H \to y_c$ and hence q, the discharge per unit width of weir, is determined using critical flow conditions. The equation then is

$$q = \sqrt{g y_c^3}$$

For $\dfrac{H}{z} > 20$, it is possible to write

$$q = \sqrt{g y_c^3} = \sqrt{g} (H + z)^{3/2} \tag{9.39}$$

9.6.6.2 Suppressed weirs

Suppressed weir (Figure 9.15) is a weir that has its breadth the same as the width of the approach channel. The nappe then contracts only in the vertical direction. The following formula for C_d has been obtained by Rehbock (1929):

$$C_d = 0.605 + 0.08 \frac{H}{z} + \frac{1}{1,000H} \tag{9.40}$$

For large z, $C_d \approx 0.61$.

9.6.6.3 V-notch (triangular weir)

V-notch (Figure 9.16) has a less variable coefficient because it has the same shape, whatever the value of H.

The same equation can be used:

$$v_2 = \left(2gh + v_1^2\right)^{\frac{1}{2}}$$

Figure 9.15 Suppressed weir.

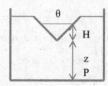

Figure 9.16 Triangular weir (V-notch).

For a V-notch, the cross-sectional area of the approach channel is much larger and therefore v_1^2 is negligible.

$$dQ = \sqrt{2gh}\,b\,dh$$

where

$$b = 2(H-h)\tan\frac{\theta}{2}$$

Therefore,

$$Q = 2\tan\frac{\theta}{2}\sqrt{2g}\int_0^H (H-h)h^{\frac{1}{2}}dh$$

$$= 2\tan\frac{\theta}{2}\sqrt{2g}\left\{\frac{2}{3}Hh^{\frac{3}{2}} - \frac{2}{5}h^{\frac{5}{2}}\right\}_0^H$$

$$= \frac{8}{15}\tan\frac{\theta}{2}\sqrt{2g}H^{\frac{5}{2}} \tag{9.41}$$

$$Q_{actual} = C_d\frac{8}{15}\tan\frac{\theta}{2}\sqrt{2g}H^{\frac{5}{2}} \quad (\theta \text{ is usually} = 30° - 90°) \tag{9.42}$$

For width of channel grater then four times the maximum width of Nappe, and for H large enough for the Nappe to clear off the notch plate, $C_d \approx 0.59$ (for water).

9.6.6.4 Broad crested weir (critical flow weir)

Broad crested weirs can have different shapes (Figure 9.17). The basic principle is that the flow over the weir is critical. Using Bernoulli equation between sections 1 and 2,

$$z + H + \frac{v_1^2}{2g} = z + d_c + \frac{v_2^2}{2g}$$

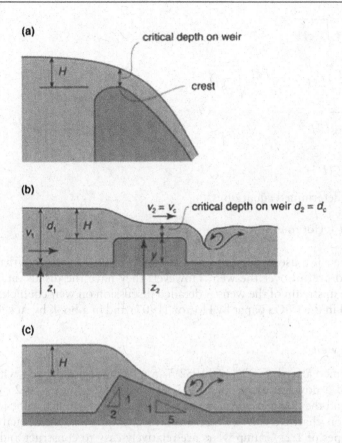

Figure 9.17 Critical depth weirs. (a) Solid weir, (b) broad-crested weir, and (c) crump weir.

If $v_1 = 0$ (or $v_1 <<<< v_2$), then

$$\frac{v_2^2}{2g} = H - d_c \Rightarrow v_2 = \sqrt{2g(H - d_c)}$$

But $d_c = \frac{2}{3}E_s = \frac{2}{3}H = \frac{2}{3}(\text{Total upstream energy})$ (assuming $v_1 <<<<< v_2$)   (9.43)

Therefore,

$$\frac{v_2^2}{2g} = H - \frac{2}{3}H = \frac{1}{3}H \Rightarrow v_2 = \sqrt{\frac{2}{3}gH}$$ (9.44)

$$Q_{ideal} = Av_2 = Ld_cv_2 = Ld_c\sqrt{2g(H - d_c)}$$

where L is the length of weir normal to flow.

The velocity v_2 is known as the critical velocity and is equal to the velocity of propagation of a surface wave. The maximum discharge under ideal conditions is therefore

$$Q_{ideal} = L\frac{2}{3}H\sqrt{\frac{2}{3}gH}$$

$$= \frac{2}{3}\sqrt{\frac{2}{3}g}LH^{3/2} = \frac{2LH}{3}\sqrt{\frac{2}{3}gH}$$

$$= \sqrt{\frac{8}{27}g}LH^{3/2}$$

$$= 1.70LH^{3/2} \tag{9.45}$$

Experimentally determined equation is

$$Q = 1.67LH^{3/2}\,(\text{for rounded corners}) \tag{9.46}$$

Broad crested weir is a useful device to measure rates of flow because of the critical conditions that would prevail over the weir. However, they have the disadvantage of having a dead water zone upstream of the weir. A detailed discussion on weir coefficients and formulas can be found in the USGS paper by Horton (1907) and in a book by Ackers et al. (1978).

9.6.6.5 Crump weir

Crump weir (named after E.S. Crump, 1952) is a triangular profile weir with an upstream slope of 1:2 and a downstream slope of 1:5. The upstream slope of 1:2 avoids sediment accumulation, and the downstream slope of 1:5 provides a stable hydraulic jump with good energy dissipation characteristics. There are also crump weirs with equal upstream and downstream slopes of 1:2. Crump weirs are relatively easy to construct and cost-effective. Much of the design and calibration work of Crump weirs have been carried out in the Hydraulic Research Station (now HR Wallingford) in Wallingford, UK.

9.6.6.6 Parshall flume

Parshall flume (Figure 9.18) is a standing wave flume and is similar to a Venturi flume for measuring flow rates in open channels. The throat is so arranged as to force the occurrence of critical flow there, followed by a short length of super-critical flow and a jump. It is a form of critical depth meter with no dead zones. The bed elevation is flat on the upstream side, which has a converging cross section, followed by an inclined throat and a diverging section.

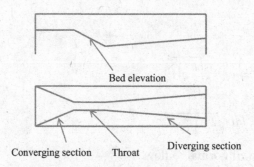

Figure 9.18 Configuration of a Parshall flume.

9.6.6.7 Rotameters

Rotameters are variable area meters that measures the flow rate by allowing the flow area to vary. A rotameter (Figure 9.19) has a tapered tube within which a float can move upwards and downwards. The float is subjected to a drag force by the moving fluid and a downward force due to its weight. The fluid moves in an upward direction lifting the float until it is balanced by its weight. The annular area between the float and the tube wall is then related to the volume flow rate.

The principle of operation is based on Bernoulli equation. Along a streamline travelling up the axis of the vertical tube gives,

$$p_a - p_b = \rho g z_b - \rho g z_a + \frac{1}{2}\rho V_b^2 - \frac{1}{2}\rho V_a^2 \tag{9.47}$$

where subscripts a and b represent the bottom and top positions of the float, V, the velocity of flow, p, the pressure, and ρ, the density.

$$\Delta p = \rho g h_f + \frac{1}{2}\rho V_b^2 \left(1 - \left(\frac{V_a}{V_b}\right)^2\right) \tag{9.48}$$

where h_f is the height of the float or the distance from the bottom to the indicator of the float, which depends on the float design. From continuity,

$$Q = A_a V_a = \overline{A_b} V_b \Rightarrow \frac{V_a}{V_b} = \frac{\overline{A_b}}{A_a}$$

Figure 9.19 Rotameter.

where $\overline{A_b} = A_b - A_f$ is the annular area between the tube and the float, which, when substituted into Bernoulli equation gives,

$$\Delta p = \rho g h_f + \frac{1}{2}\rho\left(\frac{Q}{\overline{A_b}}\right)^2\left(1-\left(\frac{\overline{A_b}}{A_a}\right)^2\right)$$

The pressure drop resulting mostly from the weight of the float is (ignoring the fluid friction caused by pressure drop)

$$\Delta p = \frac{\overline{W}_{\text{Float}}}{A_{\text{Float}}} = \frac{W_{\text{Float}} - B_{\text{Float}}}{A_{\text{Float}}} = \frac{V_f\left(\rho_f - \rho\right)g}{A_f}$$

where the subscript f represents the float, V_f is the volume, A_f is the cross-section area and ρ_f is the density of the float.

$$Q = \overline{A_b}\sqrt{\frac{2\left(\Delta p - \rho g h_f\right)}{\rho\left(1-\left(\frac{\overline{A_b}}{A_a}\right)^2\right)}} = \overline{A_b}\sqrt{\frac{2g\left(\frac{V_f(\rho_f - \rho)}{A_f} - \rho h_f\right)}{\rho\left(1-\left(\frac{\overline{A_b}}{A_a}\right)^2\right)}} \qquad (9.49)$$

Advantages of rotameters include cost-effectiveness and works without any external power source. The tube is calibrated to give the flow rate values.

9.7 FLOW MEASUREMENT IS NATURAL STREAMS

9.7.1 Use of tracers

A tracer can be any substance that can be detected when mixed with water. There should not be any chemical reaction between the tracer and water. Tracers include chemical tracers such as dyes, chlorides (NaCl) etc., and radioactive tracers, which are usually isotopes (since 1958).

Isotopes are chemically the same but have different radioactive properties and atomic weights. They can be classified into two types:

- Environmental isotopes e.g. Tritium, deuterium, oxygen-18, carbon-14, silicon-32. Deuterium and tritium are found in all waters in proportions which are determined by their abundances on each. Deuterium is a common isotope of hydrogen; Oxygen-18 is an isotope of Oxygen-16; Tritium H3 is produced by cosmic radiation; Carbon-14, which has a half-life of 5568 years is used for groundwater studies.
- Artificial isotopes are isotopes artificially produced by bombarding stable elements with nuclear particles such as α-particles, such as for example in nuclear reactions, nuclear bombs, cyclotrons, etc. They include

- Br-82 (half life 35.7 h)
- Na-24 (half life 15 h)
- I-131 (half life 8.05 days)
- Au-198 (half life 64.8 h)

A good tracer

- should mix with water (solubility high)
- be capable of determination of concentration in water
- be non-toxic
- should not be unduly absorbed
- be stable (chemically)
- should have a very low background concentration.

9.7.2 Radioactive decay

Radioactive decay follows the relationship

$$N = N_0 e^{-\lambda t} \tag{9.50}$$

where N is the concentration of the material at time t and N_0 is its initial concentration. The advantage of radioisotopes is that they are easy to be detected, and the disadvantage is safety.

9.7.3 Dilution methods (constant rate of injection)

Dosing must be continued at a constant predetermined rate until the concentration of the chemical at the sampling point is constant. The applicable equations are

$$q c_0 + Q c_1 = (Q + q) c_2 \tag{9.51}$$

where
q – rate of injection (say m^3/s)
Q – rate of flow in stream (m^3/s)
c_0 – concentration of chemical in dosing solution
c_1 – concentration of chemical naturally occurring in the stream water
c_2 – concentration of chemical in water at a sampling point
Therefore,

$$Q = \left[\frac{c_0 - c_2}{c_2 - c_1} \right] q$$

Since $c_0 \gg c_2$,

$$Q = \left[\frac{c_0}{c_2 - c_1} \right] q \approx \frac{c_0}{c_2} q \quad \text{since } c_2 \gg c_1 \tag{9.52}$$

9.7.4 Continuous sampling method (sudden injection)

A known volume V of the dosing solution is added to the stream as rapidly as possible. Samples are taken at regular intervals of time, and the chemical concentration is determined to draw a concentration-time curve. From continuity,

$$c_1 V_{\bar{1}} = \int c \, dV \qquad (9.53)$$

where $V_{\bar{1}}$ – volume of tracer injected
c_1 – concentration of tracer
The total volume 'V' which passes a given point is related to the constant flow rate 'Q' by

$$V = Qt$$

or

$$dV = Q \, dt$$

Therefore,

$$c_1 V_{\bar{1}} = Q \int c \, dt$$

$$Q = \frac{c_1 V_{\bar{1}}}{\int c \, dt} \qquad (9.54)$$

If two stations are considered, the applicable equations are

$$(c_0 - c_1) V = \int_0^T (c_2 - c_1) \, dt$$

Again since $c_0 \gg c_1$,

$$Q = \frac{V c_0}{\int_0^T (c_2 - c_1) \, dt} \qquad (9.55)$$

9.7.5 Mixing length

Mixing length has several interpretations. The widely accepted definition, according to Prandtl's mixing length theory, assumes that a disturbance retains its random velocity over a mixing length before mixing with the surrounding fluid. It is given by

$$u_* = l \frac{d\bar{u}}{dy} \qquad (9.56)$$

where u_* is the shear velocity, l the mixing length and $\bar{u}$ the mean velocity.

In tracer studies, the effect of dispersion is very important. In particular, many, if not all, mathematical models to predict longitudinal dispersion are valid only after an initial period of time (or distance) known as the initial period, or sometimes known as the convective period. Most studies on this topic are based on laboratory experiments and the results so obtained are not always applicable to field conditions. The dispersion coefficient, which is one of the key parameters in tracer studies, is estimated by a number of methods such as the change of moments method based on Fickian diffusion which assumes a Gaussian distribution of concentration and the routing method (Fischer, 1968) which requires a large amount of field data, and a time-dependent dispersion coefficient based on laboratory data (Jayawardena and Lui, 1983). The convective or initial period can be translated to a length which may be taken as the mixing length. More details of tracer studies and dispersion can be found in a book by Fischer et al. (1979) in which empirical relationships for the mixing length have been suggested. They include the length for complete mixing L given as

$$L = \frac{0.1\bar{u}W^2}{\varepsilon_t} \text{ (Eq. 5.10 in Fischer et al., 1979)} \tag{9.57a}$$

for a tracer introduced at the centreline, and

$$L = \frac{0.4\bar{u}W^2}{\varepsilon_t} \text{ (Eq. 5.10a in Fischer et al., 1979)} \tag{9.57b}$$

for a tracer introduced from a side, where $\bar{u}$ is the mean velocity of the stream, W is the width of the stream and ε_t the eddy diffusivity in the transverse direction.

Example 9.1

The flow in an open channel of $0.25\,\text{m}^2$ cross-sectional area goes over a sharp-crested rectangular weir 500 mm wide with a head of 150 mm. Calculate the discharge assuming that

    a. the approach velocity in the channel is negligible, and
    b. the approach velocity in the channel is NOT negligible.

FIRST APPROXIMATION: ASSUMING THAT THE VELOCITY OF APPROACH IS ZERO

Discharge formula for a rectangular sharp-crested weir is

$$Q = \frac{2}{3}C_D\sqrt{2g}bH^{3/2} = \frac{2}{3}\times 0.65 \times \sqrt{2\times 9.81}\times 0.5 \times (0.15)^{3/2} = 0.05575\,\text{m}^3/\text{s}.$$

$$\text{Velocity of approach} = \frac{\text{Discharge}}{\text{Area}} = \frac{0.05575}{0.25} = 0.223\,\text{m/s}$$

From Bernoulli equation,

$$H_1 + \frac{V_1^2}{2g} \Rightarrow 0.150 + \frac{0.223^2}{2g} = 0.1525$$

SECOND APPROXIMATION: TOTAL HEAD WITH THE APPROACH VELOCITY

$$Q = \frac{2}{3}C_D\sqrt{2g}bH^{3/2} = \frac{2}{3}\times 0.65 \times \sqrt{2\times 9.81}\times 0.5 \times (0.1525)^{3/2} = 0.05715\,\text{m}^3/\text{s}.$$

This procedure could be repeated until the difference between two successive approximations becomes insignificant.

Example 9.2

A cylindrical water tank mounted horizontally on a truck has a length of 5 m and a diameter of 2 m. It is fitted with an orifice of diameter 50 mm at its bottom. Assuming that the orifice has a coefficient of discharge of 0.8, calculate the time taken to lower the water level to one-quarter of its full value.

(a) When the water level in the cylindrical tank is between full and half-full position,

$$Q = -A\frac{dh}{dt}$$

$$A = L \times 2\sqrt{r^2 - h^2}$$

where r is the radius and h is the depth measured from the half-full position. Also,

$$Q = C_D a \sqrt{2g(h+r)} \quad \text{where } a \text{ is the area of the orifice.}$$

Therefore,

$$Q = C_D a \sqrt{2g(h+r)} = -2L\sqrt{r^2 - h^2}\,\frac{dh}{dt}$$

$$dt = -\frac{2L}{C_D a \sqrt{2g}}\left(\frac{r^2 - h^2}{r+h}\right)^{1/2} dh$$

$$t_1 = -\frac{2L}{C_D a \sqrt{2g}}\int_r^0 (r-h)^{1/2}\, dh = \frac{2L}{C_D a \sqrt{2g}}\frac{2}{3}r^{3/2}$$

(b) When the water level in the cylindrical tank is between half-full and empty position,

$$Q = -A\frac{dh}{dt}$$

$$A = L \times 2\sqrt{r^2 - (r-h)^2} = 2L\left(2rh - h^2\right)^{1/2}$$

$$Q = C_D a \sqrt{2gh}$$

Therefore,

$$Q = C_D a \sqrt{2gh} = -2L\left(2rh - h^2\right)^{1/2}\frac{dh}{dt}$$

$$dt = -\frac{2L}{C_D a \sqrt{2g}}(2r - h)^{1/2}\, dh$$

$$t_2 = -\frac{2L}{C_D a \sqrt{2g}}\int_r^{r/2}(2r - h)^{1/2}\, dh$$

$$= \frac{2L}{C_D a \sqrt{2g}}\frac{2}{3}r^{3/2}\left(\left(\frac{3}{2}\right)^{3/2} - 1\right)$$

$$= \frac{2L}{C_D a \sqrt{2g}}\frac{2}{3}r^{3/2}(0.837)$$

Data given: $L = 5\,\mathrm{m}$; $r = 1\,\mathrm{m}$; $C_D = 0.8$; and $a = \dfrac{\pi}{4}(0.05)^2 = 0.001963\,\mathrm{m}^2$

$$t_1 = \frac{2 \times 5}{0.8 \times 0.001963\sqrt{2 \times 9.81}} \times \frac{2}{3} \times 1 = 958\ \mathrm{s}$$

$$t_2 = t_1 \times 0.837 = 802\ \mathrm{s}$$

$$T = t_1 + t_2 = 1,760\ \mathrm{s}$$

REFERENCES

Ackers, P., White, W. R., Perkins, J. A., and Harrison, A. J. M. (1978): *Weirs and Flumes for Flow Measurement*, John Wiley & Sons, Chichester, 327 pp.

Crump, E. S. (1952): A new method of gauging streamflow with little efflux by means of a submerged weir of triangular profile, *Proceedings of the Institution of Civil Engineers*, vol. 1, pp. 223–242.

Faxen, H. (1922): Die Bewegung einer starren Kugel längs der achse eines mit zäher Flussigkeit gefüllten Rohres, *Arkiv för Matematik, Astronomi och Fysik*, Stockholm 17, no. 27, pp. 1–28.

Fischer, H. B. (1968): Dispersion prediction in natural streama, *Journal of the Sanitary Engineering Division*, Proceedings of the American Society of Civil Engineers, vol. 94, pp. 927–944.

Fischer, H. B., List, E. J., Koh, R. C. Y., Imberger, J., and Brooks, N. (1979): *Mixing in Inland Coastal Waters*, Academic Press, New York, 483 pp.

Francis, A. W. (1933): Wall effect in falling ball method for viscosity, *Physics*, vol. 4, pp. 403–406.

Fulmer, E. I. and Williams, J. C. (1936): A method for the determination of the wall correction for the falling sphere viscometer, *Journal of Physical Chemistry*, vol. 40, pp. 143–149.

Horton, R. E. (1907): Weir experiments, coefficients and formulas, Revision of paper no. 150; United States Geological Survey, Department of the Interior, Water Supply and Irrigation Paper No. 200; Series, M., General Hydrographic Investigations, 24, Washington, 195 pp.

Jayawardena, A. W. and Lui, P. H. (1983): A time-dependent dispersion model based on Lagrangian correlation, *Hydrological Sciences Journal*, vol. 28, no. 4, pp. 455–473, doi: 10.1080/02626668309491988.

Ladenburg, R. (1907): Uber den Einfluß von Wanden auf die Bewegung einer Kugel in einer reibenden, *Flussigkeit, Ann Physik*, vol. 4, no. 23, pp. 447–458.

Lamb, H. (1906): *Hydrodynamics*, University Press, Cambridge. p. 634.

Rehbock, T. (1929): Discussion of "Precise weir measurements". In: E. W. Schoder and K. B. Turner (eds.), *Transactions of the American Society of Civil Engineers*, New York, vol. 93, p. 1143.

von Mises, R. (1917): Berechnung von Ausfluss und Uberfullzahlen, *Z. ver Deuts. Ing*, vol. 61, p. 447.

Williams, J. C. and Fulmer, E. I. (1938): The evaluation of yield value with sphere viscometer, *Journal of Applied Physics*, vol. 9, pp. 760–764. doi: 10.1063/1.1710388.

Chapter 10

Pipe flow

10.1 INTRODUCTION

Flow in pipes is important for many reasons. The studies of turbulence and boundary layer in pipes have led to a better understanding of turbulence and boundary layer in general. Since pumps and turbines are always connected by pipes, the estimation of head losses is important in pipe systems. Boundary shear (stress) is important in gas flow. Heat and fluid transfer in engineering is always via ducts. Water supply and drainage are always via pipes and conduits. Overland conveyance of oil over long distances is also via pipes.

Most pipes are circular in cross section, but there are applications where the cross sections are not circular. The main difference between pipe (or conduit) flow and open channel flow is that, in the former case, the fluid inside the pipe is at a pressure greater than atmospheric in most cases, whereas in open channel flow, the pressure at the open water surface is always at atmospheric pressure.

The key dimensions in pipe flow are the length, the diameter of the pipe, the hydraulic diameter if not circular and the hydraulic mean depth (cross-sectional area divided by the wetted perimeter).

Laminar flow exists for Reynolds numbers up to about 2,300 $\left(Re = \dfrac{VD}{v}; \text{ where } D \text{ is the} \right.$ hydraulic diameter, which is four times the hydraulic radius$\left. \vphantom{\dfrac{VD}{v}}\right)$. For most engineering applications, it is taken as 2,000.

10.2 MOMENTUM EQUATION FOR FULLY DEVELOPED FLOW

Fully developed flow refers to the condition when the boundary shear stress τ_0, the pressure gradient $\dfrac{\partial p}{\partial x}$, and the velocity u have attained their maximum values. The momentum flux for a fully developed flow is

$$\beta \rho \frac{\pi D^2}{4} V \cdot V$$

where V is the average velocity, D the diameter of pipe, ρ the density of fluid and β the momentum correction factor, given by

$$\beta = \frac{\int u^2 \, dA}{V^2 A}$$

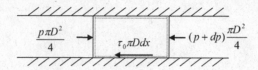

Figure 10.1 Forces acting on a fluid element in pipe flow.

Considering a control volume inside a pipe (Figure 10.1), the momentum equation can be written as

$$p\frac{\pi D^2}{4} - (p+dp)\frac{\pi D^2}{4} - \tau_0 \pi D dx = \frac{\pi D^2}{4}\frac{d}{dx}\left(\rho\beta V^2\right)dx$$

$$\Rightarrow -\frac{dp}{dx} = \frac{4\tau_0}{D} + \rho V^2 \frac{d\beta}{dx} + \beta\rho V\frac{dV}{dx} + \beta V\frac{d}{dx}(\rho V)$$

(10.1)

which, for fully developed incompressible flow (steady state, no acceleration), simplifies to (since $\beta V\frac{d}{dx}(\rho V)$, $\frac{d\beta}{dx}$, and $\frac{dV}{dx}$ are each equal to zero for fully developed flow, and the last term in Eq. 10.1 is zero by continuity consideration).

$$-\frac{dp}{dx} = \frac{\Delta p}{L} = \frac{4\tau_0}{D}$$

(10.2)

Equation 10.2 implies that the pressure decreases with increasing x, and the pressure drop is used in overcoming wall shear stress. This relationship depends on the Reynolds number. For laminar flow, $\Delta p \propto V$, and for turbulent flow, $\Delta p \propto V^a$ with $1.7 < a < 2$. This relationship (Eq. 10.2) can also be obtained as follows.

Considering a cylindrical fluid element of radius r and length L the forces acting are the pressure difference force and the wall shear force. Since the flow is steady and uniform, there is no acceleration, and hence no inertia force. Then, the force balance equation is

$$(p_1 - p_2)\pi r^2 = 2\pi r L\tau_0 \Rightarrow \frac{\Delta p}{L} = \frac{2\tau_0}{r} = \frac{4\tau_0}{D}$$

The velocity profile and shear distribution in laminar flow is shown in Figure 10.2.

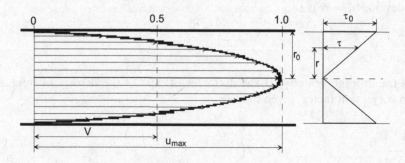

Figure 10.2 Velocity profile and shear distribution in laminar flow.

For developing flow, at the entrance of a pipe $\beta = 1$ and approaches $\beta = 4/3$ for fully developed laminar flow in circular pipes. For turbulent flow, $\beta \approx 1.03$. Therefore, for developing flow, Eq. 10.1 becomes (since $D^2V_1 = D^2V_2 \Longrightarrow V_1 = V_2$; therefore, $\dfrac{dV}{dx} = 0$ and $\dfrac{d}{dx}(\rho V) = 0$)

$$-\frac{dp}{dx} = \frac{4\tau_0}{D} + \rho V^2 \frac{d\beta}{dx} \tag{10.3}$$

When $\dfrac{d\beta}{dx}$ tends to zero, Eq. 10.3 will be the same as Eq. 10.2 for fully developed flow.

For compressible flow $\dfrac{d\beta}{dx} \approx 0$, but V varies with x because of density variations. Therefore, τ_0 and $\dfrac{dp}{dx}$ will vary with x, giving

$$-\frac{dp}{dx} = \frac{4\tau_0}{D} + \beta \rho V \frac{dV}{dx} \tag{10.4}$$

10.3 DARCY–WEISBACH EQUATION AND FRICTION FACTOR USING DIMENSIONAL ANALYSIS

The change of pressure head per unit length, $\dfrac{\Delta p}{L}$, can be obtained using dimensional analysis. It can be expressed as

$$\frac{\Delta p}{L} = f(D, k, V, \rho, \mu) \tag{10.5}$$

where D is the diameter of the pipe, k is the pipe roughness, V is the average velocity, and ρ and μ are the density and the viscosity of the fluid, respectively.

There are six ($n = 6$) variables with three ($m = 3$) fundamental dimensions, M, L and T. Therefore, according to **Buckingham's π theorem** there will be at least one set of ($n - m$) independent dimensionless groups in a dimensional analysis.

Each of these will consist of m ($= 3$) variables in common and are called repeating variables. The rules for the selection of repeating variables are

- They must include among them, all fundamental quantities
- Dependent variables should not be used as repeating variables
- For a fluid system, the most significant groups will result if the repeating variables are chosen so that one is a geometrical characteristic, one is a fluid property and one is a flow characteristic.

After the repeating variables are chosen, each one of the remaining variables is included in these groups. In this case, the repeating variables can be taken as D, ρ and V (average velocity). Therefore,

$$\pi_1 = D^{a_1} \rho^{b_1} V^{c_1} \left(\frac{\Delta p}{L} \right) \tag{10.6a}$$

$$\pi_2 = D^{a_2} \rho^{b_2} V^{c_2} k \tag{10.6b}$$

$$\pi_3 = D^{a_3} \rho^{b_3} V^{c_3} \mu \tag{10.6c}$$

Substituting the fundamental quantities M, L and T for each, it can be shown that

$$\pi_1 = \frac{\Delta p}{L} \frac{D}{\rho V^2} \quad \text{(Twice the friction factor, } f) \tag{10.7a}$$

$$\pi_2 = \frac{k}{D} \quad \text{(Relative roughness of a pipe)} \tag{10.7b}$$

$$\pi_3 = \frac{\mu}{VD\rho} \left(\frac{1}{\text{Reynolds No.}} \right) \tag{10.7c}$$

Using the friction factor f, Eq. 10.7a can be written as

$$\frac{\Delta p}{L} = \frac{4f}{D} \frac{\rho V^2}{2} \tag{10.8a}$$

This form of the head loss equation is called the **Darcy–Weisbach** equation. It can also be written as

$$\Delta p = h_f \rho g \quad \text{or} \quad h_f = 4f \left(\frac{L}{D} \right) \frac{V^2}{2g} \tag{10.8b}$$

Note: The relationship that π_1 is twice the friction factor f is based on the values of f given in the English system. In the US, the value of f is four times that defined by the above relationships. With their values of friction factor, the Darcy–Weisbach equation becomes

$$\frac{h_f}{L} = \frac{f'V^2}{2Dg} \tag{10.8c}$$

where $f' = 4f$.

Also, from dimensional analysis, $\pi_1 = F(\pi_2, \pi_3) \Rightarrow f = F(\pi_2, \pi_3)$, or,

$$f = F(\text{Relative roughness, Reynolds Number}) \tag{10.9}$$

The friction factor f can be obtained analytically (Hagen–Poiseuille and Darcy–Weisbach relationships) for laminar flow, and semi-empirically for turbulent flow.
From Eq. 10.2,

$$\tau_0 = \frac{\Delta p}{L} \frac{D}{4}$$

Substituting for $\frac{\Delta p}{L}$ from Eq. 10.8, the boundary shear stress can be expressed as

$$\tau_0 = \frac{4f}{D} \frac{\rho V^2}{2} \frac{D}{4} = \frac{f\rho V^2}{2} \tag{10.10}$$

The shear velocity is defined as

$$u_* = \sqrt{\frac{\tau_0}{\rho}} = V\sqrt{\frac{f}{2}} \tag{10.11}$$

For circular pipes, the characteristic length is D; for non-circular pipes it is the hydraulic diameter (not radius!) $D_h = \dfrac{4A}{P}$; hydraulic diameter is the same as the geometrical diameter.

With the US system, Eqs. 10.10 and 10.11, respectively, become

$$\tau_0 = \frac{f'}{D}\frac{\rho V^2}{2}\frac{D}{4} = \frac{f'\rho V^2}{8} \tag{10.12}$$

and

$$u_* = \sqrt{\frac{\tau_0}{\rho}} = V\sqrt{\frac{f'}{8}} \tag{10.13}$$

10.4 CRITICAL REYNOLDS NUMBER

The transition from laminar to turbulent flow takes place when the Reynolds number is around 2,000. The critical velocity for this Reynolds number is

$$u_{\text{crit}} = \frac{2{,}000\,\nu}{D}$$

For water, $\nu = 1.15 \times 10^{-6}$ m²/s; therefore, for a pipe of 25 mm diameter, $u_{\text{cirt}} = 92$ mm/s. This is a very small velocity compared with the velocities that are normally encountered in pipe flow. If the velocity is to be greater, a smaller diameter pipe would be needed. However, a pipe of diameter smaller than 25 mm is not of any practical value.

For oil, the kinematic viscosity ν is about 200 times that of water. Therefore, for a pipe of 25 mm diameter, $u_{\text{cirt}} = 18.4$ m/s. This means that the transition takes place when the velocity is around 18.4 m/s, which is a very high velocity. Therefore, all oil flows can be considered as laminar.

10.5 LAMINAR FULLY DEVELOPED FLOW

Laminar fully developed flow is attained usually after a length of approximately 100 pipe diameters (Figures 10.3 and 10.4).

10.5.1 Circular pipes

For laminar flow,

$$\tau = \mu\frac{\partial u}{\partial y} = \mu\frac{du}{dy} = -\mu\frac{du}{dr} \quad (\text{because } y = R - r; \ dy = -dr) \tag{10.14}$$

where R is the radius of the pipe, and y is the distance from the boundary to any radius r.

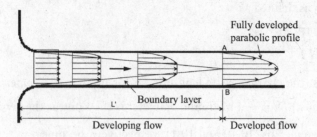

Figure 10.3 Velocity profile development of the boundary layer along a pipe in laminar flow.

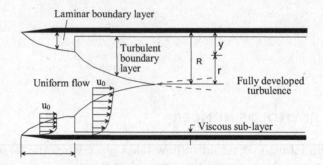

Figure 10.4 Development of turbulent fully developed boundary layer in a pipe (not to scale).

Therefore,

$$\frac{du}{dr} = -\frac{\tau}{\mu} = -\frac{\Delta p}{L}\frac{r}{2\mu} \text{ (substituting from Eq. 10.2)}$$

Integration and substitution of boundary condition $u = 0$ at $r = R$ gives

$$u = \frac{1}{4\mu}\frac{\Delta p}{L}\left(R^2 - r^2\right) \tag{10.15}$$

The flow rate Q can then be calculated as

$$Q = \int_0^R 2\pi ru\,dr = \frac{\pi}{2\mu}\frac{\Delta p}{L}\int_0^R r\left(R^2 - r^2\right)dr = \frac{\pi R^4}{8\mu}\frac{\Delta p}{L} \tag{10.16}$$

The velocity profile given in Eq. 10.15 has parabolic variation and the flow is called **Hagen–Poiseuille flow** in honour of the German engineer G. H. L. Hagen (1797–1884) who carried out experiments of laminar water flow in circular pipes and published in 1839, and the French physician J. L. M. Poiseuille (1799–1869) who independently carried out experiments of water flow in glass capillary tubes to study blood flow in veins and published in 1840. However, his results are not applicable to blood flow since the walls of veins are not rigid and that blood is not a Newtonian fluid with a constant viscosity.

From Eq. 10.16, the pressure gradient can be expressed as

$$\frac{\Delta p}{L} = \frac{8\mu Q}{\pi R^4}; \quad \text{or,}\ \frac{\Delta p}{L} = \frac{128\mu Q}{\pi D^4} \tag{10.17}$$

This equation can also be obtained by integrating the simplified Navier–Stokes equation in cylindrical co-ordinates, which is

$$\frac{dp}{dx} + \rho g \sin\theta = \mu\left(\frac{d^2 u}{dr^2} + \frac{1}{r}\frac{du}{dr}\right) = \frac{\mu}{r}\frac{d}{dr}\left(r\frac{du}{dr}\right)$$

with two boundary conditions $u = 0$ at $r = R$ and $\frac{du}{dr} = 0$ at $r = 0$ (Potter and Wiggert, 1991; p 263).

From Eq. 10.15, the maximum velocity at $r = 0$ is

$$u_{\max} = \frac{1}{4\mu}\frac{\Delta p}{L}R^2 \tag{10.18}$$

Therefore,

$$\frac{u}{u_{\max}} = 1 - \left(\frac{r}{R}\right)^2 \tag{10.19}$$

$$\text{Average velocity } V = \frac{Q}{A} = \frac{\int u\,dA}{A} = \frac{1}{A}\int_0^R (2\pi r\,dr)u$$

Substituting for u from Eq. 10.19 and integrating,

$$V = \frac{u_{\max}}{2}$$

Therefore,

$$\frac{\Delta p}{L} = \frac{8\mu V}{R^2} \quad \text{or} \quad \frac{128\mu Q}{\pi D^4} \quad \text{(from Eq.10.18)} \tag{10.20}$$

Equations 10.20 and 10.8 give

$$4\frac{f}{D}\frac{\rho V^2}{2} = \frac{8\mu V}{R^2} \Rightarrow f = \frac{16\mu}{VD\rho} = \frac{16}{Re_D} \tag{10.21a}$$

The corresponding equation with the US friction factor is

$$f' = \frac{64\mu}{VD\rho} = \frac{64}{Re_D} \tag{10.21b}$$

The relationship given by Eq. (10.21) has been experimentally verified and is valid for smooth and rough pipes for Reynolds numbers up to about 2,000. This equation is sometimes given as

$$c_f = \frac{64}{Re} \tag{10.21c}$$

where $c_f\,(= 4f)$ is the local skin friction coefficient given by

$$c_f = \frac{\tau}{\frac{1}{2}\rho V^2}$$ (10.21d)

10.6 TURBULENT FULLY DEVELOPED FLOW

In laminar flow, the fluid particles in one layer stay in the same layer and the layers slide without any crossing of particles from one layer to the adjacent layer. Thus, the flow is smooth and regular. On the other hand, in turbulent flow, the fluid particles jump from one layer to the adjacent layer and mix randomly. Eddies and swirls play significant roles in turbulent flow. Osborne Reynolds was the first to observe the difference between laminar and turbulent flows through his experimental studies. Turbulent flow is necessarily three-dimensional and unsteady. The transition from laminar to turbulent takes place when the Reynolds number exceeds about 2,300, but in most practical problems, it is taken as 2,000. The velocity profile for laminar flow is parabolic with the mean velocity equal to half the maximum velocity, and it can be determined theoretically as well as experimentally. Velocity profiles in turbulent flows can only be determined by experiments, and there are several empirical equations to describe the turbulent velocity profile.

10.6.1 Velocity profiles

In turbulent flow, the velocity profile is flatter than that for laminar flow (Figure 10.5). The higher the Reynolds number, the flatter the velocity profile. This is because of the radial component of the velocity. This means steep velocity gradients at the boundary and larger boundary shear stress (for the same Reynolds number).

Turbulent flow is not as simple as laminar flow. Using statistical representation of turbulence, several investigators have simplified the analysis of turbulence by assuming the turbulence to be homogeneous and isotropic (for example, Taylor, 1937; von Karman, 1921; Kolmogorov 1941; among others). See also Chapters 14 and 15.

Total shear stress can be expressed as a sum of molecular shear stress and a turbulent shear stress as

$$\tau = \mu\frac{d\bar{u}}{dy} + \eta\frac{d\bar{u}}{dy} = (\mu + \eta)\frac{d\bar{u}}{dy}$$

where η is referred to as the turbulent viscosity or eddy viscosity. Unlike the molecular viscosity μ, the eddy viscosity is not a fluid property. It depends upon the flow conditions such as the degree of turbulence, type of flow, etc.

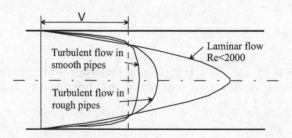

Figure 10.5 Velocity profiles for laminar and turbulent flows.

Reynolds (1883) showed that turbulent shear stress in two-dimensional flow could be expressed as

$$\tau = -\rho \overline{u'v'} \tag{10.22a}$$

where according to Reynolds (1895), decomposition, $u' = u - \bar{u}$; $v' = v - \bar{v}$ and $\bar{u}$ and $\bar{v}$ are the average velocities.

Prandtl (1925) suggested that

$$u' \propto l \frac{d\bar{u}}{dy}$$

where ℓ is the mixing length in the transverse direction. It is also assumed that $v' \propto u'$ and y is defined as the distance normal to the flow from the boundary. Then, $v' \propto l \dfrac{d\bar{u}}{dy} \Rightarrow \rho u'v' = \rho \overline{u'v'}$

Therefore, turbulent shear stress can be expressed as (Eq. 10.22a)

$$\tau_{\text{turbulent}} = \rho \ell^2 \left| \frac{d\bar{u}}{dy} \right| \frac{d\bar{u}}{dy} \tag{10.22b}$$

with the proportionality factor included in ℓ. Thus, the eddy viscosity can be expressed as

$$\eta = \rho l^2 \frac{d\bar{u}}{dy} \tag{10.22c}$$

Substituting $\ell = \kappa y$ (near the boundary the mixing length varies as y), the boundary shear stress is

$$\tau_0 = \rho \kappa^2 y^2 \left(\frac{du}{dy} \right)^2 \tag{10.23}$$

which can be simplified to

$$\sqrt{\frac{\tau_0}{\rho}} = \kappa y \frac{du}{dy} \Rightarrow u_* = \kappa y \frac{du}{dy} \Rightarrow \frac{dy}{y} = \frac{\kappa}{u_*} du$$

which when integrated gives

$$\frac{u}{u_*} = \frac{1}{\kappa} \ell n(y) + \text{constant} \tag{10.24a}$$

where u_* is the shear velocity defined as $u_* = \sqrt{\dfrac{\tau_0}{\rho}}$

Since $u = u_{\max}$ when $y = R$

$$\frac{u_{\max} - u}{\sqrt{\tau_0/\rho}} = \frac{u_{\text{mas}} - u}{u_*} = \frac{1}{\kappa} \ell n \left(\frac{R}{y} \right) \tag{10.24b}$$

Experimentally, $\kappa = 0.36$ ($\cong 0.4$) (von Karman constant)

Equation 10.24 is called the **universal velocity profile** or the **logarithmic velocity profile**. The constant in Eq. 10.24 has to be experimentally determined. With the experimental values obtained by Nikuradse (1932) and Reichardt (1943), Eq. 10.24 takes the form

$$\frac{\bar{u}}{u_*} = 2.5\ln\frac{u_*}{v}y + 5.5 \tag{10.25a}$$

or

$$\frac{\bar{u}}{u_*} = 5.75\log\frac{u_*}{v}y + 5.5 \tag{10.25b}$$

Von Karman assumed that turbulent velocity fluctuations vary with a length that he considered to vary with $\dfrac{du}{dy}\bigg/\dfrac{d^2u}{dy^2}$ and $\dfrac{du}{dy}$. He also assumed turbulent shear to vary with $\rho\ell^2\left(\dfrac{du}{dy}\right)^2$ as did Prandtl. Thus, the von Karman analysis shows

$$u_* = -\frac{\kappa\left(\dfrac{du}{dy}\right)^2}{\dfrac{d^2u}{dy^2}} \tag{10.26}$$

which, upon integration, gives

$$\frac{u_{max} - u}{u_*} = -\frac{1}{\kappa}\left[\ln\left(1 - \sqrt{1 - \frac{y}{R}}\right) + \sqrt{\frac{y}{R}}\right] \tag{10.27}$$

where $\kappa = 0.36$. This fits experimental data quite well. Prandtl's and von Karman's concepts are not valid at pipe centre line and pipe boundary where the flow is laminar (laminar sub-layer; viscous sub-layer).

10.6.2 Flow regimes in viscous flow

In viscous flow, three flow regimes can be identified:

a. **Hydraulically smooth region:** roughness elements submerged within the sub-layer

$$0 \le \frac{u_*k}{v} \le 5 \tag{10.28a}$$

b. **Transitional region:** roughness elements partly submerged within the sub-layer and partly projecting outside the sub-layer

$$5 \le \frac{u_*k}{v} \le 70 \tag{10.28b}$$

c. **Hydraulically rough region:** all roughness elements project outside the sub-layer

$$\frac{u_*k}{v} > 70 \tag{10.28c}$$

10.6.3 Empirical equations for friction factor for smooth pipes

Empirical friction formulae include those of

$$\textbf{Blasius}: f = \frac{0.079}{\left(Re_D\right)^{\frac{1}{4}}}\left(3{,}000 \le Re \le 10^5\right) \text{ or}, f' = \frac{0.316}{\left(Re_D\right)^{\frac{1}{4}}} \text{ in US system} \tag{10.29}$$

This is called **Blasius law** of pipe friction for smooth pipes. It is obtained by assuming $\frac{u}{u_{max}} = \left(\frac{y}{R}\right)^{\frac{1}{7}}$ outside the sub-layer, but not extending to the centre line. For high Reynolds numbers, the power varies from $\frac{1}{6} - \frac{1}{10}$ (Schlichting, 1979). This variation of the power (in the power law) suggest a formula of the form

$$\frac{1}{\sqrt{f}} = 2\log\left(Re_D\sqrt{f}\right) - 0.8 \tag{10.30}$$

which is called **Prandtl's law** of pipe friction for smooth pipes. **Nikuradse** has experimentally verified this for Reynolds numbers up to 3.4×10^6.

$$\textbf{Moody}: f = 0.001375\left[1 + \left(20{,}000\frac{k}{D} + \frac{10^6}{Re}\right)^{\frac{1}{3}}\right] \tag{10.31}$$

for $4{,}000 < Re < 10^7$ and for $\frac{k}{D}$ up to 0.01. This is an approximate formula. It is less reliable than the Moody diagram.

$$\textbf{Nikuradse}: \frac{1}{\sqrt{f}} = 2\log\left(\frac{Re\sqrt{f}}{2.51}\right) \tag{10.32}$$

$$\textbf{Colebrook}: \frac{1}{\sqrt{f}} = 1.8\log\left(\frac{Re}{6.9}\right) \tag{10.33}$$

This is an approximated form of Nikuradse's equation.

10.6.4 Empirical equations for friction factor for rough pipes

For fully rough conditions, the friction factor f is independent of Re_D. Empirical equations include those of

$$\textbf{von Karman}: \frac{1}{\sqrt{f}} = 2\log\left(\frac{3.71D}{k}\right) \tag{10.34}$$

$$\textbf{Colebrook}: \frac{1}{\sqrt{f}} = -2\log\left(\frac{2.51}{Re\sqrt{f}} + \frac{k}{3.71D}\right) \tag{10.35}$$

This equation is applicable for all pipes. For smooth pipes ($k = 0$) it reduces to the smooth pipe equation, and for large Re, it reduces to the rough pipe equation.

$$\text{Nikuradse}: \frac{1}{\sqrt{f}} = 2 \log\left(\frac{Re}{k}\right) + 1.74 \tag{10.36}$$

$$\text{Colebrook and White}: \frac{1}{\sqrt{f}} = 1.74 - 2 \log\left(\frac{2k}{D} + \frac{18.7}{Re_D \sqrt{f}}\right) \tag{10.37}$$

This equation is applicable to all flow regimes, and it is equivalent to Eq. 10.38 for large Reynolds numbers and to Eq. 10.29 for smooth pipes ($k = 0$).

$$\text{Prandtl (simplified form)}: \frac{1}{\sqrt{f}} = C + 2 \log D \tag{10.38}$$

where $C = 9.14 - 2 \log\left(\dfrac{10^4 \, kV + 1}{V}\right)$ and k is in mm.

$$\text{Prandtl and Nikuradse}: f = \frac{1}{\left(2 \log \dfrac{D}{2k} + 1.74\right)^2} \tag{10.39}$$

For commercial pipes, the exact roughness k is not known. It is also difficult to determine their shape and spacing. Therefore, for commercial pipes the equivalent sand-grain roughness is used. Such results are satisfactory for engineering calculations.

It is important to note that

- For laminar flow, f depends on Re only and is not affected by k.
- For hydraulically smooth pipes, f for turbulent flow also depends on Re only.
- For hydraulically rough pipes, f depends only on $\dfrac{k}{D}$ and not on Re for large Re.
- For smooth-rough transition zone, f depends on $\dfrac{k}{D}$ and Re.

With the empirical data obtained from Nikuradse's measurements on artificially roughened pipes, the above equations can be shown in a single set of graphs with friction factor f varying as a function of Re and $\dfrac{k}{D}$.

10.7 MOODY DIAGRAM (MOODY, 1944)

The friction factor f depends upon the velocity, pipe diameter, density of fluid, viscosity of fluid and the roughness height. By dimensional analysis, it can be shown that (Eq. 10.9)

$$f = F\left(\frac{\rho V D}{\mu}, \frac{k}{D}\right) = F(Re, \text{Relative roughness})$$

This relationship leads to the Moody diagram, which gives the variation of the friction factor with the Reynolds number, originally proposed by **Stanton**, and modified by Moody, and therefore called the Moody diagram. It enables the various pipe flow parameters to be linked by a series of curves. In this diagram, a single line on the left-hand side corresponds to laminar flow ($Re < 2,000$) where f varies according to

$$f = \frac{16}{Re} \text{ or } f = \frac{64}{Re}$$

It is independent of the roughness of the pipe. For turbulent flow, the curves are based on Nikuradse's experiments. For each value of $\frac{k}{d}$, there is a different curve. For large Re (turbulent region), f is independent of Re. It is also important to note that the friction factors used in some textbooks (US origin) are four times those used in some other textbooks (UK origin). This difference can also be seen in the corresponding Moody diagrams. Care should, therefore, be taken when using these values in the appropriate head loss equation. Figure 10.6 shows the Moody diagram from Glasgow College of Nautical Studies.

10.8 TYPICAL PIPE FLOW PROBLEMS

Typical pipe flow problems can be categorized into three types:

Type I: Given Q, L, D, ν, k Find h_f
Type II: Given h_f, L, D, ν, k Find Q
Type III: Given h_f, L, Q, ν, k Find D

In all cases, it is necessary to use the Darcy–Weisbach equation, the continuity equation and the Moody diagram. Very often, an iterative approach is needed.

10.9 MINOR LOSSES IN PIPES

Losses at entrance, exit, bends, contractions, expansions and pipe fittings, etc. are called minor losses. They are usually expressed as a velocity head in the form

$$\left(h_L = k_L \frac{V^2}{2g} \right) \tag{10.40}$$

where k_L is the loss coefficient that needs to be experimentally determined, except for sudden expansions. For a sudden expansion from A_1 to A_2, the loss coefficient is

$$k_L = \left(1 - \frac{A_1}{A_2} \right)^2 \tag{10.41}$$

If A_2 is very large, such as, for example, when a pipe is discharging into a reservoir, $k_L = 1$, implying that the entire kinetic energy is lost. Loss coefficients can also be expressed in terms of the equivalent pipe length L_e as

$$k_L \frac{V^2}{2g} = \frac{4fL_e V^2}{2gD} \Rightarrow k_L = L_e \frac{4f}{D}; \text{or as } L_e = \frac{k_L D}{4f} \tag{10.42}$$

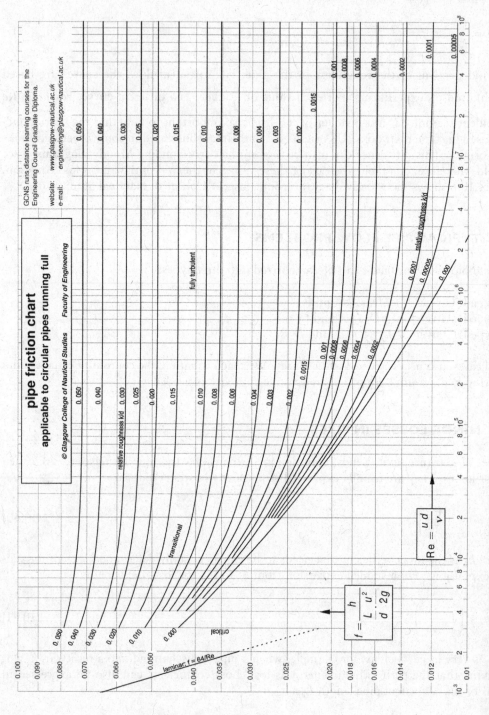

Figure 10.6 Moody diagram. (Courtesy of City of Glasgow College, Faculty of Nautical & STEM.)

Table 10.1 Approximate Loss Coefficients

Fitting	Loss Coefficient k_L
Globe valve (wide open)	10.0
Gate valve (wide open)	0.19
Gate valve (three quarters open)	1.15
Gate valve (half open)	5.6
Gate valve (quarter open)	24.0
Pump foot valve	1.5
90° elbow (threaded)	0.9
45° elbow (threaded)	0.4
Side outlet of T junction	1.8
Close return bend	2.2
Swing check valve (wide open)	2.5
Angle valve (wide open)	5.0

Source: Streeter, V.L., Wylie, E.B. and Bedford, K.W., *Fluid Mechanics* (Ninth Edition), McGraw Hill, New York, 1998; Massey B., revised by Ward-Smith, J., *Mechanics of Fluids* (Eighth Edition), Taylor & Francis, Oxford, UK, 2006.

For short lengths, minor losses may be substantially larger than the frictional losses, and for long lengths, they may be insignificant. For a sudden contraction, it can be shown that (Streeter et al., 1998; p. 300)

$$\left(h_L = \left(\frac{1}{C_c} - 1 \right)^2 \frac{V^2}{2g} \right) \tag{10.43}$$

where C_c is the coefficient of contraction.

Entrance losses when a flow takes place from a square opening is usually taken as $0.5 \frac{V^2}{2g}$ and $(0.01 - 0.05) \frac{V^2}{2g}$ for well-rounded entrances. If the pipe discharges into a reservoir, the entire kinetic energy is lost.

Other minor losses due to pipe fittings are given either as $h_L = \kappa_L \frac{V^2}{2g}$, or in terms of an equivalent pipe length. They are experimentally determined and given in the form of table such as Table 10.1:

In general, the Bernoulli equation for pipe flow can be modified as follows:

$$\frac{p_1}{\rho g} + \frac{V_1^2}{2g} + z_1 = \frac{p_2}{\rho g} + \frac{V_2^2}{2g} + z_2 + \frac{4fLV_2^2}{2gD} + k_L \frac{V_2^2}{2g} \tag{10.44}$$

The minor losses should be added together.

Example 10.1

Determine the size of galvanized steel pipe needed to carry water a distance of 180 m at 85 l/s with a head loss of 9 m.

$$h_f = 9 = \frac{4f(180)}{D} \left(\frac{0.085}{\pi D^2/4} \right)^2 \frac{1}{2(9.81)} \Rightarrow D^5 = 0.0478f$$

$$Re = \frac{VD}{v} = \frac{0.085}{\pi D^2/4} \frac{D}{1.14 \times 10^{-6}} = \frac{9.49 \times 10^4}{D}$$

For galvanized steel, $k = 0.15$ mm.

Try $f = 0.006$. Then $D = 0.1956$ and $Re = 4.85 \times 10^6$; $\frac{k}{D} = 0.00077$.

For these values of Re and $\frac{k}{D}$, Moody diagram gives $f = 0.00475$.

Substituting back, $D = 0.1867$ and $Re = 5.08 \times 10^5$; $\frac{k}{D} = 0.00080$.

For these values, $f = 0.0048$, which is close enough to the previous value. This value can be further improved. However, commercially available pipes have fixed sizes, and hence, the next largest fixed size is usually used.

Example 10.2

Water flows under gravity from an upper reservoir to a lower reservoir B through a 2-km long pipe of diameter 0.15 m. The water levels of the two reservoirs remain constant, and the discharge through the pipe is 0.011 m³/s. It is now required to increase this discharge by 30%. The proposed solution is to install a new pipe parallel to the existing pipe but only over half the length (that is, 1 km long). Determine a suitable diameter for the new pipe. Take the friction factors of all pipes at 0.032.

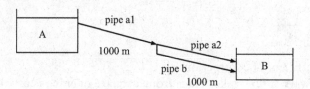

Original case → pipes $a_1 + a_2$ → $Q = 0.011 \, \text{m}^3/\text{s}$

$$h_f = \frac{8flQ^2}{\pi^2 g d^5} = \frac{8 \times 0.032 \times 2{,}000 \times 0.011^2}{\pi^2 \times 9.81 \times 0.15^5} = 8.426 \, m = \Delta H$$

New case → add pipe b → $Q = 0.011 \times 1.3 = 0.0143 \, \text{m}^3/\text{s}$

Consider pipe a_1 and pipe a_2. Their total head loss is ΔH.

$$\Delta H = h_{f,a_1} + h_{f,a_2} = \frac{8 \times 0.032 \times 1{,}000 \times 0.0143^2}{\pi^2 \times 9.81 \times 0.15^5} + \frac{8 \times 0.032 \times 1{,}000 \times Q_{a_2}^2}{\pi^2 \times 9.81 \times 0.15^5} = 8.426$$

Thus, $Q_{a_2} = 0.00621 \, \text{m}^3/\text{s}$

Consider pipe a_2 and pipe b. They have the same head loss, $h_{f,b}$.

$$\text{pipe } a_2 \rightarrow h_{f,b} = \frac{8flQ^2}{\pi^2 g d^5} = \frac{8 \times 0.032 \times 1{,}000 \times Q_{a_2}^2}{\pi \times 9.81 \times 0.15^5}$$

$$\text{pipe } a_2 \rightarrow h_{f,b} = \frac{8flQ^2}{\pi^2 g d^5} = \frac{8 \times 0.032 \times 1{,}000 \times Q_b^2}{\pi \times 9.81 \times d_b^5} \rightarrow \frac{Q_b}{Q_{a_2}} = \left(\frac{d_b}{0.15}\right)^{2.5}$$

Since $Q_b + Q_{a_2} = Q_{a_1} = Q$, therefore, $Q_b = 0.0143 - 0.00612 = 0.00818$.

Thus, $d_b = 0.15 \times \left(\dfrac{0.0818}{0.0612} \right)^{0.4} = 0.168$ m.

REFERENCES

Kolmogorov, A. N. (1941): The local structure of turbulence in incompressible viscous fluid for very large Reynolds numbers, *Proceedings of the USSR Academy of Sciences (in Russian)*, vol. 30, pp. 299–303. Translated into English: Kolmogorov, A. N. (July 8, 1991). Translated by Levin, V. (1991): The local structure of turbulence in incompressible viscous fluid for very large Reynolds numbers, *Proceedings of the Royal Society A*, vol. 434, pp. 9–13, doi: 10.1098/rspa.1991.0075. Archived from the original (PDF) on September 23, 2015.

Moody, L. F. (1944): Friction factors for pipe flows, Transactions of the ASME November.

Nikuradse, J. (1932): Gesetzmassigkeiten der turbulenten Stromung in glatten Rohren, *Ver Dtsch Ing Forschungsh*, vol. 356, pp. 1–36.

Reichardt, H. (1943): NACA Technical Memorandum No 1047.

Reynolds, O. (1883): An experimental investigation of the circumstances which determine whether the motion of water shall be direct or sinuous, and of the law of resistance in parallel channels, *Philosophical Transactions of the Royal Society of London*, vol. 174, pp. 935–982. Published by: The Royal Society Stable, http://www.jstor.org/stable/109431.

Reynolds, O. (1895): On the dynamical theory of incompressible viscous fluids and the determination of the criterion, *Philosophical Transactions of the Royal Society of London, Series A*, vol. 186, p. 123. (Reynolds decomposition to mean and fluctuating).

Schlichting, H. (1979): *Boundary Layer Theory*, (7th Edition), McGraw-Hill, New York, p. 817.

Streeter, V. L., Wylie, E. B. and Bedford, K. W. (1998): *Fluid Mechanics*, WCB McGraw-Hill, Boston, MA, p. 740.

Taylor, G. I. (1937): The statistical theory of isotropic turbulence, *Journal of the Aeronautical Sciences*, vol. 4, no. 8, pp. 311–315. doi: 10.2514/8.419.

ADDITIONAL READING

Blasius, H. (1908): Grenzschichten in Flussigkeiten mit kleiner Reibung, *Zeitschrift fur Mathematik und Physik*, vol. 56, no. 1, pp. 1–37. (NACA Technical Memorandum No. 1256).

Blasius, H. (1913): Das Ähnlichkeitsgesetz bei Reibungsvorgängen in Flüssigkeiten. *Mitt. Forsch. Gebiete Ing. Insbes. Laboratorien Tech. Hochsch.* 131:1–41.

Colebrook, C. F. (1938–1939): Turbulent flow in pipes, with particular reference to the transition region between smooth and rough pipe laws, *Journal Institution of Civil Engineers, London*, vol. 11, pp. 133–156.

Massey, B., revised by Ward-Smith, J. (2006): Mechanics of Fluids, (Eighth Edition), Taylor & Francis, Oxford, UK.

Nikuradse, J. (1942): *Laminare Reibungsschichten an der Langssangestromen Platte*, Monograph, Zentrale f wiss. Berichtswegung, Berlin.

Potter, M. C. and Wiggert, D. C. (1991): *Mechanics of Fluids*, Prentice Hall, Upper Saddle River, NJ, p. 692.

Prandtl, L. (1925): Bericht uber Untersuchungen zur ausgebildeten Turblenz, *Zeitschrift für Angewandte Mathematik*, vol. 5 no. 2, p. 136

von Karman, T. (1921): Uber laminare und turbulente Reibung, *Zeitschrift für Angewandte Mathematik*, vol. 1, pp. 233–252. (English translation: On laminar and turbulent friction, NACA TM 1092).

Chapter 11

Pipe networks

11.1 INTRODUCTION

In many conveyance systems, several pipes are connected together to form a network of pipes. The network may have loops as well as open ends. The determination of the head losses in each branch of such a system is more difficult than in a single pipe. It is also important to ensure that the pressure within the network does not become negative. If the pressure line coincides with the axis of the pipe, then the gauge pressure is zero (atmospheric). If it is above, the pressure is greater than atmospheric, and if it is below, the pressure is less than atmospheric. When the gauge pressure is negative, it is equivalent to a siphon. If the gauge pressure falls below 1 bar (-100 kN/m²), the flow would stop because this pressure corresponds to a perfect vacuum. In actual fact, the flow would stop much earlier, at about -75 kN/m², which corresponds to about 7.64 m below the pipe. This is the practical suction limit.

11.2 PIPE SYSTEMS

11.2.1 Pipes in series

For pipes arranged in series (Figure 11.1), the pressure is the same. The velocity varies according to the pipe size. The total head is the sum of the individual head losses in each pipe.

11.2.2 Pipes in parallel

The principle is that at any point in the pipe system, there can be only one head. Therefore, at point 1, the head is (Figure 11.2)

$$\frac{p_1}{\rho g} + \frac{V_1^2}{2g} + z_1.$$

Similarly, at point 2, the head is

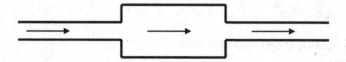

Figure 11.1 Pipes in series.

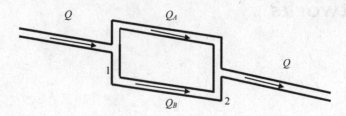

Figure 11.2 Pipes in parallel.

$$\frac{p_2}{\rho g} + \frac{V_2^2}{2g} + z_2.$$

and is independent of the path from point 1 to point 2. Therefore, for steady state,

$$(h_L)_{\text{Path } A} = (h_L)_{\text{Path } B} \tag{11.1}$$

The continuity condition states that

$$Q = Q_A + Q_B \tag{11.2}$$

The solution of these two equations enables the determination of head losses.

11.3 PIPE NETWORKS

The principles are the same as before. Therefore, the net head loss round any loop must be zero, and the net flow into any junction must be zero. It is, however, almost impractical to attempt to solve pipe networks analytically. Trial and error methods are very tedious and time-consuming. Methods of successive approximations are therefore employed. Hardy Cross Method (Cross, 1936) is one in which the head losses are balanced by correcting assumed flows, or flows are balanced by correcting assumed heads. The former type is more common.

11.3.1 Hardy Cross method

In the Hardy Cross method, the frictional losses are assumed to be exponential functions of discharge. i.e., $h_f = kQ^n$, in general, but mostly, $h_f = kQ^2$. The procedure is as follows.
For any pipe, assume Q_0 as the initial flow. Then,

$$Q = Q_0 + \delta Q \tag{11.3}$$

where Q is the correct flow, Q_0, the assumed flow, and δQ is the correction. For each pipe

$$h_f = kQ^n = k(Q_0 + \delta Q)^n \tag{11.4a}$$

which for small δQ may be written as a series

$$h_f = kQ^n = k(Q_0 + \delta Q)^n = k\left(Q_0^n + nQ_0^{n-1}\delta Q + \cdots\right) \tag{11.4b}$$

For a loop,

$$\sum h_f = \sum kQ|Q|^{n-1} = \sum kQ_0 |Q_0|^{n-1} + \delta Q \sum kn |Q_0|^{n-1} = 0 \qquad (11.5)$$

The absolute value is used to take account of the direction of summation around a loop. For each loop,

$$\delta Q = -\frac{\sum kQ_0 |Q_0|^{n-1}}{\sum kn |Q_0|^{n-1}} \qquad (11.6)$$

In applying this correction, the directional sense is very important. Usually, it is added to the flow in the clockwise direction and subtracted from the flow in the anti-clockwise direction. Some standard guidelines for the procedure are as follows:

- Assume the best distribution of flows that satisfy continuity by careful visual examination of the network
- For each pipe in a simple loop, calculate the net head loss $\sum h_f$ and δQ, and add the correction algebraically to each flow in the loop.
- Repeat the procedure for all other simple loops.
- Repeat the above steps until the corrections needed are smaller than a prescribed small value.

When $n = 2$, as in most situations, Eq. 11.6 simplifies to

$$\delta Q = -\frac{\sum kQ_0 |Q_0|}{2 \sum k |Q_0|} \qquad (11.7)$$

In the second type in which the balancing of flow is done, an error in the head at a junction is assumed. If h'_j is the correct head, then, the assumed head h_j is related to h'_j by

$$h'_j = h_j - \delta h \qquad (11.8)$$

For pipe 1,

$$h_f = h_A - h_j = k_1 Q_1^2 \qquad (11.9)$$

The value of Q_1, obtained by the above equation will be in error by δQ.
    For the correct head and flow,

$$h_A - h'_j = k_1 (Q_1 - \delta Q_1)^2 \qquad (11.10)$$

Substituting for h_j from Eq. 11.8 in equation $h_A - h_j = k_1 Q_1^2$,

$$h_A - (h'_j + \delta h) = k_1 Q_1^2 \qquad (11.11)$$

Substituting Eq. 11.10 in Eq. 11.11,

$$\delta h = -k_1\left(2Q_1\delta Q_1\right) = -2k_1Q_1\delta Q_1 = -2h_1\frac{\delta Q_1}{Q_1} \ \left(\text{because } h_1 = k_1Q_1^2\right) \tag{11.12}$$

This error can be written in terms of the flow errors in all other pipes:

$$\delta h = 2h_1\frac{\delta Q_1}{Q_1} = 2h_2\frac{\delta Q_2}{Q_2} = \cdots \tag{11.13}$$

Therefore,

$$\frac{Q_1}{h_1}\delta h = 2\delta Q_1; \quad \frac{Q_2}{h_2}\delta h = 2\delta Q_2; \quad \frac{Q_3}{h_3}\delta h = 2\delta Q_3 \cdots \tag{11.14}$$

$$\Rightarrow \left(\frac{Q_1}{h_1} + \frac{Q_2}{h_2} + \cdots\right)\delta h = 2\left(\delta Q_1 + \delta Q_2 + \cdots\right)$$

$$\Rightarrow \delta h = \frac{2\sum \delta Q}{\sum \dfrac{Q}{h}} \tag{11.15}$$

which should be added or subtracted from the assumed had.

11.3.2 Branched pipes: several reservoirs feeding into a junction

The basic principles involved are that the continuity must be satisfied at the junction, there can only be one value of head at any point, and that the friction equation must be satisfied for each pipe.

It is obvious that water will flow from the reservoir with the highest free surface elevation towards the junction and that water will flow into the reservoir with the lowest free surface elevation from the junction. It should, however, be noted that sometimes the direction of flow may not be quite evident. In such cases, it is assumed and checked for its physical justifiability. The piezometric head at the junction, h_j, determines the direction of flow.

Referring to Figure 11.3,

$$\text{Pipe } a : h_{f_1} = h_1 - h_j = k_aQ_a^2 \tag{11.16a}$$

$$\text{Pipe } b : h_{f_2} = h_2 - h_j = k_bQ_b^2 \tag{11.16b}$$

$$\text{Pipe } c : h_{f_3} = h_3 - h_j = k_cQ_c^2 \tag{11.16c}$$

$$h_1 - h_j = h_{f_1} \tag{11.17a}$$

$$h_j - h_2 = h_{f_2} \tag{11.17b}$$

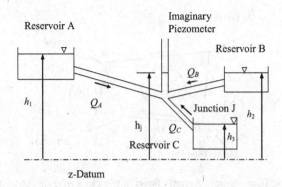

Figure 11.3 Pipes meeting at a junction.

$$h_j - h_3 = h_{f_3} \qquad\qquad\qquad (11.17c)$$

$$Q_A = Q_B + Q_C \qquad\qquad\qquad (11.17d)$$

There are four equations and four unknowns, h_j, Q_A, Q_B, and Q_C. The head loss h_f is a function of Q. The solution is usually obtained by trial and error.

If the direction of flow in pipe B is different, the corresponding equations would be

$$h_1 - h_j = h_{f_1} \qquad\qquad\qquad (11.18a)$$

$$h_2 - h_j = h_{f_2} \qquad\qquad\qquad (11.18b)$$

$$h_j - h_3 = h_{f_3} \qquad\qquad\qquad (11.18c)$$

$$Q_A + Q_B = Q_C \qquad\qquad\qquad (11.18d)$$

11.3.3 Pipe network: a ring network

A ring network is found in the distribution of water supply to a number of nodal points. There is one or more supply and extraction points in the network. A common problem is to find the discharges in the branches of the ring network (Figure 11.4):

The criterion to be used is that the head loss around a loop is zero: $\sum h_f = 0$; or

$$(H_A - H_B) + (H_B - H_C) + (H_C - H_D) + (H_D - H_A) = 0 \qquad (11.19)$$

The procedure involves first assuming all the discharges, Q's (with directions), which satisfy the inflow and outflow conditions at the nodes, computing the head losses in all branches using the head loss equation $h_f = KQ^2$ and adding all losses around the loop. If the discharges are correct, then $\sum h_f$ should be zero.

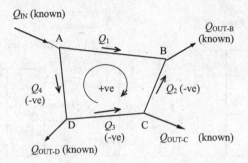

Figure 11.4 A ring network. (Convention: Assume flows in the clockwise direction as +ve. Head loss in the clockwise direction follows the sign of Q.)

When the assumed discharges are incorrect, $\sum h_f \neq 0$, but equal to Δh_f (an error). This error in total head loss around the loop is due to the accumulation of error in each branch:

$$\sum h_f = \sum \Delta h_f = \delta h_{f,AB} + \delta h_{f,BC} + \delta h_{f,CD} + \delta h_{f,DA} \tag{11.20}$$

By $\delta h_f = 2h_f \dfrac{\delta Q}{Q}$, $\displaystyle \sum h_f = \left(\frac{2h_{f,AB}}{Q_1} + \frac{2h_{f,BC}}{Q_2} + \frac{2h_{f,CD}}{Q_3} + \frac{2h_{f,DA}}{Q_4} \right) \delta Q,$

from which $\delta Q = \dfrac{1}{2} \dfrac{\sum h_f}{\sum h_f/Q}$ $\tag{11.21}$

It means that if $\sum h_f$ around the ring is +ve, the estimated Q's are each too high by the amount δQ and smaller values should be used. The iteration equation for refining the discharges in the pipes is

$$Q = Q - \frac{1}{2} \frac{\sum h_f}{\sum h_f/Q} \left(\sum h_f \text{ around the ring should be zero.} \right) \tag{11.22}$$

11.4   QUASI-STEADY FLOW IN PIPES

Quasi-steady implies Q varies with time, but the acceleration of the fluid and the forces are negligible. Examples include the case of a pipe discharging from a reservoir with a surface area A very much larger than the cross-sectional area of the pipe a. When quasi-steady conditions are attained,

$$h = \frac{4fLV^2}{2gD} + \frac{k_L V^2}{2g} \tag{11.23}$$

Continuity equation requires that

$$-A\frac{dh}{dt} = aV \Rightarrow -A\frac{dh}{aV} = dt \Rightarrow dt = \frac{-A\,dh}{a\left\{\dfrac{4fL}{2gD} + \dfrac{k_L}{2g}\right\}^{-\frac{1}{2}} h^{\frac{1}{2}}}$$

$$\int_0^t dt = -\int_{h_1}^{h_2} \frac{A}{a}\left\{\frac{2g}{4fL/D} + k_L\right\}^{\frac{1}{2}} h^{-\frac{1}{2}}dh \tag{11.24}$$

If A is constant, Eq. 11.24 can be integrated. If, however, A varies with h, direct integration may not be possible, and resort should be made to use numerical methods.

11.4.1 Two reservoirs connected by a pipe

The rate of flow depends on the difference between the surface elevations of the two reservoirs. The following equations can be written:

$$Q = -A_1\frac{dh_1}{dt} = A_2\frac{dh_2}{dt} \tag{11.25}$$

$$\frac{dh}{dt} = \frac{d}{dt}(h_1 - h_2) = \frac{dh_1}{dt} - \frac{A_1}{A_2}\left(-\frac{dh_1}{dt}\right) = \frac{dh_1}{dt}\left(1 + \frac{A_1}{A_2}\right) \tag{11.26}$$

From continuity,

$$A_1\left(-\frac{dh_1}{dt}\right) = av = a\left\{\frac{2g}{4fL/D + k_L}\right\}^{\frac{1}{2}} h^{-\frac{1}{2}}\left(1 + \frac{A_1}{A_2}\right) \tag{11.27}$$

Thus, $$\frac{dh}{dt} = -\frac{a}{A_1}\left(\frac{2g}{4fL/D + k}\right)^{1/2} h^{-1/2}\left(1 + \frac{A_1}{A_2}\right) \tag{11.28}$$

This can be integrated for constant A_1 and A_2 as in Eq. 11.24.

Example 11.1

When water is pumped through a pipe (pipe A) at a flow rate of 2.0 m³/s, the head loss is 30 m. When water is pumped through another pipe (pipe B) at a flow rate of 1.4 m³/s, the head loss is 25 m. Assume that the friction factors of the pipes do not vary with the discharges through them.

  a. If pipes A and B are connected in series and water is pumped through them jointly at a total flow rate of 1.5 m³/s, what is the resulting head loss?
  b. What is the resulting head loss if pipes A and B are connected in parallel, and water of the same total flow rate of 1.5 m³/s is pumped through them? What will be the discharge through pipe A in this case?

$$h_f = \frac{4fL\bar{u}^2}{2gD} = \frac{32fLQ^2}{\pi^2 gD^5} = KQ^2$$

Pipe A: $h_f = KQ^2 \rightarrow K_A = 30/2.0^2 = 7.5$

Pipe B: $\rightarrow K_B = 25/1.4^2 = 12.755$

Pipes A and B in series $\rightarrow$ same $Q, h_f = h_{f_A} + h_{f_B} = (7.5 + 12.755) \times 1.5^2 = 45.57\,\text{m}$

Pipes A and B in parallel $\rightarrow$ same h_f, $Q = Q_A + Q_B$

$$K_A Q_A^2 = K_B Q_B^2 \rightarrow Q_A/Q_B = (K_B/K_A)^{0.5} = (12.755/7.5)^{0.5} = 1.304$$

$$Q = Q_A + Q_B = 1.5 \rightarrow (1.304 + 1)Q_B = 1.5 \rightarrow Q_B = 0.651; Q_A = \underline{0.849\,\text{m}^3/\text{s}}$$

$$h_f = 7.5 \times 0.849^2 = \underline{5.405\,\text{m}}$$

Example 11.2

Three pipes from three reservoirs meet at a common junction. The pipe and reservoir data are given in the following table. If it is observed that there is no flow in pipe b between the junction and reservoir B, calculate the elevation of reservoir C which has the lowest elevation.

Reservoir level (m above datum)	Pipe	Diameter (mm)	Length (m)	Friction factor
A: 250	a	500	1,500	0.004
B: 150	b	100	500	0.008
C: unknown	c	600	1,250	0.005

Since no flow from B to J, $H_J = H_B = 150$,

$$\text{Pipe } a: h_f = \frac{4fL\bar{u}^2}{2gD} = \frac{32fLQ^2}{\pi^2 gD^5} = KQ^2 \Rightarrow K_a = \frac{32fL}{\pi^2 gD^5} = \frac{32 \times 0.004 \times 1,500}{\pi^2 \times 9.81 \times 0.5^5} = 63.457$$

$$Q_a = (h_a/K_a)^{0.5} = [(250 - 150)/63.457]^{0.5} = 1.255$$

Pipe $c: Q_c = Q_a = 1.255$

$$K_c = \frac{32fL}{\pi^2 gD^5} = \frac{32 \times 0.005 \times 1250}{\pi^2 \times 9.81 \times 0.6^5} = 26.565$$

$$h_c = KQ^2 = 26.565 \times 1.255^2 = 41.84 = H_J - H_C$$

$$H_C = 150 - 41.84 = \underline{108.2\,\text{m}}$$

REFERENCE

Cross, H. (1936): Analysis of flow in networks of conduits or conductors, Engineering Experiment Station, University of Illinois Bulletin 286, pp 3–29.

Chapter 12

Fluid machinery

12.1 INTRODUCTION

Fluid machinery can be broadly classified into two types; namely, pumps and turbines. Pumps convert mechanical energy into fluid energy, thereby increasing the energy of the fluid (increase of pressure). Turbines convert the fluid energy into mechanical energy, thereby decreasing the energy of the fluid.

Compressor is also a pump. Its primary function is to increase the pressure of the gas. Fan and Blower are machines used only for causing the movement of gas.

12.2 TYPES OF PUMPS

There are various types of pumps; some have only historical importance. Examples include.

12.2.1 Chain or bucket pumps

This is the simplest type of pump that uses gravity lift. It has a valve which opens when it goes below the water line and closes when lifted above the water line.

12.2.2 Static or positive displacement pumps

Positive displacement type functions by changes of the volume occupied by the fluid within the machine. They create an expanding cavity on the suction side and a contracting cavity on the delivery side. The operation of a positive displacement type machine depends only on mechanical and hydrostatic principles. Only a few principles of fluid dynamics are involved. Positive displacement pumps are more suitable for low flow high-pressure situations, and their efficiency is relatively independent of the pressure. They are preferred for moving viscous fluids such as oil, asphalt, etc. There are several types of positive displacement pumps.

12.2.2.1 Reciprocating type

They are usually of piston–cylinder combination with inlet and exit valves involving suction and compression. Work done by pump is $\int pdv$, integrated over a complete cycle. If only one cylinder is working, the output will be fluctuating. If there are several cylinders, a fairly steady outflow can be maintained. They can also be single acting or double acting. In a single acting pump, the fluid is displaced in either the forward or backward stroke. In a double acting pump, the fluid is displaced in both forward and backward strokes.

12.2.2.2 Diaphragm pumps

These are particularly useful for handling sewage, acids, etc. as the moving parts of the pumps are not in contact with the fluid in motion. Typical speeds are in the range of 50–70 strokes/min. The heart is a diaphragm type pump. These pumps are self-priming.

12.2.2.3 Rotary pumps

In this type, the fluid is moved without a change in velocity in the spaces between inter-meshing rotors or gears. They create high pressure on the delivery side. Delivery is steady compared to a reciprocating pump. They are used for moving high-viscosity fluids such as oil, polymers, asphalts, etc.

If the valves in a positive displacement type pump were closed, pressure will build up and the casing will burst.

12.2.2.4 Screw pumps

A screw pump uses two or more intermeshing screws to increase the pressure of fluid, including solids and move them. The screws take in fluid from the inlet side and push it out from the outlet side while increasing its pressure. The pump system can be of single-screw type, two- screw type or three- screw type. Single-screw type, which is also known as the Archimedes screw, has limited capacity and is used mainly to move water and wastewater including sewage. In the two- screw type, the screws are not in contact with each other, thereby minimizing wear and tear and extending their life. In the three-screw type, the driving screw is in contact with the other two to create the pressure. The disadvantage of the three-screw type is that the narrow space between the screws does not allow solid material to be moved. Thus, they can only be used to move relatively clean liquids.

12.2.2.5 Gear pumps

Gear pumps consist of gears that are arranged with the teeth meshed. The gears rotate in opposite directions so that they pull fluid into the spaces between the gear teeth and the pump casing. The fluid is finally released through the pump discharge due to the movement of the teeth. A fairly constant fluid flow is maintained by smaller teeth, while bigger teeth will produce a more pulsating fluid flow pattern.

12.2.2.6 Hydraulic ram

Hydraulic ram can be used when there is a large volume of water flowing at a pressure higher than necessary. Some of the dynamic pressure energy of the main body of water is used to increase the static pressure on part of the supply. For the hydraulic ram to function, there must be a source of water at an elevation higher than where the ram is located. The water is allowed to run through a pipeline towards the ram, which has a valve to let the water in. When the flow reaches its maximum velocity, the inlet valve is suddenly closed, resulting in a build-up of pressure due to the inertia of the flow. The build-up pressure then opens an exit valve forcing the fluid to flow into the delivery pipe. The pressure in the ram falls, and the entry valve opens allowing more water to flow in and build up the pressure. This cycle is repeated. The water wheel used in some parts of the world in ancient times operates on the same principle.

12.2.3 Rotodynamic type (turbomachines)

In this type of machines, there is a transformation between pressure and velocity heads. There is a significant velocity change across the pump. If the valves in a rotodynamic machine were closed, there will only be an increase in temperature. The casing will not burst. There are three common rotodynamic type pumps; namely, centrifugal pumps, axial flow pumps and mixed flow pumps. They are used mainly for high-flow and low-pressure situations.

12.3 TYPES OF TURBINES

There are two main categories of turbines; namely, impulse turbines and reaction turbines. Pelton wheel is an example of an impulse turbine, whereas Francis and Kaplan turbines are reaction type of turbines.

In both types, fluid passes through a runner having blades. The momentum of the fluid in the tangential direction is changed and so a tangential force on the runner is produced. The runner rotates and does work. The energy in the fluid is reduced.

In impulse turbines, there is no change in static pressure across the runner which is open and at atmospheric pressure. The energy change is from kinetic to mechanical. In reaction turbines, static pressure decreases as the fluid passes through the runner which is enclosed.

For any turbine, the fluid energy is in the form of pressure. In water turbines, it is the difference between the upper and lower water levels that gives this pressure. In steam turbines, it is the steam pressure produced by heat. In gas turbine, it is the gas pressure produced by chemical energy.

12.4 CENTRIFUGAL PUMP

The flow in a centrifugal pump is radially outwards. They are so called because the centrifugal force or the variation of pressure due to rotation is an important factor in their operation. A rotating impeller provides the energy to the fluid in the form of a velocity head which is converted into a pressure head as the fluid leaves the pump. There is a pronounced change in the radius from inlet to outlet. Centrifugal pumps operate efficiently for water and other liquids with low viscosity. They are not suitable for high-viscosity fluids and high pressure. Multi-stage pumps can be used to deliver high pressure.

A centrifugal pump consists of an impeller rotating inside a casing. There are centrifugal turbines, pumps and compressors. The flow is usually towards the larger radius for a pump and radially inwards for a turbine.

Centrifugal pumps are broadly divided into two classes; namely, the diffuser type and the volute type. A diffuser type pump is one in which the impeller is surrounded by a diffuser containing stationary guide vanes. There is a lowering of kinetic energy and, therefore, an increase of pressure energy. Volute type pumps do not have diffuser vanes, but instead, have diverging spiral casings so proportioned to reduce the velocity of flow and to convert some of the velocity head to static pressure head. The spiral is called the volute. The rotating impeller converts the mechanical energy to kinetic energy and the volute, or the diffuser then converts some of the kinetic energy to pressure energy. The faster and larger the impeller, the greater the amount of energy transferred to the fluid. The diffuser type is more bulky and expensive. But the gain in efficiency is about 80% compared with 75%–80% for volute type. They are not suitable for viscous fluids and high-pressure applications.

In a crude form, the centrifugal pump can be thought of as a paddle wheel rotating in a chamber. When the paddle wheel is rotated inside a closed chamber, there will be an increase in pressure from the centre to the circumference. If the water is assumed to rotate at the same speed as the impeller the peripheral velocity of which is u, the pressure difference between the centre and the circumference would be $\frac{u^2}{2g}$. That means the height h to which water will rise is given by

$$h = \frac{u^2}{2g} \qquad (12.1)$$

If the height of the tube is less than h, water will flow out.

Centrifugal pumps can be self-priming or non-self-priming. Initially, when a centrifugal pump is started, it sucks in air and water resulting in the pump chamber to be filled with a mixture of air and water. Because of the density difference, the water gets to the bottom of the pump chamber while the air fills the top part. A self-priming pump mixes the air and water to create a fluid with pumping properties like those of water. The pump then gets rid of air and moves the water only. Self-priming pumps need to have the pump chamber (casing) full of water. They have a water reservoir above or in front of the impeller. Standard (non-self-priming) centrifugal pumps need the suction lines to be flooded or the pump placed below the water line (submersible pumps).

In centrifugal pumps, fluid enters the 'eye' of the impeller with little or no whirl velocity (component of velocity in the direction of rotation of the impeller and therefore tangential) and leaves the casing with increased pressure and velocity. The outward flow has a substantial whirl component. The volute, which is a passage with increasing cross section, serves to reduce the velocity of fluid leaving the pump. They also have fixed guide vanes, also known as stationary diffusers, to reduce the velocity of the fluid at exit.

The velocity triangles and the basic equations are the same for pumps and turbines. Centrifugal pumps do not usually have any guide vanes at inlet. The fluid, therefore, approaches without appreciable whirl giving rise to a right-angled velocity triangle at the inlet. If the design conditions are not met, the direction of the relative velocity R_1 will not coincide with that of the blade. Then, the fluid changes direction abruptly and results in loss of energy.

For ideal conditions (steady-state uniform flow),

$$v_{w_1} = 0 \qquad (12.2)$$

Therefore, the work done on the fluid per unit mass is $u_2 v_{w_2}$, which is independent of inlet radius. The whirl velocities at inlet and outlets are denoted by v_{w_1} and v_{w_2}. In terms of an equivalent head (energy per unit weight), this will be

$$E = \frac{u_2 v_{w_2}}{g} \qquad (12.3)$$

which is also known as the Euler head.

12.4.1 Manometric head

Manometric head H_m is the difference in piezometric heads across the pump. It is the difference in head that would be recorded by a manometer connected to the inlet and outlet flanges of the pump. It represents the increase in fluid energy by the pump.

Manomeric efficiency $\eta_m = \dfrac{\text{Energy out per unit time}}{\text{Energy in per unit time}}$

$= \dfrac{\text{Increase in fluid energy per unit time}}{\text{Energy imparted by impeller per unit time}}$

$$= \frac{gH_m}{u_2 v_{w2}} \qquad (12.4)$$

This gives a measure of the effectiveness in converting the impeller energy to fluid energy. The velocity heads at inlet and outlet are usually small. If these are taken into account, an overall efficiency can be defined as follows:

Overall efficiency, $\eta_0 = \dfrac{\text{Power out}}{\text{Power in}} = \dfrac{\rho g Q H}{\text{Shaft power in}} \qquad (12.5)$

where H is the total head across the pump, including velocity heads. Overall efficiency is usually less than the manometric efficiency because of friction in bearings, etc.

12.4.2 Velocity diagrams

At inlet, under ideal conditions, the velocity diagram is a right-angled triangle since the whirl velocity at inlet is zero. However, in practice, there will be a small non-zero whirl velocity at inlet. At the outlet, the velocity triangle consists of impeller velocity (u_2), the velocity of fluid relative to the blade (R_2), and the absolute velocity of fluid (V_2) as shown in the velocity triangles in Figure 12.1.

12.4.3 Performance of a pump — characteristic curves

The performance of a pump depends (among other things) on the outlet angle of the impeller blades, ϕ_2, the value of which determines the shape of the velocity diagram. It may be forward facing ($\phi_2 > 90°$), radial ($\phi_2 = 90°$), or backward facing (($\phi_2 < 90°$). In the second case, there is no whirl slip. Then,

$$\phi_2 = \beta_2$$

β_2 is the angle between the tangent and the direction of the relative velocity of the fluid. The direction of R_2 is the same as that of the blade angle at the outlet.

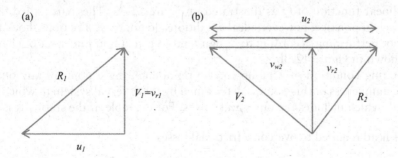

Figure 12.1 Inlet (ideal case) (a) and outlet (b) velocity diagrams.

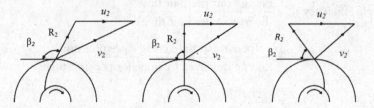

Figure 12.2 Outlet velocity triangles as a function of the outlet blade angle.

Forward facing ($\beta > 90°$) Radial ($\beta = 90°$) Backward facing ($\beta < 90°$)

The values of v_{w_2} are different for the three cases shown in Figure 12.2. Therefore, for the same value of u_2, the work done would be different for different values of ϕ_2.

Referring to the velocity triangles (Figure 12.1),

$$v_{w_2} = u_2 - v_{r_2} \cot \beta_2 = \left(u_2 + \frac{v_{r_2}}{\tan(\pi - \beta_2)} \right) \tag{12.6}$$

where v_{r_2} is the radial component of the exit velocity. Therefore, the increase in energy per unit weight is

$$\frac{u_2 v_{w_2}}{g} = \frac{u_2}{g}\left(u_2 - v_{r_2} \cot \beta_2 \right)$$

$$= \frac{u_2}{g}\left(u_2 - \frac{Q}{A_2} \cot \beta_2 \right) \tag{12.7}$$

where Q is the flow rate through the pump and A_2 is the outlet area normal to the radial direction which is the difference between the peripheral area of impeller and the area occupied by blades.

Since $u = \omega r$, the ideal increase in energy as a head, H may be written in the form

$$H = C_1 N^2 - C_2 Q N \tag{12.8}$$

where C_1, C_2 are constants and N is the speed of rotation.

If N is constant,

$$H = C_1' - C_2' Q \tag{12.9}$$

which is a linear function of Q as illustrated in Figure 12.3a. These are called the characteristic curves. In practice, however, ideal conditions do not exist. Friction, shock losses and the pump not performing according to design tend to give a non-linear set of characteristic curves as shown in Figure 12.3b.

The operating condition of a pump can be represented by a point in any one of these curves. The uniqueness of this point is determined by the external system to which the pump is connected, which includes the pipes and valves. For example, if the static lift is h,

$$H = h + \text{head required to overcome frictional losses}$$

$$= h + kQ^2 \left(h_f \alpha V^2 \right) \tag{12.10}$$

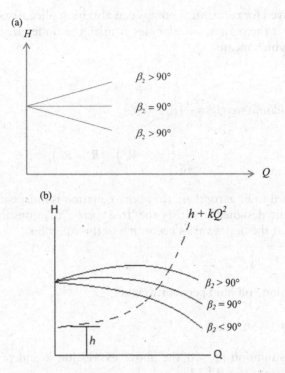

Figure 12.3 (a) Characteristic curves under ideal conditions for different outlet blade angles. (b) Characteristic curves under normal conditions for different outlet blade angles.

The intersection of this curve with the characteristic curve gives the values of H and Q at the operating condition (Figure 12.3b).

Note: Impellers with backward facing angles are preferred because they give the smallest value of V_2 and hence less dissipation of energy in the volute.

12.4.4 Whirl slip

In the previous analysis, it was assumed that $\phi_2 = \beta_2$. This is so if the volute is perfectly designed and if the impeller had a large number (infinite) of blades. In practice, the actual number of blades is limited (usually 6–12 for backward facing type and up to 60 for forward facing type). The effect of this is to distort the direction of flow of fluid relative to the blade. The deviation of the angle $(\phi_2 - \beta_2)$ can be as high as 15° in some designs. Such deviations from ideal conditions lead to a non-zero 'whirl slip'.

12.5 AXIAL FLOW PUMP (PROPELLER PUMP)

The flow in this type is largely parallel to the axis of rotation. There are also axial flow turbines (Kaplan turbine), fans and compressors.

The axial flow pump (propeller pump) rotates within a cylindrical casing with a clearance which is made as small as possible. In the outlet side of the impeller, a set of stationary guide vanes is usually fitted – the purpose being to reduce the whirl velocity. Guide vanes on the inlet side are usually not provided because they do not improve the pump performance significantly.

The equations derived for centrifugal pumps can also be applied to axial flow pumps. The main difference is that the peripheral velocities at inlet and outlet are the same (same radii at inlet and outlet), which means

$$u_1 = u_2$$

Work done on fluid/unit weight $= \dfrac{u}{g}\left(v_{w_2} - v_{w_1}\right)$

$$= \frac{1}{2g}\left\{\left(V_2^2 - V_1^2\right) - \left(R_2^2 - R_1^2\right)\right\} \qquad (12.11)$$

Usually, v_{w_1} is assumed to be zero; then, the above equation is independent of the radius.

Impellers are usually designed to satisfy the 'free vortex' relationship which implies that the velocity of whirl at the outlet varies according to the equation

$$v_{w_2} r = \text{Constant} \qquad (12.12)$$

Therefore, the work done on fluid per unit weight

$$\omega\, r\left(v_{w_2} - v_{w_1}\right) = \omega\left(v_{w_2} r_2 - v_{w_1} r_1\right) \qquad (12.13)$$

With the further assumption $v_{w_1} = 0$ the above expression is independent of the radius because $v_{w_2} r_2$ is a constant (Eq. 12.12).

Note: These equations are based on the assumption that uniform conditions exist at inlet and outlet. This is not so for positions close to the impeller because the blades are so widely spaced.

Axial-flow pumps are high capacity, low head type.

12.6 MIXED FLOW (DIAGONAL FLOW) PUMP

This type is intermediate between the radial flow type and the axial flow type. The flow may go along the conical surfaces of revolution.

In all these types, there must be a rotating member, usually called a rotor or an impeller which does the work on the fluid and one or more stationary members called a stator. Volute and guide vanes are to guide the flow before and after the impeller.

In summary, positive displacement type pumps have the following characteristics:

- Unsteady delivery
- Clearance between moving and stationary parts is very small. Therefore water containing gravel or other solid matter cannot be pumped.
- If the valves are closed, the casing will burst.
- Bulkier than rotodynamic type for the same design.

Rotodynamic type pumps, on the other hand, have the following characteristics:

- Steady delivery
- Clearance large
- If the valves are closed, the kinetic energy will be converted into heat energy
- Less bulky.

12.7 PUMP SYSTEMS

When two or more similar pumps are connected in series, the resulting effects are the same as that obtained by having a number of impellers on a single shaft. For a given capacity, the total head is the sum of the head for each pump or stage.

For two or more pumps in parallel, the capacities will be the sum of the individual capacities while the head will remain the same.

12.8 IMPULSE TURBINES: PELTON WHEEL (AFTER LESTER A. PELTON, 1829–1908)

Pelton wheel is a tangential flow machine used in high head systems, particularly in hydro-electric schemes in mountain areas. The rotor consists of a number of buckets attached to its periphery. One or more nozzles are mounted so that they discharge jets along a tangent to the circle through the centreline of the buckets (Figure 12.4).

The buckets are split in half at the centre, and the incoming jet divides into two equal halves upon impinging. The notch in the outer rim of each bucket prevents the jet from the preceding bucket being intercepted too soon.

The maximum change of momentum – and hence the maximum force – would be obtained if the fluid is deflected through 180°. But in practice, this deflection is around 165° to prevent the fluid, leaving one bucket striking the back of the next.

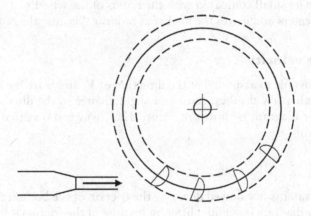

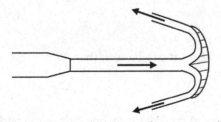

Figure 12.4 Jet and buckets in a Pelton wheel.

The pressure of the fluid after it has left the nozzle is constant at atmospheric. Normally, the Pelton wheel rotates in a vertical plane, i.e., in a horizontal shaft, having only two jets placed symmetrically. In horizontal wheels more number of jets (up to 6) can be used.

12.8.1 Basic equations and velocity vectors

The Pelton wheel has a number of velocities:

V_1: Absolute velocity of the jet before striking the bucket
V_2: Absolute velocity of jet after striking the bucket
ω: Angular velocity of wheel
r: Radius at the centreline of buckets
u: Absolute velocity of bucket at radius r ($u = \omega r$); $u_1 = u_2 = u$
R_1: Velocity of the incoming jet relative to bucket
R_2: Velocity of the outgoing jet relative to bucket
θ: Angle through which fluid is deflected by bucket
C_v: Coefficient of velocity for the nozzle (≈ 0.97)
Q: Discharge rate
ρ: Density of fluid

Jet velocity $V_1 = C_v\sqrt{2gH}$, where H is the net head (after allowing for frictional losses).
In the basic equations, the following assumptions are made:

- Steady-state, uniform flow
- Radius of the jet small compared with the radius of the wheel r
- All the fluid enters and leaves the bucket at radius r (because the radius of jet is small)

12.8.2 Relative velocity

At the inlet, it is possible to assume that the direction of V_1 and u are the same although the wheel turns through a few degrees, causing a slight change in the direction of the bucket. The velocity vector diagram is shown in Figure 12.5, which shows that the velocities are co-linear in the ideal case.

$$\vec{R_1} = \vec{V_1} - \vec{u} \tag{12.14}$$

At the exit, if the water is not slowed down by the friction of the bucket, $R_2 = R_1$. However, at exit, the velocity diagram is slightly different because of the frictional losses in the bucket (although the bucket surfaces are smooth it is not possible to eliminate all friction losses) and losses at the splitting edge of the bucket as this edge cannot have zero thickness. Therefore it is possible to write

$$R_2 = kR_1 \, (k < 1)$$

From the velocity triangle (Figure 12.5), it can be seen that

$$\vec{V_2} = \vec{R_2} + \vec{u} \tag{12.15}$$

Due to the symmetry, it is necessary to consider only one half of the bucket.

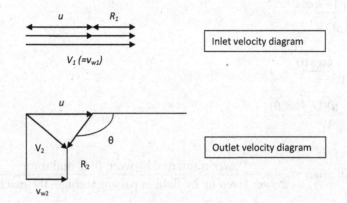

Figure 12.5 Inlet and outlet velocity diagrams for a Pelton wheel.

12.8.3 Velocity of whirl

The velocity of whirl, v_w, is the component of V in the direction of motion of the bucket. The direction of whirl is the direction tangential to the periphery of the runner. Therefore, at exit

$$v_{w_2} = u - R_2 \cos(\pi - \theta) \qquad (12.16)$$

This is assuming that the direction of u is positive.

At the inlet, $v_{w_1} = V_1$ (because V_1 is in the tangential direction).

Therefore, the change in velocity of whirl is given by

$$V_1 - (u - R_2 \cos(\pi - \theta)) = V_1 - u + R_2 \cos(\pi - \theta)$$

$$= R_1 + R_2 \cos(\pi - \theta)$$

$$= R_1(1 - k \cos\theta) \quad \text{because } R_2 = k R_1 \qquad (12.17)$$

Changes in momentum in the direction in which the wheel is rotating (in the peripheral direction) only need to be considered. Therefore, the whirl velocity only needs to be considered.

Change in momentum in the peripheral direction $= \rho Q R_1 (1 - k \cos\theta)$

$$= \text{force driving the wheel}$$

Torque, $T = \rho Q R_1 r (1 - k\cos\theta) \qquad (12.18)$

Power, $P = T\omega = \rho Q R_1 r \omega (1 - k\cos\theta) = \rho Q R_1 u (1 - k\cos\theta) \qquad (12.19)$

The energy imparted to the wheel is in the form of kinetic energy of the jet and is given by $(1/2)\rho Q V_1^2$ per unit time. Therefore,

Wheel efficiency, $\eta_w = \dfrac{\text{Power out}}{\text{Power in}}$

$$\eta_w = \frac{\rho Q R_1 u (1 - k \cos \theta)}{\frac{1}{2} \rho Q V_1^2}$$

$$= \frac{2u R_1 (1 - k \cos \theta)}{V_1^2} \qquad (12.20)$$

$$= \frac{2u (V_1 - u)(1 - k \cos \theta)}{V_1^2}$$

$$\text{Hydraulic efficiency} = \frac{\text{Power tranferred between fluid and rotor}}{\text{Power given up by fluid in passing through the machine}}$$

$$\eta_b = \frac{\rho Q R_1 u (1 - k \cos \theta)}{\rho g Q H} = \frac{\rho Q \{V_1 - (u - R_2 \cos(\pi - \theta))\} r \omega}{\rho Q g H} \qquad (12.21)$$

$$\text{Overall efficiency} = \text{Mechanical } \eta \times \text{Hydraulic } \eta \qquad (12.22)$$

The difference between η_b and η_w is the effect of the coefficient of velocity C_v because the hydraulic head is not completely converted into kinetic energy. Therefore, the hydraulic efficiency is less than the wheel efficiency.

The wheel efficiency indicates the efficiency with which the wheel converts the fluid energy into mechanical energy. Part of this energy is lost in friction in shaft bearings, and therefore the energy available at the shaft is less than the above. There are also losses due to air friction (aerodynamic). When all these are taken into account, the overall efficiency comes down to about 85%–90% for large machines.

It should be noted that the various frictional losses increase with increasing speed u. The peak theoretical efficiency occurs when $\frac{u}{V_1} = 0.5$. (This can be checked by making $\frac{\partial \eta_w}{\partial u} = 0$.)

This ratio is termed the speed ratio. In practice, it is about 0.46 (Figure 12.6).

In the design of Pelton wheels, care is taken to ensure that the machine operates at about the maximum efficiency for varying output conditions. Pelton wheels are always coupled with electrical generators on the same shaft. These have varying electrical outputs

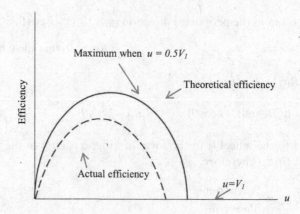

Figure 12.6 Efficiency curves.

depending upon the demand. Therefore, the shaft power should be adjusted to varying output demands. This can be done either by varying the speed u (i.e. ω), or by varying the discharge Q. Change of angular velocity is not permitted because it would alter the frequency of the electrical output. The latter can be achieved either by varying the discharge velocity V_1 or by varying the area of the jet.

Varying the discharge velocity is not allowed because it would alter the speed ratio, which will affect the efficiency. Hence, the only parameter that could be varied is the jet area. This is done by a spear valve that can control the area of flow.

There are other factors to be considered in the design of Pelton wheels, such as the ratio of bucket width to jet diameter and the ratio of wheel diameter to jet diameter, and there are optimal values for these. If the bucket width is too small, the fluid is not smoothly deflected, resulting in energy losses in turbulence. If it is too large, the frictional losses along the surfaces will be too high. The optimum is about 4–5. If the wheel diameter is too small, the bucket spacing will be too crowded. If it is too large, it becomes bulky and the whole system gets too big. The optimum value is about 10.

The jet velocity is controlled by the available head. The bucket velocity for maximum efficiency should satisfy the relationship $\dfrac{u}{V_1} = 0.46$. Since $u = \omega r$, r, the radius of the pitch circle of the buckets may be calculated for a given shaft speed from the equation

$$\omega r = 0.46 V_1$$

Therefore, for a given head the shaft speed or the radius of pitch circle may be determined from one another.

Required power output determines the volume rate of flow, Q since

$$P = \rho g H Q \eta_0 \tag{12.23}$$

where η_0 is the overall efficiency of the turbine. Therefore,

$$Q = \frac{P}{\rho g H \eta_0} \tag{12.24}$$

12.9 REACTION TURBINES

12.9.1 Francis turbine

The basic difference between reaction and impulse turbines is that there is a change in pressure when the fluid goes through the runner. Francis turbine is one of the common types of reaction turbines and has been developed by the U.S. engineer J.B. Francis (1815–1892). It is also called the radial flow or inward flow turbine. Radial flow is turned into axial flow from the centre of the runner. There is no whirl at exit. It can be thought of as a reverse centrifugal pump. It has a spiral casing through which fluid enters and its cross-sectional area decreases along the fluid path in such a way as to keep the fluid velocity constant in magnitude. The flow rate decreases along the spiral casing as fluid enters the runner.

From the spiral volute, the fluid passes between stationary guide vanes to the runner. The stationary guide vanes deflect the fluid by the desired angle according to design.

Blades of Francis turbines resemble airfoils. As water flows over a blade, there will be high pressure on one side and low pressure on the other side giving rise to a lift force. The blades have bucket type shape towards the outlet, thereby producing an impulse force before leaving the runner. Both impulse force and lift force make the runner rotate. Guide vanes at the entrance to the runner convert part of the pressure energy to kinetic energy. They also reduce the swirl of inflow into the runner.

When the fluid passes through the runner, the angular momentum is changed by the blades. At the exit, it has very little or no whirl velocity (compare with a centrifugal pump which has the same conditions at the inlet).

12.9.2 Kaplan turbines

Developed by Austrian engineer V. Kaplan (1876–1934), the Kaplan turbine is equivalent to an axial flow pump in reverse. The flow enters through a spiral casing with decreasing cross-sectional area to ensure uniform velocity as the fluid flows along the periphery. Before the runner, the radial flow is turned into axial flow between the guide vanes and the runner.

12.9.3 Mixed flow turbines

In mixed flow turbines, the flow enters radially and leaves with a substantial axial flow component of velocity. All reaction turbines run full of fluid. Usually, they are mounted on vertical shafts, and the motion is in a horizontal direction. They are used on low head high-flow situations, with heads of the order of about 15–30 m. Efficiencies are around 90% for large machines.

For any reaction turbine, the total head is given by (Figure 12.7)

$$H = \text{Total head at inlet} - \text{Total head at outlet}$$

$$= \frac{P_c}{\rho g} + \frac{V_c^2}{2g} + z_c - \frac{V_E^2}{2g} \tag{12.25}$$

The subscripts c and E refer to the inlet to the turbine and the outlet to the tail race. Effective head across any machine is the difference of head between the inlet and outlet. Therefore, the elevation difference between the centreline of the turbine and the outlet water level z_c is

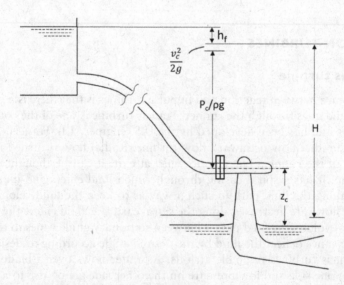

Figure 12.7 Components of heads in a reaction turbine.

included. This does not occur in the case of a Pelton wheel as the fluid enters and leaves at atmospheric pressure.

12.10 BASIC EQUATIONS FOR ROTODYNAMIC MACHINES

The basic assumptions are that the flow is steady state and that all velocities at inlet and outlet are uniform. The velocities involved are the same as for the Pelton wheel. Suffixes 1 and 2 indicate the inlet and outlet conditions.

12.10.1 Relative velocity

Relative velocity refers to the velocity of the fluid relative to the blade. It can be expressed as

Fluid relative to blade = Velocity of fluid − velocity of blade

or

$$\vec{R} = \vec{V} - \vec{u} \tag{12.26}$$

Motion in the circumferential direction only needs to be considered because the impeller (or runner) moves in that direction only. The momentum change in this direction only contributes to forces that do work. There may be momentum changes in other directions, but the corresponding forces have no moments about the axis of rotation. The torque about a given fixed axis is equal to the rate of change of angular momentum about that axis, and angular momentum is equal to the moment of momentum. Torque can be expressed as

$$T = \int dm \left(v_{w_1} r_1 - v_{w_2} r_2 \right) \tag{12.27}$$

Assuming uniform and steady conditions at inlet and outlet, the torque exerted on the runner by the fluid is

$$T = \rho Q \left(v_{w_1} r_1 - v_{w_2} r_2 \right) \tag{12.28}$$

where Q is the volume rate of flow through the runner (or impeller as the case may be).

The torque on the runner must be equal to the difference between the angular momentum of the fluid entering the runner per unit time and the angular momentum of the fluid leaving the runner per unit time. The conditions will be reversed for a pump.

The above equation is sometimes known as Euler's equation. It is applicable regardless of changes of density or components of velocity in other directions. The path taken by the fluid is of no significance. It involves inlet and outlet conditions only. It is independent of losses by turbulence, friction between fluid and blades and changes in temperature as these factors of course would affect the outlet conditions and are therefore already implicitly accounted for.

The torque available at the shaft of a turbine is less than above due to friction in bearings, etc.

Power developed at the runner $= T\omega =$ work done by the fluid per unit time

$$= \rho Q \left(v_{w_1} r_1 - v_{w_2} r_2 \right) \omega$$

Since $u = \omega r$ (implying $u_1 = \omega r_1$ and $u_2 = \omega r_2$),

$$\text{Work done per unit weight} = \frac{\rho Q \left(v_{w_1} u_1 - v_{w_2} u_2 \right)}{\rho g Q}$$

$$= \frac{\left(v_{w_1} u_1 - v_{w_2} u_2 \right)}{g}$$

(12.29)

This expression has the units of 'head' and is sometimes called 'Euler head'. It is also called the energy transfer per unit weight, E. This is the head imparted by the fluid to the runner (or by the impeller to the fluid in the case of a pump).

12.10.2 Velocity diagrams

The relative velocity at inlet, R_1, is in line with the inlet edge of the blade (Figure 12.8). This is the ideal condition. Fluid enters smoothly. If they are not in line, there will be turbulence, eddy formation, etc. resulting loss of energy. Therefore, for all rotodynamic machines, the correct alignment of blades with the velocities relative to them is very important.

For a turbine, the angle α_1 (Figure 12.8), which defines the direction of the absolute velocity of fluid, is determined by setting the guide vanes, which are usually pivoted. For each setting of this angle, there is only one shape of the inlet velocity diagram which gives the ideal conditions. The angle of R_1 is determined by the geometry of the vector diagram.

At the outlet, the direction of R_2 is determined by the outlet blade angle and the geometry of the vector diagram then determines the velocity V_2. The fluid at the exit has some kinetic energy, $\dfrac{V_2^2}{2g}$.

For maximum efficiency, V_2 must be a minimum. For a given rate of flow, V_2 is a minimum when it is normal to u_2 (Figure 12.8). However, this is not achievable, and a small velocity of whirl exists in practice. If v_{w_2} is zero, the Euler head is given by

$$\text{Euler head}, E = \frac{u_1 v_{w_1}}{g}$$

(12.30)

From the velocity vector diagram, it is easy to see the following relationships (trigonometric):

$$R_1^2 = u_1^2 + V_1^2 - 2u_1 V_1 \cos \alpha_1$$

$$= u_1^2 + V_1^2 - 2u_1 v_{w_1} \left(\text{because } V_1 \cos \alpha_1 = v_{w_1} \right)$$

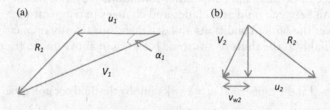

Figure 12.8 Inlet (a) and outlet (b) velocity diagrams for a reaction turbine.

Therefore,

$$u_1 v_{w_1} = \frac{u_1^2 + V_1^2 - R_1^2}{2} \tag{12.31a}$$

Similarly,

$$u_2 v_{w_2} = \frac{u_2^2 + V_2^2 - R_2^2}{2} \tag{12.31b}$$

Therefore, work done per unit weight (Euler head) E, is

$$E = \frac{1}{2g} \left\{ \left(V_1^2 - V_2^2 \right) + \left(u_1^2 - u_2^2 \right) - \left(R_1^2 - R_2^2 \right) \right\} \tag{12.32}$$

The first term represents pressure change from centrifugal effects, the second term the change in the kinetic energy of fluid, and the third term the pressure change due to relative kinetic energy.

In a radial flow or mixed flow machine all three terms are effective. In an axial flow machine, fluid enters and leaves at the same radius. Therefore, $u_1 = u_2$, and

$$E = \left\{ \frac{V_1^2 - V_2^2}{2g} - \frac{R_1^2 - R_2^2}{2g} \right\} \tag{12.33}$$

Gas turbines and compressors are designated on the above parameter.

For a radial flow turbine, E is positive. This is easily seen if the flow is inward.

$$u_1 > u_2 \text{ and } R_2 > R_1 \left(\text{small area at } R_2 \right)$$

For a pump it will be the opposite.

12.11 REVERSIBLE MACHINES (PUMP TURBINES)

Pump turbines operate as pumps in one direction of rotation and as turbines in the other direction of rotation. They were used mainly in pumped storage hydro-electric schemes since the early 1930s. Surplus power available during off-peak hours is used to pump water from a lower level to a higher level which is released during peak hours to generate electricity. All three types of machines (radial, axial and mixed) are used.

12.12 DYNAMIC SIMILARITY AND SPECIFIC SPEED

If the principles of dynamic similarity are applied, it is possible to predict the performance of one machine from the results of tests on another geometrically similar machine, and also of the same machine under different operating conditions.

Geometrical similarity and kinematic similarity are essential for dynamic similarity. Geometrical similarity must apply to all significant parts of the machine; the impeller, entrance and discharge tubes, etc. So is kinematic similarity; corresponding velocities should have a constant ratio. Machines that are geometrically similar belong to the same homologous series.

12.12.1 Dimensional analysis

D – Diameter of impeller (rotor)	L
Q – Volume rate of flow through the machine	L^3T^{-1}
N – Speed of rotation	T^{-1}
H – Difference in head across the machine	L
g – Weight per unit mass	LT^{-2}
P – Power transferred between the machine and the fluid	ML^2T^{-3}
ρ – Density of fluid	ML^{-3}
μ – Viscosity of fluid	$ML^{-1}T^{-1}$

For geometrical similarity, the length ratio is constant; for kinematic similarity, the velocity ratio is constant; and for dynamic similarity, the force ratio is constant. These dimensionless ratios can be determined using Buckingham's π theorem.

There are eight variables. However, the term 'g' appears only as $\dfrac{V^2}{2g}$ in the transformation of static head to velocity head and vice versa. The effect of 'g' is negligible and, therefore, can be replaced by 'gH' rather than using 'g' and 'H'. This also eliminates the difference between D and H as a length parameter.

The same conclusions may be arrived at by considering the pressure differences instead of H.

The number of separate variables then is 7; D, Q, N, gH, P, ρ and μ. The number of basic variables is 3; M, L and T. Therefore, in a dimensional analysis, there must be four independent dimensionless functions (7–3 = 4).

Considering D, N and ρ as the recurring[1] variables, it is possible to write

$$\pi_1 = D^{a_1} N^{b_1} \rho^{c_1} Q \tag{12.34a}$$

$$\pi_2 = D^{a_2} N^{b_2} \rho^{c_2} gH \tag{12.34b}$$

$$\pi_3 = D^{a_3} N^{b_3} \rho^{c_3} \mu \tag{12.34c}$$

$$\pi_4 = D^{a_4} N^{b_4} \rho^{c_4} P \tag{12.34d}$$

Substituting the fundamental quantities, M, L and T, the indices can be determined, and, it can be shown that (see also Chapter 8)

$$\pi_1 = \frac{Q}{ND^3} \quad \pi_2 = \frac{gH}{N^2 D^2} \quad \pi_3 = \frac{\mu}{ND^2 \rho} \quad \pi_4 = \frac{P}{N^3 D^5 \rho} \tag{12.35}$$

12.12.2 Significance of the parameters

- All lengths of the machines are proportional to D
- All areas of the machines are proportional to D^2

---

[1] (i) Repeating variables must include among them all fundamental quantities.

- Average velocity of fluid is proportional to $\dfrac{Q}{D^2}$ which can be considered as a typical velocity of fluid.
- Peripheral velocity of the impeller is proportional to ND.

Therefore,

$$\pi_1 = \frac{Q}{ND^3} = \frac{Q/D^2}{ND} \quad \propto \quad \frac{|\text{Fluid velocity}|}{|\text{Blade velocity}|} \tag{12.36}$$

π_1 is sometimes known as the 'discharge number' or 'flow coefficient' and must be the same for kinematic similarity (similarity of velocity vector diagrams).

If $\dfrac{Q}{D^2}$ is taken as the typical velocity, the ratio

$$\frac{\pi_1}{\pi_3} = \frac{Q}{ND^3} \frac{ND^2\rho}{\mu} = \frac{Q}{D^2} \frac{D\rho}{\mu} = \frac{VD\rho}{\mu} = Re \tag{12.37}$$

represents the Reynolds number. If π_1 is held the same for kinematic similarity, then π_3 is proportional to the reciprocal of the Reynolds number.

In the operating range of speeds in hydraulic machines, the flow is highly turbulent and therefore the Reynolds number is not significant and π_3 can be disregarded. Then,

$$\frac{\pi_4}{\pi_1\pi_2} = \frac{P}{D^5N^3\rho} \frac{ND^3}{Q} \frac{N^2D^2}{gH} = \frac{P}{\rho gHQ} \tag{12.38}$$

The term ρgHQ represents the energy imparted to the fluid in going through the pump per unit time, and the ratio $\dfrac{P}{\rho gHQ}$ represents the hydraulic efficiency of a turbine (reciprocal of the hydraulic efficiency for a pump). It is now possible to write

$$\phi_1\left(\frac{Q}{ND^3}, \frac{gH}{N^2D^2}, \frac{P}{\rho N^3 D^5}\right) = 0 \tag{12.39}$$

$$\Rightarrow \phi_1\left(\pi_1, \pi_2, \pi_4\right) = 0$$

or, in a different arrangement of π's (because the π functions can be multiplied, divided and/or powered),

$$\phi_2\left(\eta, \frac{gH}{N^2D^2}, \frac{P}{\rho N^3 D^5}\right) = 0 \Rightarrow \phi_2\left(\frac{\pi_4}{\pi_1\pi_2}, \pi_2, \pi_4\right) = 0 \tag{12.40}$$

The graphs of these parameters obtained from model tests can be applied to any machine in the same homologous series.

For a turbine, the quantities of interest are N, P and H. A dimensionless number independent of the size of the turbine (represented by D) referred to as the specific speed parameter can be obtained as follows:

$$K_n = \frac{\pi_4^{\frac{1}{2}}}{\pi_2^{\frac{5}{4}}} = \frac{NP^{\frac{1}{2}}}{\rho^{\frac{1}{2}}(gH)^{\frac{5}{4}}} \tag{12.41}$$

For a particular value of this parameter, the flow conditions of all turbines belonging to the same homologous series will be the same.

Alternatively, for a pump, the quantities of interest are N, H and Q. The corresponding dimensionless parameter can therefore be defined as

$$K_n = \frac{\pi_1^{1/2}}{\pi_2^{3/4}} = \frac{Q^{1/2}}{N^{1/2}D^{3/2}} \frac{N^{3/2}D^{3/2}}{(gH)^{3/4}} = \frac{NQ^{1/2}}{(gH)^{3/4}} \tag{12.42}$$

It can be thought of as the speed of a geometrically similar hypothetical pump operating under a unit head and unit discharge.

Since 'ρ' and 'g' are constants, they can be taken out from K_n and another parameter can be defined as follows:

$$N_s = \frac{NP^{1/2}}{H^{5/4}} \text{ for a turbine} \tag{12.43a}$$

and

$$N_s = \frac{NQ^{1/2}}{H^{3/4}} \text{ for a pump} \tag{12.43b}$$

This parameter N_s is called the specific speed and is not dimensionless. Specific speed is not a speed; it is somewhat of a shape factor. For a pump, the three important dimensionless parameters are

$$\text{Dimensionless flow parameter, } \pi_1 = \frac{Q}{ND^3} \tag{12.44a}$$

$$\text{Dimensionless head parameter, } \pi_2 = \frac{gH}{N^2D^2} \tag{12.44b}$$

$$\text{Dimensionless speed parameter, } K_n = \frac{NQ^{1/2}}{(gH)^{3/4}} \tag{12.44c}$$

The effect of the shape of the impeller on the specific speed is that for radial flow, impellers have lower values of K_n and N_s, and for axial flow, impellers have higher values of K_n and N_s.

Usually, with decreasing K_n values, the pump sizes increase. Therefore, higher K_n values are preferred. The volute shape also affects K_n and N_s.

Typical values of the specific speed parameter K_n (Massey, 1980, Table 14.1) are (usually expressed as per runner or jet)

- Pelton wheel (single jet) 0.015–0.024
- Pelton wheel (two jets) 0.022–0.033
- Francis turbine (<370 m) 0.055–0.37
- Kaplan turbine (<60 m) 0.3–0.75.

12.12.3 Other useful relationships

For dynamic similarity, π_1 is a constant. If D is kept constant, then $\dfrac{Q}{N}$ is a constant. Therefore, Q is proportional to N. From π_2 and π_4 it can be seen that for constant ρ and g,

$$Q \propto N \tag{12.45a}$$

$$H \propto N^2 \tag{12.45b}$$

$$P \propto N^3 \tag{12.45c}$$

Similarly, if N is kept constant,

$$Q \propto D \tag{12.46a}$$

$$H \propto D^2 \tag{12.46b}$$

$$P \propto D^5 \tag{12.46c}$$

These are called the affinity laws for pumps. They allow the prediction of performance characteristics of a pump at one speed from knowledge at another speed.

The performance characteristics (power output, efficiency and the flow rate) of a turbine are usually plotted against N as the independent variable. The head is kept constant. These parameters can be converted into their dimensionless forms by manipulating with the π functions.

$$\pi_2^{-1/2} = \frac{ND}{(gH)^{1/2}} \tag{12.47a}$$

$$\frac{\pi_1}{\pi_2^{1/2}} = \frac{Q}{D^2(gH)^{1/2}} \tag{12.47b}$$

$$\frac{\pi_4}{\pi_2^{3/2}} = \frac{P}{\rho D^2(gH)^{3/2}} \tag{12.47c}$$

Since ρ and g are constants, they are usually omitted from the parameters. Sometimes, D is also omitted. Then, the resulting ratios are

$$\frac{N}{N^{1/2}}; \quad \frac{Q}{H^{1/2}}; \quad \frac{P}{H^{3/2}}$$

which are referred to as the unit power, unit flow and the unit speed, respectively. Only their numerical values are considered. Since turbines are always coupled to electricity generators and therefore to maintain the frequency of generation of electricity, they are usually run at constant speed.

12.13 CAVITATION

Cavitation is caused by

- Vapourisation of the liquid and/or release of dissolved air at low pressure. When the velocities and elevations are high, pressures can reach low values, according to Bernoulli's equation. This occurs when the pressure is lowered to the vapour pressure at the temperature considered.
- Movement of the vapour into a high-pressure region. The bubbles are carried along by the flow.
- Collapse of the bubbles (vapour cavities) due to high pressure. Vapour is condensed to liquid again. When the cavity collapses, liquid from all directions rushes to the centre of gravity of the cavity giving rise to very high local pressure.
- Release of energy and pressure wave of high intensity results in
  a. Pitting and erosion of metal surfaces
  b. Noise and vibration of machine
  c. Loss of energy and efficiency

Although cavitation may not be formed on solid surfaces, pressure waves from nearby cavities can affect them. These pressures, intense in magnitude, act only for a short time. But they act repeatedly at high frequencies, causing the metal to fail by fatigue.

In reaction turbines, the point of minimum pressure is the outlet end of a runner blade on the leading side. For flow between such a point and the outlet tailrace (where the pressure is atmospheric), the energy equation is

$$\frac{p_{\min}}{\rho g} + \frac{V^2}{2g} + z = \frac{p_{\mathrm{atm}}}{\rho g} + h_L \quad \text{(frictional losses in draft tube)} \tag{12.48}$$

If V is large, $p_{\min}$ is small. Therefore, V (outlet velocity) should be kept to a minimum:

$$\frac{V^2}{2g} - h_L = \frac{p_{\mathrm{atm}} - p_{\min}}{\rho g} - z \tag{12.49}$$

From this equation, if

$$\sigma_c = \frac{1}{H}\left(\frac{p_{\mathrm{atm}}}{\rho g} - \frac{p_{\min}}{\rho g} - z\right) \tag{12.50}$$

is a certain fraction of the net head across the machine, H, then, for cavitation not to occur, $p_{\min} >$ vapour pressure of liquid, or, $\sigma > \sigma_c$ where

$$\sigma = \frac{1}{H}\left(\frac{p_{\mathrm{atm}}}{\rho g} - \frac{p_{vp}}{\rho g} - z\right) \tag{12.51a}$$

which is

$$\left(\frac{p_{\mathrm{atm}} - p_{vp}}{\rho g} - z\right)\frac{1}{H} > \left(\frac{p_{\mathrm{atm}} - p_{\min}}{\rho g} - z\right)\frac{1}{H} \tag{12.51b}$$

The above parameter is known as Thoma's cavitation parameter (German Engineer, Dietrich Thoma (1881–1943)).

If z or H is increased, σ is reduced. To check whether cavitation occurs or not σ is first calculated and compared with σ_c given for the particular machine. σ_c may be computed for a particular machine design operating under its design conditions.

This expression can also be used to determine the maximum elevation of the machine above the tailrace for cavitation not to occur.

$$z_{max} = \frac{p_{atm} - p_{vp}}{\rho g} - \sigma_c H \tag{12.52}$$

If H is high, z is low, sometimes even below the tail water level.

For pumps,

$$\sigma_c = \frac{1}{H}\left(\frac{p_{atm}}{\rho g} - \frac{p_{min}}{\rho g} - z - h_L \right) \tag{12.53}$$

For turbines, for cavitation not to occur $\sigma > \sigma_c$ where

$$\sigma = \frac{1}{H}\left(\frac{p_{atm}}{\rho g} - \frac{p_{vp}}{\rho g} - z - h_L \right) \tag{12.54}$$

For σ to be large, z should be as small as possible which means that the pump may have to be installed below the water level to avoid cavitation.

Example 12.1

The impeller of a centrifugal pump has an outer diameter of 250 mm and an effective outlet area of 0.017 m². The outlet blade angle is 32°. The diameters of suction and discharge openings are 150 and 125 mm, respectively. At 24.2 rev/s and discharge 0.03 m³/s, the pressure heads at suction and discharge openings were respectively 4.5 m below and 13.3 m above atmospheric pressure, the measurement points being at the same level. The shaft power was 7.76 kW. Water enters the impeller without shock or whirl. Assuming that the true outlet whirl component is 70% of the ideal, determine the overall efficiency and the manometric efficiency based on the true whirl component.

$D = 250$ mm
$A = 17{,}000$ mm²
$\phi = 32°$
$N = 24.2$ rev/s
$Q = 0.03$ m³/s
$H_s = -4.5$ m
$H_d = 13.3$ m
$P = 7.76$ kW

$$\text{Manometric efficiency} = \frac{gH_m}{u_2 V_{w2}}$$

$H_m\,(\text{manometric head}) = 13.3 - (-4.5) = 17.8$

$u_2 = \omega r_2 = (24.2) \times (2\pi) \times 0.125$

$\quad = 19.0$ m/s

$$V_{r_2} = \frac{Q}{A} = \frac{0.03}{17,000 \times 10^{-6}} = 1.765 \text{ m/s}$$

$$Vw_2 = u_2 - \frac{Vr_2}{\tan\phi_2}$$

$$= 19 - \frac{1.765}{\tan 32°} = 19 - 2.825$$

$$= 16.175$$

True whirl component $= 0.70 \times 16.175$

$$= 11.322$$

Manometric efficiency $= \dfrac{9.82 \times 17.8}{19.0 \times 11.322} = 0.8126 = 81.3\%$

Overall efficiency $= \dfrac{\text{Power given to fluid}}{\text{Shaft power}} = \dfrac{\rho g Q H}{P}$

where H is the overall gain in head

$$H = \left(P_d + \frac{V_d^2}{2g} \right) - \left(P_s + \frac{V_s^2}{2g} \right)$$

$$V_s = \frac{Q}{A_S} = \frac{0.03}{\dfrac{\pi}{4} \times 0.150^2} = 1.679 \text{ m/s}$$

$$V_d = \frac{Q}{A_d} = \frac{0.03}{\dfrac{\pi}{4} \times 0.125^2} = 2.445 \text{ m/s}$$

$$\frac{V_s^2}{2g} = 0.1469$$

$$\frac{V_d^2}{2g} = 0.3047$$

Therefore,

$$H = 17.8 + 0.3047 - 0.1469 = 17.9578 \text{ m}$$

Overall efficiency $= \dfrac{1,000 \times 9.82 \times 17.9578 \times 0.03}{7.76 \times 1,000} = 68.1\%$

Example 12.2

In a hydro-electric scheme, a number of Pelton wheels are to be used under the following conditions: total output required 30 MW; gross head 245 m; speed 6.25 rev/s; two jets per wheel; C_v of nozzle 0.97; maximum overall efficiency (based on conditions immediately before the nozzles) 81.5%; dimensionless specific speed not to exceed 0.022 rev/jet; head lost due to friction in pipeline not to exceed 12 m. Calculate

a. the number of wheels required
b. the diameters of the jets and wheels
c. the hydraulic efficiency if the blades deflect the water through 165° and reduces its relative velocity by 15%, and
d. the percentage of the input power which remains as kinetic energy of the water at discharge.

Output power $= 30 \times 10^6$ W

Gross head $= 245$ m

Friction loss $= 12$ m

Net head $= (245 - 12) = 233$ m

$N = 6.25$ rev/s

$\eta_0 = 0.815$

$K_n = 0.022$

Dimensionless specific speed $K_n = \dfrac{NP^{\frac{1}{2}}}{\rho^{\frac{1}{2}}(gH)^{\frac{5}{4}}}$

$0.022 = \dfrac{6.25 \times P^{\frac{1}{2}}}{\sqrt{1{,}000} \times (9.81 \times 233)^{\frac{5}{4}}} \Rightarrow P = 3.09 \times 10^6$ W

where P is the power generated per jet

$P = 3.09 \times 10^6$ W

Total power $= 30 \times 10^6$ W

No. of wheels $= \dfrac{30}{2 \times 3.09} = 5$

Overall efficiency $= \dfrac{\text{Power out}}{Q\rho gH}$

$0.815 = \dfrac{30 \times 10^6}{Q \times 1{,}000 \times 9.81 \times 233} \Rightarrow Q = 16.1 \ \text{m}^3/\text{s}$

Velocity at inlet $V_1 = C_v \sqrt{2gH}$

$= 0.97\sqrt{2 \times 9.81 \times 233}$

$= 65.58 \ \text{m/s}$

Total area of jets $= \dfrac{Q}{V_1} = \dfrac{16.10}{65.58} = 0.245 \ \text{m}^2$

Aea of one jet $= \dfrac{0.245}{10} = 0.0245\,\text{m}^2$.

$$\dfrac{\pi d^2}{4} = 0.0245$$

$$d = 0.177\,\text{m} = 177\,\text{mm}.$$

Assume a speed ratio of 0.46

$$\dfrac{u}{V_1} = 0.46$$

$$u = 0.46 \times 65.58 = 30.17\,\text{m/s}.$$

$$u = \omega r$$

$$r = \dfrac{30.17}{2\pi \times 6.25} = 0.7682\,\text{m}$$

$$D = 1.536\,\text{m}$$

Hydraulic efficiency $= \dfrac{\text{Power tranferred between fluid and rotor}}{\text{Power given up by fluid in passing through the machine}}$

$$= \dfrac{\rho Q\{v_1 - (u - R_2 \cos(\pi - \theta))\}\, r\omega}{\rho Q g H}$$

where the velocities are as shown in the vector diagram and H is the difference in head across the machine.

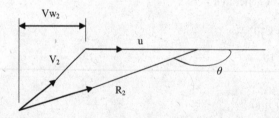

$$R_2 = kR_1$$

$$= 0.85 R_1$$

$$= 0.85 (v_1 - u)$$

$$= 0.85 (65.58 - 30.17)$$

$$= 0.85 \times 35.41$$

$$= 30.1\,\text{m/s}$$

$$\eta_b = \frac{\rho Q R_1 \left\{ 1 + k \cos 15 \right\} u}{\rho Q g H}$$

$$= \frac{R_1 \left\{ 1 + k \cos 15 \right\} u}{gH}$$

$$= \frac{35.41 \times 30.17 \times 1.821}{9.81 \times 233}$$

$$= 85.1\%$$

From the velocity triangle,

$$V_2^2 = u^2 + R_2^2 - 2uR_2 \cos 15$$

$$= 30.17^2 + 30.10^2 - 2 \times 30.17 \times 30.10 \cos 15$$

$$= 61.89$$

$$\text{K.E. at exit} = \frac{1}{2}(\rho Q)V_2^2$$

Input power $= \rho g Q H$

Percentage of input power remaining as K.E. $= \dfrac{V_2^2}{2gH} \times 100$

$$= \frac{61.89^2}{2 \times 9.81 \times 233} \times 100$$

$$= 1.35\%$$

REFERENCE

Massey, B. S. (1980): *Mechanics of Fluids*, (4th Edition), Van Nostrand Reinhold Company Ltd., England, p. 543.

Chapter 13

Applications of basic fluid flow equations

13.1 INTRODUCTION

There are many problems in nature that involve principles of fluid mechanics. Solutions to such problems are obtained by applying the governing equations of fluid flow or their approximations. The three basic equations involved are the continuity equation, the momentum equation and the energy equation. Assumptions and/or simplifications necessary to apply the basic equations to certain types of practical problems include incompressibility, steady state condition, ideal fluid, and sometimes reducing the dimensionality of the problem. In this chapter, how the governing equations of fluid flow are applied to some typical practical problems in nature are highlighted.

13.2 KINETIC ENERGY CORRECTION FACTOR

When flows in open channels or pipes are considered, it is generally assumed to be one-dimensional with an average velocity at each section. The kinetic energy is $\dfrac{V^2}{2g}$ per unit weight, but is different from $\displaystyle\int_{\text{cross section}} \dfrac{v^2}{2g}$. Therefore, a correction factor, α, is used for $V^2/2g$ so that $\dfrac{\alpha V^2}{2g}$ is the average kinetic energy for unit weight, passing the section.

Referring to Figure 13.1, the kinetic energy passing δA per unit time is

$$\int \frac{v^2}{2}(\rho v dA)$$

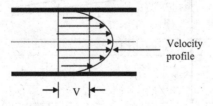

Figure 13.1 Velocity profile in pipe flow.

where $\rho v dA$ is the mass.

Therefore,

$$\alpha \frac{V^2}{2} \rho VA = \int_A \frac{v^2}{2} (\rho v dA)$$

which leads to

$$\alpha = \frac{1}{A} \int_A \left(\frac{v}{V}\right)^3 dA \qquad (13.1)$$

The Bernoulli equation then becomes

$$z + \frac{p}{\rho g} + \alpha \frac{V^2}{2g} = \text{Constant}$$

For laminar flow in pipes $\alpha = 2$; for turbulent flows in pipes, α varies from 1.01 to 1.10 and is usually ignored.

All terms in the Bernoulli equation are available energy, and, for real fluids flowing through a system, the available energy decreases in the downstream direction. It is the energy that it is available for doing work, as in hydropower generation.

A plot of available energy along the direction of flow is called the energy grade line, whereas a plot of the two terms $z + \frac{p}{\rho g}$ is called the hydraulic grade line (Figure 13.2).

13.3 MOMENTUM CORRECTION FACTOR

Similar to the kinetic energy correction factor, a correction factor has to be applied to the momentum to account for the cross-sectional variation of velocity.

Considering a control volume (Figure 13.3) separated by two cross sections with θ_1 and θ_2 as the directions of the velocities to the horizontal, the momentum equation gives

$$F_x = m V_2 \cos \theta_2 - m V_1 \cos \theta_1 \Rightarrow \propto V^2$$

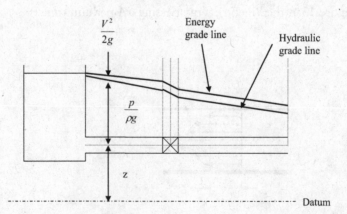

Figure 13.2 Energy and hydraulic grade lines.

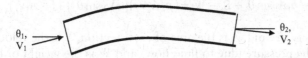

Figure 13.3 Control volume for pipe flow.

where
m = mass rate through section = $\rho_1 A_1 V_1 = \rho_2 A_2 V_2$
ρ = density of fluid
V = average velocity
A = areas of cross sections (suffixes refer to the two sections)

The relationship of the momentum when using the average velocity V and the point velocity in a cross section can be expressed in terms of β, as

$$\int_{\text{cross section}} \rho v^2 \, dA = \beta \rho V^2 A$$

where β, the momentum correction factor is given as

$$\beta = \frac{1}{A} \int_{\text{cron section}} \left(\frac{v}{V}\right)^2 dA \tag{13.2}$$

For laminar flow in straight circular tubes $\beta = \frac{4}{3}$; for uniform flow $\beta = 1$; always $\beta \geq 1$.

13.4 FORCES ON A PIPE BEND CARRYING A FLUID

In the following examples, all fluids will be assumed to be frictionless, implying no shear forces.

Referring to the control volume bounded by the pipe surface and the two sections 1 and 2 (Figure 13.4), with B_x and B_y as the resultant of the forces acting on the fluid within the control volume in the x and y directions, the momentum equations in the two directions give

$$x: \quad p_1 A_1 \cos\theta_1 + B_x - p_2 A_2 \cos\theta_2 = (\rho_2 A_2 V_2) V_2 \cos\theta_2 - (\rho_1 A_1 V_1) V_1 \cos\theta_1 \tag{13.3a}$$

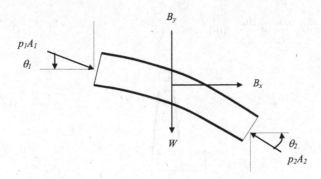

Figure 13.4 Forces acting on fluid in a pipe bend.

$$y: \quad p_2A_2 \sin\theta_2 - p_1A_1 \sin\theta_1 + B_y - W = (p_2A_2V_2)(-V_2 \sin\theta_2) - (p_1A_1V_1)(-V_1 \sin\theta_1) \quad (13.3b)$$

where ρ is the density of fluid, A is the area of cross section, V is the velocity (average across the section), p is the pressure (due to fluid flow) and W is the weight of fluid in the control volume with the subscripts 1 and 2 indicating the two cross sections.

The two unknowns, B_x and B_y, can therefore be expressed as

$$B_x = V_2^2 \rho_2 A_2 \cos\theta_2 - V_1^2 \rho_1 A_1 \cos\theta_1 - p_1A_1 \cos\theta_1 + p_2A_2 \cos\theta_2 \quad (13.4a)$$

$$B_y = -V_2^2 \rho_2 A_2 \sin\theta_2 - p_2A_2 \sin\theta_2 + W - p_1A_1 \sin\theta_1 - p_1A_1V_1^2 \sin\theta_1 \quad (13.4b)$$

The forces exerted by the fluid on the bend will be equal and opposite in direction. It should be emphasized that the change of momentum should be the difference between the final momentum and the initial momentum. From continuity,

$$\rho_1 V_1 A_1 = \rho_2 V_2 A_2 = m(\text{mass per unit time})$$

Then,

$$B_x = m(V_2 \cos\theta_2 - V_1 \cos\theta_1) - p_1A_1 \cos\theta_1 + p_2A_2 \cos\theta_2 \quad (13.5a)$$

$$B_y = -m(V_2 \sin\theta_2 + V_1 \sin\theta_1) - p_2A_2 \sin\theta_2 + W - p_1A_1 \sin\theta_1 \quad (13.5b)$$

13.5 NATURAL CONVECTION – CHIMNEY FLOW DYNAMICS

The equation of motion for the flow of gases through a chimney can be written as follows.

The momentum equation in the vertical direction with w as the velocity in the vertical direction is (Figure 13.5)

$$F = \frac{d}{dt}(mw) = m\frac{dw}{dt} = m\left\{w\frac{\partial w}{\partial z} + \frac{\partial w}{\partial t}\right\} \quad (13.6)$$

If the flow is steady, $\dfrac{\partial w}{\partial t} = 0$, and the partial derivatives can be replaced by the total derivatives. Therefore, $\dfrac{\partial w}{\partial z} = \dfrac{dw}{dz}$, and,

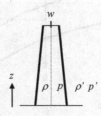

Figure 13.5 Control volume for chimney flow.

$$F = mw \frac{dw}{dz}$$ (13.7)

For static equilibrium inside the chimney (Figure 13.6),

$$F_z = pdA - \left(p + \frac{dp}{dz} dz \right) dA - \rho g dz dA = 0$$

$$= -\left(\frac{dp}{dz} + \rho g \right) dz dA = 0 \Rightarrow \frac{dp}{dz} = -\rho g$$ (13.8)

But, for dynamic equilibrium, $F_z = mw \dfrac{dw}{dz}$. Substituting back,

$$-\left(\frac{dp}{dz} + \rho g \right) dz dA = \rho dz dA w \frac{dw}{dz}$$

$$\Rightarrow \frac{1}{\rho} \frac{dp}{dz} + g + w \frac{dw}{dz} = 0$$ (13.9)

$$\text{or } \ w \frac{dw}{dz} = -\frac{1}{\rho} \frac{dp}{dz} - g$$

Note that $\dfrac{dp}{dz} < 0$, i.e. pressure decreases with increasing height, and that the density of the gas inside the chimney is less than that of air outside because of its higher temperature.

The above equation assumes that there is no buoyant force acting on the fluid element.

If there is a buoyant force, which exists when there is a density difference between the parcel of gas and the ambient air, the equation of motion is (assuming the air outside has a density ρ' $(> \rho)$)

$$pdA - \left(p + \frac{dp}{dz} dz \right) dA - \rho g dz dA + \rho' g dz dA = \rho dz dA w \frac{dw}{dz}$$

$$\Rightarrow -\frac{1}{\rho} \frac{dp}{dz} + \frac{(\rho' - \rho)}{\rho} g = w \frac{dw}{dz}$$ (13.10a)

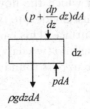

Figure 13.6 Control volume for static equilibrium.

If $\dfrac{dp}{dz}$ is known, this equation can be integrated. This is the equation that is applicable when the warm air from the chimney leaves it. It can also be seen that the net force is in the upward direction. Therefore, w increases in the upward direction.

i.e. $\dfrac{dw}{dz} > 0$ (This can be proved)[1]

This type of motion due to insufficient gravity forces is termed natural convection. With this theory, it is possible to determine a relationship between the chimney height and the chimney draft velocity at its upper end.

The outside pressure (p') gradient may be obtained from static equilibrium considerations.

$$p'dA - \left(p' + \frac{dp'}{dz} dz \right) dA - \rho' g dA dz = 0 \Rightarrow \frac{dp'}{dz} = -\rho' g \tag{13.11}$$

If it is assumed that $\dfrac{dp}{dz} = \dfrac{dp'}{dz}$ (i.e., the static pressure gradient is the same as the dynamic pressure gradient), then

$$w \frac{dw}{dz} = -\frac{1}{\rho}(-\rho' g) + \left(\frac{\rho' - \rho}{\rho} \right) g = \left\{ 2 \frac{\rho'}{\rho} - 1 \right\} g \tag{13.12}$$

which can be integrated as

$$\frac{w^2}{2} = \left(2 \frac{\rho'}{\rho} - 1 \right) gz + \text{constant}$$

At $z = 0$, $w = 0$; therefore,

$$w = \left\{ 2 \left(2 \frac{\rho'}{\rho} - 1 \right) gz \right\}^{\frac{1}{2}} \tag{13.13}$$

The analysis is approximate because the flow is not entirely frictionless and that the flow is not entirely one dimensional.

13.6 STAGNATION POINT

Referring to Figure 9.5 (Chapter 9) in which the flow past a cylinder is considered, the streamline closest to the leading edge must divide into two halves, one going along the upper surface and the other along the lower surface. This can only happen if the flow is momentarily brought to rest at S before dividing into two halves. The point S is a stagnation point. It occurs every time a fluid has to divide to pass round an object.

---

[1] Since $\dfrac{dp}{dz} < 0$, $-\dfrac{dp}{dz} > 0$, and $\rho' - \rho > 0$, therefore $\dfrac{dw}{dz} > 0$ (Eq. 13.10a)

13.7 PITOT TUBE – AN AIRSPEED INDICATOR

Pitot tube is an example of the application of Bernoulli's equation and is used to measure air (or fluid) speed in many instants, e.g. in aircrafts. The principle is that when fluid particles pass through a blunt obstacle, they move from one side to the other, leaving a stagnation point where the velocity is zero (see also Chapter 9).

This is done at a constant level, so that the potential energy remains constant.

$$\frac{p_1}{\rho g} + \frac{V_1^2}{2g} + z = \frac{p_2}{\rho g} + \frac{V_2^2}{2g} + z \tag{13.14}$$

If $V_2 = 0$, then

$$V_1 = \left[(p_2 - p_1) \frac{2g}{\rho g} \right]^{1/2} \tag{13.15}$$

The pressure difference is measured by a manometer as h, and density corrections are required at different levels.

13.8 IMPACT OF JETS

When a jet of fluid strikes a vane, the momentum of the fluid is transferred to the vane. If the vane is fixed, the jet gets deflected resulting in momentum change. In the case of a moving vane, energy transfer can take place from the fluid to the vane or vice versa. The following assumptions are needed to determine the forces on the vane:

- Friction between jet and vane ignored,
- Uniform velocity throughout the jet,
- Elevation difference between the ends of vane ignored,
- Steady-state, two-dimensional, frictionless fluid,
- Gravity ignored.

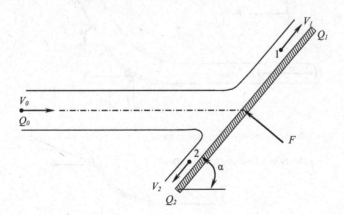

Figure 13.7 Jet impinging on a fixed vane.

Let a be the cross-sectional areas of jet normal to its direction of flow, V_0, the uniform speed of jet, a_1 and a_2, the cross-sectional areas of the deflected streams with velocities V_1 and V_2, respectively (Figure 13.7).

Consider a control volume that includes the jet and the vane as shown in Figure 13.8, continuity equation for steady state gives

$$\rho a V_0 = \rho a_1 V_1 + \rho a_2 V_2 \tag{13.16}$$

From Bernoulli's equation,

$$\frac{V^2}{2} + gz + \frac{p}{\rho} = \text{constant} \tag{13.17}$$

If gravity is ignored,

$$\frac{V^2}{2} + \frac{p}{\rho} = \text{constant.} \tag{13.18}$$

If the pressure is atmospheric all along the jet, then $p = p_0$. Therefore, V is a constant and from the continuity equation,

$$a = a_1 + a_2 \tag{13.19}$$

Because the fluid is frictionless, the force exerted by the fixed wall will be normal to the wall. It can however be resolved into horizontal and vertical components.

From momentum equation normal to the vane

$$F = (\rho a V)V \sin\alpha = \rho a V^2 \sin\alpha \tag{13.20}$$

The force F is independent of how a is divided into a_1 and a_2, but they can be determined by applying the momentum equation parallel to the vane as

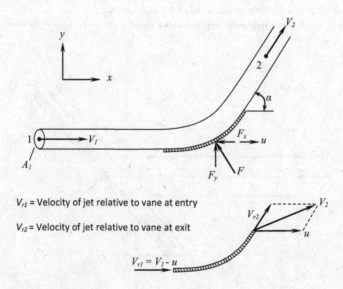

V_{r1} = Velocity of jet relative to vane at entry

V_{r2} = Velocity of jet relative to vane at exit

$V_{r1} = V_1 - u$

Figure 13.8 Fixed and moving vanes.

$$0 = (\rho a_1 V)V - (\rho a_2 V)V - (\rho a V)V \cos\alpha \Rightarrow a_1 - a_2 = a\cos\alpha \qquad (13.21)$$

For real fluids, the tangential force may be calculated if the final fluid velocity is known.

13.8.1 Fixed vanes

The theory of turbo-machines is based on the relationship between jets and vanes. Energy is transferred from the fluid to the moving vanes or vice versa. When a free jet impinges on a fixed vane, the jet is deflected and the momentum is changed. With the following assumptions, the force on the vane by the jet (or its reaction) can be obtained (Figure 13.8):

- Fluid is frictionless,
- Jet impinges the vane tangentially without a shock,
- Friction between jet and vane ignored,
- Uniform velocity throughout the jet,
- Elevation difference between the ends of the vane ignored (no gravity).

Applying Bernoulli's equation,

$$\frac{p_0}{\rho} + \frac{V_1^2}{2} + gz(=0) = \frac{p_0}{\rho} + \frac{V_2^2}{2} + gz(=0) \Rightarrow V = \text{constant} \qquad (13.22)$$

This applies to fixed vanes only.
Force exerted on the vane:

$$-F_x = (\rho A V)V\cos\alpha - (\rho A V)V \quad \left(\text{Because} \quad V_1 = V_2 = V\right) \qquad (13.23a)$$

$$F_y = (\rho A V)V\sin\alpha \qquad (13.23b)$$

13.8.2 Moving vanes

No work is done by the fluid on the vane (or vice versa) when the vane is fixed. When the vanes can be displaced, work is done by the fluid (or vice versa). The analysis is similar.

The problem of a moving vane can be reduced to that of a fixed vane by superimposing a velocity equal and opposite to that of vane on the vane and the fluid. Once the vane is brought to rest, the rest of the problem is the same as that for a fixed vane. Then,

$$-F_x = \left[\rho A(V-u)\right](V-u)\cos\alpha - \rho A(V-u)(V-u) \qquad (13.24a)$$

$$F_y = \rho A(V-u)^2 \sin\alpha. \qquad (13.24b)$$

These equations are for a single vane. For a series of vanes, they become

$$-F_x = \rho Q(V-u)\cos\alpha - \rho Q(V-u) \qquad (13.25a)$$

$$F_y = \rho Q(V-u)\sin\alpha \qquad (13.25b)$$

where Q is the rate of flow through the vanes.

Pelton wheel (see also Chapter 12) is an example of moving vanes. The relative velocities in the x-direction are $(V - u)$ at entry and $(V - u)\cos\alpha$ at exit. The force components are

$$F_x = \rho A(V - u)^2 (1 - \cos\alpha) = \rho Q(V - u)(1 - \cos\alpha) \tag{13.26a}$$

$$F_y = \rho Q(V - u)\sin\alpha \tag{13.26b}$$

F_x is a maximum when $(1 - \cos\alpha)$ is a maximum., i.e. when $\cos\alpha = -1$, or $\alpha = 180°$. In practice, however, this is not possible because it interferes with other vanes (blades). The optimum value is about 165°.

Power transferred to the wheel at 100% efficiency = Force × velocity

$$= \rho A(V - u)^2 (1 - \cos\alpha) u$$

$$= \rho Q(V - u)(1 - \cos\alpha) u \quad \text{Watts.} \tag{13.27}$$

13.9 LOSSES DUE TO SUDDEN CHANGES OF CROSS SECTIONS

The following assumptions are made in the derivations:

- Shear forces on the walls between the two sections ignored,
- Uniform velocity over the cross sections,
- Gravity ignored,
- Pressure distribution at the enlarged section uniform.

Referring to Figure 13.9 and applying the momentum equation gives

$$p_1 A_1 + p_x (A_2 - A_1) - p_2 A_2 = (\rho A_2 V_2) V_2 - (\rho A_1 V_1) V_1 \tag{13.28}$$

Assuming that p_x, the pressure in the stagnation just after the expansion equals p_1,

$$p_1 A_2 - p_2 A_2 = \rho \left\{ A_2 V_2^2 - A_1 V_1^2 \right\} \tag{13.29a}$$

or

$$p_1 - p_2 = \rho \left\{ V_2^2 - \frac{A_1}{A_2} V_1^2 \right\} \tag{13.29b}$$

Bernoulli's equation gives

$$\frac{p_1}{\rho g} + \frac{V_1^2}{2g} + z = \frac{p_2}{\rho g} + \frac{V_2^2}{2g} + z + h_L \tag{13.30}$$

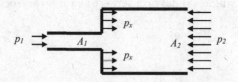

Figure 13.9 Sudden enlargement.

where h_L =head lost due to change of cross section. Here, kinetic energy correction factor is assumed to be unity. Therefore,

$$h_L = \frac{p_1 - p_2}{\rho g} + \frac{V_1^2 - V_2^2}{2g}$$

or

$$\frac{p_1 - p_2}{\rho g} = h_L + \frac{V_2^2 - V_1^2}{2g} \tag{13.31}$$

Equating the two expressions for $\frac{p_1 - p_2}{\rho g}$ (Eqs. 13.29b and 13.31),

$$h_L + \frac{V_2^2 - V_1^2}{2g} = \frac{\rho}{\rho g} \left\{ V_2^2 - \frac{A_1 V_1^2}{A_2} \right\} \tag{13.32a}$$

or

$$h_L = \frac{1}{g} \left\{ V_2^2 - \frac{A_1 V_1^2}{A_2} \right\} - \frac{1}{2g} \left\{ V_2^2 - V_1^2 \right\} \tag{13.32b}$$

But from continuity,

$$V_1 A_1 = V_2 A_2 \Rightarrow \frac{A_1}{A_2} = \frac{V_2}{V_1}$$

Therefore,

$$h_L = \frac{1}{g} \left\{ V_2^2 - V_2 V_1 - \frac{1}{2} V_2^2 + \frac{1}{2} V_1^2 \right\} = \frac{1}{2g} \left\{ V_2^2 - 2 V_2 V_1 + V_1^2 \right\}$$

$$= \frac{1}{2g} \left\{ V_2 - V_1 \right\}^2 = \frac{1}{2g} (V_1 - V_2)^2 \propto V^2 \tag{13.33a}$$

or

$$h_L = k \frac{V^2}{2g} \tag{13.33b}$$

The losses are usually given as a velocity head or as an equivalent pipe length.

13.10 FORCES ON A SOLID BODY IN A FLOWING FLUID

The momentum equation may be used to determine the force exerted on a solid body by fluid flowing past it. The body exerts an equal and opposite force on the fluid, and this force corresponds to a change of momentum of the fluid. Referring to Figure 13.10 and a control volume excluding the plate,

$$\text{Mass flow crossing } 12 = \int_1^2 \rho u_0 \, dy \qquad (13.34\text{a})$$

$$\text{Momentum transfer across } 12 = \int_1^2 \rho u_0^2 \, dy \qquad (13.34\text{b})$$

$$\text{Mass flow crossing } 23 = \int_2^3 \rho v \, dx \qquad (13.34\text{c})$$

$$\text{Momentum transfer across } 23 = \int_2^3 \rho v u_0 \, dx \qquad (13.34\text{d})$$

$$\text{Mass flow across } 34 = \int_3^4 \rho u \, dy \qquad (13.34\text{e})$$

$$\text{Momentum transfer across } 34 = \int_3^4 \rho u^2 \, dy \qquad (13.34\text{f})$$

Total rate of change of momentum = Momentum leaving through 23 and 34

− momentum entering through 12

Mass flow entering through 12 = Mass flow leaving through 23 and 34

Force exerted on the fluid by the body in the x-direction

= Rate of change of momentum in the x-direction

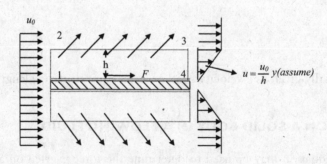

Figure 13.10 Forces on a solid body.

Since it is symmetrical about the plate, only one half needs to be considered.

$$F\,(\text{on one side}) = \int_{CS} (\rho V)V dA$$

If it is assumed that there is no pressure gradient along the plate, and the momentum equation is written in a horizontal direction, the shear force F_x exerted by the plate on the fluid is given by

F_x = Rate of change of momentum in the control volume

$$= \int_3^4 (\rho u)u dy + (\text{mass rate across } 23)u_0 - \int_1^2 (\rho u_0)u_0\, dy$$

$$= \int_0^b \rho u^2 dy - \int_0^b \rho u_o^2 dy + \dot{m}u_0 \tag{13.35}$$

The third term in the expression is the rate of momentum passing through the control surface 23. Across the surface 23, there is a velocity normal to it as well as one parallel to it. The normal component is small, and the parallel component is very nearly equal to u_0.

Mass rate flowing across 23 = Mass rate entering across 12 – Mass rate leaving 34

$$= \rho u_0 h - \rho u_0 \frac{h}{2} = \rho u_0 \frac{h}{2}$$

Therefore,

$$F_x = \int_0^b \rho u^2 dy - \int_0^b \rho u_0^2 dy + \rho u_0^2 \frac{h}{2}$$

$$= \int_0^b \rho \frac{u_0^2}{h^2} y^2 dy - \rho u_0^2 h + \rho u_0^2 \frac{h}{2}$$

$$= \rho u_0^2 \frac{h}{3} - \rho u_0^2 h + \rho u_0^2 \frac{h}{2}$$

$$= -\frac{1}{6} \rho u_0^2 h \tag{13.36}$$

The negative sign indicates that it is a retarding force. This force is also called the drag force. In general, any velocity variation function may be used. The basic equation is

$$F_x = \int_4^3 \rho u^2 dy + \dot{m}u_0 - \int_1^2 \rho u_0^2\, dy \tag{13.37}$$

where $\dot{m}$ and u will depend upon the velocity distribution ($u = u(y)$).

13.11 HYDRAULIC JUMP

Under certain conditions of flow, a fast moving flow stream with a free surface can suddenly change into a slow moving flow with increased depth. This phenomenon is known as a hydraulic jump and is frequently encountered in open channel hydraulics (see also Chapter 17). The rapidly varying flow expands and converts part of its kinetic energy into potential energy by way of increased elevation. It is accompanied by a loss of energy. The nature of the jump depends upon the flow conditions, and the basic equations can be used to determine the losses due to the jump and the relationships between different depths. Referring to Figure 13.11, the relevant equations are

(i) Continuity equation

$$A_1 V_1 = A_2 V_2$$

(ii) Momentum equation

$$\rho g \frac{y_1^2}{2} - \rho g \frac{y_2^2}{2} = (\rho A_2 V_2) V_2 - (\rho A_1 V_1) V_1$$

(iii) Bernoulli's equation (energy equation)

$$\frac{p_0}{\rho g} + \frac{V_1^2}{2g} + y_1 = \frac{p_0}{\rho g} + \frac{V_2^2}{2g} + y_2 + h_L$$

These three equations together with the relationships $A_1 = y_1$, and $A_2 = y_2$, will enable a solution of the three unknowns V_2, y_2 and h_L as follows:

$$\frac{\rho g}{2}\left(y_1^2 - y_2^2\right) = \rho A_2 V_1^2 \frac{A_1^2}{A_2^2} - \rho A_1 V_1^2$$

$$\Rightarrow \frac{\rho g}{2}(y_1 - y_2)(y_1 + y_2) = \rho V_1^2 \frac{y_1^2}{y_2} - \rho y_1 V_1^2$$

$$\Rightarrow \frac{g}{2}(y_1 - y_2)(y_1 + y_2) = \frac{V_1^2}{y_2} y_1 (y_1 - y_2) \tag{13.38}$$

If $y_1 \neq y_2$,

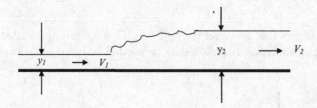

Figure 13.11 Hydraulic jump.

$$\frac{g}{2}(y_1 + y_2)y_2 - V_1^2 y_1 = 0$$

$$\Rightarrow y_2^2 + y_1 y_2 - \frac{2V_1^2}{g} y_1 = 0$$

$$\Rightarrow y_2 = -\frac{y_1}{2} \pm \frac{1}{2}\sqrt{\left\{y_1^2 + 4\left(\frac{2V_1^2}{g}\right)y_1\right\}}$$

$$= -\frac{y_1}{2} + \sqrt{\left(\frac{y_1}{2}\right)^2 + \frac{2V_1^2 y_1}{g}} \tag{13.39}$$

The depths y_1 and y_2 are known as the conjugate depths. Similarly, solving the third equation,

$$h_L = (y_1 - y_2) + \frac{V_1^2}{2g} - \frac{V_2^2}{2g}$$

Substituting $V_2 = V_1 \frac{A_1}{A_2} = V_1 \frac{y_1}{y_2}$ in the momentum equation,

$$V_1^2\left(y_2 \frac{y_1^2}{y_2^2} - y_1\right) = \frac{g}{2}\left(y_1^2 - y_2^2\right)$$

$$\Rightarrow V_1^2\left(y_1^2 - y_1 y_2\right) = \frac{g}{2} y_2\left(y_1^2 - y_2^2\right) \Rightarrow V_1^2 = \frac{gy_2(y_1 + y_2)}{2y_1}$$

Substituting back,

$$h_L = (y_1 - y_2) + \frac{gy_2(y_1 + y_2)}{2y_1(2g)} - \frac{gy_2(y_1 + y_2)}{2y_1} \frac{y_1^2}{y_2^2(2g)}$$

$$\Rightarrow (y_1 - y_2) + \frac{(y_1 + y_2)(y_2^2 - y_1^2)}{4y_1 y_2}$$

$$\Rightarrow (y_1 - y_2) + \frac{(y_1 + y_2)(y_2 - y_1)(y_2 + y_1)}{4y_1 y_2}$$

$$\Rightarrow \frac{(y_1 - y_2)}{4y_1 y_2}\left\{4y_1 y_2 - (y_1 + y_2)^2\right\} \tag{13.40a}$$

$$h_L = -\frac{(y_1 - y_2)(y_1 - y_2)^2}{4y_1 y_2} = \frac{(y_2 - y_1)^3}{4y_1 y_2} \tag{13.40b}$$

13.12 DRAG, LIFT, FLOW SEPARATION, WAKE AND TERMINAL VELOCITY

13.12.1 Drag

When a fluid flows past a solid body, it produces a force that tends to drag it in the direction of flow. It acts in a direction opposite to the direction of relative flow velocity. It is known as drag and takes two forms, skin friction drag and form drag. In unsymmetrical flow, there can also be a lift. The lift force acts in a direction normal to the direction of flow and not necessarily in the vertical direction. The drag force is affected by cross-section area (form drag) and by surface smoothness (surface drag).

Skin friction drag sometimes referred to as surface drag, refers to the frictional resistance due to viscous effects between the solid surface and the fluid layer in contact with the solid surface. Skin friction drag occurs in solid bodies which have longer dimensions in the direction of flow compared to their heights normal to the direction of flow. In thin objects (height to length ratio small), the skin friction drag dominates. The shear stress acting between the solid surface and the moving fluid layer next to the solid surface is known as the 'wall shear stress', and the corresponding force is known as 'wall shear force'. From the surface of contact, the velocity of the fluid varies within the boundary layer until it reaches the free stream velocity according to some velocity profile. Outside the boundary layer, the velocity is assumed to be constant and equal to the free stream velocity. Skin drag force is due to forces tangential to the surface whereas form drag is due to forces normal to the surface. The force that propels the body against the drag force is known as the thrust which in automobiles, ships, airplanes, etc. is provided by the engine.

13.12.2 Lift

Lift force acts in a direction normal to the relative motion of the fluid but not necessarily in a vertical direction. It is created by the pressure difference on the upper and lower sides of an object due to fluid flow. Since velocity is inversely proportional to pressure according to Bernoulli equation, a higher velocity will generate a lower pressure and vice versa. An airplane wing exerts a force on air and deflects it downward. According to Newton's third law, the air exerts an upward reaction force creating lift. The larger the wing, the more air it can deflect and thus more lift. It also depends upon the velocity. The lift coefficient is defined as

$$\text{Lift coefficient, } C_L = \frac{\text{Lift force}}{\frac{1}{2}(\text{Dynamic pressure})(\text{Projected Area})} = \frac{F_L}{\frac{1}{2}\rho u_0^2 A} \tag{13.41}$$

For an airplane wing, the projected area is the product of the span of the wing and the mean chord. Since form drag is due to pressure only, the drag force if the body is inclined at an angle is defined as

$$F_D = \int_A p\cos\theta\, dA \tag{13.42}$$

Drag coefficient is dimensionless and therefore independent of the size of the body and the shape of the body. The pressure coefficient is defined as

$$C_P = \frac{p - p_0}{\frac{1}{2}\rho u_0^2} \tag{13.43}$$

where p is the pressure acting at any point on the surface, and p_0 is the static pressure in the undisturbed flow. It can also be related to the velocity differences between the undisturbed and disturbed flows by using Bernoulli equation:

$$\frac{p_0}{\rho g}+\frac{u_0^2}{2g}=\frac{p}{\rho g}+\frac{u^2}{2g}\Rightarrow C_P=1-\frac{u^2}{u_0^2} \tag{13.44}$$

Because of the difficulty in calculating the drag forces theoretically, they are usually obtained experimentally using the drag coefficient. For slow moving fluids ($Re < 1$) the drag force due to pressure differences and shear stresses for a sphere can be expressed as

$$\text{Drag}=C_D\frac{1}{2}\rho u_0^2 A;\quad\text{where } C_D=\frac{24}{Re}\quad\text{(Stokes flow)} \tag{13.45}$$

The Re in this case is defined with respect to the sphere diameter as the characteristic length.

13.12.3 Flow separation

Normally, in a flow, the shear stress caused by the viscosity of the fluid has a retarding effect on the flow. This is normally offset by a negative pressure gradient, i.e. the pressure decreasing in the downstream direction. A negative pressure gradient is known as a favourable pressure gradient, which enables the flow to continue. However, sometimes, the pressure gradient becomes positive and the condition is known as adverse pressure gradient. Under this condition the velocity decreases, and if the adverse pressure gradient persists, the velocity can become zero and even negative. When the velocity becomes negative, flow reversal takes place. A flat plate placed normal to the flow will lead to flow separation whereas the flow will be smooth if it is placed longitudinally along the flow direction.

One of the serious effects of an adverse pressure gradient is separation of flow. At the point of flow separation, the wall shear stress becomes zero. From the point of flow separation, the shear stress becomes negative and the flow becomes recirculating. The flow does not follow the contour of the body and hence the use of the term separation.

Sometimes, the flow may get re-attached to the contour of the body. Re-attachment is the opposite of separation. This usually happens in turbulent flow. In an areofoil, if the separation occurs towards the trailing edge, re-attachment may not occur. In this situation, the separated region merges with the wake resulting in loss of lift.

13.12.4 Wake

Wake is the flow region downstream of the body where the fluid flow is disturbed by the body. Depending on the body shape and smoothness of the surface, separation of flow can take place. If the body is smooth and streamlined, the pressure drag is small and the total drag consists of mainly the friction drag. Such a body is called a streamlined body. On the other hand, if the body shape causes substantial flow separation, the wake is large and the normal pressure drag is greater than the friction drag. Such a body is called a bluff body, which experiences more pressure drag compared to a streamlined body.

Skin drag force = Wall shear stress × wetted area

$$\text{Drag coefficient} = \frac{\text{Drag force}}{\text{Dynamic pressure} \times \text{wetted area}}$$

$$= \frac{\text{Drag force}}{\frac{1}{2}\rho u_0^2 (\text{wetted area})} \tag{13.46}$$

where u_0 is the undisturbed uniform velocity. Dynamic pressure is the pressure resulting from converting kinetic energy of flow to pressure energy. For a thin plate inclined at an angle θ to the horizontal, the drag force may be expressed as

$$F_D = \int_A \tau_w \cos\theta \, dA \tag{13.47}$$

where τ_w is the wall shear stress.

Form drag (sometimes referred to as pressure drag) applies to bodies with heights greater than their lengths. It is called form drag because it depends on the form of the body. There is positive pressure on the upstream side and negative pressure on the downstream side. At the stagnation points (both upstream and downstream) the velocity is zero, and there is a corresponding increase in pressure. The drag coefficient due to form drag is defined as

$$\text{Drag coefficient} = \frac{\text{Drag force}}{\text{Dynamic pressure} \times \text{projected area}}$$

$$= \frac{\text{Drag force}}{\frac{1}{2}\rho u_0^2 (\text{Projected area})} \tag{13.48}$$

The projected area refers to the area normal to the direction of flow (area exposed to flow).

13.12.5 Terminal velocity

When a body falls under gravity, a point is reached at which the gravity force (weight) is balanced by the drag force. Then there will be no acceleration and the velocity attains a steady value referred to as the terminal velocity. Small particles settling in liquids are modelled on this principle. Falling sphere viscometer is one example of the application of the principle of terminal velocity. The force balance equation is that the buoyant weight is equal to the viscous resistance force based on Stokes' law which is applicable for $Re < 0.2$. It is of the form

$$\frac{\pi D^3 (\rho_s - \rho) g}{6} = 3\pi D \mu V \tag{13.49}$$

where V is the terminal velocity of a sphere of diameter D and density ρ_s falling through a fluid of density ρ and viscosity μ. This equation assumes that the fluid is of infinite medium. From Eq. 13.49, the terminal velocity is given by

$$V = \frac{(\rho_s - \rho) g D^2}{18\mu} \tag{13.50}$$

The falling sphere method of determining the viscosity of a fluid is based on this equation, although the experiment is carried in a confined space.

For $0.2 < Re < 500$ the flow is called Allen flow and the following expression gives the empirical relationship between the drag coefficient and the Reynolds number.

$$C_D = 18.5 Re^{-0.6} \tag{13.51}$$

Drag coefficient as given by Eq. 13.43 is

$$C_D = \frac{\text{Drag force}}{\frac{1}{2}\rho u_0^2 (\text{Projected area})} = \frac{\dfrac{\pi D^3 (\rho_s - \rho) g}{6}}{\dfrac{1}{2}\rho V^2 \dfrac{\pi D^2}{4}} = \frac{8 Dg(\rho_s - \rho)}{6\rho V^2} \tag{13.52}$$

Equating Eq. 13.51 with Eq. 13.52,

$$\frac{8 Dg(\rho_s - \rho)}{6\rho V^2} = 18.5\left(\frac{\rho V D}{\mu}\right)^{-0.6} \tag{13.53}$$

The terminal velocity V can be calculated from Eq. 13.53 as

$$V = \left(\frac{8}{6 \times 18.5}\frac{D^{1.6}g(\rho_s - \rho)}{\rho^{0.4}\mu^{0.6}}\right)^{\frac{1}{1.4}} \tag{13.54}$$

For $500 < Re < 10^5$, which is referred to as Newton flow, C_D takes a constant value of 0.44. Substituting $C_D = 0.44$ in Eq. 13.52 gives

$$V = \sqrt{\frac{3.03 Dg(\rho_s - \rho)}{\rho}} \tag{13.55}$$

Example 13.1

Calculate the drag force on each side of a thin smooth plate of $2\,\text{m} \times 1\,\text{m}$ with the length parallel to the flow with a velocity of $30\,\text{m/s}$. Assume the density of fluid to be $800\,\text{kg/m}^3$ and viscosity to be $0.008\,\text{kg/ms}$. The drag coefficient for smooth surfaces may be assumed to be $0.074(\text{Re})^{-1/5}$.

$$Re = \frac{\rho U_0 l}{\mu} = \frac{800 \times 30 \times 2}{0.008} = 6 \times 10^6.$$

$$C_D = 0.074(6 \times 10^6)^{-1/5} = 0.00326$$

$$\text{Dynamic pressure} = \frac{1}{2}\rho U_0^2 = 0.5 \times 800 \times 30 \times 30 = 360 \text{ kPa}$$

$$\text{Wall shear stress} = C_D \times \text{Dynamic pressure} = 0.00326 \times 360 = 1.173 \text{ kPa}$$

$$\text{Drag force} = \text{Wall shear stress} \times \text{wetted area} = 1.173(2 \times 1) = 2.347 \text{ kN}$$

Example 13.2

A cylinder 80 mm in diameter and 200 mm long is placed in a flow field with a velocity of 0.5 m/s. The axis of the cylinder is normal to the direction of flow. The density of the fluid is 800 kg/m³. The drag force is measured and found to be 30 N.

i. Calculate the drag coefficient
ii. Calculate the velocity at a point where the pressure is 96 Pa above ambient pressure.

Projected area $= 0.08 \times 0.2 = 0.016 \text{ m}^2$

Dynamic pressure $= (1/2)\rho V^2 = 0.5 \times 800 \times 0.25 = 100$

$$\text{Drag coefficient} = \frac{\text{Drag force}}{\text{Dynamic pressure} \times \text{projected area}} = \frac{30}{100 \times 0.016} = 18.75$$

From Bernoulli equation,

$$\frac{p - p_0}{\rho g} = \frac{u_0^2 - u^2}{2g} \Rightarrow u = 0.1 \text{ m/s}$$

Example 13.3

Determine the force required to keep a pipe bend in place given the following data:
At entry: $r_1 = 25$ cm, $\theta_1 = 45°$, $p_1 = 8.5$ kPa; At exit: $r_2 = 15$ cm, $\theta_2 = 30°$, $p_2 = 5.83$ kPa; Fluid weight $= 2$ N; Flow rate $= 50$ l/s.

$$x: \quad p_1 A_1 \cos\theta_1 + B_x - p_2 A_2 \cos\theta_2 = \rho Q(V_2 \cos\theta_2 - V_1 \cos\theta_1)$$

$$y: \quad p_1 A_1 \sin\theta_1 - p_2 A_2 \sin\theta_2 + B_y - W = \rho Q(V_2 \sin\theta_2 - V_1 \sin\theta_1)$$

From the data given,

$$p_1 A_1 = 8,500\pi(0.25)^2 = 1,669 \text{ N}; \quad p_2 A_2 = 5,830\pi(0.15)^2 = 412.1 \text{ N}$$

$$V_1 = \frac{Q}{A_1} = 0.255 \text{ m/s}; \quad V_2 = \frac{Q}{A_2} = 0.707 \text{ m/s}$$

Substituting,

$$B_x = -801.5 \text{ N}; \quad B_y = -963.3 \text{ N}$$

Equal and opposite force components will be needed to keep the pipe bend in place.

Chapter 14

Turbulence

14.1 INTRODUCTION

In the context of fluid mechanics, there are several definitions of turbulence but none quite precise. The earliest definition is perhaps after Osborne Reynolds (1883), who described it as a kind of 'sinuous motion'. Subsequently, Taylor (1937) and von Kármán (1937) defined it as 'Turbulence is an irregular motion which in general makes its appearance in fluids, gaseous or liquid, when they flow past solid surfaces or even when neighbouring streams of the same fluid flow past or over one another'. In the book by Hinze (1959), he defines it as 'Turbulent fluid motion is an irregular condition of flow in which the various quantities show a random variation with time and space coordinates, so that statistically distinct average values can be discerned'. Some characteristics of turbulence include three-dimensional unsteady behaviour, disorder, chaotic, seemingly random but deterministic, non-repeatability, sensitive to initial conditions, extremely large range of space and time scales, and enhanced mixing and dissipation.

Almost all types of flows in nature are turbulent, some to a lesser degree than others. Laminar flows occur in rare, controlled conditions only. Turbulent flow in nature include the mixing of milk when stirred in a coffee cup, blood flow in arteries due to complexity of geometry and pulsatality of the flow, flow around a wall in a channel, boundary layer, rotation and vorticity, smoke coming out of a chimney, waves, flow of a fluid around an object, and, motion of an object such as an automobile, train, ship, airplane, rocket etc. in a fluid (liquid and/or gas). In large scale systems, such as in storms and cyclones, the Coriolis force needs to be considered. The dissipation of energy is characterized by the diffusion and mixing of mass (matter), momentum, and energy at a rapid rate compared to laminar flows. The physical complexity of the problem increases when the effects of compressibility and combustion are taken into consideration, and in multi-phase analysis. Turbulence is an area of fluid mechanics which is not fully understood at the present time, but useful analysis and predictions can still be made using physical intuition, dimensional analysis, numerical techniques and empirical methods. Turbulence, though seemingly random, is fully deterministic and obeys the laws of conservation of mass, momentum and energy and hence the Navier–Stokes equations. The dependent variables involved in fluid mechanics include velocities, pressures, and temperatures which appear to fluctuate with the independent variables in space and time. Such fluctuations can be irregular, chaotic and usually unpredictable. The dependent variables in turbulent flow are usually represented as the sum of a mean value and a fluctuating component and their variations are expressed in terms of statistical parameters (or moments) such as the variance, kurtosis, etc. Turbulent flow is necessarily non-linear with Reynolds numbers (or other similar dimensionless numbers) exceeding critical values.

14.2 OVERVIEW OF THE HISTORICAL DEVELOPMENT OF THE UNDERSTANDING OF TURBULENCE

The first scientist to embark on turbulent flow studies through experiments on pipe flow was Osborne Reynolds (1883). In his pipe flow experiments, he observed the flow pattern to change from regular motion to irregular or turbulent flow when the Reynolds number exceeded a certain critical value. Reynolds number $\left(Re = \dfrac{VD\rho}{\mu} = \dfrac{VD}{\nu} \right)$ increases with increasing size of the object characterized by D, increasing velocity of flow past the object characterized by V, increasing density of the fluid ρ and decreasing viscosity of the fluid, μ. Reynolds also introduced the concept of turbulent stress which together with Reynolds number now forms the basis of modern day understanding of turbulence. He also introduced the concept of describing the instantaneous values of state variables as a sum of a mean and a fluctuating component, which is known and widely used as Reynolds' decomposition. Boussinesq (1877) hypothesized that turbulent stresses are linearly proportional to the large scale mean strain rates. Prandtl (1925) introduced the mixing length theory and the logarithmic velocity profile near a solid wall, Taylor (1921, 2000) introduced the idea of presenting turbulence in statistical terms such as correlations, Fourier transforms and power spectra, as well as the concept of mixing length, and subsequently (1935–1936) the statistical theory of turbulence. Richardson (1922) introduced the concept of energy cascade, which was followed by Kolmogorov (1941), who postulated that the statistics of small scales are isotropic and uniquely determined by the length scale, l, the kinematic viscosity, ν, and an average rate of kinetic energy dissipation per unit mass, ε. Kolmogorov subsequently presented a statistical approach of describing turbulence. Lorenz (1963) proposed a link between chaos and turbulence.

14.3 SOME DEFINITIONS

- **Turbulent (eddy) viscosity:** Eddy viscosity, the constant of proportionality between turbulent (Reynolds) stresses and mean (large-scale) strain rate, is analogous to physical (molecular) viscosity in Newton's law of viscosity.

- **Turbulent intensity:** Turbulent intensity, I, is defined as $I = \dfrac{\overline{u'}}{U}$, where $\overline{u'}$ is the root-mean-square of the fluctuating component of velocity and U is the mean velocity. $\overline{u'}$ is computed from

$$\overline{u'} = \sqrt{\frac{1}{3}\left(u'^2 + v'^2 + w'^2 \right)} = \sqrt{\frac{2}{3}k} \tag{14.1}$$

  where k is the turbulent kinetic energy per unit mass. The mean velocity U is calculated as

$$U = \sqrt{\overline{u}^2 + \overline{v}^2 + \overline{w}^2} \tag{14.2}$$

  where $\overline{u}, \overline{v}, \overline{w}$ are the time-averaged values of u, v, w, the instantaneous values of velocities in three mutually perpendicular directions.
- **Turbulence strength:** Turbulence strength is the root mean square of the fluctuating component of velocity defined as

$$u_{\text{rms}} = \sqrt{u'(t)^2} \tag{14.3}$$

- **Turbulent kinetic energy:** Turbulence kinetic energy per unit mass, k, is the kinetic energy calculated with turbulent fluctuating velocities.
- **Turbulent rate of dissipation:** Turbulence energy dissipation rate per unit mass, ε, is the rate at which turbulence energy is converted to thermal energy by viscous effects on small scales.
- **Reynolds stress:** Reynolds stress is the component of stress in a flowing fluid due to turbulence. It arises from the averaging procedure used in deriving the Reynolds Averaged Navier–Stokes (RANS) equations.
- **Mixing length:** Mixing length is the distance over which a hypothesized turbulent eddy retains its identity.
- **Energy cascade:** Energy cascade refers to the transfer of kinetic energy from large macroscopic scales of motion, through successively smaller scales, ending with viscous dissipation and conversion to heat (thermal energy).
- **Vorticity:** Vorticity is a mathematical entity that is not a measurable physical quantity. It gives a measure of the local rotation and is defined as

$$\text{Vorticity, } \Omega = \frac{\text{Circulation}}{\text{Area}} = \frac{\partial v}{\partial x} - \frac{\partial u}{\partial y} \tag{14.4}$$

Circulation is a scalar integral quantity, whereas vorticity is a vector field. It can be considered as the amount of force that pushes along a closed boundary or path such as a circle. Circulation and vorticity are the two primary measures of rotation in a fluid. Circulation is a macroscopic measure of rotation for a finite area of the fluid, whereas vorticity is a microscopic measure of rotation at any point.

Circulation is defined as the line integral of velocity along a closed path as

$$\Gamma = \oint q_s \, ds \tag{14.5}$$

where q_s denotes the component of velocity along an element of the fluid. It is the Curl[1] of the velocity field.

$$\text{Circulation } \Gamma = u\delta x + \left(v + \frac{\partial v}{\partial x} \delta x \right) \delta y - u\delta y - \left(u + \frac{\partial u}{\partial y} \delta y \right) \delta x$$

$$= \frac{\partial v}{\partial x} \delta x \delta y - \frac{\partial u}{\partial y} \delta y \delta x \tag{14.6}$$

For irrotational flow, both circulation and vorticity are zero.

---

[1] Curl is the cross product of a vector field defined as

$$\nabla \times U = \begin{vmatrix} j & j & k \\ \dfrac{\partial}{\partial x} & \dfrac{\partial}{\partial y} & \dfrac{\partial}{\partial z} \\ u & v & w \end{vmatrix} = \left(\frac{\partial w}{\partial y} - \frac{\partial v}{\partial z} \right) i + \left(\frac{\partial u}{\partial z} - \frac{\partial w}{\partial x} \right) j + \left(\frac{\partial v}{\partial x} - \frac{\partial u}{\partial y} \right) k$$

If $\nabla \cdot U = 0$, then the vector U is called solenoidal

If $\nabla \times U = 0$, then U is irrotational.

14.4 PHYSICAL PROCESS

Turbulence can be generated by frictional forces at boundaries such as in pipe flow, and, flow past solid bodies, etc. which are identified as 'wall turbulence'. The flow of layers of fluids flowing one over the other at different velocities is identified as 'free turbulence'. When there is a gradient in the mean velocity in the direction of motion due to viscosity, which is associated with a shear stress, it is called 'shear-flow turbulence'. Some initial perturbations such as variations in flow, roughness in boundaries, changes in fluid density and external mechanical vibrations, etc. are necessary to start turbulence.

The physical process of turbulence is three dimensional, unsteady, contains stretching of vortices that increases the intensity of turbulence, stirring, mixing, diffusion and dissipation. Kinetic energy is reduced by viscosity. Mixing is necessary for chemical reactions and heat transfer, but the downside is that increased mixing of momentum causes increased friction and thereby increasing the power required to pump the flow or the drag force on a moving object. Momentum transfer takes place at microscopic scale due to viscosity as well as at a macroscopic scale due to advection.

Turbulence in flow is caused by excessive kinetic energy in some parts of flow which tend to overcome the effects of fluid resistance due to viscosity. As a result, it is easier for turbulence to occur in less viscous fluids than in high viscous fluids. The basic features of turbulent flow are unsteadiness, irregularity, diffusivity, rotationality and dissipation. Irregularity comes from the chaotic nature of turbulent flow. Diffusivity comes from mixing and is characterized by a turbulent diffusion coefficient which is different from the molecular diffusion coefficient, and which is much larger than the latter. Rotationality comes from the vortex formation in the flow. Turbulent flows have non-zero vorticity, which tend to stretch along the flow direction, forming a cascade of eddies from larger sizes to smaller sizes. Energy dissipation takes place at the lower scale, where the kinetic energy is transformed into internal energy (heat). The large eddies are usually unstable and eventually break up into smaller eddies while transferring their kinetic energy to smaller eddies in the form of cascades until small-scale eddies reach a sufficiently small length scale where the viscosity of the fluid can dissipate the kinetic energy into internal energy (heat).

Turbulence causes mixing at a faster rate than molecular diffusion. Advective transport is much greater than diffusive transport. If kinematic viscosity is small, advective non-linear transport is dominant; if not molecular diffusion of momentum is dominant. Turbulence is always associated with high Reynolds numbers and includes low momentum diffusion, high momentum advection and rapid variation of pressure and velocity spatially and temporally. At low Reynolds numbers, external disturbances are usually dampened by the viscous resistance tending to resemble laminar flow.

Vortex stretching mechanism transfers fluctuating energy and vorticity to smaller and smaller scales via non-linear interactions until velocity gradients become so large that the energy is converted to heat. The stretching is in the direction of motion, implying that vortices will become thinner in the direction normal to flow. Continued turbulence requires a continuous supply of energy to make up for the loss of energy.

Turbulence consists of eddies (vortices) of different sizes and strengths which change with time thereby giving a random appearance. The characteristic lengths of such eddies (represented by the diameters) and the average orbital velocities are important measures of turbulence. For different sizes of eddies, these two variables change with time. By assuming homogeneous and isotropic conditions, the velocities of the same scale eddies are considered to have the same value. In other words, it is assumed that the velocity is a function of the length scale.

Energy is supplied by external forces, and turbulence is suppressed by viscosity. Flow of energy is from large scale to smaller scales in a cascade manner up to the smallest small scale when the energy is dissipated by viscosity. The Reynolds numbers for large scale eddies are high since the viscous effects are negligible. At the large-scale end, the turbulence is driven by inertial forces. At the lowest end of the eddy size spectrum, viscous effects come into force, Reynolds number becomes small and energy dissipation begins to take place. The process of passing down the energy from the largest scale to the smallest scale sequentially is known as the energy cascade, a concept suggested by Richardson (1922).

14.5 SCALES (ORDER OF MAGNITUDE OF THE PARAMETERS) OF TURBULENCE

There are many scales of turbulence because of the existence of different sizes of eddies. Large eddies contain most of the turbulent kinetic energy, and their scales are of the same size as the confinement of the flow. Energy is transferred from the large eddies to small eddies in a cascade until they are dissipated at the lowest scale and converted to thermal energy. The difficulty in turbulence modelling is to capture the contributions from all the scales. The computational effort is proportional to Re_L^3 where Re_L is the Reynolds number with the characteristic length equal to the largest length scale L. It is also reported that the total number of degrees of freedom needed to account for all scales is of the order of $Re_l^{9/4}$ for three-dimensional flows for a single time step and Re_l^2 for two-dimensional flows (McDonough, 2007, p. 51; Lesieur, 1997, p. 375) where Re_l is the Reynolds number based on the integral length scale l.

In turbulent flow, there exists a broad spectrum of eddy sizes. The largest turbulent length scale L is a physical quantity that describes the size of the large eddies that contain the energy in turbulent flow. It can be related to the physical size of the problem and therefore assigning a value to it is relatively easy. It should not be larger than the physical size of the problem. The largest scale extends from boundary to boundary in a confined flow. In river flows, the depth is very much smaller than the width, which is very much smaller than the length. In atmospheric motions, height is very much smaller than the horizontal distance. It is also important to recognize the difference between longitudinal and transverse eddies as they may rotate in the vertical direction (about a horizontal axis), or in a horizontal direction (about a vertical axis). Most eddies are, however, very much smaller in size.

There are four length scales widely used in describing turbulence flows. They are the large scale L, the integral scale l, the Taylor micro scale, λ and the Kolmogorov small scale, η. These are briefly described below.

14.5.1 Large scale

For the calculation of Reynolds number, the characteristic velocity U can be the mean velocity or the free stream velocity, and the characteristic length L which is the physical size of the problem such as for example the pipe diameter, depth of flow, chord length in aerodynamics, etc. The Reynolds number is then

$$Re_L = \frac{UL}{v} \tag{14.7}$$

The convective time scale is

$$t_c = \frac{L}{U}, \tag{14.8}$$

and the diffusive time scale is

$$t_d = \frac{L^2}{v}. \tag{14.9}$$

The ratio of these two possible time scales is

$$\frac{t_d}{t_c} = \frac{L^2}{v}\frac{U}{L} = \frac{UL}{v} = Re_L \tag{14.10}$$

Re_L can therefore be interpreted as the ratio of the diffusive (microscopic) time scale to the convective (macroscopic) time scale, in addition to the interpretation as the ratio of the inertia force to the viscous force. If Re_L is large, the diffusive time is very much longer than the convective time, and hence the diffusive transport effects are negligible.

14.5.2 Integral scale

The integral scale is taken as a fraction of the largest scale. Taylor (2000) defined this length scale using a statistical approach, which is also sometimes referred to as the Eulerian length scale, as

$$l = \int_0^\infty R_y \, dy \tag{14.11}$$

where R_y is the spatial correlation of velocities of two points separated by a distance y. The upper limit of the above integration can be replaced by the eddy size since the correlation will be zero outside the eddy. This is similar to the definition of the Lagrangian length scale. Similarly, the integral time scale can be defined as

$$t = \int_0^\infty R_\tau \, d\tau \tag{14.12}$$

where R_τ is the temporal correlation of velocities of two points separated by a temporal distance τ. Here, the velocity needed for calculating the autocorrelation is the turbulent fluctuating velocity u', which is different from the fluctuating component defined in the RANS equation and the Large Eddy Simulation (LES) equation. It is somewhere between RANS u' and LES u'. In complex turbulent flows, l is a function of space and time.

Integral scale Reynolds number is

$$Re_l = \frac{|u'|l}{v} \tag{14.13}$$

where u' is defined as

$$|u'| = k^{1/2} = \left\{ \frac{1}{3}(u'^2 + v'^2 + w'^2) \right\}^{1/2} \tag{14.14}$$

which is sometimes referred to as turbulent intensity (see Eq. 14.1).

14.5.3 Taylor micro-scale

Taylor micro-scale length is given as (Lesieur, 1997, p. 185)

$$\lambda = \left(\frac{\nu \left\langle \left| u' \right|^2 \right\rangle}{\varepsilon} \right)^{1/2} \qquad (14.15)$$

and the time scale as

$$t_\lambda = \frac{\lambda}{\left| u' \right|} \qquad (14.16)$$

and the Reynolds number as

$$\mathrm{Re}_\lambda = \frac{\left| u' \right| \lambda}{\nu}. \qquad (14.17)$$

14.5.4 Kolmogorov small scale

Small-scale flows have relatively small time scales. It is reasonable to assume that small-scale motions are statistically independent of large-scale turbulence. With this assumption the small-scale length η depends only on the rate at which energy is supplied and on the kinematic viscosity of the fluid. The rate at which energy is supplied at the large scale should be equal to the rate of dissipation of energy at the small scale provided there is no significant accumulation or destruction of energy. Therefore, the rate of energy supply should be equal to the rate of energy dissipation, ε. This is what is referred to as Kolmogorov's universal equilibrium theory applicable to small-scale turbulence.

Based on dimensional analysis, the rate of kinetic energy dissipation per unit mass passed down from the largest scale via the cascade in functional form is (Davidson, 2015, p. 25)

$$\sim \frac{u'^2}{(l / u')} = \frac{u'^3}{l} \qquad (14.18)$$

The rate of energy dissipation ε per unit mass is given by

$$\varepsilon = \nu S_{ij} S_{ij} \qquad (14.19)$$

where S_{ij} is the rate of strain associated with the smallest eddies. The rate of strain using Einstein notation[2] can be expressed as

$$S_{ij} = \frac{1}{2} \left(\frac{\partial u_i}{\partial x_j} + \frac{\partial u_j}{\partial x_i} \right) \qquad (14.20)$$

---

[2] $y = c_i x_i = \sum_{i=1}^{3} c_i x_i = c_1 x_1 + c_2 x_2 + c_3 x_3$ Rules: Repeated indices are implicitly summed over; each index can appear up to two in any term; each term must contain identical non-repeated indices.

Therefore, from a dimensional point of view, ϵ may be expressed in functional form as

$$\varepsilon \sim v\left(\frac{v}{\eta}\right)^2 \tag{14.21}$$

where v is the small-scale velocity. Equating these two functional forms gives

$$\frac{u^3}{l} \sim v\frac{v^2}{\eta^2}. \tag{14.22}$$

Combining Eqs. 14.18, 14.21 and 14.22, and the fact that the Reynolds number at this scale is of the order of unity, the following relationships can be obtained (Davidson, 2015, p. 25):

$$\eta \sim l\,Re^{-3/4}, \quad \text{or,} \quad \eta = \left(\frac{v^3}{\varepsilon}\right)^{1/4} \tag{14.23}$$

$$v_\eta \sim u\,Re^{-1/4} \quad \text{or,} \quad v_\eta = (v\varepsilon)^{1/4} \tag{14.24}$$

where Re is the Reynolds number for the large-scale eddies given as $Re = \frac{ul}{v}$, and the time scale from the velocity and length scales by

$$t_\eta = \left(\frac{v}{\varepsilon}\right)^{1/2} \tag{14.25}$$

From these, it can be seen that the Reynolds number in the Kolmogorov small scale is $Re_\eta = \frac{\eta v}{v} = 1$, implying small-scale motion is highly viscous.

Substituting for η and v, it can also be seen that

$$\frac{l}{\eta} \sim Re_l^{3/4} \tag{14.26}$$

$$\frac{\lambda}{\eta} \sim Re_\lambda^{1/2} \tag{14.27}$$

Typical magnitudes of these length scales in laboratory experiments have been reported to be of the orders of 4 cm for the integral scale, 2 mm for the Taylor micro-scale and 0.1 mm for the Kolmogorov scale (Lesieur, 1997, p. 185).

An approximate relationship between the integral length scale l and the physical size of the confinement (largest scale) is (http://jullio.pe.kr/fluent6.1/help/html/ug/node178.htm)

$$l = 0.07D \tag{14.28}$$

where D is the diameter for circular pipes and hydraulic diameter D_h for non-circular pipes. The hydraulic diameter is given as

$$D_h = 4\frac{\text{Cross sectional area}}{\text{Wetted perimeter}} \tag{14.29}$$

which, for circular pipes, is equal to the pipe diameter. The factor of 0.07 is based on the maximum value of the mixing length in fully developed turbulent pipe flow. It is normally taken as a percentage of the characteristic dimension of the problem. For fully developed pipe flow, it is taken as 3.8% of the hydraulic diameter, which is the same as the geometrical diameter (https://www.cfd-online.com/Wiki/Turbulence_length_scale). It can also be determined using hot-wire anemometry and other velocity measurements (O'Neill et al., 2004; El-Gabry et al., 2014).

There are also other empirical relationships to calculate the integral length scale. For example, in $k-\varepsilon$ models, it is calculated as (Versteeg and Malalasekera, 2007, p. 70)

$$l = \frac{k^{3/2}}{\varepsilon} \tag{14.30}$$

where k is the turbulence kinetic energy per unit mass and ε is the turbulence kinetic energy dissipation rate per unit mass. In fully developed confined flows such as in pipes, l is limited by the size of the confinement, since turbulent eddies cannot exceed the size of the confinement. Another relationship for estimating l, such as in some CFD codes, is (Versteeg and Malalasekera, 1995, p. 72)

$$l = C_\mu^{3/4} \frac{k^{3/2}}{\varepsilon} \tag{14.31}$$

where C_μ is a model constant normally taken as 0.09.

The Kolmogorov micro-scale (Kolmogorov, 1941) is based on the hypothesis that for very high Reynolds numbers, the small-scale turbulence is isotropic. In general, the large scales of a flow are not isotropic, since they are determined by the particular geometrical features of the boundaries. Kolmogorov also hypothesized that the statistics of small scales are universal (implying independent of flow geometry) and uniquely determined by two parameters, the kinematic viscosity ν, and the rate of energy dissipation ε. They are the same for all turbulent flows at high Reynolds numbers. For large Reynolds numbers, the largest and the smallest length scales can differ by several orders of magnitude.

The Taylor micro-length scale lies somewhere between the largest and the smallest length scales, and the region covered within these two scales is known as the inertial sub-range. In this scale, the viscosity affects the dynamics of turbulent flow, whereas for length scales greater than the Taylor micro-scale, the dynamics are not strongly affected by viscosity. Taylor's micro-length scale length, λ, is proportional to $Re_l^{-1/2}$, whereas the Kolmogorov length scale is proportional to $Re_l^{-3/4}$.

The three length scales can be represented as

$$l \propto \frac{k^{3/4}}{\varepsilon} \quad \text{for integral length scale,} \tag{14.32}$$

$$\lambda \propto \left(\frac{15\nu u'^2}{\varepsilon} \right)^{1/2} \quad \text{for the Taylor micro-length scale, and,} \tag{14.33}$$

$$\eta \propto \left(\frac{\nu^3}{\varepsilon} \right)^{1/4} \quad \text{for the Kolmogorov length scale} \tag{14.34}$$

For large Reynolds numbers, $l \geq \lambda \geq \eta$.

The ratios of these scales can therefore be expressed as

$$\frac{\text{Taylor microscale}}{\text{Integral scale}} = \frac{\lambda}{l} \sim \sqrt{15} Re_l^{-1/2} \tag{14.35}$$

$$\frac{\text{Kolmogorov micro scale}}{\text{Taylor micro scale}} = \frac{\eta}{\lambda} \sim \frac{Re_l^{-1/4}}{\sqrt{15}} \tag{14.36}$$

$$\frac{\text{Kolmogorov micro scale}}{\text{Integral scale}} = \frac{\eta}{l} \sim Re_l^{-3/4} \tag{14.37}$$

$$\frac{Re_\lambda}{Re_l} = \frac{\lambda}{l} \tag{14.38}$$

14.6 GOVERNING EQUATIONS OF MOTION

The Navier–Stokes equations form the basis of all fluid flows. They are based on the assumption that the fluid is a continuum and that all the fields of interest such as pressure, velocity, density and temperature are differentiable. They are derived using the basic laws of physics, namely the conservation of mass, momentum and energy which are defined for a system but which can be transformed to a control volume. For an incompressible Newtonian fluid, the continuity and momentum equations respectively take the form

$$\frac{\partial u}{\partial x} + \frac{\partial v}{\partial y} + \frac{\partial w}{\partial z} = 0 \tag{14.39}$$

$$\rho\left(\frac{\partial u}{\partial t} + u\frac{\partial u}{\partial x} + v\frac{\partial u}{\partial y} + w\frac{\partial u}{\partial z}\right) = -\frac{\partial p}{\partial x} + \mu\left(\frac{\partial^2 u}{\partial x^2} + \frac{\partial^2 u}{\partial y^2} + \frac{\partial^2 u}{\partial z^2}\right) + F_x \tag{14.40a}$$

$$\rho\left(\frac{\partial v}{\partial t} + u\frac{\partial v}{\partial x} + v\frac{\partial v}{\partial y} + w\frac{\partial v}{\partial z}\right) = -\frac{\partial p}{\partial y} + \mu\left(\frac{\partial^2 v}{\partial x^2} + \frac{\partial^2 v}{\partial y^2} + \frac{\partial^2 v}{\partial z^2}\right) + F_y \tag{14.40b}$$

$$\rho\left(\frac{\partial w}{\partial t} + u\frac{\partial w}{\partial x} + v\frac{\partial w}{\partial y} + w\frac{\partial w}{\partial z}\right) = -\frac{\partial p}{\partial z} + \mu\left(\frac{\partial^2 w}{\partial x^2} + \frac{\partial^2 w}{\partial y^2} + \frac{\partial^2 w}{\partial z^2}\right) + F_z \tag{14.40c}$$

where u, v, w represent the instantaneous velocity components in three mutually perpendicular axes x, y and z; p, the pressure; μ, the viscosity and F_x, F_y and F_z, the body forces per unit volume in the three mutually perpendicular directions. In abbreviated form using Einstein notation, the Navier–Stokes equation can be written as

$$\rho\left\{\frac{\partial u_i}{\partial t} + \frac{\partial(u_j u_i)}{\partial x_j}\right\} = -\frac{\partial p}{\partial x_i} + \frac{\partial}{\partial x_j}\left(\mu\frac{\partial u_i}{\partial x_j}\right) \tag{14.41}$$

These equations (without the body force term) under certain assumptions can also be written as

$$\frac{\partial \rho}{\partial t} + \nabla \cdot (\rho u) = 0 \Rightarrow \nabla \cdot u = 0 \text{ for incompressible fluids} \tag{14.42}$$

$$\rho \frac{Du}{Dt} = -\nabla p + \mu \nabla^2 u \tag{14.43}$$

where

$$\frac{Du}{Dt} = \frac{\partial u}{\partial t} + u\nabla \cdot u \tag{14.44}$$

Different flow conditions lead to more simplified forms of these equations. For example, for one-dimensional irrotational incompressible flows with $F=0$, the equation becomes

$$\rho \frac{\partial u}{\partial t} + \rho u \cdot \nabla u = -\nabla p + \mu \nabla^2 u \tag{14.45}$$

For low Reynolds numbers, when the inertia terms are small and can be ignored, it becomes

$$\rho \frac{\partial u}{\partial t} = -\nabla p + \mu \nabla^2 u \tag{14.46}$$

For low Reynolds numbers with negligible inertia terms at low pressures, it becomes

$$\rho \frac{\partial u}{\partial t} = \mu \nabla^2 u \text{ which is the diffusion equation.} \tag{14.47}$$

For high Reynolds numbers, viscous effects can be ignored, which leads to the Euler equation for inviscid fluids:

$$\rho \frac{\partial u}{\partial t} + \rho u \cdot \nabla u = -\nabla p \tag{14.48}$$

With no pressure force, it becomes

$$\rho \frac{\partial u}{\partial t} + \rho u \cdot \nabla u = \mu \nabla^2 u \tag{14.49}$$

For steady incompressible flow,

$$\frac{\partial u}{\partial t} = 0 \tag{14.50}$$

Using the Einstein notation, the Navier–Stokes equation can be written as

$$\frac{\partial u_i}{\partial t} + \frac{\partial (u_i u_j)}{\partial x_j} = -\frac{1}{\rho} \frac{\partial p}{\partial x_i} + \frac{1}{\rho} \frac{\partial (2\mu S_{ij})}{\partial x_j} \tag{14.51}$$

where S_{ij} is as defined by Eq. 14.20.

The momentum equation can also be written as

$$\frac{\partial u_i}{\partial t} + u_j \frac{\partial u_i}{\partial x_j} = -\frac{1}{\rho} \frac{\partial p}{\partial x_i} + v \frac{\partial^2 (u_i)}{\partial x_i \partial x_j} \tag{14.52}$$

The general Navier–Stokes equations do not have any analytical solution. However, they can be solved numerically under various assumptions and using various numerical methods.

14.7 DIRECT NUMERICAL SIMULATION (DNS)

Direct numerical simulation (DNS) refers to solving the Navier–Stokes equations numerically by resolving all scales from the largest energy providing scale down to the smallest scale of viscous dissipation. The approach does not require any turbulence model to describe turbulent stresses. No physical assumptions and/or simplifications over and above those inherent in the Navier–Stokes equations are needed. The resulting four equations (continuity+momentum equations in three directions) for the four unknowns, u, v, w and p can be numerically solved. For compressible fluids, the density ρ and the temperature T must also be included. It can be thought of as a reasonable substitute to the exact analytical solution (although an exact solution does not exist) of Navier–Stokes equations. DNS gives the velocity and pressure fields from which other quantities of interest can be estimated. The downside of this approach is that it requires resolutions from the largest scale to the dissipating smallest scale, and the computations required are of the order of Re^3 or higher. Since the applicable Reynolds numbers are of the order of 10^4, the computational resource requirements are prohibitively high, particularly for problems involving high Reynolds numbers. To fully resolve all scales in the Navier–Stokes equations, there are restrictions on the grid spacing, Δx, Δy, Δz and the time step Δt. For relatively simple problems in simple domains with low Reynolds numbers, the DNS approach has been a valuable tool for studying various aspects of turbulence including statistical aspects. For problems where the primary interest is to determine the statistical properties of the large scale, numerical simulations with the spatial grid size not greater than three times the Kolmogorov length scale is sufficient. i.e., $\Delta x, \Delta y, \Delta z \leq 3\eta$ (Moin and Mahesh, 1998). For example, if the Kolmogorov length scale η is of the order of $0.1\,\text{mm}$ as in the case of laboratory experiments, $\Delta x, \Delta y, \Delta z \leq 0.3$ mm. To model the turbulence covering all scales in a hypothetical flow domain $1 \times 1 \times 1\,\text{m}^3$ would require $\dfrac{1 \times 1 \times 1}{0.3 \times 0.3 \times 0.3 \times 10^{-9}} \approx 4 \times 10^{10}$ grid points. For high Reynolds numbers, the DNS method can be used for simple flow problems with simple domains discretized with equal spacing. For high accuracy in DNS, spectral methods are preferred compared to finite difference methods as it is necessary to solve Poisson's equation to determine the pressure gradient.

The first attempts to carry out DNS were in the 1970s for wall-bounded flows followed by flows in fully developed plane and curved channels (Orszag and Patterson, 1972). They were for incompressible homogeneous flows in the longitudinal direction. The results have been mainly qualitative until the late 1980s when quantitative analysis of plane and curved channel flows have been carried out. A review of such studies and subsequent studies can be found in an article by Moin and Mahesh (1998).

Numerical problems in the DNS approach include the choice of the numerical method such as for example, finite volume method, finite difference methods, finite element methods, hybrid methods, etc., the imposition of boundary and initial conditions, time marching scheme (explicit or implicit schemes when finite differencing methods are employed), spatial discretization, among others. The time marching scheme is mostly explicit due to the high resolution needed.

In the DNS approach, the energy dissipation should be entirely due to viscosity. Some numerical schemes sometimes introduce certain types of artificial dissipation which should be avoided.

14.8 REYNOLDS AVERAGED NAVIER-STOKES (RANS) EQUATIONS

The Navier–Stokes equation (Eq. 14.40) can only be solved numerically at a very high computer cost to determine $u(x, y, z, t)$.

Even then, solutions can only be obtained for relatively simple problems with low Reynolds numbers. Instead of attempting to obtain instantaneous spatial values of the velocity field, it is more meaningful to seek the statistical characteristics of the velocity field, which engineers are more interested in. This is possible using an approach proposed by Reynolds in which the system variables are expressed as the sum of a time-averaged mean component and a fluctuating component. This approach is called Reynolds decomposition which leads to the RANS equations.

14.8.1 Derivation of RANS equation

The Navier–Stokes equations for an unsteady three-dimensional incompressible Newtonian fluid flow using Einstein notation are

$$\frac{\partial u_i}{\partial x_i} = 0 \tag{14.53}$$

$$\frac{\partial u_i}{\partial t} + u_j \frac{\partial u_i}{\partial x_j} = F_i - \frac{1}{\rho}\frac{\partial p}{\partial x_i} + v \frac{\partial^2 (u_i)}{\partial x_j \partial x_j} \tag{14.54}$$

Using Reynolds decomposition,

$$u(x,y,z,t) = \bar{u}(x,y,z) + u'(x,y,z,t) \tag{14.55a}$$

$$v(x,y,z,t) = \bar{v}(x,y,z) + v'(x,y,z,t) \tag{14.55b}$$

$$w(x,y,z,t) = \bar{w}(x,y,z) + w'(x,y,z,t) \tag{14.55c}$$

$$p(x,y,z,t) = \bar{p}(x,y,z) + p'(x,y,z,t) \tag{14.55d}$$

where $\bar{u}, \bar{v}, \bar{w}, \bar{p}$ are time-averaged values of u, v, w and p which are therefore independent of time. They are respectively defined as

$$\bar{u} = \frac{1}{T}\int_0^T u\,dt; \quad \bar{v} = \frac{1}{T}\int_0^T v\,dt; \quad \bar{w} = \frac{1}{T}\int_0^T w\,dt; \quad \bar{p} = \frac{1}{T}\int_0^T p\,dt \tag{14.56}$$

implying, their time derivatives are zero; i.e.

$$\frac{\partial \bar{u}}{\partial t} = \frac{\partial \bar{v}}{\partial t} = \frac{\partial \bar{w}}{\partial t} = \frac{\partial \bar{p}}{\partial t} = 0 \tag{14.57}$$

$$\bar{\bar{u}} = \bar{u}; \quad \overline{u'} = 0; \quad \overline{v'} = 0; \quad \overline{w'} = 0; \quad \overline{p'} = 0 \tag{14.58}$$

When these new variables are substituted into the Navier–Stokes equations, a set of equations known as the RANS equations is obtained. These equations are statistically equivalent to the original Navier–Stokes equations but contain more unknowns than the number of equations due to the non-linearity of the terms of the equations. This is known as the closure problem in turbulence. The RANS equations are derived as follows.

The continuity equation using Reynolds decomposition is

$$\frac{\partial \bar{u}}{\partial x} + \frac{\partial u'}{\partial x} + \frac{\partial \bar{v}}{\partial y} + \frac{\partial v'}{\partial y} + \frac{\partial \bar{w}}{\partial z} + \frac{\partial w'}{\partial z} = 0 \qquad (14.59)$$

Time-averaged continuity equation is (over-bar indicates time averaging)

$$\overline{\frac{\partial \bar{u}}{\partial x} + \frac{\partial u'}{\partial x} + \frac{\partial \bar{v}}{\partial y} + \frac{\partial v'}{\partial y} + \frac{\partial \bar{w}}{\partial z} + \frac{\partial w'}{\partial z}} = 0 \qquad (14.60)$$

Time averaging follows the following rules:

$$\overline{\frac{\partial \bar{u}}{\partial x}} = \frac{\partial \bar{u}}{\partial x}; \quad \overline{\frac{\partial u'}{\partial x}} = 0 \qquad (14.61)$$

Therefore, Eq. 14.59 when time-averaged becomes

$$\frac{\partial \bar{u}}{\partial x} + \frac{\partial \bar{v}}{\partial y} + \frac{\partial \bar{w}}{\partial z} = 0 \qquad (14.62)$$

The time-dependent Navier–Stokes equation ignoring the body force term is

$$\frac{\partial u}{\partial t} + u\frac{\partial u}{\partial x} + v\frac{\partial u}{\partial y} + w\frac{\partial u}{\partial z} = -\frac{1}{\rho}\frac{\partial p}{\partial x} + \frac{\mu}{\rho}\left(\frac{\partial^2 u}{\partial x^2} + \frac{\partial^2 v}{\partial y^2} + \frac{\partial^2 w}{\partial z^2}\right) \qquad (14.63)$$

Time-averaged Navier–Stokes equation in the x-direction is obtained first by transforming the convective terms on the left-hand side of Eq. 14.63 as follows:

$$u\frac{\partial u}{\partial x} + v\frac{\partial u}{\partial y} + w\frac{\partial u}{\partial z} = \frac{\partial}{\partial x}\left(u^2\right) + \frac{\partial}{\partial y}(uv) + \frac{\partial}{\partial z}(uw) - u\left(\frac{\partial u}{\partial x} + \frac{\partial v}{\partial y} + \frac{\partial w}{\partial z}\right)$$

$$= \frac{\partial}{\partial x}\left(u^2\right) + \frac{\partial}{\partial y}(uv) + \frac{\partial}{\partial z}(uw) \qquad (14.64)$$

Substituting the transformed convective terms and the Reynolds decomposition in the Navier-Stokes equation gives

$$\rho\left\{\frac{\partial}{\partial t}(\bar{u}+u') + \frac{\partial}{\partial x}(\bar{u}+u')^2 + \frac{\partial}{\partial y}(\bar{u}+u')(\bar{v}+v') + \frac{\partial}{\partial z}(\bar{u}+u')(\bar{w}+w')\right\}$$

$$= -\frac{\partial}{\partial x}(\bar{p}+p') + \mu\left\{\frac{\partial^2(\bar{u}+u')}{\partial x^2} + \frac{\partial^2(\bar{v}+v')}{\partial y^2} + \frac{\partial^2(\bar{w}+w')}{\partial z^2}\right\} \qquad (14.65)$$

Taking the time averages,

$$\frac{\partial\left(\overline{u^2}\right)}{\partial x} = \frac{\partial\overline{(\bar{u}+u')^2}}{\partial x} = \frac{\partial\overline{(\bar{u}^2 + 2\bar{u}u' + u'^2)}}{\partial x} = \frac{\partial\left(\bar{u}^2 + 0 + \overline{u'^2}\right)}{\partial x} \qquad (14.66a)$$

$$\frac{\partial\overline{(uv)}}{\partial y} = \overline{\frac{\partial}{\partial y}\left[(\bar{u}+u')(\bar{v}+v')\right]} = \frac{\partial}{\partial y}(\bar{u}\,\bar{v} + \overline{u'v'}) \qquad (14.66b)$$

$$\frac{\partial(uw)}{\partial z} = \frac{\partial}{\partial z}\left[(\bar{u}+u')(\bar{w}+w')\right] = \frac{\partial}{\partial z}\left(\bar{u}\,\bar{w} + \overline{u'w'}\right) \tag{14.66c}$$

The basis of the derivation of RANS equations is that the mean has to be time-independent. i.e. $\frac{\partial \bar{u}}{\partial t} = 0$.

Therefore, the first term on the left-hand side of Eq. 14.63 using Reynolds decomposition is

$$\frac{\partial u}{\partial t} = \frac{\partial \bar{u}}{\partial t} + \frac{\partial u'}{\partial t} = 0 + 0 = 0,\ \text{because the time averaged value of}\ u' = 0 \tag{14.67}$$

The mean value of a spatial derivative of a variable is equal to the corresponding spatial derivative of the mean value of that variable, i.e.

$$\overline{\frac{\partial u_i}{\partial x_j}} = \frac{\partial \bar{u}_i}{\partial x_j}; \quad \overline{\frac{\partial u'_i}{\partial x_j}} = \frac{\partial \bar{u}_i}{\partial x_j} \tag{14.68}$$

Equation 14.68 is also a basis of the derivation of RANS equations.

Since the time averages of u', v', w' are each zero, the time averages of $\bar{u}u'$, $\bar{u}v'$, $\bar{u}w'$ are also zero. Taking time averages,

$$\rho\left\{\frac{\partial(\bar{u}\,\bar{u})}{\partial x} + \frac{\partial\left(\overline{u'u'}\right)}{\partial x} + \frac{\partial(\bar{u}\,\bar{v})}{\partial y} + \frac{\partial\left(\overline{u'v'}\right)}{\partial y} + \frac{\partial(\bar{u}\,\bar{w})}{\partial z} + \frac{\partial\left(\overline{u'w'}\right)}{\partial z}\right\}$$

$$= -\frac{\partial \bar{p}}{\partial x} + \mu\left(\frac{\partial^2 \bar{u}}{\partial x^2} + \frac{\partial^2 \bar{u}}{\partial y^2} + \frac{\partial^2 \bar{u}}{\partial z^2}\right) \tag{14.69}$$

which can also be written as

$$\rho\left\{\bar{u}\frac{\partial(\bar{u})}{\partial x} + \bar{v}\frac{\partial(\bar{u})}{\partial y} + \bar{w}\frac{\partial(\bar{u})}{\partial z}\right\} = -\frac{\partial \bar{p}}{\partial x} + \mu\left(\frac{\partial^2 \bar{u}}{\partial x^2} + \frac{\partial^2 \bar{u}}{\partial y^2} + \frac{\partial^2 \bar{u}}{\partial z^2}\right)$$

$$-\rho\left\{\frac{\partial\left(\overline{u'u'}\right)}{\partial x} + \frac{\partial\left(\overline{u'v'}\right)}{\partial y} + \frac{\partial\left(\overline{u'w'}\right)}{\partial z}\right\} \tag{14.70}$$

Upon rearranging,

$$\rho\left[\frac{\partial\left(\bar{u}^2\right)}{\partial x} + \frac{\partial(\bar{u}\,\bar{v})}{\partial y} + \frac{\partial(\bar{u}\,\bar{w})}{\partial z}\right] = -\frac{\partial \bar{p}}{\partial x} + \mu\left(\frac{\partial^2 \bar{u}}{\partial x^2} + \frac{\partial^2 \bar{u}}{\partial y^2} + \frac{\partial^2 \bar{u}}{\partial z^2}\right)$$

$$-\left[\frac{\partial\left(\rho\overline{u'^2}\right)}{\partial x} + \frac{\partial\left(\rho\overline{u'v'}\right)}{\partial y} + \frac{\partial\left(\rho\overline{u'w'}\right)}{\partial z}\right] \tag{14.71}$$

The second term on the right-hand side of this equation represents the molecular viscous stresses, while the last term in the square brackets represents the Reynolds (turbulent) stresses.

Equation 14.71 can also be written as

$$\rho\left[\frac{\partial(\bar{u}^2)}{\partial x} + \frac{\partial(\bar{u}\,\bar{v})}{\partial y} + \frac{\partial(\bar{u}\,\bar{w})}{\partial z}\right]$$

$$= -\frac{\partial \bar{p}}{\partial x} + \left[\frac{\partial}{\partial x}\left(\mu \frac{\partial \bar{u}}{\partial x} - \rho\overline{u'^2}\right) + \frac{\partial}{\partial y}\left(\mu \frac{\partial \bar{u}}{\partial y} - \rho\overline{u'v'}\right) + \frac{\partial}{\partial z}\left(\mu \frac{\partial \bar{u}}{\partial z} - \rho\overline{u'w'}\right)\right] \qquad (14.72)$$

In abbreviated form, it can also be written as

$$\rho\frac{D\bar{u}_i}{Dt} = -\frac{\partial \bar{p}}{\partial x_i} + \mu\Delta\bar{u}_i - \rho\frac{\partial(\overline{u_i'u_j'})}{\partial x_j} \qquad (14.73)$$

The last term in Eq. 14.73 is the Reynolds stress.

$$\mu\Delta\bar{u}_i - \rho\frac{\partial\left(\overline{u_i'u_j'}\right)}{\partial x_j} = \mu\frac{\partial}{\partial x_j}\left(\frac{\partial u_i}{\partial x_j}\right) - \rho\frac{\partial}{\partial x_j}\left(\overline{u_i'u_j'}\right)$$

$$= \frac{\partial}{\partial x_j}\left(\mu\frac{\partial u_i}{\partial x_j} - \rho\overline{u_i'u_j'}\right) = \frac{\partial}{\partial x_j}\left(\tau_{ij}\right) \qquad (14.74)$$

The general RANS equation is

$$\rho\left(\frac{Du_i}{Dt}\right) = -\frac{\partial p}{\partial x_i} + \frac{\partial}{\partial x_j}\left(\tau_{ij}\right) + F_i \qquad (14.75)$$

where

$$\tau_{ij} = \mu\frac{\partial u_i}{\partial x_j} + \rho\left\{v_t\left(\frac{\partial u_i}{\partial x_j} + \frac{\partial u_j}{\partial x_i}\right) - \frac{2}{3}k\delta_{ij}\right\} \qquad (14.76)$$

The terms $\mu\frac{\partial \bar{u}}{\partial x}$, $\mu\frac{\partial \bar{u}}{\partial y}$, $\mu\frac{\partial \bar{u}}{\partial z}$ are the normal stresses due to molecular viscosity; the terms $\rho\overline{u'^2}$, $\rho\overline{u'v'}$, $\rho\overline{u'w'}$ inside the square brackets on the right-hand side of Eq. 14.72 are known as the Reynolds stresses. They are the components of stresses in a flowing fluid due to turbulence. The divergence of the Reynolds stresses with a negative sign represents convective momentum transfer due to the turbulent eddy motion.

Similar equations can be written for the other two directions. In summary, the RANS equations in the three directions are

$$\rho\left(\frac{\partial \bar{u}}{\partial t} + \bar{u}\frac{\partial \bar{u}}{\partial x} + \bar{v}\frac{\partial \bar{u}}{\partial y} + \bar{w}\frac{\partial \bar{u}}{\partial z}\right) = F_x - \frac{\partial \bar{p}}{\partial x} + \mu\nabla^2\bar{u} - \rho\left[\frac{\partial}{\partial x}\left(\overline{u'u'}\right) + \frac{\partial}{\partial y}\left(\overline{u'v'}\right) + \frac{\partial}{\partial z}\left(\overline{u'w'}\right)\right]$$

$$\rho\left(\frac{\partial \bar{v}}{\partial t} + \bar{u}\frac{\partial \bar{v}}{\partial x} + \bar{v}\frac{\partial \bar{v}}{\partial y} + \bar{w}\frac{\partial \bar{v}}{\partial z}\right) = F_y - \frac{\partial \bar{p}}{\partial y} + \mu\nabla^2\bar{v} - \rho\left[\frac{\partial}{\partial x}\left(\overline{u'v'}\right) + \frac{\partial}{\partial y}\left(\overline{v'v'}\right) + \frac{\partial}{\partial z}\left(\overline{v'w'}\right)\right]$$

$$\rho\left(\frac{\partial \bar{w}}{\partial t} + \bar{u}\frac{\partial \bar{w}}{\partial x} + \bar{v}\frac{\partial \bar{w}}{\partial y} + \bar{w}\frac{\partial \bar{w}}{\partial z}\right) = F_z - \frac{\partial \bar{p}}{\partial z} + \mu\nabla^2\bar{w} - \rho\left[\frac{\partial}{\partial x}\left(\overline{u'w'}\right) + \frac{\partial}{\partial y}\left(\overline{w'v'}\right) + \frac{\partial}{\partial z}\left(\overline{w'w'}\right)\right]$$

$$(14.77)$$

The third terms on the right-hand side of the above equations are, respectively,

$$\mu\left(\frac{\partial^2 \bar{u}}{\partial x^2}+\frac{\partial^2 \bar{u}}{\partial y^2}+\frac{\partial^2 \bar{u}}{\partial z^2}\right),\ \mu\left(\frac{\partial^2 \bar{v}}{\partial x^2}+\frac{\partial^2 \bar{v}}{\partial y^2}+\frac{\partial^2 \bar{v}}{\partial z^2}\right),\ \text{and,}\ \mu\left(\frac{\partial^2 \bar{w}}{\partial x^2}+\frac{\partial^2 \bar{w}}{\partial y^2}+\frac{\partial^2 \bar{w}}{\partial z^2}\right) \tag{14.78}$$

which, in abbreviated form using Einstein notation, is of the form

$$\rho\left[\frac{\partial \bar{u}_i}{\partial t}+\frac{\partial}{\partial x_j}\left(\bar{u}_j \bar{u}_i\right)\right]=\frac{\partial}{\partial x_j}\left[\mu\left(\frac{\partial \bar{u}_i}{\partial x_j}\right)-\rho\overline{u_j' u_i'}\right]-\frac{\partial \bar{p}}{\partial x_i} \tag{14.79}$$

This equation can be written in different forms. For example,

$$\rho\frac{D\bar{u}_i}{Dt}=F_i-\frac{\partial \bar{p}}{\partial x_i}+\mu\Delta^2 \bar{u}_i-\rho\left(\frac{\partial\left(\overline{u_i' u_j'}\right)}{\partial x_j}\right) \tag{14.80a}$$

$$\rho\frac{D\bar{u}_i}{Dt}=F_i-\frac{\partial \bar{p}}{\partial x_i}+\mu\frac{\partial}{\partial x_j}\left(\frac{\partial u_i}{\partial x_j}\right)-\rho\left(\frac{\partial\left(\overline{u_i' u_j'}\right)}{\partial x_j}\right) \tag{14.80b}$$

$$\rho\frac{D\bar{u}_i}{Dt}=F_i-\frac{\partial \bar{p}}{\partial x_i}+\frac{\partial}{\partial x_j}\left(\tau_{ij}\right) \tag{14.80c}$$

$$\frac{\partial \bar{u}_i}{\partial t}+\bar{u}_j\frac{\partial \bar{u}_i}{\partial x_j}=-\frac{1}{\rho}\frac{\partial \bar{p}}{\partial x_i}+\frac{1}{\rho}\frac{\partial \tau_{ij}}{\partial x_j} \tag{14.80d}$$

where

$$\tau_{ij}=\mu\frac{\partial u_i}{\partial x_i}-\rho\overline{u_i' u_j'} \tag{14.80e}$$

Reynolds stress has six components as shown below:

$$\tau_{ij}'=\tau_{ji}'=-\rho\overline{u_i' u_j'}=\rho\begin{vmatrix} \overline{u'^2} & \overline{u'v'} & \overline{u'w'} \\ \overline{v'u'} & \overline{v'^2} & \overline{v'w'} \\ \overline{w'u'} & \overline{w'v'} & \overline{w'^2} \end{vmatrix} \tag{14.81}$$

The diagonal terms in this equation represent the normal stresses, whereas the other terms are the shear stresses. In turbulent flows, the shear stresses are dominant, and they contribute to the transport of mean momentum:

$$\tau_{ij}=\overline{u_i}\,\overline{u_j}-\overline{u_i u_j} \tag{14.82}$$

As can be seen, the RANS equations add extra unknowns making the equations insolvable since the number of unknowns is more than the number of equations. This is known as the 'closure problem in turbulence'. Additional conditions are therefore needed to make the RANS equations solvable. Several approximate methods, commonly known as turbulence models, have been proposed in the past. Among them, the models (or hypotheses) briefly described in Chapter 15 are quite popular.

Although RANS equations are generally applicable to situations where the mean flow is statistically stationary, the method can be extended to unsteady mean flow (statistically non-stationary) conditions too. The resulting equations are known as Unsteady Reynolds Average Averaged Navier–Stokes (URANS) equations.

REFERENCES

Boussinesq, J. (1877): Essai sur la théorie des eaux courantes, *Mémoires présentés par divers savants à l'Académie des Sciences*, vol. 23, no. 1, pp. 1–680.

Davidson, P. A. (2015): *Turbulence: An Introduction for Scientists and Engineers*, (2nd Edition), Oxford University Press, Oxford, UK, 630 pp.

El-Gabry, L. A., Thurman, D. R. and Poinsatte, P. E. (2014): Procedure for determining turbulence length scales using hotwire anemometry, *NASA/TM—2014–218403*, December 2014, p. 10. https://ntrs.nasa.gov/search.jsp?R=20150000733 2018-07-04T01:02:30+00:00Z.

Hinze, J. O. (1959): *Turbulence: An Introduction to its Mechanism and Theory*, McGraw Hill Book Company, Inc, New York, Toronto, London, p. 586.

Kolmogorov, A. N. (1941): The local structure of turbulence in incompressible viscous fluid for very large Reynolds numbers, *Proceedings of the USSR Academy of Sciences (in Russian)*, vol. 30, pp. 299–303. Translated into English: Kolmogorov, A. N. (July 8, 1991), Translated by Levin, V.: The local structure of turbulence in incompressible viscous fluid for very large Reynolds numbers, *Proceedings of the Royal Society A*, vol. 434, pp. 9–13. doi: 10.1098/rspa.1991.0075. Archived from the original (PDF) on September 23, 2015.

Lesieur, M. (1997): *Turbulence in Fluids*, (3rd Revised and Enlarged Edition), Kluwer Academic Publishers, Dordrecht/Boston/London, p. 515.

Lorenz, E. N. (1963): Deterministic non-periodic flow, *Journal of the Atmospheric Sciences*, vol. 20, pp. 130–141.

McDonough, J. M. (2007): *Introductory Lectures on Turbulence, Physics, Mathematics and Modeling*, Departments of Mechanical Engineering and Mathematics, University of Kentucky, Kentucky, p. 174.

Moin, P. and Mahesh, K. (1998): *Annual Review of Fluid Mechanics*, Palo Alto, Santa Clara, CA, vol. 30, 539. pp. 1–33.

O'Neill, P. L., Nicolaides, D., Honnery, D. and Soria, J. (2004): Autocorrelation functions and the determination of integral length with reference to experimental and numerical data, *In 15th Australasian Fluid Mechanics Conference*, The University of Sydney, Sydney, Australia, 13–17 December 2004, pp. 1–4.

Orszag, S. A and Patterson, G. S., Jr. (1972): Numerical simulation of three-dimensional homogeneous isotropic turbulence, *Physical Review Letters*, vol. 28, no. 2. doi: 10.1103/Phys Rev Lett.28.76.

Prandtl, L. (1925): Bericht u¨ber Untersuchungen zur ausgebildeten Turbulenz, *Zeitschrift für Angewandte Mathematik und Mechanik*, vol. 5, pp. 136–139.

Reynolds, O. (1883): An experimental investigation of the circumstances which determine whether the motion of water shall be direct or sinuous, and of the law of resistance in parallel channels, *Philosophical Transactions of the Royal Society of London*, vol. 174, pp. 935–982, Published by: The Royal Society Stable. http://www.jstor.org/stable/109431.

Richardson, L. F. (1922): *Weather Prediction by Numerical Process*, Cambridge University Press, Cambridge. pp. xii+236.

Taylor, G. I. (1921): Diffusion by continuous movements. *Proceedings of the London Mathematical Society*, vol. 20, pp. 196–211.

Taylor, G. I. (1937): The statistical theory of isotropic turbulence, *Journal of the Aeronautical Sciences*, vol. 4, no. 8, pp. 311–315. doi: 10.2514/8.419.

Taylor, G. I. (2000): Statistical theory of turbulence (with notes by Kelly Hendrickson). https://www.seas.harvard.edu/brenner/taylor/handouts/taylor_stats/taylor_stats.html.

Versteeg, H. K. and Malalasekera, W. (1995): *An Introduction to Computational Fluid Dynamics: The Finite Volume Method*, Pearson Education, New York. 257 pp.

von Kármán, Th. (1937): The fundamentals of the statistical theory of turbulence, *Journal of the Aeronautical Sciences*, vol. 4, no. 4, pp. 131–188. doi: 10.2514/8.350.

Chapter 15

Turbulence modelling

15.1 INTRODUCTION

In view of the prohibitively high computing resources needed for direct numerical solution of the governing Navier–Stokes equations, several alternative approaches of modelling turbulence have emerged in recent years. They can be broadly classified into two types: models based on the Reynolds Averaged Navier–Stokes (RANS) equations and the Large Eddy Simulation (LES) approach. The RANS type models can be further sub-divided into several categories. They include

- Mixing length model (zero equation model)
- Spalart–Allmaras model (one equation model)
- k-ε type models (two equations models)
- Reynolds stress model (seven equations model)
- Algebraic stress model.

The number of equations indicates the additional number of partial differential equations that need to be solved. The first three are also known as first-order models, and they are based on the analogy between laminar and turbulent flows. The last two types belong to second-order models, which do not depend upon the Boussinesq assumption. The RANS equations describe the mean flow, but velocity fluctuations still appear in the convective acceleration term in the Navier–Stokes equation which is non-linear. This term in the equation is referred to as the Reynolds stress. It has to be expressed in terms of mean velocities and pressures without any reference to the fluctuating parts. This is the closure problem in turbulence. RANS approach is not predictive as it involves modelling the Reynolds stress.

In LES models, large eddies are explicitly calculated while small eddies are modelled using a subgrid scale. They are based on space-filtered equations and are time dependent.

15.2 BOUSSINESQ HYPOTHESIS

Many turbulence models use the Boussinesq (1877) hypothesis which can be expressed as

$$\tau_{ij} = \mu \varepsilon_{ij} = \mu \left(\frac{\partial u_i}{\partial x_j} + \frac{\partial u_j}{\partial x_i} \right) \text{ for molecular viscous stresses} \qquad (15.1a)$$

and

$$\tau_{ij} = -\rho\overline{u_i'u_j'} = \mu_t\left(\frac{\partial\overline{u}_i}{\partial x_j} + \frac{\partial\overline{u}_j}{\partial x_i} - \frac{2}{3}\frac{\partial\overline{u}_k}{\partial x_k}\delta_{ij}\right) - \frac{2}{3}\rho k\delta_{ij} \tag{15.1b}$$

for turbulent viscous stresses, which for incompressible fluids becomes

$$-\rho\overline{u_i'u_j'} = \mu_t\left(\frac{\partial\overline{u}_i}{\partial x_j} + \frac{\partial\overline{u}_j}{\partial x_i}\right) - \frac{2}{3}\rho k\delta_{ij} \tag{15.1c}$$

and, which is also written as

$$-\rho\overline{u_i'u_j'} = 2\mu_t S_{ij} - \frac{2}{3}\rho k\delta_{ij} \tag{15.1d}$$

where S_{ij} is the mean rate of strain, defined as $S_{ij} = \frac{1}{2}\left(\frac{\partial\overline{u}_i}{\partial x_j} + \frac{\partial\overline{u}_j}{\partial x_i}\right)$, k is the turbulent kinetic energy per unit mass, δ_{ij} is the Kronecker delta function and μ_t is the eddy viscosity ($ML^{-1}T^{-1}$). The kinematic eddy (turbulent) viscosity v_t is defined as $v_t = \frac{\mu_t}{\rho}$ (L^2T^{-1}), which can be estimated by a number of methods. Perhaps the mixing length model may be the simplest.

Although the eddy viscosity is not homogeneous, it is assumed to be isotropic. The objective of turbulence modelling is to predict the eddy viscosity μ_t. The Boussinesq hypothesis is approximate but used in many turbulence models.

15.3 FIRST-ORDER MODELS

15.3.1 Mixing length model

Mixing length model does not use a new partial differential equation. It is sometimes referred to as a zero equation model or algebraic model. Using dimensional analysis, the eddy viscosity is expressed as a product of a turbulent velocity scale and a length scale as

$$v_t(L^2T^{-1}) \propto V(LT^{-1})l(L) \tag{15.2}$$

Assuming that the turbulent velocity scale can be expressed as

$$V \propto l\frac{\partial\overline{u}}{\partial y} \tag{15.3}$$

the eddy viscosity can be then be expressed as

$$v_t = l^2\left|\frac{\partial\overline{u}}{\partial y}\right| \tag{15.4}$$

which is Prandtl's (1925) mixing length model. This model is simple and easy to use and works well for simple flows. However, it is unable to take care of variations in the turbulent length scale, flow separation and circulation. It only calculates mean flow and turbulent shear stress. It is not used in most commercial CFD software as there are better models available. Typical values of mixing lengths based on the study by Rodi (1980) are given by Versteeg and Malalasekera (1995; Table 3.3). For example, for jets it is $0.09L$ where L is the jet half width. Turbulent Reynolds stress is given by

$$\tau_{xy} = \tau_{yx} = -\rho \overline{u'v'} = -\rho l^2 \left|\frac{\partial \bar{u}}{\partial y}\right| \frac{\partial \bar{u}}{\partial y}. \tag{15.5}$$

In general, the stress tensor consists of nine components in three dimensions of which the three diagonal terms are the normal stresses whereas the remaining six are the shear stresses as shown below:

$$\tau_{ijk} = \begin{bmatrix} \tau_{xx} & \tau_{xy} & \tau_{xz} \\ \tau_{yx} & \tau_{yy} & \tau_{yz} \\ \tau_{zx} & \tau_{zy} & \tau_{zz} \end{bmatrix} \tag{15.6}$$

Similarly, the rates of strain components are given as

$$e_{ijk} = \begin{bmatrix} e_{xx} & e_{xy} & e_{xz} \\ e_{yx} & e_{yy} & e_{yz} \\ e_{zx} & e_{zy} & e_{zz} \end{bmatrix} \tag{15.7}$$

The decomposition rule gives the relationship

$$\varepsilon_{ijk}(t) = E_{ijk} + e'_{ijk} \tag{15.8}$$

where $\varepsilon_{ijk}(t)$ is the instantaneous value of the rate of strain, E_{ijk} the mean value and e'_{ijk} the fluctuating part.

15.3.2 Spalart–Allmaras model (one-equation model)

This model (Spalart and Allmaras, 1992) solves a single partial differential conservation equation for eddy viscosity. It contains convective and diffusive transport terms as well as production and dissipation terms. It is accurate and economical to use but not suitable for flow separation, recirculation and free shear flows. It has been designed for aerodynamic flows and for boundary layers subject to adverse pressure gradients. Reynolds stress is expressed as

$$-\overline{u'_i u'_j} = 2\nu_t S_{ij} \tag{15.9}$$

15.3.3 k-ε Type models (two equation models)

This type of model focusses on the mechanisms that affect the turbulent kinetic energy, which can be expressed as a mean and a fluctuating component according to Reynolds decomposition. Two equations, one for the turbulent kinetic energy per unit mass k and the other for the rate of dissipation of kinetic energy per unit mass ε, need to be solved to establish a relationship for eddy viscosity. It can be considered as an improvement to the mixing length model. One important limitation in k-ε models is that the eddy viscosity is assumed to be isotropic. The mean (K), the instantaneous value $k(t)$ and fluctuating components of kinetic energy (k) can be expressed as

$$K = \frac{1}{2}\left(\bar{u}^2 + \bar{v}^2 + \bar{w}^2\right) \tag{15.10}$$

$$k = \frac{1}{2}\left(\overline{u'^2} + \overline{v'^2} + \overline{w'^2}\right) \tag{15.11}$$

$$k(t) = K + k \tag{15.12}$$

The equation for mean kinetic energy K takes the form (Versteeg and Malalasekera, 1995, p. 68)

$$\frac{\partial}{\partial t}(\rho K) + \operatorname{div}(\rho K \bar{u}) = \operatorname{div}\left(-\bar{p}\bar{u} + 2\mu \bar{u} E_{ij} - \rho \bar{u}\overline{u_i'u_j'}\right) - 2\mu E_{ij}.E_{ij} + \rho \overline{u_i'u_j'} E_{ij} \tag{15.13}$$

In this equation, the first term on the left-hand side represents the rate of change of kinetic energy, the second term the divergence[1] of the transport of kinetic energy by convection, the first term on the right-hand side the divergence of the sum of the transport of kinetic energy by pressure, viscous stresses and Reynolds stresses, the second term on the right-hand side the rate of dissipation of K and the last term on the right-hand side the turbulence production. Here E_{ij} is the mean rate of deformation tensor.

The equation for turbulent kinetic energy k takes the form (Versteeg and Malalasekera, 1995, pp. 68–71)

$$\frac{\partial}{\partial t}(\rho k) + \operatorname{div}(\rho k \bar{u}) = \operatorname{div}\left(-\overline{p'u'} + 2\mu \overline{u'e_{ij}'} - \rho \frac{1}{2}\overline{u_i'u_i'u_j'}\right) - 2\mu \overline{e_{ij}'e_{ij}'} - \rho \overline{u_i'u_j'} E_{ij} \tag{15.14}$$

In this equation, the first term on the left-hand side is the rate of change of k, the second term on the left-hand side is the divergence of the transport of k by convection, the first term on the right-hand side is the divergence of the transport of k by pressure, viscous stresses and Reynolds stresses, the second term on the right-hand side the rate of dissipation of k and the last term the turbulence production. Here e_{ij}' is the fluctuating component of the rate of deformation tensor.

Equation 15.14 contains too many unknowns. For practical purposes, the following simplified equation is used (Launder and Spalding, 1974):

$$\frac{\partial}{\partial t}(\rho k) + \frac{\partial}{\partial x_i}(\rho k \bar{u}) = \frac{\partial}{\partial x_j}\left[\frac{\mu_t}{\sigma_k}\frac{\partial k}{\partial x_j}\right] + 2\mu_t E_{ij} \cdot E_{ij} - \rho \varepsilon \tag{15.15}$$

In this equation, the first term on the left-hand side represents the rate of change of k, the second term the convective transport, the first term on the right-hand side the diffusive

---

[1] Divergence $\nabla = \dfrac{\partial}{\partial x} + \dfrac{\partial}{\partial y} + \dfrac{\partial}{\partial z}$; $\nabla^2 = \dfrac{\partial^2}{\partial x^2} + \dfrac{\partial^2}{\partial y^2} + \dfrac{\partial^2}{\partial z^2}$ = Laplacian

transport, the second term on the right-hand side the rate of production and the last term the rate of destruction. The coefficient σ_k is the Prandtl number which is typically taken as unity.

Similarly, for turbulent dissipation ε per unit mass, an exact equation that has many unknowns can be derived. The simplified equation used in standard k-ε models (Launder and Spalding, 1974) is

$$\frac{\partial}{\partial t}(\rho\varepsilon) + \text{div}(\rho\varepsilon\bar{u}) = \frac{\partial}{\partial x_j}\left[\frac{\mu_t}{\sigma_\varepsilon}\frac{\partial\varepsilon}{\partial x_j}\right] + C_{1\varepsilon}\frac{\varepsilon}{k}2\mu_t E_{ij}\cdot E_{ij} - C_{2\varepsilon}\rho\frac{\varepsilon^2}{k} \qquad (15.16)$$

The terms in this equation have meanings similar to those in the equation for k. Typical values of the coefficients obtained by data fitting for a wide range of turbulent flows are: σ_ε (also a Prandtl number) = 1.3; $C_{1\varepsilon}$ = 1.44; $C_{2\varepsilon}$ = 1.92. These two simplified equations lead to what is commonly referred to as the standard k-ε model. The rate of dissipation per unit mass ε is given as

$$\varepsilon = 2v_t\overline{e'_{ij}e'_{ij}} \qquad (15.17)$$

The eddy viscosity then is calculated from k and ε as

$$\mu_t = \rho C_\mu \frac{k^2}{\varepsilon}; \quad C_\mu = 0.09 \qquad (15.18)$$

Finally, the Reynolds stresses in Eq. 15.14 are calculated from

$$-\rho\overline{u'_iu'_j} = \mu_t\left(\frac{\partial\bar{u}_i}{\partial x_j} + \frac{\partial\bar{u}_j}{\partial x_i}\right) - \frac{2}{3}\rho k\delta_{ij} = 2\mu_t E_{ij} - \frac{2}{3}\rho k\delta_{ij}.$$

which is the same as Eq. 15.1 with $\delta_{ij} = 1$, if $i = j$, and, $\delta_{ij} = 0$ if $i \neq j$.

The k-ε models have their advantages as well as disadvantages. Advantages include ease of implementation, stable calculations and reasonable results. On the other hand, they are not suitable for swirling flows, flow separation and axisymmetric flows, and valid only for fully turbulent flows. Discussion of the pros and cons of the k-ε models is well documented elsewhere (e.g., Davidson, 2015, pp. 169–174). They are also not suitable for flows with large adverse pressure gradients. Improvements made to k-ε models include k-ε RNG, k-ε realizable, k-ω, algebraic stress and non-linear models. Most such models have k as one of the variables and other choices for the remaining variable.

In k-ω models, where ω is the specific dissipation, a variable that determines the scale of turbulence, the kinematic eddy viscosity is given as

$$v_t = \frac{k}{\omega} \qquad (15.19)$$

and the corresponding equations for k and ω are respectively given as (Wilcox, 1988)

$$\frac{\partial k}{\partial t} + \bar{u}_j\frac{\partial k}{\partial x_j} = \tau_{ij}\frac{\partial\bar{u}_i}{\partial x_j} - \beta^*k\omega + \frac{\partial}{\partial x_j}\left[(v+\sigma^*v_t)\frac{\partial k}{\partial x_j}\right] \qquad (15.20)$$

and

$$\frac{\partial \omega}{\partial t} + \bar{u}_j \frac{\partial \omega}{\partial x_j} = \alpha \frac{\omega}{k} \tau_{ij} \frac{\partial \bar{u}_i}{\partial x_j} - \beta \omega^2 + \frac{\partial}{\partial x_j}\left[(v + \sigma v_t)\frac{\partial \omega}{\partial x_j}\right] \qquad (15.21)$$

These two equations have too many closure coefficients. The values of the closure coefficients given by Wilcox are $\alpha = \dfrac{5}{9}$; $\beta = \dfrac{3}{40}$; $\beta^* = \dfrac{9}{100}$; $\sigma = \dfrac{1}{2}$; $\sigma^* = \dfrac{1}{2}$; $\varepsilon = \beta^* \omega k$.

The standard k-ε model (Launder and Spalding, 1974) is semi-empirical. The k-equation is exact but simplified for the standard k-ε model; the ε equation uses physical reasoning but lacks mathematical justification. In Eqs. 15.20 and 15.21, the molecular kinematic viscosity v is small compared to kinematic eddy viscosity v_t and can be neglected for fully turbulent flows.

Mixing length model and k-ε models both assume that the eddy viscosity is isotropic. This is one of the limitations of these two types. The latter includes the effects of turbulent transport properties by mean flow and diffusion as well as production and destruction of turbulence. Most CFD software use k-ε models.

15.4 SECOND-ORDER MODELS

15.4.1 Reynolds stress model (RSM)

This is a second-order model and is the most complex turbulence model and perhaps the most exact representation of the Reynolds stresses. The k-ε type models assume isotropy which may not always be satisfied. There may be directional changes in the Reynolds stresses which are not accounted in either the mixing length model or the family of k-ε models. The Reynolds stress equation model can handle the anisotropic conditions of the Reynolds stresses. The modelling approach is based on the work of Launder et al. (1975). The governing equation for the transport of Reynolds kinematic stress takes the form (Versteeg and Malalasekera, 1995, pp. 75)

$$\frac{DR_{ij}}{Dt} = P_{ij} + D_{ij} - \varepsilon_{ij} + \Pi_{ij} + \Omega_{ij} \qquad (15.22)$$

The first term in this equation is the rate of change of $R_{ij}\left(= \overline{u_i' u_j'}\right)$ and the transport of R_{ij} by convection; the first term on the right-hand side the rate of production of R_{ij}; the second term the transport by diffusion, the third term the rate of dissipation, the fourth term the transport of R_{ij} due to turbulent pressure–strain interactions and the last term the transport of R_{ij} due to rotation.

This equation describes six partial differential equations – one each for the six independent Reynolds stress components. Commercial CFD software use different forms to describe the terms in the above equation. For example (Versteeg and Malalasekera, 1995, pp. 76–77),

$$P_{ij} = -\left(R_{im}\frac{\partial \bar{u}_j}{\partial x_m} + R_{jm}\frac{\partial \bar{u}_i}{\partial x_m}\right) \qquad (15.23)$$

$$D_{ij} = \frac{\partial}{\partial x_m}\left(\frac{v_t}{\sigma_k}\frac{\partial R_{ij}}{\partial x_m}\right) = \mathrm{div}\left(\frac{v_t}{\sigma_k}\mathrm{grad}(R_{ij})\right) \qquad (15.24)$$

with $v_t = C_\mu \dfrac{k^2}{\varepsilon}$; $C_\mu = 0.09$; $\sigma_k = 1.0$

$$\varepsilon_{ij} = \frac{2}{3}\varepsilon\delta_{ij} \tag{15.25}$$

$$\Pi_{ij} = -C_1 \frac{\varepsilon}{k}\left(R_{ij} - \frac{2}{3}k\delta_{ij}\right) - C_2\left(P_{ij} - \frac{2}{3}P\delta_{ij}\right) \quad \text{with} \quad C_1 = 1.8; \quad C_2 = 0.6 \tag{15.26}$$

$$\Omega_{ij} = -2\omega_k\left(R_{jm}e_{ikm} + R_{im}e_{jkm}\right) \tag{15.27}$$

with $e_{ijk} = 1$ if i, j and k are different and in cyclic order, $e_{ijk} = -1$ if i, j, k are different and in anti-cyclic order and $e_{ijk} = 0$ if any two indices are the same.

$$k = \frac{1}{2}(R_{11} + R_{22} + R_{33}) = \frac{1}{2}\left(\overline{u_1'^2} + \overline{u_2'^2} + \overline{u_3'^2}\right) \tag{15.28}$$

The six equations for the Reynolds stress transport together with a model equation for dissipation rate ε make this a seven equation model. Details can be found in Launder et al. (1975). The dissipation equation is that proposed by Hanjalic and Launder (1972) and is of the form

$$\frac{D\varepsilon}{Dt} = \frac{\partial}{\partial x_i}\left(C_\varepsilon \frac{k}{\varepsilon}\overline{u_iu_j}\frac{\partial \varepsilon}{\partial x_j}\right) + C_{\varepsilon 1}\frac{P_\varepsilon}{k} - C_{\varepsilon 2}\frac{\varepsilon^2}{k} \tag{15.29}$$

with $C_\varepsilon = 0.15$; $\quad C_{\varepsilon 1} = 1.44$; $\quad C_{\varepsilon 2} = 1.92$.

15.4.2 Algebraic stress model

This model which is also a second-order model is derived by removing the convective and diffusive transport terms from the RSM model, resulting in a set of algebraic equations. Removal of these two terms leads to a substantial saving in computational cost and in some cases appears to be reasonably accurate (Naot and Rodi, 1982; Demuren and Rodi, 1984). The algebraic stress model takes the form

$$R_{ij} = \overline{u_i'u_j'} = \frac{2}{3}k\delta_{ij} + \left(\frac{C_D}{C_1 - 1 + \dfrac{P}{\varepsilon}}\right)\left(P_{ij} - \frac{2}{3}P\delta_{ij}\right)\frac{k}{\varepsilon} \tag{15.30}$$

with adjustable constants C_D and C_1. This leads to a set of six algebraic simultaneous equations for the six Reynolds stress components which can be solved for known values of k and ε. Some CFD software use $C_D = 0.55$ and $C_1 = 2.2$ for swirling flows.

Criteria for model assessment include the level of description, completeness, cost and ease of use, range of applicability and accuracy. Not all models are applicable to all types of flows. DNS is complete, but the cost is high. Mixing length model is incomplete. Accuracy depends upon the numerical scheme used.

15.5 LARGE EDDY SIMULATION (LES)

The idea of large eddy simulation (LES) was first proposed by Joseph Smagorinsky (1963) in his highly mathematical and detailed paper that describes atmospheric air currents using the primitive equations. The primitive equations in this context include the conservation of mass equation, the momentum equation represented by the Navier–Stokes equation and a thermal energy equation. They are a set of non-linear partial differential equations that have been used to describe large scale (global) atmospheric flow. In the Navier–Stokes equations, the vertical motion is assumed to be much smaller than the horizontal motion and that the depth of fluid layer is much smaller than the radius of the earth. The method was first explored by Deardorff (1970). The approach has been applied to a wide variety of engineering problems, including combustion, acoustics and simulations of the atmospheric boundary layer (Mason, 1994). The LES falls between the DNS and the RANS approaches and has better accuracy than the latter. The downside is the excessive computer resource requirement although LES requires far less computing resources than the DNS approach. It is reported that the relative comparison of the computing required for DNS is of the order of Re^3, for LES of the order of Re^2 and for RANS of the order of Re (McDonough, 2004, 2007). However, the accuracy depends upon two issues – the filtering process and the numerical solution. LES is more meaningful because of a number of factors. For example, mass, momentum and energy are transported mainly by large eddies; they are problem dependent and the geometry and boundary conditions play significant roles whereas small scale eddies are less dependent on the geometry and are generally isotropic and therefore universal; the chance of finding a universal turbulence model for small scale eddies is much higher. It is also important to note that the LES models must run for a sufficiently long time to attain stable conditions, thereby adding up to computing cost. The method is focussed on isotropy and fully developed turbulent flows.

LES requires modelling from parts of the large scale to the beginning of the dissipation scale. The LES procedure generally converges to the DNS solution as the discretization size gets smaller and smaller, and thereby implicitly converging to the Navier–Stokes solutions. It is a three-dimensional transient modelling approach. The procedure consists of decomposition of the flow variables into large and small scale components using a technique known as filtering. It removes a range of small scales from the solution to the Navier–Stokes equation, thereby reducing the computational cost. The smallest length scales are the most computationally demanding to resolve. This operation is equivalent to a low-pass filter, meaning filtering out the scales with high frequency. It can also be thought as simulating the moving averages of the flow field variables which smooth out the high-frequency components. The Navier–Stokes equations are then filtered, and the decomposition parts are substituted into the non-linear terms to construct the 'unclosed' terms to be modelled. Large eddies are retained and solved. This implies ignoring the small scale part. Resolving the large eddies only permits a much coarser mesh and longer time steps compared to the DNS approach but a finer and shorter time steps compared to the RANS approach.

The derivation of the filtered Navier–Stokes equations is analogous to the derivation of the RANS equations. In the RANS equations, the Reynolds stresses are modelled whereas in LES, the small scale motions are modelled. LES resolves the large scale turbulence and, therefore, suitable for problems with large scale turbulence, such as when flow passes over a bluff body. Since the small scale turbulence, which takes much of the computational cost, is not resolved in LES, the overall computational cost is significantly less than that for DNS. The basic procedure for LES involves filtering, derivation of the filtered Navier–Stokes equations for the resolved-scale variables that are as close as possible to the Navier–Stokes equations, picking a suitable 'closure' model for the small scale unresolved turbulence and solving the filtered Navier–Stokes equations numerically. In choosing a model for the subgrid scale stresses, it is important to ensure that the subgrid scale stresses can be calculated from the resolved scales. The model may also have some adjustable parameters. The filtering uses a decomposition procedure of the form

$$u(x,t) = \tilde{u}(x,t) + u'(x,t) \qquad (15.31)$$

where u is the velocity field, $\tilde{u}(x,t)$ is the spatially filtered velocity field at a given time t which will be resolved and $u'(x,t)$ is the residual velocity field which is not resolved and modelled at the subgrid scale. Although the notation is similar to that used in the RANS equations, $\tilde{u}(x,t)$ and $u'(x,t)$ in this procedure have different meanings from the corresponding notations used in the RANS equations. The RANS equations are time independent whereas the LES equations are both space and time dependent. For example,

$$\tilde{\tilde{u}} \neq \tilde{u}, \quad \text{and} \quad \text{hence} \quad \bar{u}' \neq 0; \quad \tilde{u}(x,y,z,t) \text{ is a random field.} \qquad (15.32)$$

The general filtering equation (Leonard, 1974) is of the form

$$\tilde{u}(x,t) = \int\limits_{\text{over the entire domain}} G(r,x)u(x-r,t)\,dr \qquad (15.33)$$

Compare with the Reynolds decomposition where

$$\bar{u}(x) = \frac{1}{T}\int\limits_0^T u(x,t), \qquad (15.34)$$

with G satisfying the condition

$$\int\limits_{\text{over the entire domain}} G(r,x)\,dr = 1, \qquad (15.35)$$

and

$$\overline{u'(x,t)} \neq 0 \quad \left(\text{Compare with Reynolds decomposition where } \overline{u'(x,t)} = 0\right)$$

The function G can take several forms. The widely used ones are the box filter, the Gaussian filter and the sinc filter (also known as sharp spectral filter defined as $\dfrac{\sin(\pi r/L)}{\pi r}$). In a one dimensional framework, they are described as follows (Pope, 2000; Table 13.2).

The Box filter in physical space is of the form

$$G(x,r) = \begin{cases} \dfrac{1}{\Delta} & \text{if } |x-r| \leq \dfrac{\Delta}{2} \\[2mm] 0 & \text{otherwise} \end{cases} \qquad (15.36)$$

In spectral space[2] (wave number space where κ is the wave number), it is

$$\hat{G}(\kappa) = \frac{\sin\left(\frac{1}{2}\kappa\Delta\right)}{\frac{1}{2}\kappa\Delta} \tag{15.37}$$

The Gaussian filter in physical space is of the form

$$G(x,r) = \left(\frac{6}{\pi\Delta^2}\right)^{1/2} \exp\left(-\frac{6(x-r)^2}{\Delta^2}\right) \tag{15.38}$$

which in the spectral space takes the form

$$\hat{G}(\kappa) = \exp\left(-\frac{\kappa^2\Delta^2}{24}\right) \tag{15.39}$$

The sharp spectral filter in physical space is

$$G(x,r) = \frac{\sin\left(\pi(x-r)/\Delta\right)}{\pi(x-r)} \tag{15.40}$$

which in the spectral space takes the form

$$\hat{G}(\kappa) = H\left(\kappa_c - |\kappa|\right) \tag{15.41}$$

where $\kappa_c = \dfrac{\pi}{\Delta}$ and Δ is a dimensionless cut-off length scale, also known as the filter width, and H is the Heaviside function ($H(x) = 1$ if $x \geq 0$ and $H(x) = 0$ if $x < 0$). Scales less than Δ or the grid spacings used in the computations are eliminated in LES. The transformation from physical space to spectral space (in other words, from time domain to wave number domain) is of the form

$$\hat{G}(\kappa) = \int G(r)e^{-i\kappa r}\, dr = 2\pi\left(\text{Fourier Transform of } G(r)\right) \tag{15.42}$$

In the spectral representation,

$$\hat{u}(\kappa) = \text{Fourier Transform of } u(x) \tag{15.43}$$

and then,

$$\hat{\bar{u}}(\kappa) = \text{Fourier Transform of } \bar{u}(x) = \hat{G}(\kappa)\hat{u}(\kappa) \tag{15.44}$$

It can be seen that filtering and Reynolds decomposition are analogous but have different interpretations. Starting from the Navier–Stokes equations,

---

[2] If the autocorrelation is a function of time interval, the transformed variable is a frequency (T^{-1}). If the autocorrelation is a function of space interval, the transformed variable is a wave number (L^{-1}). The latter is referred to as the spectral space.

$$\frac{\partial u_i}{\partial x_i} = 0 \tag{15.45}$$

$$\frac{\partial u_i}{\partial t} + \frac{\partial (u_i u_j)}{\partial x_j} = -\frac{1}{\rho}\frac{\partial p}{\partial x_i} + v\frac{\partial^2 u_i}{\partial x_j \partial x_j}, \tag{15.46}$$

filtering the momentum equation is as follows:

$$\overline{\frac{\partial u_i}{\partial t}} + \overline{\frac{\partial (u_i u_j)}{\partial x_j}} = -\overline{\frac{1}{\rho}\frac{\partial p_i}{\partial x_i}} + v\overline{\frac{\partial^2 u_i}{\partial x_j \partial x_j}} \tag{15.47}$$

which can be written as

$$\frac{\partial \overline{u}_i}{\partial t} + \frac{\partial (\overline{u_i u_j})}{\partial x_j} = \frac{1}{\rho}\frac{\partial \overline{p}}{\partial x_i} + v\frac{\partial^2 \overline{u}_i}{\partial x_j \partial x_j} \tag{15.48}$$

This is on the assumption that filtering and differentiation are commutative. This equation has a non-linear filtered advective term $\overline{u_i u_j}$ posing as the main problem in LES. This term requires knowledge of the unfiltered velocity field which is unknown and therefore needs to be modelled. It causes interaction between large and small scales preventing separation of scales. Since unfiltered u_i, u_j are unknown, they are substituted as

$$\frac{\partial \overline{u}_i}{\partial t} + \frac{\partial (\overline{u}_i \overline{u}_j)}{\partial x_j} = -\frac{1}{\rho}\frac{\partial \overline{p}}{\partial x_i} + v\frac{\partial^2 \overline{u}_i}{\partial x_j \partial x_j} - \left(\frac{\partial (\overline{u_i u_j})}{\partial x_j} - \frac{\partial (\overline{u}_i \overline{u}_j)}{\partial x_j}\right) \tag{15.49}$$

Leonard (1974) proposed the replacement of this non-linear term by introducing a residual stress tensor τ_{ij}^r (analogous to the Reynolds stress) as follows to get over this problem:

$$\overline{u_i u_j} = \tau_{ij}^r + \overline{u}_i \overline{u}_j \tag{15.50}$$

With the above decomposition and substitution, the filtered Navier–Stokes equations for incompressible fluids take the form

$$\frac{\partial \overline{u}_i}{\partial x_i} = 0 \tag{15.51}$$

$$\frac{\partial \overline{u}_i}{\partial t} + \frac{\partial}{\partial x_j}(\overline{u}_i \overline{u}_j) = -\frac{1}{\rho}\frac{\partial \overline{p}}{\partial x_i} + v\frac{\partial}{\partial x_j}\left(\frac{\partial \overline{u}_i}{\partial x_j} + \frac{\partial \overline{u}_j}{\partial x_i}\right) - \frac{\partial \tau_{ij}^r}{\partial x_j}$$

$$= -\frac{1}{\rho}\frac{\partial \overline{p}}{\partial x_i} + 2v\frac{\partial \overline{S}_{ij}}{\partial x_j} - \frac{\partial \tau_{ij}^r}{\partial x_j} \tag{15.52}$$

where

$$\overline{S}_{ij} = \frac{1}{2}\left(\frac{\partial \overline{u}_i}{\partial x_j} + \frac{\partial \overline{u}_j}{\partial x_i}\right) \text{ is the rate of strain tensor.} \tag{15.53}$$

This form of filtered Navier–Stokes equation is in the physical space, but it can also be written in the wave number space. Unresolved scales are generally represented by the eddy viscosity model of the form

$$\tau_{ij}^r = 2\rho v_r \bar{S}_{ij} + \frac{1}{3}\delta_{ij}\tau_{kk}^r \tag{15.54}$$

where v_r is the eddy viscosity of residual motion. The residual stress is decomposed into three terms as follows:

$$\tau_{ij}^r = L_{ij} + C_{ij} + R_{ij} \tag{15.55}$$

where L_{ij} is the Leonard tensor representing interactions among large scales, C_{ij} is the Clark tensor representing the cross-scale interactions between large and small scales and R_{ij} is similar to Reynolds stress and represents the interactions among sub-filter (subgrid) scales. Modelling (which is not resolved) is done using subgrid scale models. This stress tensor interacts with all scales making it challenging for modelling.

The subgrid scale Reynolds stresses are given as

$$\tau_{ij}^r = -2v_t S_{ij} \tag{15.56}$$

with v_t as the Smagorinsky turbulent (or eddy) kinematic viscosity given as

$$v_t = l_S^2 \bar{S} \tag{15.57}$$

where l_S is the Smagorinsky length scale and $\bar{S}$ is a characteristic filtered rate of strain defined as

$$\bar{S} = \sqrt{2\bar{S}_{ij}\bar{S}_{ij}}. \tag{15.58}$$

Smagorinsky (1963) defined the eddy viscosity as

$$v_r = C_s^2 L^2 \left(2\bar{S}_{ij}\bar{S}_{ij}\right)^{1/2} \tag{15.59}$$

where C_s is called the Smagorinsky coefficient. In Eq. 15.59, L is chosen so that $\bar{u}$ contains bulk of the energy-containing eddies, and L lies in the inertial range. Averaging is done over the length L. This model seems to work well for isotropic and free shear turbulence.

There are other subgrid scale models, dynamic models (e.g., Germano et al., 1991) and mixed models (e.g., Armenio and Piomelli, 2000; Zang et al., 1993). Many turbulence models use the Boussinesq hypothesis which relates the stress to strain in the form as given in Eq. 15.1:

$$\tau_{ij} = \mu\varepsilon_{ij} = \mu\left(\frac{\partial u_i}{\partial x_j} + \frac{\partial u_j}{\partial x_i}\right) \text{ for molecular viscous stress}$$

and

$$\tau_{ij} = -\rho\overline{u_i'u_j'} = \mu_t\left(\frac{\partial \bar{u}_i}{\partial x_j} + \frac{\partial \bar{u}_j}{\partial x_i}\right) \text{ for turbulent viscous stress.}$$

When these are substituted into the filtered Navier–Stokes equations, the resulting equation is of the form

$$\frac{\partial \bar{u}_i}{\partial t} + \frac{\partial}{\partial x_j}\left(\bar{u}_i \bar{u}_j\right) = -\frac{1}{\rho}\frac{\partial \bar{p}}{\partial x_i} + v\frac{\partial}{\partial x_j}\left(\frac{\partial \bar{u}_i}{\partial x_j} + \frac{\partial \bar{u}_j}{\partial x_i}\right)$$

$$= -\frac{1}{\rho}\frac{\partial \bar{p}}{\partial x_i} + 2v\frac{\partial \bar{S}_{ij}}{\partial x_j} + \frac{\partial}{\partial x_j}\left(\left(v + v_t\right)\frac{\partial \bar{u}_i}{\partial x_j}\right) \qquad (15.60)$$

This form of filtered Navier–Stokes equation is in the physical space, but it can also be written in the wave number space. Unresolved scales are generally represented by the eddy viscosity model shown by Eq. 15.54.

Historical development of LES, which started with the work of Smagorinsky (1963), is well documented (e.g., Pope, 2000; Chapter 13; McDonough, 2004, 2007; among others). Examples of LES applications include isotropic turbulence that uses the wave number space with the pseudo-spectral method, isotropic turbulence that uses the physical space with finite difference method, and free shear flow that uses uniform rectangular grid.

LES generally accounts for about 80% of the kinetic energy in the flow field (Pope, 2000, p. 560).

REFERENCES

Armenio, V. and Piomelli, U. (2000): A lagrangian mixed subgrid-scale model in generalized coordinates, *Flow, Turbulence and Combustion*, vol. 65, pp. 51–81.

Boussinesq, J. (1877): Essai sur la théorie des eaux courantes, *Mémoires présentés par divers savants à l'Académie des Sciences*, vol. 23, no. 1, pp. 1–680.

Davidson, P. A. (2015): *Turbulence: An Introduction for Scientists and Engineers*, (2nd Edition), Oxford University Press, Oxford, UK, 630 pp.

Deardorff, J. W. (1970): A numerical study of three-dimensional turbulent channel flow at large Reynolds numbers, *Journal of Fluid Mechanics*, vol. 41, no. 2, pp. 453–480.

Demuren, A. O. and Rodi, W. (1984): Calculation of turbulence-driven secondary motion in non-circular ducts, *Journal of Fluid Mechanics*, vol. 140, pp. 189–222.

Germano, M., Piomelli, U, Moin, P. and Cabot, W. (1991): A dynamic sub-grid scale eddy viscosity model, *Physics of Fluids*, A, vol. 3, no. 7, pp. 1760–1765. doi: 10.1063/1.857955.

Hanjalic, K. and Launder, B. E. (1972): Reynolds stress model of turbulence and its application to thin shear flows, *Journal of Fluid Mechanics*, vol. 52, no. 4, pp. 609–638.

Launder, B. E. and Spalding, D. B. (1974): The numerical computation of turbulent flows, *Computer Methods Applied Mechanical Engineering*, vol. 3, pp 269–289.

Launder, B. E., Reece, G. J. and Rodi, W. (1975): Progress in the development of Reynolds stress turbulence closure, *Journal of Fluid Mechanics*, vol. 68, no. 3, pp. 537–566.

Leonard, A. (1974): Energy cascade in large-eddy simulations of turbulent fluid flows, *Advances in Geophysics*, vol. 18A, pp. 237–248. doi: 10.1016/S0065-2687(08)60464-1.

Mason, P. J. (1994): Large eddy simulation: A critical review of the technique, *Quarterly Journal of Royal Meteorological Society*, vol. 120, pp. 1–26.

McDonough, J. M. (2004, 2007): Introductory lectures on turbulence, Physics, Mathematics and Modeling, Departments of Mechanical Engineering and Mathematics, University of Kentucky, p. 174.

Naot, D. and Rodi, W. (1982): Numerical simulation of secondary currents in channel flow, *Journal of Hydraulic Division, ASCE*, vol. 108, no. HY8, pp. 948–968.

Pope, S. B. (2000): *Turbulent Flows*, Cambridge University Press, Cambridge, UK, 771 pp.

Prandtl, L. (1925): Bericht u¨ber Untersuchungen zur ausgebildeten Turbulenz, *Zeitschrift für Angewandte Mathematik und Mechanik*, vol. 5, pp. 136–139.

Rodi, W. (1980): *Turbulent Models and Their Applications in Hydraulics: A State of the Art Review*, IAHR, Delft, The Netherlands.

Smagorinsky, J. (1963): General circulation experiments with the primitive equations, *Monthly Weather Review*, vol. 91, no. 3, pp. 99–164.

Spalart, P. R. and Allmaras, S. R. (1992): A one-equation turbulence model for aerodynamic flows, AIAA Paper 92–439.

Versteeg, H. K. and Malalasekera, W. (1995): *An Introduction to Computational Fluid Dynamics: The Finite Volume Method*, Pearson Education, London.257 pp.

Wilcox, D. C. (1988): Re-assessment of the scale-determining equation for advanced turbulence models, *AIAA Journal*, vol. 26, no. 11, pp. 1299–1310.

Zang, Y., Street, R. L. and Koseff, J. R. (1993): A dynamic mixed subgrid-scale model and its application to turbulent recirculating flows, *Physics of Fluids A*, vol. 5, pp. 3186–3196.

Chapter 16

Computational fluid dynamics

16.1 INTRODUCTION

With the advent of high-speed computers, many fluid flow problems that were intractable in the early days can now be solved using numerical methods. Examples of fluid flow analysis include aerodynamics, hydrodynamics, combustion in internal combustion engines and gas turbines, hydraulic machines, chemical processes, wind engineering, marine engineering, hydrology, oceanology, meteorology, blood flow, etc. Among such a vast range of topics that can be handled using computational fluid dynamics, or CFDs, those that are relevant to civil engineers include hydrodynamics, hydraulic machines, hydrology, meteorology (for weather forecasting), marine engineering (forces and effects of forces on off-shore structures), wind engineering (wind loading on buildings and other structures) and, environmental engineering (pollutant transport processes). Prior to the CFD era, the only other option available for understanding complex fluid flow problems was by experimental analysis, which was expensive, time consuming and with limitations in carrying out experimental work on certain types of flow problems such as when carrying out experiments using hazardous materials.

All CFDs need, a set of governing equations, a domain of analysis for the problem including boundary and initial conditions, a scheme of discretization of the domain and a numerical scheme to solve the governing equations. For fluid flows, the usual governing equations are the Navier–Stokes equations or their modified forms depending on the problem. The domain depends upon the problem, and in most CFDs, it is three-dimensional in space and time-dependent. Discretization of the domain depends upon the numerical scheme to be used and may be uniform for simple flow problems and non-uniform for problems that would have steep gradients. The last step in CFDs is the solution scheme upon which the accuracy, reliability, applicability, etc. depend. In this context, it is important to ensure that the results of the numerical solution satisfy the conditions of convergence, consistence, and stability. Convergence implies that the numerical solution approaches the exact solution as the discretization (grid size) tends to zero. Consistency implies that the algebraic equations that need to be solved are equivalent to the original governing equations as the discretization approaches zero. Stability implies that the round-off errors that accumulate during repeated calculations dampen to zero as the number of calculations increase.

The widely used numerical methods in CFDs are the finite volume method (FVM), the finite difference method, the finite element method and hybrid methods, with each type having its own pros and cons. All CFDs give results that are approximate because they solve the difference equations of the governing differential equations in a finite number of locations within the domain of interest. Instead of a continuous solution of the governing equations, what CFDs produce are solutions at a finite number of points within the domain.

In all CFD software, the approaches followed include a pre-processing stage during which the domain of analysis of the problem is identified and discretized to form a mesh of cells.

They may be uniform or non-uniform, and may involve a combination of geometrical shapes. This can be achieved by using some CAD software. The next stage involves identifying the physical problem and expressing it in a mathematical form leading to the governing equations. They may be one, two or three dimensional in space and steady or transient. Some simplifications of the governing equations may sometimes be necessary to make them solvable, followed by the imposition of the boundary and initial conditions. The original or the simplified governing equations subject to the boundary and initial conditions are then solved numerically using any one of the numerical methods described below. Finally, the results of the numerical solutions are displayed graphically and numerically using some post-processing package for visualization and interpretation.

16.2 GOVERNING EQUATIONS

The governing equations are based on the laws of conservation of mass, momentum and energy. They can be expressed in differential form or in integral form. For flow problems, the conservation of mass leads to the continuity equation and the conservation of momentum leads to the Navier–Stokes equations. They are respectively given as follows:

$$\frac{\partial \rho}{\partial t} + \frac{\partial(\rho u)}{\partial x} + \frac{\partial(\rho v)}{\partial y} + \frac{\partial(\rho w)}{\partial z} = 0 \tag{16.1}$$

$$\rho \frac{Du}{Dt} = -\frac{\partial p}{\partial x} + \rho g_x + \mu\left(\frac{\partial^2 u}{\partial x^2} + \frac{\partial^2 u}{\partial y^2} + \frac{\partial^2 u}{\partial z^2}\right) + \frac{\mu}{3}\frac{\partial}{\partial x}\left(\frac{\partial u}{\partial x} + \frac{\partial v}{\partial y} + \frac{\partial w}{\partial z}\right) \tag{16.2a}$$

$$\rho \frac{Dv}{Dt} = -\frac{\partial p}{\partial y} + \rho g_y + \mu\left(\frac{\partial^2 v}{\partial x^2} + \frac{\partial^2 v}{\partial y^2} + \frac{\partial^2 v}{\partial z^2}\right) + \frac{\mu}{3}\frac{\partial}{\partial y}\left(\frac{\partial u}{\partial x} + \frac{\partial v}{\partial y} + \frac{\partial w}{\partial z}\right) \tag{16.2b}$$

$$\rho \frac{Dw}{Dt} = -\frac{\partial p}{\partial z} + \rho g_z + \mu\left(\frac{\partial^2 w}{\partial x^2} + \frac{\partial^2 w}{\partial y^2} + \frac{\partial^2 w}{\partial z^2}\right) + \frac{\mu}{3}\frac{\partial}{\partial z}\left(\frac{\partial u}{\partial x} + \frac{\partial v}{\partial y} + \frac{\partial w}{\partial z}\right) \tag{16.2c}$$

where ρ is the density of fluid, u, v and w are the velocity components in the x, y and z directions which are orthogonal to each other, p is the pressure, μ is the dynamic viscosity of the fluid and g_x, g_y and g_z are acceleration due to gravity in the three directions.

For scalar quantities such as concentration, temperature, etc., the governing equation may be of the form

$$\frac{\partial c}{\partial t} + u\frac{\partial c}{\partial x} = \frac{\partial}{\partial x}\left(D\frac{\partial c}{\partial x}\right) + kc \pm Q_s \tag{16.3}$$

where c is the concentration (or any equivalent scalar quantity), D is a diffusion or dispersion coefficient, k is a decay coefficient and Q_s is a source term.

16.3 DOMAIN DISCRETIZATION

The physical domain of interest can be one, two or three dimensional. It can be discretized into a regular grid or an irregular grid. A regular grid is easy for problem formulation and solution and suitable for problems where the gradients of the scalar quantity of interest are mild. On the other hand, in problems where there are sharp and steep gradients, irregular grids with finer discretisations in regions where there are steep gradients should be used.

16.4 NUMERICAL SOLUTION

16.4.1 Finite volume method (control volume approach)

The Finite Volume Method FVM is a numerical method to solve differential or partial differential equations which can be written in the divergence form. The governing equations are recast in a conservative form and solved over control volumes. Gauss theorem is applied to transform the volume integrals into surface integrals across the boundaries, thereby integrating the differential of the dependent variable inside the control volume into surface integrals of the fluxes of the dependent variable across the boundary of the control volume. The basis of this approach is that all mass that would leave the control volume must pass the boundary of the control volume at some time. The FVM, which was first used by McDonald (1971) and McCormack and Paullay (1972), is now one of the widely used numerical techniques in CFDs. Details of the method with examples are described in the book by Versteeg and Malalasekera (1995).

The steps involved in using FVM consist of discretizing the solution domain into a number of non-overlapping control volumes which can have arbitrary shapes. FVM works with control volumes and not nodes, and therefore any unstructured grid can be used with a free choice of shapes and sizes. Structured meshes consist of two sets of lines and are used in finite difference methods (FDM). Unstructured meshes in FVM can be arbitrary combinations of triangles and quadrilaterals in two dimensional domains, and tetrahedrons, hexahedra or pyramids in three dimensional domains.

The discretization can belong to the cell-centered scheme in which the variables are stored at the centroids of the cells (or control volumes) (Figure 16.1b), or the cell-vertex scheme in which the variables are stored at the grid points (Figure 16.1a). The control volume in the latter case can be the union of all cells sharing the grid point, or some volume centered around the grid point. The grids can be structured or unstructured, thereby making the method more flexible than the FDM. Cell centered is more convenient and widely used.

The FVM is based on the approximate solution of integral form of conservation equations. Integrals are evaluated at nodal points. Key features of FVM are that they are conservative locally as well as globally, equivalent to Galerkin's method of Finite Element Method

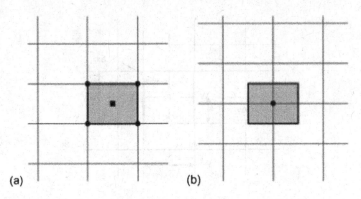

(a)   (b)

Figure 16.1 Discretization schemes in FVM: (a) cell-vertex scheme, (b) centroid scheme.

(FEM) with the weighting function equal to 1, and that the method is widely used in CFD software as it satisfies the conservation principles. The starting governing equations are integrated over the control volume to give a discretized equation at the nodal point. Integration of the governing equation over the CV gives a discretized equation at its nodal points. The variation of the properties between nodal points is usually described by linear interpolation functions. This practice is quite common in finite difference methods. Higher order interpolation functions can also be used but the formulation becomes more complicated and that there is no guarantee that the use of higher order interpolation functions will lead to more accurate results.

The steps involved in using the FVM consist of grid generation, discretization and solution of the resulting equations. Grid generation is done by dividing the domain into discrete control volumes and placing a nodal point at the centroid of each control volume resulting in each node been surrounded by a control volume. In the grid generation step, the compass notation is used to identify the geometry of the control volume and the positioning of the nodal points. This notation is quite general and used in all CFD software. In a two-dimensional domain, the nodes to the north, south, east and west of a nodal point P are denoted by N, S, E and W, respectively (Figure 16.2). The faces of the control volume are denoted by lower case letters as n, s, e and w depending on whether the surface is to the north, south, east or west. The distances between the nodes P and N, P and S, P and E and P and W, are respectively denoted as δx_{PN}, δx_{PS}, δx_{PE} and δx_{PW}. The distances between P and the faces n, s, e and w, are denoted by δx_{pn}, δx_{ps}, δx_{pe} and δx_{pw}, respectively. The control volume widths can be expressed in terms of the above distances. If a three-dimensional domain is considered, another set of notations and distances need to be added in the third direction.

Discretization scheme should satisfy conservativeness, boundedness and transportiveness. Conservativeness implies that the conservative laws for each control volume when added together should satisfy the global conservation laws. Numerical discretization should satisfy this requirement. Central differencing satisfies conservativeness, whereas quadratic interpolation does not. Boundedness implies that in the absence of sources, the values of the scalar quantity described in the governing equation should lie between the values at the boundaries. A diagonally dominant matrix is desired for this condition to be satisfied. Transportiveness describes the influence of the upstream nodes on the downstream nodes. This is related to the Peclet number. For pure diffusion for which the Peclet number is zero, the isolines are circular, implying the same influences from upstream as well as from downstream. When the Peclet number increases, the isolines tend to become ellipses implying more influence from upstream than from downstream. When the Peclet number is very high, there is hardly any influence from downstream.

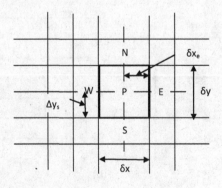

Figure 16.2 Grid generation and compass notation used in FVM.

16.4.1.1 Governing equations

The general transport equation can be described as

$$\frac{\partial(\rho\phi)}{\partial t} + \text{div}(\rho\phi u) = \text{div}(D\,\text{grad}\,\phi) + S_\phi \tag{16.4}$$

where ϕ is the scalar variable such as the concentration, temperature or electric current. The first term on the left-hand side of this equation represents the rate of change of ϕ and the second term the rate of flow of ϕ out of fluid element (convective term). On the right-hand side, the first term represents the rate of increase of ϕ due to diffusion where D is the diffusion coefficient in the case of a diffusion problem but will have a different property for other transport problems, and the last term represents the rate of increase of ϕ due to source. This equation, which states that the flux is proportional to the concentration gradient, follows Fick's law for diffusion, Fourier law for heat conduction and Ohm's law for electric current. The FVM starts with this type of equation.

The key step in FVM is the integration of the above equation over a one-, two- or three-dimensional control volume.

$$\int_{CV}\frac{\partial(\rho\phi)}{\partial t}\,dV + \int_{CV}\text{div}(\rho\phi u)\,dV = \int_{CV}\text{div}(D\,\text{grad}\,\phi)\,dV + \int_{CV}S_\phi\,dV \tag{16.5}$$

The volume integral in Eq. 16.5 can be reduced to a surface integral using Gauss divergence theorem which is

$$\int_{CV}\text{div}(u\,dV) = \int_A n\cdot u\,dA \ (u \text{ is a vector and } n \text{ is normal to the surface } dA) \tag{16.6}$$

Applying Gauss theorem to the governing equation gives,

$$\frac{\partial}{\partial t}\left(\int_{CV}\rho\phi\,dV\right) + \int_A n\cdot(\rho\phi u)\,dA = \int_A n\cdot(D\,\text{grad}\,\phi)\,dA + \int_{CV}S_\phi\,dV \tag{16.7}$$

The first term will be zero for a steady-state problem. For a time-dependent problem it will be

$$\int_{\Delta t}\frac{\partial}{\partial t}\left(\int_{CV}\rho\phi\,dV\right)dt + \int_{\Delta t}\int_A n\cdot(\rho\phi u)\,dA\,dt$$
$$= \int_{\Delta t}\int_A n\cdot(D\,\text{grad}\,\phi)\,dA\,dt + \int_{\Delta t}\int_{CV}S_\phi\,dV\,dt \tag{16.8}$$

Steady state problems lead to elliptic type equations. Transient problems lead to parabolic or hyperbolic type equations which lead to marching problems. Steady-state problems need only the boundary conditions whereas transient problems need initial as well as boundary conditions. Most transient flow problems are of the parabolic type.

For a one-dimensional steady-state transport problem described by

$$\frac{d}{dx}\left(D\frac{d\phi}{dx}\right) + S = 0 \tag{16.9}$$

the discretized equation at its nodal point, obtained by integrating the governing equation over a control volume, takes the form

$$\int_{\Delta V} \frac{d}{dx}\left(D\frac{d\phi}{dx}\right)dV + \int_{\Delta V} S\,dV = \left(DA\frac{d\phi}{dx}\right)_e - \left(DA\frac{d\phi}{dx}\right)_w + \bar{S}\Delta V = 0 \qquad (16.10)$$

where A is the cross-sectional area of the face of the control volume, ΔV is the volume of the control volume and $\bar{S}$ is the average value of S over the control volume. Subscripts e and w represent the east and west boundaries (or faces) of the control volume. To evaluate the diffusion coefficient D and the gradient $\frac{d\phi}{dx}$ at the interfaces in terms of the values at the nodal points, some kind of interpolation function that will describe the variation of the scalar quantity within the control volume is needed. Linear interpolation, which leads to the central differencing scheme, appears to be the obvious choice. Then,

$$D_w = \frac{D_W + D_P}{2}; \quad D_e = \frac{D_P + D_E}{2} \qquad (16.11)$$

and

$$\left(DA\frac{d\phi}{dx}\right)_e = D_e A_e \frac{\phi_E - \phi_P}{\delta x_{PE}}; \quad \left(DA\frac{d\phi}{dx}\right)_w = D_w A_w \frac{\phi_P - \phi_W}{\delta x_{WP}} \qquad (16.12)$$

Substituting these into Eq. 16.12, gives

$$D_e A_e \left(\frac{\phi_E - \phi_P}{\delta x_{PE}}\right) - D_w A_w \left(\frac{\phi_P - \phi_W}{\delta x_{WP}}\right) + S_P \phi_P = 0 \qquad (16.13)$$

Writing similar equations for all the control volumes, assembling them, and incorporating the boundary conditions lead to a set of simultaneous equations which can be solved by any of the method available for the unknown ϕ values at each nodal point.

For transient problems, such as the one shown below, the formulation is as follows:

The procedure is the same as for steady-state formulation except for the time integration part which is illustrated for the simple case of a one-dimensional problem described below:

$$\rho u \frac{\partial c}{\partial t} = \frac{\partial}{\partial x}\left(D\frac{\partial c}{\partial x}\right) + S \qquad (16.14)$$

Integrating over the control volume,

$$\int_t^{t+\Delta t}\int_{CV} \rho u \frac{\partial c}{\partial t}\,dV\,dt = \int_t^{t+\Delta t}\int_{CV}\frac{\partial}{\partial x}\left(D\frac{\partial c}{\partial x}\right)dV dt + \int_t^{t+\Delta t}\int_{CV} S\,dV\,dt \qquad (16.15)$$

which, according to the discretized control volume, may be written as

$$\int_w^e\left[\int_t^{t+\Delta t}\rho u\frac{\partial c}{\partial t}\,dt\right]dV = \int_t^{t+\Delta t}\left[\left(DA\frac{\partial c}{\partial x}\right)_e - \left(DA\frac{\partial c}{\partial x}\right)_w\right]dt + \int_t^{t+\Delta t}\bar{S}\Delta V dt \qquad (16.16)$$

The left-hand side of this equation may be written as

$$\int_{w}^{e}\left[\int_{t}^{t+\Delta t} \rho u \frac{\partial c}{\partial t} dt\right] dV = \rho u \left(c_P - c_P^i\right) \Delta V \tag{16.17}$$

where c_P^i is the concentration at time 't'.

Applying central differencing to the right-hand side of the equation,

$$\rho u \left(c_P - c_P^i\right) \Delta V = \int_{t}^{t+\Delta t}\left[\left(D_e A \frac{c_E - c_P}{\delta x_{PE}}\right) - \left(D_w A \frac{c_P - c_W}{\delta x_{WP}}\right)\right] dt + \int_{t}^{t+\Delta t} \bar{S} \Delta V \, dt \tag{16.18}$$

It is now necessary to assume some relationship for the variation of c_P, c_E and c_W with time. The time integral can be calculated using the c value at time t, or $t + \Delta t$, or as a weighted sum of the values at t and $t + \Delta t$. The weighting parameter lies between 0 and 1 and depending upon its value, the formulation will lead to explicit or implicit schemes. The final formulation will be of the form

$$\rho u \left(\frac{c_P - c_P^i}{\Delta t}\right) \Delta x = \theta\left[\frac{D_e\left(c_E - c_P\right)}{\delta x_{PE}} - \frac{D_w\left(c_P - c_W\right)}{\delta x_{WP}}\right]$$

$$+ (1-\theta)\left[\frac{D_e\left(c_E^i - c_P^i\right)}{\delta x_{PE}} - \frac{D_w\left(c_P^i - c_W^i\right)}{\delta x_{WP}}\right] + \bar{S}\Delta x \tag{16.19}$$

The weighting parameter $\theta = 0$ leads to an explicit scheme, $\theta = 0.5$ leads to the Crank–Nicolson scheme and $\theta = 1$ leads to the fully implicit scheme.

16.4.2 Finite difference method

The Finite Difference Methods FDM solve differential or partial differential equations by approximating the derivative terms by difference terms. All finite difference formulations are based on the Taylor series expansion which may be written as

$$F(x + \Delta x) = F(x) + \Delta x F'(x) + \frac{(\Delta x)^2}{2!} F''(x) + \frac{(\Delta x)^3}{3!} F'''(x) + \cdots \tag{16.20}$$

$$F(x - \Delta x) = F(x) - \Delta x F'(x) + \frac{(\Delta x)^2}{2!} F''(x) - \frac{(\Delta x)^3}{3!} F'''(x) + \cdots \tag{16.21}$$

Ignoring terms containing $O(\Delta x)$ and higher, Eqs. 16.20 and 16.21 give

$$F'(x) = \frac{F(x + \Delta x) - F(x)}{\Delta x} \quad \text{Forward differencing} \tag{16.22}$$

$$F'(x) = \frac{F(x) - F(x - \Delta x)}{\Delta x} \quad \text{Backward differencing} \tag{16.23}$$

Subtracting Eq. (16.21) from Eq. (16.20) and ignoring terms of $O(\Delta x)^2$ and higher

$$F'(x) = \frac{F(x + \Delta x) - F(x - \Delta x)}{2\Delta x} \quad \text{Central Differencing} \tag{16.24}$$

Addition of the two equations gives the second derivative

$$F''(x) = \frac{F(x + \Delta x) - 2F(x) + F(x - \Delta x)}{(\Delta x)^2} - \frac{(\Delta x)^2}{12} F'''(x) + \cdots \tag{16.25}$$

Ignoring $(\Delta x)^2$ and higher order terms, above equation simplifies to

$$F''(x) = \frac{F(x + \Delta x) - 2F(x) + F(x - \Delta x)}{(\Delta x)^2} \tag{16.26}$$

Thus, the accuracy of all finite difference schemes depends upon the step size, Δx. The forward and backward differencing schemes have trailing first-order terms and are therefore called **first-order approximations**. The central differencing scheme has trailing second-order terms and is therefore called **second-order approximation**. In a first-order scheme, the error would decrease by a factor of **two** if the mesh size (Δx) is halved. In a second-order scheme, the error would be reduced by a factor of **four** when the mesh size is halved. It does not, however, say anything about the magnitude of the error. A second-order scheme may not necessarily be capable of modelling a process more accurately than a first-order approximation.

16.4.2.1 Explicit methods

In explicit methods, the unknown value is expressed explicitly in terms of a combination of known values. Therefore, solutions can be obtained in a recursive manner.

16.4.2.2 Implicit methods

In implicit methods, the unknown value is related to other unknown values as well as known values. Solutions cannot be obtained in a recursive manner. Simultaneous solutions are therefore needed.

16.4.3 Finite element method

The Finite Element Method FEM is a powerful numerical technique which can be used to obtain approximate solutions to certain types of partial differential equations that describe field phenomena. It is basically similar to the more widely used finite difference method and has the added advantage of being extremely flexible when used to solve field problems. Although the FEM was originally developed for solving complex problems in structural dynamics, it has now found a wide range of applications in many areas of science and engineering. The application of the method to flow problems is relatively new, and much of its potential uses in this area, is yet to be explored.

To obtain a FEM solution to a field problem, it is necessary to have

- a discretized domain within which the solution has to be obtained,
- a set of constitutive equations that describe the field phenomenon with appropriate boundary and initial conditions, and
- a computer algorithm to carry out the solution on a computer.

16.4.3.1 General principles

The FEM has been developed on the basis that a continuum can be represented by an assemblage of subdivisions called finite elements. These elements are considered interconnected at joints called nodes. Piecewise approximations are made by assuming the unknown function to vary within each finite element in terms of their nodal values according to some interpolation functions. In some literature, the term shape function, displacement function and displacement model are used synonymously. Generally, the interpolation functions are assumed to be polynomials of a given degree.

The FEM was first used for solving complex problems in structural dynamics (Turner et al., 1956; Clough, 1960; Zienkiewicz, 1971). In recent years the method has been widely used for solving a vast range of field problems in many areas of science and engineering. Complex boundary conditions, local variations in the frame of reference of the constitutive equations and heterogeneous material properties offer no intractable difficulties when using the FEM.

The theory, development, applications and limitations of the general FEM are readily available elsewhere (Zienkiewicz, 1971; Oden, 1972; Desai and Abel, 1972). For a given problem, which is often described by a partial differential equation, three widely used approaches are the variational method, the method of weighted residuals and the boundary element method.

16.4.3.2 Variational method

Using principles of calculus of variations, a solution to a field problem may sometimes be obtained by solving a mathematically equivalent problem. The equivalent system often consists of finding the extrema of a functional leading to a Raleigh–Ritz type of solution. The system must satisfy the following conditions in order that a solution may be obtained by a variational method (Forsythe and Wasow, 1960, p. 169; Finlayson and Scriven, 1966):

- The system be linear
- The operation be self-adjoint.

An operator L is said to be self-adjoint if

$$(L\phi, \psi) = (L\psi, \phi) \tag{16.27}$$

where $\phi, \psi \; \varepsilon \; \Omega$, and $(L\phi, \psi) = \text{inner product} = \int_{\Omega} L(\phi) \cdot \psi d\Omega$

This is analogous to the symmetry of a matrix operation. For a linear self-adjoint problem defined by an equation of the form

$$L(\phi) = f \tag{16.28}$$

where f is the driving function, the equivalent minimizing functional is given by

$$F(\phi) = (L(\phi), \phi) - 2(\phi, f) \tag{16.29}$$

where $F(\phi)$ is the minimizing functional and $(,)$ indicate the inner product (Oden, 1972, p. 118; Hutton and Anderson, 1971). When the functional F has some physical meaning, such as total potential energy or total kinetic energy, the governing field equations can be derived using variational calculus. The term $2(\phi, F)$ is a functional of a single independent function $\phi = \phi(x)$, i.e. $F = F(\phi, \phi_x)$ where $\phi_x = \dfrac{\partial \phi}{\partial x}$. Then

$$\delta F = \frac{\partial F}{\partial \phi} \delta\phi + \frac{\partial F}{\partial \phi_x} \delta\phi_x \tag{16.30}$$

Total variation $\displaystyle\int_\Omega \delta F \cdot d\Omega = 0$ implies $\displaystyle\int_\Omega \left(\frac{\partial F}{\partial \phi} \delta\phi + \frac{\partial F}{\partial \phi_x} \delta\phi_x \right) d\Omega = 0$.

Using integration by parts, this equation can be written as

$$\int_\Omega \left[\frac{\partial F}{\partial \phi} - \frac{d}{dx}\left(\frac{\partial F}{\partial \phi_x} \right) \right] \delta\phi\, d\Omega + \left[\frac{\partial F}{\partial \phi_x} \delta\phi \right]_B = 0 \tag{16.31}$$

The last term in Eq. 16.31 is applicable only at the boundary B of the domain.

By equating each term of this equation to zero, the Euler equations for the field problem may be obtained. They are

$$\frac{\partial F}{\partial \phi} - \frac{d}{dx}\left(\frac{\partial F}{\partial \phi_x} \right) = 0 \tag{16.32}$$

$$\left[\frac{\partial F}{\partial \phi_x} \delta\phi \right]_B = 0 \tag{16.33}$$

This principle can be extended to functionals of several independent variables. The condition $\left[\dfrac{\partial F}{\partial \phi_x} \right]_B = 0$ is called a natural boundary condition. If it is satisfied, it is called a free boundary condition. If not,

$$\delta\phi = 0; \quad \text{i.e. } \phi = \phi_B$$

which is called a forced boundary condition. Minimization of the functional F is carried out by assuming the function ϕ to vary in a prescribed manner within each finite element. If

$$\phi = \sum \phi_i N_i \tag{16.34}$$

where N_i's are the interpolation functions and ϕ_i's are the nodal values of the unknown function, then

$$\frac{\partial F}{\partial \phi} = \begin{bmatrix} \dfrac{\partial F}{\partial \phi_i} \\[2mm] \dfrac{\partial F}{\partial \phi_i} \\[2mm] \cdot \\ \cdot \end{bmatrix} = 0 \tag{16.35}$$

This matrix equation is a system of simultaneous equations for each element, the number of equations being governed by the type of element used. This process is repeated for all the elements to obtain the final global system of equations, which is solved subject to given boundary conditions.

Although the approach is for problems involving self-adjoint operators, those involving non-self-adjoint operators have been rendered self-adjoint by the use of a transformation (Guymon, 1970; Guymon et al., 1970). However, the functional obtained by the transformations has been later proven to be non-unique (Smith et al., 1973). The overall matrix for ϕ (in structural dynamics, this matrix is known as the stiffness matrix) is symmetric for self-adjoint problems.

16.4.3.3 Method of weighted residuals – Galerkin's method

One of the most attractive features of the method of weighted residuals is that it can be applied to non-linear and non-self-adjoint problems. Of the many approximations used in this approach, Galerkin's approximation seems to give the best results (Zienkiewicz, 1971, p. 40). The method of weighted residuals, in general, consists of substituting an approximate solution to the constitutive equations and making the weighted residuals (errors) vanish. In mathematical notation, if

$$(\phi)^a = \sum N_i \phi_i = (N)(\phi)^\theta \tag{16.36}$$

then

$$\int_\Omega \left\{ L(\phi)^a - f \right\} W d\Omega = 0 \tag{16.37}$$

where $(\phi)^a$ is an approximate solution to Eq. 16.37 (in this case L is not restricted), (N) is a vector of interpolation functions (functions of the co-ordinates only), and W is any function of the co-ordinates and is known as the weighting function.

By choosing the number of weighting functions W_i's equal to the number of unknowns (ϕ^a)'s in the domain Ω, a set of simultaneous equations for $(\phi)^a$ can be obtained from Eq. 16.37. Different approximations are possible depending upon the way the weighting functions W_i's are defined within an element. The case where the weighting function is made equal to the interpolation function is known as Galerkin's method. Further mathematical treatment of Galerkin's method and the method of weighted residuals in general are well documented by several authors (Hutton and Anderson, 1971; Finlayson and Scriven, 1966; Crandall, 1956).

One disadvantage of Galerkin's method is that higher order continuity has to be satisfied when defining interpolation functions. However, in many problems, higher order derivatives can often be reduced by using integration by parts. If this reduction can be accomplished in the general form, then there is no restriction on the choice of interpolation functions and that continuity requirements of the reduced order of integrals only need to be satisfied. Natural or derivative type boundary conditions are automatically generated in the process of reducing the order of the integral. Forced boundary conditions are simply substituted into the final matrix equation before a solution is attempted. When the problem is linear and self-adjoint, the variational approach and Galerkin's approach lead to the same formulation.

16.4.3.3.1 Linear interpolation functions

A local co-ordinate system is required to define the interpolation functions and to carry out the integration over each element. It is best to use the natural co-ordinate system which permits the specification of a point within the element by a set of dimensionless numbers whose magnitude never exceeds unity (Desai and Abel, 1972; Oden, 1972). This system is used in defining the interpolation functions.

If v is the variable in the governing equation to be determined, its variation within an element can be represented as

$$v = Ax + B \tag{16.38}$$

where x is the local co-ordinate and A and B are constants. It can be shown that

$$A = \frac{v_j - v_i}{2}; \quad B = \frac{v_j + v_i}{2} \tag{16.39}$$

where v_j and v_i are the values of v at nodes j and i. Therefore,

$$v = \frac{1-x}{2} v_i + \frac{1+x}{2} v_j \tag{16.40}$$

In matrix notation, $(v) = \begin{pmatrix} N_i & N_j \end{pmatrix} \begin{bmatrix} v_i \\ v_j \end{bmatrix}$ \tag{16.41}

where

$$N_i = \frac{1}{2}(1-x); \quad N_j = \frac{1}{2}(1+x) \tag{16.42}$$

Similarly,

$$(w) = \begin{pmatrix} N_i & N_j \end{pmatrix} \begin{bmatrix} w_i \\ w_j \end{bmatrix}$$

$$(v) = \begin{pmatrix} N_i & N_j \end{pmatrix} \begin{bmatrix} v_i \\ v_j \end{bmatrix}$$

16.4.3.3.2 Quadratic interpolation functions

In general, the local variations of the various functions within an element may be approximated to any degree of accuracy by assuming higher order polynomials for interpolation functions. But polynomials of degree higher than three have rarely been used for finite element analysis. Various factors such as element continuity requirements, computing costs, accuracy of data, etc. limit the extent to which higher order interpolations functions can be used. For quadratic variations,

$$v = Ax^2 + Bx + C \tag{16.43}$$

Three nodes are required to determine A, B and C uniquely. For convenience, the third node is assumed to be located midway between the two end nodes. If v_i, v_j and v_k denote the nodal values of v at nodes i, j and k, then it can be shown that

$$A = \frac{1}{2}\left(v_i + v_j - 2v_k\right) \tag{16.44a}$$

$$B = \frac{1}{2}\left(-v_i + v_j\right) \tag{16.44b}$$

$$C = v_k \tag{16.44c}$$

Substituting back, Eq. 16.43 can be written as

$$v = v_i N_i + v_j N_j + v_k N_k \tag{16.45}$$

or in matrix notation,

$$v = \begin{bmatrix} N_i \ N_j \ N_k \end{bmatrix} \begin{bmatrix} v_i \\ v_j \\ v_k \end{bmatrix} \tag{16.46}$$

where

$$N_i = \frac{1}{2}x(x-1) \tag{16.47a}$$

$$N_j = \frac{1}{2}x(x+1) \tag{16.47b}$$

$$N_k = \left(1-x^2\right) \tag{16.47c}$$

Similar expressions can be written for other variables too.

16.5 PROBLEMS ASSOCIATED WITH NUMERICAL METHODS

16.5.1 Consistency

Consistency implies that the numerical equation approaches the continuum equation as the mesh size decreases to zero.

16.5.2 Convergence

If the numerical solution approaches the exact solution as the mesh size decreases to zero, then the scheme is said to be convergent. It means that the difference equation approaches the differential equation as the mesh size tends to zero. It is difficult to prove this condition except for problems which have closed form solutions.

16.5.3 Stability

In a numerical solution, computations can only be done up to a finite number of significant digits resulting in **round-off errors**. If these round-off errors do not build up as the number of computations increase, then the scheme is said to be stable. Stability depends on the type of scheme used. A necessary condition for the stability of explicit methods is

$$0 < \Delta t < \frac{1}{2} \frac{S}{T} \frac{(\Delta x)^2 (\Delta y)^2}{(\Delta x)^2 + (\Delta y)^2}$$

(This is for the two-dimensional equation for flow through porous media.)

In this condition, $\Delta t > 0$, which is trivial, is also a necessary condition. This means that it is not possible to work back in time. Although a smaller Δt is preferred from a stability point of view, it has the opposing effect of introducing more round-off errors because more computations will be needed to reach the same time level.

16.5.4 Accuracy

Accuracy implies that the numerical solution does not deviate too much from the true exact solution. It depends on

- the mesh size,
- finite difference approximation (whether backward, forward or central differencing), and
- type of problem (e.g., presence of areas of sharp variations of the function to be evaluated).

In general, the error is proportional to the square of the mesh size.

16.6 CFD SOFTWARE

The beginning of CFDs coincides with the advent of digital computers and the first numerical computations of fluid flow based on the Navier–Stokes equations were carried out in the 1950s at Los Alamos National Laboratory in USA. The development and use of commercial software began in the 1960s followed by developments in Boeing and NASA in the 1970s. Since then, there have been rapid developments in both the methods of analysis as well as the computing power of computers. Now there are over 200 CFD commercial software packages in circulation. The choice for the user depends upon the problem to be analyzed and the associated costs. Generally, CFD software can be classified into five categories as OPEN SOURCE software, OPEN WRAPPER software, CAD integrated software, Specialty software and Comprehensive packages. Open SOURCE software allows the user to access the source code, change and improve under license. OPEN Wrappers have bundled additional software such as pre- and post-processers. Examples include Visual-CFD, Caedium, HELYX and SimFlow. CAD integrated software include SolidWorks and Autodesk Inventor, and they include meshing and post-processing tools. They are catered mainly to product designers. Specialty software are primarily for specialized flow problems. Examples include CONVERGE which has the flexibility to accommodate moving meshes, multi-phase flow, etc. Other software in this category include FloTHERM, FINE/Marine, 6Sigma, EXA, XFlowCFD, CFX, etc. Comprehensive Packages include the widely used and the most popular Fluent and Star-CCM+ as well as the less widely used COMSOL CFD module and Altair's AcuSolve. OPEN SOURCE codes are cost-effective with increasing usage and reliable as they are used, scrutinized and improved by the developers as well as by skilled users.

16.6.1 ANSIS

ANSIS has a range of software of which ANSYS FLUENT and ANSYS CFX are the most popular and widely used ones. There are other ANSYS software for specific types of analysis, such as for aerodynamics, simulating aircraft icing and custom made ones for various industrial design analysis. They cover a broad range of fluid flow analysis including turbulence modelling, cavitation, turbo-machinery, multi-phase flow, fluid structure interaction, combustion, heat transfer, and, reacting flows under steady-state and transient conditions. The package GAMBIT does the mesh generation for using FLUENT and CFX. The graphical outputs of ANSYS CFX or ANSYS FLUENT can show how the fluid flow variables evolve with time.

16.6.2 PHOENICS

PHONICS is a commercial software package that can be used for the analysis of fluid flow, heat and mass transfer, chemical reactions, combustion, etc. It uses the FVM in which the original partial differential equations are converted to algebraic equations by discretizing the solution domain into a finite number of control volumes. The finite volume equations are obtained by integrating the governing partial differential equations. The numerical procedure used in PHOENICS is the FVM with discretization assumptions for the transient, convection, diffusion and source terms. Fully implicit backward difference scheme is used for the transient terms and central differencing is used for the diffusion terms. The software has the facility to use higher order convection schemes consisting of 5 linear schemes and 12 non-linear schemes. The linear schemes include CDS, QUICK, linear upwind and cubic upwind. Non-linear schemes include SMART, H-QUICK, UMIST, SUPERBEE, MINMOD, OSPRE, MUSCL and van-Leer harmonic.

REFERENCES

Clough, R. W. (1960): The finite element in plane stress analysis, *Proceedings of the 2nd ASCE Conference on Electronic Computation*, Pittsburgh, PA.

Crandall, S. H. (1956): *Engineering Analysis: A Survey of Numerical Procedures*, McGraw Hill Book Co., New York, 417 pp.

Desai, C. S. and Abel, J. F. (1972): *Introduction to the Finite Element Method, a Numerical Method for Engineering Analysis*, Van Nostrand Reinhold Co. New York, 477 pp.

Finlayson, B. A. and Scriven, L. E. (1966): The method of weighted residuals: A review, *Applied Mechanics Review*, vol. 19, no. 9, pp. 735–748.

Forsythe, G. E. and Wasow, W. R. (1960): *Finite Difference Methods for Partial Differential Equations*, John Wiley & Sons Inc., New York, 444 pp.

Guymon, G. L. (1970): A finite element solution of the one-dimensional diffusion-convection equation, *Water Resource Research*, vol. 6, no. 1, pp. 204–210.

Guymon, G. L., Scott, V. H. and Herrmann, L. R. (1970): A general numerical solution of the two-dimensional diffusion-convection equation by the finite element method, *Water Resources Research*, vol. 6, no. 6, pp. 1611–1617.

Hutton, S. G. and Anderson, D. L. (1971): Finite element method, a Galerkin approach, *Proceedings of the ASCE Journal of Engineering Mechanics Division*, vol. 97, no. EM5, paper 8448, pp. 1503–1519.

McCormack, R. W. and Paullay, A. J. (1972): Computational efficiency achieved by time splitting of finite difference operators, AIAA paper, pp 72–154, San Diego, CA.

McDonald, P. W. (1971): The computation of transonic flow through two-dimensional gas turbine cascades, ASME paper 71-GT-89. p. 7.

Oden, J. T. (1972): *Finite Elements of Non-Linear Continua*, McGraw Hill Book Co., New York. p. 432.

Smith, I. M., Farraday, R. V. and O'Connor, B. A. (1973): Rayleigh-Ritz and Galerkin finite elements for diffusion-convection problems, *Water Resources Research*, vol. 9, no. 3, pp. 593–606.

Turner, M. J., Clough, R. W., Martin, H. C. and Topp, L. J. (1956): Stiffness and deflection analysis of complex structures, *Journal of Aeronautical Sciences*, vol. 23, no. 9, pp. 805–823.

Versteeg, H. K. and Malalasekera, W. (1995): *An Introduction to Computational Fluid Dynamics: The Finite Volume Method*, Longman Scientific & Technical, New York. 257 pp.

Zienkiewicz, O. C. (1971): *The Finite Element Method in Engineering Science*, McGraw Hill Book Co., London. 521 pp.

Chapter 17

Open channel flow

17.1 INTRODUCTION

Open channel flow refers to flows in which the surface of flow is at atmospheric pressure. They may be natural or artificial by origin, prismatic or non-prismatic by geometry (prismatic channels have uniform cross section and constant slope). Examples include rivers, canals, flumes, sewers running partly full, etc. Forces acting are the gravity (driving force) where the Froude number is important and the viscous shear at the boundary (retarding force) where the Reynolds number is important. Therefore, both Re[1] and Fr are important in open-channel flows.

The variables involved in describing open channel flows include the depth h, normal to surface, the pressure at the bed ($= h \cos \theta \cong h$ vertical), the velocity V, the bed slope S_0 ($S_0 = \tan \theta \cong \sin \theta$) and the frictional slope S_f (S_f = slope of the energy line).

17.2 TYPES OF FLOW

17.2.1 Uniform or non-uniform

Uniform flow has the surface profile parallel to the bed. Velocity need not be constant over the entire cross section, but maybe the same at corresponding points along the length.

$$\frac{\partial V}{\partial x} = 0 \Rightarrow \frac{\partial h}{\partial x} = 0; \quad \frac{\partial A}{\partial x} = 0 \left(\frac{\partial V}{\partial y} \text{ and } \frac{\partial V}{\partial z} \text{ need not be zero} \right)$$

17.2.2 Steady or unsteady

For steady flow, $\frac{\partial}{\partial t} = 0$, and for unsteady flow, $\frac{\partial}{\partial t} \neq 0$.

17.2.3 Gradually varied or rapidly varied

For gradually varied flow, $\frac{\partial}{\partial x}$ is small, and for rapidly varied flow, $\frac{\partial}{\partial x}$ is large.

---

[1] $Re = \dfrac{\text{Inertia force}}{\text{Viscous force}}$ $Fr = \dfrac{\text{Inertia force}}{\text{Gravity force}}$ $We = \dfrac{\text{Inertia force}}{\text{Surface tension force}}$ (Weber No.)

17.2.4 Tranquil or rapid

For tranquil flow, the Froude number $Fr < 1$, which means that a small disturbance can travel upstream and downstream. Froude number has the same ratio as the ratio of the velocity to the velocity of a wave due to a small disturbance. For rapid flow, Froude number $Fr > 1$, which means a small disturbance cannot travel upstream. It will be washed downstream.

17.2.5 Laminar or turbulent

For laminar flow, Reynolds number is small (<about 600), and the flow is controlled by viscous forces. For turbulent flow, Reynolds number is large (>about 600) and the flow is controlled by inertia forces. The characteristic length for the Reynolds number is the hydraulic mean depth (radius), R. Most practical problems are turbulent.

In laminar flows, the fluid particles in one layer stay in the same layer and the layers slide without any crossing of particles from one layer to the next layer. Momentum transfer between fluid layers is by molecular exchange. Shear stress between fluid layers is due to viscosity only according to Newton's law of viscosity. The velocity profile in laminar flows is parabolic and the maximum velocity in the cross section is twice the cross sectional mean velocity.

In turbulent flow the fluid particles jump from one layer to the adjacent layer and mix. Eddies and swirls exist in turbulent flow. Turbulent flows are three-dimensional and time-dependent. Momentum transfer between fluid particles takes place due to molecular activity and due to particle motion. Shear stress between fluid layers is due to viscosity and turbulence. There is significant mixing due to turbulence. Different velocity profiles have been proposed for turbulent flows.

17.2.6 Critical

Critical flow occurs when the Froude number is equal to 1. It also means that the specific energy is a minimum.

17.2.7 Sub-critical

Sub-critical flows refer to flows where the Froude number is less than 1. It also means the flow is slow and the depth of flow is relatively high. Sub-critical flows are also identified as tranquil flows.

17.2.8 Super-critical

Super-critical flows refer to flows where the Froude number is greater than 1. It also means that the flow is fast and the depth of flow is relatively low. Super-critical flows are also identified as rapid flows.

17.3 FLOW ENERGY

17.3.1 Total energy

The total energy, E, expressed as a head consists of the sum of pressure energy, kinetic energy and potential energy as given by

$$E = \frac{p}{\rho g} + \alpha \frac{V^2}{2g} + z \qquad (17.1a)$$

where α is a kinetic energy correction factor, sometimes called the Coriolis coefficient.

This is a constant along a streamline when friction is negligible. If the streamlines are parallel and straight, the pressure is hydrostatic. Otherwise, the curvature will introduce acceleration normal to the streamlines resulting in differences in pressures in addition to hydrostatic pressures. Since $p = h\rho g$, the total energy expressed as a height becomes

$$E = h + \alpha \frac{V^2}{2g} + z \qquad (17.1b)$$

17.3.2 Energy gradient (frictional slope, S_f)

The energy gradient is the slope of the total energy line. For uniform flow, it will be parallel to the bed slope, thus making $S_0 = S_f$.

$$S_f = \frac{dE}{ds}$$

17.4 STEADY UNIFORM FLOW

This is the simplest of all open channel flows. All variables are invariant with time and distance along the flow direction. There can, however, be velocity variations across a cross section. Uniform flow can occur as sub-critical or super-critical. A certain minimum length is necessary for the flow to attain uniform conditions. In this transition region, the flow is accelerating. At the critical slope, the flow is undulating although on average it may be considered as uniform. Normal depth is the depth of uniform flow.

Most empirical equations for the mean velocity of uniform turbulent flow in open channels are of the form

$$V = CR^x S^y \qquad (17.2)$$

where V is the mean velocity, R is the hydraulic radius, S is the slope of the energy line, x and y are indices and C is a factor of flow resistance, which is a function of V, R, roughness, viscosity, etc. There are two commonly used equations to calculate the cross-sectional average velocity.

17.4.1 Chezy's equation (1769)

Considering a control volume of fluid, the weight of the fluid in the direction of flow is balanced by the resistance force (momentum equation in the steady uniform flow condition)

$$W \sin \alpha = \tau_0 P \Delta x \qquad (17.3)$$

$$A \Delta x \rho g \sin \alpha = \tau_0 P \Delta x$$

$$\tau_0 = \frac{A}{P} \rho g \sin \alpha = R \rho g S_0. \qquad (17.4)$$

The boundary shear stress in terms of the dynamic pressure can be expressed as

$$\tau_0 = C_D \frac{1}{2} \rho V^2 = f \frac{1}{2} \rho V^2 \tag{17.5}$$

Therefore,

$$f\rho V^2 = 2\rho g R S_0 \Rightarrow V = \left(\frac{R S_0 2g}{f} \right)^{\frac{1}{2}} = C \left(R S_0 \right)^{\frac{1}{2}} \tag{17.6}$$

where $C = \left(\dfrac{2g}{f} \right)^{\frac{1}{2}}$. C has the dimension of $[g]^{\frac{1}{2}}$.

17.4.2 Manning's equation (1889)

$$V = \frac{1}{n} R^{\frac{2}{3}} S_0^{\frac{1}{2}} \tag{17.7a}$$

In English units, Manning's equation is given as

$$V = \frac{1.49}{n} R^{\frac{2}{3}} S_0^{\frac{1}{2}} \tag{17.7b}$$

Comparing Chezy's equation with Manning's equation, the relationship between Chezy's 'C' and Manning's 'n' can be expressed as

$$C = \frac{1}{n} R^{\frac{1}{6}} \tag{17.8a}$$

In English units, it can be expressed as

$$C = \frac{1.49}{n} R^{\frac{1}{6}} \tag{17.8b}$$

Equation (17.7) is also called the Strickler equation (in Europe), and $1/n$ is the Strickler coefficient. Strictly speaking, n has a dimension of $[TL^{-1/3}]$, but it makes no sense to have a surface roughness dependent on time. Therefore, the equation is considered as a numerical formula applicable to a particular set of units only; in this case, SI units.

Manning's 'n' is a function of

- Surface roughness
- Vegetation
- Channel irregularity
- Channel alignment
- Silting and scouring
- Obstructions
- Size and shape of channel
- Stage and discharge
- Seasonal changes
- Bed load and suspended load.

Values of n vary from about 0.01 to about 0.12. Typical values for some surfaces are

- Cast iron conduits ~ 0.013
- Concrete trowel finish ~ 0.013
- Streams (no vegetation) ~ 0.03

A more detailed table of Manning's roughness coefficient 'n' based on field and laboratory investigations from various sources has been compiled by Chow (1959, pp. 110–113).

17.5 OPTIMUM CHANNEL CROSS SECTION

Discharge is a function of hydraulic mean depth R which is the cross-sectional area of flow divided by the wetted perimeter ($Q = AV = f$ (hydraulic mean depth, R)). For a given area A, n and S, Q is a maximum when R is a maximum, or, when P is a minimum. Therefore, an optimum cross section is one which has a minimum wetted perimeter.

The best closed section is a circle running full and a semi-circle when running partially full. The best polygonal section is one circumscribed about a semi-circle for free surface flow, or about a circle for closed flow.

For a given number of sides, a regular polygon or half polygon is better than an irregular one, and one with a larger number of sides is better than one with a smaller number of sides.

17.5.1 Prismatic shapes

For non-rectilinear shapes, the semi-circle has the maximum R. For a trapezoidal section,

$$A = bh + \frac{h^2}{\tan \alpha} \Rightarrow b = \frac{A}{h} - \frac{h}{\tan \alpha} \tag{17.9}$$

$$P = b + \frac{2h}{\sin \alpha} \tag{17.10}$$

$$R = \frac{A}{P} = \frac{A}{\left[\dfrac{A}{h} - \dfrac{h}{\tan \alpha} + \dfrac{2h}{\sin \alpha} \right]} \tag{17.11}$$

This is a maximum when the denominator is a minimum; i.e. when its partial derivative (with respect to h or α) is zero.

17.5.1.1 With respect to h

$$-\frac{A}{h^2} - \frac{1}{\tan \alpha} + \frac{2}{\sin \alpha} = 0$$

$$\Rightarrow A = h^2 \left\{ \frac{2}{\sin \alpha} - \frac{1}{\tan \alpha} \right\} \tag{17.12}$$

Substituting this value in the expression for R gives

$$R = \frac{h^2 \left\{ \dfrac{2}{\sin \alpha} - \dfrac{1}{\tan \alpha} \right\}}{h \left\{ \dfrac{2}{\sin \alpha} - \dfrac{1}{\tan \alpha} \right\} - \dfrac{h}{\tan \alpha} + \dfrac{2h}{\sin \alpha}}$$

(17.13)

$$R = \frac{h \left\{ \dfrac{2}{\sin \alpha} - \dfrac{1}{\tan \alpha} \right\}}{2 \left\{ \dfrac{2}{\sin \alpha} - \dfrac{1}{\tan \alpha} \right\}} = \frac{h}{2}$$

Special case: Rectangular section, $\alpha = 90°$

$$R = \frac{h}{2}$$

$$A = bh = h^2 \left\{ \frac{2}{\sin 90} - \frac{1}{\tan 90} \right\}$$

$$= 2h^2$$

Therefore, $b = 2h$.

17.5.1.2 With respect to α

If differentiation is carried out with respect to α, it can be shown that R is a maximum when $\alpha = 60°$, i.e. a hexagonal channel half full is the optimum.

17.5.2 Closed circular conduits running partly full

The cross-sectional area of flow, A, and the wetted perimeter, P, can be expressed as functions of the radius, r and the angle the water surface subtends at the centre, 2θ, as follows:

$$A = r^2\theta - 2\left(\frac{1}{2} r \sin\theta \, r \cos\theta \right)$$

(17.14)

$$= r^2 \left(\theta - \frac{1}{2} \sin 2\theta \right)$$

and

$$P = r(2\theta)$$

(17.15)

The velocity and discharge may be calculated using Manning's equation.

Since the depth of flow, h, is a function of θ, the variables A, P, R, V and Q can be expressed as fractions of the corresponding values when the conduit is flowing full. It can be shown that the maximum flow occurs not when flowing full, but when $\dfrac{h}{D}$ is about 0.94

(D is the diameter of the pipe). The maximum value of the velocity occurs when $\dfrac{h}{D}$ is about

0.81. Both these conditions can be obtained by differentiating Q and V with respect to θ. It can be seen that the maximum velocity and maximum discharge are both greater than their corresponding full flow values.

17.6 SPECIFIC ENERGY

The specific energy E_s (first introduced by Bahkmeteff (1912)) is defined as

$$E_s = h + \frac{V^2}{2g} = h + \frac{1}{2g}\left[\frac{Q}{A}\right]^2 \tag{17.16}$$

In the definition of total energy, $(h+z)$ is the height of the free surface above a datum. Specific energy is measured with respect to the bed of the channel. It may increase, remain constant or decrease.

Of the three variables involved, E_s, h and Q, any one depends on the other two. The minimum value of E_s occurs when

$$\frac{\partial E_s}{\partial h} = 1 + \left(\frac{Q^2}{2g}\right)\left(-\frac{2}{A^3}\frac{\partial A}{\partial h}\right) = 0 \tag{17.17}$$

As $\delta h \Rightarrow 0, \dfrac{\delta A}{\delta h} \Rightarrow \dfrac{\partial A}{\partial h} \Rightarrow B$, because $\delta A = B \delta h$.

Therefore,

$$\frac{\partial E_s}{\partial h} = 1 - \left(\frac{Q^2}{gA^3}\right)B = 0, \text{ giving } Q^2 = \frac{gA^3}{B}$$

$$\tag{17.18}$$

$$\Rightarrow \frac{gA}{B} = \frac{Q^2}{A^2} = V^2$$

If $\dfrac{A}{B}$ is the mean depth of flow, $\bar{h}$ $(\neq R)$, then

$$V = \left(g\bar{h}\right)^{\frac{1}{2}} \tag{17.19}$$

This condition, which corresponds to minimum specific energy, is known as the critical condition. The corresponding depth, h_c, and velocity, V_c, are called the critical depth and critical velocity, respectively (The critical velocity is the velocity at which the depth of flow is critical). The specific energy diagram is shown in Figure 17.1.

If a rectangular channel of cross-sectional area, bh and discharge q per unit width is considered (i.e. $Q = qb$)

$$E_s = h + \frac{1}{2g}\left(\frac{Q}{A}\right)^2 = h + \frac{1}{2g}\left(\frac{qb}{hb}\right)^2 = h + \frac{1}{2g}\left(\frac{q}{h}\right)^2 \tag{17.20}$$

and critical condition occurs when $\dfrac{\partial E_s}{\partial h} = 0$.

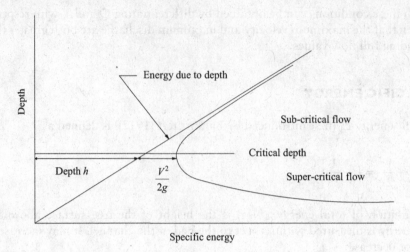

Figure 17.1 Specific energy diagram.

$$\frac{\partial E_s}{\partial h} = 1 + \left(\frac{q^2}{2g}\right)\left(-\frac{2}{h^3}\right) = 0$$

$$\Rightarrow q^2 = gh^3$$

$$\Rightarrow h^3 = \frac{q^2}{g} \qquad (17.21)$$

$$\Rightarrow h = \left(\frac{q^2}{g}\right)^{1/3}, \text{ or } h_c = \left(\frac{q^2}{g}\right)^{1/3}$$

Therefore, the minimum (critical) specific energy, E_{sc} is given as

$$E_{sc} = h_c + \left(\frac{1}{2g}\right)\frac{gh_c^3}{h_c^2} = \frac{3}{2}h_c \qquad (17.22)$$

The critical depth corresponds to minimum specific energy. For any other value of E_s, there will be two depths. These are known as the alternate depths (or conjugate depths). For a given value of q, when h is small, $\frac{v^2}{2g}$ and E_s will be large. On the other hand, when h is large, $\frac{v^2}{2g}$ is small and E_s is asymptotically equal to h. As q increases, h_c also increases and therefore, for large values of q the h vs. E_s curves tend to move away from the origin.

Since any one of the three variables depends on the other two, the condition $\frac{\partial q}{\partial h} = 0$ corresponds to maximum q:

$$E_s = h + \frac{1}{2g}\left(\frac{q}{h}\right)^2 \Rightarrow q^2 = 2gh^2\left(E_s - h\right)$$

Therefore,

$$2q\frac{\partial q}{\partial h} = 2g\left(2E_s h - 3h^2\right)$$

and

$$\frac{\partial q}{\partial h} = 0 \text{ when } 2E_s h = 3h^2 \text{ i.e. when } h_c = \frac{2}{3}E_{sc} \text{ or } E_s = \frac{3}{2}h_c \qquad (17.23)$$

Equation (17.23) is the same as Eq. (17.22). Therefore, at the critical depth, the discharge is a maximum for a given specific energy and the specific energy is a minimum for a given discharge. At critical conditions, $q^2 = gh_c^3$. Since $q = Vh$, the following relationship can be written:

$$\left(V_c h_c\right)^2 = gh_c^3 \Rightarrow V_c = \left(gh_c\right)^{\frac{1}{2}} \qquad (17.24)$$

The critical slope is the slope at which critical uniform flow is attained. This is possible in long open channels if the slope is suitable.

Using Chezy's equation,

$$u_c = C\sqrt{RS_c} = \sqrt{gh_c} \text{ gives the critical slope}$$

Using Manning's equation,

$$u_c = \frac{1}{n}R^{\frac{2}{3}}S_c^{\frac{1}{2}} \text{ also gives the critical slope.}$$

The critical slope determines whether a physical slope is mild or steep. If the physical slope is less than the critical slope, it is considered as mild and if it is greater than the critical slope, it is considered as steep. Since critical slope is a function of depth, which is a function of flow rate, different flow rates have different critical slopes. A channel which has a mild slope for one flow rate may have a steep slope for another flow rate.

The critical velocity is also the velocity of propagation of a small surface wave in shallow water. When $V < V_c$, it is possible for a surface wave to propagate upstream and downstream. When $V > V_c$, it can propagate downstream only. When $V = V_c$, a standing (or stationary) wave is formed. This also means that when $V < V_c$, downstream effects can influence upstream flow behaviour.

When $V < V_c$, the flow is called tranquil or sub-critical flow. When $V > V_c$, the flow is called rapid or super-critical flow. The ratio $\dfrac{V}{V_c}\left(=\dfrac{V}{\sqrt{gh_c}}\right)$ has the same form as the Froude number Fr, which is defined as the ratio of inertia forces to gravity forces. Therefore, critical, sub-critical and super-critical flow can be identified with the Froude number as well.

- For critical conditions $Fr = 1$
- For sub-critical conditions $Fr < 1$
- For super-critical conditions $Fr > 1$.

17.7 HYDRAULIC JUMP

When rapid flow takes place such as for example through a sluice gate, it cannot be sustained indefinitely without a sufficiently steep bed slope. For a given discharge, the uniform flow depth is determined by friction in the cross section and the bed slope.

$$Q = AV = f\left(h, S_0, n\right) \Rightarrow h = f_1\left(Q, S_0, n\right) \Rightarrow h = f_2\left(S_0, n\right)$$

For a mild slope, this uniform flow depth is greater than h_c, and hence the flow is tranquil. A gradual transition from rapid flow to tranquil flow is not possible, as can be seen from the specific energy diagram. If a smooth transition is to take place, it should follow the specific energy curve which means that E_s must decrease with increasing h. But E_s can decrease only up to E_{sc} because thereafter E_s is increasing again. This is not possible, because if a horizontal flow plane is considered, the specific energy will be the same as the total energy when the datum is at the bed of the channel, and total energy cannot increase in the direction of flow. Therefore, a discontinuity occurs in the flow when changing from rapid to tranquil. This is known as the hydraulic jump. The specific energy diagram is not valid at the jump. Most hydraulic jumps that can be seen in waterways are stationary jumps, but there are a few moving hydraulic jumps which are also referred to as tidal bores in certain tidal rivers in some parts of the world. Hydraulic jumps are characterized by high turbulence, air entrainment and rolling effect.

17.7.1 Stationary hydraulic jump

Stationary hydraulic jumps (Figure 17.2) are sometimes referred to as standing waves, which occur when a super-critical flow encounters an obstruction or a sudden change of slope. After the jump, the flow becomes sub-critical with an increased depth of flow and reduced velocity. Some energy is lost in the jump. They can be seen at the bottom of spillways and when water discharges from a sluice gate.

Considering a horizontal rectangular channel, the following equations can be written:

Continuity: $Q = A_1V_1 = A_2V_2 \Rightarrow h_1V_1 = h_2V_2$ (17.25)

Momentum: $p_1A_1 - p_2A_2 = \rho A_2V_2^2 - \rho A_1V_1^2$

$$\Rightarrow \frac{1}{2}\rho gh_1^2 - \frac{1}{2}\rho gh_2^2 = \rho h_2V_2^2 - \rho h_1V_1^2$$ (17.26)

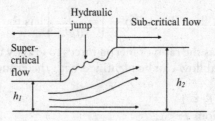

Figure 17.2 Hydraulic jump.

Energy: $\dfrac{p_1}{\rho g}+\dfrac{V_1^2}{2g}+z_1=\dfrac{p_2}{\rho g}+\dfrac{V_2^2}{2g}+z_2+h_L$

$$\Rightarrow h_1+\frac{V_1^2}{2g}=h_2+\frac{V_2^2}{2g}+h_L \qquad (17.27)$$

Since $q=h_1V_1=h_2V_2\Rightarrow V_2=\dfrac{q}{h_2};\quad V_1=\dfrac{q}{h_1}$

Substituting the relationships

$q=h_1V_1=h_2V_2$

$\Rightarrow V_2=\dfrac{q}{h_2};\quad V_1=\dfrac{q}{h_1}$

in the momentum equation yields

$\dfrac{1}{2}\rho gh_1^2-\dfrac{1}{2}\rho gh_2^2=\rho h_2V_2^2-\rho h_1V_1^2\Rightarrow\dfrac{1}{2}gh_1^2-\dfrac{1}{2}gh_2^2=h_2V_2^2-h_1V_1^2$

$\Rightarrow\dfrac{1}{2}g\left(h_1^2-h_2^2\right)=\dfrac{q^2}{h_2}-\dfrac{q^2}{h_1}$

$\Rightarrow\dfrac{1}{2}g(h_1-h_2)(h_1+h_2)=q^2\dfrac{h_1-h_2}{h_1h_2}\quad\Rightarrow h_1+h_2=\dfrac{2q^2}{gh_1h_2}=2Fr_1^2\dfrac{h_1^2}{h_2}\qquad (17.28a)$

$\Rightarrow\left(\dfrac{h_2}{h_1}\right)^2+\left(\dfrac{h_2}{h_1}\right)-2Fr_1^2=0$

$\Rightarrow\dfrac{h_2}{h_1}=\dfrac{1}{2}\left(\sqrt{1+8Fr_1^2}-1\right)$

It can also be shown that

$$\Rightarrow\frac{h_1}{h_2}=\frac{1}{2}\left(\sqrt{1+8Fr_2^2}-1\right) \qquad (17.28b)$$

The head loss h_L can be shown to be

$$h_L=\frac{(h_2-h_1)^3}{4h_1h_2} \qquad (17.29)$$

The strength of the hydraulic jump increases as the Froude number increases.

17.7.2 Circular hydraulic jump

A circular hydraulic jump can be seen when a vertical jet of water impinges on a horizontal plane surface. It is also a type of stationary hydraulic jump.

17.7.3 Moving hydraulic jump – tidal bore

A moving hydraulic jump, or tidal bore, occurs when a tide travels upstream against the direction of flow of a tidal river or a narrow bay. A moving hydraulic jump can also be formed as a cascade of roll waves or undulating waves moving downstream and overtaking a shallower downstream flow of water. Some conditions necessary for a tidal bore to be formed include a fairly shallow river, a narrow outlet to the sea and a reasonably high tidal range of the order of about 6 m.

The world's largest tidal bore is in the Qiantang River in Hangzhou, China. It is reported to have 9 m high tidal waves travelling upstream at speeds as fast as 40 km/h with a roar that can be heard hours before it moves up the river. The bore has become a famous tourist attraction for surfers and surfing events, as well as for onlookers.

The Amazon River in Brazil and the Orinoco River in Venezuela are known to have tidal bores up to 4 m high moving upstream at up to 21 km/h. Locally the bore in the Amazon River is known as *pororoca*. Another well-known tidal bore can be observed in the River Severn which flows out into Bristol Channel in the UK. The Severn Estuary is reported to have the third largest tidal range in the world, amounting to some 15 m. Only the Bay of Fundy and Ungava Bay, both in Canada have higher tidal ranges. The highest tidal bore in the Severn River in Maryland is reported to be about 2.8 m. Other notable tidal bores include that in Petitcodiac River in North America with over 2 m high waves; Bono, Kampang River at Meranti Bay; Pelalawan, in Indonesia with waves over 6 m travelling upstream to about 40–130 km; Batang River in Malaysia, locally known as benak; Ganges–Brahmaputra Rivers in India and Bangladesh; Indus River in Pakistan; and Sittaung River in Myanmar.

Factors that contribute to a high tidal bore include low freshwater outflow, low atmospheric pressure, wind speed and direction, narrowing river section and the season. Timing of tidal bores is not regular. They can occur daily, like in the Batang River in Malaysia, known locally as *benak*. *Pororoca* in the Amazon River occurs during spring tides which coincide with the times of new moon and full moon.

Tidal bores are characterized by high degree of turbulence, mixing, sediment erosion beneath the bore front and the banks as well as rumbling noise. The rumbling noise can be heard far away because of the low frequency sound.

17.8 ROUGHNESS PARAMETER

Several empirical equations are available for the estimation of the roughness parameter C in Chezy's equation. They depend upon the units used. Typical equations include the following.

17.8.1 Ganguillet and Kutter (1869)

$C = f(S, n, R)$; n is Kutter's coefficient of roughness   (17.30a)

In English units, it takes the form

$$C = \frac{41.65 + \dfrac{0.00281}{S} + \dfrac{1.811}{n}}{1 + \left(41.65 + \dfrac{0.00281}{S}\right)\left(\dfrac{n}{\sqrt{R}}\right)}$$ (17.30b)

17.8.2 Bazin (1897)

$$C = f(R, m); \; m \text{ is the coefficient of roughness} \tag{17.31a}$$

In English units, it takes the form

$$C = \frac{157.6}{1 + \dfrac{m}{\sqrt{R}}} \tag{17.31b}$$

The coefficient of roughness m in this equation is different from Kutter's n in Eq. 17.30. For example, smooth surfaces have a value of $m = 0.11$.

17.8.3 Powell (1950)

$$C = f(C, R, \text{Re}, \varepsilon); \; \varepsilon \text{ is the channel roughness parameter.} \tag{17.32a}$$

In English units, it takes the form

$$C = -42\log\left(\frac{C}{4\,\text{Re}} + \frac{\varepsilon}{R}\right) \tag{17.32b}$$

This is an implicit formula.

In compound channels, the equivalent roughness parameter is estimated by dividing the flow area into several parts with wetted perimeters $P_1, P_2, P_2 \ldots$ etc. with associated roughness coefficients n_1, n_2, n_3 etc (if Manning's equation is used). Suggested equations include

$$n = \left(\frac{\sum\limits_{i=1}^{N} P_i n_i^{3/2}}{P}\right)^{2/3} \tag{17.33}$$

$$n = \left(\frac{\sum\limits_{i=1}^{N} P_i n_i^{2}}{P}\right)^{1/2} \tag{17.34}$$

$$n = \left(\frac{P R^{5/3}}{\sum\limits_{i=1}^{N} \dfrac{P_i R_i^{5/3}}{n_i}}\right) \tag{17.35}$$

where P is the total wetted perimeter. Total discharge in a compound channel can therefore be expressed as

$$Q = \left(\sum\limits_{i=1}^{N} \frac{A_i R_i^{2/3}}{n_i}\right) S_0^{1/2} \tag{17.36}$$

if the same slope is assumed for all parts of the compound channel.

17.9 CONVEYANCE OF A CHANNEL SECTION

The conveyance of a channel is a function of the cross-sectional area of flow and the mean velocity, and can be expressed as

$$Q = AV = ACR^\alpha S^\beta = KS^\beta \tag{17.37}$$

where

$$K = CAR^\alpha \tag{17.38}$$

The term K is known as the conveyance of the channel section. It is a measure of the carrying capacity of the section.

$$Q \propto K$$

When Chezy's equation is used, $\beta = 0.5$,

$$Q = K\sqrt{S} \Rightarrow K = \frac{Q}{\sqrt{S}}, \text{and also } K = CAR^{1/2} \tag{17.39}$$

When Manning's equation is used,

$$K = \frac{Q}{\sqrt{S}}, \text{and } K = CAR^{2/3} = \frac{1}{n} AR^{2/3} \tag{17.40}$$

The term $AR^{2/3}$ is called the section factor for uniform flow computation which can be expressed as

$$AR^{2/3} = nK = \frac{nQ}{\sqrt{S}} \tag{17.41}$$

For a given flow condition with n, Q and S given, there is only one value of $AR^{2/3}$ which implies that there is only one value of depth for maintaining uniform flow provided that $AR^{2/3}$ always increases with increasing depth. This is true in most cases. This depth is the normal depth.

When n and S are known at a section, there can be only one value for Q (subject to the above proviso) and this value is the normal discharge.

To simplify the calculation, dimensionless curves showing the function $AR^{2/3}$ vs. h have been prepared for rectangular, trapezoidal and circular sections. In circular sections, the curve is turning back. Unlike in rectangular or trapezoidal sections, the flow and velocity do not always increase with increasing depth. This is due to the convergence of the side slopes at the top.

The maximum velocity occurs when $h = 0.81h_0$. The maximum discharge occurs when $h = 0.938h_0$. Mathematically, this can be obtained by making

$$\frac{d\left(AR^{2/3}\right)}{dy} = 0$$

Also, since $V = \dfrac{1}{n} R^{2/3} S^{1/2}$, maximum velocity is obtained when $\dfrac{d\left(R^{2/3}\right)}{dy} = 0$ for given n and S in both cases.

Furthermore, when $h > 0.81 h_0$, it is possible to have two different depths for the same discharge; one $> 0.938 h_0$ and the other $< 0.938 h_0$.

Similarly, when $h > 0.5 h_0$, it is possible to have two different depths for the same velocity, one above and the other below $0.81 h_0$.

The above results are based on the assumption that Manning's n is constant (or independent of h). In fact, n increases by as much as 25% as the depth is reduced from h to $0.25 h$. This effect causes the actual maximum discharge and velocity to take place at depths of about $0.97 h_0$ and $0.94 h_0$.

When n is constant, the velocity at half depth is equal to the velocity at full depth. When n is varying, the velocity at half depth is approximately equal to 0.8 times the velocity at full depth.

Example 17.1

For a circular pipe not running full, find relationships for the ratio of depth to diameter for maximum velocity and for maximum discharge based on Chezy's equation for velocity

$$\text{Area of flow} = r^2\theta - r^2 \sin\theta \cos\theta = r^2\left(\theta - \frac{1}{2}\sin 2\theta\right)$$

$$\text{Wetted perimeter} = 2r\theta$$

For maximum velocity, R should be a maximum because $V = C\sqrt{RS}$

$$\frac{\partial}{\partial}\left(\frac{A}{P}\right) = \frac{1}{P^2}\left(P\frac{dA}{d\theta} - A\frac{dP}{d\theta}\right) = 0 \Rightarrow P\frac{dA}{d\theta} = A\frac{dP}{d\theta}$$

$$2r\theta r^2(1 - \cos 2\theta) = r^2\left(\theta - \frac{1}{2}\sin 2\theta\right)2r$$

$$\theta(1 - \cos 2\theta) = \left(\theta - \frac{1}{2}\sin 2\theta\right) \Rightarrow 2\theta = \tan 2\theta \Rightarrow 2\theta = 257.5°$$

$$\text{Depth of flow} = r - r\cos\theta = r(1 - \cos\theta) = r(1 + 0.626) = 1.626r$$

$$\frac{\text{Depth}}{\text{Diameter}} = \frac{1.626r}{2r} = 0.81$$

For maximum discharge,

$$Q = AV = AC\sqrt{RS} = C\sqrt{\frac{A^3}{P}S}$$

Q will be a maximum when $\dfrac{A^3}{P}$ is a maximum.

$$\frac{d}{d\theta}\left(\frac{A^3}{P}\right) = \frac{1}{P^2}\left(3PA^2\frac{dA}{d\theta} - A^3\frac{dP}{d\theta}\right) = 0 \Rightarrow 3P\frac{dA}{d\theta} - A\frac{dP}{d\theta}$$

Substituting

$$3(2r\theta)r^2(1-\cos 2\theta) = r^2\left(\theta - \frac{1}{2}\sin 2\theta\right)2r \Rightarrow 3\theta(1-\cos 2\theta) = \theta - \frac{1}{2}\sin 2\theta$$

$$\Rightarrow 4\theta - 6\theta\cos 2\theta + \sin 2\theta = 0 \Rightarrow 2\theta = 308° \Rightarrow \theta = 154°$$

Depth for maximum discharge $= r(1-\cos\theta) = 1.9r$

$$\frac{\text{Depth}}{\text{Diameter}} = \frac{1.9r}{2r} = 0.95$$

Example 17.2

A sill is to be constructed on the bed of a rectangular channel 6 m wide laid on a slope of 0.0004. The channel carries a discharge of 24 m³/s. Neglecting losses and assuming that the material of the channel has a Manning's 'n' of 0.0113,

   i. Determine the critical depth of flow
  ii. Determine the minimum sill height to create critical flow
 iii. Sketch the variation of the upstream depth and the depth over the sill as a function of the sill height and describe the possible flow profiles in the vicinity of the sill.

$Q = 24 \text{ m}^3/\text{s}$; $BW = 6 \text{ m}$; $S_0 = 0.0004$; $n = 0.0113$

Normal depth of flow is given by

$$V = \frac{1}{n}R^{2/3}S_0^{1/2} \Rightarrow \frac{Q}{A} = \frac{1}{0.0113}R^{2/3}S_0^{1/2} \Rightarrow \frac{24}{6y_n} = \frac{1}{0.0113}\left(\frac{6y_n}{6+2y_n}\right)^{2/3}(0.0004)^{1/2}$$

By trial and error $y_n = 2.0$ m.

Neglecting the elevation difference between the sill and a point just upstream, Bernoulli equation gives

$$y_1 + \frac{V_1^2}{2g} = y_2 + \frac{V_2^2}{2g} + z$$

where z is the height of the sill.

$$V_1 = \frac{Q}{A_1} = \frac{24}{6y_1} = \frac{4}{y_1}; \quad V_2 = \frac{4}{y_2}$$

Therefore,

$$y_1 + \frac{4^2}{2gy_1^2} = y_2 + \frac{4^2}{2gy_2^2} + z \tag{17.42}$$

Critical depth of flow is given by

$$y_c = \sqrt[3]{\frac{q^2}{g}} = \sqrt[3]{\frac{4^2}{9.81}} = 1.177 \text{ m}$$

At just the critical condition, the depth over the sill will be y_c and the upstream depth will be normal.

Therefore,

$$y_n + \frac{4^2}{2gy_n^2} = y_c + \frac{4^2}{2gy_c^2} + z_c \Rightarrow 2 + 0.2039 = 1.177 + 0.5887 + z_c \Rightarrow z_c = 0.438 \text{ m.}$$

As the sill height increases from zero to the critical sill height z_c, the depth at the sill y_2 is given by Eq. 17.42 with $y_1 = y_n$. After the critical sill height, y_2 will always be y_c and the upstream depth y_1 will no longer be normal.

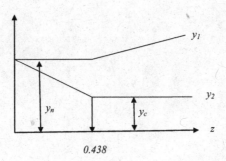

0.438

REFERENCES

Bahkmeteff, B. A. (1912): *O Neravnomernom Dvizhenii Zhidkosti v Otkrytom Rusle ("Varied flow in open channel")*, St. Petersburg, Russia.

Chow, V. T. (1959): *Open Channel Hydraulics*, McGraw-Hill Inc., New York, 680 pp.

Chapter 18

Gradually varied flow in open channels

18.1 INTRODUCTION

Gradually varied flow refers to flows where the flow parameters vary along the length of the flow plane gradually. Steady conditions are still prevailing. To distinguish from uniform flow conditions, the depth of flow is assigned the variable y instead of h.

If $\dfrac{dy}{dx}$ is positive, the depth increases.

If $\dfrac{dy}{dx}$ is negative, the depth decreases.

If $\dfrac{dy}{dx}$ is zero, the depth remains constant (uniform flow).

18.2 GOVERNING EQUATIONS

For a rectangular channel, the total energy E as given by

$$E = z + y + \frac{V^2}{2g} \tag{18.1}$$

In terms of the discharge per unit width q, Eq. 18.1, may be written as

$$E = z + y + \frac{q^2}{2gy^2} \tag{18.2}$$

$$\frac{dE}{dx} = \frac{dz}{dx} + \frac{dy}{dx} - \left(\frac{2q^2}{2gy^3}\right)\frac{dy}{dx} \tag{18.3}$$

Using the relationships $S_0 = -\dfrac{dz}{dx}$ and $S_f = -\dfrac{dE}{dx}$, Eq. 18.3 simplifies to

$$-S_f = -S_0 + \frac{dy}{dx}\left(1 - \frac{q^2}{gy^3}\right) \tag{18.4}$$

319

Therefore,

$$\frac{dy}{dx} = \frac{(S_0 - S_f)}{(1 - q^2/gy^3)} = \frac{(S_0 - S_f)}{(1 - Fr^2)} \tag{18.5a}$$

which for non-prismatic channels, may be written in the form

$$\frac{dy}{dx} = \frac{(S_0 - S_f)}{(1 - Q^2 B / gA^3)} = \frac{(S_0 - S_f)}{(1 - Fr^2)} \tag{18.5b}$$

Because $S_f = \dfrac{V^2 n^2}{R^{4/3}}$ (from Manning's equation), $V = \dfrac{q}{y}$, and that $R = y$ for a wide rectangular

channel (if the width of the channel is greater than ten times the depth of flow, the channel is assumed as a wide channel)

$$S_f = \frac{\left(\dfrac{q}{y}\right)^2 n^2}{y^{4/3}} = \frac{q^2 n^2}{y^{10/3}} \tag{18.6}$$

Therefore,

$$\frac{dy}{dx} = S_0 \frac{\left(1 - \dfrac{S_f}{S_0}\right)}{\left(1 - \dfrac{q^2}{gy^3}\right)} = S_0 \frac{\left(1 - \dfrac{q^2 n^2}{S_0 y^{10/3}}\right)}{\left(1 - \dfrac{q^2}{gy^3}\right)} \tag{18.7}$$

In terms of the normal depth of flow y_n and the critical depth of flow y_c

$$S_0 = \frac{V^2 n^2}{R^{4/3}} = \frac{\left(\dfrac{q}{y_n}\right)^2 n^2}{y_n^{4/3}} = \frac{q^2 n^2}{y_n^{10/3}} \tag{18.8}$$

and

$$\frac{q^2}{gy^3} = \frac{\left(\dfrac{q^2}{g}\right)}{y^3} = \frac{y_c^3}{y^3} = \left(\frac{y_c}{y}\right)^3 \tag{18.9}$$

Therefore, from Eq. 18.7,

$$\frac{dy}{dx} = S_0 \frac{1 - \left(\dfrac{q^2 n^2}{y^{10/3}}\right)\left(\dfrac{y_n^{10/3}}{q^2 n^2}\right)}{1 - \left(\dfrac{y_c}{y}\right)^3} \tag{18.10}$$

$$\frac{dy}{dx} = S_0 \frac{1 - \left(\dfrac{y_n}{y}\right)^{10/3}}{1 - \left(\dfrac{y_c}{y}\right)^{3}} \tag{18.11}$$

The sign of $\dfrac{dy}{dx}$ depends on the relative magnitudes of y, y_c and y_n. Equation 18.5 can be numerically integrated as follows:

$$\left(S_0 - S_f\right)dx = \left(1 - \frac{q^2}{gy^3}\right)dy$$

$$x = \int dx = \int_{y_1}^{y_2} \frac{\left(1 - \dfrac{q^2}{gy^3}\right)}{\left(S_0 - S_f\right)}\,dy \tag{18.12}$$

If a non-rectangular section is considered, V in Eq. 18.1 is replaced by $\dfrac{Q}{A}$. Then, $\dfrac{d\left(\dfrac{V^2}{2g}\right)}{dx}$ becomes $-\dfrac{2Q^2}{2gA^3}\dfrac{dA}{dx}$, and $\dfrac{dA}{dx}$ can be replaced by $B\dfrac{dy}{dx}$ because $dA = Bdy$. Then, $\dfrac{Q^2 B}{gA^3}$ can be written as $\dfrac{V^2 B}{gA}$, which in turn can be written as $\dfrac{V^2}{g\bar{h}}$, where $\bar{h} = \dfrac{A}{B}$. The term $\dfrac{V^2}{g\bar{h}}$ has the same form as Fr^2, and is of the same form as $\dfrac{q^2}{gy^3}$ in Eq. 18.5.

18.3 METHODS OF COMPUTATION

18.3.1 Graphical and numerical integration method

In this method, the depth of flow is specified and the distance is calculated:

$$x_2 - x_1 = \int_{x_1}^{x_2} dx = \int_{y_1}^{y_2} \frac{dx}{dy}\,dy = \text{area under the } \frac{dy}{dx} \text{ vs, } y \text{ curve}$$

The ratio $\dfrac{dx}{dy}$ is calculated from the slope $\dfrac{dy}{dx}$ obtained from Eq. 18.5.

18.3.2 Direct step method

In this method also, the depth is specified and the distances are calculated. It is suitable for prismatic channels.

In terms of the specific energy, Eq. 18.2 can be written as

$$E = z + E_s \Rightarrow \frac{dE}{dx} = \frac{dz}{dx} + \frac{dE_s}{dx} \tag{18.13}$$

$$-S_f = -S_0 + \frac{dE_s}{dx}$$

$$\frac{dE_s}{dx} = S_0 - S_f$$

(18.14)

which, for numerical integration, can be written as

$$\Delta x = \frac{\Delta E_s}{\left(S_0 - S_f\right)}$$

(18.15)

This is the basis of the computations in the direct step method.

The computations may be carried in tabular form as follows:

y	V	$\frac{V^2}{2g}$	E_s	ΔE_s	S_f	$\bar{S}_f$	Δx	x
y_1				—			—	—
y_2								

18.3.3 Standard step method

This method is suitable for non-prismatic channels. The distances are specified, and the depth is calculated by trial and error. When the total energy at one section is known, the energy at another section at a given distance is adjusted until they match.

The standard step method starts at a control section where the depth and velocity are known. Depending upon whether the water surface profile from the control section is sub-critical or super-critical, the computation has to move either upstream or downstream. For example, if the control section is a dam or spillway, the computation needs to be carried out in the upstream direction whereas if the control section is a sluice gate, the computation has to be carried out in the downstream direction. For the computation of a backwater curve, the iterative procedure should start at the downstream control and move upstream.

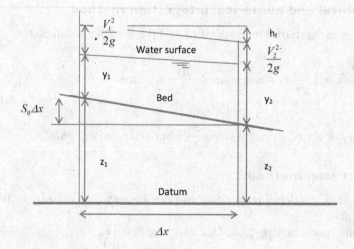

Figure 18.1 Definition sketch for the standard step method.

Referring to Figure 18.1, the water surface elevations at the two sections can be expressed as

$$H_1 = S_0 \Delta x + y_1 + z_1 \qquad (18.16a)$$

and

$$H_2 = y_2 + z_2 \qquad (18.16b)$$

where y_2 is the assumed value of water depth in Section 2. The average friction head loss between the two sections can be expressed as

$$h_f = \bar{S}_f \Delta x = \frac{1}{2}\left(S_{f_1} + S_{f_2}\right)\Delta x \qquad (18.17)$$

The friction slopes S_{f_1} and S_{f_2} can be estimated using Manning's or Chezy's equation. If Manning's equation is used,

$$S_{f_1} = \frac{V_1^2 n}{R_1^{4/3}}, \text{ and } S_{f_2} = \frac{V_2^2 n}{R_2^{4/3}}$$

The velocities V_1 and V_2 can be calculated using the continuity relationship $V = \dfrac{Q}{A}$. Water area at Section 1 is known because y_1 is known. The water area at Section 2 for the assumed value of y_2 can be calculated. The energy balance equation can then be expressed as

$$H_1 + \frac{V_1^2}{2g} = H_2 + \frac{V_2^2}{2g} + h_f \qquad (18.18)$$

This is the basic equation for the standard step method. If the assumed value of y_2 is correct, the above equation will be satisfied. If not, the trial value y_2 should be adjusted until Eq. 18.18 is satisfied within reasonable tolerance limits.

The steps involved in the procedure can be summarized as follows:

- Assume the downstream depth y_2
- Calculate the downstream water surface elevation as $H_2 = z_2 + y_2$. Since y_2 is assumed, A_2 and R_2 can be calculated, and the velocity V_2 is calculated using the discharge and the flow area.
- Calculate friction slopes at Sections 1 and 2, and their mean friction slope using

$$S_{f_1} = \frac{V_1^2 n}{R_1^{4/3}}; \; S_{f_2} = \frac{V_2^2 n}{R_2^{4/3}}; \; \bar{S}_f = 0.5\left(S_{f_1} + S_{f_2}\right)$$

- Calculate the frictional head loss as $h_f = \bar{S}_f \Delta x$

- Calculate the total head at Section 2 as $H_2 + \dfrac{V_2^2}{2g} + h_f$

- Compare H_2 from the last step with the total head at Section 1
- Repeat until they agree within a given tolerance level.

18.3.4 Method of direct integration

$$\frac{dy}{dx} = S_0 \frac{1 - \left(\dfrac{y_n}{y}\right)^N}{1 - \left(\dfrac{y_c}{y}\right)^M} \tag{18.19}$$

where the exponents N and M have different values for different cross-sectional shapes. For example, $N=3.33$; $M=3$ for a wide rectangular section.

Using the substitution $u = \dfrac{y}{y_n}$ (hence $dy = y_n du$), the following equation can be written:

$$dx = \frac{y_n}{S_0} \left\{ 1 - \frac{1}{1 - u^N} + \left(\frac{y_c}{y_n}\right)^M \left(\frac{u^{N-M}}{1 - u^N}\right) \right\} du \tag{18.20}$$

Proof: Replacing y by (uy_n) and dy by $y_n du$ in Eq. 18.19,

$$y_n du = S_0 \left[\frac{1 - \left(\dfrac{1}{u}\right)^N}{1 - \left(\dfrac{y_c}{uy_n}\right)^M} \right] dx$$

Therefore,

$$dx = \frac{y_n}{S_0} \left[\frac{1 - \left(\dfrac{y_c}{y_n}\right)^M \dfrac{1}{u^M}}{1 - \dfrac{1}{u^N}} \right] du$$

$$= \frac{y_n}{S_0} \left[\frac{1}{1 - \dfrac{1}{u^N}} - \left(\frac{y_c}{y_n}\right)^M \frac{1}{u^M} \frac{1}{1 - \dfrac{1}{u^N}} \right] du$$

$$= \frac{y_n}{S_0} \left[1 + \frac{\dfrac{1}{u^N}}{1 - \dfrac{1}{u^N}} - \left(\frac{y_c}{y_n}\right)^M u^{N-M} \frac{1}{u^N - 1} \right] du$$

$$= \frac{y_n}{S_0} \left[1 - \frac{1}{1 - u^N} + \left(\frac{y_c}{y_n}\right)^M u^{N-M} \frac{1}{1 - u^N} \right] du$$

This is then integrated for the length x of the flow profile. The exponents N and M are assumed constant within the range of limits of integration.

Integrating

$$x = \frac{y_n}{S_0} \left\{ u - \int_0^u \frac{du}{1-u^N} + \left(\frac{y_c}{y_n}\right)^M \int_0^u \left(\frac{u^{N-M}}{1-u^N}\right) du \right\} \qquad (18.21a)$$

The integral $\int_0^u \frac{du}{1-u^N} = F(u,N)$ is called the varied flow function.

It is given in the form of tables in most textbooks on open channel flow (e.g. Chow, 1959, Appendix 18.1A).

$\int_0^u \left[\frac{u^{N-M}}{1-u^N}\right] du$ can also be expressed as a varied flow function by the substitution

$$v = u^{N/J} \text{ and } J = \frac{N}{(N-M+1)}$$

Then,

$$\frac{N}{J} = N-M+1; \quad dv = \left(\frac{N}{J}\right) u^{(N/J)-1} du = \left(\frac{N}{J}\right) u^{N-M} du$$

Therefore,

$$u^{N-M} du = \frac{J}{N} dv; \text{ and, } u^N = v^{(J/N)N} = v^J$$

$$\int \frac{J}{N} dv \frac{1}{1-v^J} = \frac{J}{N} \int \frac{dv}{1-v^J} = \frac{J}{N} F(v,J)$$

Collecting terms

$$x = \frac{y_n}{S_0} \left\{ u - F(u,N) + \left[\frac{y_c}{y_n}\right]^M \frac{J}{N} F(v,J) \right\} \qquad (18.21b)$$

The advantage of this method is that intermediate points have no effect on the final result. Just two points are sufficient for the calculation.

18.3.5 Backwater profile

When the flow in a mildly sloping channel is dammed by a hydraulic structure, the water surface profile will be parallel to the pool surface at the dam and will attain the upstream normal depth of flow asymptotically. The resulting water surface profile is known as the backwater profile. It is a $M1$ profile. Backwater flow computations must start from the downstream control section and follow upstream until the sub-critical normal depth of flow is attained.

18.4 CLASSIFICATION OF SURFACE PROFILES

There are three zones of flow that can be identified with respect to the relative magnitudes of the depth of flow y, the critical depth y_c and the normal y_n. The normal depth and critical depth can be represented by two lines parallel to the bed, and their relative positions can change depending on whether the flow plane is steep or not. The actual depth of flow may take any position with respect to these two lines.

Zone 1: $y > y_n$ and y_c and y is above the upper line;

$$S_0 > S_f; Fr < 1; \frac{dy}{dx} > 0. \text{ (y increases with increasing x)}$$

Zone 2: y is between y_n and y_c and y is between the two lines

$$S_0 < S_f; Fr < 1; \frac{dy}{dx} < 0. \text{ (y deceases with increasing x)}$$

Zone 3: $y < y_n$ and y_c and y is below the lower line.

$$S_0 < S_f; Fr > 1; \frac{dy}{dx} > 0. \text{ (y increases with increasing x)}$$

There are also five slopes that can be identified with respect to the relative magnitude S_0, S_f and the horizontal:

- Horizontal slope $S_0 = 0$
- Mild slope $S_0 < S_f$ or $y_n > y_c$
- Steep slope $S_0 > S_f$ or $y_n < y_c$
- Critical slope $S_0 = S_f$
- Adverse slope $S_0 < 0$

Thus there are 15 possible surface profiles for different combinations of zones and slopes. Three of these are non-existent ($H1$, $C2$ and $A1$). Continuous profile occurs only in one zone.

It can also be seen that when

$y \to y_n, \dfrac{dy}{dx} \to 0$, and the flow profile will be asymptotic to the normal depth line

$y \to y_c, \dfrac{dy}{dx} \to \infty$, and the flow profile will cross the critical depth line normally

$y \to 0, \dfrac{dy}{dx} \to \infty$ if Manning's equation is used

$y \to \infty, \dfrac{dy}{dx} \to S_0$, and the flow profile will be asymptotic to a horizontal line

M – Profiles : $S_0 < S_f$ and $y_n > y_c$

M1: Backwater curve; sub-critical flow; $y > y_n$ and y_c; $\dfrac{dy}{dx} > 0$

$$\text{As } y \to y_n, S_0 \to S_f \text{ and } \frac{dy}{dx} \to 0$$

$$\text{When } y \gg y_n, \frac{dy}{dx} \to S_0 \text{ (i.e. horizontal)}$$

M2: Drawdown curve; sub-critical flow; $y_n > y > y_c$

Occurs when the bottom of the channel at the downstream end is submerged in a reservoir to a depth less than the normal depth, or at a free overall

As $y \to y_n$, $\dfrac{dy}{dx} \to 0$ (Asymptotic to normal depth line)

As $y \to y_c$, $\dfrac{dy}{dx} \to \infty$ (Sharp drawdown near critical depth)

M2: is produced when a mild slope changes to a steep slope, and at a drop structure.
M3: Backwater curve; super critical flow; y_n and $y_c > y$.

Profile in a stream below a sluice and profile after the change in bottom slope from steep to mild; terminates in a hydraulic jump.

As $y \to 0$, $\dfrac{dy}{dx} \to S_0 \left(\dfrac{y_n}{y_c} \right)^3$

For low values of y, M3 is asymptotic to an imaginary straight line.

As $y \to y_c$, $\dfrac{dy}{dx} \to \infty$ (the profile approaches critical depth line perpendicularly)

S – Profiles: $S_0 > S_f$ and $y_c > y_n$
   S1: Backwater curve; sub-critical flow; $y > y_c$ and y_n
   Begins with a jump and becomes tangential to horizontal pool surface; flow behind a dam in a steep channel and in a steep canal emptying into a pool of high elevation.

When $y \gg y_n$, $\dfrac{dy}{dx} \to S_0$ (profile becomes horizontal)

As $y \to y_c$, $\dfrac{dy}{dx} \to \infty$

This profile is not common. It transforms super-critical flow to sub-critical flow.
   S2: Drawdown curve; super-critical flow; $y_c > y > y_n$
   Occurs on the downstream side of an enlargement of a channel section. S2 is asymptotic to the normal depth line. Observed when slow flow is transformed into fast flow and when flow enters from a reservoir to a channel.
   S3: Backwater curve; super-critical flow; $y_c > y_n$ and y
   Occurs on the steep slope side as the channel slope changes from steep to less steep. Also, below a sluice with the depth of the entering flow less than the normal depth on a steep slope.
   Profile approaches normal depth asymptotically from a shallower depth.

H – Profiles: $S_0 = 0$ and $y_n \to \infty$
   H1: Non-existent because $y_n \to \infty$
   H2: Drawdown curve; sub-critical flow; $y_n > y$ and y_c
   H3: Backwater curve; super-critical flow; $y_n > y_c > y$ (terminates in a jump)

H2 and H3 are the limiting cases of M2 and M3.
   C – Profiles: $S_0 = S_f$ and $y_n = y_c$
Critical profiles are rare in natural channels.

   C1: Backwater curve; sub-critical flow; $y > y_c = y_n$
      Asymptotic to horizontal

C2: Parallel to bed; uniform critical flow; $\dot{y} = y_c = y_n$
C3: Backwater curve; super-critical flow; $y_c = y_n > y$

A – Profiles: $S_0 < 0$
A1: Non-existent; $y > y_n > y_c$; y_n is not real
A2: Drawdown; sub-critical; $y_n > y > y_c$; similar to $H2$
A3: Backwater; super-critical; $y_n > y_c > y$; similar to $H3$

18.4.1 Discharge problem

Discharge is always not known before longitudinal profiles are plotted. However, it can be calculated by trial and error. For example, consider the discharge from a lake into a mild channel, which has a sluice gate control at a downstream point:

Assume Q; then y_c and y_n can be calculated. Hence, the profiles from the controls can be calculated. The profile from upstream of control arrives at the lake with certain y and V, from which the specific energy E can be calculated. Verify whether this E agrees with the specific energy available at the lake exit, which is the elevation of the lake level above the datum.

18.4.2 Effect of a 'choke'

Sometimes, contractions are produced by bridge piers and culverts. First consider a case in which the contraction does not act as a 'choke'. In other words, it is not so severe to affect the flow. In a mild channel, the flow then remains subcritical.

If the contraction width is reduced, there will be a point when critical conditions are attained. This is the threshold of 'choking' condition. The contraction is then a control.

If the contraction is further narrowed,

- Flow within the contraction still remains critical.
- Flow cannot change to supercritical because at the same width, depths can change only along the constant q specific energy curve.
- The discharge per unit width q must increase, and therefore, the critical depth will also increase. Within the contraction and upstream, the specific energy will increase. $M1$ profile will be produced.
- As long as the $M1$ profile does not reach the source, the flow is unaffected by the choking which only produce local disturbances.

18.5 DELIVERY OF A CANAL CONNECTING TWO RESERVOIRS

18.5.1 Sub-critical flow

When the levels of the two reservoirs are different, the discharge in the canal for varying levels is called the delivery of the canal. There are three possible variables; namely, the upstream depth, the downstream depth and the discharge.

18.5.1.1 Upstream end depth y_1 held constant

The downstream level can vary, causing a variation in the discharge as a function of the downstream end depth y_2. A plot of the discharge as a function of the downstream reservoir level is called the delivery curve. Several flow conditions are possible:

- **Uniform flow**

$$y_1 = y_2 = y_n; \quad Q = Q_n = K_n \sqrt{S_0} \tag{18.22}$$

Conveyance K is given by $\dfrac{1}{n} A R^{2/3}$ when Manning's equation is used

- **Flow with maximum discharge**

$$y_2 = y_c; \quad Q = Q_c \tag{18.23}$$

When y_2 is at critical, the discharge is a maximum. For sub-critical flow, y_2 cannot be less than y_c. The head between the two reservoirs is also a maximum when $y_2 = y_c$. If the downstream reservoir level falls below y_c, then a free overfall will take place.

The discharge is determined by trial and error. Start with Q_n and increase the discharge keeping $y_2 = y_c$. The discharge that makes y_1 equal to the given upstream depth is the Q_{max}.

- **M1 profile** $y_2 > y_n > y_c$

  The upper limit of this case is when the two reservoir levels become equal. The discharge then is zero. The lower limit occurs when the M1 profile coincides with uniform flow line. A delivery curve is obtained by assuming a discharge less than Q_n and then computing y_2. This gives the upper part of the curve.

- **M2 profile** $y_c < y_2 < y_n$

  The lower limit is the critical flow line. Delivery curve is obtained in the same way as for the M1 profile. i.e., Assume $Q < Q_{max}$ and obtain y.

  From the delivery curve, it can be seen that the curve is very steep from the position of normal depth to critical depth. This means that there is hardly any significant difference in the normal flow discharge and the critical flow discharge. Assuming that the end of the limiting M2 profile ends at a depth of $0.99y_n$, and that the corresponding length L' is less than L, Q_c can be taken to be equal to Q_n. This also means that the length of the canal is long.

18.5.1.2 Downstream end depth y_2 kept constant

The delivery curve will be a function of y_1

- **Uniform flow**

$$y_1 = y_2 = y_n; \quad Q = Q_n = K_n \sqrt{S_0} \tag{18.24}$$

- **Flow with maximum discharge** $y_1 > y_n; y_2 = y_c$

  When y_1 increases to y_{max} which will make y_2 critical, the discharge becomes a maximum. $y_1 > y_{max}$ is not considered because it will increase the downstream end depth y_2. The discharge is equal to critical discharge at the downstream end, which violates constant y_2 condition.

- **M1 profile** $y_1 < y_n$

  The discharge is less than Q_n, and the lower limit is when the levels of the reservoirs are equal. i.e. when $y_1 = y_2$. In this limiting case, the discharge is zero.

- **M2 profile** $y_n < y_1 < y_{max}$

  The discharge is less than Q_{max}, but greater than Q_n.

18.5.1.3 Constant discharge

The discharge is kept constant while the levels of the two reservoirs are allowed to fluctuate.

- **Uniform flow** $y_1 = y_2 = y_n$; $Q = Q_n = K_n \sqrt{S_0}$
- **M1 profile** $y > y_n$
  For depths greater than y_n, **M1** profile is produced. The upper limit of the M_1 profile is a horizontal line with $y_2 = y_1 + S_0 L$. As the limiting case approaches, the head decreases and hence the velocity decreases. But the discharge is maintained constant because of the increase in the cross-sectional area of flow.
- **M2 profile** $y < y_n$
  The lower limit is when the $M2$ profile approaches the critical depth flow profile at the downstream end.

18.5.1.4 Constant discharge curves

The curves of constant discharges can be shown in the $y_1 - y_2$ axes. Three characteristic features of this diagram are

- **N-line**: The normal depth line; $y_1 = y_2 = y_n$, 45° to y_1 axis or y_2 axis
- **Z-line**: $y_2 = y_1 + S_0 L$: The upper limit of $M1$ Profile
- **C-curve**: Critical depth curve; there is no y_2 less than y_c for a given discharge. Hence, the Q-constant curve terminates at the C-curve.

18.5.2 Super-critical flow

Super-critical flow occurs when the canal is steep. Steep canals are usually short; for example, spillways, chutes, etc. The flow in such short steep canals is usually unsteady and therefore is outside the scope of this topic.

The discharge in a steep canal is governed from upstream controls. The upstream section becomes a critical section. Flow profiles depend on the tail-water conditions.

18.5.2.1 Tail-water level less than the downstream end depth

Flow is unaffected by the tail-water condition. The flow profile is of the $S2$ type.

18.5.2.2 Tail-water level greater than the downstream end depth

The downstream end water depth is raised by the tail-water level. The flow profile is of the $S1$ type producing a hydraulic jump. The flow upstream of the jump is not affected by the tail water.

As the tail-water level increases, the jump will move upstream until the normal depth is reached and thereafter will move upstream gradually with decreasing height. The height of the jump approaches zero as it reaches the upstream end critical depth.

18.6 FLOW PASSING ISLANDS

When a river channel passes by an island, the flow gets divided into two routes. The division of the main flow along the two channels can be obtained using gradually varied flow computations. Normally the flow is assumed to be sub-critical, and therefore, the computations should be carried out upstream from the downstream control point. The flow is assumed to be divided in varying ratios and the flow profiles calculated from the downstream end. The depth of flow at the point of division must be the same when computed along both canals when the ratio is correct.

If the flow is super-critical, the computations should be carried out in the downstream direction, starting from the upstream control point. In both cases, the flow is assumed to be uniform. The method of computation is a trial and error procedure.

Example 18.1

Water discharges at the rate of $15 \, m^3/s$ through a sluice gate mounted on a 3 m wide rectangular channel of slope 0.006. The depth of water upstream of the gate is 3 m, the gate opening is 1 m and the coefficient of contraction is 0.6. At a downstream point, the flow is normal with a depth of 2.5 m. Determine as accurately as possible the location of the hydraulic jump that will form. Point out any possible sources of errors in the method. Assume a value of 0.015 for Manning's 'n'.

$$Q = 15 \, m^3/s; \text{ therefore } q = \frac{15}{3} = 5 \, m$$

Critical depth $y_c = \sqrt[3]{\dfrac{q^2}{g}} = 1.366 \, m$

$$V_2 = \frac{15}{3 \times 2.5} = 2 \, m/s$$

Sequent depths are given by

$$y_1 = \frac{y_2}{2}\left(-1 \pm \sqrt{1 + 8Fr_2^2}\right)$$

$$Fr = \frac{V}{\sqrt{gy}} \Rightarrow Fr_2 = \frac{V_2}{\sqrt{9.81 \times 2.5}} = \frac{2}{\sqrt{9.81 \times 2.5}} = 0.404$$

Therefore,

$$y_1 = \frac{2.5}{2}\left(-1 \pm 1.518\right) = 0.648 \, ms$$

The location of the hydraulic jump is where the depth of flow is y_1, i.e. 0.648 m. The depth at the gate opening is 1 m, and with a coefficient of contraction of 0.6, the depth at the Vena Contracta is 0.6 m. Using the continuity equation and the gradually varied flow relationship,

$$y_1 = 0.6 \, m; \quad y_2 = 0.648 \, m; \quad P_1 = 4.2 \, m; \quad P_2 = 4.296 \, m;$$

$$R_1 = \left(\frac{A}{P}\right)_1 = \frac{3 \times 0.6}{4.2} = 0.428 \, m; \quad R_2 = \left(\frac{A}{P}\right)_2 = \frac{3 \times 0.648}{4.296} = 0.452 \, m$$

$$V_1 = \frac{Q}{A_1} = \frac{15}{3 \times 0.6} = 8.333 \, m/s; \quad V_2 = \frac{15}{3 \times 0.648} = 7.716 \, m/s;$$

Specific energy at Vena Contracta $E_1 = 0.6 + \dfrac{8.333^2}{2g} = 4.139$

Specific energy at the start of hydraulic jump $E_2 = 0.648 + \dfrac{7.716^2}{2g} = 3.682$

$$S_{f_1} = \frac{V_1^2 n^2}{R_1^{4/3}} = \frac{8.333^2 \times 0.015^2}{R_1^{4/3}} = 0.0484$$

$$S_{f_2} = \frac{V_2^2 n^2}{R_2^{4/3}} = \frac{7.716^2 \times 0.015^2}{R_2^{4/3}} = 0.0386$$

$$\overline{S_f} = 0.5\left(S_{f_1} + S_{f_2}\right) = 0.0435$$

$$S_0 = 0.006$$

Distance between the Vena Contracta and the start of hydraulic jump
$$\Delta x = \frac{E_2 - E_1}{S_0 - \overline{S_f}} = 12.19\,\text{m}.$$

Example 18.2

A 4 m wide rectangular channel of bed slope 1:250 carries a flow of 4 m³/s. Using the single-step method, determine the distance between an upstream section of depth 1 m and a downstream section of depth 0.9 m. Assume a value of 0.015 for Manning's 'n'.

For gradually varied flow single-step method gives

$$\Delta x = \frac{E_2 - E_1}{S_0 - \overline{S_f}}$$

Bed width = 4 m; $Q = 4\,\text{m}^3/\text{s}$; $S_0 = \dfrac{1}{250} = 0.004$

For y=1.0 m:

$$A = 4\,\text{m}^2; \quad P = 6\,\text{m}; \quad R = \frac{4}{6} = \frac{2}{3}\,\text{m}$$

$$R^{4/3} = 0.5824; \quad S_f = \frac{V^2 n^2}{R^{4/3}} = 3.863 \times 10^{-4}$$

$$V = \frac{Q}{A} = \frac{4}{4} = 1\,\text{m/s}; \quad E = y + \frac{V^2}{2g} = 1 + 0.05096 = 1.05096$$

For y=0.9 m:

$$A = 4 \times 0.9 = 3.6\,\text{m}^2; \quad P = 5.8\,\text{m}; \quad R = \frac{3.6}{65.8} = 0.6207\,\text{m}$$

$$R^{4/3} = 0.5294; \quad S_f = \frac{V^2 n^2}{R^{4/3}} = 5.246 \times 10^{-4}$$

$$V = \frac{Q}{A} = \frac{4}{3.6} = 1.111\,\text{m/s}; \quad E = y + \frac{V^2}{2g} = 0.9 + 0.0629 = 0.9629$$

$$E_2 - E_1 = -0.08806;$$

$$\overline{S}_f = 0.5(3.863 + 5.246) \times 10^{-4} = 4.554 \times 10^{-4}$$

Therefore, $\Delta x = 24.84\,\text{m}$.

18.1A APPENDIX

18.1.1A Super-Critical *flow profile* (M3)

When water discharges through a sluice gate, the resulting flow is super-critical. It is also a form of backwater curve because the gradient of the water surface profile is positive. The following Fortran program and the example illustrate the procedure.

Calculate the water surface profile for the following input conditions:

Discharge = 15 m³/s
Channel cross section is rectangular with a bed width of 3 m
Bed slope = 0.006
Upstream depth of flow = 0.6 m
Manning's '*n*' = 0.015
Kinetic energy correction factor = 1.1

```
C
C       THIS PROGRAM COMPUTES THE WATER SURFACE PROFILE (M3)IN A
C       RECTANGULAR OR TRAPEZOIDAL OPEN CHANNEL OF BEDWIDTH 'BW'
C       WHEN THE UPSTREAM DEPTH, DISCHARGE RATE, BED SLOPE, SIDE
C       SLOPES (VERTICAL/HORIZONTAL) AND MANNING'S 'N' ARE GIVEN.
C
        DIMENSION D(100),A(100),P(100),R(100),V(100),VH(100),SF(100),
       CSFF(100),SFD(100),X(100),E(100),DELTAE(100)
        REAL MN
        OPEN(5,FILE=' ')
        OPEN(6,FILE=' ')
C
C       ISECT = 1, TRAPEZOIDAL; ISECT = 0, RECTANGULAR
C       Q     = DISCHARGE
C       BW    = BED WIDTH
C       D0    = UPSTREAM INITIAL DEPTH
C       DELTAD= DEPTH INCREMENT
C       MN    = MANNING'S N
C       LIMIT = MAXIMUM NO. OF DEPTH INCREMENTS
C
        READ(5,*) Q,BW,SS,D0,S0,MN,DELTAD,LIMIT,ISECT
        WRITE(6,15) Q,BW,D0
        WRITE(6,25) S0,MN
   15 FORMAT(5X,'INPUT DATA :   DISCHARGE      = ',F9.3,'   CUMEC',/,
       1      5X,'               BED WIDTH      = ',F9.3,' M',/,
       2      5X,'               UPSTREAM DEPTH = ',F9.3,' M')
   25 FORMAT(5X,'               BED SLOPE      = ',F9.3,/,
       1      5X,'               MANNINGS N     = ',F9.3,/)
        IF(ISECT.EQ.1) WRITE(6,35) SS
   35 FORMAT(5X,'               SIDE SLOPES    = ',F9.3,/)
```

```
      I=1
      X(I)=0.0
      D(I)=D0
      WRITE(6,66)
      WRITE(6,115)
 66 FORMAT(//)
115  FORMAT(9X,'DEPTH',4X,'VELOCITY',2X,'V**2/2G',5X,'ES',8X,' SF',
     16X,'S0-SFF',7X,'X',/)
 30 CONTINUE
      IF (ISECT.EQ.0) GO TO 40
      A(I)=BW*D(I)+ D(I)*D(I)/SS
      P(I)=BW+2.0*D(I)*SQRT(1.0+1.0/SS**2)
      GO TO 70
 40 A(I)=BW*D(I)
      P(I)=BW+2.0*D(I)
 70 CONTINUE
      R(I)=A(I)/P(I)
      V(I)=Q/A(I)
      VH(I)=(V(I)**2)/(2.0*9.81)*1.1
      SF(I)= (V(I)**2)*(MN**2)/(R(I)**1.333)
      E(I)=VH(I)+D(I)
      IF(I.EQ.1) GOTO 50
      SFF(I)=0.5*(SF(I)+SF(I-1))
      SFD(I)=S0-SFF(I)
C      IF(SFD(I).LT.0.0) GOTO 100
      DELTAE(I)=E(I)-E(I-1)
      DELTAX=DELTAE(I)/(SFD(I))
      X(I)=X(I-1)+DELTAX
      WRITE(6,55) I,D(I),V(I),VH(I),E(I),SF(I),SFD(I),X(I)
 55 FORMAT(I3,2X,7E10.3)
      GOTO 60
 50 WRITE(6,65) I,D(I),V(I),VH(I),E(I),SF(I)
 65 FORMAT(I3,2X,5E10.3)
 60 I=I+1
      D(I)=D(I-1)+DELTAD
      IF(I.GT.LIMIT) GOTO 100
      GO TO 30
100 STOP
      END
```

```
15.0 3.0 1.5 0.6 0.006 0.015 0.005 20 0
      INPUT DATA :  DISCHARGE       =     15.000 CUMEC
                    BED WIDTH       =      3.000M
                    UPSTREAM DEPTH  =       .600M
                    BED SLOPE       =       .006
                    MANNINGS N      =       .015
```

	DEPTH	VELOCITY	V**2/2G	ES	SF	S0-SFF	X
1	.600E+00	.833E+01	.389E+01	.449E+01	.483E-01	---	0
2	.605E+00	.826E+01	.383E+01	.443E+01	.472E-01	-.418E-01	.142E+01
3	.610E+00	.820E+01	.377E+01	.438E+01	.460E-01	-.406E-01	.283E+01
4	.615E+00	.813E+01	.371E+01	.432E+01	.449E-01	-.395E-01	.425E+01
5	.620E+00	.806E+01	.365E+01	.427E+01	.439E-01	-.384E-01	.567E+01

6	.625E+00	.800E+01	.359E+01	.421E+01	.429E-01	-.374E-01	.709E+01
7	.630E+00	.794E+01	.353E+01	.416E+01	.419E-01	-.364E-01	.851E+01
8	.635E+00	.787E+01	.348E+01	.411E+01	.409E-01	-.354E-01	.994E+01
9	.640E+00	.781E+01	.342E+01	.406E+01	.400E-01	-.344E-01	.114E+02
10	.645E+00	.775E+01	.337E+01	.401E+01	.391E-01	-.335E-01	.128E+02
11	.650E+00	.769E+01	.332E+01	.397E+01	.382E-01	-.326E-01	.142E+02
12	.655E+00	.763E+01	.327E+01	.392E+01	.374E-01	-.318E-01	.156E+02
13	.660E+00	.758E+01	.322E+01	.388E+01	.365E-01	-.309E-01	.171E+02
14	.665E+00	.752E+01	.317E+01	.383E+01	.357E-01	-.301E-01	.185E+02
15	.670E+00	.746E+01	.312E+01	.379E+01	.350E-01	-.293E-01	.199E+02
16	.675E+00	.741E+01	.308E+01	.375E+01	.342E-01	-.286E-01	.214E+02
17	.680E+00	.735E+01	.303E+01	.371E+01	.335E-01	-.278E-01	.228E+02
18	.685E+00	.730E+01	.299E+01	.367E+01	.328E-01	-.271E-01	.243E+02
19	.690E+00	.725E+01	.294E+01	.363E+01	.321E-01	-.264E-01	.257E+02
20	.695E+00	.719E+01	.290E+01	.360E+01	.314E-01	-.258E-01	.272E+02

18.1.2A Calculation procedure

Column 1: Cross-section no.
Column 2: Assumed depths (upstream depth given as 0.6 m; subsequent downstream depths decreased by 0.005 m
Column 3: Velocity = Q/A (A = Bed width x depth)
Column 4: Velocity head $1.1(V^2/2g)$
Column 5: Specific energy E_s = depth+velocity head
Column 6: Friction slope $S_f = (V^2 n^2 / R^{4/3}) - (R = A/P; P = $ Bed width $+ 2 \times$ depth)
Column 7: $S_0 - $ mean S_f
Column 8: distance $\Delta x = \Delta E_s / (S_0 - $ mean $S_f)$

REFERENCE

Chow, V. T. (1959): *Open Channel Hydraulics*, McGraw Hill, Inc., New York, 680 pp.

Chapter 19

Rapidly varying flows

19.1 INTRODUCTION

In steady uniform flows and gradually varied flows, the streamlines are parallel or assumed to be parallel making the acceleration negligible, and the pressure is hydrostatic or assumed to be hydrostatic. When there are sharp curvatures in streamlines, centrifugal effects come into play with the acceleration playing an important role, and the hydrostatic pressure assumption becomes invalid. Also, when there are discontinuities in the flow profile such as for example in a hydraulic jump, changes in the cross section of flow in size and shape, changes in flow direction etc., the flow cannot be analysed using the approaches used for steady and gradually varied flows which assume parallel streamlines. Typical examples where rapidly varying flows in open channels take place include mountainous streams, rivers during periods of high floods, spillway chutes, conveyance channels, sewer systems, and inlet and outlet works. The analysis of such flows is complicated due to difficulties such as formation of roll waves, air entrainment and cavitation. Instabilities may also develop if the Froude number exceeds a critical value giving rise to roll waves or slug flow. Shock waves, standing waves and large surface disturbances also come under rapidly varied flows. Much of the developments in rapidly varying flows are from experimental analysis.

Rapidly varying flows have been analysed using either the Boussinesq (1877) assumption which is a linear vertical velocity variation from zero at the boundary to the maximum at the surface that uses momentum principles, or, the Fawer (1937) assumption which is an exponential vertical velocity variation from zero at the boundary to the maximum at the water surface that uses energy principles.

The distances involved in rapidly varying flows are relatively short, and therefore, losses due to shear at the boundary are ignored. The same applies when there are sudden changes in slope. Due to sudden change of geometry of flow, there can be flow separation, eddy formation, swirling effects, etc. In such cases, the kinetic energy correction factor α and the momentum correction factor β used in the energy equation in terms of the average velocities can be significantly different from unity, usually greater than unity.

There are no general solutions for rapidly varied flows. Solutions to specific cases can be obtained using numerical techniques. Empirical information is generally used for design purposes. One-dimensional analysis is not valid for rapidly varying flows. Three dimensional or depth averaged two-dimensional formulations are generally used. Many approximate results can, however, be obtained by using a simple one-dimensional model along with experimentally determined coefficients.

Many open channel flow measuring devices are based on principles associated with rapidly varied flows. Examples include sharp-crested weirs, broad crested weirs, critical flow flumes and sluice gates.

19.2 CONSERVATION LAWS

19.2.1 Continuity equation

Unlike in uniform flows, the mean values of velocities cannot be used in conservation laws. The velocity and pressure distributions at a section should be known. For example, in an abrupt drop in the channel, the continuity equation may be written as

$$Q = \int_A v \cdot dA \quad \text{(The dot product takes care of the orthogonality of } v \text{ and } A) \tag{19.1}$$

which may be expressed as

$$Q = \int_A v \, dA \quad \text{(If } v \text{ is orthogonal to } A) \tag{19.2}$$

where the velocity v is normal to cross-sectional area dA. Considering three sections one before, one at the change, and one after the abrupt change as sections 1, 2 and 3, for steady flow, and assuming the velocities at Sections 1 and 3 to be uniform and the pressure is hydrostatic

$$Q_1 = Q_3 \Rightarrow V_1 A_1 = V_3 A_3 \tag{19.3}$$

where the V's represent the cross-sectional mean velocities. At the abrupt cross-sectional change (Section 2), the flow separates and Eq. 19.2 should be used.

19.2.2 Momentum equation

$$\text{Rate of change of momentum in the } x\text{-direction} = \rho \int u(u \, dA) \tag{19.4}$$

Similarly, in the other two directions they can be expressed as $\rho \int v(v \, dA)$ and $\rho \int w(w \, dA)$

Using the momentum correction factor β,

$$\text{Rate of momentum flux} = \rho Q(\beta u) \tag{19.5}$$

This is valid for Sections 1 and 3. For Section 2, Eq. 19.4 should be used.

To determine the force acting on a section, the pressure and velocity distributions should be known. They are usually obtained by measurements.

19.2.3 Energy equation

The energy flux is given by

$$P = \rho g \iint \left(z + \frac{p}{\rho g} + \frac{v^2}{2g} \right) v \, dA \tag{19.6}$$

because the total energy H cannot be expressed in terms of average velocity and hydrostatic pressure according to

$$H = z + y + \frac{\alpha V^2}{2g} \tag{19.7}$$

Evaluation of the integral in Eq. 19.6 requires the velocity and pressure distributions at the section, which are normally obtained from experimental observations. Problems can be solved by using the continuity equations and energy principles if the energy losses are known. However, if losses are unknown, the momentum principle must be used.

19.3 EXAMPLES OF RAPIDLY VARYING FLOWS

19.3.1 Expansions

Expansions and contractions are considered under sub-critical flow conditions. Considering three sections, one before the expansion, one at the expansion and one after the expansion with subscripts 1, 2 and 3, and assuming that

$$E_2 = E_1; \quad F_{x_1} = F_{x_2}; \quad y_1 = y_2$$

Henderson (1966) has shown that

$$E_1 - E_3 = \frac{V_1^2}{2g}\left(\left(1 - \frac{B_1}{B_2}\right)^2 + 2Fr_1^2\left(B_2 - B_1\right)\frac{B_1^3}{B_2^4}\right) \tag{19.8}$$

where E_1, E_3, B_1, B_2 and Fr_1, respectively, are the specific energies, widths of channel sections, and Froude number at the sections indicated by the subscripts which can be approximated to

$$H_L = \frac{\left(V_1 - V_3\right)^2}{2g} \tag{19.9}$$

where H_L is the head loss in sudden expansions. For gradual expansions, the head loss is less than the above. A suggested equation (Formica, 1955) is of the form

$$H_L = 0.3\frac{\left(V_1 - V_3\right)^2}{2g} \quad \text{for 1:4 tapered expansions.} \tag{19.10}$$

19.3.2 Contractions

Contractions may be sudden or gradual and may be due to change of channel width or raising or lowering the channel bed or a combination. Generally, the head losses in contractions are less than in expansions. Empirically, the following equations, among other equations, have been suggested (Formica, 1955):

$$H_L = 0.23\frac{V_3^2}{2g} \quad \text{for square edged contractions, and} \tag{19.11}$$

$$H_L = 0.11 \frac{V_3{}^2}{2g} \quad \text{for rounded contractions} \tag{19.12}$$

19.3.3 Spillways

The function of a spillway is to release excess water during flood seasons to ensure the safety of the dam. Spillways can be broadly classified as controlled or uncontrolled. In controlled spillways, there are gates that can be opened and closed, depending on the reservoir water level. Closed position of gates allows a larger storage in the reservoir, whereas opened position allows the release of excess waters from the reservoir. Gates can be vertical, radial or drum type. Uncontrolled spillways can be the dam itself or separate open channel type spillways near the dam, chute spillways, stepped spillways, bell mouth spillways (also known as morning glory spillways) where the spillway is located inside the reservoir and siphon spillways where the driving force is the pressure difference between the inlet and the outlet. Since the outflow from a spillway carries a significant amount of kinetic energy, provision should be made to dissipate at least part of that energy at the downstream end of the spillway. Devices to dissipate excess kinetic energy include steps in the spillway itself, flip buckets at the base of the spillway, ski jumps that can direct the spilled water horizontally and finally into a stilling basin or two ski jumps that can direct the water to collide with each other. A stilling basin with friction blocks at the base of the spillway is provided to further dissipate the energy and to prevent erosion of the discharge channel.

Theoretically, the projectile principle can be used to derive the governing equation for flow over a weir type spillway. The governing equations then are

$$x = v_0 t \cos\theta; \quad y = -v_0 t \sin\theta + \frac{1}{2} g t^2 + C' \tag{19.13}$$

where v_0 is the velocity at the point where $x = 0$ and θ is the angle of inclination of the velocity v_0 to the horizontal, and C' is the value of y at $x = 0$. Here the horizontal component of the velocity remains constant. Eliminating t,

$$\frac{y}{H} = A\left(\frac{x}{H}\right)^2 + B\frac{x}{H} + C + D \tag{19.14}$$

where

$$A = \frac{gH}{2v_0^2\cos^2\theta}; \quad B = -\tan\theta; \quad C = \frac{C'}{H}; \quad D = \frac{T}{H}$$

and T is the vertical thickness of the nappe. The values of the constants A, B, C and D have been established by the United States Bureau of Reclamation (USBR). However, Eq. 19.13 has been experimentally shown to be invalid when $\frac{x}{H}$ is less than about 0.5, thereby invalidating the projectile assumption used in deriving it. Instead, empirical equations under the assumption that the nappe is at atmospheric pressure are widely used. They are of the form

$$Q = C_D L H^{3/2} \tag{19.15}$$

where Q is the flow rate over the spillway, and L is the effective length of the spillway, which is a kind of weir. The effective length L is given as

$$L = L' - 0.1NH \tag{19.16}$$

where L' is the measured length of the spillway and N is the number of contractions. The discharge coefficient C_D is empirically given by a number of equations. For example,

$$C_D = 0.611 + 0.075\frac{H}{h} + \frac{0.36}{H\sqrt{\frac{\rho g}{\sigma} - 1}} \quad \text{(Rehbock,1929)} \tag{19.17}$$

$$C_D = 0.611 + 0.08\frac{H}{h} \quad \text{(Henderson, 1966)} \tag{19.18}$$

where h is the height of weir, H is the measured head above the crest and σ is the surface tension of the liquid. In using these equations, it is important to ensure that the nappe is at atmospheric pressure. It can be seen that C_D becomes equal to 0.611 as h becomes large. For very low weirs, Henderson (1966, p. 177) suggests the following formula:

$$C_D = 1.06\left(1 + \frac{h}{H}\right)^{3/2} \tag{19.19}$$

19.3.4 Hydraulic jump

A hydraulic jump occurs when a supercritical flow changes to sub-critical flow. Details of this topic have been discussed in Chapter 17. The type of hydraulic jump very much depends on the upstream Froude number. These can be broadly classified into five types as shown below. When the Froude number is less than one, a hydraulic jump cannot occur.

- $1 < Fr < 1.7$: Undular jump with the water surface having undulations of decreasing size.
- $1.7 < Fr < 2.5$: Weak jump with the water surface remaining smooth.
- $2.5 < Fr < 4.5$: Oscillating jump with the water jet oscillating from bottom to top.
- $4.5 < Fr < 9$: Steady jump with energy loss between 45% and 70%.
- $Fr > 9$: Strong jump with rough downstream water surface and energy loss up to 85%.

19.3.5 Weirs

Assuming that the velocity profile upstream of the weir plate is uniform and that the pressure within the nappe is atmospheric, and that the fluid flows horizontally over the weir plate with a non-uniform velocity profile, the discharge over the weir can be expressed as a function of the head over the weir with an empirically established coefficient of discharge. The flow rate equations depend upon the type of weir. They are of the form

$$Q = C_D \frac{2}{3}\sqrt{2g}LH^{3/2} \quad \text{(Sharp-crested rectangular)} \tag{19.20}$$

$$Q = C_D \frac{8}{15}\tan\frac{\theta}{2}\sqrt{2g}H^{5/2} \quad \text{(Sharp-crested triangular)} \tag{19.21}$$

$$Q = C_D L\sqrt{g}\left(\frac{2}{3}\right)^{3/2} H^{3/2} \quad \text{(Broad crested)} \tag{19.22}$$

A sharp-crested trapezoidal weir can be considered as a combination of a rectangular and a triangular weir. In these equations, C_D is an empirically determined coefficient of discharge, H is the head over the weir, L is the effective length of the weir and θ is the subtended angle of the triangular weir.

19.3.6 Underflow gates

Underflow gated structures can be vertical, radial or drum type. They act as control structures in spillways or at the entrances of canals or rivers from a lake. When the fluid issues as a supercritical jet with the free surface at atmospheric pressure, the flow can be considered as free outflow. Then, the flow rate is expressed as

$$q = C_D a \sqrt{2 g y_1} \tag{19.23}$$

where q is the flow rate per unit width, a is the height from the base to the lower edge of the plate and y_1 is the upstream depth. The coefficient of discharge is a function of the coefficient of contraction and the ratio of the upstream depth to the gate opening a.

$$C_D = f\left(C_C, \frac{y_1}{a}\right) = f\left(\frac{y_2}{a}, \frac{y_1}{a}\right); \ (y_2 \text{ is the downstream depth}) \tag{19.24}$$

The outflow may be drowned or controlled by some downstream condition.

19.3.7 Culverts

Another example of rapidly varying flows is the inlet zone of a culvert, which has a contraction zone where the streamlines contract as the water enters and an expansion zone immediately downstream of the contraction. The head loss as the water passes through the inlet zone is dependent on the velocity inside the culvert as well as its geometry. Fish passage can be hindered due to high velocities in the contraction zone of culverts. A software package called 'FishXing' approximates the maximum contraction velocity by assuming the calculated entrance head loss is converted entirely to kinetic energy.

19.3.8 Plunging flow

Water leaving an inclined culvert has both horizontal and vertical velocity components. Ignoring the frictional effects, the projectile equation can be used to describe the path of water plunging down the culvert. FishXing has the following equations:

$$H = V_{out}(\sin\theta)t - \frac{1}{2}gt^2 \tag{19.25}$$

$$L = V_{out}(\cos\theta)t \tag{19.26}$$

where H and L are the vertical and horizontal plunge distances, V_{out} is the exit water velocity at the culvert outlet, θ is the exit angle of the plunging water, and t is the time taken for the plunging water to fall from the culvert outlet to the pool downstream. Since H is known, L and t can be calculated.

19.3.9 Surges

A surge in an open channel is caused by the sudden change of flow, resulting in a change of depth of flow which may be increasing or decreasing. Examples of surges include flow resulting from a dam breach, storm surges resulting from extreme weather coupled with high tide, waves resulting from tsunami, surface waves resulting from landslides into reservoirs and flood waves during periods of heavy rainfall. The surges can travel upstream or downstream, and the velocity of the wave is known as the celerity which is different from the velocity of flow.

The celerity or speed of propagation of the wave relative to the water is given by $\sqrt{gy}$ where y is the water depth. Therefore, the velocity of the wave relative to a stationary observer is $\sqrt{gy} \pm v$. A positive surge leads to an increase in depth as the wave propagates, whereas a negative surge leads to a decrease in the depth as the wave propagates. In negative surges, the celerity decreases with increasing time and are dispersive. When a sluice gate is opened, positive surges move downstream whereas when the sluice gate is closed, positive surges move upstream. There are six possible types of surges with

$$v_w > v_1 > v_2; \quad v_w > v_2 > v_1; \quad v_2 > v_1 > v_w; \quad v_1 > v_2 > v_w; \quad v_2 > v_1 > v_w; \quad v_1 > v_2 > v_w$$

where the subscripts 1 and 2 correspond to the upstream and downstream water velocities and the subscript w corresponds to the surge velocity. For small disturbances, the conditions upstream and downstream are approximately the same, and the surge velocity is known as the celerity.

19.3.9.1 Upstream positive surges

An upstream positive surge is created by the sudden closure of a gate resulting in a rapid change in flow rate. This causes a wave front to travel upstream. The governing equations are

$$A_1(v_1 + c) = A_2(v_2 + c) \quad \text{(Continuity)} \tag{19.27}$$

$$gA_1\bar{y}_1 - gA_2\bar{y}_2 + A_1(v_1 + c)(v_1 - v_2) = 0 \quad \text{(Momentum)} \tag{19.28}$$

where the suffixes 1 and 2 refer to the upstream and downstream conditions and the over-bar refers to the depth at the centre of area. From these two equations in which the frictional effects are neglected, it can be shown that (Pandey, 2015)

$$c = \left(gA_2 \frac{A_2\bar{y}_2 - A_1\bar{y}_1}{A_1(A_2 - A_1)} \right)^{1/2} - v_1 \tag{19.29}$$

For the special case of a rectangular channel where $A = by$; $\bar{y} = \dfrac{y}{2}$,

$$c = \left(\left(\frac{gy_2}{2} \right) \frac{(y_2^2 - y_1^2)}{y_1(y_2 - y_1)} \right)^{1/2} - v_1 = \left(\left(\frac{gy_2}{2} \right) \frac{(y_2 + y_1)}{y_1} \right)^{1/2} - v_1 \tag{19.30}$$

The hydraulic jump is a stationary wave. Making $c = 0$ in Eq. 19.30 gives

$$v_1^2 = \frac{gy_2}{2} \frac{(y_2 + y_1)}{y_1} \Rightarrow \frac{2v_1^2 y_1}{g} = y_2^2 + y_2 y_1 \tag{19.31}$$

Using the definition of Froude number $Fr = \dfrac{v_1}{\sqrt{gy_1}}$, Eq. 19.31 can be written as

$$y_2 = \frac{y_1}{2}\left(\sqrt{1+8Fr_1^2} -1\right) \tag{19.32}$$

which is the relationship between the upstream and downstream depths of a hydraulic jump.

19.3.9.2 Downstream positive surges

They occur at the downstream of a gate when it is suddenly opened, resulting in an instantaneous increase in the flow rate and the flow depth. As before, using the continuity and momentum equations, it can be shown (Pandey, 2015) that

$$c = \frac{gy_1}{2y_2}\left(y_1 + y_2\right)^{1/2} + v_2 \tag{19.33}$$

19.3.9.3 Upstream negative surges

An upstream negative surge may be caused by the sudden opening of a gate. In this case, it is necessary to calculate the speeds of the wave crest and the wave trough because the wavefront flattens as it travels along the channel due to the wave having a greater velocity at the top than at the bottom.

It can be shown (Pandey, 2015) that the celerity at the crest is

$$c_1 = \sqrt{gy_1} - v_1, \tag{19.34}$$

and at the trough,

$$c_2 = 3\sqrt{gy_2} - 2\sqrt{gy_1} - v_1 \tag{19.35}$$

19.3.9.4 Downstream negative surges

As before, it has been shown (Pandey, 2015) that

$$c_1 = 3\sqrt{gy_1} - 2\sqrt{gy_2} + v_2 \tag{19.36}$$

$$c_2 = \sqrt{gy_2} + v_2 \tag{19.37}$$

In this context, crest refers to the highest depth of the wave and trough the lowest.

19.3.10 Dam break problem

Dam break problem is a surge that can be simulated by the instantaneous removal of a dam. This causes a positive surge in the positive x-direction, which is assumed to be in the upstream direction and a negative surge in the negative x-direction. A positive surge has an increasing depth whereas a negative surge has a decreasing depth. Hydraulically, the dam break problem can be considered as equivalent to the instantaneous opening of a sluice gate. In the simplified analysis, the bed slope of the downstream channel and friction effects are ignored. It is also assumed that the downstream channel has a rectangular cross section. A sluice gate operation can have moving water in both upstream and downstream sides, whereas a dam break problem will have still water on the upstream side initially. The downstream channel can be a dry bed or with a finite known depth. The dry bed condition is analysed first.

The dam break problem or sudden opening of a gate is equivalent to a surge that will have a zero velocity at the beginning and the wave moving downstream until it attains a steady state. The celerity of the wave is given as

$$c = 3\sqrt{gy} - 2\sqrt{gy_1} \tag{19.38}$$

and the distance travelled by the wave at time t is given by

$$x = ct = \left(3\sqrt{gy} - 2\sqrt{gy_1}\right)t \tag{19.39}$$

which is of parabolic shape and tangential to the channel bed at the wave front. At the original position of the dam ($x = 0$), the depth of water $y = \dfrac{4y_1}{9}$, where y_1 is the upstream water depth and is constant. The water velocity at this point is $\dfrac{2c_0}{3}$ where c_0 is the speed of the receding wave front on the upstream side.

The case with a downstream water depth is analysed by Henderson (1966, pp. 307–309). In his analysis, the downstream wave front terminates at the downstream water surface instead of the dry channel bed, thereby creating a discontinuity as it is not asymptotic to the water surface.

19.4 GOVERNING EQUATIONS FOR UNSTEADY NON-UNIFORM FLOW

Rapidly varying flows in general can be described by the three-dimensional governing equations for unsteady non-uniform flows as

$$\frac{\partial u}{\partial x} + \frac{\partial v}{\partial y} + \frac{\partial w}{\partial z} = 0 \tag{19.40}$$

which is the continuity equation for an incompressible fluid in 3 dimensions, and,

$$\frac{\partial u}{\partial t} + u\frac{\partial u}{\partial x} + v\frac{\partial u}{\partial y} + w\frac{\partial u}{\partial z} = g_x - \frac{1}{\rho}\frac{\partial p}{\partial x} + \frac{\mu}{\rho}\left(\frac{\partial^2 u}{\partial x^2} + \frac{\partial^2 u}{\partial y^2} + \frac{\partial^2 u}{\partial z^2}\right) \tag{19.41}$$

$$\frac{\partial v}{\partial t} + u\frac{\partial v}{\partial x} + v\frac{\partial v}{\partial y} + w\frac{\partial v}{\partial z} = g_y - \frac{1}{\rho}\frac{\partial p}{\partial y} + \frac{\mu}{\rho}\left(\frac{\partial^2 v}{\partial x^2} + \frac{\partial^2 v}{\partial y^2} + \frac{\partial^2 v}{\partial z^2}\right) \tag{19.42}$$

$$\frac{\partial w}{\partial t} + u\frac{\partial w}{\partial x} + v\frac{\partial w}{\partial y} + w\frac{\partial w}{\partial z} = g_z - \frac{1}{\rho}\frac{\partial p}{\partial z} + \frac{\mu}{\rho}\left(\frac{\partial^2 w}{\partial x^2} + \frac{\partial^2 w}{\partial y^2} + \frac{\partial^2 w}{\partial z^2}\right) \tag{19.43}$$

which are the momentum equations in the three directions.

These equations have no general solutions. However, numerical solutions can be obtained under certain simplified assumptions. One simplification is to reduce the three-dimensional equations to two-dimensional depth averaged equations. The continuity equation can be written as

$$\int_{Z_b}^{Z} \frac{\partial u}{\partial x} dz + \int_{Z_b}^{Z} \frac{\partial v}{\partial y} dz + w(Z) - w(Z_b) = 0 \tag{19.44}$$

where Z and Z_b refer to the z coordinates of the water surface and the bed.

$$\int_{Z_b}^{Z} \frac{\partial u}{\partial x} dz = \frac{\partial}{\partial x} \int_{Z_b}^{Z} u \, dz - u(Z) \frac{\partial Z}{\partial x} + u(Z_b) \frac{\partial Z_b}{\partial x} \tag{19.45}$$

$$\int_{Z_b}^{Z} \frac{\partial v}{\partial y} dz = \frac{\partial}{\partial y} \int_{Z_b}^{Z} v \, dz - v(Z) \frac{\partial Z}{\partial y} + v(Z_b) \frac{\partial Z_b}{\partial y} \tag{19.46}$$

$$w(Z) = \frac{DZ}{Dt} = \frac{\partial Z}{\partial t} + u(Z) \frac{\partial Z}{\partial x} + v(Z) \frac{\partial Z}{\partial y} \tag{19.47}$$

$$w(Z_b) = \frac{DF_b}{Dt} = u(Z_b) \frac{\partial Z_b}{\partial x} + v(Z_b) \frac{\partial Z_b}{\partial y} \tag{19.48}$$

where $F_b = Z_b(x, y) - z = 0$ if the channel bed is rigid in which case $Z_b(x, y)$ represents the z co-ordinate of bed.

Substituting

$$\frac{\partial Z}{\partial t} + \frac{\partial (\bar{u} d)}{\partial x} + \frac{\partial (\bar{v} d)}{\partial y} = 0 \tag{19.49}$$

where $\bar{u} = \frac{1}{d} \int_{Z_b}^{Z} u \, dz; \quad \bar{v} = \frac{1}{d} \int_{Z_b}^{Z} v \, dz; \quad d = Z - Z_b$ measured normal to the bed.

Assuming vertical acceleration is negligible $\frac{Dw}{Dt} = 0; \mu \nabla^2 w = 0$, the momentum equation in the z-direction becomes

$$g_z - \frac{1}{\rho} \frac{\partial p}{\partial z} = 0 \tag{19.50}$$

which when integrated gives

$$p = \rho g_z (z - Z) \tag{19.51}$$

Similarly,

$$-\frac{1}{\rho}\frac{\partial p}{\partial x} = g_z\frac{\partial Z}{\partial x}; \quad -\frac{1}{\rho}\frac{\partial p}{\partial y} = g_z\frac{\partial Z}{\partial y} \tag{19.52}$$

After substitution and simplifying, the depth-averaged equations become (Chaudhry, 1993, p. 350)

$$\frac{\partial(\bar{u}d)}{\partial t} + \frac{\partial(\bar{u}^2d)}{\partial x} + \frac{\partial(\overline{uv}d)}{\partial y} = \left(g_x + g_z\frac{\partial d}{\partial x}\right)d - \frac{g}{C^2}\bar{u}\sqrt{\bar{u}^2 + \bar{v}^2} \tag{19.53}$$

$$\frac{\partial(\bar{v}d)}{\partial t} + \frac{\partial(\overline{uv}d)}{\partial x} + \frac{\partial(\bar{v}^2d)}{\partial y} = \left(g_y + g_z\frac{\partial d}{\partial y}\right)d - \frac{g}{C^2}\bar{v}\sqrt{\bar{u}^2 + \bar{v}^2} \tag{19.54}$$

The simplified equations also do not have analytical solutions but can be solved numerically under certain assumptions.

Example 19.1

A tidal river has a flow depth of 3 m and a velocity of 1 m/s. As a result of intercepting a tidal bore, the flow depth suddenly increases to 4 m. Determine the speed of propagation of the tidal bore upstream and the flow velocity behind the bore.

In this example, the surge is moving upstream. $y_1 = 3$ m; $y_2 = 4$ m.

$$\text{The wave velocity } v_w = v_1 - \sqrt{\frac{gy_2}{2y_1}(y_1 + y_2)}$$

$$= 1.0 - \sqrt{\frac{9.81\times4}{2\times3}(3+4)} = -5.766 \text{ m/s}$$

$$\text{The flow velocity } v_2 = v_w + \sqrt{\frac{gy_1}{2y_2}(y_1 + y_2)}$$

$$= -5.766 + \sqrt{\frac{9.81\times3.0}{2\times4.0}(3.0+4.0)} = -0.69 \text{ m/s}$$

Example 19.2

A sluice gate installed in a rectangular channel 2.5 m wide carrying a discharge of 10 m³/s at a depth of 2 m is suddenly closed. Determine the depth and velocity of the resulting surge.

$$y_1 = 2\text{ m}; \quad v_1 = \frac{10}{2\times2.5} = 2\text{ m/s}$$

$$y_2 = ? \quad v_2 = 0$$

Continuity equation: $y_1(v_w + 2) = y_2(v_w + 0) \Rightarrow v_w = \frac{2\times2}{y_2 - 2} = \frac{4}{y_2 - 2}$

Momentum equation: $c = \left(\left(\dfrac{gy_2}{2} \right) \dfrac{(y_2 + y_1)}{y_1} \right)^{1/2} - v_1$

$$c = \left(\left(\dfrac{9.81y_2}{2} \right) \dfrac{(y_2 + 2)}{2} \right)^{1/2} - 2 = \dfrac{4}{y_2 - 2}$$

By trial and error, $y_2 \approx 3\,\mathrm{m}$

Velocity of surge $= v_w = \dfrac{4}{3-2} = 4\,\mathrm{m/s}$

Height of surge $= y_2 - y_1 = 3 - 2 = 1\,\mathrm{m}$.

REFERENCES

Boussinesq, J. (1877, 1878): Essae sur la theorie des eaux courantes (Essay on the theory of water flow), *Memoires presents par divers savants, a F Academie des Sciences, Paris*, vol. 23, pp. 1–680; vol. 24, no. 2.

Chaudhry, M. H. (1993): *Open Channel Flow*, Prentice Hall, Englewood Cliffs, NJ, 483 p.

Fawer, C. (1937): Etude de quelgues ecoilemeals permanents a filets courbes (study of some permanent flows with curved filaments), Thesis, Universite de Lausanne, Laussanne, Switzerland.

Formica, G. (1955): Esperienze preliminary sulle perdite di carico nei anali dovute a cambiameni di sezione (Preliminary tests on head losses in channels due to cross sectional changes), *L'Energia Elletrica, Milan*, vol. 32, no. 7, p. 554.

Henderson, F. M. (1966): *Open Channel Flow*, MacMillan, New York, 522 pp.

Pandey, B. R. (2015): Open channel surges, *Journal of Advanced College of Engineering and Management*, vol. 1, pp. 35–43.

Rehbock, T. (1929): Discussion of "Precise weir measurements". In: E. W. Schoder and K. B. Turner (eds.), *Transactions of the American Society of Civil Engineers*, New York, vol. 93, p. 1143.

Chapter 20

Hydraulics of alluvial channels

20.1 INTRODUCTION

The topics on open channel flows discussed so far assume that the channel geometry is invariant. Such channels are usually artificially made by lining the bed and the banks with non-erodible materials such as concrete and/or masonry. In nature, however, rivers and canals are formed in loose sedimentary material, which allows the shape and the alignment to shift and adjust. The loose sedimentary material is known as alluvium, which consists of gravel, sand, silt and debris. The flow in the channel can erode the bed and banks, and the eroded materials which are usually known as sediments are carried downstream. Some of the sediments carried by the flow get deposited on the bed and banks as the flow energy becomes weaker. The two complementary processes of erosion and deposition take place until sediment equilibrium is attained, or until the sediments are transported to a reservoir or to the sea as in some rivers. Such rivers and channels formed on loose sedimentary material are called alluvial channels. They have the ability to change their geometry and alignment.

In addition to the sediments eroded from the channel itself, sources of sediments include mountain catchments, weathered sedimentary rocks as well as construction sites. They are dislodged by the impact of rain and carried towards the river channel by overland flow. Once the sediments reach the flowing channel, they may be transported in suspension if the sediments are fine and light, or as bed load along the bed layer.

The transported sediments finally get deposited in river banks, river beds, reservoirs and oceans. The total amount of sediments transported to the oceans is approximately 14×10^9 ton/year of solids and a further 4×10^9 ton/year of dissolved solids. This is equivalent to 100 ton/km²/year. UN-FAO estimates that the global loss of productive land by erosion is about 5–7 million hectares per year.

This chapter focuses on describing the sediment properties, modes and mechanics of erosion, transport and deposition.

20.2 TYPES OF SEDIMENTS

Sediments can be broadly classified as cohesive soils and non-cohesive soils. In cohesive soils, the weight is insignificant and, once eroded, may behave like non-cohesive soils in the transport process. Non-cohesive soils, sometimes referred to as alluvial soils, have no cohesive forces between particles, but the weight is a dominant force.

20.3 SEDIMENT PROPERTIES

Sediment properties include the size of sediments characterized by different indicators and classified according to certain standards. Table 20.1 shows the classification of sediments according to British Standard 1377 (1975). A similar American Society for Testing Materials (ASTM) classification has been proposed by the sub-committee on sediment terminology of the American Geophysical Union (Vanoni, 1975: p. 20, Table 20.1).

20.3.1 Particle size

Indicators of particle size include

- **Sieve diameter:** Sieve size that just passes
- **Sedimentation diameter:** Diameter of a sphere of the same density and the same fall velocity in the same fluid at the same temperature as the given particle – a rather difficult definition
- **Nominal diameter:** Diameter of a sphere of equal volume
- Triaxial dimensions, a, b and c with c being the shortest
- Shape factor, $\dfrac{c}{\sqrt{ab}}$
- Phi-index scale, ϕ – Transforms the diameters into integers.

Table 20.1 Sediment classification

Type of particle	Particle size
Very fine clay	0.24–0.5 μm
Fine clay	0.5–1.0 μm
Medium clay	1.0–2.0 μm
Coarse clay	2.0–4.0 μm
Very fine silt	4.0–8.0 μm
Fine silt	8.0–16.0 μm
Medium silt	16.0–31.0 μm
Coarse silt	31.0–62.0 μm
Very fine sand	62.0–125.0 μm
Fine sand	125.0–250.0 μm
Medium sand	250.0–500.0 μm
Coarse sand	0.5–1.0 mm
Very coarse sand	1.0–2.0 mm
Very fine gravel	2.0–4.0 mm
Fine gravel	4.0–8.0 mm
Medium gravel	8.0–16.0 mm
Coarse gravel	16.0–32.0 mm
Very coarse gravel	32.0–64.0 mm
Small cobbles	64.0–128.0 mm
Large cobbles	128.0–256.0 mm
Small boulders	256.0–512.0 mm
Medium boulders	512.0–1,024 mm
Large boulders	1,024.0–2,048.0 mm
Very large boulders	2,048.0–4,096 mm

20.3.2 Grain size distribution

Grain size distribution is a histogram plot of sediment size (x-axis) vs. percentage finer (y-axis), or, as a cumulative distribution with sediment size (x-axis) and cumulative percentage finer (y-axis). For most sediment mixtures, such plots resemble a normal distribution, which will plot as a straight line if the probability scale is used instead of the arithmetic scale. The mean, median and mode in a normal distribution will be the same and correspond to the 50% cumulative frequency value. In a normal distribution, approximately 68%, 95% and 99.7%, respectively, of all values fall within one, two and three standard deviations from the mean value. If the distribution is skewed, the mean will be halfway between the 15.9% and 84.1% values and the median will be at the 50% value. If the distribution is plotted on a log-normal probability scale, the geometric mean will correspond to the 50% size. The grain size distribution is characterized by several indicators:

- **Geometric mean:** d_g is the geometric mean of the sizes corresponding to 15.9% and 84.1% probabilities, and is given as

$$d_g = \left(d_{15.9}d_{84.1}\right)^{1/2} \tag{20.1a}$$

- **Geometric standard deviation**

$$\sigma_g = \left\{\frac{d_{84.1}}{d_{15.9}}\right\}^{1/2} \text{ or, } \sigma_g = \frac{d_{84.1}}{d_{50}} = \frac{d_{50}}{d_{15.9}} \tag{20.1b}$$

- **Natural mean diameter:** $\bar{d}$ (arithmetic mean of the size distribution)

$$\bar{d} = \frac{\sum d_i \Delta p_i}{\sum \Delta p_i} = \frac{\sum d_i \Delta p_i}{100} = \sum f_i d_i \tag{20.1c}$$

$$\sigma = \left\{\frac{\sum (d_i - \bar{d})^2 \Delta p_i}{\sum \Delta p_i}\right\}^{1/2} \tag{20.1d}$$

where f_i is the percentage frequency in each class, and Δp_i is the percentage weight corresponding to size d_i. Also, $\bar{d}$ and d_g are related to each other by

$$\bar{d} = d_g e^{0.5\ln^2 \sigma_g} \tag{20.1e}$$

- **Effective grain size:** d_e (many definitions). For example,

$$d_e = \frac{1}{9}\left(d_{10} + d_{20} + d_{30} + \ldots + d_{90}\right) \tag{20.1f}$$

- **Governing mean grain size:** d_m (or, average size, d_a)

$$d_m = \frac{\displaystyle\sum_0^{100\%} \bar{d}_i \Delta p_i}{\displaystyle\sum_0^{100} \Delta p_i} \tag{20.1g}$$

where $\overline{d_i} = \dfrac{1}{2}(d_i + d_{i+1})$, and, Δp_i is the percentage weight of size d_i.

- **Sorting coefficient:** c_s

$$c_s = \left\{ \frac{d_{75}}{d_{25}} \right\}^{1/2} \quad \text{or} \quad s_c = \frac{d_{90}}{d_{10}} \tag{20.1h}$$

- **Sorting index:** S_i

$$S_i = \frac{1}{2} \left\{ \frac{d_{84}}{d_{50}} + \frac{d_{50}}{d_{10}} \right\} \tag{20.1i}$$

- **Uniformity coefficient:** u_c

$$u_c = \frac{d_{60}}{d_{10}} \quad \left(\text{Hazen, 1904} \right) \tag{20.1j}$$

$$u_c = \frac{\displaystyle\sum_{0}^{50} \Delta p_i d_i}{\displaystyle\sum_{0}^{100} \Delta p_i} \quad \text{(Kramer, 1935)} \tag{20.1k}$$

- ϕ: **Index scale**

$$\phi = \log_2 d = -\frac{\log(d)}{\log(2)} \quad d \text{ in mm} \tag{20.1l}$$

For example

D (mm)	8	4	2	1	0.5	0.25	0.125
ϕ		-3	-2	-1	0	1	2	3

- **Sample size for analysis (ASTM)**

$$M = 0.082 b^{1.5} \tag{20.1m}$$

where M is in kg and b is the maximum intermediate triaxial size in mm.
- **Shape factor,** S_F (several definitions)

$$S_F = \frac{c}{\sqrt{ab}} \tag{20.1n}$$

- **Criteria for treating sediments as uniform:** (from the hydraulic point of view)

$$\frac{d_{95}}{d_5} < 4 \text{ or } \sigma_g < 1.35$$

20.3.2 Fall velocity

Fall velocity refers to the velocity with which a sediment particle will settle in water. It can also be considered as the terminal velocity of a particle moving in a fluid. It is a function of the density difference between the particle and the fluid, the viscosity of the fluid, the surface roughness of the sediment particles, size and shape of sediment particles, the concentration of sediments in the fluid and the level of turbulence. Terminal velocity is attained when the drag force is balanced by the submerged weight. The drag force, F_D, is given by

$$F_D = C_D \frac{\rho w^2}{2} A \qquad (20.2)$$

where C_D is the drag coefficient, w is the downward (fall) velocity, ρ is the density of fluid and A is the projected area of the particle. The drag coefficient depends upon the body shape, Reynolds' number, $\frac{wd}{v}$, Mach number; $\frac{w}{\sqrt{\kappa_s/\rho}}$, the bulk modulus of elasticity, κ_s, Froude number, $\frac{w}{\sqrt{gd}}$, surface roughness and free-stream turbulence. For equilibrium (when there is no acceleration), the drag force is balanced by the submerged weight:

$$C_D \frac{\rho w^2}{2} A = \frac{\pi d^3}{6} g(\rho_s - \rho) \qquad (20.3)$$

where ρ_s is the density of the sediments. Stokes' drag force is given by

$$F_D = 3\pi d\mu w \qquad (20.4)$$

which is valid for $Re < 0.1$, but used for Re up to about 1. Stokes' law is applicable to silt and clay particles that have very small fall velocities. They are not found in the bed layer and are carried downstream as wash load.

From Eqs. 20.2 and 20.4,

$$C_D = \frac{24}{Re} \qquad (20.5)$$

and from Eqs. 20.3 and 20.4,

$$w = \frac{d^2}{18\mu} g(\rho_s - \rho) \qquad (20.6)$$

For $Re \leq 2$ (in some literature, $Re \leq 5$), Oseen (1927) obtained the approximate relationship

$$C_D = \frac{24}{Re}\left(1 + \frac{3}{16} Re\right) \qquad (20.7)$$

The solution obtained by Goldstein (1929) takes the form

$$C_D = \frac{24}{Re}\left(1 + \frac{3}{16} Re - \frac{19}{1280} Re^2 + \ldots\right) \qquad (20.8)$$

Empirically, for $Re < 800$, Schiller and Naumann (1933) obtained the following relationship:

$$C_D = \frac{24}{Re}\left(1 + 0.15 Re^{0.687}\right) \qquad (20.9)$$

If C_D is known, the fall velocity w is given by

$$w = \left\{ \frac{1}{C_D} \frac{4}{3} gd(s_s - 1) \right\}^{1/2} \tag{20.10}$$

where s_s is the specific gravity of sediments.

Empirically, for quartz sands at 20°C, the average fall velocities seem to fit the following equations:

$$w(\text{mm/s}) = 663d^2; \, d \text{ in mm } \text{ for } d < 0.15 \text{ mm}$$

$$w(\text{mm/s}) = 134.5d^{0.52}; \, d \text{ in mm } \text{ for } d > 1.5 \text{ mm}$$

Effective fall velocity is given by

$$w = \frac{\sum p_i w_{pi}}{\sum p_i} \tag{20.11}$$

where p_i is the weight size and w_{pi} is the fall velocity of size i.

20.4 MODES OF TRANSPORT AND BED FORMATION

Transport of sediments depends upon the type of sediments which can be classified as bed material load and wash load. Bed material load can be further sub-divided into bed load and suspended load. Wash load refers to particle sizes smaller than about 0.05 mm carried away as suspended load. Bed load is moved by rolling, sliding and saltation (jumping), whereas suspended load is moved in suspension caused by turbulence.

20.4.1 Bed forms

Depending upon the sediment size, channel geometry and type of flow, different bed forms can take place. They can be broadly classified as

- **Plane bed**: Bed surface has no elevations or depressions larger than the largest grain size of bed material
- **Ripples**: Small bed forms with wavelengths < 30 cm and heights < 5 cm; sediment sizes < 0.6 mm and low Froude numbers $\ll 1$
- **Bars**: Bed forms having lengths of the same order as the channel width and heights comparable to the mean depth of flow
- **Dunes**: Bed forms larger than ripples but smaller than bars. Profile out of phase with the water surface profile
- **Transition**: Intermediate between dunes and plane bed forms
- **Antidunes (Standing waves)**: Bed and water surface profiles are in phase. Sand waves and water waves move upstream
- **Chutes and pools**: Large elongated mounds of sediments that occur in relatively steep slopes.

20.5 THRESHOLD OF MOVEMENT

20.5.1 Incipient motion

The condition at which the particles just begin to move as the flow rate increases is called the critical condition or condition of incipient motion. It can also be viewed either as the minimum shear stress needed to move a given particle or as the largest grain size that can be moved by a given shear stress. There are a number of criteria to describe this condition all of which are empirical as the physics is too complicated.

20.5.2 Forces acting on a particle of sediment on a flat bed

As shown in Figure 20.1, the forces acting on a particle of sediment on a flatbed consist of hydrostatic forces which tend to balance on either side of the control volume, hydrodynamic forces, resistance (frictional) forces and submerged weight. Hydrodynamic forces include the drag force[1] which can be expressed as

$$F_D = c_1 \tau_0 d^2 \tag{20.12}$$

where τ_0 is a bed shear stress ($= \tau_c$ at the point of incipient motion) and c_1 is a constant, and a lift force normal to the flow, which is usually ignored. The frictional force can be expressed as

$$F_R = \mu R = \tan\phi \, W_s \cos\theta \tag{20.13}$$

where μ is the coefficient of friction, ϕ is the friction angle (angle of repose) and θ is the angle of the longitudinal (bed) slope of the channel. The submerged weight is expressed as

$$W_s = c_2(\rho_s - \rho)gd^3 \tag{20.14a}$$

where c_2 is a constant. For a spherical particle, it is given by

$$W_s = \frac{1}{6}\pi(\rho_s - \rho)gd^3 \tag{20.14b}$$

20.5.3 Critical shear stress, τ_c

Critical shear stress refers to the shear stress at the point of incipient motion. It can be expressed by considering the equilibrium of forces as

$$F_D + W_s\sin\theta = \mu R = \tan\phi \, W_s \cos\theta \tag{20.15}$$

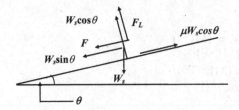

Figure 20.1 Forces acting on a particle of sediment on a flat bed with longitudinal slope angle θ.

---

[1] Fluid forces consist of skin friction forces by viscous shear acting on the surface and pressure forces (Drag and Lift) due to pressure differences along the surface.

(μ here and in Eq. 20.15 is the coefficient of friction which should not be confused with viscosity.)

from which, using Eqs. 20.12–20.14,

$$\tau_c = c(\rho_s - \rho)gd \cos\theta(\tan\phi - \tan\theta) \tag{20.16}$$

where

$$c = \frac{c_2}{c_1}$$

If the longitudinal slope angle is zero, Eq. 20.16 simplifies to

$$\tau_c = c(\rho_s - \rho)gd \tan\phi \tag{20.17a}$$

or

$$\frac{\tau_c}{gd(\rho_s - \rho)} = c\tan\phi \tag{20.17b}$$

The critical shear stress can also be expressed empirically as

$$\tau_c = 4 \text{ N/m}^2 \text{ for quartz sand upto 2 mm} \tag{20.18a}$$

$$\tau_c = \frac{10^{-4}}{6}g(\rho_s - \rho)\frac{d}{M} \text{ N/m}^2 \quad (\text{Kramer, 1935}) \tag{20.18b}$$

where d is the arithmetic mean diameter in mm and M is the uniformity coefficient defined by Kramer, and as

$$\tau_c = 0.285\left|\left(\frac{\rho_s - \rho}{\rho}\right)\frac{d}{M}\right|^{1/2} \quad (\text{USWES equation}) \tag{20.19}$$

20.5.4 Dimensional analysis

The critical shear stress can also be expressed in terms of other independent variables using dimensional analysis. The variables involved are

- Depth of flow, y_n (L)
- Bed slope, S_0 (0)
- Gravity, g (LT^{-2})
- Density of sediments, ρ_s (ML^{-3})
- Density of water, ρ (ML^{-3}/L^3)
- Viscosity of water, μ (ML^{-1}T^{-1})
- Sediment particle size, d (L)

Using Buckingham's π theorem, it can be shown that

$$\frac{u_* d}{v} = F\left(\frac{\rho u_*^2}{y_s^* d}, \frac{y_n}{d}, \frac{\rho_s}{\rho}\right) \tag{20.20}$$

where

$$u_* = \sqrt{\frac{\tau_c}{\rho}}; \quad y_s^* = g(\rho_s - \rho); \quad v = \frac{\mu}{\rho}$$

and $\dfrac{\rho u_*^2}{y_s^* d}$ is known as the Shields parameter (Shields, 1936).

Eq. 20.20 is equivalent to the expression

$$Re_* = F\left(\text{Shield's parameter, Relative roughness, Relative density}\right)$$

The significant parameters τ_c, $g(\rho_s - \rho)$, d and v are contained in the two dimensionless quantities Re_* and Shields' parameter (also called the entrainment function). Therefore, it is possible to write

$$\frac{u_* d}{v} = f\left(\frac{\rho u_*^2}{y_s^* d}\right), \text{ or, } \left(\frac{\rho u_*^2}{y_s^* d}\right) = \phi\left(\frac{u_* d}{v}\right) \tag{20.21a}$$

which is equivalent to

$$\frac{\tau_c}{gd(\rho_s - \rho)} = \phi\left(\frac{u_* d}{v}\right) \tag{20.21b}$$

20.5.5 Shields' analysis

Determination of the critical shear stress from Eq. 20.21 requires the function ϕ which has to be obtained from experimental data. Shields (1936) obtained this function in the form of a diagram, which is widely known as the Shields diagram for incipient motion. It depends on the geometry of the channel and sediments, and each point on the Shields' curve corresponds to incipient condition. The curve is similar to the curves of the friction factor f vs. Reynolds number for pipe flow. The straight line part of the curve (Figure 20.2) corresponds to the case when $\delta' \ggg d$ where δ' is the laminar sub-layer thickness. This condition means that the particles are completely submerged in the laminar sub-layer. The sagging part of the curve corresponds to the case when $\delta' = O(d)$. Under this condition, turbulent eddies disturb the laminar sub-layer. This occurs when $5 < Re_* < 40$. The steady part of the curve corresponds to the case when $\delta' \lll d$. Under this condition, the laminar sub-layer is destroyed by the particles protruding into it. The flow is turbulent, and the Shields' parameter is independent of the Re_*. This happens when the Shields' parameter is about 0.056, and Re_* is about 400. At this value of Shields' parameter, which corresponds to incipient motion condition, the following relationships can be established:

$$\tau_0 = \tau_c \Rightarrow \rho gRS = \tau_c^* \rho g(s_s - 1)d \Rightarrow 0.056\rho g(s_s - 1)d \tag{20.22}$$

$$d = 10.8RS \approx 11RS \tag{20.23a}$$

This is the limiting particle size for a stable channel with given R and S.

For a wide channel, $R = y$, and therefore the limiting particle size is given by

$$d = 11yS \tag{20.23b}$$

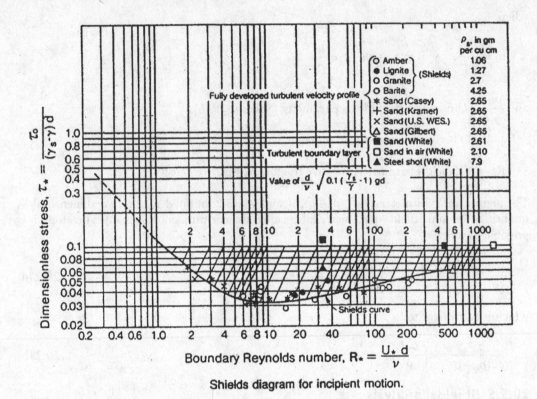

Shields diagram for incipient motion.

Figure 20.2 Shields' diagram. (Reproduced with permission from ASCE.)

In Shields' diagram, τ_c can only be determined implicitly. To avoid this difficulty, American Society of Civil Engineers (ASCE) Sedimentation manual gives a third parameter, which enables explicit determination of τ_c. It is of the form

$$\frac{d}{\nu}\left(0.1gd\left(\frac{\rho_s - \rho}{\rho}\right)\right)^{1/2} \tag{20.24}$$

which plots as a set of parallel lines on the Shields' diagram.

20.5.6 White's analysis

Using laboratory data, for high-speed flow with $\mathrm{Re}_* > 3.5$, the critical shear stress obtained by White (1940) is given by

$$\tau_c = \eta \frac{\pi}{6}(\rho_s - \rho)gd\cos\theta(\tan\phi - \tan\theta) \tag{20.25a}$$

where η is a packing coefficient. Under this condition, the surface drag $<<<$ form drag. Surface drag is due to skin friction whereas form drag is due to pressure. Form drag acts through the centroid of the particle. For a horizontal bed, the equation becomes

$$\tau_c = \eta \frac{\pi}{6}(\rho_s - \rho)gd\tan\phi \tag{20.25b}$$

For $Re_* > 3.5$, a turbulence factor T_f, which is the ratio of the instantaneous shear stress to the mean shear stress, is introduced into the equation. Then,

$$\tau_c = \frac{1}{T_f} \eta \frac{\pi}{6} (\rho_s - \rho) gd \tan \phi \tag{20.25c}$$

For open channel flow with $Re_* > 3.5$ (turbulent flow) with $T_f = 4.0$, $\eta = 0.4$ and $\tan \phi = 1.0$,

$$\frac{\tau_c}{gd(\rho_s - \rho)} = 0.052 \tag{20.26}$$

which is in reasonable agreement with Shields' results.

For low-speed flow with $Re_* < 3.5$, the critical shear stress is given by

$$\tau_c = \beta \eta \frac{\pi}{6} (\rho_s - \rho) gd \cos \theta (\tan \phi - \tan \theta) \tag{20.27}$$

where β is a factor introduced into the equation to account for the eccentricity of the line of action of the surface drag force. Under this condition, the surface drag >>>> form drag. Surface drag force acts with an eccentricity from the centroid of the particle. For a horizontal bed, the equation becomes

$$\tau_c = \beta \eta \frac{\pi}{6} (\rho_s - \rho) gd \tan \phi \tag{20.28}$$

By observation, it has been found that $\beta \eta = 0.34$. Then, for $Re_* < 3.5$,

$$\frac{\tau_c}{gd(\rho_s - \rho)} = 0.18 \tan \phi \tag{20.29}$$

which is not in agreement with Shields' results in which Shields' parameter is inversely proportional to Re_* for $Re_* < 3.5$ (straight line part).

20.5.7 Critical shear stress on a transversely sloping bed

Figure 20.3 shows the forces acting on a particle of sediment on a transversely sloping bed. For equilibrium,

$$F_D^2 + (W_s \sin \gamma)^2 = (\mu W_s \cos \gamma)^2 \Rightarrow F_D = W_s \cos \gamma \tan \phi \left(1 - \frac{\tan^2 \gamma}{\tan^2 \phi}\right)^{1/2} \tag{20.30}$$

At the channel bed, $\gamma = 0$, and therefore,

$$(F_D)_b = W_s \tan \phi \tag{20.31}$$

The tractive force ratio, K, is given as

$$K = \frac{(F_D)_s}{(F_D)_b} = \left(1 - \frac{\sin^2 \gamma}{\sin^2 \phi}\right)^{1/2} \tag{20.32}$$

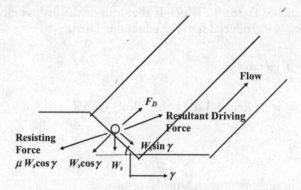

Figure 20.3 Forces acting on a sediment particle on a transverse slope.

20.6 CHANNEL ARMOURING AND DEGRADATION

When the channel sediment supply from upstream is less than the transport capacity of the channel, erosion from the channel bed takes place until a new sediment equilibrium is established. The result is a degradation of the channel. Finer materials are first transported, and at a faster rate than coarser materials. The channel will, as a result, be left with a coarse streambed. The coarsening process will stop once the finer materials are completely covered by coarser material. The streambed is then called 'armoured' and the coarse layer of sediments is called the 'armour layer'. Because of the variation of bed material size, more than one armouring layers may be necessary to protect the streambed.

The armouring layer thickness as proposed by Yang (1996) is (Figure 20.4)

$$Y_a = Y - Y_d \tag{20.33}$$

where Y_a is the thickness of the armouring layer, Y is the depth from the original streambed to bottom of the armouring layer and Y_d is the depth from the original streambed to the top of the armouring layer. This is also equal to the depth of degradation. It can also be expressed as (Yang, 1996)

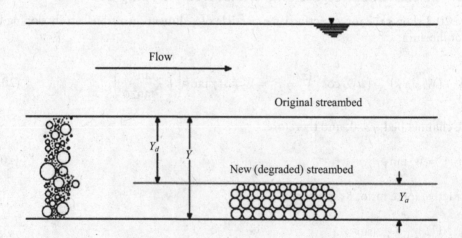

Figure 20.4 Definition sketch for armouring.

$$Y_a = (\Delta p) Y \tag{20.34}$$

where Δp is the fraction of material larger than the armouring size.
From these two equations,

$$Y_d = Y_a \left(\frac{1}{\Delta p} - 1 \right) \tag{20.35}$$

The design armouring layer thickness varies with the size of the armouring material. Usually, 2–3 particle diameters, or about 15 cm, whichever is small, is used. The procedure is as follows:

- Compute the mean particle size $(\bar{d})$ from a number of different empirical equations. For example, Shields' diagram, Yang's method, USBR method, Meyer-Peter and Muller (1948) method, etc.
- Assume three layers of armouring material, the thickness of the armouring layer is equal to $3\,\bar{d}$
- Assume a value for Δp, (20%–50%), the fraction of bed material larger than $\bar{d}$
- Hence the depth of degradation Y_d.

The reverse process is called aggradation when the stream bed elevation begins to increase. In a reservoir, this process is normally referred to as siltation which forms a delta on the downstream part of the reservoir. This is caused by excessive sediment supply from catchments and/or tributaries or reduction of water flow, which can occur when sediment free water is diverted, for example, for water supply schemes to provide potable water. It can also be due to a reduction of the slope of the stream as a result of, for example, the construction of a dam downstream. Examples of natural rivers affected by aggradation can be found in Japan, China, etc. where the river bed elevations have become higher than the surrounding ground levels.

Example 20.1

Determine the limiting particle size for Shields' stress to be independent of Reynolds number.

$$Re_c^* = \frac{u_* d}{v} \quad \text{and} \quad u_* = \sqrt{\frac{\tau_0}{\rho}}$$

$$\tau_* = \frac{\tau_0}{\rho g d (s_s - 1)} = 0.056$$

Substituting $v = 1.306 \times 10^{-6}\,\text{m}^2/\text{s}$; $s_s = 2.65$ and $g = 9.81$; $d = 6.7\,\text{mm}$.
Therefore, the grain size must exceed 6.7 mm for the Shields' stress to be independent of Reynolds number.

Example 20.2

River sand has a mean diameter of 0.2 mm. What is the fall velocity of this sand in water at 15°C? Assume particles are spheres with a specific gravity of 2.65. The kinematic viscosity of water at 15°C is $1.14 \times 10^{-6}\,\text{m}^2/\text{s}$.

The fall velocity is given by

$$w = \left\{ \frac{1}{C_D} \frac{4}{3} g d (s_s - 1) \right\}^{1/2}$$

$$= \left\{ \frac{1}{C_D} \frac{4}{3} 9.81 \times 0.2 \times 10^{-3} (2.65 - 1) \right\}^{1/2} = \frac{0.0657}{\sqrt{C_D}}$$

Assume C_D, calculate w, Re and C_D. Repeat until the assumed C_D agrees with the computed one. Computation of C_D can be done using

$$C_D = \frac{24}{Re} \left\{ 1 + \frac{3}{16} Re \right\} \text{ for } Re \text{ up to about 5.}$$

The graphical relationship can be used for the entire range of Re.
    Some assumed values are shown below:

Assumed C_D	w (m/s)	Re	Computed C_D
10	0.0208	3.645	6.92
7	0.0248	4.35	5.76
6	0.0268	4.70	5.30
5	0.0294	5.16	4.80
4.8	0.0299	5.26	4.725
4.7	0.0303	5.317	4.672 (OK)

The graphical relationship for fall velocity with particle size and shape factor can be used. Approximately, it reads as velocity = 3 cm/s.

REFERENCES

Hazen, A. (1904): On sedimentation, *Transactions of ASCE*, vol. 53, pp. 45–71.

Goldstein, S. (1929): The steady flow of viscous fluid past a fixed spherical obstacle at small Reynolds Number, *Proceedings of the Royal Society of London, Series A*, vol. 123, pp. 225–235.

Kramer, H. (1935): Sand mixtures and sand movement in fluvial models, *Transactions of ASCE*, vol. 100, pp. 798–873.

Meyer-Peter, E. and Muller, R. (1948): Formula for bed-load transport, Proceedings, International Association for Hydraulic Research, Second meeting, Stockholm, pp. 39–64.

Oseen, C. (1927): *Hydrodynamik, Chapter 10*, Akademische Verlagsgessellschaft, Leipzig.

Schiller, L. and Naumann, A. (1933): Uber die grundlegenden Berechnungen bei der Schwerkraftaufbereitung, *Zeitschriftd Vereines Deutscher Inge*, vol. 77, pp. 318–321.

Shields, A. (1936): *Anwendung der Ähnlichkeitsmechanik auf die Geschiebebewegung*, Preussische Versuchanstalt für Wasserbau und Schiffbau, Mitteilungen, no. 26, Berlin, 25 p.

Vanoni, V. A. ed. (1975): *Sedimentation Engineering*, ASCE Task Committee for the Preparation of the Manual on Sedimentation of the Sedimentation Committee of the Hydraulic Division, Reston, VA, 745 pp.

White, C. M. (1940): The equilibrium of grains on the bed of an alluvial channel, *Proceedings of the Royal Society of London, Series A*, vol. 174, pp. 332–338.

Yang, C. T. (1996): *Sediment Transport – Theory and Practice*, McGraw Hill Companies, Inc., New York, 396 pp.

Chapter 21

Channel stability analysis

21.1 INTRODUCTION

Alluvial channels change their geometry and alignment as a result of varying flow conditions, erosion, transport and deposition of sediments. A stable channel is one in which there is no erosion and deposition. Lane (1953) defined a stable channel as 'A stable channel is an unlined channel for carrying water, the banks and bed of which are not scoured by the moving water and in which objectionable deposits of sediments do not occur'.

A channel is in regime when the average dimensions and mean level do not show a trend in time.

21.2 VARIABLES INVOLVED IN CHANNEL DESIGN

21.2.1 Flow variables

- Water discharge, Q
- Sediment discharge, Q_s
- Average velocity of flow, V_0 (for no silting, no scour)
- Depth of flow D, or hydraulic radius, R
- Frictional slope, S_f.

21.2.2 Sediment variables

- Particle size
  - Mean diameter
  - Geometric standard deviation σ_g
- Density difference, $(\rho_s - \rho)$.

21.2.3 Fluid variables

- Density, ρ
- Dynamic viscosity, μ
- Gravity, g (not really a fluid variable but affects fluid flow).

21.2.4 Channel geometrical variables

- Average bed width, B or Wetted perimeter, P
- Cross-sectional shape factor, S_f.

For natural channels carrying sand as the sediment and water as the fluid, σ_g, $(\rho_s-\rho)$, ρ, and g can be taken as constants. Therefore, the design problem involves the determination of V_0, D (or R), B and Q_s for a given Q, d and S_f. Sometimes, the given variables may be different from those given above.

21.3 CONDITIONS TO BE SATISFIED

21.3.1 Continuity

$$Q = AV_0$$
$$= BDV_0 \left(\text{or } PRV_0 \right) \tag{21.1}$$

21.3.2 Resistance law

$$V_0 = f\left(R,\ S_f,\ \text{Resistance parameter} \right) \tag{21.2}$$

If the resistance parameter is unknown, an additional equation such as Strickler's equation (shown below) will be necessary:

$$\left[n = 0.01312 \left(d_{75} \right)^{1/6} \right] \tag{21.3}$$

21.3.3 Sediment transport law

$$Q_s = f(\tau_0,\ d,\ B,...) \tag{21.4}$$

21.3.4 Relationship for stable channel width B or wetted perimeter P

A channel section can be deep and narrow or shallow and wide. See also Section 21.4.2.1.

21.3.5 Condition for bank stability

To make the design procedure simple, some of the above conditions are ignored. For example, if the sediment transport rate is not taken into consideration, condition 21.3.3 is omitted (regime method of Kennedy and Lacey assume this simplification). For non-erodible banks, condition 21.3.5 is ignored.

21.4 METHODS OF STABLE CHANNEL DESIGN

21.4.1 Method of maximum permissible velocity

This is one of the oldest methods, and the design is for the greatest velocity which will not cause erosion or deposition. Design formulae are based on velocities measured in stable channels in various places and under various flow conditions. Empirical data are transformed into equations linking permissible velocities to flow parameters, soil type and geometry. Among them,

- Fortier and Scobey (1926) using field survey for depths up to approximately 1 m gave a typical set of permissible velocities as shown in Table 21.1 (original units are in feet, pounds, seconds).
- Hjulstrom (1935), based on the study of the movement of uniform material, obtained a set of graphs (mean sediment size vs. velocity, reproduced on p. 26, Yang, 1996), generally referred to as the Hjulstrom curves, which demarcates the zones of erosion, transport and deposition. The movement of bed material is influenced by the bottom velocity which is difficult to measure. It is therefore replaced by the average flow velocity. The effect of the ratio of depth to particle size $\left(\dfrac{D}{d}\right)$ is not taken into account in this study which is valid for depths not less than about 1 m.

When the actual conditions are different from the conditions given in Table 21.1 (flow depths > 1 m and for colloidal sediments), some adjustments must be made to the permissible velocities given.

A correction factor for the variation of flow depth has been derived by Mehrota (1983) on the assumption of the same critical tractive force.

$$\tau_c = \rho g R_1 S_{01} = \rho g R_2 S_{02} \tag{21.5}$$

which gives

$$\frac{S_{02}}{S_{01}} = \frac{R_1}{R_2} \tag{21.6}$$

Table 21.1 Maximum permissible velocities for different materials

Material	n	Clear water		Water with colloidal silts	
		V (m/s)	τ_0 (N/m²)	V (m/s)	τ_0 (N/m²)
Fine sand, colloidal	0.020	0.457	1.293	0.762	3.591
Sandy loam, non-colloidal	0.020	0.533	1.771	0.762	3.591
Silt loam, non-colloidal	0.020	0.601	2.298	0.914	5.267
Alluvial silts, non-colloidal	0.020	0.601	2.298	1.067	7.182
Ordinary firm loam	0.020	0.762	3.591	1.067	7.182
Volcanic ash	0.020	0.762	3.591	1.067	7.182
Stiff clay, very colloidal	0.025	1.143	12.449	1.524	22.025
Alluvial silts colloidal	0.025	1.143	12.449	1.524	22.025
Shales and hardpans	0.025	1.829	32.080	1.829	32.080
Fine gravel	0.020	0.762	32.080	1.524	15.322
Graded loam to cobbles when non-colloidal	0.030	1.143	18.194	1.524	31.601
Graded silts to cobbles when colloidal	0.030	1.219	20.588	1.676	38.304
Coarse gravel, non-colloidal	0.025	1.219	14.364	1.829	3.080
Cobbles and shingles	0.035	1.524	43.571	1.676	52.668

where S_{01} and S_{02} are channel slopes and R_1 and R_2 are hydraulic radii. From Manning's equation,

$$V_1 = \frac{1}{n} R_1^{2/3} S_{01}^{1/2} \text{ and } V_2 = \frac{1}{n} R_2^{2/3} S_{02}^{1/2}$$

which gives

$$\frac{V_2}{V_1} = \left(\frac{R_2}{R_1}\right)^{2/3} \left(\frac{S_{02}}{S_{01}}\right)^{1/2} \tag{21.7}$$

if n is independent of depth.

The correction factor k is given by (combining with Eq. 21.6)

$$k = \frac{V_2}{V_1} = \left(\frac{R_2}{R_1}\right)^{2/3} \left(\frac{S_{02}}{S_{01}}\right)^{1/2} = \left(\frac{R_2}{R_1}\right)^{2/3} \left(\frac{R_1}{R_2}\right)^{1/2} = \left(\frac{R_2}{R_1}\right)^{1/6} \tag{21.8}$$

In these equations, V_1, R_1 (or D_1) refer to the conditions given in Table 21.1, and V_2, R_2 (or D_2) refer to the actual conditions.

For coarse materials (non-cohesive), the graphical relationship can be used.

Approximate guidelines for corrections are as follows:

- When the depth > 1 m, velocity is increased by 0.15 m/s
- When coarse sediments are present, velocity is decreased by 0.15 m/s
- When discharges are extremely high for short durations, velocities may be increased by as much as 30%
- Yang (1973), using laboratory data from various sources, obtained a dimensionless velocity defined as

$$\frac{V_{cr}}{w} = \frac{2.5}{\log\left(\frac{u_* d}{v}\right) - 0.06} + 0.66 \text{ for } 1.2 < \frac{u_* d}{v} < 70 \tag{21.9}$$

and

$$\frac{V_{cr}}{w} = 2.05 \text{ for } \frac{u_* d}{v} \geq 70 \tag{21.10}$$

where w is the terminal fall velocity and u_* is the shear velocity. His results are shown graphically (Reynolds number based on shear velocity u_* vs. dimensionless critical velocity; Yang, 1996: p. 30).

- Neill (1967, 1968) related the mean erosion velocity to sediment size, specific gravity and depth of flow as

$$\frac{V_p^2}{(s_s - 1)gd} = 2.5 \left(\frac{d}{D}\right)^{-0.20} \tag{21.11}$$

where V_p is the permissible mean velocity or the mean erosion velocity.

The method of maximum permissible velocity is based more on experience than on theory.

21.4.2 Regime method (empirical)

The regime method aims to maintain a channel in regime. The design criteria have all been obtained from observations made in the field. Many such observations were made in India, Pakistan and Egypt where a large number of irrigation canals have been in existence for many thousands of years. The regime method started with an equation proposed by Kennedy (1895) using data collected from the Bari Doab canal system in Punjab. Lacey (1929) improved Kennedy's work with more extensive data, including the data used by Kennedy in Bari Doab and additional data from the lower Chenab and Madras canals. The design approach is to relate D, B, S_0 and V empirically to the discharge Q and particle size d. In other words, D, B, S_0, $V = f(Q, d)$.

21.4.2.1 Kennedy's approach (1895)

The general equation for permissible velocity is expressed as

$$V_0 = aD^b \tag{21.12}$$

where a and b are empirical constants. For the data collected from the Bari Doab canal system, Eq. 21.12 in SI units is

$$V_0 = 0.548D^{0.64} \tag{21.13}$$

For regions that have different sediment sizes to those found in the Bari Doab canal system, Eq. 21.13 is modified as

$$V = 0.548mD^{0.64} \tag{21.14}$$

where m is the critical velocity ratio $\dfrac{V}{V_0}$ defined as

$$m = \frac{\text{velocity of channel under study}}{\text{velocity of channel having the same sediment size as Bari Doab}} \tag{21.15}$$

Equation 21.15 gives a critical velocity ratio. The critical velocities can be found from graphs giving its variation with particle size (e.g. Yang, 1996: p. 26), as well as with shear velocity Reynolds number (Yang, 1996: p. 30). For sand coarser than those found in the Bari Doab system, $m > 1$, whereas for sand finer than those found in the Bari Doab system, $m < 1$.

Equations 21.13 and 21.14 together with Chezy's (or Manning's) equation can be used to solve for the bed width B, and S_f provided one of the variables is given.

In all these equations, the flow described corresponds to average conditions, and hence uniform flow conditions prevail. Therefore, the friction slope S_f can be replaced with bed slope S_0 or simply slope S. The method allows the choice of narrow and deep (small B, large D) sections as well as wide and shallow (large B, small D) sections. This is because one of the two variables can be determined only when the other is assumed.

Because of this unsatisfactory condition, Kennedy introduced the ratio $\dfrac{B}{D}$ which must be specified. He suggested $\dfrac{B}{D} = 3.5 - 7.0$ for channels with Q ranging from 0.28 to 28 m^3/s. To use Kennedy's approach, Q, S, m and a roughness parameter must be known. The critical velocity ratio comes from knowledge of the bed material size. The roughness parameter can be Manning's 'n' or Kutter's 'N' if Chezy's equation is used. The procedure is as follows:

- Compute velocity from Eqs. 21.12 or 21.13 for an assumed D
- Determine area of cross section A from Q and velocity
- With A and an assumed side slopes, compute the bed width and R for an assumed value of $\dfrac{B}{D}$
- Compute velocity using Manning's or Chezy's equation with the values of R, S, roughness parameter and m.
- Compare with the velocity obtained in the first step
- If compared well, the design is satisfactory. Otherwise, repeat the procedure with a new D.

The procedure described above is constrained. It has been later realized that a stable channel can adjust the width, the depth and the slope independently to a given set of conditions.

21.4.2.2 Lacey's approach (1929)

Lacey's approach is based on Kennedy's work as well as consolidating it with new data. The basic empirical equation has the same format as Eq. 21.12, but with R as the independent variable. It is of the form

$$V_0 = aR^b \tag{21.16}$$

which with Lacey's parameters takes the form

$$V_0 = 0.646R^{0.5} \tag{21.17}$$

He also introduced a silt factor f defined as

$$f = \left(\frac{a'}{0.646}\right)^2 \tag{21.18}$$

Eliminating a' from Eqs. 21.16, 21.17 and –21.18,

$$V_0 = 0.646\,(fR)^{0.5} \tag{21.19}$$

or, alternatively

$$R = 2.4\frac{V_0^2}{f} \tag{21.20}$$

Lacey's silt factors are (as given in p. 212, Raudkivi, 1990; for example)

Coarse sand 1.56 – 1.44

Medium sand 1.31

Standard Bari Doab silt 1.00

Lacey also obtained the following additional equations:

$$Af^2 = 134.2\,V_0^5 \text{ for } 0.46 < Af^2 < 278.7 \tag{21.21}$$

$$Qf^2 = 134.2\,V_0^6 \text{ or } 0.30 < V_0 < 1.22 \tag{21.22}$$

The mean velocity may be obtained from Manning's equation with 'n' given as

$$n = 0.0225f^{0.2} \tag{21.23}$$

For a given Q, and an assumed cross-sectional shape, the procedure is as follows:

- Assume a value for f
- Compute V_0 by Eq. 21.22
- Compute cross-sectional area A from $Q = A\,V_0$
- Compute hydraulic mean radius by Eq. 21.20

Dividing Eq. 21.21 by Eq. 21.20,

$$Pf = 56\,V_0^3 \tag{21.24}$$

Combining Eqs. 21.22 and 21.24, and eliminating f,

$$P = 4.8326\,Q^{0.5} \tag{21.25}$$

For a stable channel, Eq. 21.25 shows that the wetted perimeter is independent of the silt size. However, in practice, it is known that the silt size affects the shape of the cross section.

The limitations of Lacey's approach are that in the formulation, no consideration is given to the sediment discharge and that the range of validity is narrow. In 1953, Lacey modified some of his original equations and added some others. The important ones are given below:

$$P = 4.832\,Q^{0.5} \tag{21.26a}$$

$$A = 2.282\,\frac{Q^{5/6}}{f^{1/3}} \tag{21.26b}$$

$$R = 0.4725\,\frac{Q^{1/3}}{f^{1/3}} \tag{21.26c}$$

$$V_0 = 0.4382\,Q^{1/6}f^{1/3} \tag{21.26d}$$

$$S = \frac{1}{3169.8}\frac{f^{5/3}}{Q^{1/6}} \tag{21.26e}$$

$$V_0 = 10.8\,R^{2/3}S^{1/3} \tag{21.26f}$$

$$n = 0.0225f^{1/4} \tag{21.26g}$$

$$f = 1.76\,d^{0.5} \text{ with } d \text{ in mm.} \tag{21.26h}$$

$$f = 281.6\,R^{1/3}S^{2/3} \tag{21.26i}$$

Similar sets of equations have been obtained by Simons and Albertson (1963a, b) using Indian and US data and by Chitale (1966, 1976) using data from the Indian subcontinent, Egypt and USA.

21.4.3 Tractive force method

The tractive force method was originally developed by the USBR under the direction of Lane (1953). The original approach assumed no upstream sediment supply, and hence stability is ensured by preventing erosion. Bed material will be stable when the prevailing maximum bed stress is less than the critical bed shear stress. The critical bed shear stress has to be determined from experimental and/or empirical studies. On side slopes, the component of the weight in the direction of side slopes also must be taken into account. The tractive force method of stable channel design involves the determination of the following:

- Distribution of shear stress along a periphery
- Limiting tractive stress for various materials
- Effect of side slopes on limiting tractive stress
- Relationship between roughness coefficient and grain size.

21.4.3.1 Distribution of shear stress on periphery of a trapezoidal channel

The shear stress, averaged over the wetted perimeter, is given by

$$\tau_0 = \rho g R S = \rho g D S \text{ for a wide channel} \tag{21.27}$$

However, it is not uniformly distributed on the bed and side slopes. The boundary shear stress varies from a maximum at the deepest point to zero just above the water's edge. For practical purposes, the maximum values in the bed and side slopes are taken as 0.97 and 0.75 times $\rho g D S$ when the bed width is four times the depth and the side slopes are 2:1, although they depend upon the values of $\dfrac{B}{D}$ and the cross-sectional shape (Figure 21.1). The Highway Research Board (1970) has obtained a set of curves, which relate the ratio of maximum shear stress to $\rho g R S$ vs. $\dfrac{B}{D}$ values for different side slopes. The distribution is

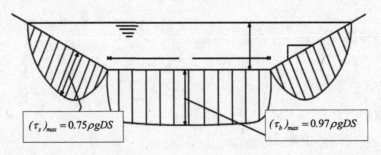

$(\tau_s)_{max} = 0.75\rho g D S \qquad (\tau_b)_{max} = 0.97\rho g D S$

Figure 21.1 Variation of tractive force on the bed and the side slopes.

determined using the power law for velocity distribution, and using membrane analogy and finite differences.

21.4.3.2 Limiting tractive stress for various materials

The limiting tractive stress can be empirically related to various materials. Lane (1955), using US data, obtained the following empirical relationship

$$\tau_c = 0.754 \, d_{75} \tag{21.28}$$

where τ_c is the limiting tractive stress in N/m^2 on horizontal bed and d_{75} is in mm. A similar relationship obtained by the Highway Research Board (1970) takes the form

$$\tau_c = 0.628 \, d_{50} \tag{21.29}$$

The original equation is $\tau_c = 4 \, d_{50}$ with τ_c in psi and d_{50} in ft which, when converted to SI units, is Eq. 21.29.

The USBR gives a graphical relationship between the critical shear stress and the mean sediment diameter.

21.4.3.3 Effect of side slopes on limiting tractive stress

A flatbed is stable if bed shear stress τ_0 is less than the critical shear stress $(t_0 \le \tau_c)$. When the component of the weight in the direction of the side slopes is taken into account, the limiting tractive stress will be less than that for an equivalent flatbed which means

$$\tau_s \le K \, \tau_c. \tag{21.30}$$

where K is the tractive force (or stress) ratio, which is given as

$$K = \frac{(F_D)_s}{(F_D)_b} = \left(1 - \frac{\sin^2 \gamma}{\sin^2 \phi}\right)^{1/2} \tag{21.31}$$

The prevailing maximum stress on side slopes depends on the side slopes and the $\dfrac{B}{D}$ ratio. The angle of repose ϕ is dependent on the size and shape of sediments. USBR has shown that ϕ needs to be considered only for coarse non-cohesive soils. For cohesive soils, gravity is negligible. Lane (1953) gives in graphical form the relationships between the angle of repose and the particle size for different shapes. Table 21.2 gives a set of values recommended by Simons (1957).

21.4.3.4 Relationship between roughness coefficient and grain size

The roughness coefficient (Manning's 'n' or Chezy's 'C') is expected to have a relationship with the particle size. Several relationships linking the Manning's 'n' to the particle size for coarse non-cohesive materials have been suggested in the literature. For example,

$$n = \frac{d_{50}^{1/6}}{21.1} \text{ with } d \text{ in m} \quad \text{(Strickler, 1923)} \tag{21.32a}$$

Table 21.2 Values of ϕ recommended by Simons (1957)

Median diameter in mm	Angle of repose in degrees		
	Crushed ledge rock	Very angular	Very rounded
0.30	32.0	31.4	29.2
1.50	34.5	32.9	29.5
3.0	36.6	33.8	29.9
15.0	40.0	37.5	32.5
30.0	40.8	39.1	34.8
150.0	42.0	41.2	38.5
300.0	42.2	41.5	39.2

$$n = \frac{d_{90}^{1/6}}{26} \quad \text{with } d \text{ in m} \quad \text{(Meyer-Peter and Muller, 1948)} \tag{21.32b}$$

$$n = \frac{d_{75}^{1/6}}{39} \quad \text{with } d \text{ in in.} \quad \text{(Lane and Carlson, 1953)} \tag{21.32c}$$

Because of the washing out of the finer material, d_{50} is difficult to be known. Eq. 21.32a has therefore been replaced by

$$n = \frac{d_{75}^{1/6}}{21.1} = 0.04739 \, d_{75}^{1/6} \tag{21.33}$$

where d_{75} refers to the channel material.

21.4.3.5 *Effect of sinuosity*

If the channel has bends, the limiting tractive stress should be reduced by a certain percentage depending on the sinuosity of the channel. The effect of sinuosity on the tractive stress and velocity as given by Garde and Ranga Raju (1977) is shown in Table 21.3.

21.5 STABLE CHANNELS CARRYING SEDIMENT-LADEN WATER

The criterion for stability is that the sediment transport capacity must be equal to the sediment supply. Then there will be no silting and no scour. For design purposes, the regime method and the tractive force method can be used. An empirical equation cited in Garde

Table 21.3 Effect of sinuosity

Sinuosity	Relative limiting tractive stress	Corresponding relative limiting velocity
Straight	1.0	1.0
Slightly sinuous	0.90	0.95
Moderately sinuous	0.75	0.87
Very sinuous	0.60	0.78

Source: Garde, R.J. and Ranga Raju, K.G., *Mechanics of Sediment Transportation and Alluvial Stream Problems,* Wiley, New Delhi, p. 305, 1977. With permission.

and Ranga Raju (1977: p. 324) for rivers which they attribute to Maddock and Leopold is of the form

$$\frac{W_s}{y_0} = 155.4 \left(\frac{\sqrt{S}/n}{(c_s w_0)^{0.395}} \right)^{3.0} Q^{0.555} \tag{21.34}$$

where W_s is the water surface width, y_0 is the hydraulic depth, w_0 is the fall velocity of sediments and c_s is the sediment concentration in ppm by weight. This equation is in SI units. It is not verified but is used as there is no other formula available. It can give a guide for channel width, q and q_s (discharges per unit width). Then, the continuity equation and a resistance equation together with a sediment transport law can be used to determine the average depth and slope.

DuBoys (1879), using the formula

$$q_s = c_d \tau (\tau - \tau_c) \tag{21.35}$$

where

$$\tau = \rho g R S = \rho g D S$$

$$\tau_c = \rho g R S_c = \rho g D S_c$$

obtained the relationship

$$q_s = c_d (\rho g)^2 y_0^2 S^2 \left(1 - \frac{S_c}{S} \right) \tag{21.36}$$

where c_d is a sediment discharge coefficient as given in (Raudkivi, 1990: p. 144)

$$c_d = \frac{0.689 \times 10^{-5}}{d^{0.75}} \text{ with } d \text{ in mm.} \tag{21.37}$$

The ratio of sediment discharge to water discharge using Chezy's equation is

$$\frac{q_s}{q} = \frac{c_d}{C} (\rho g)^2 y_0^{1/2} S^{3/2} (1 - \frac{S_c}{S}) \tag{21.38}$$

where C is the Chezy coefficient. The critical slope given by Straub (1950) is

$$S_c = 0.00025 \left(\frac{\bar{d} + 0.8}{y_0} \right) \text{ with } \bar{d} \text{ in mm.} \tag{21.39}$$

The slope S can be expressed by Manning's equation. Hence, the equilibrium velocity can be determined. Any bed load function can be used in the above approach.

21.6 OPTIMAL SHAPE FOR STABLE CHANNEL

The optimal shape and cross section refers to the minimum stable cross section. It depends upon the soil type and should have the least earthwork, least width and maximum velocity.

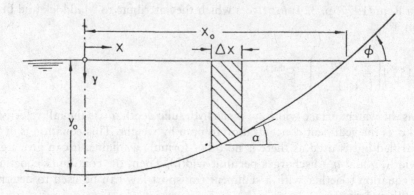

Figure 21.2 Cross sectional shape for a stable channel.

The commonly used cross section is trapezoidal. The design value of the boundary shear stress occurs over only part of the perimeter. Glover and Florey (1951) developed the following equations (Figure 21.2):

$$\rho g y S \Delta x = \frac{\tau \Delta x}{\cos \alpha} \Rightarrow \tau = \rho g y S \cos \alpha \tag{21.40}$$

If τ_{max} is the maximum shear stress at $y = y_0$, then

$$\frac{\tau}{\tau_{max}} = \frac{y}{y_0} \cos \alpha \tag{21.41}$$

At the threshold of motion, $\tau_{max} = \tau_0$
   Comparing Eq. 21.41 with Eq. 21.31,

$$\frac{y}{y_0} = \left(1 - \frac{\tan^2 \alpha}{\tan^2 \phi}\right)^{1/2} \tag{21.42}$$

Since $\tan \alpha = \dfrac{dy}{dx}$, Eq. 21.42 can be written as

$$\left(\frac{y}{y_0}\right)^2 = 1 - \frac{\left(\dfrac{dy}{dx}\right)^2}{\tan^2 \phi} \tag{21.43}$$

or

$$\left(\frac{dy}{dx}\right)^2 + \left(\frac{y}{y_0}\right)^2 \tan^2 \phi = \tan^2 \phi \tag{21.44}$$

$$\frac{dy}{\sqrt{1 - \left(\dfrac{y}{y_0}\right)^2}} = \tan \phi \, dx \tag{21.45}$$

which, upon integration, gives

$$y_0 \cos^{-1}\left(\frac{y}{y_0}\right) = x \tan\phi \qquad (21.46a)$$

or

$$\frac{y}{y_0} = \cos\left(\frac{x \tan\phi}{y_0}\right) \qquad (21.46b)$$

Thus, a stable channel has the shape of a cosine curve.

Half width x_0 corresponds to $y = 0$, or, $\cos^{-1}\left(\frac{y}{y_0}\right) = \cos^{-1}(0) = \frac{\pi}{2}$, which is when

$$x_0 = \frac{\pi y_0}{2\tan\phi} \qquad (21.47a)$$

Surface width, $B = 2x_0 = \dfrac{\pi y_0}{\tan\phi}$ $\qquad (21.47b)$

It can also be shown that

$$A = \frac{2y_0^2}{\tan\phi} \qquad (21.48)$$

$$P = \frac{2y_0}{\sin\phi} \int_0^{\pi/2} \left(1 - \sin^2\phi\cos^2\alpha\right)^{1/2} d\theta \qquad (21.49)$$

where $\theta = \dfrac{x\tan\phi}{y_0}$.

The $\cos^2\theta$ term in the integral can be replaced by $\sin^2\theta$ without changing the result. Then,

$$P = \frac{2y_0}{\sin\phi} \int_0^{\pi/2} \left(1 - \sin^2\phi\sin^2\alpha\right)^{1/2} d\theta \qquad (21.50)$$

This integral (elliptic integral of the second kind) can be found in mathematical tables (For example, p. 159, *Engineering Mathematics Handbook* by J. J. Tuma, McGraw-Hill Professional; 4th Edition, 1997). The procedure can be summarized as follows:

- For given d, find ϕ form Table 21.2 and n from Eq. 21.33
- Find B as a function of y_0 from Eq. 21.47b
- Find A as a function of y_0 from Eq. 21.48
- Find P as a function of y_0 from Eq. 21.50
- Find V $(= Q/A)$ as a function of y_0
- Substitute V above in Manning's equation and solve for y_0
- Determine y–x relationship from Eq. 21.46

Note: The USBR (1951) has given the following relationships to determine the stable channel cross sections (Simons and Senturk, 1992: p. 442):

$$R = \frac{y_0 \cos\phi}{E(\sin\phi)} \tag{21.51}$$

where

$$E(\sin\phi) \cong \frac{\pi}{2}\left(1 - \frac{1}{4}\sin^2\phi\right) \tag{21.52}$$

Hence, P can be determined approximately.

Example 21.1

Design a trapezoidal channel to carry $20\,\text{m}^3/\text{s}$ along a slope of 0.0016 in a terrain where the bed material consists of non-colloidal coarse gravel and pebbles (very rounded) with $d_{50} = 32\,\text{mm}$ and $d_{75} = 40\,\text{mm}$.

From Lane's equation (Eq. 21.28),

$$\tau_c = 0.754\, d_{75} = 30.16\,\text{N/m}^2$$

From Strickler's equation (Eq. 21.3),

$$n = 0.01312(d_{75})^{1/6} = 0.0243$$

For $d_{50} = 32\,\text{mm}$, very rounded, the angle of repose (as recommended by Simons (1957)) is 34.8°. Side slopes should be flatter than the angle of repose. Choose side slopes 2:1 which gives an angle $\alpha = 26.6°$.

$$K = \left(1 - \frac{\sin^2\alpha}{\sin^2\phi}\right)^{1/2} = 0.620 = \frac{\tau_s}{\tau_b}$$

Assume $\dfrac{B}{D} = 7$.

From graphs of $\dfrac{\tau(\max)}{\rho g R S}$ vs. $\dfrac{B}{D}$ for maximum shear stress on bed and on side slopes

$$\frac{\tau_s(\max)}{\rho g R S} = 1.0; \quad \frac{\tau_b(\max)}{\rho g R S} = 1.26$$

$$\frac{K\tau_b}{\rho g R S} = 1.0; \quad \frac{30.16}{\rho g R S} = 1.26$$

$$\frac{0.62 \times 30.16}{\rho g R S} = 1.0; \quad \frac{30.16}{\rho g R S} = 1.26$$

$$\Rightarrow R = 1.19\,\text{m}; \quad R = 1.525\,\text{m}$$

The lower value governs the design.

Cross-sectional area $A = BD + 2D^2 = 7D^2 + 2D^2 = 9D^2$

$$V = \frac{1}{n}R^{2/3}S^{1/2} = \frac{1}{0.0243}1.19^{2/3} \times 0.0016^{1/2} = 1.848\,\text{m/s}$$

$$Q = AV \Rightarrow 20 = 9D^2 \times 1.848 \Rightarrow D = 1.096 \approx 1.1\,\text{m}$$

$$B = 7D = 7.676\,\text{m} \approx 7.7\,\text{m}.$$

Check whether these dimensions can carry the required discharge.

$$Q = AV = 9 \times 1.1^2 \times 1.848 = 20.12\,\text{m}^3/\text{s} > 20\,\text{m}^3/\text{s}$$

Hence, the design is ok.

Example 21.2

A wide river carrying a flow of $50\,\text{m}^3/\text{s}$ at a depth of $2\,\text{m}$ with an average velocity of $0.75\,\text{m/s}$ on a slope of 0.0005. The bed material consists of particles $1-2\,\text{mm}$ in diameter. Examine whether the particles will move or not.

For a wide channel, $R = D$.

Using Shields' criterion for incipient motion, the dimensionless stress is given by

$$\tau_* = \frac{\tau_c}{gd(\rho_s - \rho)} = \frac{\rho g D S}{\rho g d \left(\dfrac{\rho_s}{\rho} - 1\right)} = \frac{DS}{d\left(\dfrac{\rho_s}{\rho} - 1\right)}$$

Using the given data $D = 2\,\text{m}$; $S = 0.0005$, and $\dfrac{\rho_s}{\rho} = 2.65$, $\tau_* = \dfrac{2 \times 0.0005}{d \times 1.65} = \dfrac{6.06 \times 10^{-4}}{d}$.

The dimensionless parameter $\dfrac{d}{\nu}\left(0.1gd\left(\dfrac{\rho_s - \rho}{\rho}\right)\right)^{1/2}$ for $d = 1\,\text{mm}$ is equal to 40.23 and corresponds to $\tau_* = 0.033$.

Therefore, $\dfrac{6.06 \times 10^{-4}}{d} = 0.033 \Rightarrow d = 18.8\,\text{mm} > 1\,\text{mm}\cdot$

Therefore $1\,\text{mm}$ particles will move.

For $2\,\text{mm}$ particles, $\dfrac{d}{\nu}\left(0.1gd\left(\dfrac{\rho_s - \rho}{\rho}\right)\right)^{1/2} = 114$, and corresponds to $\tau_* = 0.042$.

Therefore, $\dfrac{6.06 \times 10^{-4}}{d} = 0.042 \Rightarrow d = 14.4\,\text{mm} > 1\,\text{mm}\cdot$

Therefore, all the particles in the bed will move.

REFERENCES

Chitale, S.V. (1966): Design of alluvial channels, Proceedings, 6th Congress of ICID, Delhi, Q20, R 17.

Chitale, S.V. (1976): Shape and size of alluvial channels, *Journal of the Hydraulic Division*, vol. 102 no.: 7, pp. 1003–1011.

DuBoys, M. P. (1879): Le Rhone et les Rivieres a Lit affouillable, *Annales de Ponts et Chausses*, vol. 18 sec.: 5, pp. 141–195.

Fortier, S. and Scobey, F. C. (1926): Permissible canal velocities, *Transactions of the ASCE*, vol. 89 paper no.: 1588, pp. 940–984.

Garde, R. J. and Ranga Raju, K.G. (1977): *Mechanics of Sediment Transportation and Alluvial Stream Problems*, Wiley, New Delhi, 618 pp.

Glover, R. E. and Florey, Q. L. (1951): Stable channel profiles, United States Department of the Interior, Bureau of Reclamation, Hydraulic Laboratory Report No. Hyd-325, 45 pp.

Highway Research Board (1970): Tentative design procedure for rip-rap lined channels, National Academy of Sciences, National Cooperative Highway Research Program Report 108.

Hjulstrom, F. (1935): *The Morphological Activity of Rivers as Illustrated by River Fyris, Chapter 3*, vol. 25, Bulletin of the Geological Institute, Uppsala.

Kennedy, R. G. (1895): The prevention of silting in irrigation canals, Proceedings of the Institution of Civil Engineers, vol. CXIX, London.

Lacey, G. (1929): Stable channels in alluvium, Proceedings of the Institution of Civil Engineers, vol. 229, pp. 259–384, Thomas Telford-ICE Virtual Library, London.

Lacey, G. (1953): Uniform flow in alluvial rivers and canals, Proceedings of the Institution of Civil Engineers, vol. 237, p. 421, Thomas Telford-ICE Virtual Library, London.

Lane, E. W. (1953): Progress report on studies on the design of stable channels of the Bureau of Reclamation, Proceedings of the ASCE, vol 79, Reston, VA.

Lane, E. W. (1955): Design of stable canals, *Transactions of the ASCE*, vol. 120, pp. 1234–1260.

Lane, E. W. and Carlson, E. J. (1953): *Some Factors Affecting the Stability of Canals Constructed in Coarse Granular Materials*, International Association for Hydraulic Research, Minneapolis, MN.

Mehrota, S. C. (1983): Permissible velocity correction factor, *Journal of the Hydraulics Division, ASCE*, vol. 109 no.: HY2, pp. 305–308.

Meyer-Peter, E. and Muller R. (1948): Formulas for bed load transport, Proceedings, Second Meeting, International Association for Hydraulic Research, Paper No. 2, pp. 39–64, Delft.

Neill, C.R. (1967): Mean-velocity criterion for scour of coarse uniform bed material, International Association of Hydraulic Research 12th Congress Proceedings, vol. 3, pp. 46–54, Fort Collins, CO.

Neill, C. R. (1968): *A Re-Examination of the Beginning of Movement for Coarse Granular Bed Materials, Report IT 68*, Hydraulic Research Station, Wallingford.

Raudkivi, A. J. (1990): *Loose Boundary Hydraulics*, (3rd Edition), Pergamon Press, Oxford. 538 pp.

Simons, D. B. (1957): Theory and design of stable channels in alluvial material, Ph.D. thesis, Department of Civil Engineering, Colorado State University.

Simons, D. B. and Albertson, M. L (1963a): Uniform water conveyance channels in alluvial material, *ASCE Journal of Hydraulic Division*, vol 86, issue 5, pp 33–71.

Simons, D. B. and Albertson, M. L.(1963b): Discussion on "Uniform water conveyance channels in alluvial materials", *Transactions, ASCE*, vol. 128, Paper No. 3399, pp. 65–105.

Simons, D. B. and Senturk, F. (1992): *Sediment Transport Technology, Water and Sediment Dynamics*, Water Resources Publications, Littleton, CO, 897 pp.

Strickler, A (1923): *Some Contributions to the Problem of the Velocity Formula and Roughness Factors for Rivers, Canals and Closed Conduits*, Mitteilungen des eidgenossischen Amtesfur Wasserwirtschaft, no.: 16, Bern (Original in German).

Straub, L. G. (1950): "Chapter XII". In: *Engineering Hydraulics*, edited by H. Rouse, Wiley & Sons, New York, p. 806.

Yang, C. T. (1973): Incipient motion and sediment transport, *Journal of the Hydraulics Division, ASCE*, vol. 90, no.: HY10, pp. 1679–1704.

Yang, C. T. (1996): *Sediment transport, Theory and Practice*, McGraw Hill Book Companies Inc., New York, 396 pp.

Chapter 22

Sediment transport and deposition

22.1 INTRODUCTION

Open channel hydraulics is an idealized version of natural rivers or channel flow. Natural rivers and unlined artificial channels carry sediments as well as water. Erosion of river or channel bed and banks, as well as deposition of eroded or transported sediments, take place in all natural channels until a sediment equilibrium state is attained. Sediment transport and deposition are, therefore, important topics in the study of hydraulics of alluvial channels. They are problems in some major rivers in the world as well as in artificially constructed irrigation and other types of conveyance canals. The starting point of sediment transport is a source of sediment which is usually a mountainous catchment or a construction site. The mechanism of sediment generation is by the kinetic energy of rainfall which dislodges soil particles from the soil surface. Transport from the source to a watercourse is by overland flow. Once the sediments reach a watercourse, transport can be in different forms assisted by the energy of the flowing water and gravity. During the transport process, new sediments may be generated by erosion, and some may get deposited in the bed and banks of the watercourse before reaching the oceans where all sediments finally get deposited. In this chapter, some details of the types of sediments, their mechanisms of transport, different approaches of estimating the rates of sediment transport, as well as methods of estimating sediment yield will be presented.

22.2 SEDIMENT CLASSIFICATION

Sediment load is classified as *bed load* and *suspended load* depending on the mode of transport. It can also be classified as *wash load* and *bed material load*.

- **Bed load**: that part of the sediment load moving in the bed or near the bed by rolling, saltating or sliding. Usually, bed load is about 5%–25% of the suspended load in natural rivers. Saltation heights are of the order of one grain size diameter.
- **Suspended load**: that part of the sediment load moving in suspension.
- **Wash load**: finest part of the sediment mixture, e.g., clay and silt. It is washed through the channel, leaving no trace. It depends only on the supply and generally not correlated to the flow characteristics.
- **Bed material load**: particles that are found in the bed material. It is usually correlated with the water discharge.

22.3 TYPES OF SEDIMENT TRANSPORT FORMULAS

Pure theoretical approaches to determining sediment transport rates are not available. Most formulas are empirical or semi-empirical. There are several approaches, and all assume

- non-cohesive sediments,
- steady state conditions, and,
- uniform flow.

22.3.1 Shear stress approach

The formulas include

- DuBoys' formula
- Shields' formula
- Einstein's formula
- Meyer-Peter-Muller formula
- Einstein-Brown formula
- Parker et al.'s formula

22.3.2 Power approach

- Engelund-Hansen formula
- Ackers-White formula
- Yang's formula

22.3.3 Parametric approach

- Colby relations

Sediment transport formulas are also classified as

- Bed load formulas
  - DuBoys
  - Shields
  - Einstein
  - Einstein-Brown
  - Meyer-Peter-Muller
  - Parker et al.
- Suspended load formulas
  - Einstein-suspended load method
- Bed material load Formulas
  - Colby relations
  - Engelund–Hansen
  - Ackers–White
  - Yang

22.4 BED LOAD FORMULAS

22.4.1 DuBoys' formula (1879)

In the development of this formula, it is assumed that uniform sediments move as a series of superimposed layers with each thickness d' of the same magnitude as the grain diameter (Figure 22.1).

If the velocities in the layers are assumed to vary linearly, then, for $(n-1)$ layers in motion, the surface layer will have a velocity $(n-1)\Delta u$ where Δu is the velocity increment between two adjacent layers. The average velocity is then

$$\frac{(n-1)\Delta u + 0}{2} = \frac{(n-1)}{2}\Delta u$$

The volume rate of bed load q_b per unit width is then given by

$$q_b = \frac{1}{2}(n-1)\Delta u\, n d' \tag{22.1}$$

where n is the number of layers, d' is the thickness of each layer and $\frac{1}{2}(n-1)\Delta u$ is the average velocity. Bed shear stress τ_o is given by

$$\tau_o = \mu(\rho_s - \rho)gnd' \tag{22.2}$$

where μ is the coefficient of friction.

The threshold for bed load is when the top layer is at the point of incipient motion. At incipient motion, $n = 1$. The critical shear stress then is

$$\tau_c = \mu(\rho_s - \rho)gd' \tag{22.3}$$

Dividing Eq. (22.2) by Eq. (22.3)

$$\frac{\tau_o}{\tau_c} = n. \tag{22.4}$$

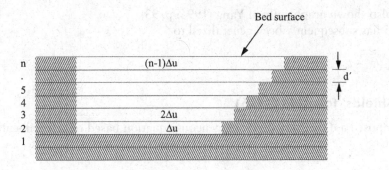

Figure 22.1 Definition sketch for DuBoys' formula.

Substituting in Eq. 22.1 gives

$$
\begin{aligned}
q_b &= \frac{1}{2}\frac{\tau_o}{\tau_c}d'\left(\frac{\tau_o}{\tau_c}-1\right)\Delta u \\
&= \frac{1}{2}d'\Delta u(\tau_0-\tau_c)\frac{\tau_o}{\tau_c^2} \\
&= c_d\tau_o(\tau_o-\tau_c)
\end{aligned}
\tag{22.5}
$$

where

$$
c_d = \frac{1}{2}\frac{d'\Delta u}{\tau_c^2}\ (\text{characteristic sediment coefficient})
\tag{22.6}
$$

With laboratory data, it has been shown that

$$
c_d = \frac{0.173}{d^{3/4}}(\text{Straub, 1935, in fps units})
\tag{22.7a}
$$

$$
c_d = \frac{0.689\times10^{-5}}{d^{3/4}}(\text{Raudkivi, 1990: p. 144, in SI units})
\tag{22.7b}
$$

and

$$
\tau_c = 0.061+0.093\,d\ \text{kg/m}^2\ \left(\text{multiply by 'g' to convert to N/m}^2\right)
\tag{22.8}
$$

where d is the uniform sediment diameter in mm (Chang, 1992: p. 135). Typical values given by Raudkivi (1990), based on Straub's (1950) data are as follows:

d (mm)	1	2	4
c_d (m^6/N^{2}s)$\times10^5$	0.689	0.405	0.243
τ_c (N/m^2)	1.535	2.443	4.311

They are also shown graphically in Yang (1996: p. 93).

Eq. 22.5 has subsequently been generalized to

$$
q_b = c_d'(\tau_o-\tau_c)^m
\tag{22.9}
$$

22.4.2 Shields' formula (1936)

Shields proposed a dimensionally homogeneous equation based on experimental data. It is of the form

$$
\frac{q_b(\gamma_s-\gamma)}{qS\gamma}=10\frac{\tau_o-\tau_c}{(\gamma_s-\gamma)d}
\tag{22.10a}
$$

where $\gamma_s = \rho_s g$, $\gamma = \rho g$, q is the water discharge rate/unit width, S is the slope, τ_o is the bed shear stress and τ_c is the critical bed shear stress, which can be obtained from Shields' diagram. In Eq. 22.10a, q_b is in weight per unit width. Any system of units may be used because it is dimensionless.

Note: Eq. 22.10a also has different versions; e.g. Chang (1992: p. 135) gives it as

$$\frac{q_b\left(\frac{\gamma_s}{\gamma}-1\right)}{qS\gamma}=10\frac{\tau_o-\tau_c}{(\gamma_s-\gamma)d}$$ (22.10b)

The factor 10 is used to account for the range of scatter in the data which fits into the above equation.

22.4.3 Einstein's bed load function (1942, 1950)

Einstein obtained his bed load function using concepts from Fluid Mechanics and Probability. Fluid forces in turbulent flow vary both in space and time. Therefore, the movement of any particle depends upon the probability at a particular time of the applied force exceeding the resisting force. He defined two dimensionless parameters, the bed load intensity ϕ and the flow intensity ψ, given as

$$\phi=\frac{q_b}{\gamma_s}\left\{\frac{\gamma}{\gamma_s-\gamma}\frac{1}{gd^3}\right\}^{1/2}$$ (22.11a)

$$\psi=\frac{\gamma_s-\gamma}{\gamma R'S}d$$ (22.11b)

where R' is the hydraulic radius with respect to grain roughness, and

$$\phi=f(\psi)$$ (22.12)

For fully rough conditions, R' is obtained from Manning–Strickler formula in the form

$$\frac{u}{u'_*}=7.66\left(\frac{R'}{d_{65}}\right)^{1/6}$$ (22.13)

where $u'_*=(gR'S)^{1/2}$ and, d_{65} is used as the roughness height k_s.

For conditions less than fully rough, Einstein (1950) proposed the following velocity profile:

$$\frac{u}{u_*}=5.75\log\left(30.2\frac{z}{\Delta}\right)$$ (22.14)

where u is the velocity at a height z from the bed (estimated as $z=0.35d_x$ from experiments; d_x is the characteristic grain size of the sediment mixture). For mean velocity, he proposed another velocity profile as

$$\frac{\bar{u}}{u_*}=5.75\log\left(12.27\frac{D}{\Delta}\right)$$ (22.15)

where D is the depth of flow, Δ is the apparent roughness of the bed surface ($= k_s/x$), k_s is the roughness height, and x is a correction factor that compensates for conditions where the channel bed is not fully rough. It is given as a function of k_s/δ where $k_s \cong d_{65}$.

From experiments, it has been shown that the velocity acting on particles must be measured at a height $0.35\,d_x$ where d_x is a characteristic grain size. Therefore, the velocity relevant to sediments is given by

$$\frac{u}{u_*} = 5.75 \log\left(30.2 \times 0.35 \frac{d_x}{\Delta} \right) = 5.75 \log\left(10.6 \frac{d_x}{\Delta} \right) \tag{22.16}$$

Hence, R' can be determined if u is known.

Einstein and Barbarosa (1952) proposed that

$$\tau_0 = \tau_0' + \tau_0'' \tag{22.17}$$

which is equivalent to the bed shear being equal to the sum of flatbed shear due to grain roughness and shear due to bed feature or form drag, and

$$\tau_0 = \rho g R S = \rho g S (R' + R'') \tag{22.18}$$

in which R' corresponds to τ_0' and R'' corresponds to τ_0''.

Also, the corresponding shear velocities

$$u_*' = \sqrt{gR'S_f}\,;\ u_*'' = \sqrt{gR''S_f} \tag{22.19}$$

with

$$u_*^2 = u_*'^2 + u_*''^2 \tag{22.20}$$

and

$$\frac{u}{u_*} = 5.75 \log\left(12.27 \frac{R'}{k_s} x \right) \tag{22.21}$$

An empirical relationship between $\dfrac{u}{u_*''}$ and the flow intensity function

$$\psi' = \frac{\rho_s - \rho}{\rho} \frac{d_{35}}{R'S_f}\ \text{(similar to Eq. 22.11b)} \tag{22.22}$$

has been obtained by Einstein.

The calculation of R' is a trial and error procedure. The steps involved are as follows:

- Assume $R' = R$
- Calculate ψ'

- Read $\dfrac{u}{u_*''}$ from Einstein's graphs

- Calculate u_*'' (knowing u, the average velocity)
- Calculate R'' from $u_*'' = \sqrt{gR''S_f}$
- Adjust R' as R' (new) $= R'$ (old) $- R_*''$
- Repeat until R' (new) $\cong R'$ (old).

The ϕ-ψ relationship is obtained from empirical data.

Because the bed load transport of grains in a sediment mixture is affected by other grains due to size variation and hiding effect, Einstein divided the bed load into several size fractions each of which is represented by the geometric mean. Sediment discharge of each fraction is computed separately, and the total sediment transport is obtained by summation:

$$q_b = \sum_{i=1}^{n} p_i q_{bi} \tag{22.23}$$

where p_i is the fraction of the total sediment mixture of size i, q_{bi} is the sediment transport rate for the sediment class i and n is the number of size fractions.

Einstein (1950) later presented more sophisticated formulas for ϕ and ψ for each size fraction. They are of the form

$$\phi_{*i} = \frac{q_{bi}}{p_i \gamma_s} \left\{ \frac{\gamma}{\gamma_s - \gamma} \frac{1}{g d_i^3} \right\}^{1/2} \tag{22.24a}$$

$$\psi_{*i} = \xi_i Y \left[\frac{\log 10.6}{\log\left(10.6 \dfrac{d_x}{\Delta}\right)} \right]^2 \frac{(\gamma_s - \gamma)}{\gamma R'S} d_i \tag{22.24b}$$

The parameter ψ_{*i} has three parts:

Part I $[\xi_i Y]$: Consists of a hiding factor ξ_i which is a function of $\dfrac{d_i}{d_x}$ where d_x is the characteristic grain diameter of the mixture (ξ_i vs. $\dfrac{d_i}{d_x}$ is given graphically). For uniform grains $\xi_i = 1$, and ξ_i increases with decreasing particle size in the mixture.

For rough bed:

$$d_x = 0.77\Delta \text{ for } \frac{\Delta}{\delta} > 1.8 \tag{22.25a}$$

For smooth bed:

$$d_x = 1.39\delta \text{ for } \frac{\Delta}{\delta} < 1.8 \tag{22.25b}$$

where Δ is the apparent roughness of the bed surface $\left(= \dfrac{k_s}{x}\right)$ and δ is the laminar sub-layer thickness.

The term Y is a correction factor for the change in the lift coefficient, which is given graphically as a function of $\dfrac{d_{65}}{\delta}$. The factors ξ_i and Y are determined experimentally.

$$\textbf{Part II} \left[\frac{\log 10.6}{\log\left(10.6 \dfrac{d_x}{\Delta}\right)} \right]^2 : \text{Accounts for the hydrodynamic lift} \tag{22.26}$$

Part III $\left[\dfrac{\gamma_s - \gamma}{\gamma R'S}\, d_i\right]$: Reciprocal of the grain shear stress (22.27)

The quantities ϕ_{*i} and ψ_{*i} are expressed as empirical functions in the form

$$1 - \frac{1}{\sqrt{\pi}} \int_{l_1}^{l_2} e^{-t^2}\, dt = \frac{A\phi_{*i}}{1 + A\phi_{*i}} \tag{22.28}$$

where

$$I_1 = -\left(\frac{1}{7}\psi_{*i} - 2\right) \tag{22.29}$$

$$I_2 = \frac{1}{7}\psi_{*i} - 2 \tag{22.30}$$

and A is a constant (= 43.5).

Eq. 22.28 is Einstein's bed load function. It is expressed as ϕ_* vs. ψ_* or ϕ_* vs. $\dfrac{1}{\psi_*}$.

Note: If the ψ values fall outside the range given in the graphs, then the ϕ values need to be computed using Eq. 22.28. The integral on the left-hand side of Eq. 22.28 may be written as

$$\frac{1}{\sqrt{\pi}} \int_{l_1}^{l_2} e^{-t^2}\, dt = \frac{1}{\sqrt{\pi}} \int_{0}^{l_2} e^{-t^2}\, dt - \frac{1}{\sqrt{\pi}} \int_{0}^{l_1} e^{-t^2}\, dt$$

$$= \frac{1}{2}\operatorname{erf}(I_2) - \frac{1}{2}\operatorname{erf}(I_1) \tag{22.31}$$

in which

$$\operatorname{erf}(x) = \frac{2}{\sqrt{\pi}} \int_{0}^{x} e^{-t^2}\, dt \tag{22.32}$$

Values of the error function $\operatorname{erf}(x)$ are given in mathematical handbooks. In the use of Eq. 22.28, it is also necessary to make use of the property that

$$\operatorname{erf}(-x) = -\operatorname{erf}(x)$$

The bed load per unit width q_{bi} for each fraction of sediment size is then calculated from Eq. 22.24a as

$$q_{bi} = \frac{\phi_{*i} p_i \gamma_s d_i^{3/2}}{\left(\dfrac{1}{s_s - 1}\dfrac{1}{g}\right)^{1/2}} \tag{22.33}$$

To use this function,

- ψ_{*i} is computed from given bed sediment characteristics and flow conditions for each size fraction i
- ϕ_{*i} is obtained from the Figure given in Technical Bulletin no. 1026 (Einstein, 1950) or Eq. 22.28
- q_{bi} is obtained from Eq. 22.24a

22.5 SUSPENDED LOAD FORMULAS

22.5.1 Theoretical basis

Diffusion analogy has been used to describe the suspended sediment transport mechanism under the following assumptions:

- Sediment concentration does not change in the longitudinal and transverse directions,
- Diffusion does not occur in the longitudinal and transverse directions,
- Diffusion occurs in the vertical direction, and
- Two-dimensional steady uniform flow conditions prevail.

The diffusion equation then is

$$w \frac{\partial c}{\partial z} = \frac{\partial}{\partial z}\left(\varepsilon_z \frac{\partial c}{\partial z} \right) \tag{22.34}$$

where c is the sediment concentration and ε_z is the diffusion coefficient in the z-direction. Integration gives

$$w_s c + \varepsilon_s \frac{\partial c}{\partial z} = \text{constant} \tag{22.35}$$

($\uparrow w = -w_s \downarrow$ (downward fall velocity); ε_s – diffusion coefficient for sediments)

$\frac{dc}{dz} = 0$ for $c = 0$ gives constant $= 0$.

Considering momentum transfer, the momentum flux is given by

$$\rho \varepsilon_m \frac{du}{dz} \tag{22.36}$$

where ε_m is the kinematic eddy viscosity or the diffusion coefficient for momentum. It can also be shown that

$$\tau = \rho \varepsilon_m \frac{du}{dz} \tag{22.37}$$

Similarly, the mass flux can be written as

$$\rho \varepsilon_s \frac{dc}{dz} \tag{22.38}$$

where ε_s is a diffusion coefficient.

In general, $\varepsilon_s \neq \varepsilon_m$, but for fine sediments, it has been proved by experiments that $\varepsilon_s = \varepsilon_m$. For coarse particles, $\varepsilon_s = \beta\varepsilon_m$ where $\beta < 1$.

For two-dimensional steady uniform flow, the shear stress τ at height z above the bed is given by

$$\tau = \rho g(D - z)S = \rho g D S - \rho g z S = \rho g D S\left(1 - \frac{z}{D}\right) = \tau_0\left(1 - \frac{z}{D}\right) \tag{22.39}$$

If the velocity profile $u(z)$ is known, ε_s can be expressed as a function of z using

$$\varepsilon_s = \beta\varepsilon_m \text{ and } \tau = \tau_0\left(1 - \frac{z}{D}\right)$$

If the logarithmic profile (Prandtl-Karman) is used,

$$\frac{\partial u}{\partial z} = \frac{u_*}{Kz} \quad \left(\text{i.e. } u = \frac{u_*}{K}\ln(z) + \text{constant}\right) \tag{22.40}$$

where K – von Karman constant (≈ 0.4).

Therefore,

$$\tau = \rho\varepsilon_m\frac{du}{dz} = \rho\varepsilon_m\frac{u_*}{Kz} \tag{22.41}$$

Substituting for τ from Eq. 22.39,

$$\tau_0\left(1 - \frac{z}{D}\right) = \rho\varepsilon_m\frac{u_*}{Kz} \tag{22.42a}$$

$$\frac{\tau_0}{\rho}\left(1 - \frac{z}{D}\right) = \varepsilon_m\frac{u_*}{Kz} \tag{22.42b}$$

Because $u_*^2 = \dfrac{\tau_0}{\rho}$

$$\varepsilon_m = Ku_*z\left(1 - \frac{z}{D}\right) \tag{22.43a}$$

or

$$\varepsilon_m = Ku_*\frac{z}{D}(D - z) \tag{22.43b}$$

Therefore,

$$\varepsilon_s = \beta\varepsilon_m = \beta Ku_*\frac{z}{D}(D - z) \tag{22.44}$$

$$\left[\varepsilon_m \text{ and } \varepsilon_s = 0 \text{ at } z = 0, \text{ or } z = D, \text{ and attain maximum values at } \frac{D}{2}\right].$$

Substituting for ε_s in Eq. 22.35,

$$\frac{dc}{c} + \frac{w_s}{\beta K u_*} \frac{D}{(D-z)z} dz = 0 \tag{22.45}$$

Putting $z_* = \dfrac{w_s}{\beta K u_*}$

$$\frac{dc}{c} + z_* \frac{D}{(D-z)z} dz = 0 \text{ which can be integrated as follows:}$$

$$\ln(c) + z_* \int \frac{D}{(D-z)z} dz = 0$$

$$\ln(c) + z_* \int \left[\frac{1}{D-z} + \frac{1}{z} \right] dz = 0$$

$$\ln(c) + z_* \left[-\ln(D-z) + \ln(z) \right] = 0$$

$$\ln(c) + z_* \ln\left(\frac{z}{D-z} \right) = 0$$

$$\ln(c) + \ln\left(\frac{z}{D-z} \right)^{z_*} = 0$$

$$\ln(c) = \ln\left(\frac{D-z}{z} \right)^{z_*}$$

Between the limits $z = a$ and $z = z$

$$\left[\ln(c) \right]_a^z = \left[\ln\left(\frac{D-z}{z} \right)^{z_*} \right]_a^z \tag{22.46a}$$

or

$$\frac{c}{c_a} = \left[\frac{D-z}{z} \frac{a}{D-a} \right]^{z_*} \tag{22.46b}$$

where c_a represents the concentration of sediments (weight of sediments per unit volume of water-sediment mixture; kg/m^3) with fall velocity w_s at a level $z = a$. This equation is called the Rouse equation (1937) and is given graphically too.

The computation of z_* requires β, K and w_s.

- For fine particles, $\beta \approx 1$
- For coarse particles, $\beta < 1$
- von Karman constant $K = 0.4$
- w_s needs to be measured in still water

Einstein and Chien (1954) used

$$z'_* = z_* \beta = \frac{w_s}{Ku_*} \qquad (22.47)$$

and β was obtained by comparing with measured suspended sediment load (given graphically in the plot of z_* vs. z'_*). This relationship is still used as there are no other similar relationships available.

22.5.2 Einstein's suspended load formula (1950)

$$q_{ss} = \int_a^D cu\,dz \qquad (22.48)$$

where q_{ss} is the suspended sediment discharge rate per unit width of channel, a is the lower limit where suspension begins and D is the depth. (For narrow channels, D should be replaced by R.)

Substituting Eq. 22.46b in Eq. 22.48 and using the relationship

$$\frac{u}{u'_*} = 5.75 \log\left(\frac{30.2z}{\Delta}\right) \quad \text{(same as Eq. 22.14)}$$

$$q_{ss} = \int_a^D c_a\left(\frac{D-z}{z}\frac{a}{D-a}\right)^{z_*} 5.75 \log\left(\frac{30.2z}{\Delta}\right)u'_* dz \qquad (22.49)$$

where the velocity profile is resulting from sediment roughness only.

Replacing 'a' and 'z' by dimensionless values,

$$A = \frac{a}{D} \text{ and } \eta = \frac{z}{D},$$

Equation 22.49 becomes

$$q_{ss} = \int_A^1 Dcu\,d\eta$$

$$= Du'_* c_a\left(\frac{A}{1-A}\right)^{z_*} 5.75 \int_A^1 \left(\frac{1-\eta}{\eta}\right)^{z_*} \log\frac{30.2\eta}{\Delta/D}\,d\eta$$

which has been transformed to the following form by Einstein:

$$= 11.6 c_a u'_* a\left[2.303 \log\left(\frac{30.2D}{\Delta}\right)I_1 + I_2\right] \qquad (22.50)$$

where

$$I_1 = 0.216 \frac{A^{z*-1}}{(1-A)^{z*}} \int_A^1 \left(\frac{1-\eta}{\eta}\right)^{z*} d\eta \tag{22.51a}$$

$$I_2 = 0.216 \frac{A^{z*-1}}{(1-A)^{z*}} \int_A^1 \left(\frac{1-\eta}{\eta}\right)^{z*} \ln(\eta)\, d\eta \tag{22.51b}$$

The values of I_1 and I_2 are given in graphical form as functions of A and $z*$ (I_2 is always negative).

Equation 22.50 can be used to calculate suspended load for different size fractions. For each size fraction, Einstein used $a = 2d_i$ as the lower limit of integration. He also defined the bed load as the sediment moving in the bed layer, which has a thickness of two grain diameters. Then, for each size fraction p_i, the suspended transport rate is $p_i q_{ssi}$.

The steps involved in using Eq. 22.50 can be summarized as follows:

For each size fraction of sediments,

- Assume the depth of flow z (or R) to cover the whole range of depths
- Compute $u* \left[= (gRS)^{0.5}\right]$
- Compute $\delta' \left[= 11.6\dfrac{\nu}{u*}\right]$
- Compute $\dfrac{k_s}{\delta'}$, with $k_s \cong d_{65}$
- Obtain x from x vs. $\dfrac{k_s}{\delta'}$ graph
- Compute $\Delta\left[= \dfrac{k_s}{x}\right]$
- Compute $z'_* \left[= \dfrac{w_s}{Ku*}\right]$ with w_s, the settling velocity known, and $K = 0.4$
- Obtain $z*$ from $z*$ vs. z'_* graphs
- Compute $A = \dfrac{a}{D}$ and $\eta = \dfrac{z}{D}$
- Obtain I_1 and I_2 for the computed values of A and η from graphs
- Substitute in Eq. 22.50 (c_a is assumed to be known).

22.5.3 Transition between bed load and suspended load

If u_b is the average velocity of the bed-material load, then the weight of particles of a given size per unit area is

$$\frac{p_i q_{bi}}{u_b} = \frac{\text{weight}}{\text{area}}$$

The concentration of sediments in the bed layer which Einstein defined as a layer of thickness of two grain diameters, i.e. $2d_i$, is (c_{ai})

$$\frac{p_i q_{bi}}{u_b (2d_i)} = \frac{\text{weight of sediments}}{\text{volume of mixture}}$$

Using a correction factor A' to allow for errors that may be caused by assuming the concentration in the bed layer to be uniform and the assumption that $u_b = u'_*$,

$$c_{ai} = A' \frac{p_i q_{bi}}{u_*(2d_i)}$$

Experimentally, A' has been found to be $\dfrac{1}{11.6}$. Therefore,

$$c_{ai} = \frac{1}{11.6} \frac{p_i q_{bi}}{u_*(2d_i)} \tag{22.52}$$

which can also be written as

$$p_i q_{bi} = 11.6 \, u'_* \, c_{ai} \, 2d_i = 11.6 \, u'_* \, c_{ai} \, a \tag{22.53}$$

Combining with Eq. 22.50,

$$p_i q_{ssi} = p_i q_{bi} \left[2.303 \log\left(\frac{30.2D}{\Delta} \right) I_1 + I_2 \right] \tag{22.54}$$

Equation 22.54 is dimensionally homogeneous. The total sediment transport q_T for a single size fraction p_i is given by

$$p_i q_T = p_i q_{bi} + p_i q_{ssi}$$

$$= p_i q_{bi} \left[1 + \left\{ 2.303 \log\left(\frac{30.2D}{\Delta} \right) \right\} I_1 + I_2 \right] \tag{22.55}$$

22.6 BED MATERIAL LOAD FORMULAS

22.6.1 Ackers–White formula (1973)

Ackers and White related the concentration of bed material load as a function of the mobility number, F_g:

$$c_s = c s_s \frac{d}{R} \left(\frac{u}{u_*} \right)^n \left(\frac{F_g}{A} - 1 \right)^m \tag{22.56}$$

where n, c, A and m are coefficients. The mobility number is given by

$$F_g = \frac{u_*^n}{\left[gd(s_s - 1) \right]^{1/2}} \left[\frac{u}{(32)^{1/2} \log\left(10 \dfrac{R}{d} \right)} \right]^{1-n} \tag{22.57}$$

They also expressed the sediment size by a dimensionless grain diameter d_g defined as

$$d_g = d \left[\frac{g(s_s - 1)}{v^2} \right]^{1/3} \tag{22.58}$$

where ν is the kinematic viscosity of water, s_s is the specific gravity of sediments, d is the median diameter of bed material, and c_s is the sediment concentration by weight $\left(= \dfrac{Q_s}{Q}\right)$.

The coefficients have been determined from about 1,000 sets of laboratory data with sediment size greater than 0.04 mm and Froude number less than 0.8. They are as follows:

Coefficient	$D_g > 60$	$60 > d_g > 1$
c	0.025	$\log c = 2.86 \log d_g - \left(\log d_g\right)^2 - 3.53$
n	0	$1 - 0.56 \log d_g$
A	0.17	$\dfrac{0.23}{\left(d_g\right)^{1/2}} + 0.14$
m	1.50	$\dfrac{9.66}{d_g} + 1.34$

22.7 SEDIMENT YIELD

Sediment yield refers to that part of the gross erosion that is transported. Gross erosion is the total amount of material eroded.

22.7.1 Method of estimation

22.7.1.1 Sediment sampling

Sediment sampling method requires the sediment rating curve, which is obtained by sampling, and the flow duration curve, which is obtained from flow records. The sediment rating curve gives the relationship between the sediment discharge and the flow discharge. The flow duration curve gives the relationship between the magnitude of flow with the percentage of the time it is equalled or exceeded. The procedure is as follows:

- Separate the flow duration curve into a certain number of class intervals,
- Compute the range of each probability interval and the mid-point of each interval,
- Determine the flow discharge from the flow duration curves,
- Using the sediment rating curve determine the sediment discharge for each flow discharge,
- Compute the expected sediment discharge as (sediment discharge) × (probability interval),
- Compute the total sediment as (number of days in the year) × (sum of the expected sediment discharges),
- Divide by the sediment area.

22.7.1.2 Universal soil loss equation

This equation is based on a statistical analysis of data from 47 locations in 24 states in USA and is of the form

$$A = RKLSCP \tag{22.59}$$

where

    A – Computed soil loss in Tons/unit area/year
    R – Rainfall factor
    K – Soil erodibility factor
    L – Slope-length factor
    S – Slope-steepness factor
    C – Crop-management factor
    P – Erosion-control practice factor

The rainfall factor R accounts for differences in rainfall intensity-duration-frequency for different locations; soil erodibility factor accounts for the susceptibility of a given soil to erosion; slope-length factor accounts for the increase in the quantity of runoff that occurs as the distance from the top of the slope increases; slope-steepness factor accounts for the increase in velocity of runoff as slope steepness increases; crop-management factor accounts for the crop rotation used. Typical values of these factors can be found in tables, maps and graphs (e.g. Yang, 1996: pp. 269–274).

22.7.1.3 Use of sediment yield equations

$$\text{Sediment yield} = \frac{\text{Sediment accumulated}}{\text{Trap efficiency}} \tag{22.60}$$

where the 'trap efficiency' is the fraction of the sediment inflow which gets deposited in the reservoir. In a different form, it is

$$\text{Trap efficiency} = \frac{\text{Sediment retained}}{\text{Sediment inflow}} \tag{22.61}$$

and is a function of the reservoir capacity – inflow ratio. For large reservoirs, it is nearly 100%, whereas for small reservoirs, it is almost insignificant. The reason being that large reservoirs retain the water for long periods of time, and therefore, there is sufficient time for the sediments (particles in suspension in particular) to settle down. In small reservoirs, water is not retained for long periods of time. An empirical graphical relationship between the trap efficiency and the capacity-inflow ratio has been obtained by Brune (1953), commonly known as Brune's curves. The capacity is measured at the mean operating level, and the inflow is the mean annual inflow into the reservoir. Trap efficiency is given as a percentage.

22.7.1.4 Use of sediment delivery ratios (SDR)

All sediment material eroded is not delivered to the downstream point. The sediment delivery ratio is defined as

$$\frac{\text{Sediment transported to a particular point}}{\text{Gross erosion in the catchment upstream of the point}}$$

and it is a function of

- Sediment source (catchment vs. channel),
- Magnitude and proximity of source (distant vs. nearby),

- Characteristics of the transport system (channel network, slope, etc.),
- Frequency, duration and intensity of the erosion producing storms,
- Texture of soil material,
- Potential for sediment deposition (presence of depressions, etc.), and,
- Catchment characteristics (area, slope and shape).

22.7.2 Where sediments get deposited

Large particles and most of the bed load are trapped in the delta while fine sediments tend to accumulate at the face of the dam.

22.7.3 Density of deposited sediments

Most sediment transport estimates give the sediment load as a weight, e.g. tons/day, tons/year, etc. To estimate the loss of storage due to sediment deposition, the sediment loads by weight must be converted to sediment loads by volume. The conversion requires the density of sediments, which depend upon

- the sediment size,
- the depth of deposit,
- the degree of submergence, and,
- the length of time the sediments have been deposited (consolidation).

An empirical equation proposed by Lane and Koelzer (1943) has found wide use in estimating density variation in time. It is of the form

$$w_t = w_1 + k \log t \tag{22.62}$$

where w_t is the density after t years, w_1 is the density after 1 year, and k is a constant for each class of sediments and operating conditions to represent consolidation

Typical values of the constants recommended by Lane and Koelzer (1943) are given in Table 22.1. All the values are in kg/m^3.

22.7.4 Sediment control

All reservoirs will someday be filled with sediments. Preventive measures can, however, extend the useful life of a reservoir. There are instances where reservoirs have been filled

Table 22.1 Typical values of the constants in Eq. 22.62

Reservoir operation	Sand 0.05 mm w_1	k	Silt 0.005–0.05 mm w_1	k	Clay <0.005 mm w_1	k
Sediments always submerged or nearly submerged	1,488	0	1,040	91.2	480	256
Normally a moderate drawdown	1,488	0	1,184	43.2	736	170
Normally a considerable drawdown	1,488	0	1,264	16.0	960	96
Reservoir normally empty	1,488	0	1,312	0	1,250	0

with sediments in very short periods of time (A reservoir on Solonian River near Osborne, Kansas was filled in 1 year). Sediment deposition may be retarded by

- selecting a site where the normal sediment inflow is low,
- increasing the size of the reservoir so that the life span is justifiable,
- adopting upstream land use and soil conservation measures such as terracing, strip cropping, contour ploughing, check dams in gullies, etc.,
- having vegetation, and
- providing facilities for silt ejection periodically.

However, it should be noted that when the sediment equilibrium is disturbed by upstream control, there will be adverse effects in the downstream reaches until a new sediment equilibrium is attained.

22.8 CONCLUDING REMARKS

Sediment transport is a topic that does not have precise theoretical treatment. Much of the information available in the literature is based on experimental and/or field observations and as such is empirical or semi-empirical. The equations presented in this chapter have been developed and tested by pioneers in this field and involves several assumptions and empirical constants which may therefore be not universally applicable. Simple site-specific regression relationships may sometimes be more applicable than the rigorous equations presented in this chapter. It should also be noted that there are many more methods and equations for estimating sediment transport rates than those presented in this chapter.

There are several empirical graphical relationships needed for the estimation of bed load and suspended load using Einstein's methods. These can be found in his classical publication 'Technical Bulletin 1026'. These need to be referred to in any application of his methods.

Example 22.1

A sediment sampling survey in a river gave the following bed load composition:

Mean particle size (mm)	Fraction by weight, p_i
0.090	0.08
0.180	0.21
0.354	0.35
0.707	0.27
1.450	0.09

Determine the bed load transport rate for sediments of mean size 0.707 mm in a river carrying a flow of 105 m³/s at a depth of 2.0 m. The river has a slope of 0.00027 and a bed width of 46 m with side slopes 2:1. Specific gravity of bed material is 2.65 and the kinematic viscosity of water in the river may be taken as 1.139×10^{-6} m²/s.

Data given: $Q = 105$ m³/s; $S = 0.00027$; $D = 2.0$ m

Area $A = BD + 2D^2 = 46 \times 2 + 8 = 100$ m²

Wetted perimeter $P = B + 2\sqrt{5}D = 54.9$ m

$$R = \frac{A}{P} = 1.82 \text{ m}$$

Roughness height $k_s = d_{65} = 0.354\,\text{mm}$.

i. Shear velocity $u_* = \sqrt{gRS} = 0.0694\,\text{m/s}$

ii. $\delta = 11.6\dfrac{v}{u_*} = 0.190 \times 10^{-3}\,\text{m} \Rightarrow \dfrac{k_s}{\delta} = 1.863$

iii. From the graph of x vs. $\dfrac{k_s}{\delta}$, $x = 1.42$ (From Technical Bulletin 1026, Einstein)

iv. $\Delta = \dfrac{k_s}{x} = 2.493 \times 10^{-4}$

v. $\dfrac{\Delta}{\delta} = 1.312 \quad (<1.8)$

vi. Therefore, $d_x = 1.39\delta = 2.641 \times 10^{-4}$

vii. $\psi_{*i}\xi_i Y \left[\dfrac{\log 10.6}{\log\left(10.6\dfrac{d_x}{\Delta}\right)}\right]\dfrac{\gamma_s - \gamma}{\gamma RS}d_i$

viii. $\dfrac{d_i}{d_x} = \dfrac{0.707}{0.2641} = 2.67 \Rightarrow \xi_i = 1.0$

ix. $\dfrac{k_s}{\delta} = 1.863 \Rightarrow Y = 0.63$

x. Substituting, $\psi_{*i} = 1.42$

xi. $\phi_{*i} = 5.0$ (From graph; From Technical Bulletin 1026, Einstein)

xii. $\phi_{*i} = \dfrac{q_{bi}}{p_i\gamma_s}\left(\dfrac{\gamma}{\gamma_s - \gamma}\dfrac{1}{gd_i^3}\right)^{1/2} \Rightarrow q_{bi} = \phi_{*i}p_i\gamma_s d_i^{3/2}\left(\dfrac{1}{\dfrac{\gamma}{\gamma_s - \gamma}g}\right) = 0.27\,\text{kg/s m}$

xiii. Total bed load transport rate $Q_{bi} = 0.27 \times 54.9 = 14.8\,\text{kg/s}$

xiv. Calculations must be repeated for all the size fractions and the total bed load
$$Q_T = \sum Q_{bi}$$

Assumptions: Diffusion in the vertical direction only; Steady two-dimensional flow; Sediment concentration varies in the vertical direction only

Example 22.2

Estimate the sediment concentration at a depth of 2 m in a wide alluvial channel 5 m having a slope of 0.00025 if a sediment sample taken at a depth of 2.5 m had a concentration of 1,000 mg/l. Assume that the bed material is uniform and has a fall velocity of 0.01 m/s, von Karman constant is 0.4 and $\beta = 1$.

Data given: $D = 5\,\text{m}$; $a = 0.25\,\text{m}$; $z = 2\,\text{m}$; $c_a = 1,000\,\text{mg/l}$; $w_s = 0.01\,\text{m/s}$; $\beta = 1$

Use Rouse equation $\dfrac{c}{c_a} = \left[\dfrac{D-z}{z}\dfrac{a}{D-a}\right]^{z_*}$

$u_* = \sqrt{gRS} = \sqrt{9.81 \times 5 \times 0.00025} = 0.11\,\text{m/s}$

$z_* = \dfrac{w_s}{\beta K u_*} = \dfrac{0.01}{1 \times 0.4 \times 0.11} = 0.227$

$$\frac{c}{c_a} = \left[\frac{D-z}{z} \frac{a}{D-a} \right]^{z*} \Rightarrow \frac{c}{1,000} = \left[\frac{5-2.0}{2.0} \frac{2.5}{5-2.5} \right]^{0.227}$$

$$\Rightarrow \frac{c}{1,000} = 1.5^{0.227} = 1.1 \Rightarrow c = 1,100 \, mg/l$$

REFERENCES

Ackers, P and White, W.R. (1973): Sediment transport: new approach and analysis, *Proceedings of the ASCE*, vol 99 no.:HY 11, pp. 2041–2060.

Brune, G. M. (1953): Trap efficiency of reservoirs, *Transactions of the American Geophysical Union*, vol. 34, no.:3, pp. 407–418.

Chang, H. H. (1992): *Fluvial Processes in River Engineering*, Krieger Publishing Co., Malabar, FL, 432 pp.

DuBoys, M.P. (1879): Le Rhone et les Rivieres a Lit affouillable, Mem Doc., Pont et Chaussees, Ser 5 vol XVIII.

Einstein, H. A. (1942): Formula for the transportation of bed-load, *Transactions of the ASCE*, vol. 107, p. 561.

Einstein, H. A. (1950): *The Bed-Load Function for Sediment Transportation in Open Channel Flows*, US Department of Agriculture, Soil Conservation Service, Washington, DC, Technical Bulletin no. 1026.

Einstein, H. A. and Barbarossa, N. L. (1952): River channel roughness, *Proceedings of the ASCE*, vol. 117, pp. 1121–1146.

Einstein, H. A. and Chien, N. (1954): *Second Approximation to the Solution of the Suspended Load Theory, MRD Sediment Series No.2*, University of California and Missouri River Division, US Corps of Engineers, Berkeley.

Lane, E. W. and Koelzer, V. A. (1943): *Density of Sediment Deposited in Reservoirs, Report 9*, U.S. Corps of Engineers, University of Iowa, St. Paul, IA.

Raudkivi, A. J. (1990): *Loose Boundary Hydraulics*, (3rd Edition), Pergamon Press, Oxford, 538 pp.

Rouse, H. (1937): Modern concepts of the mechanics of turbulence, *Transactions of the ASCE*, vol. 102, p. 4630.

Shields, A. (1936): *Application of Similarity Principles and Turbulence Research to Bed Load Movement*, California Institute of Technology, Pasadena (Translated from German).

Straub, L. G. (1935): Missouri River Report, US Army Corps of Engineers, House Document 238, Appendix XV, 73rd US Congress Second Session, p. 1156. Also see Vanoni (1975) pp. 58–61.

Straub, L. G. (1950): "Chapter XII". In: *Engineering Hydraulics*, edited by H. Rouse, Wiley & Sons, New York, p. 794.

Yang, C. T. (1996): *Sediment transport, Theory and Practice*, McGraw Hill Book Companies Inc., New York, 396 pp.

Chapter 23

Environmental hydraulics

23.1 INTRODUCTION

Water, soil and air constitute the physical environment. Human activities as well as natural changes in any one of these will induce changes in the other. Environmental Hydraulics can be considered as the domain of science that investigates the physical, chemical and biological changes taking place in the water environment with the objective of protecting and improving the quality of the environment. Fluid Mechanics, Hydrodynamics and Hydraulics are branches of science that describe the motion of water at different scales. They form the backbone of environmental hydraulics.

Pollution of the water environment can be due to natural causes, as well as due to human activities such as the discharge of domestic pollutants, industrial wastes, agricultural wastes and pesticides, and, accidental discharges such as toxic chemicals and oil from ocean going vessels and other modes of transportation. The degree of pollution is measured by the concentration of the pollutant (e.g. mg/l) and when the concentration of the pollutants exceeds a certain threshold, the water becomes harmful to the users.

Types of wastes include organic waste, inorganic waste, sediments, thermal, pathogenic bacteria, heavy metals (include lead, mercury, cadmium), trace metals (include iron, magnesium, lithium, zinc, copper, chromium, nickel, cobalt, vanadium, arsenic, molybdenum, manganese, selenium), synthetic organic chemicals and radioactive waste.

Such wastes finally end up in the oceans but on their way to the oceans, they also pollute the water bodies that convey them to their final destination. Concentrations are reduced by dilution as a result of mixing as well as due to biological and chemical reactions. Natural environment has a certain degree of self-purifying capacity and when it is exceeded, the concentrations will reach undesirable levels which in certain situations may become toxic to living organisms. The self-purifying capacity depends upon physical factors such as the size of the receiving water body, dispersion and sedimentation characteristics, biological factors such as natural degradation and decomposition and chemical factors such as oxidation and reduction. The parameters of importance are dissolved oxygen, *E. coli*, ionized ammonia and total inorganic nitrogen. The safe limits of such parameters are set by the local authorities who generally follow the World Health Organization (WHO) standards.

In this chapter, the mechanics of the processes that govern the mixing and transport of solutes in a solvent is described from the points of view of fluid mechanics and hydraulics. Some material in this chapter is reproduced from the author's earlier book (Jayawardena, 2014). A brief description of eutrophication in coastal water bodies is also given.

23.2 SOME DEFINITIONS

i. **Advection:** Horizontal transport by an imposed current system, e.g. as in rivers and coastal waters.

ii. **Convection:** Vertical transport induced by hydrostatic instability, e.g. parcel of warm air moving upwards.

iii. **Diffusion (molecular):** Scattering of particles by random molecular motion. Can be described by Fick's law.

iv. **Diffusion (turbulent):** Scattering of particles by turbulent motion. Analogous to molecular diffusion but with 'eddy' diffusion coefficient in place of molecular diffusion coefficient. Eddy diffusion coefficient >>> molecular diffusion coefficient.

v. **Shear flow:** Advection of fluid at different velocities at different positions.

vi. **Dispersion:** Scattering of particles by the combined effect of shear flow and transverse diffusion.

vii. **Mixing:** Diffusion or dispersion.

viii. **Concentration:** Mass of tracer per unit volume; $c = \text{Limit}_{\Delta v \to 0} \dfrac{\Delta M}{\Delta V}$; In general $c = f(x, y, z, t)$.

ix. **Time average of** c: $\bar{c}_t = \dfrac{1}{T} \displaystyle\int_{t_0}^{t_0 + T} c(x, y, z, t)\, dt$ (function of x, y, z, t_0, T).

x. **Spatial average of** c: $\bar{c}_v(x_0, y_0, z_0, t) = \dfrac{1}{V} \displaystyle\iiint_{\Delta V} c(x, y, z, t)\, dV$ and is a function of t.

xi. **Flux average of** c: $\displaystyle\int_A cu\, dA = \bar{c}_f \int_A u\, dA = \bar{c}_f Q$, or, $c_f(t) = \dfrac{\displaystyle\int_A cu\, dA}{Q}$.

23.3 FICKIAN DIFFUSION (FOR MOLECULAR DIFFUSION)

According to Fick's law, in a quiescent fluid, the mass flux of solute crossing a unit area per unit time in a given direction is proportional to the concentration gradient of the solute in that direction. In a one-dimensional framework, it is of the form

$$q = -D \frac{\partial c}{\partial x} \tag{23.1}$$

where q is the solute mass flux rate, c is the concentration and D is the diffusion coefficient ($L^2 T^{-1}$). Mass flux rate is the mass of a solute crossing a unit area per unit time in a given direction. Considering a control volume of unit cross-sectional area normal to the direction of flow, the continuity equation can be written as

$$q - \left(q + \frac{\partial q}{\partial x} \delta x\right) = \frac{\partial c}{\partial t} \cdot 1 \cdot \delta x \text{ which simplifies to} \tag{23.2a}$$

$$\frac{\partial c}{\partial t} + \frac{\partial q}{\partial x} = 0 \left(\text{This equation is independent of the transport process}\right) \tag{23.2b}$$

Substituting Fick's law (Eq. 23.1), the governing Fickian diffusion equation becomes

$$\frac{\partial c}{\partial t} = D \frac{\partial^2 c}{\partial x^2} \tag{23.2c}$$

23.4 SOLUTION OF THE FICKIAN DIFFUSION EQUATION

Solution of the Fickian diffusion equation depends upon the boundary and initial conditions. For example,

i. Initial slug of mass M introduced at time $t = 0$ at $x = 0$.

Since the diffusion equation is linear, the solution to this case can be used as a basic solution from which solutions to other complex initial and boundary conditions may be obtained. This case can be compared with the Instantaneous Unit Hydrograph IUH which is the solution to an instantaneous rainfall excess of unit magnitude. The solution can be obtained by any of the several mathematical techniques available of which dimensional analysis is one.

Concentration $c(x, t)$ can only be a function of M, D, x, and t. Since the process is linear, the concentration is proportional to mass of slug and has the dimension of ML^{-1} in a one-dimensional framework. Therefore,

$$c = \frac{M}{\text{Characteristic length}}$$

Since D has the dimension of L^2T^{-1}, $\sqrt{Dt}$ is a suitable characteristic length. Therefore,

$$c(x, t) = \frac{M}{\sqrt{4\pi Dt}} f\left(\frac{x}{\sqrt{4Dt}}\right) \tag{23.3}$$

The factors 4π and 4 are added for convenience and that Eq. 23.3 gives only the form of the solution to the diffusion equation (Eq. 23.2c). The basic solution, obtained by reducing Eq. 23.2c, which is a partial differential equation, to an ordinary differential equation can be shown to be

$$c(x, t) = \frac{M}{\sqrt{4\pi Dt}} e^{\left[\frac{-x^2}{4Dt}\right]} \tag{23.4}$$

ii. Initial slug of mass M introduced at time $t = 0$ at $x = \xi$. i.e. $c(x, 0) = M\delta(x - \xi)$ where δ is the Dirac Delta function.

This means at $t = 0$, $c = c_0$ at $x = \xi$, and $c = 0$ for $x \neq \xi$

The solution to Eq. 23.2c, then is

$$c(x, t) = \frac{M}{\sqrt{4\pi Dt}} e^{\left[\frac{-(x-\xi)^2}{4Dt}\right]} \tag{23.5}$$

iii. Instead of an initial slug assume an initial condition of the form

$$c(x, 0) = f(x) \text{ for } -\infty < x < \infty$$

This can be considered as a series of separate slugs which diffuse independently. Approximating the function by a series of slugs as shown below the mass contained in each slug is

$$M = f(\xi)d\xi \tag{23.6}$$

Concentration at (x, t) resulting from the slug at ξ of magnitude $f(\xi)$ and width $d\xi$ is

$$c(x, t) = \frac{f(\xi)d\xi}{\sqrt{4Dt}} e^{\left[\frac{-(x-\xi)^2}{4Dt}\right]} \tag{23.7}$$

Therefore, the total concentration at (x, t) due to all the slugs can be obtained by superposition:

$$c(x, t) = \int_{-\infty}^{\infty} \frac{f(\xi)}{\sqrt{4\pi Dt}} e^{\left[\frac{-(x-\xi)^2}{4Dt}\right]} d\xi \tag{23.8}$$

Special cases of Eq. 23.8:
   When $f(x)$ is given as $c(x, 0) = 0$ for $x < 0$, and $c(x, 0) = c_0$ for $x > 0$
   The solution in this case is

$$c(x, t) = \int_{-\infty}^{\infty} \frac{c_0}{\sqrt{4\pi Dt}} e^{\left[\frac{-(x-\xi)^2}{4Dt}\right]} d\xi$$

which can be transformed by substituting $u = \dfrac{x - \xi}{\sqrt{4Dt}}$ to

$$c(x, t) = \frac{c_0}{\sqrt{\pi}} \int_{-\infty}^{x/\sqrt{4Dt}} e^{-u^2} du = \frac{c_0}{\sqrt{\pi}} \left[\frac{\sqrt{\pi}}{2} + \int_{0}^{x/\sqrt{4Dt}} e^{-u^2} du \right] \tag{23.9}$$

$$= \frac{c_0}{\sqrt{\pi}} \left[1 + erf\left(\frac{x}{\sqrt{4Dt}} \right) \right]$$

where

$$erf(z) = \frac{2}{\sqrt{\pi}} \int_{0}^{z} \exp(-\xi^2) d\xi \tag{23.10}$$

iv. Boundary condition specified as a function of time, i.e. concentration is expressed as mass added per time.
   a. For a single slug of mass M at $x = 0$, the concentration is given by Eq. 23.4 which is a Gaussian distribution. i.e.

$$c(x, t) = \frac{M}{\sqrt{4\pi Dt}} e^{\left[\frac{-x^2}{4Dt}\right]}$$

b. For a continuous injection at the rate of $\dot{M}$ is equivalent to a series of slugs of amounts $\dot{M}\,\delta t$ where $\delta t \to 0$. The concentration for this case is

$$c(x, t) = \int\limits_{-\infty}^{t} \frac{\dot{M}(\tau)}{\sqrt{4\pi D(t-\tau)}} e^{\left[\frac{-x^2}{4D(t-\tau)}\right]} d\tau \qquad (23.11)$$

which is the convolution integral.

This is analogous to the convolution integral for the instantaneous unit hydrograph (IUH), which is

$$y(t) = \int\limits_{-\infty}^{t} x(\tau)\,h(t-\tau)\,d\tau$$

where $x(\tau) = \dot{M}$, and, $h(t) = \dfrac{M}{\sqrt{4\pi Dt}} e^{\left[\frac{-x^2}{4Dt}\right]}$.

c. Combining (a) and (b) is equivalent to a distributed input of mass $m(x, t)$ where m has units of mass/unit length/unit time.

The solution, obtained by superposition is as follows:

$$c(x, t) = \int\limits_{-\infty}^{t}\int\limits_{-\infty}^{x} \frac{m(\xi, \tau)}{\sqrt{4\pi D(t-\tau)}} \exp\left[\frac{-(x-\xi)^2}{4D(t-\tau)} d\xi d\tau\right] \qquad (23.12)$$

This is the general solution to Eq. 23.2c.

23.5 ADVECTIVE DIFFUSION

The transport by the mean motion of the fluid is referred to as advection. Advective diffusion is the combined effect of advection and diffusion. Assuming laminar flow conditions and constant and isotropic D, the mass flux can be expressed as

$$q = uc + \left(-D\frac{\partial c}{\partial x}\right) \qquad (23.13)$$

where the first term on the right-hand side represents the advective flux and the second term the diffusive flux. The corresponding governing equation will then be

$$\frac{\partial c}{\partial t} + \frac{\partial(uc)}{\partial x} = D\frac{\partial^2 c}{\partial x^2} \qquad (23.14)$$

23.6 TURBULENT DIFFUSION (MACROSCOPIC NEGLECTING MOLECULAR DIFFUSION)

In a three-dimensional framework, the continuity equation is

$$\frac{\partial c}{\partial t} + \frac{\partial}{\partial x}(cu) + \frac{\partial}{\partial y}(cv) + \frac{\partial}{\partial z}(cw) = 0 \qquad (23.15)$$

Variables expressed as sums of a mean value and a fluctuating component

$$c = \bar{c} + c'$$

$$u = \bar{u} + u'; \quad v = \bar{v} + v'; \quad w = \bar{w} + w'$$

where

$$\bar{u} = \frac{1}{T} \int_t^{t+T} u\, dt; \bar{v} = \frac{1}{T} \int_t^{t+T} v\, dt; \bar{w} = \frac{1}{T} \int_t^{t+T} w\, dt; \bar{c} = \frac{1}{T} \int_t^{t+T} c\, dt \tag{23.16}$$

$$\frac{\partial}{\partial t}(\bar{c} + c') + \frac{\partial}{\partial x}(\bar{c} + c')(\bar{u} + u') + \frac{\partial}{\partial y}(\bar{c} + c')(\bar{v} + v') + \frac{\partial}{\partial z}(\bar{c} + c')(\bar{w} + w') = 0 \tag{23.17}$$

Since the mean values of the fluctuating components will be zero

(i.e. $\bar{c'} = 0; \bar{u'} = 0; \bar{v'} = 0; \bar{w'} = 0$)

$$\frac{\partial}{\partial t}(\bar{c}) + \bar{u}\frac{\partial}{\partial x}(\bar{c}) + \bar{v}\frac{\partial}{\partial y}(\bar{c}) + \bar{w}\frac{\partial}{\partial z}(\bar{c}) + \bar{c}\left(\frac{\partial\bar{u}}{\partial x} + \frac{\partial\bar{v}}{\partial y} + \frac{\partial\bar{w}}{\partial z}\right)$$

$$+ \frac{\partial}{\partial x}(\overline{u'c'}) + \frac{\partial}{\partial y}(\overline{v'c'}) + \frac{\partial}{\partial z}(\overline{w'c'}) = 0 \tag{23.18}$$

Continuity equation is

$$\frac{\partial\bar{u}}{\partial x} + \frac{\partial\bar{v}}{\partial y} + \frac{\partial\bar{w}}{\partial z} = 0 \tag{23.19}$$

Therefore,

$$\frac{\partial}{\partial t}(\bar{c}) + \bar{u}\frac{\partial}{\partial x}(\bar{c}) + \bar{v}\frac{\partial}{\partial y}(\bar{c}) + \bar{w}\frac{\partial}{\partial z}(\bar{c}) + \frac{\partial}{\partial x}(\overline{u'c'}) + \frac{\partial}{\partial y}(\overline{v'c'}) + \frac{\partial}{\partial z}(\overline{w'c'}) = 0 \tag{23.20}$$

The net convection of mass due to turbulent fluctuations (by analogy with Fick's law) is given by

$$\overline{u'c'} = -D_x\frac{\partial\bar{c}}{\partial x}; \overline{v'c'} = -D_y\frac{\partial\bar{c}}{\partial y}; \overline{w'c'} = -D_z\frac{\partial\bar{c}}{\partial z} \tag{23.21}$$

where D_x, D_y and D_z are coefficients of turbulent diffusion in the x, y and z directions, respectively.

Simplifying

$$\frac{\partial}{\partial t}(\bar{c}) + \bar{u}\frac{\partial}{\partial x}(\bar{c}) + \bar{v}\frac{\partial}{\partial y}(\bar{c}) + \bar{w}\frac{\partial}{\partial z}(\bar{c}) = \frac{\partial}{\partial x}\left(D_x\frac{\partial\bar{c}}{\partial x}\right) + \frac{\partial}{\partial y}\left(D_y\frac{\partial\bar{c}}{\partial y}\right) + \frac{\partial}{\partial z}\left(D_z\frac{\partial\bar{c}}{\partial z}\right) \tag{23.22}$$

It is to be noted that $D_{\text{Tubulent}} \ggg D_{\text{Molecular}}$.

Typical time and length scales for molecular diffusion are 10^{-12} s and 10^{-7} cm, respectively, whereas for turbulent diffusion they are of the same order as observations.

In a one-dimensional case, taking values of all parameters across the section

$$\bar{u} = \frac{1}{A} \int_0^A u \, dA \qquad (23.23)$$

$$\bar{c} = \frac{1}{A} \int_0^A c \, dA \qquad (23.24)$$

with $\dfrac{\partial \bar{c}}{\partial y} = \dfrac{\partial \bar{c}}{\partial z} = 0$; and, $v, w = 0$.

(In the above integrals, A can sometimes be replaced by h, the depth.)
The longitudinal dispersion equation then is

$$\frac{\partial c}{\partial t} + u \frac{\partial c}{\partial x} = \frac{1}{A} \frac{\partial}{\partial x} \left(AD \frac{\partial c}{\partial x} \right) \qquad (23.25)$$

Taylor (1954) obtained the same equation assuming that $c' \lllless \bar{c}$ and that advection and diffusion balance each other.

By including a first-order decay term and a source/sink term, the longitudinal dispersion equation can be expanded to

$$\frac{\partial c}{\partial t} + u \frac{\partial c}{\partial x} = \frac{1}{A} \frac{\partial}{\partial x} \left(AD \frac{\partial c}{\partial x} \right) - kc \pm L_a \qquad (23.26a)$$

where k is the decay coefficient and L_a is the rate of addition of material in $ML^{-3}T^{-1}$ (usually in mg/l/day) at the point of discharge.

In a different form, the equation may also be written as

$$\frac{\partial (Ac)}{\partial t} + \frac{\partial (Auc)}{\partial x} = \frac{\partial}{\partial x} \left(AD \frac{\partial c}{\partial x} \right) - kAc \pm Q_s \qquad (23.26b)$$

23.7 LONGITUDINAL DISPERSION

Longitudinal dispersion is the process by which a solute introduced into a solvent dilutes by the combined action of molecular diffusion and cross-sectional velocity variation. It is different from molecular diffusion which takes place due to molecular activity and is considered on a microscopic scale such as for instance in chemical and biological reactions. It is an important topic in Environmental Engineering. Knowledge of the dispersion characteristics is necessary for estimating the waste receiving capacity and the distribution of pollutants in a water body.

Longitudinal dispersion is treated in accordance with Fick's (1855) law of diffusion which is an extension to Fourier's law of heat conduction (see Section 23.3). By combining Fick's law with the continuity equation, which is dependent on the type of process by which the solute is transported, it is possible to obtain the partial differential equation that describes the process. The governing equation and the solutions to specific input and boundary conditions are described in Section 23.4.

In this type of diffusion, the flow is assumed to be laminar and therefore the molecular diffusion coefficients have constant values in all directions. This is called the advective diffusion equation, or simply the diffusion equation. Laminar flow in the environment is uncommon. Most flows in nature are turbulent. The Fickian diffusion coefficient for laminar flow is then replaced by an analogous coefficient called turbulent mixing coefficient ε_x. The analogous equation for turbulent flow can then be written in terms of ε_x as

$$\frac{\partial c}{\partial t} = \varepsilon_x \frac{\partial^2 c}{\partial x^2} \tag{23.27}$$

It is possible to show that the coefficient ε_x can be compared with D since the mass flux can be expressed as

$$\overline{uc} = -\varepsilon_x \frac{\partial \overline{c}}{\partial x} \left(\text{analogous to Eq. 23.1} \right) \tag{23.28}$$

where $\overline{u}$ and $\overline{c}$ are the time averaged values of u and c, respectively. Thus, the flux is proportional to the concentration gradient. The coefficient ε_x is called the Fickian turbulent diffusion coefficient or eddy diffusivity.

23.7.1 Some characteristics of turbulent diffusion

Since the scale of distances for turbulent diffusion is very much greater than that for molecular diffusion, the former is more effective in dispersing on a macro scale. Turbulence is associated with eddies, large and small, with scales varying with the scale of the diffusion process.

Turbulent flow is something that cannot be expressed in exact terms. It has a random component in its behaviour and therefore the characteristics of turbulent motion can only be described in statistical terms. This applies to the solute which mixes with the solvent in turbulent motion. By assuming further that the turbulent process is stationary and homogeneous which implies that the variance of the velocity remains constant in time and space, it is possible to define statistical parameters of concentration such as mean concentration, variance of concentration – time curve, variance of concentration – distance curve, etc. These parameters define the overall character of turbulent diffusion.

23.7.2 Shear flow dispersion

Shear flow dispersion refers to the dispersion due to changes in velocity across a cross section. If the separation of particles by advection is greater than that caused by molecular diffusion, then, it is called laminar shear flow.

In this type of flow also, the mass transport in the streamwise direction is proportional to the concentration gradient in the streamwise direction which is identical to the result obtained for molecular diffusion. The difference being that in shear flow dispersion, both advection and diffusion are considered together whereas in ordinary molecular diffusion, only molecular diffusion is considered. The diffusion coefficient is therefore replaced by a bulk transport coefficient or dispersion coefficient. It expresses the diffusive property of the velocity distribution and is generally known as the longitudinal dispersion coefficient. The corresponding governing equation is

$$\frac{\partial \overline{c}}{\partial t} + \overline{u} \frac{\partial \overline{c}}{\partial x} = D \frac{\partial^2 \overline{c}}{\partial x^2} \tag{23.29}$$

in which the velocity and concentration values are cross-sectionally averaged, i.e.

$$\bar{u} = \frac{1}{A} \int_A u \, dA \qquad\qquad (23.30)$$

$$\bar{c} = \frac{1}{A} \int_A c \, dA \qquad\qquad (23.31)$$

where A is the cross-sectional area of flow.

Extending to three dimensions, the governing equation based on the law of conservation of mass can be written as the following partial differential equation:

$$\frac{\partial c}{\partial t} + u \frac{\partial c}{\partial x} + v \frac{\partial c}{\partial y} + w \frac{\partial c}{\partial z} = \frac{\partial}{\partial x}\left(\varepsilon_x \frac{\partial c}{\partial x}\right) + \frac{\partial}{\partial y}\left(\varepsilon_y \frac{\partial c}{\partial y}\right) + \frac{\partial}{\partial z}\left(\varepsilon_z \frac{\partial c}{\partial z}\right) \qquad (23.32)$$

where c is the time-averaged concentration $[ML^{-3}]$, u, v, w are the time-averaged velocities in the x, y, and z directions, respectively $[LT^{-1}]$, ε_x, ε_y and ε_z are the turbulent mixing coefficients in the x, y, and z directions, respectively $[L^2T^{-1}]$, and t is the time $[T]$. The terms on the left-hand side of the equation correspond to advection along streamlines while the terms on the right-hand side correspond to turbulent diffusion between streamlines. Molecular diffusion which takes place at a microscopic scale is small compared with turbulent diffusion and therefore ignored in the above formulation.

In this equation, a solute introduced into a turbulent flow field is considered to be transported by two mechanisms: advection by the time-averaged velocities and diffusion by turbulent fluctuations. In deriving the governing equation, it is assumed that the instantaneous variations are averaged over a time period long enough to average out short-time fluctuations, but short enough to consider long-term changes in the time-averaged values.

There is no general analytical solution of Eq. 23.32. Very often, assumptions are made to make the equation tractable. The lateral and transverse velocities, v and w, are very much small in comparison to the longitudinal velocity and can be neglected without causing too much error. The turbulent mixing coefficients ε_x, ε_y and ε_z are also assumed to be constants in which case the equation becomes linear.

23.7.3 Taylor's approximation

Taylor (1953, 1954), by studying uniform flow through long straight circular pipes, argued that in a co-ordinate system that moves with the average velocity of flow, the spreading of a solute should follow Fick's law of diffusion (Eq. 23.1) which when combined with the continuity equation gives

$$\frac{\partial \bar{c}}{\partial t} = D \frac{\partial^2 \bar{c}}{\partial \xi^2}, \qquad\qquad (23.33)$$

where $\xi = x - \bar{u}t$, $\bar{u}$ and $\bar{c}$ are the cross-sectional means as defined by Eqs. 23.30 and 23.31. Taylor made further assumptions and approximations to Eq. 23.32 to obtain what is generally known as the longitudinal dispersion equation. Among them, he assumed

$$c = \bar{c} + c' \qquad\qquad (23.34a)$$

and

$$u = \bar{u} + u' \qquad\qquad (23.34b)$$

where c' and u' are the deviations of the concentration and velocity from their respective mean values. He further assumed that $c' = c'(y, z)$, the longitudinal transport due to turbulence, $\frac{\partial}{\partial x}\left(\varepsilon_x \frac{\partial c}{\partial x}\right)$, is negligible, the rate of change of concentration at a point moving with the mean velocity $\frac{\partial c}{\partial t} + \bar{u}\frac{\partial c}{\partial x}$ is also negligible and that the lateral and transverse velocities, v, w are zero. With these drastic assumptions, Eq. 23.32 becomes

$$u'\frac{\partial \bar{c}}{\partial \xi} = \frac{\partial}{\partial y}\left(\varepsilon_y \frac{\partial c'}{\partial y}\right) + \frac{\partial}{\partial z}\left(\varepsilon_z \frac{\partial c'}{\partial z}\right) \tag{23.35}$$

with $\frac{\partial c'}{\partial y} = \frac{\partial c'}{\partial z} = 0$ at the boundary. This is Taylor's simplified form of the governing equation.

He obtained an expression for the dispersion coefficient (given below by Eq. 23.39) using a solution for the above equation that assumes universal velocity distribution in pipes.

23.7.4 Turbulent mixing coefficients

Of the three turbulent mixing coefficients, ε_x in the longitudinal direction is assumed to be zero. In the y direction for the vertical transport of momentum (and assuming Reynolds analogy), ε_y can be derived theoretically for turbulent flow in an infinitely wide open channel since the distribution of velocity and shear are known. Assuming a logarithmic velocity distribution, it has been shown (Fischer, 1966) that ε_y takes the form

$$\varepsilon_y = \kappa\left(1 - \frac{y}{h}\right)\left(\frac{y}{h}\right)hu_* \tag{23.36}$$

where κ is the von Karman constant, and h is the depth of flow, and u_* is the shear velocity $\left(= \sqrt{\frac{\tau_0}{\rho}}, \tau_0 \text{ is the boundary shear stress and } \rho \text{ is the density of fluid}\right)$. Elder (1959), also using the logarithmic velocity distribution, has obtained the following expressions for the lateral mixing coefficient, ε_z as

$$\varepsilon_z = 0.23\,hu_* \tag{23.37}$$

There are also several other expressions for the turbulent mixing coefficients derived under different assumptions.

23.8 DISPERSION COEFFICIENT AND METHODS OF ITS ESTIMATION

The dispersion coefficient is a dominant parameter in the dispersion equation. A reasonable estimate of the highly variable dispersion coefficient is a pre-requisite to the solution of the longitudinal dispersion equation. Various attempts have been made by various investigators to find a relationship between the dispersion coefficient and the flow parameters. Some findings contradict others. The first such attempt was by Taylor (1953, 1954) who carried out laboratory experiments in long straight circular pipes under laminar and turbulent flow conditions. His studies have become the basis of analysis by many investigators subsequently. Taylor, by assuming universal velocity distribution within a circular section, obtained the following relationship:

$$D = \frac{r^2 \bar{u}^2}{4.8K} \qquad (23.38)$$

where D is the dispersion coefficient, r is the pipe radius, $\bar{u}$ is the discharge velocity and K is the coefficient of molecular diffusivity. His finding was for laminar flow in pipes and demonstrates the opposing effects of the advective velocity $\bar{u}$ and the diffusivity K.

He extended his analysis to turbulent pipe flow and obtained an equation for the effective dispersion coefficient by using two approaches: universal velocity distribution in a pipe, and Reynolds analogy.[1] Using these two assumptions, Taylor (1954) worked out an equation for an effective dispersion coefficient as

$$D = 10.06\, r u_* \qquad (23.39)$$

He further allowed for the longitudinal turbulent diffusion by assuming that the coefficient of longitudinal diffusion is equal to the coefficient of lateral diffusion (longitudinal diffusion is normally negligible compared to advection). The effect of longitudinal diffusion has been found to be

$$D' = 0.052\, r u_* \qquad (23.40)$$

Therefore, the corrected value of the longitudinal dispersion coefficient after allowing for longitudinal diffusion is

$$D = (10.06 + 0.052)\, r u_* = 10.1\, r u_* \qquad (23.41)$$

Taylor (1953) gave a similar expression $D = 8.98\ r u_*$ without proof.

Experiments conducted by Taylor have yielded values of dispersion coefficient very close to the above provided an initial time $T = \dfrac{15r}{u_*}$ has elapsed. But dispersion at which this time is attained is seldom encountered in practice.

The next important contribution to dispersion prediction is due to Elder (1959) who obtained a similar expression for the dispersion coefficient for open channel flow assuming a logarithmic velocity distribution and Reynolds analogy in the depth-wise direction in an infinitely wide channel. The relationship he obtained is

$$D = 5.93\, h u_* \qquad (23.42)$$

where h is the depth of flow. Both these expressions (Eqs. 23.41 and 23.42) have been verified experimentally.

The difference between the constant terms in Eqs. 23.41 and 23.42 is partly attributed to the different velocity distributions assumed. If Taylor's empirical velocity distribution (universal) is used for open channel flow, Eq. 23.42 becomes

$$D \approx 11.7\, h u_* \qquad (23.43)$$

On the other hand, if Elder's logarithmic distribution is used for pipe flow, the result is

$$D = 5.84\, r u_* \qquad (23.44)$$

---

[1] Transfers of matter, heat and momentum by turbulence are analogous.

These theoretical predictions have not been always accurate when they are applied to natural streams. Fischer (1967) reported that the dispersion coefficient for natural streams varied from 50 to 700 Ru_* where R is the hydraulic radius. He pointed out that the concept of Fickian diffusion is applicable only after an initial period of time since the injection of a tracer. During this initial period, the dispersion of the tracer cloud is dominated by advection. Fischer's results can be summarized as follows.

During the diffusive period (i.e. after the initial period of time) using Taylor's analysis, Fischer (1966, 1968) derived the following expression for the dispersion coefficient assuming that dispersion in natural streams is caused primarily by the transverse velocity variations (Elder considered the vertical velocity variations). The separation between zones of differing velocities is much greater in the lateral direction than in the vertical direction since for practically all natural streams the widths are many times greater than the depths:

$$D = -\frac{1}{A}\int_0^W q'(z)\,dz \int_0^z \frac{1}{e_z h(z)}\,dz \int_0^z q'(z)\,dz \qquad (23.45a)$$

in which

$$q'(z) = \int_0^{h(z)} u'(y, z)\,dy \qquad (23.45b)$$

where $u'(y,z)$ is the fluctuating component of the velocity at any point in the cross section, $h(z)$ is the depth of flow at any point in the cross section, W is the channel width, y is the Cartesian coordinate in the depth-wise direction and e_z is the turbulent mixing coefficient in the transverse (z) direction which has to be experimentally determined. Elder (1959) has found it to be equal to $0.23hu_*$ which has been shown to be satisfactory by other investigators as well. Velocity is measured as a function of the cross-sectional position, and $u(z)$ and $u'(z)$ are depth averaged.

Equations 23.45a and b require only the channel geometrical data, cross-sectional distribution of downstream relative velocity, u', and u_*. $(u' = u - \bar{u}; u$ is the actual velocity at any point in the x-direction.)

With extensive field data, the dispersion coefficient can be estimated by approximating Eq. 23.45a to a summation as

$$D = -\frac{1}{A}\sum_{k=2}^n q'(k)\Delta z \left[\sum_{j=2}^k \frac{\Delta z}{\varepsilon_{z_j} h_j}\left(\sum_{i=1}^{j-1} q'_i \Delta z\right)\right] \qquad (23.46)$$

in which
$q'_i = \frac{1}{2}(h_i + h_{i+1})u'_i$, u_i is the mean velocity in the ith vertical slice, $u'_i = u_i - \bar{u}$, $\bar{u}$ is the mean velocity of the cross section, h_i is the depth at the beginning of the ith slice, Δz is the width of the vertical slice, u_* is the shear velocity $\left(= \sqrt{gRS_f}\right)$, R is the hydraulic radius and S_f is the slope of the energy grade line. An example application illustrating the procedure for computation of some of the above methods is given by Fischer (1968).

This method requires cross-sectional velocity profile as well as cross-sectional geometry, and therefore is data intensive. Subsequently, Fischer (1975) gave a simplified form of the above equation which takes the form

$$D = I \frac{u'^2 l^2}{\varepsilon_y} \tag{23.47}$$

where I is a non-dimensional integral, whose value depends on the velocity distribution, and l is the distance from the point of maximum velocity to the most distant bank. Using the values that Fischer used ($I = 0.07$; $u'^2 = 0.2\,\bar{u}^2$; $l = 0.7W$; and the mean transverse turbulent diffusion coefficient $\varepsilon_y = 0.6\,hu_*$), Eq. 23.47 becomes

$$D = 0.011 \frac{\bar{u}^2 W^2}{hu_*}. \tag{23.48}$$

Using a similar approach, Liu (1977) developed the following equation in which the parameter ψ relates to the channel cross-sectional profile and transverse velocity distribution:

$$D = \psi \frac{u^2 W^2}{hu_*}; \psi = 0.18 \left(\frac{u_*}{u} \right)^{3/2} \tag{23.49}$$

Seo and Cheong (1998) also developed an equation using hydraulic data that can be more easily obtained from natural rivers. Their equation is of the form

$$D = 5.915\,hu_* \left(\frac{W}{h} \right)^{0.620} \left(\frac{u}{u_*} \right)^{1.428}. \tag{23.50}$$

Other equations that have been derived include that by Deng et al. (2001) who used direct integration of the triple integral of Eq. 23.45a, and that by Kashefipour and Falconer (2002) who used dimensional and regression analysis. Their equations respectively are as follows:

$$D = \frac{0.15\,hu_*}{8\varepsilon_{t_0}} \left(\frac{W}{h} \right)^{5/3} \left(\frac{u}{u_*} \right)^2, \tag{23.51}$$

$$D = hu \frac{u}{u_*} \left[7.428 + 1.775 \left(\frac{W}{h} \right)^{0.620} \left(\frac{u}{u_*} \right)^{0.572} \right]. \tag{23.52}$$

A review as well as a comparison of their application is given by Ahsan (2008). Under field conditions, the dispersion coefficient is reported to have values ranging from about 4.6 to 670 m²/s, with an exceptionally high value of 1500 m²/s for Missouri River (Fischer et al., 1979, Table 5.3).

23.8.1 Change of moment method

A theoretically exact method of determining the dispersion coefficient is the change of moment method which is based on the properties of the diffusion equation. Under field conditions, however, the concentration profile has long tails, and therefore, it is difficult to calculate the moments. Dispersion coefficient by the change of moment method is given as

$$D = \frac{1}{2} \frac{d\sigma_\xi^2}{dt}. \tag{23.53}$$

where σ_ξ^2 is the variance of the concentration-distance profile. This equation is valid during the diffusive period regardless of the shape of the concentration profile during the convective period. Because it is difficult to measure concentration-distance profiles, the spatial variance can be converted into a time variance for uniform steady flow by the transformation

$$\sigma_\xi^2 = \bar{u}^2 \sigma_t^2 \tag{23.54}$$

where σ_t^2 is the variance of the concentration-time profile at a fixed point. Therefore, by taking measurements at two fixed points, the dispersion coefficient can be estimated as

$$D = \frac{1}{2}\bar{u}^2 \frac{\sigma_{t_2}^2 - \sigma_{t_1}^2}{\bar{t}_2 - \bar{t}_1} \tag{23.55}$$

where the subscripts 1 and 2 refer to the two measuring points, and $\bar{t}$ refers to the time of passage of the centroid of the tracer cloud at the station.

23.8.2 Routing method

The routing method (Fischer, 1966, 1968) of estimating the dispersion coefficient consists of matching the measured concentration-time profile with a predicted concentration-time profile for an assumed value of the dispersion coefficient until the mismatch between the two profiles as measured by the sum of squared differences is a minimum. Using the properties of the underlying linear theory in the bulk diffusion process, the principle of superposition is used to obtain the concentration distribution profile at any given time if the profile at some initial time is known. If at $t = t_0$, the initial concentration profile $c = c_0(\xi, t_0)$, then at any subsequent time, the concentration-distance profile is given by

$$c(\xi, t) = \int_{-\infty}^{\infty} c_0(\xi', t_0) \frac{e^{-\frac{(\xi - \xi')^2}{4D(t-t_0)}}}{\sqrt{4\pi D(t - t_0)}} d\xi' \tag{23.56}$$

However, in practice, it is difficult to measure the concentration-distance profile in the field. Therefore, the concentration-distance profile is obtained from the concentration-time profile by the following transformation which is approximate:

$$c(\xi, \bar{t}_0) = c(x_0, t) \tag{23.57a}$$

where

$$\xi = \bar{u}(\bar{t}_0 - t) \tag{23.57b}$$

Then the distance integration of Eq. 23.56 becomes the following time integration:

$$c(x_1, t) = \int_{-\infty}^{\infty} c(x_0, \tau) \frac{\exp\left[\frac{-\left(\bar{u}(\bar{t}_1 - t) + \bar{u}(\tau - \bar{t}_0)\right)^2}{4D(\bar{t}_1 - \bar{t}_0)}\right]}{\sqrt{4\pi D(\bar{t}_1 - \bar{t}_0)}} \bar{u}\, d\tau \tag{23.58}$$

In Eq. 23.58, $c(x_1, t)$ is the concentration-time profile at station x_1, $c(x_0, \tau)$ is the concentration-time profile at station x_0, $\bar{t}_0$ and $\bar{t}_1$, respectively, are the mean times of passage of the tracer cloud past stations x_0 and x_1, and $\bar{u}$ is the mean velocity of flow. In deriving Eq. 23.58, $c(\xi, t)$ is replaced by $c(x_1, t)$, ξ by $\bar{u}(\bar{t}_1 - t)$, ξ' by $\bar{u}(\bar{t}_0 - \tau)$, $c_0(\xi', t_0)$ by $c(x_0, \tau)$, $(t - t_0)$ by $(\bar{t}_1 - \bar{t}_0)$ and $d\xi$ by $\bar{u}\, d\tau$.

The routing method assumes that no dispersion takes place while the tracer cloud passes through the measuring station. In actual practice, the concentration profiles at two stations are measured, and the upstream concentration profile is used as an input and the corresponding concentration profile at a downstream station is determined for an assumed value of D. It is then compared with the measured concentration profile at the downstream station, and the dispersion coefficient is adjusted iteratively until the two concentrations match within a certain specified tolerance.

All the methods proposed hitherto for estimating the dispersion coefficient are valid only after the convective period has elapsed. An expression derived by Fischer (1967) for the convective period has been reported to underestimate when applied to natural streams (Beltaos, 1980; Liu and Cheng, 1980). Attempts to explain this discrepancy has been made by proposing non-Fickian models in which the analogy between the one-dimensional dispersion and diffusion equations is not used (Mcquivey and Keefer, 1976; Sabol and Nordin, 1978), and modified Fickian models using the concept of dead zone storage effects (Thackston and Schnelle, 1970; Valentine and Wood, 1977). It is argued that the presence of dead zones, channel irregularities and non-uniformities merely prolong the time required to attain Fickian behaviour. In the latter type of models, the analogy between the one-dimensional dispersion equation and the diffusion equation is used with an extra term added to the dispersion equation to account for dead zone storage effects. Liu and Cheng (1980), using the same modified Fickian concept, dispensed with this extra term by introducing a time-dependent dispersion coefficient. However, the practical application of these methods to natural streams is limited.

23.8.3 Time-dependent dispersion coefficient

Following the approach proposed by Liu and Cheng (1980) and embodying the arguments put forward by Sullivan (1971), Jayawardena and Lui (1983) proposed a time-dependent dispersion model which has been founded upon a large number of laboratory tests. Their findings in which the dispersion coefficient is considered as a function of time are given by

$$D = D_F \left\{ 1 - \frac{T}{t} \left(1 - e^{-\frac{t}{T}} \right) \right\} = D_F f(t'). \tag{23.59}$$

In this equation, D is the time-dependent dispersion coefficient, D_F is the Fickian dispersion coefficient, T is the Lagrangian time scale, which is usually considered as a measure of the longest time during which on average a particle persists in motion, and $t' = \dfrac{t}{T}$ is the dimensionless time.

Through a large number of laboratory experiments, the function $f(t')$ has been empirically obtained and consists of three parts: a linear part for small t, a transitional part for medium t and an asymptotic part for large t. In this context, small t for practical purposes correspond to $\dfrac{t}{T} \leq 0.1$; intermediate t correspond to $0.1 \leq \dfrac{t}{T} \leq 20$, and large t correspond to $\dfrac{t}{T} > 20$. When $\dfrac{t}{T} > 20$, D lies between $0.95\, D_F$ and D_F. The function $f(t')$ is shown in Figure 23.1.

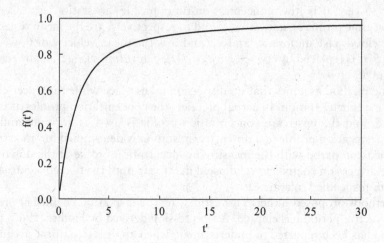

Figure 23.1 Function $f(t')$ of Eq. 23.59.

A field application of the above approach has been carried out for four narrow streams in Hong Kong using sodium chloride and Rhodamine-B as tracers (Lui and Jayawardena, 1983). The results, when compared with the dispersion coefficient estimated by the routing method agree to within a factor of about 2. Dispersion prediction is considered as satisfactory if they differ by a factor of less than 4 (Fischer et al., 1979).

23.9 TIME SCALES – DIMENSIONLESS TIME

There is one common limitation for all the methods described above. It was assumed that cross-sectional mixing is complete which was a pre-requisite for applying the one-dimensional dispersion equation. They are applicable after an initial period of time commonly referred to as the convective period. It can also be identified as the mixing period, or when converted into an equivalent length, the mixing length. The 'time scale' is a characteristic time which is used to obtain dimensionless time from real time. Two types of time scales have been defined for this purpose: the Eulerian time scale and the Lagrangian time scale. Eulerian time scale is obtained by expressing the time required for cross-sectional mixing using the Eulerian co-ordinate system whereas the Lagrangian time scale is obtained by tracking the movement of a fluid particle using the Lagrangian co-ordinate system.

23.9.1 Eulerian time scale

Cross-sectional mixing is important because it is the mechanism for establishing steady state cross-sectional concentration distribution. From dimensional analysis, it is possible to work out an approximate time required for mixing in the vertical direction. The problem is similar to heat conduction in a rod with sudden addition of heat at one end. The thermal diffusivity and eddy diffusivity are analogous and have dimensions of $L^2 T^{-1}$. By dimensional analysis,

$$T \propto \frac{d^2}{\varepsilon_y}; \quad \text{or,} \quad T = \alpha \frac{d^2}{\varepsilon_y} \tag{23.60}$$

where d is the depth, ε_y is the eddy diffusivity in the vertical direction and T is the Eulerian time scale. In a generalized form Eq. 23.60 can be written as

$$T = \frac{l^2}{\varepsilon} \tag{23.61}$$

where l is a characteristic length and ε is the eddy diffusivity. This equation can be used in the vertical or transverse direction. More precisely, the Eulerian time scale can be defined as

$$T_E = \int_0^\infty R_E(\tau)\, d\tau \tag{23.62}$$

where

$$R_E(\tau) = \frac{[u_t u_{t+\tau}]}{u_t^2} \tag{23.63}$$

The Eulerian correlation coefficient $R_E(\tau)$ gives the correlation between the values of velocity, or any other variable quantity, at a fixed point in the flow field at two different times separated by a time τ which is referred to as the lag. The term $[u_t u_{t+\tau}]$ is the product of u_t and $u_{t+\tau}$ averaged over a large number of trials. The Eulerian time scale can be considered as an approximate measure of the longest connection in the turbulent behaviour of $u(t)$. In a similar manner, it is possible to define an Eulerian length scale as

$$L_E = \int_0^\infty F(\xi)\, d\xi \tag{23.64}$$

where $F(\xi)$ represents the correlation between velocities at two points separated by a distance ξ. Using Taylor's assumptions, which are justified for a turbulent field which is homogeneous and stationary,

$$\frac{\partial}{\partial t} = -\bar{u}\frac{\partial}{\partial x} \tag{23.65}$$

from which it can be seen that

$$L_E = \bar{u} T_E \tag{23.66}$$

$$\text{Dimensionless time} = \frac{\text{Real time}}{\text{Time scale}}$$

Using this approach, the Eulerian time scale for lateral mixing is given by

$$T_E = \frac{l^2}{\varepsilon_z} = \frac{l^2}{0.23\, R u_*} \tag{23.67}$$

($\varepsilon_z = 0.23\, hu_*$ has been obtained by Elder (1959) experimentally; h is replaced by R, the hydraulic radius, and u_* the shear velocity.)

In this equation, l is a characteristic length (normally taken as half the width for symmetrical channels, and the distance between the thread of maximum velocity and the furthest bank for non-symmetrical channels).

23.9.2 Lagrangian time scale

The Lagrangian time scale is defined as

$$T_L = \int\limits_0^\infty R_L(\tau)\,d\tau \tag{23.68}$$

where R_L is the Lagrangian autocorrelation function. The prefix 'Lagrangian' is used to mean that the velocity is that following the mean motion of a particular particle, and not a velocity at a fixed point. (Recall the Lagrangian and Eulerian description of fluid flow.) The Lagrangian autocorrelation function is defined as

$$R_L(\tau) = \frac{[u_t u_{t+\tau}]}{u_t^2} \tag{23.69}$$

where $R_L(\tau)$ is the correlation coefficient between the velocity u of a particle at any given instant and that of the same particle after an interval of time τ which is referred to as the lag. The Lagrangian time scale T_L is considered as a measure of the longest time during which, on average, a particle persists in a motion in a given direction. In other words, it is a measure of how long the particles take to lose memory of the initial velocity. This statistical representation of dispersion was first put forward by Taylor (1922).

The distance travelled by a particle in time t can be given as

$$x(t) = \int\limits_0^t u(t)\,dt + x(0) \tag{23.70}$$

where u is the instantaneous velocity of a particle with zero mean. To proceed further it is also assumed that the turbulent flow is homogeneous in space and time. The distance travelled by a marked particle with starting time t_0 during the time interval t can be expressed as

$$x(t_0 + t) = \int\limits_0^t u(t_0 + t')\,dt' \tag{23.71}$$

Therefore, the variance of x can be expressed as

$$\left[x(t)^2\right] \approx 2\left[u^2\right]t T_L \tag{23.72}$$

This implies that for long times the variance of x increases linearly. In this context, long time means time long enough for the Lagrangian velocity fluctuations to be not correlated with each other. Differentiating Eq. 23.72,

$$\frac{d}{dt}\left[x(t)^2\right] = 2\left[u^2\right]T_L.$$

The turbulent mixing coefficient is now defined as

$$\varepsilon_* = \frac{1}{2}\frac{d}{dt}\left[x(t)^2\right] = \left[u^2\right]T_L. \tag{23.73}$$

In a similar manner, the Lagrangian length scale L_L can be expressed as

$$L_L^2 = \left[u^2\right]T_L^2 \tag{23.74}$$

This gives the order of magnitude of the distance a fluid particle will travel before it loses the memory of its initial position. From Eq. 23.72, the real length scale L is approximately

$$L^2 \approx 2\left[u^2\right]tT_L \tag{23.75}$$

For $t > T_L$, it is equivalent to saying $L^2 > 2L_L^2$, implying that the real length scale > 2 (Lagrangian length scale), i.e. the size of the dispersing cloud > 2 (distance over which turbulent motions are correlated).

Since the dispersion coefficient, $D = \dfrac{1}{2}\dfrac{d}{dt}\left[x^2\right]$ for the diffusion equation, in terms of the equations above,

$$D = \frac{1}{2}\,2\left[u^2\right]T_L = \left[u^2\right]T_L \tag{23.76}$$

In open channel flow, the mean turbulence is sufficiently small compared with the deviations within the cross section of time-averaged velocity. The following approximation can therefore be used:

$$[u^2] = \overline{u'^2} \tag{23.77}$$

where $u' = u - \bar{u}$, and, $\overline{u'^2}$ is the squared spatial velocity deviation from the cross-sectional mean.

Therefore, the dispersion coefficient based on the diffusive transport method can be estimated using the Lagrangian time scale as

$$D = \overline{u'^2}T_L \tag{23.78}$$

23.9.3 Time scale for two-dimensional flows

For a two-dimensional flow with logarithmic velocity profile down an infinitely wide plane, both the above scales can be calculated. For Eulerian scale, the characteristic length is h and the average value of vertical mixing coefficient is

$$\overline{\varepsilon_y} = \frac{\kappa}{6}hu_* \text{ where } \kappa \text{ is the von Karman constant } (= 0.41). \tag{23.79}$$

Therefore, $T_E = \dfrac{l^2}{\varepsilon_y} = \dfrac{6}{\kappa}\dfrac{h}{u_*}$ $\qquad\qquad\qquad\qquad$ (23.80)

Averaging the velocity distribution given by the logarithmic law gives

$$\overline{u'^2} = \left(\frac{u_*}{\kappa}\right)^2 \tag{23.81}$$

Elder's (1959) result, as originally derived including von Karman constant, is

$$D = \frac{0.404}{\kappa^3} h u_*. \tag{23.82}$$

Inserting these values in $D = \overline{u'^2} T_L$,

$$T_L = \frac{0.404}{\kappa^3} h u_* \left(\frac{\kappa}{u_*}\right)^2 = \frac{0.404}{\kappa} \frac{h}{u_*} \tag{23.83}$$

Therefore, the relationship between Eulerian and Lagrangian time scales is

$$T_E = 14.8 T_L \tag{23.84}$$

$$\left(\text{or,} \ \frac{\kappa}{0.404} T_L = \frac{\kappa}{6} T_E \Rightarrow T_E = \frac{6}{0.404} T_L = 14.8 T_L \right)$$

23.9.4 Time scales in natural streams

In natural streams, lateral mixing is more important than vertical mixing for Eulerian time scale. The important length is the distance over which mixing must take place to establish a uniform distribution. Characteristic length is defined as the distance between the thread of maximum velocity and the furthest distant bank. Therefore,

$$T_E = \frac{l^2}{\varepsilon_z} = \frac{l^2}{0.23 \, R u_*} \tag{23.85}$$

If $T_E = 14.8 \, T_L$, the Lagrangian time scale is given by

$$T_L = \frac{0.30 \, l^2}{R u_*} \tag{23.86}$$

Therefore,

$$D = 0.30 \, \overline{u'^2} \, \frac{l^2}{R u_*} \tag{23.87}$$

Dimensionless time $t' = \dfrac{t}{T_L}$. $\tag{23.88}$

Fischer (1967) using experimental data classified the dimensionless time t' into three categories as follows:

- $0 < t' < 3$: Convective period and Taylor's theory is not applicable
- $3 < t' < 6$: Transition period; variance growth is nearly linear but one-dimensional dispersion equation is not applicable
- $t' > 6$: Diffusive period and the one-dimensional dispersion equation is applicable. Taylor's dispersion theory is applicable only during the dispersive period when the normally skewed concentration profile approaches a Gaussian distribution.

The condition $t' > 6$ implies $L > 1.8 \dfrac{l^2}{R} \dfrac{\bar{u}}{u_*}$ where L can be considered as the mixing length.

23.10 NUMERICAL SOLUTION OF THE LONGITUDINAL DISPERSION EQUATION

To understand the consequences of accidental discharges of pollutants into a water body as well as to understand possible environmental hazards under various scenarios, it is necessary to predict how the pollutant is dispersed as it is transported downstream. Such an attempt must start with the governing equations including the boundary and initial conditions, simplifications and assumptions, estimation of the parameters of the processes involved, a geometrical framework upon which a solution is sought and a numerical scheme to carry out the solution as there are no analytical solutions for most real-world problems. The one-dimensional dispersion equation which governs the spreading of a non-conservative pollutant introduced into a water course can be written as (Eq.23.26b)

$$\frac{\partial}{\partial t}(A\bar{c}) + \frac{\partial}{\partial x}(\bar{u}A\bar{c}) = \frac{\partial}{\partial x}\left(DA \frac{\partial \bar{c}}{\partial x} \right) - kA\bar{c} + Q_s$$

where A is the cross-sectional area of flow, k is a decay constant (assuming first-order decay) and Q_s is a source term. Other variables in this equation have the same meanings as those in Eq. 23.29. The relevant boundary conditions can be written as

$$\bar{c} = \bar{c}_0, \text{ at } x = 0 \text{ for } t > 0 \tag{23.89a}$$

and

$$\frac{\partial \bar{c}}{\partial x} = 0, \text{ at } x = L \text{ for } t > 0 \tag{23.89b}$$

in which $\bar{c}_0$ may be either constant or time varying, and L is the length of the reach. Some initial condition is also necessary. The problem involves the solution of a parabolic type partial differential equation with two independent variables, x and t. It can be done using the finite difference method or the finite element method or their combinations.

23.10.1 Finite difference method (see Chapter 16)

23.10.2 Finite element method (see also Chapter 16)

The finite element method has the advantage of been more flexible than the finite difference method although the formulation is more difficult. Widely used approaches of finite element formulation include the variational method, the method of weighted residuals of which Galerkin's method is a special case, boundary element methods and adaptive methods. Each type has its own pros and cons, but Galerkin's method by far has the widest application. General details of the finite element method can be found in several textbooks (e.g. Zienkiewicz, 1971, 1977; Oden and Carey, 1983; among others). Specific details relevant to the solution of the dispersion equation can be found in Price et al. (1968), Smith et al.

(1973), Guymon (1970) and Jayawardena and Lui (1984), among others. Since the dispersion equation is of parabolic type, the time domain is open and a marching solution using finite differences in time and finite elements in space is normally carried out.

An application of a numerical procedure using finite elements in space with the Galerkin's formulation, and finite differences in time with the Crank-Nicolson scheme, in which the results have been compared with analytical solutions for specific assumptions, and subsequently applied to laboratory data as well as field data is given by Jayawardena and Lui (1984). In the finite element formulation in this application, linear and quadratic interpolation functions have been used. Their comparisons have been made against the analytical solutions for the following cases:

For the case with constant parameters without the decay and source terms, the governing equation simplifies to (same as Eq. 23.29).

$$\frac{\partial \overline{c}}{\partial t} + \overline{u}\frac{\partial \overline{c}}{\partial x} = D\frac{\partial^2 \overline{c}}{\partial x^2}$$

subject to the boundary and initial conditions

$$\overline{c}(0, t) = \overline{c}_0, \text{ for } t > 0 \tag{23.90a}$$

$$\overline{c}(\infty, t) = 0, \text{ for } t > 0 \tag{23.90b}$$

$$\overline{c}(x, 0) = 0, \text{ for all } x. \tag{23.90c}$$

An analytical solution for this case (Ogata and Banks, 1961) is given as

$$\overline{c}(x, t) = \frac{1}{2}\overline{c}_0 \left\{ erfc\left[\frac{x - \overline{u}t}{2\sqrt{Dt}}\right] + \exp\left[\left(\frac{\overline{u}x}{D}\right)\right] erfc\left[\frac{x + \overline{u}t}{2\sqrt{Dt}}\right] \right\} \tag{23.91}$$

where $erfc$ is the complementary error function ($= 1 - erf$), and the error function erf is defined as

$$erf(z) = \frac{2}{\sqrt{\pi}} \int_0^z e^{-t^2} dt \tag{23.92}$$

For the case with the decay term, the governing equation becomes

$$\frac{\partial \overline{c}}{\partial t} + \overline{u}\frac{\partial \overline{c}}{\partial x} = D\frac{\partial^2 \overline{c}}{\partial x^2} - k\overline{c} \tag{23.93}$$

subject to the boundary and initial conditions

$$\overline{c}(0, t) = \overline{c}_0, \text{ for } 0 \le t \le \delta_t \text{ and } \overline{c}(0, t) = 0, \text{ for } t > \delta_t \tag{23.94a}$$

$$\overline{c}(x, 0) = 0, \text{ for all } x. \tag{23.94b}$$

where δ_t is a certain time interval. An analytical solution for this case (Kinzelbach and Yang, 1981) is given as

$$\bar{c}(x, t) = \frac{1}{2}\bar{c}_0 \left\{ \begin{array}{l} \exp\left[\sqrt{\dfrac{\bar{u}^2}{4D}+k}\ \dfrac{x}{\sqrt{D}}\right] erfc\left[\dfrac{x}{2\sqrt{Dt}}+\sqrt{\dfrac{\bar{u}^2}{4D}+k}\ \sqrt{t}\right] \\[4mm] + \exp\left[-\sqrt{\dfrac{\bar{u}^2}{4D}+k}\ \dfrac{x}{\sqrt{D}}\right] erfc\left[\dfrac{x}{2\sqrt{Dt}}-\sqrt{\dfrac{\bar{u}^2}{4D}+k}\ \sqrt{t}\right] \end{array} \right\} \exp\left(\dfrac{\bar{u}x}{2D}\right)$$

$$-\frac{1}{2}\left\{ \exp\left[\sqrt{\dfrac{\bar{u}^2}{4D}+k}\ \dfrac{x}{\sqrt{D}}\right] erfc\left[\begin{array}{l} \dfrac{x}{2\sqrt{D(t-\delta_t)}}+\sqrt{\dfrac{\bar{u}^2}{4D}+k}\ \sqrt{t-\delta_t}+e\dot{x}p\left[-\sqrt{\dfrac{\bar{u}^2}{4D}+k}\ \dfrac{x}{\sqrt{D}}\right] \\[4mm] erfc\dfrac{x}{2\sqrt{D(t-\delta_t)}}-\sqrt{\dfrac{\bar{u}^2}{4D}+k}\ \sqrt{t-\delta_t} \end{array} \right] \right\} \exp\left(\dfrac{\bar{u}x}{2D}\right)\theta(t-\delta_t)$$

$$(23.95)$$

where $\theta(t-\delta_t) = 0$, for $t \le \delta_t$ and, unity for $t > \delta_t$.

Comparisons of the finite element-finite difference solutions made with these analytical solutions were found to give very good agreements (Jayawardena and Lui, 1984). Stability criteria have been examined with reference to the Peclet number, p (sometimes referred to as the Reynolds cell number), defined as $p = \dfrac{\bar{u}\Delta x}{D}$ and the Courant number, c, defined as $c = \dfrac{\bar{u}\Delta t}{\Delta x}$ where Δx and Δt are the space and time increments. Successful application of the numerical procedure for laboratory and field data for four natural streams in Hong Kong is reported by Lui and Jayawardena (1983).

23.10.3 Moving finite element method

Numerical solution of the convective dispersion equation encounters difficulties when the problem is convection dominated. The normally parabolic type partial differential equation becomes almost hyperbolic causing the numerical solution with a fixed grid to oscillate and overshoot (Price et al., 1968; Ehlig, 1977; Lam, 1977; Jensen and Finlayson, 1978, 1980). Criteria for non-oscillatory solutions are normally expressed in terms of the Peclet number p, and the Courant number c. To guarantee non-oscillatory solutions using finite differences centred in space or linear finite elements, the restriction on the Peclet number is given by Jensen and Finlayson (1978) as $p < 2$, and hence the mesh spacing Δx must satisfy the condition

$$\Delta x < \frac{2D}{\bar{u}} \qquad\qquad (23.96a)$$

Using finite elements for discretization in space and centred finite differences for discretization in the time domain, Ehlig (1977) gives the restriction on the Courant number to prevent oscillations as $c < 0.1$. Using linear finite elements in space and a Crank–Nicholson finite difference approximation in time (a Crank–Nicholson Galerkin method) Jayawardena and Lui (1984) found that the condition $c < 0.1$ to be over-restrictive and that in practice the condition $c < 1$ is sufficient. The latter criterion gives the restriction on the size of the time step Δt to yield non-oscillatory numerical solutions as

$$\Delta t < \frac{\Delta x}{\bar{u}}. \qquad\qquad (23.96b)$$

Combining these two criteria (23.96a and b) the time increment Δt for non-oscillatory solutions of Eq. 23.29 using linear finite elements in space and the Crank–Nicholson scheme in time should satisfy the condition

$$\Delta t < \frac{2D}{\bar{u}^2} \tag{23.96c}$$

When the dispersion process is convection dominated, it is well known that numerical solution with standard fixed mesh finite elements or finite differences requires large numbers of elements and small time steps in order to satisfy criteria for stability and to give oscillation-free accurate solutions. For such problems, numerical solution with fixed mesh methods is consequently inefficient and computationally expensive. For example, in the hypothetical case of $\bar{u} = 1$ m/s and $D = 0.001$ m²/s with $0 < x < 90$ m, $t = 60$ s, the above criteria for stability will require $\Delta x < 0.002$ m and $\Delta t < 0.002$ s, implying 4.5×10^4 finite elements in space and 3×10^4 time steps (Johnson and Jayawardena, 1998). This is certainly computationally expensive in terms of computational time and storage requirements. Failing to meet these criteria will make the solution oscillate.

A method to overcome this problem has been proposed by Johnson and Jayawardena (1998). They introduce the moving finite element (MFE) method, a moving mesh method, which has several advantages. In the MFE method, the finite element approximation is based on a mesh of nodes which move in time. A comprehensive account of MFE theory and applications is given by Baines (1994), and the particular application to the dispersion process by Johnson and Jayawardena (1998). Numerical solutions with highly accurate resolution of steep moving fronts can be generated using the MFE method using only a small number of elements and time steps greater than those afforded by fixed mesh methods. In the special case of linear dispersion, a highly efficient, explicit, direct solution procedure, requiring no matrix inversions has been presented.

23.11 DISPERSION THROUGH POROUS MEDIA

The equation for transport of solute in a saturated porous medium is derived using the law of conservation of mass under the following assumptions:

- The porous medium is homogeneous and isotropic
- The medium is saturated and the flow is at steady state
- Darcy's law is applicable

According to Darcy's law, the solute will be advected by the average linear velocity. If this is the only transport mechanism, the solute will move as a plug. But, due to hydrodynamic dispersion, which is caused by variations in the microscopic velocity within each pore channel and from one channel to another, mixing of the solute will also take place.

Therefore, to describe the transport mechanism using macroscopic parameters yet taking into account microscopic mixing, it is necessary to introduce a second mechanism of transport in addition to advection. In a porous medium, the average linear velocity $\bar{u} = \dfrac{u}{n}$, where u is the Darcy velocity and n is the porosity of the medium. The mass of solute per unit volume of porous medium is equal to 'nc' where c is the concentration of solute (ML^{-3}) which is in mass of solute per unit volume of solution. For a homogeneous medium, n is a constant. Therefore,

$$\frac{\partial}{\partial x}(nc) = n\frac{\partial c}{\partial x} \tag{23.97}$$

Transport by advection $= \bar{u}ncdA$

Transport by dispersion $= nK\frac{\partial c}{\partial x}dA$

The dispersion coefficient K is related to the dispersivity α and the diffusion coefficient D by the relationship

$$K = \alpha\bar{u} + D \tag{23.98}$$

where α is the dispersivity which is a characteristic property of the porous medium (L) and D is the coefficient of molecular diffusion for the solute in the porous medium (L^2T^{-1}).

The form of the dispersive component of the transport by dispersion is analogous to Fick's law. The total mass transport/unit area

$$q = \bar{u}nc - nK\frac{\partial c}{\partial x} \tag{23.99}$$

(The negative sign indicates that the transport is towards the zone of lower concentration.) Conservation of mass for a conservative solute gives

$$qdA - \left(q + \frac{\partial q}{\partial x}\delta x\right)dA = n\frac{\partial c}{\partial t}dA\delta x$$

which simplifies to

$$-\frac{\partial q}{\partial x} = n\frac{\partial c}{\partial t} \tag{23.100}$$

Substituting

$$-\left\{\frac{\partial}{\partial x}(\bar{u}nc) - \frac{\partial}{\partial x}\left(nK\frac{\partial c}{\partial x}\right)\right\} = n\frac{\partial c}{\partial t}$$

which simplifies to

$$-\bar{u}\frac{\partial c}{\partial x} + \frac{\partial}{\partial x}\left(K\frac{\partial c}{\partial x}\right) = \frac{\partial c}{\partial t} \tag{23.101}$$

In a homogeneous medium K is a constant in space. Therefore,

$$K\frac{\partial^2 c}{\partial x^2} - \bar{u}\frac{\partial c}{\partial x} = \frac{\partial c}{\partial t} \tag{23.102}$$

Extending to three dimensions,

$$K_x\frac{\partial^2 c}{\partial x^2} + K_y\frac{\partial^2 c}{\partial y^2} + K_z\frac{\partial^2 c}{\partial z^2} - \left(\bar{u}\frac{\partial c}{\partial x} + \bar{v}\frac{\partial c}{\partial y} + \bar{w}\frac{\partial c}{\partial z}\right) = \frac{\partial c}{\partial t} \tag{23.103}$$

In some applications, the one-dimensional direction is taken as a curvi-linear coordinate in the direction of flow along a flow line. The transport equation then is

$$K_l\frac{\partial^2 c}{\partial l^2} - \left(\bar{u}_l\frac{\partial c}{\partial l}\right) = \frac{\partial c}{\partial t} \tag{23.104}$$

where l is the coordinate direction along the flow line. If u and K vary spatially, then

$$\frac{\partial}{\partial x}\left(K\frac{\partial c}{\partial x}\right) - \frac{\partial}{\partial x}(uc) = \frac{\partial c}{\partial t} \qquad (23.105)$$

23.12 EUTROPHICATION

A water body can be classified as oligotrophic, mesotrophic or eutrophic. Oligotrophic implies that the water body does not receive sufficient nutrients to sustain aquatic life and eutrophic implies that the water body receives too much nutrients. Mesotrophic is somewhere in between. Too much nutrients also have their negative effects. The nutrients are mainly nitrogen and phosphorus which may be released into a water body as point sources such as for example through sewage outfalls, or as non-point sources from the surrounding catchments and agricultural fields via runoff. Excessive nutrients (carbon, nitrogen and phosphorus) cause excessive growth of algae which depletes the oxygen in the water body. When the algae die bacterial degradation of the biomass causes oxygen consumption and thereby oxygen depletion in the water body. The plants beneath the algal bloom cannot survive because the algae block the sunlight necessary for photosynthesis. The dead plant matter decomposes using the dissolved oxygen in the water body. The water body then cannot support life and fish-kill is a common occurrence in such water bodies. Other forms of reduction of biodiversity can also take place.

Eutrophication is a natural process, but in recent years, it has accelerated due to human activities. Contribution by human activities, known as cultural eutrophication, leads to a continuous increase in nutrients, mainly nitrogen and phosphorus until it exceeds the capacity of the water body to self-purify. Increased concentrations of algae (microscopic organisms that live in aquatic environments and use photosynthesis to produce their biomass using energy from sunlight, just like plants) and microscopic organisms on the surface prevent sunlight penetration for photosynthesis of underwater aquatic life as well as oxygen absorption. Excessive algal growth, or algal bloom, can lead to change in the colour of the water to green, blue-green, red or brown, depending on the type of algae. Excessive eutrophication can also cause fish-kills in fish farms. Some algae produce toxins which can harm other living organisms when released into surrounding waters.

Algae or phytoplankton manufacture their own food through photosynthesis using solar energy as the energy source and carbon from carbon dioxide. Factors such as solar radiation, nutrients, water temperature, predation, disease, wind speed and tidal flushing control their growth. For algal growth, phosphorus is the limiting nutrient which controls its growth. Since phosphorus is not naturally present in water bodies and in the atmosphere, the growth of algae can be considered as due to cultural eutrophication. In the summer months, the total phosphorus concentration is highly correlated with chlorophyll 'a' concentration. Control of cultural eutrophication can therefore be achieved by reducing the amount of phosphorous inputs to the water body. The main source of phosphorus is waste waters from agricultural runoff and septic flows from sewage. Other indicators of eutrophic conditions include cyanobacterial (blue-green algae) blooms, loss of benthic invertebrates, secchi-disc depth less than 2 m and chlorophyll 'a' concentration exceeding 10 µg/l.

23.12.1 Red tides

Red tide is a term used to refer to a type of harmful algal blooms (HABs), which are caused by certain species of phytoplankton containing photosynthetic pigments that vary in colour from green to brown to red. When red tide appears, the colour of the water takes a reddish appearance and hence the use of the word red. It has no relationship with astronomical tides. Algal blooms are concentrations of aquatic microorganisms such as protozoa, dinoflagellates and diatoms. In particular, dinoflagellates, a microscopic marine life form with two long slender appendages, are the organisms responsible for red tides. When conditions are favourable, these dinoflagellates reproduce at an accelerated rate creating a red tide 'bloom'. Harmful algal blooms occur worldwide and natural cycles can vary regionally and cause negative effects on coastal ecosystems. The presence of algal blooms, some of which are toxic, affects marine life and is aesthetically unpleasant.

Consequences of red tide occurrences include increased incidences of fish kills, loss of desirable fish species, reductions in harvestable fish and shellfish, increased biomass of phytoplankton, toxic or inedible phytoplankton species, increases in blooms of gelatinous zooplankton, increased biomass of benthic and epiphytic algae, changes in macrophyte species composition and biomass, decreases in water transparency (increased turbidity), colour, smell and water treatment problems, dissolved oxygen depletion, etc.

Historically, occurrences of red tide have taken place in British Columbia, Canada, SW Florida, USA, New England, USA, Port Moresby, Papua New Guinea, Sabah, Malaysia, Northern California, USA, Gulf of Mexico, Philippines, Netherlands, Japan, Hong Kong, among other places.

23.12.2 Prediction of harmful algal blooms

From the point of view of protecting fish farmers and other users of coastal waters it would always be useful to know when there will be a harmful algal bloom. The economic damage resulting from fish kills due to occurrences of HAB in certain coastal waters have been significant. For example, during the past few decades, massive fish kills due to algal blooms and consequent anoxia have occurred in some of the marine fish culture zones in Hong Kong (Lee et al., 1991a, b) and that in April 1998, a devastating red tide resulted in the worst fish kill in Hong Kong's history destroying over 80% (3,400 tons) of fish stocks, with an estimated loss of more than HK$312 million (Dickman, 1998). Thus, there is a need to develop a forecasting system that can give a timely warning to fish farming community of an impending algal bloom.

With the present state of knowledge of the physical, chemical and biological processes responsible for the development of a harmful algal bloom, it is difficult if not impossible to go for a process based model that can predict with reasonable accuracy when a bloom will occur and how serious it would be. The alternative thus is to go for a data driven approach that will enable a reasonable and timely prediction. The first step in such an approach is to identify the measurable variables involved in the algal production process. The dependent variable would obviously be the concentration of harmful algae and the independent variables could be the factors that influence the growth of harmful algae. Harmful algae species include *Alexandrium*, *Chattonella*, *Gymnodinium* and *Heterosigma*.

It is known that the concentration of chlorophyll 'a' which is present in all types of algae is an indicator of algal biomass and therefore can be considered as the dependent variable. The independent variables include the antecedent values of chlorophyll 'a' as well as a number of other environmental variables such as solar radiation, temperature, wind speed, dissolved oxygen, ammonia nitrogen, nitrite nitrogen, nitrate nitrogen, total inorganic nitrogen, total nitrogen, orthophosphate phosphorus, total phosphorus, etc., each of which may have different contributions to the growth of algal biomass. The frequency of measurement of such data is another factor.

One of the data driven approaches that can be used and that has been used by several investigators is the artificial neural network (ANN) approach where a priori knowledge of the process dynamics is not required. Details about ANN's can be found in several publications (e.g. Haykin, 1994) and specific applications for algal bloom prediction using different input parameters can be found in several other publications (e.g. Yabunaka et al., 1997; Barciela et al., 1999; Maier et al., 1998) with different lead times ranging from 1 to 4 weeks.

In a similar data driven approach using a three-layer feed-forward neural network with backpropagation learning algorithm to predict the occurrence of HAB's in the coastal waters of Hong Kong using bi-weekly water quality data in Tolo Harbour, a semi-enclosed bay in the northeastern coastal waters of Hong Kong, and weekly phytoplankton species *Skeletonema* abundance data in Lamma Island, it has been found that the most influencing independent input variables were solar radiation, total inorganic nitrogen (TIN), time lagged chlorophyll-a (Chl-a), phosphorus (PO4), dissolved oxygen (DO), secchi-disc depth, water temperature, rainfall, wind speed and tidal range (Lee et al., 2003). The same study also concluded that using time lagged chlorophyll 'a' only as the input variable produced the optimal network in terms of the prediction accuracy, network complexity, amount of data needed and the time taken to train the network suggesting a non-linear autoregressive type of dynamics. A similar study that employed a vector autoregressive time series model with exogenous variables (VARX) which allows the variables in the data set to be modelled jointly over present and past time periods using both daily and two hour chlorophyll 'a' data with water temperature and solar radiation as exogenous variables showed that it has the advantage that useful information can be obtained for a better understanding of the underlying mechanism of algal bloom in relation to ecological parameters (e.g. nutrients, temperature, solar radiation and wind speed) (Lui et al., 2007).

23.13 GENERAL PURPOSE WATER QUALITY MODELS

Several water quality models that can simulate many different constituents are now available for general use. Among them, are the Enhanced Stream Water Quality Model (QUAL2E), and its more recent version QUAL2K, both developed by the United States Environmental Protection Agency (USEPA), Water Quality Analysis Simulation Program (WASP), and the One Dimensional Riverine Hydrodynamic and Water Quality Model (EPD-RIV1).

23.13.1 Enhanced Stream Water Quality Model (QUAL2E)

The Enhanced Stream Water Quality Model (QUAL2E) is applicable to well-mixed, dendritic streams. It simulates the major reactions of nutrient cycles, algal production, benthic and carbonaceous demand, atmospheric re-aeration and their effects on the dissolved oxygen balance. The model assumes that the major transport mechanisms, advection and dispersion, are significant only along the longitudinal direction of flow. It can predict the following 15 water quality constituent concentrations:

- Dissolved Oxygen
- Biochemical Oxygen Demand
- Temperature
- Algae as Chlorophyll 'a'
- Organic Nitrogen as 'N'
- Ammonia as 'N'
- Nitrite as 'N'
- Nitrate as 'N'
- Organic Phosphorus as 'P'

- Dissolved Phosphorus as 'P'
- Coliform bacteria
- Arbitrary Non-conservative Constituent
- Three Conservative Constituents.

It is intended as a water quality planning tool for developing total maximum daily loads (TMDLs) and can also be used in conjunction with field sampling for identifying the magnitude and quality characteristics of nonpoint sources. By operating the model dynamically, the user can study diurnal dissolved oxygen variations and algal growth. However, the effects of dynamic forcing functions, such as headwater flows or point source loads, cannot be modelled with QUAL2E. The model assumes that the stream flow and waste inputs are constant during the simulation time periods. QUAL2EU is an enhancement that allows users to perform three types of uncertainty analyses: sensitivity analysis, first-order error analysis, and Monte Carlo simulation. QUAL2K is an enhanced version of QUAL2E that takes into account the following:

- Unequally spaced river reaches and multiple loadings and abstractions in any reach.
- Two forms of Carbonaceous BOD (slowly oxidizing and rapidly oxidizing) to represent organic Carbon as well as non-living particulate organic matter.
- Anoxia by reducing oxidation reactions to zero at low oxygen levels.
- Sediment-water interactions.
- Bottom algae.
- Light extinction.
- pH.
- Pathogens.

The Windows interface provides input screens to facilitate preparing model inputs and executing the model. It also has help screens and provides graphical viewing of input data and model results. More details of the software can be found at the website http://www.epa.gov/OST/QUAL2E_WINDOWS, and in: 'The Enhanced Stream Water Quality Models QUAL2E and QUAL2E-UNCAS: Documentation and User's Manual'. (EPA 600/3-87-007). NTIS Accession Number: PB87 202 156.

23.13.2 Water Quality Analysis Simulation Program (WASP)

This program, which is based on the work of several researchers, can carry out dynamic compartment modelling of aquatic systems including the water column as well as the benthos. It can analyse a number of pollutant types in one, two or three dimensions. The program can also be linked to hydrodynamic and sediment transport models. The pollutants it can handle include

- Nitrogen
- Phosphorus
- Dissolved Oxygen
- Biochemical Oxygen Demand
- Sediment Oxygen Demand
- Algae
- Periphyton
- Organic chemicals
- Metals

- Mercury
- Pathogens
- Temperature.

More information about WASP can be found at the website: http//www.epa.gov/athens/wwqtsc/html/wasp.html.

23.13.3 One-Dimensional Riverine Hydrodynamic and Water Quality Model (EPD-RIVI)

This is a system of programs that performs one-dimensional (cross-sectionally averaged) hydraulic and water quality simulations. The hydrodynamic model is first applied, and the results are then used as inputs to the water quality model. The model can simulate the following state variables:

- Dissolved oxygen
- Temperature
- Nitrogenous Biochemical Oxygen Demand (NBOD)
- Carbonaceous Oxygen Demand (CBOD)
- Phosphorus
- Algae
- Iron
- Manganese
- Coliform bacteria
- Two arbitrary constituents.

More information about EPD-RIV1 can be found at the website: http//www.epa.gov/athens/wwqtsc/html/epd-riv1.html.

Example 23.1

A stretch of a river $ABCDE$ receives effluents at points B, C and D. At the upstream end A, the flow is 50 m³/s and the background concentration of the effluent is 50 mg/l. The segments AB, BC, CD and DE are each 1 km long and the river may be assumed to have an average velocity of 0.5 m/s. The inflows of effluents are as follows:

At B: $Q = 15$ m³/s; $c = 250$ mg/l
l
At C: $Q = 10$ m³/s; $c = 50$ mg/l
At D: $Q = 10$ m³/s; $c = 100$ mg/l

Determine the concentrations of the effluents at the points B, C, D and E and sketch the concentration profiles assuming

    i. the effluent is conservative, and
    ii. the effluent is bio-degradable with a reaction coefficient of 2/day (Example Figure 23.1).

    a. Conservative pollutant

$$cQ = c_u Q_u + c_e Q_e$$

From A to B, $c = 50$ mg/l

$$c_{bc} = \frac{50 \times 50 + 15 \times 250}{50 + 15} = 96.15 \text{ mg/l}$$

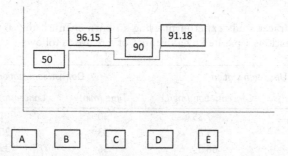

Example Figure 23.1 Concentration profile for a conservative pollutant.

$$c_{cd} = \frac{96.15 \times 65 + 10 \times 50}{75} = 90 \text{ mg/l}$$

$$c_{de} = \frac{90 \times 75 + 10 \times 100}{85} = 91.18 \text{ mg/l}$$

b. Non-conservative pollutant

$$c = c_0 e^{-Kx/u}$$

At A, $c = 50$ mg/L.
At B, upstream $c = 50e^{-Kx/u} = 50 \times 0.955 = 47.7$ mg/l.
Downstream of B, c_0 is given by

$$c_0(50 + 15) = 47.7 \times 50 + 15 \times 250 \Rightarrow c_0 = 94.38 \text{ mg/l}$$

At C, upstream $c = 94.38 \times 0.955 = 90.14$ mg/l
Downstream of C, c_0 is given by

$$c_0(50 + 15 + 10) = 90.14 \times 65 + 10 \times 50 \Rightarrow c_0 = 84.79 \text{ mg/l}$$

At D, upstream $c = 84.79 \times 0.955 = 80.97$ mg/l.
Downstream of D, c_0 is given by

$$c_0(50 + 15 + 10 + 10) = 80.97 \times 75 + 10 \times 100 \Rightarrow c_0 = 83.21 \text{ mg/l}$$

At E, upstream $c = 83.21 \times 0.955 = 79.5$ mg/l (Example Figure 23.2).

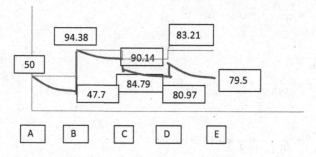

Example Figure 23.2 Concentration profile for a non-conservative pollutant.

Example 23.2

In a laboratory tracer study experiment using an open channel, the concentration-time profiles at two stations separated by a distance of 5 m are as follows:

Upstream station		Downstream station	
Time (min)	Concentration (mg/L)	Time (min)	Concentration (mg/L)
0	7.35	30	1.0
3	35.8	33	20.0
6	66.5	36	39.0
9	86.2	39	63.0
12	102.2	42	79.4
15	108.6	45	87.0
18	109.4	48	102.3
21	74.2	51	96.4
24	51.5	54	78.6
27	39.3	57	67.7
30	22.4	60	51.0
33	16.5	63	46.9
36	11.8	66	25.7
39	5.7	69	16.6
42	4.2	72	7.03
45	1.0	75	5.4

Estimate the dispersion coefficient of the channel.

c. Statistics of the given data

At the upstream station: $\sum c\bar{t} = 12{,}030; \quad \sum c\bar{t}^2 = 247{,}500$

At the downstream station: $\sum c\bar{t} = 39{,}280; \quad \sum c\bar{t}^2 = 2{,}034{,}000$

$$D = \frac{1}{2}\bar{u}^2 \frac{\sigma_{t_2}^2 - \sigma_{t_1}^2}{\bar{t}_2 - \bar{t}_1}$$

$$\bar{u} = \frac{x_2 - x_1}{\bar{t}_2 - \bar{t}_1}$$

$$\bar{t}_1 = \frac{\int ct\, dt}{\int c\, dt} = \frac{12{,}030}{738.5} = 16.3 \text{ min}$$

$$\bar{t}_2 = \frac{\int ct\, dt}{\int c\, dt} = \frac{39{,}280}{783.9} = 50.1 \text{ min}$$

$$\bar{u} = \frac{x_2 - x_1}{\bar{t}_2 - \bar{t}_1} = \frac{5}{50.1 - 16.3} = 0.1479 \text{ m/min}$$

$$\sigma_{t_1}^2 = \frac{\int ct^2\,dt}{\int c\,dt} - \overline{t_1}^2 = \frac{247,500}{738.5} - 16.3^2 = 69.4$$

$$\sigma_{t_2}^2 = \frac{\int ct^2\,dt}{\int c\,dt} - \overline{t_1}^2 = \frac{2,034,000}{783.9} - 50.1^2 = 84.7$$

$$D = \frac{1}{2} \times 0.1479^2 \times \frac{15.3}{33.8} = 0.00495 \ \text{m}^2/\text{min}$$

REFERENCES

Ahsan, N. (2008): Estimating the coefficient of dispersion for a natural stream, *Proceedings of World Academy of Science, Engineering and Technology*, vol. 34, pp 131–134, ISSN 2070-3740.

Baines, M. J. (1994). *Moving Finite Elements*, Oxford University Press, Oxford, 240 pp.

Barciela, R. M., Garcia, E. and Fernandez, E. (1999): Modelling primary production in a coastal embayment affected by upwelling using dynamic ecosystem models and artificial neural networks, *Ecological Modelling*, vol. 120, pp. 199–211.

Beltaos, S. (1980): Longitudinal dispersion in rivers, *Journal of the Hydraulic Division*, vol. 106 no.:HY1, pp. 151–172.

Deng, Z.-Q., Singh, V. P. and Bengtsson, L. (2001): Longitudinal dispersion coefficient in straight rivers, *Journal of Hydraulic Engineering*, vol. 127 no.:11, pp. 919–927.

Dickman, M. D. (1998): Hong Kong's worst red tide, In: Lee, J.H.W., Jayawardena, A.W., Wang, Z.Y. (Eds.), Proceedings of the 2nd International Symposium on Environmental Hydraulics, Balkema, Netherlands, pp. 641–645.

Ehlig, C. (1977). Comparison of numerical methods for solutions of the one-dimensional diffusion-convection equation in one and two-dimensions, In: Gray, W. G. Pinder, G. F. and Brebbia, C. A. (Eds.), *Finite Elements in Water Resources*, Pentech Press, London, pp. 1.91–1.102.

Elder, J. W. (1959): The dispersion of marked fluid inturbulent shear flow, *Journal of Fluid Mechanics*, vol. 5, pp. 544–560.

Fick, A. (1855): Poggendorf's Annalen, *der Physik und Chemie*, vol. 94, pp. 59–86. and (in English) *Philosophical Magazine*, vol. 10 no.:S4, pp. 30–39 (1855).

Fischer, H. B. (1966): *Longitudinal Dispersion in Laboratory and Natural Streams*, W. M. Keck Laboratory of Hydraulics and Water Resources Division of Engineering and Applied Science, California Institute of Technology, Pasadena, CA, Report No. KH-R-12, June 1966, . 250 pp.

Fischer, H. B. (1967): The mechanics of dispersion in natural streams, *Journal of the Hydraulics Division*, vol. 93 no.:HY6, pp. 187–216.

Fischer, H. B. (1968): Dispersion prediction in natural streams, *Journal of the Sanitary Engineering Division*, vol. 1968 no.:SA5, pp. 927–943.

Fischer, H. B. (1975): Simple method for predicting dispersion in streams, Discussion by R.S. McQuivey and T.N. Keefer, *Journal of Environmental Engineering Division*, vol. 101 no.:3, pp. 453–455.

Fischer, H. B., List, E. T., Imberger, J. and Brooks, N. H. (1979): *Mixing in Inland and Coastal Waters*, Academic Press, New York, 136 pp.

Guymon, G. L. (1970). A finite element solution of the one-dimensional diffusion-convection equation, *Water Resources Research*, vol. 6 no.:1, 204–210.

Haykin, S. (1994): *Neural Networks: A Comprehensive Foundation*, Prentice Hall, New York, 696 pp.

Jayawardena, A. W. (2014): *Environmental and Hydrological Systems Modelling*, Taylor & Francis Group, CRC Press, London, 536 pp.

Jayawardena, A. W. and Lui, P. H. (1983): A time dependent dispersion model based on Lagrangian correlation, *Hydrological Sciences Journal*, vol. 28 no.:4, pp. 455–473.

Jayawardena, A. W. and Lui, P. H. (1984): Numerical solution of the dispersion equation using a variable dispersion coefficient: method and applications, *Hydrological Sciences Journal*, vol. 29 no.:3, pp. 293–309.

Jensen, O. K. and Finlayson, B. A. (1978): Solution of the convection-diffusion equation using a moving co-ordinate system, *Proceedings of the Second International Congress on Finite Elements in Water Resources*, vol. 4, pp. 21–32.

Jensen, O. K. and Finlayson, B. A. (1980): A moving co-ordinate solution of the transport equation, *Water Resources Research*, vol. 30, pp. 9–18.

Johnson, I. W. and Jayawardena, A. W. (1998): Efficient numerical solution of the dispersion equation using moving finite elements, *Finite Elements in Analysis and Design*, vol. 28 no.:3, pp. 241–253.

Kashefipour, S. M. and Falconer, R. A. (2002): Longitudinal dispersion coefficients in natural channels, *Water Research*, vol. 36, pp. 1596–1608.

Kinzelbach, W. K. H. and Yang, R. J. (1981): On finite difference methods for solving the one dimensional pollutant transport equation of a river, *Chinese Journal of Environmental Science*, vol. 2 no.:1, pp. 25–30 (in Chinese).

Lam, D. C. L. (1977): Comparison of finite element and finite difference methods for nearshore advection-diffusion transport models, In: Gray, W. G. Pinder, G. F. and Brebbia, C. A. (Eds.), *Finite Elements in Water Resources*, Pentech Press, London, pp. 1.115–1.129.

Lee, J. H. W., Wu, R. S. S., Cheung, Y. K. and Wong, P. P. S. (1991a): Dissolved oxygen variations in marine fish culture zone. *Journal of the Environmental Engineering*, vol. 117 no.:6, pp. 799–815.

Lee, J. H. W., Wu, R. S. S. and Cheung, Y. K. (1991b): Forecasting of dissolved oxygen in marine fish culture zone. *Journal of the Environmental Engineering*, vol. 117 no.:6, pp. 816–833.

Lee, J. H. W., Huang, Y., Dickman, M. D. and Jayawardena, A. W. (2003): Neural network modelling of coastal algal blooms, *Ecological Modelling*, vol. 159, pp. 179–201.

Liu, H. (1977): Predicting dispersion coefficient of stream, *Journal of Environmental Engineering Division*, vol. 103 no.:1, pp. 59–69.

Liu, H. and Cheng, H. D. A. (1980): Modified Fickian model for predicting dispersion, *Journal of Hydraulics Division*, vol. 106 no.:HY6, pp. 1021–1040.

Lui, G. C. S., Li, W. K., Leung, K. M. Y., Lee, J. H. W. and Jayawardena, A. W. (2007): Modelling algal blooms using vector autoregressive model with exogeneous variables and long memory filter, *Ecological modelling*, vol. 200 nos.:1–2, pp. 130–138.

Lui, P. H. and Jayawardena, A. W. (1983): Application of a time dependent dispersion model for dispersion prediction in natural streams, *Hydrological Sciences Journal*, vol. 28, no.:4, pp. 475–483.

Maier, H. R., Dandy, G. C., Burch, M. D. (1998): Use of artificial neural networks for modelling cyanobacteria *Anabaena* spp. in the River Murray, South Australia. *Ecological Modelling*, vol. 105, pp. 257–272.

McQuivey, R. S. and Keefer, T. N. (1976): Convective model of longitudinal dispersion, *Journal of Hydraulics Division*, vol. 102 no.:HY10, pp. 1409–1437.

Oden, J. T. and Carey, G. F. (1983): *Finite Elements – Mathematical Aspects*, vol. IV, Prentice Hall, Englewood Cliffs, NJ, 195 pp.

Ogata, A. and Banks, R. B. (1961): *A Solution of the Differential Equation of Longitudinal Dispersion in Porous Media*. US Geological Survey Professional Paper 411-A. US Government Printing Office, Washington DC.

Price, H. S., Cavendish, J. C. and Varga, R. S. (1968): Numerical methods of higher order accuracy for diffusion-convection equations, *Journal of the Society of Petroleum Engineers*, vol. S3, pp. 293–303.

Sabol, G. V. and Nordin, C. F. (1978): Dispersion in rivers as related to storage zones, *Journal of Hydraulics Division*, vol. 104 no.:HY5, pp. 693–708.

Seo, I. W. and Cheong, T. S. (1998): Predicting longitudinal dispersion coefficient in natural streams, *Journal of Hydraulic Engineering*, vol. 124 no.:1, pp. 25–32.

Smith, I. M., Farraday, R. V. and O'Connor, B. A. (1973): Rayleigh-Ritz and Galerkin finite elements for diffusion-convection problems, *Water Resources Research*, vol. 9 no.:3, pp. 593–600.

Sullivan, P. J. (1971): Longitudinal dispersion within a two-dimensional turbulent shear flow, *Journal of Fluid Mechanics*, vol. 49, pp. 551–576.

Taylor, G. I. (1922): Diffusion by continuous movements, *Proceedings of the London Mathematical Society A*, vol. 20, pp. 196–211.

Taylor, G. I. (1953): Dispersion of soluble matter in solvent flowing through a tube, *Proceedings of the Royal Society of London A*, vol. 219, pp. 186–203.

Taylor, G.I. (1954): The dispersion of matter in turbulent flow through a pipe, *Proceedings of the Royal Society of London*, vol. 223, pp. 446–468.

Thackston, E. L. and Schnelle, K. B. (1970): Predicting effects of dead zones on stream mixing, *Journal of Sanitary Engineering Division*, vol. 96 no.:SA2, pp. 319–331.

Valentine, E. M. and Wood, I. R. (1977): Longitudinal dispersion with dead zones, *Journal of Hydraulics Division*, vol. 103 no.:HY9, pp. 975–990.

Yabunaka, K., Hosomi, M. and Murakami, A. (1997): Novel application of a back-propagation artificial neural network model formulated to predict algal bloom. *Water Science and Technology*, vol. 36 no.:5, pp. 89–97.

Zienkiewicz, O. C. (1971, 1977): *Finite Elements Method in Engineering Science*, McGraw Hill, London, 570 pp.

Chapter 24

Major hydraulic structures in the world

24.1 INTRODUCTION

Major hydraulic structures can be broadly classified into canals, storm surge barriers and dams. Some major canals are used as modes of transportation whereas some are used for transferring water from places of abundance to places with shortage. Storm surge barriers are built to protect coastal regions from combinations of high tide, waves caused at times of extreme weather and fluvial flow from rivers. Dams are built to produce hydroelectricity, for flood protection and for irrigation and water supply. All major dams are multi-purpose since constructing a single purpose dam does not justify the huge capital expenditures involved.

24.2 CANALS

24.2.1 Grand Canal

The Grand Canal (also known as the Beijing-Hangzhou Grand Canal, Da Yunhe, Jing-Hang Yunhe) is the oldest and longest canal or artificial river in the world. It is also a UNESCO World Heritage Site since June 2014, as well as a marvel of hydraulic engineering. It starts from Beijing, links the Yellow River, Qiantang River, Huai River, Wei River, Hai River and the Yangtze River and passes through the city of Tianjin and the provinces of Hebei, Shandong, Jiangsu and Zhejiang, and ends up in the city of Hangzhou. Running in a north-south direction when most natural rivers in China run from west to east, this artificial river connects two great rivers in the world. The earliest parts of the canal dates back to the 5th century BC, but the various sections were combined during the Sui dynasty (581–618 AD). Millions of labourers and soldiers were employed in the construction of the canal.

The canal is 1,776 km long, 30–61 m wide, 0.6–4.6 m deep with its highest elevation of 42 m in the mountains of Shandong. The elevation of the canal bed varies from 1 m below sea level at Hangzhou to 27 m at Beijing. It is not a single continuous man-made canal but a series of canals linking several existing artificial or natural canals, lakes, rivers and ship locks. The main purpose of constructing the Grand Canal was to transport grains from south to north using barges. It also served as an internal communication route in China. It was considered as the economic lifeline for internal trade within China. Currently, the canal is navigable only from Hangzhou to Jining, a city located in the southwest of Shandong Province. The route has seven sections – Jiangnan Canal (Hangzhou to Zhenjiang), Inner Canal (Yangtze River to Huaian), Middle Canal (Huaian to Weishan Lake), Lu Canal, also called Shandong Canal (Weishan Lake to Linqing), South Canal (Linqing to Tianjin), North Canal and Tonghui River (Tianjin to Beijing). The Grand Canal is currently used as the

Eastern Route of the South-North Water Transfer Project. Some details of the Grand Canal are given in Table 24.1 and an approximate profile in Figure 24.1.

Because of the elevation difference along the route of the canal, locks have been built to facilitate navigation from one water level to another water level. The type of locks used, known as 'pound locks', were invented during the Sung dynasty in 984 AD to help raise or lower the water level in a canal. A pound lock has a chamber with gates at both ends that control the level of water in the pound. A 'pound' in this context is the level stretch of water between two locks (also known as a reach). Nowadays, 'pound' locks are used in many canal/river systems in the world, replacing the earlier type of gates with a single gate known as 'flash locks (also known as staunch locks)'. At each lock, the water level differs by about 1.2 m and to raise the water level by 42 m required 24 locks. The canal also has 60 bridges across.

Table 24.1 Details of Grand Canal

Section	Distance (km)	Intervals (km)	Ground level (m)	Water level (m)	Depth of canal (m)
Beijing	0		36	34	
Tonghui River		29			3
Tongzhou (Bai River)		124			3–8
Tianjin	153		8	7	
Yu River		386			3–10
Linqing	539		36	35	
Huitong River		113			3
Yellow River (North side)	652		38	35	
Yellow River (South side)	652		38	42	
Hui Tong River, Qing Ji Du and Ji zhou River		69			4–4.3
Nan wang Zhen summit	721		52	42	
Jizhou River		21			4–7
Jining	742		40	35	
Huan Gong Gou		143			3
Linjia Ba (South end of lakes today)	885		36	35	
Huan Gong Gou		250			3–10
Huaiyin	1,135		18	16	
Shanyang Yundao		186			4–8
Yangtze River (North side)	1,321		5	0	12–15
Yangtze River (South side)	1,361		5	0	12–15
Jiangnan River		47			4
Danyang	1,408		17	2	
Jiannan River		298			4
Hangzhou	1,706		4	0	

Source: Ronan, C.A., *The Shorter Science and Civilization in China, An Abridgement of Joseph Needham's Original Text*, Cambridge University Press, New York, 364 pp., 1995. With permission.

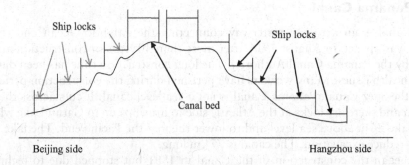

Figure 24.1 Approximate profile of Grand Canal (not to scale).

24.2.2 Suez Canal

Two of the well-known waterways that connect seas are the Suez Canal and the Panama Canal. The Suez Canal connects the Red Sea and the Mediterranean Sea, whereas the Panama Canal connects the Atlantic Ocean and the Pacific Ocean.

The Suez Canal, connecting Port Said in the north and Suez in the south, located in Egypt and 163 km (193.3 km including the southern and northern access canals) long and 300 m wide is a sea level canal because the water levels at the two ends (Gulf of Suez in Red Sea and the Mediterranean Sea) are approximately the same. It can accommodate ships up to a draft of about 19 m. The construction of Suez Canal began officially on April 25, 1859, and opened 10 years later on November 7, 1869 at a cost of some $100 million. Initially, construction was carried out using Egyptian manual labour (some forced) but towards the end, many mechanical devices were used for digging and shaping the canal. It is one of the world's busiest shipping lanes. A large percentage of Europe's energy needs are transported from Middle East oil fields via the Suez Canal. It shortens the sea voyage by about 7,000 km. Until recently, most of the Suez Canal was not wide enough for two ships to pass side by side. Passing bays at the Ballah Bypass and the Great Bitter Lake help to alleviate this problem. Typically, it would take 12–16 h to pass through the canal. The low speed helps to prevent bank erosion by waves.

The Canal has been closed for shipping twice due to wars. On July 26, 1956, Gamal Abdel Nassar, the President of Egypt at that time nationalized the Suez Canal as a result of the United States and the United Kingdom withdrawing their support for the construction of Aswan dam in Egypt. On October 29, 1956, Israel invaded Egypt and Britain and France followed on grounds that the passage through the canal was to be free. Egypt retaliated by sinking 40 ships leading to what was known as the Suez Crisis. The Suez Crisis ended in November 1956 as a result of UN intervention by arranging a truce among the United States, the United Kingdom, Egypt and Israel. Again, in 1967, the Canal was closed for commercial shipping during the Six-Day War between Israel and its Arab neighbours. The Suez Canal is currently owned and operated by the Suez Canal Authority of Egypt. It has also been closed several more times due to conflicts between Egypt and Israel.

Suez Canal is not only of strategic importance and national pride for Egypt but also a substantial source of revenue. In recent years with about 50 ships passing through the canal every day, the annual toll charges run into several billions of dollars. In August, 2015, a new 35 km expansion running parallel to the main channel was opened, enabling two-way transit through the canal. It is now called the 'New Suez Canal'. The new depth and width of the canal including the parallel sections range from 23–24 m to 205–225 m, respectively. A railway also runs parallel to the canal on the west bank.

24.2.3 Panama Canal

Panama Canal is an artificial waterway connecting the Atlantic Ocean and the Pacific Ocean. It was opened in August 1914 after more than 30 years of construction and is now managed by the Panama Canal Authority. The long construction time has been due to technical and health issues. Many workers have perished during the construction period.

Unlike the Suez Canal, Panama Canal is not a sea level canal. It cuts across the Isthmus of Panama and there are locks at the Atlantic side to lift ships up to Gatun Lake which is an artificial lake 26 m above sea level and to lower them at the Pacific end. The lake has been created to reduce excavation. The canal is 77 km long.

France began the construction of the Canal in 1881 but stopped due to technical and worker health problems. The mortality rate was very high due to tropical diseases such as malaria and yellow fever. The United States took over the project in 1904 and completed and opened the Canal on August 5, 1914. The total casualty cost has been over 25,000. The Canal reduced the sea voyage by 12,875 km and takes about eleven and half hours to pass through. Following a treaty between the United States and Panama, the Panama Canal Authority, an institution answerable to the Panamanian Government, took full control of the Canal on December 31, 1999. With over 14,000 vessels passing through annually, the Panama Canal is one of the main sources of revenue for Panama. The toll charge for the passage through Panama Canal is of the order of $450,000 for large ships. The lowest toll paid has been only 36 cents by an adventurer who swam along the Canal from one end to the other in 1928.

The profile of the Panama Canal (Figure 24.2) consists of an approach channel in Limon Bay on the Atlantic side which connects to the Gatun locks at a distance of 11 km. At Gatun, ships are lifted by a series of three locks by a height of 26 m to Gatun Lake. The lake is formed by Gatun dam across Chagres River. The channel through the lake varies in depth from about 14–26 m and extends for about 37 km to Gamboa from where the Gaillard Cut (also known as Culebra Cut or Corte de Culebra in Spanish) begins. The channel through the cut extends to a length of about 13 km with an average depth of 13 m up to Pedro Miguel Locks. The locks lower vessels by 9 m to Miraflores Lake, at an elevation of 16 m above sea level. After passing through the channel for about 2 km, vessels are lowered to sea level by a system of two-stepped locks at Miraflores. The final segment consists of a dredged approach channel about 11 km long to join the Pacific Ocean. The original lock chambers were 320 m long, 33.5 m wide and 12.56 m deep constraining the size of the ships that could pass through. The new lock chambers, opened for commercial traffic on June 26, 2016, are 426.72 m long, 54.86 m wide and 18.29 m deep enabling larger ships to pass.

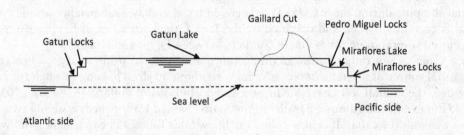

Figure 24.2 Sketch of the profile of Panama Canal (not to scale).

24.2.4 South–North Water Transfer Project

The South-to-North Water Diversion Project, aimed at transferring $44.8 \times 10^9 \, \text{m}^3$ of water annually from the Yangtze River Basin in southern China to the Yellow River Basin in arid northern China via three canal systems, is the world's largest and most expensive water transfer project in the world. The three canal systems are known as the Eastern Route, the Central Route and the Western Route.

24.2.4.1 Eastern Route

The Eastern Route generally follows the course of the Grand Canal. The quantity of water transferred annually will increase from 8.9×10^9 to 10.6×10^9 to $14.8 \times 10^9 \, \text{m}^3$ as the project progresses. These quantities are only fractions of the annual flow in Yangtze River. The completed Eastern Route is 1,152 km long and is equipped with 23 pumping stations with a power capacity of 454 megawatts. Pumping stations are needed because of the topography of the region. Eastern route will transfer $1 \times 10^9 \, \text{m}^3$ annually to Tianjin and is not expected to supply Beijing which is to be transferred by the Central Route. An important component of the Eastern Route is a tunnel under Yellow River in Shandong Province consisting of two 9.3 m diameter horizontal tunnels 70 m below the bed of Yellow River.

24.2.4.2 Central Route

The central, or middle route, runs a distance of 1,264 km starting from Danjiangkou Reservoir on the Han River, a tributary of the Yangtze River, to Beijing and Tianjin. It was completed in December 2014 initially providing $9.5 \times 10^9 \, \text{m}^3$ of water annually and is expected to increase to about $12–13 \times 10^9 \, \text{m}^3$ by the year 2030. In dry years the annual water transfer will be at least $6.2 \times 10^9 \, \text{m}^3$, with a 95% guarantee.

To let the water flow by gravity, this route required raising the crest elevation of Danjiangkou dam from 162 to 176.6 m above mean sea level thereby raising the reservoir water level from 157 to 170 m above mean sea level. A major challenge in this route was the construction of two tunnels under the Yellow River.

Construction on the central route began in 2004. By 2008, the 307 km-long northern stretch of the central route has been completed. Water in that stretch of the canal came from reservoirs in Hebei Province and not from Danjiangkou Reservoir. As a result, farmers and industries in Hebei Province had to cut back on their consumption to allow transfer of water to Beijing. Danjiangkou Reservoir is the source of two-thirds of Beijing's tap water and a third of its total supply.

24.2.4.3 Western Route

The western route was intended to transfer water from three tributaries of Yangtze River near the Bayankala Mountains to Qinghai, Gansu, Shaanxi, Shanxi, Inner Mongolia and Ningxia provinces. There are also plans to divert about $200 \times 10^9 \, \text{m}^3$ of water annually from the upstream sections of Mekong (Lancang River), Yarlung Zangbo (called Brahmaputra further downstream) and Salween (Nu River) to Yangtze River and Yellow River and to the dry areas of northern China. If implemented, this route is likely to affect downstream water users in India, Bangladesh, Myanmar, Lao, Cambodia, Thailand and Vietnam.

The western route aims to divert water from the headwaters of Yangtze River to the head-waters of Yellow River. This requires dams and tunnels to cross the Tibetan Plateau and Western Yunnan Plateau. Although designed to transfer $3.8 \times 10^9 \, \text{m}^3$ of water annually from the three tributaries (Tongtian, Yalong and Dadu Rivers) of Yangtze to northwest China, the feasibility of this route is still being studied and a firm decision to go ahead is yet to be made.

24.3 STORM SURGE BARRIERS

Storm surge barriers are structures constructed at river estuaries and tidal inlets to prevent high sea water due to high tide and wind waves under extreme weather conditions. They can be closed dams, dams with openings that can be opened and closed when the need arises or gated structures. Countries that have built storm surge barriers include the United Kingdom (Thames Barrier), the Netherlands (Eastern Scheldt Barrier, Maeslant Barrier, Hollandse IJssel Barrier, Hartel Barrier, Ramspol Barrier and Haringvliet Barrier), Italy (Venice Mose Barrier), Russia (St. Petersburg Flood Protection Barrier), Germany (Eider Barrier) and the United States (New Orleans Barrier). Of these, the Netherlands is vulnerable to more storm surges than any other country due to its topography and location. It is a geographically low-lying country, with about 20% of its area located below sea level and about 21% of its population living below sea level and with about 50% of its land lying less than one meter above sea level. The Netherlands with two-thirds of its land vulnerable to flooding utilizes a combination of dykes, embankments and storm surge barriers along the water front and river estuaries to prevent flooding from storm surges from the sea. Figure 24.3 shows the

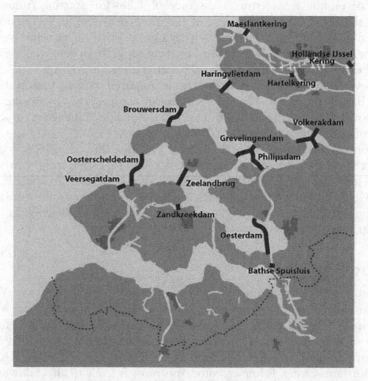

Figure 24.3 Storm surge barriers in the Netherlands. (*Source*: Reproduced with permission from http://www.deltawerken.com/modules/mediagallery/popup.php?id=1485&style_root=/home/deltawerken.com/public_html/styles/blauw&style_root_http=http://www.deltawerken.com/styles/blauw&language=en.)

locations of the storm surge barriers in the Netherlands. Most of the related construction works have been carried out by Delta Works (*Deltawerken* in Dutch) under the direction of the Directorate-General for Public Works and Water Management (*Rijkswaterstaat* in Dutch).

In determining the safe inner water level to operate the storm water barriers, it is also important to take into account not only the storm surge but also the tidal effect as well as the river runoff. Under normal conditions, the water level in the estuary is not regulated to allow navigation and tidal exchange for ecosystem functioning.

24.3.1 Oosterschelde storm surge barrier

The Oosterschelde storm surge barrier, also known as Eastern Scheldt Barrier, constructed over a period of a decade at a cost some 2.5 billion Euros, was officially opened by Queen Beatrix of the Netherlands on October 4, 1986. It is designed to withstand a storm of 4,000 year return period and the barrier is expected to last for 200 years. It is about 8 km long and is a special dam that connects the Zeeland islands of Schouwen-Duiveland and Noord-Bevelandk, and is the largest of 13 ambitious Delta Works series of dams and storm surge barriers. This part of the Delta Works was built to protect the Zeeland region from the sea after the North Sea Flood in 1953.

It is a moveable barrier that could be closed in emergencies. The moveable part is nearly 3 km long divided over three sluice gate-type openings. Colossal piles support doors that move up and down to open and close the barrier. It has 65 colossal pillars, separated by 62 sluice gates that are roughly 42 m wide and 6–12 m high and each weighing between 260 and 480 tons. Each pillar is 35–38.75 m high and weighs 18,000 tons. They were constructed in a dry dock, lifted after flooding the area using a fleet of special ships, and placed in their final positions. The pillars are hollow and filled with sand after positioning them along the length of the dam.

The gates are normally kept open and closed once a year on average in adverse weather conditions. The barrier is designed to protect Netherlands from North Sea flooding. The normal open condition creates the minimum disturbance to tidal movements and the local marine ecosystem. According to Dutch law, the gates can be completely closed only when the water level is at least three meters above normal sea level.

24.3.2 Maeslantkering storm surge barrier

The Maeslantkering storm surge barrier which is a moveable storm surge barrier is the world's second largest movable structure, built across the New Waterway (Nieuwe Waterweg, a canal that connects to River Rhine forming the artificial mouth of the River Rhine) that connects Rotterdam with North Sea. The barrier has water-filled doors which act like floating pontoons allowing them to sink. When the situation returns to normal the water inside the pontoons is pumped out and the pontoons are stored in a dry location. The main objective of the Maeslantkering storm surge barrier is to protect Rotterdam harbour.

The construction of the barrier started in 1991 and on May 10, 1997, Queen Beatrix officially opened the Maeslantkering storm surge barrier. The barrier is designed to withstand a storm with a return period of 10,000 years.

The barrier consists of two 22 m high and 210 m long steel gates welded to 237 m long steel trusses. The arms which weigh 6,800 tons each transmit the forces exerted on the gates while closed, to a single joint at the rear of each gate. During opening and closing of the gates, this ball and socket type joint allows free movement of the gates under the influence of water, wind and waves. This joint which has a diameter of 10 m and weighing 680 tons is the largest of its kind in the world.

Other storm surge barriers in the Netherland include IJessel, completed in 1958 with two vertical lift gates each 80 m wide and one sluice, Haringvlset sluices, completed in 1971 with 7×2 gates each 60 m wide and one sluice, Hartel, completed in 1997 with two gates 98 and 49 m wide and one sluice, Ramspol, completed in 2002 with three gates each 80 m wide and an inflatable dam.

24.3.3 Thames barrier

The Thames Barrier has a length of 520 m across River Thames and is located near Woolwich in Southeast London. It has been built to protect central London from tidal and storm surge flooding from North Sea and river flooding in some areas of west London. The barrier divides the river into four 61 and two 30 m spans. The 30 m spans are to allow for navigation. There are also four non-navigable spans between nine concrete piers and the two abutments. The barrier has ten steel gates, each weighing 3,300 tons that can be raised and lowered into position across the river. The flood gates are hollow and of circular cross section and filled with water when submerged and emptied when raised above the water. When raised, the gates are as tall as a five-storey building. The site at New Charlton for the barrier has been chosen because of the straight longitudinal section of the river there and the underlying chalk was strong enough to support the barrier. Construction work started in 1974 and the Barrier was officially opened by Queen Elizabeth II on May 8, 1984. As of October 2019, the Thames Barrier has been closed 186 times of which 99 were to protect from tidal flooding and 87 to protect from combined tidal/river flooding.

24.3.4 Venice Mose barrier

Mose is an acronym for 'Modulo Sperimentale Elettromeccanico' (in Italian), or 'Experimental Electromechanical Module' (in English), and refers to the biblical figure Moses who parted the Red Sea to enable the Israelites to flee to safety from Egypt. Following a flood resulting from a 1.94 m high tide on November 4, 1966 that submerged Venice, Chioggia and other built-up areas in the lagoon, Venice declared a Special Law to safeguard the city by building an effective sea defence system as a national priority. That was the beginning of the idea of constructing the MOSE Barrier.

The project was designed in 1984 and construction began in 2003 which should have been completed by 2016 but the project has been slowed down by corruption and cost over-runs. In November 2019, the city suffered its worst flooding since 1966. It is expected that the Barrier will be operational by the end of 2021.

Venice is connected to the sea at three entrances called *bocche di porto* (meaning port mouth). They are Lido, Malamocco and Chioggia and the 78 mobile gates of the barrier are located at these three inlets. When the tide from the Adriatic Sea reaches 1.1 m the gates will be raised isolating Venice from the sea during high tides and lowered when the tide subsides. The barrier is 1.7 km long.

When the gates are resting in their lowered position under normal tidal conditions, they are full of water. During high tide, the water inside the gates is expelled using compressed air causing them to rise up until they emerge above the water thereby preventing the tide entering the lagoon. When the tide subsides, the gates are again filled with water to allow them to sink to their resting positions.

24.3.5 Inner Harbor Navigation Canal Lake Borgne surge barrier

One of the deadliest natural disasters to hit the United States in recent history was the Category 5 Hurricane Katrina during August 23–31, 2005. It devastated New Orleans, Louisiana, and Gulf Coast of Mississippi causing an estimated 1833 fatalities resulting from a 5.5 m storm surge that overtopped floodwalls and levees along the Mississippi River. Responding to this unprecedented disaster, the US Army Corps of Engineers (USACE) undertook to construct the largest design and built civil engineering project in its history. The outcome of the project is the Inner Harbor Navigation Canal Lake Borgne Surge Barrier.

The barrier, located about 20 km east of downtown New Orleans, is constructed near the confluence of and across the Gulf Intracoastal Waterway (GIWW) and the Mississippi River Gulf Outlet (MRGO) near New Orleans. Construction aimed at protecting New Orleans and the vicinity from storm surge coming from the Gulf of Mexico and Lake Borgne for a designed 100 year return period began on May 9, 2009 and completed in June 2013. The barrier consists of three gates that allow vessels to pass through the 3 km long, 8.5 m high concrete barrier wall and a floodwall closure of the Mississippi River Gulf Outlet. It was first utilized on August 29, 2012 for the first time on the seventh anniversary of Hurricane Katrina to protect New Orleans from Hurricane Isaac.

24.3.6 Saint Petersburg Stormwater Barrier

Saint Petersburg Stormwater Barrier is a series of dams 25.4 km long extending from Lomonosov northward to Kotlin Island and turning eastward towards Cape Lisly Nos near Sestroretsk. It is intended to protect Saint Petersburg which is a low-lying city prone to regular flooding from storm surges, by isolating Neva Bay from the Gulf of Finland. The dam is designed to protect the city from a storm surge up to 5 m above mean sea level. It stands 8 m above sea level and spans the Gulf of Finland with Kotlin Island at the centre. It also allows two large openings to allow the passage of ships. The openings can be closed during times of storm surges.

Construction of the barrier started in 1980 but was temporarily suspended in 1987 due to public concerns about negative environmental impacts. After completion of an EIA during 2001–2002, work slowly resumed and the barrier was completed and commissioned in August 2011. Construction also stalled due to Russian political and economic upheavals of the 1990s.

The barrier, which is 25.4 km long and located across the estuary of the Neva River, consists of 11 rock and earthfill embankment type dams separated by sluices or channel openings. Other components of the barrier include six 24 m wide gated sluice complexes, a 200 m wide main navigation channel to allow for ships of 100,000 DWT capacity which can be closed with two sector gates, a secondary navigation channel 110 m wide to allow for smaller ships which can be closed with a vertical gate, navigation channels for shipping to approach the two navigation openings, a road constructed on the embankment dams, a road tunnel under the main navigation channel and a viaduct with a lift bridge above the secondary navigation channel. The gates in the main navigation opening are each 110 m wide and 22 m high. They are held by arms 132 m radius and pivoted at a ball and socket type joint that allows free movement. Each gate weighs 2650 tons. On the secondary navigation opening, the gates are 110 m long, 11.8 m deep and each weighs 2,600 tons. The layout of the barrier can be seen at https://www.themoscowtimes.com/2019/02/21/st-petersburgs-dam-is-holding-back-floods-for-now-a64066.

A unique problem in the construction and operation of the barrier is the extreme winter where the sea ice can be more than 50 cm thick, and on average, temperatures fall below zero for 175 days in a year. In Saint Petersburg, the temperature has plunged to −39°C in December 1987.

24.3.7 Eider Barrage

The Eider Barrage (*Eidersperrwerk*), Germany's largest coastal protection structure, is located at the mouth of the river Eider near Tönning on Germany's North Sea coast. The barrage is 4.9 km long, lies 8.5 m above sea level and 7 m above the average high tide. It comprises of two separate rows of five gates, each 40 m long that allow the water of the Eider River to flow into North Sea during ebb tide and vice versa during high tide. The protection is equivalent to that of having a double dyke. A 75 m long 14 m wide lift lock nearby allows passage of ships into North Sea from the adjacent harbour. Construction work started in 1967 and was celebrated as the structure of the century when it was opened on March 20, 1973.

24.4 DAMS

There are an estimated 845,000 dams in the world. China has about 23,000 large dams followed by the United States having about 9,200 large dams. These two countries are followed by India, Japan and Brazil in that order. In this context, a large dam is a dam with a height between 5 and 15 m and impounding more than $3 \times 10^6 \text{m}^3$ as adopted by the International Commission on Large Dams (ICOLD). They are built to control the flow of water in rivers and other water courses. A dam is an essential component of almost all types of water resources projects. The structure of a dam can be of concrete gravity type, concrete arch type, buttress type, rockfill embankment type and earthfill embankment type. Purposes of a dam include generation of hydroelectricity, storage of water for irrigation, water supply and industry needs etc., navigation, flood control and recreation. The size of a dam can be expressed in terms of a number of parameters including the hydroelectricity generating capacity, the height of the dam, the capacity and surface area of the impounding reservoir, the length of the dam, the spillway capacity, the cost of building the dam, and, the number of people relocated as a result of building the dam. A dam can be for a single purpose or for multiple purposes. All large dams are multi-purpose type. In terms of hydroelectricity generating capacity, the Three Gorges Dam, a concrete gravity dam across Yangtze River in China takes the top place in the world with an installed generating capacity of 22,500 MW. In terms of the height, the Jinping dam, a concrete arch dam across Yalong River, also in China, with a height of 305 m takes the top place while the Kariba dam, an arch dam in Zimbabwe, with a storage capacity of $180.5 \times 10^9 \text{m}^3$ takes the top place in terms of storage capacity. Hoover dam in the United States was the highest and had the largest hydropower generating capacity in the world at the time it was built in 1936. Bhakra dam in India was considered as the 'temple of modern India' when it was built. Some technical details of the ten largest dams in the world based on the hydroelectricity generating capacity, the height and the storage capacity are given in Tables 24.2–24.4. The information for compiling these Tables is taken from websites in public domains. As can be seen in these tables, there are more large dams in China than in any other country.

   In recent years, the rate of construction of new large dams has gone down mainly as a result of opposition from environmentalists. Construction of large dams also leads to mass relocation of people that lead to many social and economic problems. The Three Gorges Dam in China has displaced over 1.2 million people. Dams are also not without controversies, specially, when built across trans-boundary rivers. Construction of large dams which

Table 24.2 Ten largest dams in the world in terms of hydropower generating capacity

Name of dam	Country	River	Type	Power (MW)	Height (m)	Length (m)	Capacity ($10^9 m^3$)	Construction period
Three Gorges	China	Yangtze	Concrete gravity	22,500	185	2,335	42	1994–2009
Itaipu (four dams)	Brazil	Parana	Earthfill, rockfill, concrete gravity, concrete wing	14,000	225	7,200	29	1960–1984
Xiluodu	China	Jinsha	Double curvature arch	13,800	285.5	700	6.5	2005–2013
Guri	Venezuela	Caroni	Arch+embankment	10,300	162	11,409	138	1963–1978
Tucurui	Brazil	Tocantins	Concrete gravity	8,370	78	12,500	45.5	1975–2012
Grand Coulee	USA	Columbia	Concrete gravity	6,809	168	1,592	11.6	1933–1942
Xinagiiaba	China	Jinsha	Concrete gravity	6,448	161	909	5.1	2006–2012
Longtan	China	Hongshui	Concrete gravity	6,426	216	894	27.3	2001–2009
Krasnoyarsk	Russia	Yenisey	Concrete gravity	6,000	124	1,065	73.3	1956–1972
Robert-Bourassa	Canada	La Grande	Embankment	5,616	162	2,835	61.7	1974–1981
Hoover	USA	Colorado	Arch	2,080	221	379	35	1931–1936

Table 24.3 Ten largest dams in the world in terms of the height

Name	Country	River	Type	Power (MW)	Height (m)	Length (m)	Capacity ($10^9 m^3$)	Construction period
Jinping	China	Yalong	Concrete arch	3,600	305	569	7.76	2005–2013
Nurek	Tajikistan	Vakhsh	Embankment	3,015	304	700	10.5	1961–1980
Xiaowan	China	Lancang	Concrete arch	4,200	292	902	15.0	2002–2010
Xiluodu	China	Jinsha	Concrete arch	13,860	285.5	700	12.7	2005–2013
Grande Dixence	Switzerland	Dixence	Concrete gravity	2,069	285	700	0.4	1950–1961
Enguri	Georgia	Enguri	Concrete arch	1,300	271.5	680	1.1	1961–1987
Yusufeli	Turkey	Coruh	Arch double curvature	540	270	490	2.13	2013–2021(est)
Vajont	Italy	Vajont	Concrete arch	–	262	169	–	Disused
Nouzhadu	China	Lancang	Embankment	5,850	261.5	608	21.8	2004–2012
Manuel Moreno Torres (Chicoasen)	Mexico	Grijalva	Embankment earthfill	2,430	261	485	1.6	1974–1980
Tehri	India	Bhagirathi	Embankment earthfill	1,000	260.5	575	4.0	1978–2006

Table 24.4 Ten largest dams in the world in terms of storage capacity

Name of dam	Country	River	Type	Power (MW)	Height (m)	Length (m)	Capacity ($10^9 m^3$)	Construction period
Kariba	Zimbabwe	Kariba	Arch	1,470	128	617	180.5	1955–1959
Bratsk	Siberia	Angara	Concrete gravity	4,500	125	1,452	169	1954–1964
Akosombo	Ghana	Volta	Rockfill embankment	768	134	700	144	1961–1966
Daniel Johnson	Canada	Peace	Multiple arch buttress	2,660	214	1,310	140	1959–1968
Guri	Venezuela	Caroni	Concrete gravity and embankment	10,300	162	11,409	138	1963–1986
Aswan High	Egypt	Nile	Rockfill	2,100	111	3,830	132	1960–1968
W.A.C. Bennet	Canada	Peace	Earthfill embankment	2,790	183	2,068	74	1961–1967
Krasnoyarsk	Russia	Yenisey	Concrete gravity	6,000	124	1,065	73	1956–1972
Zeya	Russia	Zeya	Concrete gravity	1,290	112	714	68	1964–1975
Robert Bourassa	Canada	La Grande	Embankment	5,616	162	2,835	62	1974–1981

cost large sums of money is also prone to abuse of power by authorities and corruption. Very recently (December 31, 2019), it has been reported that the Ethiopian Federal Attorney General charged 50 people with corruption and abuse of office in relation with the construction of the Grand Ethiopian Renaissance Dam, that is being built across the Blue Nile River (https://www.occrp.org/en/daily/11369-ethiopia-50-charged-with-graft-in-nile-dam-project). The Grand Ethiopian Renaissance Dam which is expected to be completed in 2022 will be Africa's largest hydroelectricity generating dam.

REFERENCE

Ronan, C. A. (1995): *The Shorter Science and Civilization in China, An Abridgement of Joseph Needham's Original Text*, Cambridge University Press, New York, 364 pp.

Chapter 25

Hydrological cycle and its principal processes

25.1 INTRODUCTION

Hydrology can be defined as the science that treats the waters of the earth, their occurrence, circulation and distribution, their physical and chemical properties and their reaction with the environment, including their relation to living things. This is the definition adopted by the US Council for Science and Technology.

Hydrology is a multi-disciplinary subject and its development and applications have come from Civil Engineering, Soil Physics, Agriculture, Forestry, Geomorphology, Geography, etc. That part of Hydrology relevant to the Civil Engineer is called Applied Hydrology or Engineering Hydrology, whereas Scientific Hydrology refers to the study of the fundamental physical, chemical and biological processes of the Hydrological cycle. In engineering hydrology, there are three basic problems:

- Measurement, recording and publication of basic data
- Analysis of these data to develop and expand fundamental theories
- Application of these theories and data to practical problems.

In scientific hydrology, it is studied under the following areas:

- **Hydrometeorology**: study of problems intermediate between the fields of hydrology and meteorology, e.g., precipitation, evaporation, etc.
- **Limnology**: study of lakes
- **Cryology**: study of snow and ice
- **Potamology**: study of surface water
- **Geo-hydrology (groundwater hydrology)**: study of groundwater
- **Oceanology**: study of the oceans.

Three main international organizations that have contributed to and promoted the understanding and development of hydrology are:

- International Association of Hydrological Sciences (IAHS) (since 1922),
- International Hydrological Decade (IHD) (1965–1974), and,
- International Hydrological Programme (IHP) (1975 onwards).

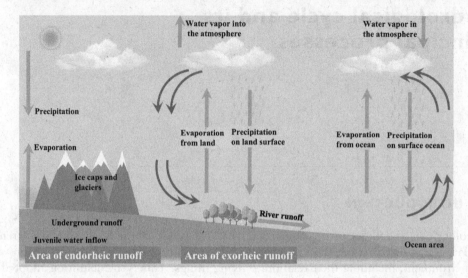

Figure 25.1 Hydrological cycle.

25.2 HYDROLOGICAL CYCLE

Earth's water resources are in a state of continuous circulation linking the atmosphere, land and the oceans. This endless process which is powered by the forces of nature is called the hydrological cycle. It is the basic framework upon which the science of hydrology is built. There are many processes that constitute the hydrological cycle, as illustrated in Figure 25.1.

25.2.1 History of the hydrological cycle

The concept of the hydrological cycle has been understood in crude form as early as 3000 BC. Egyptians learned to harness the waters of the NILE, measure the rise and fall of the river circa 3000 BC. Descriptive hydrology was conceived by Plato and Aristotle circa 400 BC. Romans constructed aqueducts without a complete knowledge of quantitative hydrology.

A reasonably clear idea of the hydrological cycle was put forward by Vitruvius circa 27–17 BC, but the concept of the hydrological cycle in its present form is only after Leonardo da Vinci (1452–1519). Since then, there have not been much significant developments in the understanding of the hydrological cycle until about 1674. Perrault in 1674 and Mariotte in 1686 independently carried out experiments on runoff and concluded that the rainfall is sufficient to produce streamflow. Other significant developments in chronological order are given in Table 25.1.

Initiatives under UNESCO IHP include

- FRIEND-Water (Flow Regimes from International Experimental and Network Data).
- GRAPHIC (Groundwater Resources Assessment under the Pressures of Humanity and Climate Change).
- G-WADI (Global Network on Water and Development Information in Arid Lands).
- HELP (Hydrology for the Environment, Life and Policy).

Table 25.1 Significant developments in hydrology

3500–3000 BC	Nileometers	Egypt
450–350 BC	Early philosophy on hydrological cycle (Plato & Aristotle)	Greece
27–17 BC	Reasonably clear philosophy on hydrological cycle (Vitruvius)	Italy
1452–1519 AD	Hydrological cycle in its present form (Da Vinci)	Italy
1674	Related rainfall to runoff (Perrault)	France
1686	Related rainfall to runoff (Mariotte)	France
1769	Chezy's equation	France
1856	Darcy's law	France
1885	Manning's equation	Ireland
1922	IASH established	
1932	Unit hydrograph theory (Sherman)	USA
1933	Infiltration theory (Horton)	USA
1941	Extreme value theory in hydrology (Gumbel)	USA
1948	Electronic analogue in flood calculations (Linsley)	USA
1950	Black box model of rainfall runoff (Sugawara)	Japan
1954	Infiltration theory (Philip)	Australia
1955	Kinematic wave theory (Lighthill and Whitham)	UK
1959	Stanford watershed model	USA
1962	Harvard water program	USA
1962	UK Institute of Hydrology established	UK
1965	International Hydrological Decade (IHD) initiated	UNESCO
1975	International Hydrological Programme (IHP) initiated	UNESCO

- IDI (International Drought Initiative).
- IFI (International Flood Initiative). Partners include the World Meteorological Organization (WMO), the United Nations University (UNU), the International Association of Hydrological Sciences (IAHS) and the International Strategy for Disaster Reduction (ISDR).
- IIWQ (International Initiative on Water Quality).
- ISARM (Internationally Shared Aquifer Resources Management).
- ISI (International Sediment Initiative).
- IWRM (Integrated Water Resources Management).
- JIIHP (Joint International Isotope Hydrology Programme).
- PCCP (From Potential Conflict to Cooperation Potential).
- UWMP (Urban Water Management Programme).
- WHYMAP (World Hydrogeological Map).

25.3 PRINCIPAL PROCESSES IN THE HYDROLOGICAL CYCLE

- Precipitation
- Evaporation and evapo-transpiration
- Interception
- Depression storage
- Infiltration
- Runoff

- Overland flow
- Interflow
- Baseflow
- Sub-surface flow
  - Flow through saturated porous media (groundwater flow)
  - Flow through partially saturated porous media (moisture flow)

These will be discussed in the chapters that follow.

Chapter 26

Hydro-meteorology

26.1 INTRODUCTION

Hydrometeorology lies between meteorology and hydrology and can be considered as the upstream component of hydrology. Meteorology is a much broader field whereas hydro-meteorology refers to the meteorological processes associated with water in solid, liquid and gaseous phases. Energy from the sun, wind, lightning, tidal and latent heat keeps the water in motion while changing phases.

In this chapter, a brief description of the atmosphere, its composition, the forces acting, the basic governing equations of motion as well as the descriptions of various weather systems, including extreme weather will be presented.

26.2 ATMOSPHERE

26.2.1 Definitions and units of measurements

26.2.1.1 Pressure

Pressure is the force exerted by the atmosphere. Units used in meteorology are millibars (1 mb = 100 N/m²). The conversion factors are as follows:

1000 mb = 1 bar = 14.7 psi = 76.0 cm of Hg = 29.53 in of Hg (Also, 10 mb = 1 kPa; 1 Atmosphere = 1013.25 mb (pressure at sea level).

In weather charts, the pressures indicated are those reduced to sea level by adding the column of air between the sea level and the point under consideration. Some typical pressure and other physical properties of the atmosphere at various altitudes are given in Table 26.1.

Table 26.1 Atmospheric properties at different altitudes

Height (m)	Pressure (mb)	Temp (°K)	Density (kg/m³)
0	1013.25	288.2	1.23
5,000	540.5	255.7	0.736
10,000	265.0	223.2	0.414
15,000	121.1	216.7	0.195
20,000	55.3	216.7	0.0889

26.2.1.2 Wind

Wind is caused by the movement of air. Wind speeds are plotted in weather maps usually in knots (1 knot = 1.15078 mph = 1.852 km/h = 0.5144 m/s). The effect of wind is categorized according to the Beaufort scale (1806), which in simplified form is given in Table 26.2.

26.2.1.3 Thermal

Thermal is a rising body of air which may carry various objects upwards at rates ranging from less than 25 mm/s to more than 30 m/s.

26.2.1.4 Air parcel or air mass

Air parcel refers to a small volume of air, which has uniform temperature, pressure, humidity, density, etc. It may expand, contract as it moves, but the matter contained within it remains constant (Figure 26.1). It is similar to the 'control volume' concept.

26.2.1.5 Convection

Convection refers to the transfer of heat (or any other property) by means of vertical motion.

26.2.1.6 Advection

Advection refers to the transfer of heat (or any other property) by means of horizontal motion.

Table 26.2 The Beaufort scale for wind effects

Force	Specifications for use on land	Equivalent mean wind speed 10m above ground (knots)	(m/s)
0	Calm, smoke rises vertically	0	0
1	Light air; wind direction shown by smoke drift, not by vanes	2	1.03
2	Light breeze; wind felt on face leaves rustle; vanes move	5	2.57
3	Gentle breeze; leaves and small twigs moving; light flags lift	9	4.63
4	Moderate breeze; dust and loose paper lift; small branches move	13	6.68
5	Fresh breeze; small leafy trees sway; crested wavelets on lakes	19	9.77
6	Strong breeze; large branches sway; telegraph wires whistle; umbrellas difficult to use	24	12.3
7	Near gale; whole trees move; inconvenient to walk against	30	15.4
8	Gale; small twigs break off; impedes all walking	37	19.0
9	Strong gale; slight structural damage	44	22.6
10	Storm; seldom experienced on land; considerable structural damage; trees uprooted	52	26.7
11	Violent storm; rarely experienced; widespread damage	60	30.8
12	Hurricane; at sea visibility is badly affected by driving foam and spray; sea surface completely white	>64	>32.9

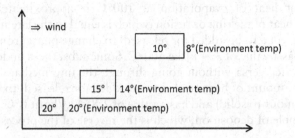

Figure 26.1 An ascending air parcel.

26.2.1.7 Cumulus

Cumulus refers to convective clouds over mountainous areas, especially when moist air covers the region. They are detached clouds and resemble a cauliflower in shape.

26.2.1.8 Lapse rate

Lapse rate refers to the temperature gradient with respect to altitude. There are three types of lapse rates in meteorology. The dry adiabatic lapse rate refers to the temperature gradient when there is no heat added or taken away from the atmospheric process. In this case, the change in temperature is caused by the change in pressure (expansion and contraction) according to

$$pv^{\gamma} = \text{constant; and, } pv = RT; \text{ Therefore, } \frac{T_1}{T_2} = \left[\frac{p_1}{p_2}\right]^{(\gamma-1)/\gamma} \text{ or, } Tv^{\gamma-1} = \text{constant} \quad (26.1)$$

A parcel of air flowing over a mountain can be adiabatic. The approximate value of the dry adiabatic lapse rate is 9.8°C/km. The environmental (ambient) lapse rate is the actual temperature gradient that exists in the environment. It can take a wide range of values. When the moist unsaturated parcel of air rises, it will at some altitude reach saturation and condensation will begin to take place. This adds the latent heat of condensation to the thermodynamic process resulting in a decrease of the lapse rate to a value of approximately 6°C/km. The lowered lapse rate is referred to as the saturated or moist lapse rate.

26.2.1.9 Latent heat

When phases change, heat must be added or taken away without any change of temperature. The processes of changing phases from solid to liquid and liquid to gas are endothermic or energy absorbing. The reverse processes of changing phases from gas to liquid and liquid to solid are exothermic or energy releasing. Latent heat of vapourization (or condensation) is the amount of heat needed to be added (or released) to change phase from liquid to vapour (or vapour to liquid) and has a value of 2,500.78 kJ/kg at 0°C. The rate of change of latent heat of evaporation with absolute temperature is equal to the difference between the specific heat at constant pressure of the vapour and the specific heat of liquid. Empirically it is given as

$$\frac{dL_{ev}}{dT} = -2.3697 \text{ kJ/kg °C} \quad (26.2)$$

Therefore, the latent heat of evaporation at 100°C is approximately 2,263.81 kJ/kg ($\cong$540 Cal/g). Latent heat of melting or fusion (which is equal to the latent heat of freezing) is the amount of heat needed to be added (or released) to change phase from solid to liquid (or liquid to solid) and has a value of 334 kJ/kg at 0°C. Some substances undergo phase changes from solid to gas or vice versa without going through the intermediate phase. Latent heat of sublimation is the amount of heat needed to be added (or released) to change phase from solid to vapour (or vapour to solid) and has a value of 2,834 kJ/kg at 0°C ($\cong$680 Cal/g). Frost formation is an example of deposition, which is the reverse of the process of sublimation.

26.2.1.10 Stability

If a parcel of air lifted to a certain height returns to its original level when released, then the condition is stable. If it remains at that height, the condition is neutral. On the other hand, if it continues to rise further when released, the condition is unstable. These three conditions can be explained with respect to the lapse rates. Instability can occur if the parcel of air is warmer, or, if the parcel of air contains more water vapour than dry air (molecular weight of water vapour is less than that of dry air in the ratio of 18:29). If the environmental lapse rate is greater than the dry adiabatic lapse rate, the atmosphere is unstable. On the other hand, if it is less than the dry adiabatic lapse rate the atmosphere is stable. It is also possible for vertical lifting to take place in a stable environment when the surface temperature is very high, e.g. over forest fires, chimneys, explosions, etc.

26.2.2 Composition in the vertical direction

The atmosphere can be sub-divided into a number of vertical regions depending upon their physical characteristics. Up to about 80 km, the constituent composition of gases is quite homogeneous whereas above 80 km the gases are stratified. Therefore, the lower 80 km or so is referred to as the homosphere while that above 80 km is referred to as the heterosphere.

26.2.2.1 Troposphere

Troposphere is the lowest layer of the atmosphere, and it contains about 75% of mass and almost all of moisture and dust of the atmosphere. All weather phenomena take place in this region. It extends to about 10–20 km. Temperature decreases in the troposphere (Figure 26.2). The top of the troposphere is known as the tropopause. Its height varies from about 15 to 20 km at the tropics to about 10 km at the poles.

26.2.2.2 Stratosphere

Stratosphere is the atmospheric shell above the top of the troposphere and below the mesopause. It is calm and with more or less constant temperature in the lower layers (Figure 26.2) and extends to about 10–50 km. The stratosphere is very stable and has very low moisture content. However, there are horizontal winds, particularly near the bottom and at higher levels in the polar regions during the winter (winds of about 300 km/h). The stratosphere is very important for life because it contains the ozone which absorbs ultra-violet radiation from the sun. The maximum ozone zone lies between 20 and 32 km. Although the thickness of the layer is high, the actual quantity of ozone is very small. If the quantity of ozone were taken down to the ground level, it would compress to a thin layer of about 1 cm by the weight of the atmosphere. Ultra-violet rays are harmful and can lead to skin cancer. The top of the stratosphere is referred to as the stratopause.

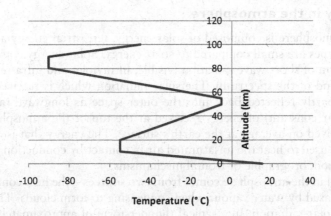

Figure 26.2 Average atmospheric temperature variation with altitude.

26.2.2.3 Mesosphere

Refers to the atmospheric shell between the top of the stratosphere and the mesopause and is about 50 km high. The top of the mesosphere is referred to as the mesopause and is at a height of about 80 km.

26.2.2.4 Thermosphere

Thermosphere refers to the atmospheric shell extending from the top of the mesosphere at about 80 km to outer space. It includes the ionosphere which is the atmospheric shell characterized by high ion density extending from about 70 km upwards and the exosphere which is the outermost portion of the atmosphere whose lower boundary is at a height of about 500 km. Temperature increases in the ionosphere and the exosphere.

The composition of the atmosphere by volume is given in Table 26.3.

Table 26.3 Principal constituents of the atmosphere

Constituent	% by volume of dry air	Concentration in ppm of air
Permanent composition		
Nitrogen	78.084	
Oxygen	20.946	
Argon	0.934	
Carbon dioxide	0.033	
Neon	0.00182	
Helium	0.000524	
Methane	0.00015	
Krypton	0.000114	
Hydrogen	0.00005	
Important variable gases		
Water vapour	0–5	
Carbon dioxide	0.034	340
Carbon monoxide		<100
Sulphur dioxide		0–1
Nitrogen dioxide		0–0.2
Ozone		0–10

26.2.3 Energy in the atmosphere

Energy in the atmosphere is composed of solar energy, terrestrial energy and tidal energy. The latter two types are small compared to solar energy. Solar energy comes from the sun mostly in the form of short wave radiation (visible, ultraviolet and infra-red rays are all at the short wave end of the spectrum). The solar radiation which is received at the surface of the earth is partly reflected back into the outer space as longwave radiation. Of the $1,380 \, W/m^2$ (solar constant) of energy received at the top of the atmosphere, only about $350 \, W/m^2$ is received on average at the earth's surface. The energy that is absorbed by the earth's surface is used to heat the unsaturated air in contact by conduction which then gets lifted by convection, orographic or frontal mechanisms.

Energy utilized in the atmosphere comes from two sources – the heat content of rising air and the heat released by water vapour when condensing to form clouds. The first source is indirectly from solar radiation. In a typical thunderstorm of approximately 5 km in diameter, there maybe 500,000 tons of condensed water. In producing these droplets, a quantity of energy equivalent to about 3.5×10^8 kWh would have been released. If the air is dry, relatively small quantities of energy are available. Energy in the atmosphere is dissipated mainly as kinetic energy in various wind systems (lightning also discharges some amount of energy). An approximate order of magnitude of the energy of various wind systems is given in Table 26.4.

Intense vortices in the atmosphere can be taken as signs that the atmosphere is unstable and has high moisture content. The total power of a system is difficult to ascertain because only part of it can be experienced at a time. By any standard, the weather systems in the atmosphere are very powerful. The amount of energy input required to develop such systems should be even larger. The difference is dissipated in overcoming friction and heating the air inside and outside the system. The energy of the systems, in general, is spread over a large area. Therefore, the destructive effect is not apparent when compared to, for example, that of a nuclear bomb. On the other hand, a tornado is concentrated around a smaller area within a radius of about 100 m and therefore the effects are explosive. At the present time, human beings are not capable of producing or modifying these systems.

26.2.4 Water vapour in the atmosphere

The amount of water vapour contained in the atmosphere (Figure 26.3) is a function of several factors such as the availability of a source of moisture, place, temperature, elevation, etc. In any location, the water vapour in the atmosphere is restricted to the lower layers despite the low density relative to dry air. The amount of water vapour contained in the atmosphere is measured by the relative humidity, which is defined as the ratio of the amount

Table 26.4 Order of magnitudes of energy in wind systems

System	Kinetic energy
Gust	<1
Dust devil[a]	10
Tornado	10^4
Thunderstorm	10^6
Hurricane	10^{10}
Cyclone	10^{11}
Nagasaki bomb	10^7
Hydrogen bomb	10^{10}

[a] In desert regions when the ground is heated to very high temperatures.

Figure 26.3 Water vapour in the atmosphere.

of moisture in the air to the amount needed to saturate the air at the same temperature. The saturation vapour pressure ranges from about 5 mb (at 0°C) to about 50 mb (at about 32°C).

26.3 ATMOSPHERIC CIRCULATION AND WIND SYSTEMS

The radiation received at the surface of the earth is non-uniform and therefore energy gradients are produced. These gradients lead to the redistribution of energy through atmospheric and oceanic circulations. The atmosphere is heated at the base and cooled at the top leading to convective motions. The atmospheric motion consists of a mean circulation and a highly variable system of smaller movements or eddies. The former determines the climate while the latter determines the weather.

26.3.1 Forces in the atmosphere

26.3.1.1 Gravitational force

Force directed towards the centre of the earth.

26.3.1.2 Pressure gradient force

Vertical: Acts upwards and balances gravity (Figure 26.4)

$$\frac{\partial p}{\partial z} \cong 100 \text{ mb/km near the ground}$$

Horizontal: Acts from high to low pressure and is important in producing wind.

$$\frac{\partial p}{\partial x} \cong 1 \text{ mb/100 km at ground level}$$

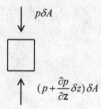

Figure 26.4 Force balance in the vertical direction.

Hydrostatic equilibrium occurs when the pressure gradient forces balance the gravity forces. Thus there is no acceleration, and the air moves with constant velocity upwards or downwards.

26.3.1.3 Frictional force

Frictional force acts in the direction opposite to that of motion. It is significant only near the ground (up to about 1 km). Shear stress due to viscosity is given by

$$\tau = \mu \frac{du}{dn} \tag{26.3}$$

26.3.1.4 Coriolis effect

Newton's law of motion is applicable to moving bodies when observed from a frame of reference which has no acceleration. Atmospheric motions are measured from a frame of reference on earth which is rotating and therefore is accelerating. Coriolis effect is a fictitious acceleration introduced into the Newtonian equation of motion to make it valid for a rotating frame of reference.

In the case of a particle moving with velocity V relative to a frame of reference rotating with angular velocity Ω about an axis normal to the direction of V, the Coriolis acceleration can be proved to be equal to

$$f = 2V\Omega \tag{26.4}$$

The above situation can be simulated by having a model train running along its track mounted on a horizontal turntable which rotates about a vertical axis. To an observer on the turntable, the centripetal acceleration is V^2/R. To an observer beside the turntable, the centripetal acceleration would appear to be $(V + \Omega R)^2/R$ where the additional velocity component ΩR is the tangential velocity due to rotation. The difference between these two components is

$$\frac{(V + \Omega R)^2}{R} - \frac{V^2}{R} = 2V\Omega + \Omega^2 R \tag{26.5}$$

The term $2V\Omega$ is the Coriolis acceleration and the term $\Omega^2 R$ (or V_t^2/R where V_t is the tangential velocity due to rotation) is the centripetal acceleration of any fixed point on the track.

Coriolis acceleration is always normal to the direction of V but may be either to the left or to the right of V, depending on the direction of rotation of the frame of reference. In the northern hemisphere, winds are deflected to the right. At a point on earth in the Northern hemisphere at latitude ϕ, the angular velocity Ω of the rotation of the earth can be resolved into two components: $\Omega \sin\phi$ along the local vertical, or z-axis, and, $\Omega \cos\phi$ along the poleward horizontal, or y-axis. The Coriolis acceleration in the Northern hemisphere is eastwards (in the x direction), or in other words, to the right of the wind direction. In the southern hemisphere, the directions are reversed. For example, for northerly winds (winds blowing from the north), v is negative, $u = 0$ and $w = 0$, where u, v and w are velocities in the East (x-axis), North (y-axis) and the local vertical (z-axis) directions, respectively. It should be noted that all the above components of Coriolis acceleration are not always important.

The components of Coriolis acceleration are

x: $-2v\Omega \sin\phi + 2w\Omega \cos\phi$

y: $2u\Omega \sin\phi$

z: $-2u\Omega \cos\phi$

26.3.2 Equations of motion

Newton's second law of motion is (Force F = mass, $m \times$ acceleration a)

$$F = m \cdot a \tag{26.6}$$

The acceleration term 'a' consists of the acceleration relative to the earth and the Coriolis acceleration:

$$a = \frac{dV}{dt} + f \tag{26.7}$$

The forces involved are pressure gradient, gravitational and frictional. In writing down the equations of motion, it is convenient to have the Coriolis acceleration as an equal and opposite force. Then, for a unit mass (Navier–Stokes equations),

$$x: \frac{du}{dt} = 2v\Omega \sin\phi - 2w\Omega \cos\phi - \frac{1}{\rho}\frac{\partial p}{\partial x} + F_x \tag{26.8a}$$

$$y: \frac{dv}{dt} = -2u\Omega \sin\phi - \frac{1}{\rho}\frac{\partial p}{\partial y} + F_y \tag{26.8b}$$

$$z: \frac{dw}{dt} = 2u\Omega \cos\phi - \frac{1}{\rho}\frac{\partial p}{\partial z} + F_z - g \tag{26.8c}$$

where F_x, F_y and F_z are the frictional forces per unit mass. The general equation of motion (Eq. 26.8a–26.8c) can be simplified to represent different scales of motion.

26.3.3 Synoptic scales of motion

The approximate order of magnitude of the various elements of the equations of motion applicable to the synoptic scale can be summarized as follows:

Length (horizontal)	L	1,000 km	10^6 m
Length (vertical)	H	10 km	10^4 m
Time	t	1 day	10^5 s
Pressure change (horizontal)	Δp	10 mb	10^3 Pa
Pressure change (vertical)	Δp	1000 mb	10^5 Pa
Air density	ρ		1 kg/m^3
Earth's angular velocity	$\Omega\ (= 7 \times 10^{-5})$		10^{-4} rad/s
Acceleration due to gravity	g		10 m/s^2
Wind speed (horizontal)	u, v		10 m/s
Wind speed (vertical)	w		10^{-1} m/s
Acceleration (horizontal)	$u/t, v/t$		10^{-4} m/s^2
Acceleration (vertical)	w/t		10^{-6} m/s^2
Coriolis acceleration	ΩV		10^{-3} m/s^2
Horizontal pressure gradient	$\Delta p/L$		10^{-3} Pa/m

Ignoring the friction terms F_x, F_y and F_z, an order of magnitude analysis gives the following:

$$x:\ \frac{du}{dt} = 2v\Omega\,\mathrm{Sin}\,\phi - 2w\Omega\,\mathrm{Cos}\,\phi - \frac{1}{\rho}\frac{\partial p}{\partial x}$$

$$\ 10^{-4}\quad 10^{-3}\qquad 10^{-5}\qquad 10^{-3}$$

$$y:\ \frac{dv}{dt} = -2u\Omega\,\mathrm{Sin}\,\phi - \frac{1}{\rho}\frac{\partial p}{\partial y}$$

$$\ 10^{-4}\qquad 10^{-3}\qquad 10^{-3}$$

$$z:\ \frac{dw}{dt} = 2u\Omega\,\mathrm{Cos}\,\phi - \frac{1}{\rho}\frac{\partial p}{\partial z} - g$$

$$\ 10^{-6}\quad 10^{-3}\qquad 10\qquad 10$$

It can also be shown that frictional forces are even smaller than the component accelerations $\frac{du}{dt}$ and $\frac{dv}{dt}$.

Neglecting higher order terms, the equations of motion simplify to

$$x:\ \frac{du}{dt} = 2v\Omega\,\mathrm{Sin}\,\phi - \frac{1}{\rho}\frac{\partial p}{\partial x} \tag{26.9a}$$

$$y:\ \frac{dv}{dt} = -2u\Omega\,\mathrm{Sin}\,\phi - \frac{1}{\rho}\frac{\partial p}{\partial y} \tag{26.9b}$$

$$z:\ -\frac{1}{\rho}\frac{\partial p}{\partial z} - g = 0 \tag{26.9c}$$

In Eqs. 26.9a and 26.9b, the LHS's are one order of magnitude smaller than the RHS's. If they are also ignored,

$$x: 2v\Omega\operatorname{Sin}\phi - \frac{1}{\rho}\frac{\partial p}{\partial x} = 0 \tag{26.10a}$$

$$y: -2u\Omega\operatorname{Sin}\phi - \frac{1}{\rho}\frac{\partial p}{\partial y} = 0 \tag{26.10b}$$

$$z: -\frac{1}{\rho}\frac{\partial p}{\partial z} - g = 0 \tag{26.10c}$$

Equations 26.10a and 26.10b are called the geostrophic equations and Eq. 26.9c the hydrostatic equation. This approximation (Eqs. 26.9a and 26.9b) is however not as good as the hydrostatic approximation shown in Eq. 26.9c.

26.3.4 Geostrophic wind (above friction layer)

For pure westerly winds (winds blowing from the west), the winds blow in straight lines and are parallel to isobars with no acceleration or frictional forces acting. The only forces then are the Coriolis and the horizontal pressure gradient forces which balance with each other (Figure 26.5). Then,

$$u > 0 \text{ and } v = 0. \ w = 0.$$

Therefore,

$$x: \frac{\partial p}{\partial x} = 0 \tag{26.11a}$$

$$y: -2u\Omega\operatorname{Sin}\phi - \frac{1}{\rho}\frac{\partial p}{\partial y} = 0 \tag{26.11b}$$

The isobars are east-west.

It can be shown that the horizontal wind speed V and the horizontal pressure gradient $\frac{\partial p}{\partial n}$ are in general related to each other by

$$\frac{1}{\rho}\frac{\partial p}{\partial n} = 2V\Omega\operatorname{Sin}\phi \tag{26.12}$$

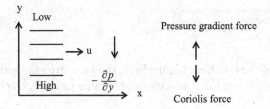

Figure 26.5 Geostrophic balance.

where

$$\left(\frac{\partial p}{\partial n}\right)^2 = \left(\frac{\partial p}{\partial x}\right)^2 + \left(\frac{\partial p}{\partial y}\right)^2 \tag{26.13}$$

and

$$V^2 = u^2 + v^2 \tag{26.14}$$

From Eq. 26.12,

$$V = \frac{1}{\rho} \frac{1}{2\Omega \sin\phi} \frac{\partial p}{\partial n} \tag{26.15}$$

The velocity which satisfies the above balance is called the **Geostrophic Wind**. This relationship however fails at low latitudes, i.e. when Sin ϕ approaches zero.

26.3.5 Gradient wind (above friction layer)

Although straight line flow is common, curved flow represents the general case in nature. When a particle moves in a circular path, it is acted on by a centrifugal force (outwards). Air particles moving in curved paths also experience centrifugal forces. Gradient winds occur in large-scale weather systems when the wind blows horizontally in a curved path parallel to isobars with no frictional forces acting. Such geostrophic departures are common in cyclonic and anticyclonic flow.

To maintain a centripetal acceleration towards the centre for a Northern hemisphere horizontal low-pressure system (cyclone), the wind speed must be such that

Pressure gradient force > Coriolis force by $\dfrac{V^2}{r}$

This is possible only if the Coriolis force < Pressure gradient force, i.e. when the wind speeds are sub-geostrophic. Balance of radial forces gives

$$\frac{V^2}{r} = \frac{1}{\rho} \frac{\partial p}{\partial r} - 2V\Omega \sin\phi \Rightarrow \frac{V^2}{r} + 2V\Omega \sin\phi - \frac{1}{\rho} \frac{\partial p}{\partial r} = 0 \tag{26.16a}$$

$$\Rightarrow V^2 + (2\Omega r \sin\phi)V - \frac{r}{\rho} \frac{\partial p}{\partial r} = 0 \tag{26.16b}$$

This is the gradient wind equation for a cyclone.

Solving quadratically for the gradient wind V,

$$V = -r\Omega \sin\phi + \sqrt{\frac{r}{\rho} \frac{\partial p}{\partial r} + (r\Omega \sin\phi)^2} \tag{26.17}$$

For anticyclonic flow in the northern hemisphere, the pressure gradient force will be directed outwards. Therefore, to maintain a net centripetal force towards the centre,

Coriolis force > Pressure gradient force by $\dfrac{V^2}{r}$

This is possible only if the pressure gradient force < Coriolis force, i.e. when the wind speeds are super-geostrophic. Balance of radial forces gives

$$\frac{V^2}{r} = \frac{1}{\rho}\frac{\partial p}{\partial r} + 2V\Omega \sin\phi \Rightarrow \frac{V^2}{r} - 2V\Omega \sin\phi - \frac{1}{\rho}\frac{\partial p}{\partial r} = 0 \qquad (26.18a)$$

$$\Rightarrow V^2 - (2\Omega r \sin\phi)\,V - \frac{r}{\rho}\frac{\partial p}{\partial r} = 0 \qquad (26.18b)$$

This is the gradient wind equation for an anticyclone.

Solving quadratically, for the gradient wind V,

$$V = r\Omega \sin\phi + \sqrt{\frac{r}{\rho}\frac{\partial p}{\partial r} + (r\Omega \sin\phi)^2} \qquad (26.19)$$

It can be seen that V from Eq. 26.19 is greater than V from Eq. 26.17.

For a given set of conditions above the friction layer the cyclonic flow will have a smaller velocity than straight line flow. Similarly, anticyclonic flow will have a velocity larger than straight line flow.

26.3.6 Small-scale motion

An approximate order of magnitude of the various elements of the equations of motion applicable to the small-scale motion can be summarized as follows:

Length (horizontal & vertical)	L	10 km	10^4 m
Minimum time scale	t		10^3 s
Pressure change (horizontal)	Δp	1 mb	10^2 Pa
Wind speed (horizontal)	u, v		10 m/s
Angular velocity of earth	Ω		10^{-4}/s
Wind speed (vertical)	w		0.5 m/s
Acceleration (horizontal)	$u/t; v/t$		10^{-2} m/s^2
Acceleration (vertical)	w/t		10^{-3} m/s^2
Pressure gradient (horizontal)	$\Delta p/L$		10^{-2} Pa/m

An order of magnitude analysis gives

$$x: \quad \frac{du}{dt} = 2v\Omega \sin\phi - 2w\Omega \cos\phi - \frac{1}{\rho}\frac{\partial p}{\partial x} + F_x$$

$$\quad 10^{-2} \quad 10^{-3} \qquad 0.5\times10^{-3} \quad 10^{-2} \qquad 10^{-5}$$

$$y: \quad \frac{dv}{dt} = -2u\Omega \sin\phi - \frac{1}{\rho}\frac{\partial p}{\partial y} + F_y$$

$$\quad 10^{-2} \quad 10^{-3} \qquad 10^{-2} \quad 10^{-5}$$

$$z:\ \frac{dw}{dt} = 2u\Omega\cos\phi - \frac{1}{\rho}\frac{\partial p}{\partial z} + F_z - g$$

$$10^{-3} \qquad 10^{-3} \qquad 10 \quad 10^{-5}\,10$$

Neglecting higher order terms,

$$x:\ \frac{du}{dt} = -\frac{1}{\rho}\frac{\partial p}{\partial x} \tag{26.20a}$$

$$y:\ \frac{dv}{dt} = -\frac{1}{\rho}\frac{\partial p}{\partial y} \tag{26.20b}$$

$$z:\ g = -\frac{1}{\rho}\frac{\partial p}{\partial z} \tag{26.20c}$$

In polar co-ordinates, Eqs. 26.20a and 26.20b transform to (v_θ is the tangential velocity)

$$\frac{v_\theta^2}{r} = \frac{1}{\rho}\frac{\partial p}{\partial r} \tag{26.21}$$

which is the equation for a forced vortex. It gives a balance of pressure gradient and centrifugal forces. Small-scale phenomena such as tornados, waterspouts are described by this equation.

In a small-scale phenomenon such as a tornado, the velocities are of the order of 50 m/s within a radius of about 100 m. The resulting pressure gradient, therefore, is of the order of 25 mb/100 m which is very powerful and destructive.

26.3.7 Jet streams

Jet streams are easterly winds (speeds > 100 knots) flowing around the entire hemisphere from west to east in the form of a meandering river. In the tropics the core of the jet stream is located at about 13 km (150 mb). In the extra tropical latitudes (20°–40°) it is located around 12 km.

26.3.8 Fronts

A front is a narrow zone of transition between air masses of contrasting physical properties. They include stationary fronts which remain stationary over a certain area, cold fronts in which the cold (denser) air is moving into warm (lighter) air, warm fronts in which warm (lighter) air is replacing the cold (denser) air by over-running, and occluded fronts in which cold fronts which travel twice as fast as warm fronts eventually catch up and merge to form an occluded front.

26.3.9 Atmospheric general circulation models (AGCMs)

Early general circulation models of the atmosphere were developed to solve the governing equations at 5°×5° latitude–longitude grid ($\cong$ 550×550 km). If a 1°×1° grid is chosen, there would be 129,600 nodes. Present-day AGCMs have a much higher resolution. For example, the Japan Meteorological Agency Global Spectral Model (JMA-GSM) uses 1,920 (longitude)×960 (latitude) grid cells corresponding to a size of approximately 20×20 km with 60 vertical layers extending up to a pressure level of 0.1 hPa (Mizuta et al., 2006). This corresponds to 1,843,200 nodal points on the surface of the earth with several degrees of freedom at each nodal point. Such computations can only be carried out with supercomputers such as the 'Earth Simulator' which is a parallel vector supercomputer system developed by the Japan Aerospace Exploration Agency (JAXA), Japan Atomic Energy Research Institute and the Japan Marine Science Technology Centre, which at one time was the fastest computer in the world. At the present time (2019), even faster computers such as 'Tianhe' and Taihulight, which are located in Guangzhou, China, can potentially handle even finer resolutions.

The governing equations (sometimes called the 'primitive equations'), or their simplified forms, together with the boundary and initial conditions, constitute the basis for AGCMs. For a solution to be obtained, the frictional force term F and the heat source/sink term Q must be expressed as functions of the unknown variables u, v, w, p, ρ and T. Any two of the three thermodynamic variables p, ρ and T would describe the system completely. The governing equations consist of the following.

26.3.9.1 Equations of motion

Equations 26.8a, 26.8b and 26.8c in the x, y and z directions, respectively.

26.3.9.2 Continuity equation

$$\frac{\partial \rho}{\partial t} + \frac{\partial}{\partial x}(\rho u) + \frac{\partial}{\partial y}(\rho v) + \frac{\partial}{\partial z}(\rho w) = 0 \tag{26.22}$$

26.3.9.3 Thermodynamic equation (first law of thermodynamics)

$$dQ = c_v dT + p dV \tag{26.23}$$

where Q is the heat source/sink per unit mass, c_v is the specific heat of air at constant volume, T is the absolute temperature, and V is the volume. Eq. 26.23 can be transformed as

$$\frac{dQ}{dt} = c_p \frac{dT}{dt} - \frac{1}{\rho}\frac{dp}{dt} \tag{26.24a}$$

where c_p is the specific heat of air at constant pressure. If Q is constant,

$$c_p \frac{dT}{dt} - \frac{1}{\rho}\frac{dp}{dt} = 0 \tag{26.24b}$$

(The earth's radius is approximately 6,370 km. Therefore, the circumference is approximately 40,000 km. Each longitude degree therefore corresponds to approximately 111 km.)

26.4 WEATHER SYSTEMS

26.4.1 Scales of meteorological phenomena

Various atmospheric phenomena have varying magnitudes both in space and time (Figure 26.6). In this figure, the end of the horizontal scale is equivalent to the earth's circumference which is about 40,000 km. Although there can be variations of an order of magnitude in the same phenomenon, the time scales give a guide to predict the scales of influence of these phenomena.

From Figure 26.6, it can be seen that it is not possible for a single thunderstorm to affect a large area such as China or the USA and that it will not last for more than a day. The most important scale for weather is the synoptic scale or weather map scale. It includes atmospheric phenomena with typical horizontal scales of 800–8,000 km.

26.5 EXTREME WEATHER

26.5.1 Monsoons

The word monsoon has its root in the Arabic word *mausim* which means season. Monsoons are characterized by reversal of wind direction between January and July by at least 120° (180° by definition). Monsoons consist of two seasonal circulations – a winter outflow from a cold continental anticyclone and a summer inflow into a continental heat low (cyclone). In other words, surface winds flowing persistently from oceans to continents in summer and just as persistently from continents to oceans in winter. The summer winds blowing from the oceans are warm and moist whereas the winter winds blowing from the continents are dry and cool. There is a corresponding change in the surface pressure gradient and prevailing weather. These definitions cover the region in the south and south-east of Asia with the south Asian mountains as a natural boundary. The region covered is approximately 35° N–25° S and 30° W–170° E. Important features of northern summer monsoons are

- **Surface pressure:** Low in land; High in oceans
- **Pressure in the upper troposphere:** High in land; Low in oceans
- **Zonal wind in the lower troposphere:** Westerlies on land; Easterlies on oceans
- **Zonal wind in the upper troposphere:** Easterlies on land; Westerlies on oceans
- **Meridional wind in the lower troposphere:** Southerly on land; Northerly on oceans
- **Tropospheric mean temperature:** Warm on land; Cold on oceans

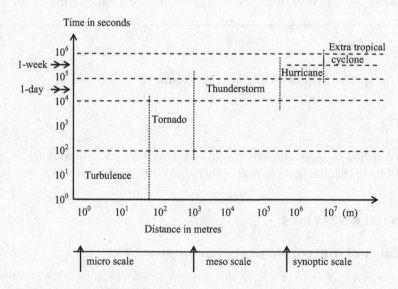

Figure 26.6 Scales of meteorological phenomena.

- **Total moisture:** Humid on land; Relatively dry on oceans
- **Rainfall:** Much larger on land than in the trade wind belt on oceans.

Monsoons bring large amounts of rainfall. World's highest recorded annual rainfall of 26,470 mm, and an average annual rainfall of about 12,000 mm have been in Cherrapunji (25° 15′ N, 91° 44′ E) in Northeast India which also has a monthly record of 9,300 mm. This rainfall is brought about by the SW monsoon. In India, 70% of the rainfall takes place during the SW monsoon (June to September). In Sri Lanka, SW monsoon which comes in summer during the period April to September is called *YALA*, and the NE monsoon which comes in winter during the period October to March is called *MAHA*.

The driving force in monsoon winds is the pressure gradient force between the large land-mass and the ocean. It can be thought of as a convective motion generated by differential heating of the land and the oceans. The swirl introduced to wind by the rotation of the earth is also a contributing factor. The differential heating is caused by the differences in the specific heats of the oceans and the landmasses. The specific heat (energy required to raise the temperature by 1°C) of water is twice that of dry soil. Solar energy received on land heats only a few metres of the earth's subsurface. Much of the energy goes into heating the air. For the oceans, it is quite the opposite. Less energy is available for heating the air. Therefore, for the same amount of energy, the temperature of dry land would be twice that of water. The effective heat capacity of the ocean is very much larger than that of land.

In summer, the rise in temperature over the oceans is less than the rise in temperature over land. The mean summer temperature over the oceans is about 5°C–10°C less than on land at the same latitude. In winter, large heat storage in the oceans leads to higher temperatures in the oceans. Westerly winds at the lower levels and easterly winds at the higher levels generate the convective motion. The reversal takes place at about 6 km altitude. Monsoon arrival is gradual and starts in June. They last from 2 to 4 months. In the Indian sub-continent, an extensive anti-cyclone dominates above the monsoon winds. In mid latitudes, the pressure gradient force and Coriolis force balance each other. At low latitudes, Coriolis force weakens, and there is no geostrophic balance.

26.5.2 Cyclones

A cyclone is any circulation around a low-pressure centre regardless of size and intensity. While rotating about the axis, they also move horizontally. They spin (or appear to spin) clockwise in the southern hemisphere and anti-clockwise in the northern hemisphere. In other words, they rotate in the same direction as the direction of rotation of the earth. The main driving force in a cyclone formation is the pressure gradient force which acts from the high-pressure to the low-pressure region.

26.5.2.1 Tropical cyclones

Tropical cyclones occur in the tropics (23° 27′ N and S). They originate around 5°–15° latitudes from the equator and are quite common in the Indian and Pacific Ocean parts of the monsoon region. Wind speeds of up to 250 knots at times have been reached. When travelling across continents, they lose energy and die down. Cyclones are usually accompanied by heavy rain. In different regions of the world, tropical cyclones have different names as shown in Table 26.5.

Any storm is a form of cyclone. Tropical cyclone is the proper generic name whereas tropical storm is a less technical term. In the Atlantic and North Pacific, tropical cyclones are called hurricanes, and in the Western North Pacific and South China Sea, they are

Table 26.5 Names used for cyclones in different regions

Country or region	Name used
North America	Hurricane
Japan, Northern China, South East Asia, North Western Pacific Ocean	Typhoon
Indian Ocean	Cyclone
Philippines	Baguios (or Bagyo)
Australia	Willey-Willys

Table 26.6 Classification of tropical cyclones in Hong Kong

Type of cyclone	Speed near the centre	Beaufort scale
Tropical depression	Up to 62 km/h	6–7
Tropical storm	63–87 km/h	8–9
Severe tropical storm	88–117 km/h	10–11
Typhoon	>118 km/h	12

called typhoons. The World Meteorological Organization classifies cyclones according to the maximum sustained wind speeds near the centre of the cyclone. In Hong Kong, the classification based on the wind speeds averaged over a period of 10 min is given in Table 26.6.

26.5.2.1.1 Structure and development of a tropical cyclone

Cyclones have a disc-like shape with a vertical scale of tens of kilometres and a horizontal scale of hundreds of kilometres. A tropical cyclone is a heat engine with energy derived from the latent heat of condensation. The power generated in a typical tropical cyclone can be of the order of 20 million MW. More details of tropical cyclones affecting Hong Kong can be found in 'Tropical cyclone annual publications' of the Hong Kong Observatory (hko.gov. hk/en/publica/pubtc.htm).

The physical processes that lead to the formation of a tropical cyclone are still not well understood. It is known that large-scale momentum surges which can provide inward eddy vorticity flux are favourable to the development of a cloud cluster to a tropical storm. After the formation of the tropical cyclone, the heating efficiency increases due to the increased vorticity associated with the system. In the Northern hemisphere, cyclones have winds in the anti-clockwise direction.

A favourable environment is necessary for a cyclone to develop into a full-scale phenomenon. It is generally known that three conditions must be satisfied for a tropical cyclone to be formed: the sea surface must be warm with temperature exceeding about 26°C, the air at low levels must converge inwards over a large area and the air flow at very high levels must be divergent so that a three-dimensional circulation can be sustained. The life span may vary from about a few days to a few weeks.

Tropical cyclones originate in the oceans with high temperatures. The air masses are lifted from the lowest layers of the atmosphere, which have about the same temperature as the sea and expanded adiabatically, resulting in condensation. Due to the high initial temperature, the air masses remain considerably warmer than the environment at least up to a level of about 12 km.

For tropical cyclones to develop, the value of the Coriolis force must be larger than a certain minimum which excludes a latitude belt of about 5°–8° on both sides of the equator.

Tropical cyclones are initiated by a pre-existing low-level disturbance (e.g., areas of bad weather and relatively low pressure) supported by the Coriolis acceleration.

A fully developed cyclone is a warm cored energy exporting system which usually remains intense for many days over the ocean. Essential to this is an extremely large surface pressure gradient near the core of the cyclone, sometimes exceeding 3 hPa/km.

The energy source for a tropical cyclone is mainly the latent heat of condensation of water vapour. Thus, if it remains over warm waters, it has a continuous source of energy and therefore can continue indefinitely.

At a given level and latitude, the pressure gradient force is balanced by the Coriolis force. The wind which acts normal to the Coriolis force is called the gradient wind. This force system is prevalent only if other forces such as friction are neglected. The result, in the Northern hemisphere, is a system of isobars in concentric circles with the gradient wind blowing counter clockwise for cyclonic systems and clockwise for anticyclonic systems. For cyclones in the Northern hemisphere, the only way the air could move in a circular path is by having a net centripetal force (directed towards the centre). This can happen when the pressure gradient force is greater than the Coriolis force. Near the ground surface, frictional forces are significant and the winds tend to get deflected inwards, thereby distorting the concentric circular isobar pattern.

26.5.2.2 Extra tropical cyclones (wavy cyclones)

Extratropical cyclones are developed in the mid-latitudes (50° S–50° N). They tend to develop whenever air masses of different properties converge such as for example when polar and tropical air masses meet in the mid-latitudes. The principal source of energy is the temperature and density differences in the two air masses. They dissipate the energy by mixing of the air masses. Winds in the region of 30–80 km/h are common.

An extratropical cyclone is essentially frontal type and has different stages of development (see also Chapter 27).

When warm air ascends over a cold wedge, adiabatic cooling leads to condensation and precipitation. As the cyclone occludes, cold air replaces warm air at low levels. The centre of gravity of the system is lowered, and the potential energy is converted to kinetic energy of the winds. Lee sides of mountains are favoured for cyclone development.

26.5.2.3 Naming of typhoons

The naming of Typhoons began in 1950 in the Atlantic Basin with names from the international phonetic alphabet of the time: Able, Baker, Charlie, and so on. From 1953, English-language female names were used. Only female names were used in the beginning because hurricanes were named after girlfriends or wives of US Army Air Corp and Navy meteorologists. Later in 1979, alternating male and female names were used. During this time, French and Spanish names have also been used. The first three male names, Bob, David and Frederick have been retired because of the sheer destruction they did in 1979 (Frederick and David) and in 1991 (Bob).

Today, The World Meteorological Organization's (WMO) Regional Association IV Hurricane Committee selects the names for Atlantic Basin and central and eastern Pacific storms. For the Atlantic Basin and the eastern Pacific, six lists of names are used, with each list used again – minus any retired names – 6 years after it was last used. The WMO's regional committee selects the names to replace those that are retired. Each year the names start with the "A" storm on that year's list, no matter how many names were used the previous year.

A storm is named when it reaches tropical storm strength with winds of 39 mph (62 km/h), and becomes a hurricane or typhoon when its wind speed reaches 74 mph (136 km/h).

Hawaiian names are used for central Pacific storms. Here, a revolving list of four sets of names is used, and each storm that forms gets the next available name on the list, regardless of the year. Similar WMO regional committees are involved in selecting names for other parts of the world, but not all nations involved go along with these names.

Nations around the western Pacific began using an entirely new system for naming typhoons in 2000. Each of the 14 nations that typhoons affect submitted a list of names for a total of 141 names. The names include animals, flowers, astrological signs and a few personal names.

Basically, there are two conventions of naming: A number based convention and a list-based convention. The former is popular in Japan while the latter is more popular in most other countries. Both conventions suffer some kind of ambiguity.

26.5.2.3.1 Number-based convention

Number-based conventions are based on the sequential number from the beginning of a typhoon season (usually from May to November). For example, Typhoon No. 15 is the 15th typhoon of the typhoon season. This kind of simplified two-digit convention like "台風 14 号" (Typhoon No. 14) is very popular in Japan, often used in the media such as newspapers and television. This name does not represent the year because at the time of using, the current year is obvious.

Japan Meteorological Agency has another official name such as "Heisei 15, Typhoon No. 14" which means that this is the 14th typhoon of Heisei 15 year, which is another system of year assigned in relation to the Emperor of Japan. This convention, however, is only used in official government documents.

Another number-based convention is to use a four-digit year + number code. For example, Typhoon 15 in the year 2011 is named T1115. A variation to this convention is to use a six-digit year + number. For example, the same Typhoon is named as T201115. This convention eliminates the problem of periodicity. The difficulty in this convention is in remembering the name.

26.5.2.3.2 List-based convention

List-based conventions are based on the list of typhoon names agreed in advance by the relevant committee of meteorological organization in the region. A new name is automatically chosen from the list upon the genesis of a typhoon. The list is defined for each basin and managed by the meteorological organization responsible for the respective basin. For example, Typhoon 200314 has a name 'Maemi', which means a cicada or a locust in North Korea, and is an Asian name chosen from the list of typhoon names for the Western North Pacific basin. The naming conventions used in different countries are given in Table 26.7.

26.5.2.3.3 Asian names

The Asian names which have been in effect since the year 2000 are a list of words submitted by individual countries to the typhoon committee, consisting of meteorological organizations of 14 countries and regions in the Asia-Pacific, that belong to the Tropical Cyclone Programme, World Meteorological Organization (WMO). This list contains words such as animals, plants and natural phenomena and does give preference to human names. Japan Meteorological Agency, for example, proposed names after constellations. Asian names are

Table 26.7 Conventions used for naming typhoons

Country	Convention used
Japan	Number-based
Philippines	List-based (only used within the country)
China	Transition from number-based to list-based convention
Korea	Primary number-based convention, and secondary list-based convention
Taiwan	List-based convention is more often used
Vietnam	Unique number-based convention only counted around Vietnam with list-based convention
United States	Only list-based convention

sorted according to the alphabetical name of countries, so typhoon names themselves are not arranged in alphabetical order.

For cyclones that originate in the Indian Ocean, the Indian Meteorological Department, since 2004, is in charge of naming cyclones from a list suggested by Bangladesh, India, the Maldives, Myanmar, Oman, Pakistan, Sri Lanka and Thailand, and the names are used in alphabetical order of country names. For example, cyclone SIDR was suggested by Oman, and cyclone NARGIS was suggested by Pakistan.

Famous ones that have female names include Typhoon KATHLEEN (194709), Typhoon IONE (194821), Typhoon KITTY (194910), Typhoon JANE (195028) and Typhoon RUTH (195115). Lists of tropical cyclones agreed by respective meteorological organizations can be found at http://en.wikipedia.org/wiki/List_of_tropical_cycone_names.

Typhoons in the Pacific area derived their names from a list arbitrarily drawn up by the Joint Typhoon Warning centre in Guam. The list consists of four series of names with 23 names arranged alphabetically. With 92 names and an average annual occurrence of about 30, each name can recur once in 3 years (or a return period of 3 years). In the early years, only female names were used, but since 1979, male names have also been introduced. The JMA, since January 1981, has undertaken the responsibility of assigning a unique code to each tropical cyclone of tropical storm intensity or above occurring within the region. The code consists of a four-digit number with the first two digits corresponding to the year and the next two digits corresponding to the sequence of the cyclone in that year.

26.5.3 Hurricanes

Hurricane derives its name from the Carib god 'Hurican' which in turn was derived from the Mayan god 'Hurakan', one of their creator gods, who blew his breath across the Chaotic water and brought forth dry land and later destroyed the men of wood with a great storm and flood. In the USA, a cyclone with wind speeds in excess of 32.6 m/s (119 km/h) is called a hurricane. Speeds of up to 90 m/s (324 km/h) have been recorded. It has a calm central area called the eye (common to all cyclones). In most cases, the surface wind speeds do not usually exceed 67 m/s (241 km/h), but they may occur over a large area. The time scale of a hurricane is of the order of a few days.

26.5.4 Thunderstorms

When the atmosphere is unstable and the moisture content is high, convective cloud development once started proceeds at a rapid rate. The cloud air, because of its buoyancy, continues rising. In a very unstable air mass (where the lapse rate is very high), the rising parcel of air becomes more and more buoyant with altitude. This is because of the temperature decrease with altitude. In some cases, the cloud air may be warmer than the environmental air up to the lower layers of the stratosphere. A cloud air ascending at the rate of perhaps 1 m/s at 1,500 m may attain speeds of 25 m/s at an altitude of 7,500 m. In this manner, small clouds become bigger and in turn develop into cumulonimbus clouds or better known as thunderstorms. These extend to altitudes of about 10–20 km. The upper limit of the growth of a thunderstorm is determined by the height of the stratosphere. This is so because the lower layers of the stratosphere are very stable, and the temperature gradient at the stratosphere is zero or negative. Once it has reached an altitude where the cloud is colder than the environment, it begins to slow down but will continue upward movement a few thousand metres because of its momentum.

When the thunderstorm is matured, the upward movement takes place at its maximum speed. Because of the growth of precipitation particles that coalesce and move downwards, there is a downward draft of equal magnitude. At this stage, heavy rain, electrical effects and gusts at the surface are common. The lifting of moist low-level air to high troposphere can take place by three mechanisms: Convectional lifting – when low-level moist air is heated by high surface temperatures caused by solar radiation; Orographic lifting – when moist air is forced up by topographical barriers such as mountain ranges; Frontal lifting – convergence of low-level air in the vicinity of cold fronts. Lightening is another feature of thunderstorms. Sometimes it cannot be seen but can be heard.

Electrons from the water droplets accumulate at the base of the cloud. This negative charge induces a positive charge on the earth's surface below the cloud. A potential gradient of about 1000 V/m occurs between the cloud and the ground. When this is too large, discharge of electrons takes place. The rapid heating of the air in the lightning path produces a violent expansion of air, which initiates a sound wave propagating outwards at the speed of sound. (Lightning travels at about 10^9 km/h, whereas sound travels at about 960 km/h.) By recording the times between seeing the flash and hearing the sound, it is possible to calculate the approximate distance to the place of lightning. Thunderstorms can affect a large area, but will not last more than a day. They bring large amounts of rain. Gustiness, falling temperatures are signs of an approaching thunderstorm.

26.5.5 Tornados

Tornados are quite common in the USA. They last only for a few minutes but with extreme force. Wind speeds are of the order of 130–180 m/s (480–640 km/h). Distances affected are of the order of 100–1,000 m. Because of the extreme low pressure, no man-made structure can survive a direct hit by a tornado. When tornados occur in water, a phenomenon known as water spout is formed.

26.5.6 Tropical depressions

These are centres of low pressure which form in the troughs. They produce deep clouds and much precipitation, mainly of the convective type. By classification, wind speeds are less than 17.4 m/s.

26.5.7 Tropical storms

These are well developed low-pressure systems surrounded by strong winds and much rain. By convention, a system qualifies as a tropical storm if winds range from 17.4 to 32.6 m/s (40–120 km/h).

26.5.8 Extreme temperatures and precipitations

Tables 26.8–26.11 show extreme temperatures and precipitation recorded and averaged over different continents. The highest recorded temperature of 57.8°C is in Africa, and the lowest recorded temperature is – 89.4°C in Antarctica. The highest average annual precipitation from 29 years of record is 13,299 mm in Lloro, Colombia whereas the lowest average annual precipitation from 29 years of record is a mere 0.762 mm in Arica in Chile, also in South America. In Asia, the highest average annual precipitation is at Mawsynram (25° 18′ N, 91° 35′ E), which is about 16 km west of Cherrapunji (re-named to its original name Sohra) on the Khasi Hills in Meghalaya state in north-eastern India. This station is also reported to have had the highest recorded annual rainfall of 26,000 mm in 1985 (https://www.worldatlas.com/articles/rainfall-records-of-the-world.html)

Table 26.8 Highest recorded temperatures in different continents

Continent	Highest temperature (°C)	Place	Elevation (m above MSL)	Date
Africa	57.8	El Azizia, Libya	112	September 13, 1922
North America	56.7	Death valley, CA	−54	July 10, 1913
Asia	53.9	Tirat Tsvi, Israel	−220	June 22, 1942
Australia	53.3[a]	Cloncurry, Queensland	190	January 16, 1889
Europe	50	Seville, Spain	8	August 4, 1881
South America	48.9	Rivadavia, Argentina	206	December 11, 1905
Oceania	42.2	Tuguegarao, Philippines	22	April 29, 1912
Antarctica	15	Vanda Station, Scott Coast	15	January 5, 1974

[a] Note: This temperature was measured using the techniques available at the time of recording, which are different from the standard techniques currently used in Australia. The most likely Australian high-temperature record using standard equipment is an observation of 50.7°C(123°F) recorded at Oodnadatta in 1960.

Table 26.9 Lowest recorded temperatures in different continents

Continent	Lowest temperature (°C)	Place	Elevation (m above MSL)	Date
Antarctica	−89.4	Vostok	3,420	July 21, 1983
Asia	−67.7	Oimekon, Russia	809	February 6, 1933
Asia	−67.7	Verkhoyansk, Russia	107	February 7, 1892
Greenland	−66.1	Northice	2,343	January 9, 1954
North America	−63	Snag, Yukon, Canada	646	Feb 3, 1947
Europe	−55	Ust'Shchugor, Russia	85	January @
South Amercia	−32.7	Sarmiento, Argentina	270	June 1, 1907
Africa	−23.8	Ifrane, Morocco	1,635	February 11, 1935
Australia	−23	Charlotte Pass, NSW	1,755	June 29, 1994
Oceania	−11.1	Mauna Kea Observatory, HI	4,198	May 17, 1979

@ Exact date unknown, lowest in the 15-year period.

26.5.8.1 Notes & additional resources

Many of the extremes reproduced in Tables 26.8–26.11 came from two sources. The first, 'Climates of the World', is an NCDC publication that lists global average temperature and precipitation information for particular locations, with highlighted global extremes.

Table 26.10 Highest average annual precipitation in different continents

Continent	Highest average annual precipitation (mm)	Place	Elevation (m above MSL)	Number of years of record
South America	13,299[ab]	Lloro, Colombia	158[c]	29
Asia	11,872[a]	Mawsynram, India	1,511	38
Oceania	11,684[a]	Mt. Waialeale, Kauai, HI	1,569	30
Africa	10,287.0	Debundscha, Cameroon	9	32
South America	8,992[b]	Quibdo, Colombia	36	16
Australia	8,636	Bellenden Ker, Queensland	1,555	9
North America	6,502	Henderson Lake, British Colombia	3.6	14
Europe	4,648	Crkvica, Bosnia-Hercegovina	1,017	22

[a] The value given is continent's highest and possibly the world's depending on measurement practices, procedures and period of record variations.
[b] The official greatest average annual precipitation for South America is 354 inches at Quibdo, Colombia. The 523.6 in. average at Lloro, Colombia [14 miles SE and at a higher elevation than Quibdo] is an estimated amount.
[c] Approximate elevation.

Table 26.11 Lowest average annual precipitation in different continents

Continent	Lowest average annual precipitation (mm)	Place	Elevation (m above MSL)	Number of years of record
South America	0.762	Arica, Chile	29	59
Africa	<2.54	Wadi Halfa, Sudan	125	39
Antarctica	20.3~	Amundsen-Scott South Pole Station	2,800	10
North America	30.48	Batagues, Mexico	4.9	14
Asia	45.72	Aden, Yemen	6.7	50
Australia	102.9	Mulka (Troudaninna), South Australia	49[a]	42
Europe	162.5	Astrakhan, Russia	14	25
Oceania	227	Puako, Hawaii, HI	1.5	13
South America	0.762	Arica, Chile	29	59
Africa	<2.54	Wadi Halfa, Sudan	125	39

[a] Approximate elevation.

The publication is available for purchase as an offline product (see NCDC Contact Information). The second publication is the updated 'Weather and Climate Extremes' (TEC-0099) published by the US Army Corp of Engineers. The report lists global extremes for various climatological parameters and presents global map inserts as well. The publication can be purchased from the National Technical Information Service.

26.5.9 Wind and water (Fung Soi; Foo sui; Feng Shui) – (風水)

All hydro-meteorological disasters are caused by wind and/or water. The two words, wind and water when combined in Chinese or Japanese languages have much more deeper meaning than the meanings of the individual words. Fung Shui (風水) is a form of geomancy (geo-divination) based on the idea that land is alive and filled with energy (Qi, or chi) (氣 is "steam (气) rising from rice (米) as it cooks). The land energy could make (productive) or break (destructive) it. The theories of Yin and Yang as well as the five element theory are some of the basic aspects of Fung Shui. In the west, it is considered as a form of interior decoration. It is an art that originated in China some 3,000 years ago. The objective is to balance the energies of any given space to ensure health and good fortune for people inhabiting it.

Yin and Yang can be thought of as the two sides of a coin. Yin cannot exist without Yang and vice versa. They have opposite effects; for example, if the earth is Yin, the sky would be Yang.

The five elements are: water, wood, fire, earth and metal.

In the productive cycle wood feeds fire which in turn feeds earth which in turn feeds metal which in turn feeds water which in turn feeds wood. In the destructive cycle, metal destroys wood, wood destroys earth, earth destroys water and water destroys metal.

Water governs pig, rat and ox in the Zodiac.

Example 26.1

Determine the cyclonic, straight line and anticyclonic velocities given the following:

$r = 200$ km (radius of curvature of air flow)

$\rho = 0.0011$ gm/cm^3

$\Omega = \dfrac{2\pi}{86,400} \cong 7.27 \times 10^{-5}$ rad/s

$\dfrac{\partial p}{\partial x} = 3$ mb/200 km

$\phi = 30° \text{N}$

The velocity V for cyclonic flow is given by Eq. 26.17 and works out to be about

10.76 m/s (38.7 km/h)

The same set of conditions for a straight-line flow (Eq. 26.15) gives a velocity of 18.75 m/s. For anticyclonic flow (Eq. 26.19) the velocity is about 25.34 m/s.

Example 26.2

If the pressure inside an aircraft suddenly drops from 700 to 400 mb, estimate the temperature inside if it was originally 22°C.

$pv^\gamma = \text{constant}$

$pv = RT$

Therefore,

$$\frac{T_1}{T_2} = \left(\frac{p_1}{p_2}\right)^{(\gamma-1)/\gamma}$$

Since $\gamma = 1.4$,

$$\frac{(273 + 22)}{(273 + t)} = \left(\frac{700}{400}\right)^{0.286}$$

$$t = \underline{-26.6^\circ C}$$

REFERENCE

Mizuta, R., Oouchi, K., Yoshimura, H., Noda, A., Katayama, K., Yukimoto, S., Hosaka, M., Kusunoki, S., Kawai, H., and Nakagawa, M. (2006): 20-km-mesh global climate simulation using JMA-GSM model – mean climate states, *Journal of the Meteorological Society of Japan*, vol 84 no.:1, pp. 165–185.

ADDITIONAL READING

Anthes, R. A., Panofsky, H. A., Cahir, J. J. and Rango, A. (1978): *The Atmosphere*, Charten E Merrill Publishing Co., Columbus, OH. 442pp

Battan, L. J. (1984): *Fundamentals of Meteorology*, Prentice Hall Inc., Englewood Cliffs, NJ, 304 pp.

Cotton, W. R. (1990): *Storms*, ASTeR Press, Fort Collins, CO, 158 pp.

Das, P. K. (1972): *The Monsoons*, Edward Arnolds, London, 162 pp.

Hanwell, J. (1980): *Atmospheric Processes*, George Allen & Unwin, London, 96 pp.

Lighthill, J. and Pearce, R. P. (Ed.) (1981): *Monsoon Dynamics*, Cambridge University Press, Cambridge, 735 pp.

Mcllveen, R. (1986): *Basic Meteorology – A Physical Outline*, Van Nostrand Reinhold (UK) Co. Ltd., New York, 457 pp.

Moran, J. M. and Morgan, M. D. (1986): *Meteorology – The Atmosphere and the Science of Weather*, Macmillan Publishing Company, New York, 557 pp.

Raudkivi, R. J. (1979): *Hydrology – An Advanced Introduction to Hydrological Processing and Modelling*, Pergamon Press, Oxford, 479 pp.

Chapter 27

Precipitation

27.1 INTRODUCTION

At any instant, the atmosphere contains about 13×10^3 km^3 of water as vapour, liquid or solid. Precipitation is the process by which this water is deposited on the earth's surface. It can take place in one of several forms:

- **Rain**: liquid precipitation
- **Snow**: solid precipitation
- **Mist**: liquid precipitation
- **Hail**: solid precipitation, but melts upon reaching the ground
- **Sleet**: solid precipitation mixed with rain
- **Dew**: liquid condensation when temperature > 0°C
- **Frost**: solid condensation when temperature < 0°C.

The effect of liquid precipitation upon the hydrological cycle is immediate whereas that of solid precipitation is slow and attenuated.

There are many stages of the precipitation process. Firstly, the air must contain the necessary water vapour.

27.2 WATER VAPOUR IN THE ATMOSPHERE

The amount of water vapour contained in the atmosphere is a function of several factors, such as the availability of a source of moisture, place, temperature, elevation, etc. In any location, the water vapour in the atmosphere is restricted to the lower layers despite the low density relative to dry air. The amount of water vapour contained in the atmosphere is measured by the relative humidity, which is defined as the ratio of the amount of moisture in the air to the amount needed to saturate the air at the same temperature.

The saturation vapour pressure (SVP) ranges from about 5 mb (at 0°C) to about 50 mb (at about 32°C).

Dew point is the temperature at which an unsaturated mass of air becomes saturated due to cooling while the pressure is kept constant (according to $pv = RT$). Therefore, if the air mass at temperature T is cooled to T_d (dew point), the corresponding SVP, e_d represents the amount of water vapour in the air. The relative humidity can then be taken as the ratio of the SVP at the dew point to the SVP at the given temperature.

An approximate relationship for the saturation vapour pressure e_s (in mb) in terms of the dew point T_d is given as[1]

$$e_s = 2.7489 \times 10^8 \exp\left(-\frac{4,278.6}{T_d + 242.79}\right) \tag{27.1}$$

where T_d is in °C. With this equation, e_s at 32°C works out to about 60 mb.

27.2.1 Precipitable water

Precipitable water is the total amount of water in a column of air. It is the maximum possible precipitation under total condensation (very rare). It, however, gives no indication of the actual precipitation because the air is always in motion, and a column when depleted of its moisture will be replaced with more moisture from adjacent columns.

Considering a column of unit area of moist air,

$$m_w = \rho_w dz \cdot 1 \tag{27.2}$$

where ρ_w is the water vapour density $\left(= \text{absolute humidity} = \dfrac{m_w}{V}\right)$ and m_w is the mass of water vapour in volume V.

$$\text{The total mass of water vapour} = \int_{z_1}^{z_2} \rho_w \, dz \tag{27.3}$$

where z_1 and z_2 are two elevations corresponding to pressures p_1 and p_2.

Since $p = -z\rho g$ (z measured positive upwards)

$$dp = -\rho g dz \tag{27.3}$$

$$\left(\rho, \text{the density of unsaturated air} = \left(\frac{m_w + m_a}{V}\right); \rho > \rho_w\right)$$

By substituting, the total mass of water vapour m_w is given by

$$m_w = -\int_{p_1}^{p_2} (\rho_w / \rho g) dp$$

$$= -\left(\frac{1}{g}\right)\int_{p_1}^{p_2} (\rho_w / \rho) dp \text{ or } \frac{1}{g}\int_{p_2}^{p_1}\left(\frac{\rho_w}{\rho}\right) dp \tag{27.5}$$

---

[1] List (1966).

The ratio (ρ_w/ρ) is called the specific humidity. It is <1.

The mass of water must then be converted to an equivalent depth.

However, the precipitable water gives no indication of the amount of rainfall expected. There is no relationship between the amount of moisture in the air over any given area and the resulting precipitation because of the continuous movement of air. Any moisture at any place precipitated is soon replaced by moisture from the surrounding.

27.3 STAGES OF PRECIPITATION

A water vapour particle undergoes various phases and physical changes before precipitation takes place. The water vapour must first be carried to upper levels where expansion and cooling take place. When the temperature has reached the dew point, condensation will take place, releasing the latent heat of condensation to an otherwise adiabatic process. Cloud formation will take place with nucleation around impurities in the water vapour. Droplets coalesce with other droplets forming raindrops, which are large enough to cause precipitation. The stages of precipitation consist of lifting, expansion and cooling, nucleation, condensation and raindrop formation. These stages are briefly described below.

27.3.1 Lifting

There are three main mechanisms by which moist air is carried upwards.

27.3.1.1 Convectional lifting

This form of lifting is quite common in the tropics. On hot days, the ground surface becomes heated, causing the air in contact to get warm and rise. The precipitation caused by convection is usually of short duration (1–2 h) and high intensities. Thunder is sometimes accompanied.

27.3.1.2 Orographic lifting

Because of topographical barriers such as mountains, moist air is sometimes forced to rise to higher levels. Usually, rainfalls produced by orographic lifting are very high. They occur when the moist laden air from the oceans strike a range of mountains. Examples include Washington and Oregon States in the USA, the Philippines, and the southern slope of the Himalayas near the Bay of Bengal in India. The world's highest annual rainfall of 12,000 mm is recorded in Cherrapunji at the foot of the Himalayas.

Air may also be forced to rise when it passes from a water body to a land area without the aid of a mountain barrier. In winter, at night, the land temperature is less than that of the oceans. The moist air carried over the land surface may lower the temperature to below dew point.

27.3.1.3 Frontal lifting

Lifting may occur as a result of the convergence of two air masses which have significantly different physical properties (for example, the **Polar Maritime** air masses (**Pm**) and the **Tropical Maritime** air masses (**Tm**); also, **Tropical maritime** (**Tm**) and **Tropical continental** (**Tc**); **Polar continental** (**Pc**) and **Tropical continental** (**Tc**); and **Continental Artic** (**A**)). The boundary between two air masses is called a frontal surface, and where the frontal surface intersects the ground is called a **Front**.

When a warm air mass moves into a cold air mass, it is called a **Warm Front**. When the cold air mass pushes under a warm air mass, it is called a **Cold Front**. In both cases, it is the warm air that is made to rise. Warm fronts have slopes of around 1:100–1:200, whereas cold fronts have slopes of around 1:50 (Figures 27.1 and 27.2).

The precipitation caused by a warm front is of long duration and less intense. There are also the signs of the arrival of a warm front such as cloud thickening. Frontal zones where the air masses do not move relative to one another are called **stationary fronts**.

Precipitation caused by cold fronts are heavy and of short duration. Arrivals of cold fronts have no warnings.

In the mid-latitudes (30°–60°) depressions often develop into wave cyclones, sometimes referred to as extratropical cyclones or occluded fronts. Occluded fronts are produced when rapidly moving cold fronts catch-up and overtake slower moving warm fronts. A cold type occluded front occurs when the air behind the front is colder than the air in front. The opposite case in which the air behind the front is warmer than the air in front is called a warm type occluded front.

27.3.1.4 Convergence lifting

Convergence lifting occurs when two air masses collide, forcing some air upward as both air masses cannot occupy the same space (Figure 27.3). This happens regularly over Florida in the USA, where air moves westward from the Atlantic Ocean and eastward from the Gulf of Mexico to collide over the Florida peninsula.

27.3.2 Expansion and cooling

When unsaturated air is carried to higher altitudes, expansion will occur due to the reduction of pressure. This expansion is adiabatic except near the earth's surface, i.e., no heat is added or taken away from the air from outside sources. However, its temperature is

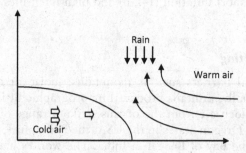

Figure 27.1 Cold front.

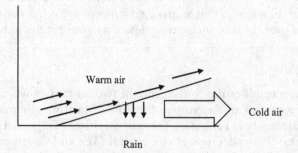

Figure 27.2 Warm front.

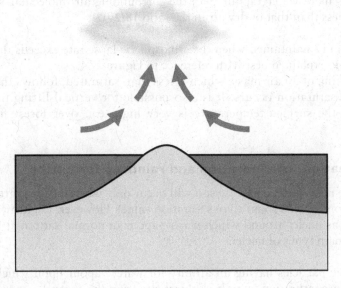

Figure 27.3 Convergence lifting.

lowered because of the heat energy that is transformed into work in the process of expansion. The rate at which the temperature decreases with altitude is called the lapse rate. There are three lapse rates in meteorology.

27.3.2.1 Dry adiabatic lapse rate, α_d

This is the lapse rate when no heat is added or taken away from the air mass. For example, a parcel of air flowing over a mountain can be adiabatic, and the temperature can fall as the pressure decreases. The temperature change in an adiabatic process is purely due to expansion and contraction. The dry adiabatic lapse rate is approximately 9.8°C/km. It can be shown that (Raudkivi, 1979: pp. 27–28)

$$\alpha_d = \left(\frac{gp}{RT} \right) \left(\frac{dT}{dp} \right) \tag{27.6}$$

27.3.2.2 Environmental (ambient) lapse rate, α_a

This is the actual lapse rate that exists in the atmosphere. It may take a wide range of values.

27.3.2.3 Saturated (moist) adiabatic lapse rate, α_s

This refers to the lapse rate when the latent heat of condensation of the saturated air is added to an otherwise adiabatic process. It occurs when moist unsaturated air becomes saturated at some elevation (due to cooling and reaching dew point). Because of the heat added (approximately 540 Cal/g), the lapse rate is lowered to a value of about 6°C/km.

  a. For an air parcel to ascend continuously, the atmospheric conditions must be unstable. This instability occurs if a. the air parcel is warmer, or b.

b. it contains more water vapour than the surrounding air (molecular weight of water vapour is less than that of dry air in the ratio 18:29).

The condition (a) is maintained when the atmospheric lapse rate exceeds the dry adiabatic lapse rate. This is explained best with reference to Figure 27.4.

Normally, lifting of an air mass, which is less than saturated, follows the dry adiabatic lapse rate until saturation occurs. It is also possible for vertical lifting in a stable environment when the surface temperature is very high, e.g. over forest fires, chimneys, explosions, etc.

27.3.3 Nucleation, condensation and raindrop formation

Normally, if the air is pure, condensation will occur only when the air is greatly supersaturated (taking more water vapour than saturation value). However, impurities present in the atmosphere act as nuclei around which water vapour in normal saturated form condense. There are two main types of nuclei:

- hygroscopic particles having an affinity for water vapour upon which condensation begins before saturation (mainly salt particles from the oceans)
- non-hygroscopic particles that need some degree of supersaturation (e.g., dust particles, smoke, ash, etc.)

Condensation nuclei range in size from a radius of about 10^{-3} µm to 10 µm. The average raindrop size is in the range of 500–4,000 µm (µm is a micron and 1 µm = 10^{-6} m).

Once cloud droplets are formed, they may grow depending on atmospheric conditions. There are several theories that explain the growth of cloud droplets. However, not all clouds produce precipitation. Small clouds on hot days disappear as a result of evaporation.

Large drops are formed by condensation of water droplets on ice crystals or by the collision of droplets with ice crystals. This means that the rain producing clouds must extend to the region where ice crystals are formed (about 5 km).

Falling crystals continue to grow both through condensation and the capture of liquid droplets. They change into rain after entering the air in which the temperature is above freezing. It is also possible that rain drops may be formed at temperatures above freezing, by the mixing of warm and cold droplets. The warm droplets evaporate and condense on cold droplets. Showers produced by this method are usually light.

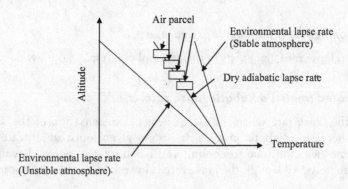

Figure 27.4 Lapse rates and stability.

27.4 MEASUREMENT OF RAINFALL

There are different types of rain gauges for different types of requirements. No single gauge is likely to satisfy all requirements. Basically, they can broadly be classified as recording type or non-recording type.

27.4.1 Non-recording gauges

Standard rain gauges are cylindrical in shape and have three components, namely, a collector, a funnel and a receiver. They are usually made out of copper. A measuring stick is also provided for measuring the depth of rain.

The US Weather Bureau gauge has the following dimensions:

- Can – 8 in. dia., 24 inches deep
- Measuring stick – 24 in. long, 3/8 in wide, 1/8 in thick
- Cross sectional area of stick = 0.1 × (receiving area)

That is, the measured depth of water is ten times the actual depth of water.

The UK Meteorological Office Mark II rain gauge has a cylindrical part with an accurately turned and bevelled brass rim to which a funnel is attached. Inside the outer case is a cylindrical inner case with a handle of brass wire, and inside the inner case is a glass bottle with a narrow neck.

The brass rim has a mean diameter of 127 mm. The cylindrical part is about 100 mm deep to minimize the amount of rain lost by splashing from the sides of the funnel.

27.4.2 Recording gauges

27.4.2.1 Tipping bucket type

Tipping bucket gauge consists of a bucket that is divided into two compartments so arranged that when one is filled, the bucket tips and brings the other into position. It is electrically connected to a recorder which records each tip of the bucket. For the gauge to record, the bucket must be full, which means that any amount of rainfall less than the bucket capacity is not recorded.

Standard buckets hold 15 g of water. This means if the gauge area is 150 cm^2, each tip records 1 mm.

27.4.2.2 Weighing type

Weighing type has a weighing scale which actuates a pen that draws a graph in the form of a mass diagram. The slope of the mass diagram gives the rate of rainfall.

27.4.2.3 Float type

Float type has a pen actuated by a float on the water surface in the receiver. Some designs have automatic siphoning arrangements. Most recording gauges can be connected by telemetry to central recording stations for immediate processing.

27.4.2.4 Interim reference precipitation gauge (IRPG)

Each country having its own standard rain gauge makes comparison of data rather difficult. Therefore the World Meteorological Organization (WMO) agreed upon an international rain gauge which incorporates both UK and US designs. Each country is expected to have at least one IRPG.

27.4.2.5 Radar rain gauge (RAdio Detecting And Ranging)

Radar rain gauge is a remotely sensed rain gauge and is being increasingly used for measuring areal rainfall. It is based upon the principle that radio waves are reflected back when hit by rain droplets. They are not reflected by clouds. These reflections, measured by the intensity of the reflected beam, are displayed on a radar screen called the plan position indicator (PPI). The direction of the target is determined by the position of the transmitting beam, and the position of the target is determined by the time between emitting a pulse and receiving its echo.

Quantitatively, the precipitation is determined by a relationship based on the correlation between the intensity of the echo signals and the precipitation rate.

27.5 ERRORS IN THE MEASUREMENT OF RAINFALL

Errors in the measurement of rainfall may be caused by any of the following factors.

27.5.1 Wind

Usually, errors are in the form of deficiencies in catch, sometimes as high as 50% (for wind speeds of about 80 km/h). Wind causes turbulence in the vicinity of the gauge which varies the rainfall fallout pattern. Since wind speed increases with increasing height above the ground, it is advantageous to keep the orifice as low as possible. But there are other sources of errors if the gauge is too low, for example, splashing effects.

It is possible to improve the catch deficiency by having wind shields.

27.5.2 Evaporation from the gauge

This type of error is particularly relevant to non-recording gauges, where the measurements are made at infrequent time intervals. Errors could be minimized by having highly reflective surfaces to insulate from heat and/or by having multiple inner containers for better insulation.

27.5.3 Placement

The placement of a rain gauge must ensure that the rainfall recorded is representative of the rainfall falling upon the area. The following conditions must be maintained for proper installation:

- The gauge orifice must be horizontal.
- The wind at the orifice level should be minimized without using any wind barriers that may disturb the normal pattern of airflow.
- The site should be flat.

- The ground surface should be grass or matted. Splashing on hard surfaces such as concrete may reach up to 1 m.
- Gauges should be separated from nearby objects by a distance of at least twice their height.

Example 27.1

On a certain day, the upper air observations using radiosonde over Hong Kong were as follows:

Pressure (mb)	Specific humidity (g/kg)	Δp (mb)	Mean (ρ_w/ρ)	$\Delta p \times \left(\dfrac{\overline{\rho_w}}{\rho}\right)$
1000	0.899			
950	0.885	50	0.892	44.6
900	0.883	50	0.884	44.2
850	0.864	50	0.874	43.7
800	0.850	50	0.857	42.8
700	0.657	100	0.753	75.3
600	0.547	100	0.602	60.2
500	0.383	100	0.465	46.5
400	0.355	100	0.369	36.9
350	0.248	50	0.302	15.1
300	0.152	50	0.200	10.0

$$\Sigma = 419.35 \text{ mb} \times \text{g/kg}$$

Calculate the depth of total precipitable water up to the 300 mb pressure level assuming total condensation.

Converting the integral in Eq. 27.5 into a summation,

$$m_w = -(1/g)\,\Sigma\,(\overline{\rho_w/\rho})\Delta p \quad \text{(per unit area)}$$

where $\overline{(\rho_w/\rho)}$ is the mean (ρ_w/ρ) over a range of pressures. It is calculated and shown in the above table. The negative sign cancels out because Δp is negative.

$$\Sigma(\rho_w/\rho)\Delta p = 419.35 \text{ (g/kg) mb} = 419.35 \times 10^{-3} \text{ mb}$$

Since 1 bar = 1,000 mb = 10^5 N/m^2 (or Pascal), 1 mb = 10^{-2} N/m^2

$$419.35 \times 10^{-3} \text{ mb} = 419.35 \times 10^{-3} \times 10^2 \text{ N/m}^2 = 0.1 \times 419.35 \text{ N/m}^2$$

Therefore, mass of precipitable water = $(0.1 \times 419.35)/9.81$ kg

$$= 4.27 \left(\text{kg/m}^2\right) \times \text{unit area} = 4.27 \text{ kg}$$

(In the derivation of the equation, a unit area of the column has been assumed.)

$$\text{Volume of mass of water} = (4.27/1,000) \text{ m}^3 = (4.27/1,000) \text{ m/m}^2$$

$$= (4.27/1,000) \ 1,000 \text{ mm/m}^2$$

$$= \underline{4.27 \text{ mm}}$$

Example 27.2

An air mass over the ocean is lifted and moved inland where it meets a mountain range of altitude 2,500 m. The temperature of the air mass at sea level is 30°C, and the dew point is 20°C. Determine

  a. the elevation at which condensation will begin to take place
  b. the temperature of the air mass at the top of the mountain range
  c. the elevation at which freezing of water vapour in the air mass will begin to take place.

The dry and saturated adiabatic lapse rates are 10°C and 5.5°C/km, respectively. Assume that the lowering of the dew point with altitude follows the saturated adiabatic lapse rate.

  a. At the condensation level, the temperature is equal to the dew point. Therefore,

$$20 - \alpha_s z = 30 - \alpha_d z = T_z \Rightarrow z(\alpha_d - \alpha_s) = 30 - 20 \Rightarrow z = \frac{10}{10 - 5.5} = 2.222 \text{ km}$$

Also, $T_z = 30 - 10 \times 2.222 = 7.78°C$

  b. From the condensation level, the temperature drop will follow the saturated lapse rate. Therefore,

$$T = T_z - \alpha_s (2,500 - 2,222) = 7.78 - \frac{5.5}{1,000} \times 278 = 6.25°C$$

  c. $0 = T_z - \alpha_s z_f \Rightarrow z_f = \frac{7.78}{5.5} = 1.415 \text{ km}$

Total altitude = 2.222 + 1.415 = 3.637 km

REFERENCES

List, R. J. (1966): *Smithsonian Meteorological Tables*, (6h Edition), Smithsonian Institution, Washington, DC.

Raudkivi, R. J. (1979): *Hydrology – An Advanced Introduction to Hydrological Processing and Modelling*, Pergamon Press, Oxford, 479 pp.

Chapter 28

Evaporation and evapo-transpiration

28.1 INTRODUCTION

Evaporation is the process by which a liquid is changed into a gas. In hydrology, it is the transfer of liquid water to water vapour. Evaporation takes place from the oceans, vegetation, bare soil and water bodies.

Water molecules are in constant motion. At high temperatures, the motion or agitation is violent and as a result, some molecules will tend to fly off into the lower layers of air. At the same time, water vapour in the lower layers will also tend to penetrate into the liquid phase. The rate of evaporation at any instant of time will depend upon the difference between the number of molecules leaving the water surface and the number of molecules entering. If it is negative, then there is condensation.

Average spacing of water molecules is about 3 Å (Å = 10^{-10}m) whereas the average spacing of water vapour molecules is about 131 Å. Thermal agitation within the molecules provides the energy for the molecules to escape the internal bonds of the liquid.

Transpiration is the process by which water vapour escapes from living plants principally from leaves and enters the atmosphere. A leaf has on average some 75–125 stomata/mm^2.

The amount of water that stays in a plant is only a small fraction of the amount of water absorbed by the root system. It depends upon the transpiration ratio which varies from crop to crop. The transpiration ratio is the ratio of the weight of water transpired during a growing season to the weight of dry matter produced. Under field conditions, it is difficult to separate transpiration from evaporation.

Typical values of transpiration ratios for some crops are

- Rice: 600–800
- Potatoes: 300–600
- Weeds: 200–1000

Evapo-transpiration refers to the combined effect of evaporation and transpiration from a given area.

Potential evapo-transpiration (PE) is the upper limit of the total losses due to evaporation and transpiration which will occur only when the supply of water is unlimited.

PE = Rainfall + Irrigation – Percolation – Runoff ± change in storage

Sometimes, **consumptive use** is defined as the sum of all the evaporation, transpiration and water stored in the plant. Empirically, the consumptive use is given as

$$u = \frac{ktp}{100} \tag{28.1}$$

where u is the monthly consumptive use (in units of depth of water), t is the mean monthly temperature, p is the monthly daytime hours given as a percentage of the year and k is a mean monthly consumptive use coefficient for each crop.

Actual evapo-transpiration is difficult to measure except for small plots or lysimeters.

28.2 FACTORS AFFECTING EVAPORATION

28.2.1 Solar radiation

When evaporation takes place, an energy equivalent of 2.43×10^6 J is absorbed by each kilogram of water. This is called the latent heat of vaporization, i.e. the energy required to change from liquid state to gaseous state [liquid to vapour – heat must be added; vapour to liquid – heat is released; solid to liquid – heat must be added; liquid to solid – heat is released].

The energy supply is from solar radiation, and it varies with latitude (location), season, weather, exposure of the surface and its reflectivity (albedo).

The amount of solar radiation received in any latitude at any time of the year depends primarily upon two factors:

- Intensity of solar radiation, which is chiefly a function of the angle at which the beam of sunlight reaches that portion of the curved earth, and,
- Duration of solar radiation or length of day compared with night.

The sun's rays do not actually change in intensity, and they are always parallel. Their varying angles, and therefore varying intensities, are only the results of the curvature of the earth's surface. Because an oblique solar ray is spread out over a larger segment of the earth's curved surface than a vertical one, it delivers less energy per unit area. Moreover, an oblique ray also passes through a thicker layer of scattering, absorbing and reflecting air which likewise greatly reduces its intensity. There is a difference in the thickness of the atmosphere that oblique and vertical solar rays have to pass.

As regards to the duration of solar radiation, obviously, the longer the period of daylight and the shorter the night, the greater the amount of solar energy received, all other conditions being equal. Figure 28.1 gives earth's average annual heat balance in percentage units. Table 28.1 gives the length of the day in various northern latitudes.

Incoming solar radiation	100 units
Absorbed by water vapour, dust and ozone	16
Absorbed by clouds	3
Backscattered by air	6 (shortwave)
Reflected by clouds	20 (shortwave)
Reflected by surface	4 (shortwave)
Absorbed by land and ocean	51
Net surface emission of long wave radiation	21 (long wave)
Absorption by water vapour	15 (long wave)
Escape into space	6 (long wave)

Net emission by water vapour, CO_2	38 (long wave)
Emission by clouds	26 (long wave)
Sensible heat flux	7
Latent heat flux	23

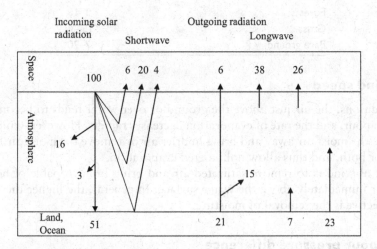

Figure 28.1 Earth's average annual heat balance in percentage units.

Table 28.1 Length of day in various northern latitudes (in hours and minutes on the 15th of each month)

Month	0°	10°	20°	30°	40°	50°	60°	70°	80°	90°
January	12:07	11:35	11:02	10:24	9:37	8:30	6:38	0:00	0:00	0:00
February	12:07	11:49	11:21	11:10	10:42	10:07	9:11	7:20	0:00	0:00
March	12:07	12:04	12:00	11:57	11:53	11:48	11:41	11:28	10:52	0:00
April	12:07	12:21	12:36	12:53	13:14	13:44	14:31	16:06	24:00	24:00
May	12:07	12:34	13:04	13:38	14:22	15:22	17:04	22:13	24:00	24:00
June	12:07	12:42	13:20	14:04	15:00	16:21	18:49	24:00	24:00	24:00
July	12:07	12:40	13:16	13:56	14:49	15:38	17:31	24:00	24:00	24:00
August	12:07	12:28	12:50	13:16	13:48	14:33	15:46	18:26	24:00	24:00
September	12:07	12:12	12:17	12:23	12:31	12:42	13:00	13:34	15:16	24:00
October	12:07	11:55	11:42	11:28	11:10	10:47	10:11	9:03	5:10	0:00
November	12:07	11:40	11:12	10:40	10:01	9:06	7:37	3:06	0:00	0:00
December	12:07	11:32	10:56	10:14	9:20	8:05	5:54	0:00	0:00	0:00

Source: Linsley, R.K., Kohler, M.A. and Paulhus, J.L.H. (1988): Hydrology for Engineers, McGraw Hill, New York, p. 9.

The effect of latitude may be simulated on a small scale within certain latitudes by changes in the direction of exposure and the degree of slope of the land. For example, the angle at which the rays of the sun strike a steep south slope is entirely different from that on a steep north slope. However, exposure is of little importance in the tropics because of the high elevation of the sun. It is of significance in the middle latitudes where the elevation is lower.

When short wave radiation reaches the earth's surface, part of it is reflected back to the atmosphere. The amount of radiation reflected depends on the albedo of the earth's surface which varies greatly for different surfaces. The values of albedo for some earth surfaces are given in Table 28.2.

Table 28.2 Albedo of some surfaces

Reflecting surface	Albedo (%)
Fresh snow	75–90
Old snow	50–70
Sand	15–20
Forests	3–10
Grass	15–30
Bare ground	7–20

28.2.2 Wind speed

In calm conditions, the air just above the ground or over water tends to become saturated with water vapour, and the rate of evaporation decreases rapidly. However, wind by convection can take the moist air away and bring in drier air or remove the moist air by turbulent dispersion, or both, and thus allow unhindered evaporation.

The effect of wind is to remove saturated air and bring in air capable of holding more water vapour immediately above the water surface. In general, the higher the wind speed, the more effective is the removal of moisture.

28.2.3 Vapour pressure difference

The rate of evaporation is strongly affected by vapour pressure deficit. The rate at which water molecules escape into the air is proportional to e_w, and, the rate at which vapour molecules escape into water is proportional to e_a, where e_w and e_a are the vapour pressures of water and air. Therefore, the rate of evaporation is proportional to $(e_w - e_a)$ which is the vapour pressure deficit.

Moist air can absorb less additional water vapour than dry air. In the evaporation process, the liquid changes to the gaseous phase. For this change, the water molecules must acquire enough energy to break through the water surface into the atmosphere. The water molecules in the atmosphere are also in continuous motion, and some return to the liquid mass. An equilibrium state is reached when the number of water molecules escaping equals that returning. At this stage, the vapour pressure concentration gradient vanishes, and the rate of diffusion goes to zero. However, if the air mass is replaced by a drier one, the process will continue indefinitely.

28.2.4 Temperature

When air temperature increase, saturation vapour pressure also increase, which means that the air can hold more water vapour. The energy content of water increases with temperature too. Hence, with increasing temperature, the rate of evaporation increases.

28.2.5 Atmospheric pressure

When atmospheric pressure is low, there are fewer molecules in the air above the water surface, tending to increase the rate of escape of water molecules from the water surface. If the effects of other factors on evaporation are equal, evaporation at a high altitude should be greater than that at a low altitude.

Because atmospheric pressure is closely related to other factors affecting evaporation, it is practically not possible to study the effect of its variation under natural conditions.

28.2.6 Quality of water

Evaporation decreases by about 1% for every 1% increase in salinity. Evaporation also decreases with increasing specific gravity. From sea water, evaporation is 2%–3% less than that from fresh water when other things are the same.

Transpiration is affected by the same atmospheric conditions that affect evaporation. However, sunlight, stage of growth of the plant and season also affect transpiration. It is restricted to day time when photosynthesis can take place.

28.3 METHODS OF MEASURING EVAPORATION

28.3.1 US weather bureau class 'A' pan (Figure 28.2)

The US Class 'A' Pan originated in the USA and is used in many parts of the world. The pan is mounted on wooden frames and placed on a flat lawn. It is filled with water, and a point gauge is used to indicate the water level. At the Observatory in Hong Kong, regular observations are made at 8.00 am every day. At each observation, an amount of water is added to the pan until the surface of the water touches the tip of the point gauge. The amount of water added represents evaporation during the past 24 h.

If there has been rain during the past 24 h, an allowance must be made for so that the evaporation is the sum of water added and the rainfall. If the rainfall is heavy, the water level in the pan may be higher than the fixed point gauge, and water has to be removed from the pan at the time of observation. Then, the evaporation is the difference between the rainfall and the water removed.

28.3.2 UK meteorological office tank

This is a tank sunk into the soil (Figure 28.3). Generally, evaporation from pans will be higher than from tanks because of the heat absorbed by the sides of the pan. The difference is represented as a pan coefficient which varies approximately between 0.5 and 0.7 (typically 0.7).

28.4 METHODS OF MEASURING EVAPO-TRANSPIRATION

28.4.1 Lysimeters

Lysimeter is an isolated block of soil upon which plants grow and evapo-transpiration takes place. By a simple water balance of the amount of water applied, the amount of outflow and the change in soil moisture, the evapo-transpiration can be worked out. They can be either weighing type or drainage type. The lysimeter used at the King's Park Meteorological Station in Hong Kong is the drainage type. Measurements are made at potential levels, i.e., that there is always sufficient water for evapo-transpiration to take place.

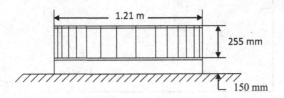

Figure 28.2 US Class A Pan.

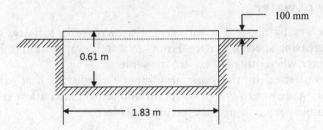

Figure 28.3 UK Standard Tank.

28.4.2 Field plots

Field plots are large lysimeters. Evapo-transpirations are obtained from a water budget calculation in which

input = rainfall + irrigation

output = surface runoff + evapo-transpiration

storage = soil moisture

28.4.3 Ground water level fluctuations

This method is applicable to situations where the plants receive the moisture from a capillary fringe above a water table. The rise and fall of the water table (Figure 28.4) which occurs during the night and the day time is monitored and converted to evapo-transpiration values by using empirical equations.

28.5 METHODS OF ESTIMATING EVAPORATION

28.5.1 Water budget method

This method is used for estimating evaporation over a long period of time, for example, annual evaporation. It is based upon the water balance in a water body (control volume) in which (Figure 28.5)

$$E = I + P - O \mp \Delta S$$

where E is the evaporation, I is the inflow, O is the outflow, ΔS is the change in storage, and P is the precipitation.

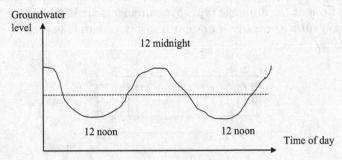

Figure 28.4 Groundwater level fluctuation with time of the day.

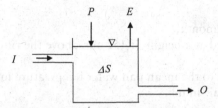

Figure 28.5 Schematic representation of the water budget method.

28.5.2 Mass transfer (aerodynamic) method

In the mass transfer or aerodynamic approach, evaporation is considered as dependent on wind speed and vapour pressure deficit in a form

$$E = f(u)(e_s - e_d) \tag{28.2}$$

where E – evaporation,
 $f(u)$ – an empirical function,
 u – wind speed,
 e_s – saturation vapour pressure corresponding to the temperature at the water surface,
 e_d – vapour pressure of the air which is equal to the SVP at the dew point.

Although this equation has an empirical structure, it can also be shown to have a physical basis (Raudkivi, 1979: p. 116). Several versions of this equation, which is based on Dalton's law of partial pressures, are available for different climatic conditions. They basically have two forms:

$$f(u) = a + bu \,(\text{Penman, 1948}), \text{ and} \tag{28.3a}$$

$$f(u) = Nu \,(\text{Harbeck and Meyers, 1986}) \tag{28.3b}$$

where a, b and N are empirical constants.

   Equation 28.2 requires the temperature of the air/water interface, which is very difficult to measure. Therefore, it has been modified to incorporate the SVP at a certain height from the water surface. If the water surface temperature is the same as the air temperature, which is rarely the case, the aerodynamic equation takes the form

$$E_a = f(u)(e_a - e_d) \tag{28.4}$$

For open water surfaces, Penman (1948) obtained the following equation, which has been calibrated using observations from a sunken evaporation tank in the UK:

$$E_a = (3.938 + 0.4893\,u)(e_a - e_d) \quad \text{mm/month} \tag{28.5}$$

where u is the wind speed measured in km/day at a height of 2 m above the water surface; e_a and e_d are in mb.

   For Hong Kong, using meteorological data measured at King's Park Meteorological station (latitude 22° 18′ 12.82″ N longitude 114° 10′ 18.75″ E) from 1957 to 1975, the aerodynamic equation takes the form (Chen, 1976)

$$E_a = (10.97 + 0.043\,u)(e_w - e_d) \tag{28.6a}$$

where

E_a – evaporation in mm/month,

u – wind speed measured at a height of 152 mm above the rim of the evaporation pan in km/day,

e_w – SVP corresponding to the mean pan water temperature in mb, and,

e_d – vapour pressure of air in mb.

This form of the equation where reference to the SVP at the mean water temperature is made is however less frequently used.

Using meteorological data from the Observatory of Hong Kong for the period 1975–1984, another mass transfer type equation obtained is of the form (Jayawardena, 1987, 1989)

$$E_a = (14.7 + 0.048\,u)(e_a - e_d) \tag{28.6b}$$

28.5.3 Energy budget method

Considering a control volume for a water (and soil) body (Figure 28.6), the energy balance equation when vertical heat exchange is negligible may be written as follows:

$$R_N = H + H_e + H_s + H_v + X_e \tag{28.7}$$

where

R_N – net radiation received at the surface of the body,

H_e – energy used for evaporation,

H_s – change in heat storage of the control volume,

H – heat transferred to the atmosphere by conduction (sensible heat),

H_v – net energy advected into the control volume,

X_e – energy absorbed by temperature, humidity changes and photosynthesis, etc.

In Eq. 28.7, X_e is small and neglected. For a soil body, there is no heat advected into the control volume. For a water body, heat advected depends upon the heat carried by the inflows and outflows. This may also be therefore neglected. Then, Eq. 28.7 becomes

$$R_N = H + H_e + H_s \tag{28.8a}$$

For a totally dry land surface, which provides no moisture for evaporation, it is reasonable to assume that the entire radiation goes into soil heat and sensible heat. Then, Eq. 28.8a becomes

$$R_N = H_s + H \tag{28.8b}$$

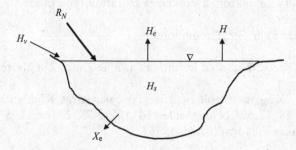

Figure 28.6 Schematic representation of the energy budget method.

The heat storage term can be estimated if the temperature distribution of the control volume is known.

On a long-term basis, the soil heat term may be assumed to be zero. Then, for a water body or a surface covered with vegetation, it is reasonable to assume that the entire radiation goes into evaporation and sensible heat. The equation then becomes

$$R_N = H_e + H \tag{28.8c}$$

28.5.3.1 Net short wave radiation

The short wave solar radiation, R_s, received at the earth's surface can either be determined from direct measurements or from indirect calculations. In Hong Kong, R_s is measured by a thermo-electric pyranometer located at the meteorological station at King's Park.

Indirect calculations are based on the solar radiation received at the outer surface of the atmosphere, I_0, which depends upon the latitude of the place and the time of the year. Values of I_0 are tabulated in most meteorological textbooks (e.g., Raudkivi, 1979: Table 4.1). But the values must be adjusted to account for losses due to absorption, reflection and scattering caused by the atmosphere. The actual radiation R_s received at the earth's surface is given by

$$R_s = I_0\left(a + b\frac{n}{N}\right) \tag{28.9}$$

where a, b are constants, n is the actual number of hours of sunshine and N is the maximum possible number of hours of sunshine.

For example, for Hong Kong (latitude $\approx 22.5°$ N), on 30th of November, $I_0 = 287.3\,\text{W/m}^2$.

A correction must then be made for the reflectivity of the receiving surface which is characterized by its albedo, r, which for water surfaces takes values ranging from 0.03 to 0.10 (Raudkivi, 1979: Table 1.4). Then, the net radiation R_N received at the evaporating surface is given by

$$R_N = R_s(1 - r) - R_B \tag{28.10}$$

where R_B is the net outgoing long wave radiation referred to as back radiation.

28.5.3.2 Net long wave radiation, R_B

A substantial amount of the absorbed radiation is reradiated by the earth as long wave radiation, particularly at night when the sky is clearer and the air dryer than during the daytime. Up to about 94% of the long wave radiation leaving the surface of the earth is absorbed by the atmosphere while the remaining 6% escapes into outer space. A part of the absorbed long wave is reradiated downwards into the earth's surface. The net long wave radiation is given by,

$$R_B = R_{LR} - R_L \tag{28.11}$$

where R_L is the downward long wave radiation from the atmosphere and R_{LR} is upward long wave radiation from the earth's surface. The ratio of these two quantities is given by

$$e = \frac{R_L}{R_{LR}} \tag{28.12}$$

where 'e' is the emissivity of the atmosphere. Therefore,

$$R_B = R_{LR}(1-e)$$

The back radiation R_B is estimated using theoretical black body radiation σT^4, modified by a function of sunshine. The net outgoing long wave radiation then is

$$R_B = 0.97 R_{LR}\left(1-e\right)\left(\alpha + \beta\,\frac{n}{N}\right) = 0.97\sigma T^4\left(1-e\right)\left(\alpha + \beta\,\frac{n}{N}\right) \tag{28.13}$$

where σ is the Stefan–Boltzmann constant $= 5.67 \times 10^{-8}$ W/m²/°K⁴; T is the absolute temperature (°K); α and β are constants; 0.97 is an average correction introduced for the water surface.

The emissivity e is given by

$$e = A + B\sqrt{e_d} \tag{28.14}$$

where A and B are empirical constants taking values ranging from 0.43 to 0.74 for A and 0.029 to 0.081 for B. In temperate climates, A is small and B is large. The sunshine function recommended by the UK Ministry of Agricultural, Fisheries and Food is $0.17 + 0.83\,\dfrac{n}{N}$ (MAFF, 1967). For Hong Kong, the coefficients A and B which gave the best estimates of E_p when compared with measured values have been determined by trial and error as 0.465 and 0.0795, respectively (Jayawardena, 1987).

The sensible heat transferred to the atmosphere depends upon the temperature difference between the air and the surface of the water. It is difficult to measure and therefore taken as a fraction of the energy used for evaporation, $H = \beta H_e$, where β is called the Bowen ratio.

28.5.4 Combination method (Penman equation)

Penman combined the theoretical concepts of the energy budget method with the empirical structure of the mass transfer method (Penman, 1948, 1950). The sensible heat transferred to the atmosphere depends upon the temperature difference between the air and the evaporating surface. The energy used up for evaporation depends upon the vapour pressure deficit. Therefore it is possible to write

$$H_e = f(u)(e_s - e_d) \tag{28.15}$$

$$H = \gamma\,f_1(u)(T_s - T_a) \tag{28.16}$$

where T_s is the temperature of the water surface, T_a is the air temperature and γ is a psychrometric constant. The two functions $f(u)$ and $f_1(u)$ are generally assumed to be the same.

In the hypothetical case when the air temperature is equal to the water surface temperature, the energy available for evaporation is given by

$$H_a = f(u)(e_a - e_d) \tag{28.17}$$

The slope of the SVP vs. temperature curve, Δ, is given by

$$\Delta = \frac{de}{dT} = \frac{(e_s - e_d)}{(T_s - T_d)} \tag{28.18}$$

If the gradients are small, this can be approximated to

$$\Delta \approx \frac{(e_s - e_d)}{(T_a - T_d)} \tag{28.19}$$

It is difficult to measure T_s, and hence the above form for Δ. Therefore,

$$H = \gamma f(u)(T_s - T_a) = \gamma f(u)\{(T_s - T_d) - (T_a - T_d)\} = \gamma f(u)\left\{\left(\frac{e_s - e_d}{\Delta}\right) - \left(\frac{e_a - e_d}{\Delta}\right)\right\}$$

$$= \frac{\gamma H_e}{\Delta} - \frac{\gamma H_a}{\Delta} \tag{28.20}$$

From Eq. 28.8a, $H_e = R_N - H - H_s$. Therefore, substituting for H,

$$H_e = R_N - \frac{\gamma H_e}{\Delta} + \frac{\gamma H_a}{\Delta} - H_s = \frac{\left\{R_N - H_s + \dfrac{\gamma H_a}{\Delta}\right\}}{\left\{1 + \dfrac{\gamma}{\Delta}\right\}} \tag{28.21}$$

or

$$\left[\frac{R_N - H_s + \dfrac{\gamma}{\Delta} H_a}{1 + \dfrac{\gamma}{\Delta}}\right]$$

In these equations, Δ corresponds to the gradient of the SVP vs. temperature curve at the air temperature. The values of the ratio (γ/Δ) are tabulated in many textbooks. Eq. (28.21) is the Penman equation. If the soil temperature does not change, $H_s = 0$; then,

$$H_e = \frac{\left\{R_N + \dfrac{\gamma H_a}{\Delta}\right\}}{\left\{1 + \dfrac{\gamma}{\Delta}\right\}} \tag{28.21a}$$

To convert the energy quantities into equivalent mass of evaporation, the latent heat of evaporation, which is given below, is used:

$$L = 2,500.78 - 2.37T_a \text{ kJ/kg} \tag{28.21b}$$

where T_a is the air temperature in °C.

28.6 METHODS OF ESTIMATING POTENTIAL EVAPO-TRANSPIRATION

There are a number of empirical equations for estimating evapo-transpiration (Thornthwaite, 1948; Penman, 1948, 1950; Blaney and Criddle, 1950; Wright, 1972; Allen and Pruitt, 1986 among others). Evapo-transpiration depends to a great extent on the crop that transpires the water. Different crops have different transpiration ratios, sometimes differing by an order of magnitude. Therefore, it is the common practice to express evapo-transpiration (potential) for plots covered with grass, which has a transpiration ratio of about 800.

All empirical equations need calibration for the region and the climatic condition that prevails in the region. It is also possible to expect some kind of correlation between evapo-transpiration and evaporation because there are many common factors that affect the two processes.

28.6.1 Thornthwaite method

The original Thornthwaite equation is of the form

$$PE = CN_m \left(\frac{10 T_m}{I} \right)^a \text{ mm/month} \tag{28.22}$$

where

I – annual heat index $\left[= \sum_{}^{12} \left(i_j^{1.514} \right) \right]$,

i_j – monthly heat index $\left(= \dfrac{T_m}{5} \right)$,

T_m – monthly mean temperature, in ^0C,

$a = 6.75 \times 10^{-7} I^3 - 7.77 \times 10^{-5} I^2 + 1.79 \times 10^{-2} I + 0.492$,

N_m – monthly sunshine adjustment factor $\left(= \dfrac{N \times ND}{360} \right)$,

ND – number of days in a month,

N – maximum possible hours of sunshine per day,

C – empirical constant.

The constant C in Thornthwaite's original equation, which has been applied to North Carolina, USA conditions, is 16. Jain and Sinai (1985) have obtained a value of 21 using data from a banana field in a semi-arid region (Jordan valley, Israel). They have then compiled a table of values of C, obtained by linear interpolation between the above two values for regions where the temperatures range from 12°C to 32°C. The original equation assumes a 12-h sunshine day. The sunshine adjustment factor accounts for the deviation of the mean duration of maximum possible sunshine hours (N), which depends upon the latitude and season.

For Hong Kong, the original Thornthwaite equation consistently over-predicts during summer months and under-predicts during the winter months. A factor of $C = 11.8$ (obtained by linear regression of measured PE vs. the remainder of the RHS of Eq. (28.22) and forced through the origin) seems to give better estimates for Hong Kong. Using linear regression forced through the origin for the data for each month, it was possible to arrive at the following monthly values for the coefficient C which gave even better estimates:

Month	January	February	March	April	May	June	July	August	September	October	November	December
C_m	29.7	29.7	16.7	14.1	11.2	11.1	11.5	9.53	10.7	13.6	19.2	23.0

Source: The following monthly values for the coefficient C which gave even better estimates (Jayawardena, 1989).

28.6.2 Penman method

The UK Ministry of Agriculture, Food and Fisheries (MAFF, 1967) modified the original Penman equation for evaporation as follows:

$$PE_P = \frac{\left\{R_{NT} + \dfrac{\gamma}{\Delta} E_{at}\right\}}{\left\{1 + \dfrac{\gamma}{\Delta}\right\}} \tag{28.23a}$$

$$E_{at} = (7.875 + 0.04893\,u)(e_a - e_d) \tag{28.23b}[1]$$

$$R_{NT} = R_s(1 - r_2) - R_B \tag{28.23c}$$

where

PE – evaporation and transpiration from the vegetated surface,

R_{NT} – net radiation received at the vegetated surface,

r_2 – albedo for grass surfaces (assumed to be 0.25).

The results of evapo-transpiration values using the above equation over-predicted the measured data for Hong Kong conditions.

The regression of measured evapo-transpiration with evaporation indicates that the former is about 0.6–0.84 times the latter. This is perhaps because evaporation values are 'pan' measurements and therefore high. Therefore, attempts were made to find a suitably modified form of the Penman equation by multiplying E_a by a factor in the form

$$E_{at} = fE_a \tag{28.24}$$

By trial and error the values of f were found to be in the region of 0.85 (Jayawardena, 1987). Marginally better results were obtained when these coefficients were estimated on a monthly basis. The monthly coefficients f_m, have a variation similar to that of the monthly coefficients C_m of the Thornthwaite equation.

28.6.3 Ramage method

Ramage (1953, 1959) using evapo-transpiration measured in Hong Kong during the period October 1951 to September 1956 obtained an equation that takes into account humidity and wind effects. His equation is of the form

$$PE_R = \sqrt{E_a} + \sqrt{e_a} - 3.5 \text{ mm/day} \tag{28.25}$$

where

$E_a = 0.35\,(e_a - e_d)(1 + 0.01u)$,

e_a – vapour pressure at the mean air temperature (mm of Hg),

u – wind speed in miles/day.

The modified form of this equation obtained by Jayawardena (1987) is of the form

$$PE_R = 10.86\left(\sqrt{E_a} + \sqrt{e_a}\right) - 84.17 \text{ mm/month} \tag{28.26}$$

where E_a is given by Eq. 28.6b.

By forcing the regressions of measured evapo-transpiration and evaporation for each month through the origin, the following simple correlation was obtained:

---

[1] The difference between Eqs. 28.23b and 28.5 is that the first term of the former is twice that of the latter.

$$PE = f'E \tag{28.27}$$

where $f' = 0.6$ for November & December; 0.7 for January, February, March, August, September & October and 0.8 for April, May, June & July.

Similar values of f' obtained by Penman (1950) for UK conditions are 0.6 for November, December, January & February; 0.7 for March, April, September & October and 0.8 for May, June, July & August. These differences in the periods of applicability are due to the differences in the climatic conditions.

An important point to note in all empirical equations is that they need to be calibrated for the conditions under which measurements are made, i.e., an equation calibrated for one geographical location cannot, in general, be applied to a different location without re-calibration. A study (Jayawardena, 1991) which analyses the reliability of some empirical equations for estimating evaporation and evapo-transpiration illustrates this point.

28.7 ENERGY AND WATER BALANCE

Table 28.3 shows the monthly energy balance estimated for Hong Kong for an average year based on meteorological data for the years 1975–1987 (Royal Observatory, Hong Kong). Columns 4–7 of Table 28.3 represent the global short wave radiation measured at the meteorological station, R_s, the net solar energy available for evaporation, R_N, the net solar energy available for evapo-transpiration, R_{NT} and the long wave back radiation, R_B respectively. Column 8 represents the sensible heat, H, transferred to the atmosphere by conduction and convection which for water bodies and surfaces covered with thin films of water should be very small in comparison to the energy used for evaporation and evapo-transpiration. It is calculated from the energy balance equation (Eq. 28.8a) with H_e obtained from the estimated values of evaporation using Penman equation and the latent heat of evaporation, L, which is given by

$$L = 2,500.78 - 2.37\,T_a \text{ kJ/kg (same as Eq. 28.21b)} \tag{28.28}$$

where T_a is the air temperature in °C (Jayawardena, 1989).

Table 28.3 Monthly energy balance for Hong Kong (based on 1975–1984 data)

Mon	Air temp (°C)	Water temp (°C)	R_s (MJ/m²)	R_N (MJ/m²)	R_{NT} (MJ/m²)	R_B (MJ/m²)	H (MJ/m²)
January	16.0	16.2	314	169	107	129.0	−28.6
February	16.3	16.7	260	157	105	89.6	−21.4
March	19.0	19.5	270	188	134	68.3	−5.62
April	22.8	23.8	344	262	193	64.3	6.43
May	25.8	27.1	435	354	267	59.0	28.3
June	28.0	29.4	483	403	306	56.0	26.8
July	29.0	30.5	571	474	360	68.1	32.8
August	28.6	30.3	518	430	327	61.3	32.3
September	27.8	28.9	464	371	278	70.6	0.70
October	25.6	25.9	441	314	226	105.0	−25.2
November	21.3	21.0	386	225	148	141.0	−49.2
December	17.8	17.3	353	175	104	161.0	−45.7
Average	23.2	23.9	403	294	213	89.4	−4.03

Source: Jayawardena (1989).

Table 28.4 Monthly water balance for Hong Kong (based on 1975–1984 data)

Month	Rainfall (mm)	Evaporation (mm)		Evapo-transpiration (mm)	
		Measured	Estimated	Measured	Estimated
January	23.1	88.4	80.4	62.7	62.3
February	41.0	76.5	72.6	63.0	62.3
March	99.0	79.7	79.0	62.2	64.7
April	187.0	99.3	104.0	85.1	84.9
May	378.0	124.0	133.0	103.0	104.0
June	335.0	146.0	154.0	123.0	123.0
July	328.0	177.0	182.0	148.0	144.0
August	422.0	157.0	164.0	114.0	114.0
September	319.0	146.0	152.0	105.0	104.0
October	177.0	144.0	139.0	108.0	107.0
November	28.1	127.0	112.0	81.4	81.5
December	9.7	109.0	89.6	66.9	66.6
Average	196.0	123.0	122.0	93.6	93.2

Source: Jayawardena (1989).

In an average year, Hong Kong receives about 403 MJ/m²month of solar radiation (equivalents to 165 mm of evaporation) of which only 293 MJ/m²month (equivalent to 120 mm of evaporation) is available for evaporation. The back radiation is about 89.5 MJ/m²month (equivalent to 36.5 mm of evaporation). The energy available for evapo-transpiration is about 213 MJ/m²month (87.2 mm equivalent evaporation). The average sensible heat is about 4 MJ/m²month from the atmosphere to the evaporating surface (1.6 mm of equivalent evaporation).

Table 28.4 shows the monthly water balance estimated for an average year based on meteorological data for the years 1975–1987 (Royal Observatory, Hong Kong). Of the 2,350 mm of rainfall received in a year, evaporation from water bodies constitutes 1,480 mm which is about 62.8%. Evapo-transpiration is about 1,120 mm which is 47.7% of the rainfall.

Example 28.1

The measurements of evaporation (in mm) made using the US Class A Pan and the UK Tank are given in the following table:

Year	UK Tank	US Class A Pan
1958	351	491
1959	536	713
1960	502	653
1961	437	586
1962	486	612
1968		621
1969		581
1970		687
1971		624
1972		568

For the period 1968–1972, measurements from US Class A Pan only were available. Estimate the volume of water lost each year in the latter period from a reservoir of surface area 1.4 km².

From a plot of tank vs. pan data, the pan coefficient can be estimated by fitting a regression line to the scatter plot. It works out to about 0.8. The rest of the estimates are given in the following table:

Year	Pan evaporation (mm)	Reservoir evaporation (mm)	Reservoir evaporation (m³)
1968	621	496.8	695,520
1969	581	464.8	650,720
1970	687	549.6	769,440
1971	624	499.2	698,880
1972	568	454.4	636,160

Example 28.2

The following data refer to meteorological observations made at the King's Park Meteorological station in Hong Kong for the month of January 1975:

Air temperature = 15.9°C; SVP at dew point, $e_d = 14.1$ mb;
$R_s = 290.73$ MJ/m²; $n = 88.9$ h; $N = 338.833$ h; $\gamma/\Delta = 1.721$.

Obtain the Penman estimate of evaporation and the Thornthwaite estimate of potential evapotranspiration

i. Penman estimate of evaporation

Let $RB1 = 0.97\sigma T^4$

$$= 0.97 \times \left(5.67 \times 10^{-8}\right)(273 + 15.9)^4 = 383.13 \text{ W/m}^2$$

Let $RB2 = \left\{1 - \left(A + B\sqrt{e_d}\right)\right\}$

$$= \left\{1 - \left(0.465 + 0.0795\sqrt{14.1}\right)\right\} = 0.2365$$

Let $RB3 = \alpha + \beta n/N$

$$= 0.17 + 0.83(88.9/338.833) = 0.3878$$

Let $RB4 = RB1 \times RB2 \times RB3$

$$= 35.14 \text{ W/m}^2$$

Radiation intensity[2] is in W/m²; energy is intensity × time {(= W/m²) s = J/m²}; Divide by 10^6 to convert to MJ/m². If ND is the number of days in the month, the conversion factor will be

$$\left(ND \times 24 \times 60 \times 60\right)/10^6 = 0.0864 \times ND$$

---

[2] Radiation intensity is also measured in Langleys (1 Langley = 1 Cal/cm² = 0.041874 MJ/m²).

Therefore, $RB = -0.0864 \times ND \times RB4$ MJ/m^2

For January, $ND = 31$, and therefore,

$$RB = 94.12 \text{ MJ/m}^2$$

Latent heat of evaporation, L, is given by

$$L = 2{,}500.78 - 2.37 \times 15.9 = 2{,}463.097 \text{ kJ/kg}$$

Let $RN1 = 0.95 R_s - RB$

$$= 0.95 \times 290.73 - 94.12 = 182.07 \text{ MJ/m}^2 \text{ month}$$

Therefore,

$$R_N = (182.07)/(2{,}463.097/1{,}000) \text{ kg/m}^2 \text{ month}$$

$$= (182.07)/(2{,}463.097/1{,}000)/1{,}000 \text{ m}^3/\text{m}^2 \text{ month}$$

$$= \{(182.07)/(2{,}463.097/1{,}000)/1{,}000\}1{,}000 \text{ mm/month}$$

$$= 73.92 \text{ mm/month}$$

(the division of 2463.097 by 1000 is to convert kJ to MJ).

Therefore,

$$E_p = \left\{R_N + (\gamma/\Delta)E_a\right\}/\left\{1 + (\gamma/\Delta)\right\}$$

$$= \{73.92 + 1.721 \times 74.35\}/\{2.721\}$$

$$= \underline{74.19 \text{ mm/month}}$$

(In the above calculation, E_a is taken from a separate calculation.)

ii. **Thornthwaite estimate of potential evapo-transpiration**

$$PE_T = 16 \, N_m \left\{10 \, T_m/I\right\}^a \text{ mm}$$

$$T = 15.9°C; \; I = \sum \left(T_m/5\right)^{1.514} = 123.34$$

$$a = 6.75 \times 10^{-7} I^3 - 7.77 \times 10^{-5} I^2 + 1.79 \times 10^{-2} I + 0.492$$

$$= 1.266 - 1.182 + 2.208 + 0.49 = 2.784$$

$$N_m = \left(n \times ND/360\right) = 338.833/360 = 0.941$$

Therefore,

$$PE_m = 16 \times 0.941 \left(10 \times 15.9/123.34\right)^{2.784}$$

$$= \underline{30.53 \text{ mm/month}}$$

REFERENCES

Allen, R. G. and Pruitt, W. O. (1986): Rational use of FAO-Blaney-Criddle method, *Journal of Irrigation and Drainage Engineering*, vol. 112 no.:2, pp. 139–155.

Blaney, H. F. and Criddle, W. D. (1950): *Determining Water Requirements in Irrigated Areas from Climatological and Irrigation Data*, SCS-TP 96, U. S. Department of Agricultural Division of Irrigation and Water Conservation, Soil Conservation Service, Washington, DC, 44 pp.

Chen, T. Y. (1976): Evaporation and evapo-transpiration in Hong Kong, Royal Observatory, Hong Kong, Technical Note No. 42, 67 pp.

Harbeck, G. E. and Meyers, J. S. (1986): Present day evaporation measurement techniques, *Proceedings of the ASCE*, vol. 96 no.:HY7, pp. 1381–1389.

Jain, P. K. and Sinai, G. (1985): Evapo-transpiration model for semi-arid regions, *Journal of Irrigation and Drainage Engineering*, vol. 111 no.:4, pp. 369–379.

Jayawardena, A. W. (1987): *Calibration of Some Empirical Equations for Evaporation and Evapotranspiration in Hong Kong, Research Report*, Department of Civil & Structural Engineering, University of Hong Kong, Pok Fu Lam, 45 pp.

Jayawardena, A. W. (1989): Calibration of some empirical equations for evaporation and evapotranspiration in Hong Kong, *Agricultural & Forest Meteorology*, vol. 47, pp. 75–81.

Jayawardena, A. W. (1991): Reliability of empirical equations for estimating evaporation and evapotranspiration, *Bulletin, Hong Kong Meteorological Society*, vol. 1 no.:2, pp. 3–10.

Linsley, R. K., Kohler, M. A. and Paulhus, J. L. H. (1988): *Hydrology for Engineers*, McGraw Hill, New York, p. 9.

Ministry of Agriculture, Fisheries and Food (1967): *Potential Transpiration*, Technical Bulletin No. 16, HMSO, London.

Penman, H. L. (1948): Natural evaporation from open water, bare soil and grass, *Proceedings of the Royal Society (London) Series A*, vol. 193, pp. 120–146.

Penman, H. L. (1950): Evaporation over the British Isles, *Quarterly Journal of Royal Meteorological Society*, vol. 77 no.:330, pp. 372–383.

Ramage, C. S. (1953): Evapotranspiration measurements made in Hong Kong, First report, Royal Observatory Technical Note, 7, 6 pp.

Ramage, C.S., 1959. Evapotranspiration in Hong Kong, A second report, *Pacific Science*, vol. XIII no.:1, pp. 81–87.

Raudkivi, A. J. (1979): *Hydrology – An Advanced Introduction to Hydrological Processes and Modelling*, Pergamon Press, London, 479 pp.

Royal Observatory, Hong Kong (1975–1989): *Meteorological Results, Part I – Surface Observations*, Royal Observatory, Hong Kong.

Thornthwaite, C. W. (1948): An approach towards a rational classification of climate, *Geographical Review*, vol. 38, pp. 55–94.

Wright, J. L. (1972): New evapo-transpiration crop coefficients, *Journal of Irrigation and Drainage Division*, 108, no.:IR2, pp. 57–64.

Chapter 29

Infiltration

29.1 INTRODUCTION

Infiltration can be defined as the process of the passage of water into the soil through the soil surface. It is generally considered as being different from percolation, which is the process of water movement within the soil. The forces which cause infiltration are due to gravity, pressure and viscosity. Infiltrated water may replenish the soil moisture storage, evaporate or become groundwater.

There are several indicators to quantify infiltration. They include the Infiltration rate which can be defined as the volume flux of water flowing into the soil per unit area. It has the units of velocity ($L^3T^{-1}L^{-2}$), the infiltration capacity which is defined as the maximum rate at which a given soil in a given condition can absorb rain as it falls (Horton, 1940) and the infiltrability which is defined as the infiltration flux resulting when water at atmospheric pressure is freely available at the soil surface.

Of these, the indicator that is more important to hydrologists is infiltration capacity which is greater than or equal to the infiltration rate. Infiltrability, a term used by Hillel (1971) is used mostly by soil scientists. The term infiltration rate can therefore be used to represent its literal meaning under any surface condition – for example, at pressures greater or less than atmospheric.

If pressure > atmospheric, infiltration > infiltrability

If pressure < atmospheric, infiltration < infiltrability

When the rate of application of water is less than the infiltrability, the process is called supply (or flux) controlled. The water then infiltrates as fast as it arrives. When the rate of application of water is greater than the infiltrability, the process is called surface (or profile) controlled.

29.2 FACTORS AFFECTING INFILTRATION

29.2.1 Meteorological factors

The main meteorological factors are the rainfall intensity, type and the time. Impact of raindrops which depend upon the rainfall intensity and type throw fine particles into suspension, thereby clogging the pores resulting in a reduction of the infiltration rate. Infiltration is usually high at first, and reaches a steady state value for longer times.

29.2.2 Geotechnical factors

- **Physical properties of the soil:** Soil physical properties such as the porosity, grain size distribution, moisture content of the soil affect the infiltration rate. For example, if the soil is initially dry, the infiltration rate would be high. On the other hand, if it is wet, the infiltration rate will be lower, and the time to attain equilibrium will also be shorter.
- **Hydraulic properties of the soil:** Soil hydraulic properties such as soil hydraulic conductivity, soil water retention characteristic and soil water diffusivity greatly affect the infiltration property of soil.
- **Soil surface condition:** If the soil surface is dry and crusty, the infiltration rates would be small until the crust is broken.
- **Sub-surface soil condition:** If the sub-surface soil has an impending layer, the infiltration rate will be small.

29.2.3 Geographical factors

- **Topography of the surface:** If the surface is steep, water would tend to flow overland faster than it is absorbed into the soil.
- **Type and density of vegetation:** The presence of roots in the sub-surface tends to increase the voids in the soils, thereby increasing the capacity of the soil to hold more water. Density of vegetation has the effect of intercepting the rainfall, thereby not distributing it evenly across the ground surface. Also, transpiration of water depletes the soil moisture storage, thereby making the soil capable of absorbing more water.

29.3 METHODS OF DETERMINING INFILTRATION

29.3.1 By measurement

Infiltration rates can be measured by infiltrometers which can be of the flooding type or the sprinkler type.

29.3.1.1 Flooding type infiltrometers

The flooding type infiltrometer consists of two concentric cylinders of approximately 300–400 mm in diameter driven into the ground to a depth of about 600 mm (Figure 29.1). The upper portion of the two cylinders projects above the ground surface to a height of about 200 mm. Water is applied to both compartments to maintain the same level, but measurements are taken only in the inner cylinder. The function of the outer cylinder is to minimize the lateral spreading of water. The volume of water added to the inner cylinder in a given time to maintain a constant level is recorded as a function of time.

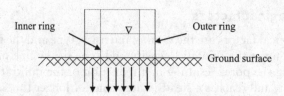

Figure 29.1 Double ring infiltrometer.

Flooding type infiltrometers give infiltration rates twice as much as the values obtained from sprinkler type infiltrometers. This is because there is no action of falling water drops. The results should therefore be applied for similar field conditions.

29.3.1.2 Sprinkler type infiltrometers

In the sprinkler type infiltrometers, rainfall is simulated by sprinklers over a test area. The surface runoff resulting from the simulated rainfall is measured, and the difference between the applied rainfall and the surface runoff gives the infiltration.

29.3.2 By estimation

29.3.2.1 Hydrograph analysis

The total loss in a storm is the difference between the volume of rainfall and the volume of direct runoff. If the total losses are assumed to be due to infiltration alone, then the hydrograph of rainfall and runoff can be used to estimate the cumulative infiltration (Figure 29.2). The gradient of the cumulative infiltration represents the infiltration rate.

The assumption that the total losses are only due to infiltration may not always be true. There are other losses such as interception losses and depression storage which cannot easily be separated from the total losses.

The advantage is that when applying the infiltration and other losses as calculated above to other storms, it is not necessary to consider them individually. The estimates by this method are more representative than those obtained by infiltrometer tests because the values obtained are for the entire catchment. One limitation is that the storm should be large enough to cover the entire catchment.

29.3.2.2 ϕ-index method

The ϕ-index is an estimate of the average value of infiltration over the period of rainfall which when added to the surface runoff makes the sum equal to the volume of precipitation (Figure 29.3). It includes any depression storage and surface retention.

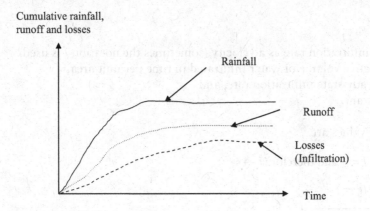

Figure 29.2 Hydrograph analysis.

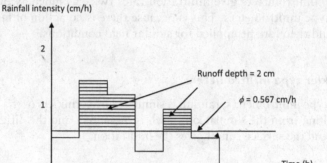

Figure 29.3 ϕ-Index method.

29.3.2.3 W-index method

The W-index method is similar to the ϕ-index method but excludes depression storage and retention. It is the average infiltration rate during the time rainfall intensity exceeds the infiltration capacity.

$$W = \frac{F}{t} = \frac{1}{t}(P - Q_s - S) \tag{29.1}$$

where F – Total infiltration,
t – Time during which rainfall exceeds infiltration capacity,
P – Total precipitation during time 't',
Q_s – Surface runoff,
S – Surface storage.

29.4 INFILTRATION EQUATIONS (EMPIRICAL)

29.4.1 Green and Ampt (1911) equation

Green and Ampt equation has some theoretical basis as well. It takes the form

$$i = i_c + \frac{b}{I} \tag{29.2}$$

where i – infiltration rate as a velocity (sometimes the notation f is used).
I – cumulative volume of water infiltrated in time per unit area,[1]
i_c – the steady state infiltration rate, and
b – a constant.

The limiting values are

At $t = 0$, $I = 0$ and therefore $i \rightarrow \infty$

as $t \rightarrow \infty$, $i \rightarrow i_c$

---

[1] $i = \dfrac{dI}{dt}$ or, $I = \int i\,dt$.

Figure 29.4 show typical shapes of some empirical equations for infiltration.
The theoretical basis can be described using Darcy's equation:

$$i = \frac{dI}{dt} = K\left(\frac{H_o - H_f}{L_f}\right) \tag{29.3}$$

where i – infiltration rate.
I – cumulative infiltration.
K – hydraulic conductivity of the transmission zone.
H_o – pressure head at entry surface.
H_f – pressure head at the wetting front.
L_f – distance from the surface to the wetting front.

If the ponding pressure is negligible and assumed to be zero (i.e., $H_o = 0$), then

$$i = \frac{dI}{dt} = -K\frac{H_f}{L_f} = K\frac{\Delta H_p}{L_f} \tag{29.4}$$

where ΔH_p is the change in pressure head from the surface to the wetting front.
If the moisture content in the wetted zone is assumed to be constant, then

$$I = L_f\Delta\theta \tag{29.5}$$

where $\Delta\theta$ is the change in moisture content of the wetted zone.
Differentiating Eq. 29.5,

$$i = \frac{dI}{dt} = \Delta\theta\frac{dL_f}{dt} \tag{29.6}$$

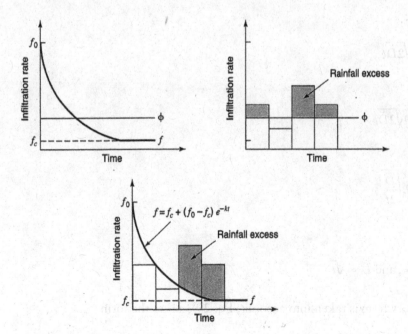

Figure 29.4 Infiltration curves.

$$= K \frac{\Delta H_p}{L_f} \quad \text{(comparing with Eq. 29.4)} \tag{29.7}$$

$$= \frac{K\Delta H_p}{I/\Delta \theta} \quad \left(\text{after substituting for } L_f \text{ from Eq. 29.5}\right) \tag{29.8}$$

$$= \frac{K\Delta\theta\Delta H_P}{I} = \frac{b}{I} \tag{29.9}$$

Thus, $i\left(= \dfrac{dI}{dt}\right)$ is inversely proportional to I.

Combining Eqs. 29.6 and 29.7 and rearranging,

$$L_f dL_f = K \frac{\Delta H_p}{\Delta \theta} dt = \tilde{D}dt \tag{29.10}$$

where $\tilde{D} = K \dfrac{\Delta H_p}{\Delta \theta}$ is considered as an effective diffusivity.

Integration of Eq. 29.10 gives,

$$\frac{L_f^2}{2} = \tilde{D}t + \text{const} (= 0) \tag{29.11}$$

Therefore,

$$L_f = \sqrt{2\tilde{D}t} \tag{29.12}$$

or

$$I = \Delta\theta\sqrt{2\tilde{D}t} \tag{29.13}$$

or

$$i = \Delta\theta\sqrt{\frac{\tilde{D}}{2t}} \tag{29.14}$$

Thus,

$$i \propto \frac{1}{\sqrt{t}}; \text{ and } L_f \propto \sqrt{t} \tag{29.15}$$

If the gravity term is taken into account, Eq. 29.3 takes the form

$$i = \frac{dI}{dt} = \frac{K(H_o - H_f + L_f)}{L_f} \tag{29.16}$$

$$= \frac{K(H_0 - H_f)}{L_f} + K \tag{29.17}$$

Substituting for L_f from Eq. 29.5,

$$i = \frac{dI}{dt} = \frac{K\Delta\theta(H_o - H)}{I} + K \tag{29.18}$$

which is of the form $i = i_c + \dfrac{b}{I}$ (Eq. 29.2).

Since $i = \Delta\theta \dfrac{dL_f}{dt}$, it is possible to write (using Eq. 29.16)

$$\Delta\theta \frac{dL_f}{dt} = \frac{K(H_0 - H_f + L_f)}{L_f} \tag{29.19}$$

Integration[2] of Eq. 29.19 gives,

$$\frac{Kt}{\Delta\theta} = L_f - (H_0 - H_f)\ln\left\{1 + \frac{L_f}{H_0 - H_f}\right\} \tag{29.20}$$

As t increases, the second term on the RHS of Eq. 29.20 increases more and more slowly in relation to the increase in L_f, so that for very large values of 't' it is possible to write,

$$L_f \approx \frac{Kt}{\Delta\theta} + \delta' \tag{29.21}$$

or

$$I = Kt + \delta \tag{29.22}$$

where δ is a constant.

The assumptions involved in these equations are as follows:

- There is a well-defined wetting front which advances at the same rate throughout the soil water system,
- The soil moisture content θ remains constant below and above the wetting front as it advances, and,
- The suction immediately below the wetting front remains constant.

29.4.2 Kostiakov (1932) equation

Kostiakov equation takes the form

$$i = Bt^{-n} \tag{29.23}$$

---

[2] $\displaystyle\int \frac{x\,dx}{A+x} = \int \frac{(A+x)\,dx}{A+x} - \int \frac{A\,dx}{A+x}.$

where B and n are empirical constants.

$i \to \infty$ as $t = 0$

$i \to 0$ as $t \to \infty$

The latter condition is somewhat unreal in vertical infiltration.

29.4.3 Horton equation (1940)

This is the most favoured empirical equation used by hydrologists. It takes the form

$$i = i_c + (i_0 - i_c) e^{-kt} \tag{29.24}$$

where i_c represents the final infiltration capacity, i_0, the initial infiltration capacity and k, the time constant.

The limiting values are

$i = i_0$ as $t = 0$

$i = i_c$ as $t \to \infty$

The difference between the infiltration capacity and the true instantaneous infiltration rate is of minor significance to hydrologists.

29.4.4 Philip's (1957) equation

Philip's equation is favoured by soil physicists. It has some theoretical basis too[3]. The equation is of the form

$$i = i_c + \frac{s}{2t^{1/2}} \tag{29.25}$$

where 's' is the **sorptivity** of the soil.

29.4.5 Holtan (1961) equation

Holtan equation is not as widely used as the others. It takes the form

$$i = i_c + a(M - I)^n \tag{29.26}$$

where 'a' is a constant, and M is the water storage capacity of the soil surface above the first impending stratum (porosity – initial moisture content), expressed in equivalent depth of water. When there is no impending stratum, the meaning of M is not clear.

$0 \le I \le M$, and, $i = i_c$ for $I > M$

[3] See Sections 29.5.4.1 and 29.5.4.2 on horizontal and vertical infiltration equations and their solutions

29.5 THEORY OF INFILTRATION

29.5.1 Physical process

The physical process of infiltration is a diffusion type of process which has many mathematical analogies. For example, the heat conduction problem and the convective diffusion problem are mathematically analogous to the infiltration problem which is one of moisture movement through partially saturated porous media.

29.5.2 The soil water system

The water in the soil can be represented by three zones as shown in Figure 29.5. At the bottom is the saturated zone in which all the voids in the soil are filled with water. The upper

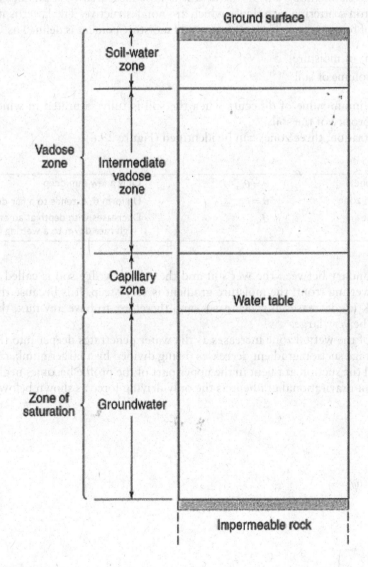

Figure 29.5 Soil water zones.

limit of this zone is the water table at which the pressure is at atmospheric. The pressure in the saturated zone is above atmospheric and is normally assumed to be hydrostatic.

The intermediate capillary water zone is also saturated because of the capillary action which pulls water from the water table upwards. The water in this zone is held by capillary forces between the soil particles and is at less than atmospheric pressure. The depth of the capillary zone depends on the type and compaction of the soil. It can range from a few centimetres in a coarse sandy soil to several metres in clay and silt.

In a soil water system, the voids are filled with air and water. The water particles are held in contact with the soil particles by the capillary forces. If the system is allowed to drain under gravity, the drainage will take place until the water content of the soil is reduced to the **field capacity**.

The state of a soil water system is measured by two parameters, namely the volumetric soil moisture content, θ, and the soil suction, ψ, which is negative pressure. The former is measured best by gravimetric methods, but in the field it is now measured by a number of methods, including neutron scattering techniques which are nondestructive. The latter is measured by various types of tensiometers. The volumetric soil moisture content is defined as

$$\theta = \frac{\text{Volume of moisture}}{\text{Bulk volume of soil}}$$

Thus, the maximum value of θ occurs when the soil is fully saturated in which case it is equal to the porosity of the soil.

During infiltration, three zones can be identified (Figure 29.6):

Saturated zone	$\theta = \theta_{sat}$	Only a few mm deep
Transmission zone	$\theta = \theta_{constant}$	Uniform θ, extends to a fair depth
Wetting zone	θ	Decreases with depth at an extremely high rate down to a wetting front

The sharp boundary between the wet soil and the relatively dry soil is called the **wetting front**. At the wetting front, the moisture gradient is very steep. It is because the hydraulic conductivity, $K(\theta)$ decreases as θ decreases, and therefore, to have any flux, the hydraulic gradient must be very large.

The length of the wetted zone increases as the water penetrates deeper into the soil. As a result, the average suction gradient decreases (being divided by a larger number). This trend continues until the suction gradient in the upper part of the profile becomes negligible, leaving the constant gravitational gradient as the only driving force as shown below:

$$q = -K \frac{\partial \phi}{\partial z} \tag{29.27a}$$

$$= -K \frac{d}{dz}(\psi - z) \tag{29.27b}$$

$$= K \tag{29.27c}$$

Therefore, the limiting value of flux = hydraulic conductivity

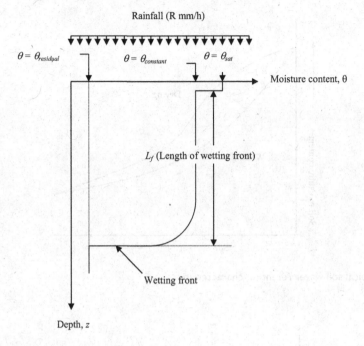

Figure 29.6 Soil water system.

29.5.3 Soil hydraulic parameters

There are three basic soil hydraulic parameters that are necessary to define the flow of moisture through partially saturated porous media. They are the soil moisture – soil suction characteristic, soil moistute – soil hydraulic conductivity characteristic and the soil moisture – soil diffusivity characteristic.

29.5.3.1 Soil suction – soil moisture content characteristic

Soil suction – soil moisture content characteristic is sometimes referred to as the retention curve. It gives the variation of the soil suction as a function of the moisture content. At the saturated end the moisture content has its maximum value, and the suction has its minimum value which is zero. At the other extreme is the dry end which is not achievable in the field and can only be attained in the laboratory after oven drying the soil for at least 24 h. The soil suction at the field dry end depends on the type of soil. The range of suctions corresponding to the range of applicable moisture contents can vary over several orders of magnitude.

An important property of this characteristic is that it depends on the history of the wetting and drying of the soil. The property known as **hysteresis** is analogous to the same property in magnetism. It can be obtained by observing the $\psi–\theta$ relationship of an initially saturated sample (**desorption**), or by observing the $\psi–\theta$ relationship of an initially dry sample (**sorption**). The two curves will be continuous but different. The moisture content on the drying curve is greater than the moisture content on the wetting curve for a given suction. There are many empirical equations that describe the soil suction-soil moisture characteristic. Some commonly used ones are given below, and the general shapes of the characteristic as the soil water content varies are given in Figure 29.7a. Typical scanning curves of drying and wetting between the boundary drying and wetting curves are shown in Figure 29.7b (after Mualem, 1976), and the typical shapes for fine and coarse textured soils are shown in Figure 29.7c.

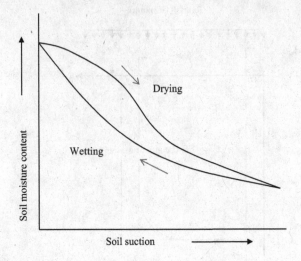

Figure 29.7a Typical soil water retention characteristic.

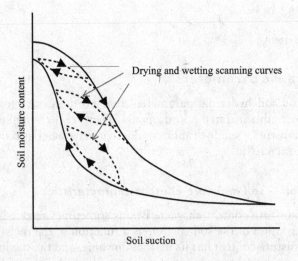

Figure 29.7b Drying and wetting scanning curves.

29.5.3.1.1 Brutsaert equation

Brutsaert (1966) equation is of the form

$$S_e = \frac{\alpha}{\alpha + |\psi|^{\beta}} \text{ or } \psi \leq 0, \text{ and } S_e = 1.0 \text{ for } \psi > 0 \tag{29.28}$$

where $S_e \, (0 \leq S_e \leq 1)$ is the effective degree of saturation, defined as

$$S_e = \frac{\theta - \theta_r}{\theta_s - \theta_r} \tag{29.29}$$

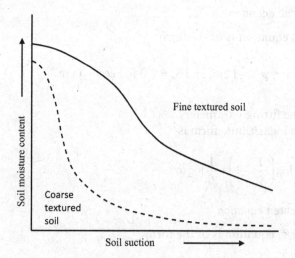

Figure 29.7c Typical soil water retention characteristic for different textured soils.

in which θ_r and θ_s are the residual and saturated soil moisture contents, and θ and ψ are the soil moisture content and soil suction, respectively. In Eq. 29.28, α and β are the parameters that should be determined.

By taking logarithms, Eq. 29.28 can be written as

$$\beta \log \psi = \log \left(\frac{1}{S_e} - 1 \right) + \log (\alpha)$$

$$\log (\psi) = \frac{1}{\beta} \log \left(\frac{1}{S_e} - 1 \right) + \frac{1}{\beta} \log (\alpha) \qquad (29.30)$$

which is a linear form for regression fitting.

29.5.3.1.2 Brooks and Corey equation

Brooks and Corey (1964) equation takes the form

$$S_e = \left(\frac{\psi_e}{\psi} \right)^{\lambda} \text{ for } \psi \leq \psi_e, \text{ and } S_e = 1.0 \text{ for } \psi > \psi_e \qquad (29.31)$$

where ψ_e is the air entry pressure, and λ is a fitting parameter.

Equation 29.31 in logarithmic form is

$$\log (\psi_e) = \frac{1}{\lambda} \log (S_e) + \log (\psi) \qquad (29.32)$$

29.5.3.1.3 Vauclin et al. equation

Vauclin et al. (1979) equation is of the form

$$S_e = \frac{\alpha}{\alpha + (\ln|\psi|)^{\beta}} \text{ for } \psi \leq -1 \text{ cm, and } S_e = 1.0 \text{ for } \psi > -1 \text{ cm} \tag{29.33}$$

where α and β are the fitting parameters.
   Equation 29.33 in logarithmic form is

$$\log\{\log(\psi)\} = \frac{1}{\beta}\log\left(\frac{1}{S_e} - 1\right) + \frac{1}{\beta}\log(\alpha) \tag{29.34}$$

29.5.3.1.4 Van Genuchten equation

Van Genuchten (1980) equation is of the form

$$S_e = \left[\frac{1}{1 + (\alpha|\psi|)^{\beta}}\right]^{\left(1 - \frac{1}{\beta}\right)} \text{ for } \psi \leq 0, \text{ and } S_e = 1.0 \text{ for } \psi > 0 \tag{29.35}$$

Equation 29.35 cannot be transformed into an explicit linear form by taking logarithms. Estimates of α and β are obtained by an iterative procedure starting from an initial set.

29.5.3.1.5 Other equations

Other forms of representation include equations of the form (Campbell, 1974; Clapp and Hornberger, 1978):

$$\psi = \psi_e\left[\frac{\theta}{\theta_s}\right]^{-b} \tag{29.36}$$

where b is a constant.

29.5.3.2 Soil hydraulic conductivity – soil moisture content characteristic

Soil hydraulic conductivity determines the capacity of the soil to conduct moisture. It is also sometimes referred to as the soil permeability. For partially saturated soils, the hydraulic conductivity depends on the degree of saturation (or the moisture content). At saturation, the hydraulic conductivity attains its maximum value asymptotically. The saturated hydraulic conductivity can be measured either in situ by field permeameters or in the laboratory using soil samples. In the latter case, the falling head permeameter and the constant head permeameter methods are commonly used. It is difficult to measure the unsaturated hydraulic conductivity and often various methods of estimation using probability concepts are used. A simple relationship used by Campbell (1974), Clapp and Hornberger (1978) is of the form

$$K = K_s\left[\frac{\theta}{\theta_s}\right]^{B} \tag{29.37}$$

where the subscript s refers to saturation conditions and B is a parameter. In this equation, K attains its saturation value when θ attains its saturation value.

29.5.3.3 Soil diffusivity — soil moisture content characteristic

Soil diffusivity is a derived parameter. It is obtained from the soil suction–moisture content and the hydraulic conductivity–moisture content relationships and defined as

$$D(\theta) = K(\theta) \frac{\partial \psi}{\partial \theta} \tag{29.38}$$

Because of the hysteresis effect of the soil suction–moisture content characteristic, the diffusivity is not a single valued function of the moisture content. At saturation, it tends to very large values. Using Eqs. 29.36 and 29.38,

$$D = bK(\theta) \left[\frac{\psi_e}{\theta_s} \right] \left[\frac{\theta_s}{\theta} \right]^{b+1} \tag{29.39}$$

29.5.4 Constitutive equations

The infiltration process is modelled by assuming that Darcy's law which is defined for the flow of water through saturated porous media is applicable to the flow of moisture through partially saturated porous media as well with some modifications. The driving force comes from the hydraulic gradients due to gravity and pressure forces. There are other forces such as osmotic and electrochemical, but they are small in comparison with the two major forces in the soil water system. The governing equations are obtained by combining the continuity equation with Darcy's law.

Considering a control volume of soil of unit cross sectional area, the continuity equation can be written as

$$-\frac{\partial q}{\partial z} = \frac{\partial \theta}{\partial t} \tag{29.40}$$

where q represents the volume flux of moisture through the control volume per unit area.

Darcy's law can be written as

$$q = -K(\theta) \frac{\partial \phi}{\partial z} \tag{29.41}$$

where $K(\theta)$ is the unsaturated hydraulic conductivity which is a function of the moisture content, thus making the relationship non-linear, and ϕ is the total potential which is the sum of the pressure head and the gravitational head:

$$\phi = \psi + z \tag{29.42}$$

Combining Eqs. 29.40, 29.41 and 29.42,

$$\frac{\partial \theta}{\partial t} = \frac{\partial}{\partial z} \left[K(\theta) \frac{\partial \phi}{\partial z} \right] \tag{29.43a}$$

$$= \frac{\partial}{\partial z} \left[K(\theta) \frac{\partial}{\partial z} (\psi + z) \right] \tag{29.43b}$$

$$= \frac{\partial}{\partial z}\left[K(\theta)\frac{\partial \psi}{\partial z}\right] + \frac{\partial K}{\partial z} \qquad (29.43c)$$

This is the basic constitutive equation of moisture movement through partially saturated porous media.

29.5.4.1 Horizontal infiltration

For horizontal infiltration, the gravity term does not exist. The only driving force is due to pressure (suction) gradient. Therefore, Eq. 29.43a simplifies to

$$\frac{\partial \theta}{\partial t} = \frac{\partial}{\partial x}\left[K(\theta)\frac{\partial \phi}{\partial x}\right] = \frac{\partial}{\partial x}\left[K(\theta)\frac{\partial \psi}{\partial x}\right] \qquad (29.44)$$

In this case, the water moves in the horizontal direction, and the soil absorbs water by suction gradients only.

For homogeneous soils, replacing $K(\theta)\dfrac{\partial \psi}{\partial \theta}$ by $D(\theta)$, Eq. 29.44 becomes

$$\frac{\partial \theta}{\partial t} = \frac{\partial}{\partial x}\left[D(\theta)\frac{\partial \theta}{\partial x}\right] \qquad (29.45)$$

The usual initial and boundary conditions are

$$\theta = \theta_i \quad \text{for } x \geq 0, \quad t = 0$$

$$\theta = \theta_0 \quad \text{for } x = 0, \quad t > 0$$

By using the Boltzmann (1894) transformation,[4] which is given as

$$\lambda(\theta) = xt^{-\frac{1}{2}} \qquad (29.46)$$

where x is the distance from the surface to the wetting front, the partial differential equation, Eq. 29.45 can be transformed into an ordinary differential equation:

$$\frac{-\lambda}{2}\frac{d\theta}{d\lambda} = \frac{d}{d\lambda}\left[D(\theta)\frac{d\theta}{d\lambda}\right] \qquad (29.47)$$

The boundary conditions then change to

$$\theta = \theta_i, \lambda \to \infty; \quad \theta = \theta_0, \lambda = 0.$$

Since $D(\theta)$ is a non-linear function of θ, Eq. 29.47 has to be solved numerically. If θ is assumed to remain constant in the transmission zone, then, from Eq. 29.46, x would be proportional to $\sqrt{t}$, and a plot of x vs. $\sqrt{t}$ would give a straight line of slope λ.

---

[4] Boltzmann transformation was originally used to transform the diffusion equation (partial differential) to an ordinary differential equation by introducing a new variable which is a combination of the two independent variables in the partial differential equation.

The cumulative infiltration, I, is given by

$$I = \int_{\theta_i}^{\theta_0} x \, d\theta \qquad (29.48)$$

where θ_i – initial moisture content.

θ_0 – final moisture content.

Substituting for x from Eq. (29.46)

$$I = \int_{\theta_i}^{\theta_0} \lambda(\theta) t^{1/2} \, d\theta = s t^{1/2} \qquad (29.49)$$

where

$$s = \int_{\theta_i}^{\theta_0} \lambda(\theta) \, d\theta = \frac{I}{t^{1/2}} \qquad (29.50)$$

This is a constant called the **sorptivity** (Philip, 1969) and is the same as that referred to in Eq. 29.25. It depends on θ_i and θ_0, and has dimensions of $LT^{-1/2}$.

By plotting I vs. $t^{1/2}$, a straight line with slope 's' can be obtained.

29.5.4.2 Vertical infiltration

For vertical infiltration, both the gravity and the suction terms apply. For the case of infiltration into a homogeneous semi-infinite medium, it is convenient to measure z positive in the downward direction. Then (by substituting $z = -z$), Eq. 29.43c takes the form

$$\frac{\partial \theta}{\partial t} = \frac{\partial}{\partial z}\left[K(\theta) \frac{\partial \psi}{\partial z} \right] - \frac{\partial K(\theta)}{\partial z} \qquad (29.51)$$

By introducing the soil diffusivity D which is defined as

$$D(\theta) = K(\theta) \frac{\partial \psi}{\partial \theta} \ (\text{see Eq. 29.38})$$

Equation 29.51 can be transformed into the following form:

$$\frac{\partial \theta}{\partial t} = \frac{\partial}{\partial z}\left[D(\theta) \frac{\partial \theta}{\partial z} \right] - \frac{\partial K(\theta)}{\partial z} \qquad (29.52)$$

Equation 29.51 can also be written as

$$C(\psi) \frac{\partial \psi}{\partial t} = \frac{\partial}{\partial z}\left[K(\psi) \frac{\partial \psi}{\partial z} \right] - \frac{\partial K(\psi)}{\partial z} \qquad (29.53)$$

where $C(\psi) = \dfrac{\partial \theta}{\partial \psi}$ (= specific moisture capacity) $\qquad (29.54)$

Equations 29.52 and 29.53 form the constitutive equations which govern the one-dimensional vertical movement of moisture through a porous medium and are usually identified as the θ-based and the ψ-based equations, respectively. Although θ and ψ are interdependent, D and K in Eq. 29.52 are usually expressed as functions of θ while C and K in Eq. 29.53 are expressed as functions of ψ.

In both Eqs. 29.52 and 29.53, the driving force consists of the suction gradient, which arises out of wetness and the gravitational gradient. Their relative magnitudes differ depending upon initial and boundary conditions.

For example, when the soil is initially dry the suction gradient is very much greater than the gravitational gradient. Then, the vertical infiltration rates approach the horizontal infiltration rates.

When $\theta \rightarrow \theta_{sat}$, $D \rightarrow \infty$, and hence the θ-based equation cannot be used. Instead, ψ-based equation is used.

As $\theta \rightarrow \theta_{sat}$, $K(\theta) \rightarrow K$ (constant), and Eq. 29.51 simplifies to

$$K\frac{\partial^2 \psi}{\partial z^2} = 0 \tag{29.55}$$

which is the **Laplace equation**.

For initially wet soils, suction gradients are small.

29.5.5 Boundary and initial conditions

There are three different boundary and initial conditions that can be applied to Eqs. 29.52 and 29.53, when describing infiltration. They are briefly defined in the following equations.

29.5.5.1 Ponded infiltration

$$\theta(z,0) = \theta_i \quad \text{for } z \geq 0; \quad t = 0 \tag{29.56a}$$

$$\theta(0,t) = \theta_0 \quad \text{for } z = 0; \quad t \geq 0 \tag{29.56b}$$

where θ_i and θ_0 are the initial and surface moisture contents, respectively (usually $\theta_0 > \theta_i$). They may be constants or functions of z or t. The most common condition in infiltration is that there is a thin layer of water available at the surface. Then, the surface moisture content is the saturated value θ_{sat} and is called ponded infiltration condition. Then,

$$\theta(0,t) = \theta_{sat} \quad \text{for } z = 0; \quad t \geq 0 \tag{29.56c}$$

29.5.5.2 Rain infiltration – low rainfall intensities

$$\theta(z,0) = \theta_i \quad \text{for } z \geq 0; \quad t = 0 \tag{29.57a}$$

$$\theta(\infty,t) = \theta_i \quad \text{for } z \rightarrow \infty; \quad t \geq 0 \tag{29.57b}$$

$$\text{Flux} = -K(\theta)\left[\frac{\partial \psi}{\partial z} - 1\right] = R \quad \text{for } z = 0; \quad t > 0 \tag{29.57c}$$

where R is the rainfall intensity.

The flux boundary condition given by Eq. 29.57c is obtained as follows:

$$q = -K(\theta)\frac{\partial \phi}{\partial z} = -K(\theta)\left[\frac{\partial \psi}{\partial z} + \frac{\partial z}{\partial z}\right];$$

Rainfall is an inward flux. Therefore, $R = -q$. Substituting for q and using the co-ordinate transformation $z = -z$,

$$-R = -K(\theta)\left[-\frac{\partial \psi}{\partial z} + 1\right] \qquad (29.57a)*$$

$$R = -K(\theta)\left[\left(\frac{\partial \psi}{\partial z} - 1\right)\right] \qquad (29.57b)*$$

Because Eq. 29.57a* can be written in the form

$$-R = K(\theta)\frac{\partial \psi}{\partial \theta}\frac{\partial \theta}{\partial z} - K(\theta) \qquad (29.57c)*$$

The condition (29.57c*) can also be written as

$$\frac{\partial \theta}{\partial z} = -\frac{R - K(\theta)}{D(\theta)} \qquad (29.57d)$$

This condition corresponds to rain infiltration and is applicable from the beginning of rainfall to the time of occurrence of incipient ponding. For low rainfall intensities $[R < K(\theta_{sat})]$, rain infiltration can continue without giving rise to ponding. As time passes, the surface moisture content approaches a limiting value θ_L.

29.5.5.3 Rain infiltration – high rain intensities

For high rainfall intensities $[R > K(\theta_{sat})]$, the ψ-based equation is preferred because the $D(\theta)$ term in the θ-based equation tends to very large values near saturation. Then,

$$\psi(z, 0) = \psi_i \quad \text{for } z \geq 0; \quad t = 0 \qquad (29.58a)$$

$$\psi(\infty, t) = \psi_i \quad \text{for } z \rightarrow \infty; \quad t \geq 0 \qquad (29.58b)$$

$$\psi(0, t) = \psi_f \geq 0 \quad \text{for } z = 0; \quad t \geq t_p \qquad (29.58c)$$

$$\text{Flux} = -K(\psi)\left[\frac{\partial \psi}{\partial z} - 1\right] = R \quad \text{for } z = 0; \quad 0 \leq t \leq t_p \qquad (29.58d)$$

where
  ψ_i – initial soil water pressure (suction),
  ψ – surface soil water pressure during ponding (hydrostatic),
  t_p – time of incipient ponding.

The physical meaning of this condition is that the rainfall intensity is exceeding the infiltration capacity of the soil, and therefore ponding of water at the surface is taking place. In Eq. 29.58c, ψ_f can be taken as zero.

29.5.6 Solutions of the equations

29.5.6.1 Ponded infiltration – linearized solution

Equation 29.52 can be linearized as follows:

$$\frac{\partial \theta}{\partial t} = D_* \frac{\partial^2 \theta}{\partial z^2} - u \frac{\partial \theta}{\partial z} \tag{28.59}$$

where $D_* (= D(\theta))$ and $u \left(= \dfrac{\partial K}{\partial \theta} \right)$ are assumed to be constants. The θ's in Eqs. 29.52 and 29.59 are identical. The relevant boundary and initial conditions are given by Eq. 29.56. Equation 29.59 is analogous to the convective dispersion equation with constant parameters for which an analytical solution has been proposed by Ogata and Banks (1961). Their solution, with variables changed to reflect the infiltration process is of the form

$$\theta = \theta_i + \frac{\theta_0 - \theta_i}{2} \left\{ erfc \frac{z - ut}{\sqrt{4D_* t}} + \exp\left(\frac{uz}{D_*} \right) erfc \frac{z + ut}{\sqrt{4D_* t}} \right\} \tag{29.60}$$

where

$$erf(x) = \frac{2}{\sqrt{\pi}} \int_0^x e^{-x^2}\, dx; \, erfc(x) = 1 - erf(x) = \frac{2}{\sqrt{\pi}} \int_x^\infty e^{-x^2}\, dx \tag{29.61}$$

and u can be written as $\dfrac{K_0 - K_i}{\theta_0 - \theta_i}$. $\tag{29.62}$

(This is also the downward rate of advance of soil wetness for large times.)

The error function $erf(x)$ varies from 0 to 1 asymptotically. It is the same as the cumulative normal probability function.

29.5.6.2 Ponded infiltration – non-linear solution

The first mathematical solution to the vertical infiltration equation for an infinitely deep soil was proposed by Philip (1957). His initial and boundary conditions were

$$\theta = \theta_i \quad \text{for } t = 0, \quad z > 0$$

$$\theta = \theta_0 \quad \text{for } t \geq 0, \quad z = 0.$$

which means that the surface value is increasing from an initial value of θ_i to a final value of θ_0 instantaneously. His solution is based on a power series expansion and takes the form

$$z(\theta, t) = f_1(\theta)t^{1/2} + f_2(\theta)t + f_3(\theta)t^{3/2} + \ldots \tag{29.63}$$

where z is the depth to any particular value of θ and $f_i(\theta)$ are calculated successively from the $K(\theta)$ and $D(\theta)$ functions.

This shows that $\theta \ \alpha \ \sqrt{t}$ for small t (same as horizontal infiltration). For large t, vertical movement of moisture approaches the constant rate u, defined in Eq. 29.62.

Similarly, Philip also showed that the cumulative infiltration, I, can be expressed as

$$I(t) = st^{1/2} + (A_2 + K_0)t + A_3t^{3/2} + A_4t^2 + \dots \tag{29.64}$$

where the A's are calculated from $K(\theta)$, $D(\theta)$ functions; and K_0 is the conductivity at $\theta = \theta_0$.

Differentiating Eq. 29.64,

$$i(t) = \frac{dI}{dt} = \frac{1}{2}st^{-1/2} + (A_2 + K_0) + \frac{3}{2}A_3t^{1/2} + \dots \tag{29.65}$$

Representing Eq. 29.65 by a two-parameter approximation (for t not too large – This is done in practice)

$$i(t) = \frac{1}{2}st^{-1/2} + A \tag{29.66a}$$

or

$$I(t) = st^{1/2} + At \tag{29.66b}$$

As $t \to \infty$, i decreases monotonically to its asymptotic value $i(\infty)$. This does not imply $A = K_0$ for small and intermediate values of time.

For large times the series is divergent, and it is possible to write Eq. 29.66 as (with A replaced by K)

$$I = st^{1/2} + Kt, \text{ or, } i = \frac{1}{2}st^{-1/2} + K \tag{29.67}$$

where K is the hydraulic conductivity of the upper layer of soil.

29.5.6.3 Rain infiltration – low intensities

For rain infiltration (flux boundary condition), there are no known analytical solutions. Numerical solutions are available for specific boundary and initial conditions. Both finite difference and finite element methods have been used (Jayawardena, 1985; Jayawardena and Kaluarachchi, 1986).

29.5.6.4 Rain infiltration – high intensities

Ponding will occur when the rainfall intensity exceeds the surface saturated conductivity value and overland flow will follow. This state is described by Eqs. 29.52 and 29.56. In the case for high rainfall intensities, solutions are only of numerical type, and a method commonly used for verifying the validity of a numerical procedure consists of comparing the depth to the air entry pressure position of the suction profile at ponding (depth of saturated zone at ponding). The analytical value of this depth, obtained by consideration of Darcy's law, is given by the following equation:

$$R = K(\theta_{sat}) \left[\frac{\psi_A - \psi_0}{B} + 1 \right] \tag{29.68}$$

where B is the depth of saturation at ponding, and ψ_0 is the surface pressure head. At $t = t_p$, $\psi_0 = 0$, and therefore

$$B(t_p) = \frac{\psi_A}{\dfrac{R}{K(\theta_{sat})} - 1} \tag{29.69}$$

Example 29.1

During a storm, 50 mm of water infiltrated into the sub-soil while maintaining 10 mm of surface ponding in a period of 2 h. The initial moisture content of the soil was 0.10, which reached a uniform value of 0.5 in the transmission zone at the end of the 2 h period. The soil suction at the end of the transmission zone is 150 mm. Determine

a. the depth to the wetting front,
b. the saturated hydraulic conductivity of the transmission zone, and,
c. the infiltration rate after 1 hour.

a. $I = L_f \Delta\theta \Rightarrow L_f = \dfrac{50}{0.4} = 125$ mm.

b. $Kt = I - (H_0 - H_f)\Delta\theta \ln\left(1 + \dfrac{I}{(H_0 - H_f)\Delta\theta}\right) = 50 - 160$

$$\times 0.4 \ln\left(1 + \frac{50}{160 \times 0.4}\right) = 50 - 36.95$$

$$Kt = 13.05 \Rightarrow K = \frac{13.05}{2} = 6.52 \text{ mm/h}.$$

c. Plot I vs. t and obtain the gradient at $t = 1$ h. Since it is easier to use I as an independent variable,

I (mm)	10	20	30	40	50
t (h)	0.109	0.398	0.828	1.369	2.00

Gradient at $t = 1$ h is $\dfrac{10}{0.541} = 18.5$ mm/h.

Example 29.2

a. Determine the ϕ-index for two rainstorms, one with a steady intensity of 25 mm/h lasting for 3 h and the other with intensities of 12.5 mm/h for the first 2 h, 35 mm/h for the next 2 h and 27.5 mm/h for the last 2 h. The volumes of direct runoff in the two cases are 45 and 115 mm, respectively.
b. Using Horton's equation with $i_0 = 18$ mm/h, $i_c = 5$ mm/h and $k = 0.5$/h determine the distribution of losses and the volumes of rainfall excesses for the storm referred to above.

a. For the steady rainfall,
Total rainfall = 75 mm
Direct runoff = 45 mm
Therefore, losses = 30 mm

$$\phi\text{-index} = \frac{30}{3} = 10 \text{ mm/h}.$$

For the intermittent rainfall,
Total rainfall = 150 mm
Direct runoff = 115 mm
Therefore, losses = 35 mm

$$\phi\text{-index} = \frac{35}{6} \approx 6 \text{ mm/h}.$$

b. Horton's equation is $i = i_c + (i_0 - i_c)e^{-kt} \Rightarrow i = 5 + 13e^{-0.5t}$.

Time (h)	i (mm/h)	Storm 1		Storm 2	
		Rainfall (mm/h)	Rainfall excess (mm/h)	Rainfall (mm/h)	Rainfall excess (mm/h)
1	12.9	25	12.1	12.5	–
2	9.78	25	15.22	12.5	2.72
3	7.90	25	17.1	35.0	27.1
4	6.76	–	–	35.0	28.24
5	6.07	–	–	27.5	21.43
6	5.65	–	–	27.5	21.85

REFERENCES

Boltzmann, L (1894): About the integration of the diffusion equation in the case of variable diffusion coefficients, *Annalen der Physik und Chemie*, vol. 53, pp. 959–964.

Brooks, R. H. and Corey, A. T. (1964): *Hydraulic Properties of Porous Media*, Hydrology Paper No. 3, Colorado State University, Fort Collins.

Brutsaert, W. (1966): Probability for pore size distributions, *Soil Science*, vol 101 no.:2, pp. 85–92.

Campbell, G. S. (1974): A simple method for determining unsaturated conductivity from moisture retention data, *Soil Science*, vol. 117, pp. 311–314.

Clapp, R. B. and Hornberger, G. M. (1978): Empirical equations for some soil hydraulic properties, *Water Resources Resources*, vol. 14 no.:4, pp. 601–604.

Green, W. H., and Ampt, G. A. (1911): Studies on soil physics, *Journal of Agricultural Science*, vol. 4 no.:1, pp. 1–24.

Hillel, D. (1971): *Soil and Water: Physical Principles and Processes*. Academic Press, New York.

Holtan, H. N. (1961): A concept for infiltration estimates in watershed engineering. USDA-ARS Bulletin 41–51, 25 pp.

Horton, R. E. (1940): The role of infiltration in the hydrologic cycle. Transactions of the AGU, 14th Annual Meeting, Washington, DC, pp. 446–460.

Jayawardena, A. W. (1985): Moisture movement through unsaturated porous media: Numerical modelling, calibration and application, Proceedings of the 21st Congress, IAHR, Aug 19–23, Melbourne, Australia, vol. 1, pp. 12–16.

Jayawardena, A. W. and Kaluarachchi, J. J. (1986): Infiltration into decomposed granite soils: Numerical modelling, application and some laboratory observations, *Journal of Hydrology*, vol. 84 no.:3–4, pp. 231–260.

Kostiakov, A. N. (1932): On the dynamics of the coefficient of water-percolation in soils and on the necessity of studying it from a dynamic point of view for purposes of amelioration, Transactions of 6th Congress of International Soil Science Society, Moscow, pp. 17–21.

Mualem, Y. (1976): A new model for predicting the hydraulic conductivity of unsaturated porous media, *Water Resources Research*, vol. 12 no.:3, pp. 513–522.

Ogata, A. and Banks, R. B. (1961): *A Solution of the Differential Equation of Longitudinal Dispersion in Porous Media, Fluid Movement in Earth Materials Geological Survey Professional Paper 411-a*, United States Government Printing Office, Washington, DC, pp. A1–A7.

Philip, J. R. (1957): The theory of infiltration: 4. Sorptivity and algebraic infiltration equations, *Soil Science*, vol. 84, pp. 257–264.

Philip, J. R., (1969). The theory of infiltration, *Advances in Hydroscience*, vol. 5, pp. 215–296.

Van Genuchten, M. T. (1980): A closed-form equation for predicting the hydraulic conductivity of unsaturated soils, *Journal of the Soil Science Society of America*, vol. 44, pp. 892–898.

Vauclin, M., Haverkamp, R. and Vachaud, G. (1979): *Resolution numerique d'une equation de diffusion non lineaire*, Presses Universitaires de Grenoble, Grenoble, 183 pp.

Chapter 30

Runoff

30.1 INTRODUCTION

Runoff is the outcome of precipitation. When the rates of rainfall (and other forms of precipitation) exceed the rates of infiltration, interception and depression storage, the difference emerges as surface runoff. The infiltrated water may also contribute to the total runoff. Components of runoff include surface runoff (overland flow), which is the component that flow over the surface of the catchment, sub-surface runoff, which is the re-emergence of infiltrated water that takes place immediately below the ground surface, and groundwater flow, or base flow. The response time of surface runoff is short, meaning that the time lag between the precipitation and resulting surface runoff is short. The response time of sub-surface runoff component is relatively long, meaning that there is a reasonable time lag between the precipitation and the resulting sub-surface runoff. It is also dependent upon the sub-surface geological and soil conditions. Groundwater flow, or base flow, is the slowest component of runoff. Replenishment takes place through percolation of infiltrated water. The graphical plot of runoff vs. time is called the hydrograph.

30.2 MEASUREMENT OF RUNOFF

Runoff is measured directly by flow measuring devices. It can also be indirectly determined by measuring the depth and velocity. In selecting a site for making measurements, it is necessary to ensure that the variation of depth does not cause a change in the cross-sectional shape, and, that the bed profile is stable.

Depth of flow is measured as stage, which is the elevation of the water surface level above a known datum. Details about stage, velocity and discharge measurements are given in Chapter 9. Rating curve gives the relationship between the stage and the discharge. Usually, it is of the form

$$Q = aH^b \tag{30.1}$$

where H is the stage (or depth), and a and b are constants. It should be re-calibrated at frequent intervals of time to ensure its validity. Extreme caution must be exercised when the rating curve has to be extrapolated.

30.3 HYDROGRAPH SHAPE

A typical hydrograph has three components, a rising segment, a crest segment and a recession segment

30.3.1 Rising segment (concentration curve)

The rising segment extends from the time of the beginning of surface runoff to the first inflexion point on the hydrograph. It depends on storm and catchment characteristics.

30.3.2 Crest segment

Crest segment refers to the part of the hydrograph between the two inflexion points (one on the rising segment and the other on the recession segment). The peak occurs in this segment.

30.3.3 Recession segment (depletion curve)

Recession segment refers to the remaining part of the hydrograph, which may or may not reduce to zero. It represents the withdrawal of water from storage after excess rainfall has ceased. It is analogous to the curve representing the draining of water from a tank. The shape of the recession curve depends on the catchment characteristics.

The recession curve can be represented by an exponentially decaying type equation of the form

$$Q_t = Q_0 e^{-kt} \qquad\qquad (30.2a)$$

where
Q_t – discharge at time 't'
Q_0 – initial discharge (at $t=0$)
K – recession constant.

This equation plots as a straight line on a semi-log scale:

$$\ln(Q_t) = \ln(Q_0) - kt \qquad\qquad (30.2b)$$

30.4 FACTORS AFFECTING THE HYDROGRAPH SHAPE

30.4.1 Climatic factors

Climatic factors dominate in determining the shape of the rising segment of the hydrograph. The main climatic factors are as follows.

30.4.1.1 Rainfall intensity and duration

The intensity governs the time to peak and the peak value whereas the duration governs the time base of the hydrograph.

30.4.1.2 Distribution of rainfall in the catchment

Depending upon where the high intensity occurs, the base length of the hydrograph may increase or decrease. For example, if the high intensity rainfall occurs near the outlet, the time base will be short. In comparison, if the high intensity rainfall occurs at the far end of the catchment, the time base would be relatively longer.

30.4.1.3 Direction of storm movement

The direction of storm movement tends to increase or decrease the peak value by decreasing or increasing the time base of the hydrograph depending upon whether the storm is moving downstream or upstream.

30.4.1.4 Type of precipitation

If the rates of rainfall excesses are high, the concentration curve would be rapidly rising. Intermittent rainfalls tend to give rather flat multi-peak hydrographs.

30.4.2 Topographical factors

30.4.2.1 Catchment area and shape

Increase in the catchment area tends to increase the base length of the hydrograph.

30.4.2.2 Stream network

Stream network affects the time to peak.

30.4.2.3 Shape of main stream and valleys

Tend to affect the velocity and hence the time to peak.

30.4.2.4 Depression storage

Reduces the peak and increases the base length.

30.4.3 Geological factors

Geological factors govern the flow of sub-surface runoff and, therefore, mainly affect the recession curve.

30.5 BASE FLOW SEPARATION

In most hydrological analysis it is necessary to compare the rainfall excess with the direct (surface) runoff. Base flow should then be separated from the total runoff. Several methods (all approximate) of which the simplest is a horizontal line from the beginning of direct runoff are available.

30.5.1 Straight line method

A straight line is drawn from the point at which direct runoff begins to the point at which direct runoff ends. The latter is obtained by a semi-log plot of total runoff vs. time and determining the point of intersection of the two segments (which should be straight lines with different slopes).

30.5.2 Fixed base length separation

In this method, the point at which direct runoff begins is the same as in the straight line method. The end point is determined on the assumption that the surface runoff always ends after a fixed interval of time T, measured after the peak of the hydrograph. The time T is expressed as an empirical function of the catchment area A, in the form

$$T = A^n \tag{30.3}$$

where n is an empirical parameter.

30.5.3 Variable slope separation

In this method, the existing recession curve is extended to a point in time just below the peak. The end point is obtained by a semi-log plot as before. The semi-log part is extended backwards to a point which is just below the second point of inflexion (on the recession curve) of the hydrograph. The two points are then joined rather arbitrarily.

30.5.4 Master depletion curve method

The procedure is based on the recession data from a number of hydrographs, which will cover a wide range of discharges and seasons. The recession curves are plotted using a semi-log scale on tracing paper with one graph for each event. On a separate master sheet, also using a semi-log scale, the recession parts are transferred from the individual sheets starting from the graph having the lowest magnitude such that all the recession parts fall upon the straight segment of the master depletion curve. An empirical equation can then be fitted to the master depletion curve.

30.6 METHODS OF ESTIMATING PEAK RUNOFF

30.6.1 Use of empirical methods – rational formula (method)

Empirical methods aim at expressing the peak runoff in terms of a few important parameters. They are still used because of the simplicity. The most widely used empirical method is the rational method.

The rational formula was first introduced by Mulvaney, an Irish Engineer, in 1851. In the USA, it is called the Kuichling formula (1889), and in the UK it is called the Lloyd–Davies formula (1906).

The rational method relates the peak runoff as a percentage of the rainfall intensity when the entire catchment is contributing to the runoff. In equation form, it is

$$Q_p = CIA \tag{30.4}$$

where
Q_p – peak runoff
C – runoff coefficient $(0 < C < 1)$
I – rainfall intensity of a storm whose duration is greater than or equal to the time of concentration of the catchment
A – catchment area.

When Q_p is expressed in m³/s; I in mm/h; A in km², $Q_p = 0.278\ CIA$.

The formula assumes that the rainfall intensity is uniform over the entire basin throughout the duration of the storm.

The time of concentration referred to in the use of the rational method is the time required for a particle of water to move from the furthest point in the catchment to its outlet. There are several empirical equations to estimate it.

30.6.1.1 Kirpich (1940) equation

$$t_c = 0.0195 L^{0.77} S^{-0.385}$$
(30.5a)

where t_c is the time of concentration (minutes), L is the maximum length of travel (metres), S is the slope ($= H/L$) where H is the difference in elevation between the furthest point in the catchment and the outlet.

The time of concentration can also be expressed as a sum of the time of entry t_e and a time of travel t_t as follows:

$$t_c = t_e + t_t$$

Time of entry is given by the following equations.

30.6.1.2 Mockus (1957) equation

$$t_e = \frac{t_L}{0.6}$$
(30.5b)

where t_e is the time of entry (overland flow concentration time or time of overland flow), and t_L is the lag time.

30.6.1.3 Ragan and Duro (1972) equation

$$t_e = \frac{6.917(nL)^{0.6}}{I^{0.4} S^{0.3}} \text{ min.}$$
(30.5c)

where n – Manning's roughness coefficient (= 0.2 for paved areas and 0.5 for grass).

Time of travel is simply the length of the channel divided by the velocity.

In addition to above, there are also equations proposed by Izzard, Kirby, etc. and charts provided by Raudkivi (1979: p. 299). There are other equations of the form

$$Q_p \propto A^N$$
(30.5c)

which is also the basis for the rational method.

30.6.1.4 Runoff coefficient

The runoff coefficient in the rational formula depends on factors such as the

- nature of the surface
- slope
- surface storage
- degree of saturation
- rainfall intensity.

Values of C are given in several textbooks and some stormwater design manuals (e.g., Hong Kong Stormwater Drainage Manual). Typical values of some surfaces are given below:

- Sandy soil 0.05–0.10
- Heavy soils flat 0.13–0.17
- Heavy soils steep 0.25–0.35
- Impervious areas 0.95.

For non-homogeneous areas, a weighted runoff coefficient can be obtained by the following equation:

$$C_w = \frac{\sum A_1 C_1}{\sum A_1} \tag{30.6}$$

30.6.2 Peak discharge as a design parameter

Peak discharges are used in the design of storm sewers, runoff pipe systems, culverts and other minor drainage structures. It is a commonly used design variable for situations where the time variation of runoff is of no significance. If the time variation is important, then the hydrograph is used as the design variable.

30.6.3 Time – area method – an extension of rational method

If the catchment is large, the times of concentration are best determined by dividing the catchment into different zones. The zoning is such that from any point on any given isochrone the runoff reaches the outlet at the same time. With the time – area distribution and assuming different runoff coefficients for different regions and times, it is possible to determine the hydrograph for a catchment.

When the catchment is non-homogeneous, it is subdivided into a number of homogeneous sub catchments and the rational method applied to each of the homogeneous sub-catchments. The practice follows two rules:

- For each inlet area, the rational method is used to compute the peak discharge
- For inlets where the runoff (drainage) is arriving from two or more sub-areas, the longest time of concentration is used to determine the design rainfall intensity. Runoff is calculated using a weighted runoff coefficient.

Example 30.1

A catchment has four sub-areas with catchment areas $A_1 = 500$ ha; $A_2 = 600$ ha; $A_3 = 800$ ha and $A_4 = 200$ ha. A_1 is closest to the outlet. The runoff coefficients and rainfall distribution are as shown below:

$$0-1\,\text{h}, \quad C = 0.5 \quad 15\,\text{mm}$$
$$1-2\,\text{h}, \quad C = 0.7 \quad 30\,\text{mm}$$
$$2-3\,\text{h}, \quad C = 0.8 \quad 30\,\text{mm}$$
$$3-4\,\text{h}, \quad C = 0.85 \quad 15\,\text{mm}$$

Assuming that the rainfall and runoff coefficients are average values for the duration of each hour, determine the resulting direct runoff hydrograph.

The computations are given in tabular form.

| Time (h) | Rainfall (mm) | Contribution from each area Q = CIA | | | | Total |
		Area A_1	Area A_2	Area A_3	Area A_4	Runoff
1	2	3	4	5	6	7
0–1	15	10.4	–	–	–	10.4
1–2	30	29.2	12.5	–	–	41.7
2–3	30	33.3	35.0	16.67	–	85.0
3–4	15	17.7	40.0	46.67	41.67	146.0
4–5			20.8	53.3	116.67	191.0
5–6				27.78	133.33	161.0
6–7					69.44	69.4

Computations for columns 3–7 are done as follows:

For A_1, hour 1, $Q_1 = 0.5 \times 15 \times 500 \ (\text{mm ha/h}) = 10.4 \ \text{m}^3/\text{s}$

hour 2, $Q_2 = 0.7 \times 30 \times 500 \ (\text{mm ha/h}) = 29.2 \ \text{m}^3/\text{s}$

hour 3, $Q_3 = 0.8 \times 30 \times 500 \ (\text{mm ha/h}) = 233.3 \ \text{m}^3/\text{s}$

hour 4, $Q_4 = 0.85 \times 15 \times 500 \ (\text{mm ha/h}) = 17.7 \ \text{m}^3/\text{s}$

For A_2, hour 1, $Q_1 = 0.5 \times 15 \times 600 \ (\text{mm ha/h}) = 12.5 \ \text{m}^3/\text{s}$

hour 2, $Q_2 = 0.7 \times 30 \times 600 \ (\text{mm ha/h}) = 35.0 \ \text{m}^3/\text{s}$

hour 3, $Q_3 = 0.8 \times 30 \times 600 \ (\text{mm ha/h}) = 40.0 \ \text{m}^3/\text{s}$

hour 4, $Q_4 = 0.85 \times 15 \times 600 (\text{mm ha/h}) = 20.8 \ \text{m}^3/\text{s}$

(For the second hour discharges are delayed by 1 h)

Similarly, for A_3 and A_4, the corresponding discharges are computed and are summed up to obtain the total effect.

Example 30.2

Determine the peak discharge from a catchment of area 20 ha for a 50-year storm given the following:

$$i = \frac{569}{(t_c + 2.7)^{0.38}} \text{ with } I \text{ in mm/h and } t_d \text{ in minutes.}$$

Length of overland flow – 100 m
  Slope of overland flow plane – 0.02
  Time of concentration for overland flow – 20 min
  Length of stream – 300 m
  Slope of stream – 0.004
  Impervious area – 10 ha with a runoff coefficient of 0.9
  Grassland area – 10 ha with a runoff coefficient of 0.4
  Stream cross section – trapezoidal with a bottom width of 2 m, maximum depth of flow of 1 m, and side slopes 1:1

Manning's 'n' for the stream section – 0.03

Will the stream overflow?

Overland time of concentration = 20 min

For the stream $A = 3\,m^2$; $P = 2 + 2\sqrt{2} = 4.828$ m; $R = \dfrac{A}{P} = 0.621$ m

Velocity $V = \dfrac{1}{n} R^{2/3} S^{1/2} = \dfrac{1}{0.03}(0.621)^{2/3}(0.004)^{1/2} = 1.535$ m/s

Time of travel $t_r = \dfrac{L}{V} = \dfrac{300}{1.535} = 195.4$ s $= 3.26$ min

Total time of concentration = 20 + 3.26 = 23.26 min

$$i = \frac{569}{(t_c + 2.7)^{0.38}} = 165 \text{ mm/h}$$

Weighted runoff coefficient $= \dfrac{10 \times 0.9 + 10 \times 0.4}{20} = 0.65$

Peak runoff $Q_p = CIA = 0.65 \times 165 \times (20 \times 10^4) \times \dfrac{10^{-3}}{3,600} = 5.96$ m^3/s

Maximum capacity of stream = $AV = 3 \times 1.535 = 4.605$ m^3/s.

Therefore, the stream will overflow.

REFERENCES

Kirpich, Z.P. (1940): Time of concentration of small agricultural watersheds, *Civil Engineering*, vol 10 no.:6, p. 362.

Kuichling, E. (1889): The relation between the rainfall and the discharge of sewers in populous districts, *Transactions of ASCE*, vol 20, pp. 1–60.

Lloyd-Davies, D.E. (1906): The elimination of storm-water from sewerage, *Minutes of the Proceedings of the Institution of Civil Engineers, London*, vol 164 no.: 1906, pp. 41–67, Part 2, E-ISSN 1753-7843.

Mockus, V. (1957): *Use of Storm and Watershed Characteristics in Synthetic Hydrograph Analysis and Application*, US Soil conservation Service, Washington, DC.

Ragan, R.M. and Duro, J.O. (1972): Kinematic wave nomograph for times of concentration, *Journal of the Hydraulics Division*, vol 98 no.:HY10, pp. 1765–1771.

Raudkivi, R.J. (1979): *Hydrology – An Advanced Introduction to Hydrological Processing and Modelling*, Pergamon Press, Oxford, 479 pp.

ADDITIONAL READING

Mulvany, W. T. (1845): Observations on regulating weirs, *Transactions of the Institution of Civil Engineers of Ireland*, vol I, pp. 83–93.

Chapter 31

Analysis and presentation of rainfall data

31.1 INTRODUCTION

Rainfall is by far the most reliable of all hydrological data. Yet it is often necessary to check for homogeneity, consistency, recording errors, transcribing errors, missing data, etc. before they are used for any rigorous analysis. Missing data are normally estimated by simple interpolation, or using isohyetal maps and interpolation between isohyets, or by comparison with data from nearby gauging stations. In the latter case, the weighted interpolation using data from a number of stations (at least three) located nearby and evenly spaced around the station with a missing record can be used as follows:

$$P = \frac{1}{N} \sum_{1}^{N} \frac{\bar{P} P_i}{\bar{P}_i} \tag{31.1a}$$

where
  N – the number of stations
  P – missing rainfall at station P
  P_i – rainfall at stations i
  $\bar{P}_i$ – long-term normal rainfall at stations i
  $\bar{P}$ – long-term normal rainfall at station P

Equation 31.1a can also be written as

$$\frac{P}{\bar{P}} = \frac{1}{N} \sum \frac{P_i}{\bar{P}_i} \tag{31.1b}$$

This method is satisfactory when there is a strong correlation between the rainfalls of the stations considered. For short-duration interpolations the mass curve may be used.

31.2 TIME ADJUSTMENTS

The time at which observations are taken is subject to variation. Usually, rainfall records list the fall, which occurred over the 24 h preceding the time of observation. For example, a 1 h storm bridging the observation time could be split into two 24 h periods, i.e. 1 h for a 48 h period. If the observer waited for the storm to last, it would show as a fall for the previous 24 h period. This type of error is applicable to non-recording gauges only.

31.3 CONSISTENCY – DOUBLE MASS CURVE

Inconsistencies may be caused by the change of observer, change of location, change of observation procedure, change in the surrounding of the gauge, change of gauge itself, etc. Statistically, inconsistency occurs when the data do not come from the same population.

One of the methods of detecting such inconsistencies is the double mass curve method. It is obtained by plotting the cumulative rainfall at the suspect gauge against the average cumulative rainfall of a number of other nearby stations which are influenced by the same meteorological conditions. The double mass curve will show a change in slope if there is any inconsistency. However, care should be taken in using this procedure by substantiating such changes by historical evidence. The new trend also must last for a reasonable length of time, say 5 years.

The method is based on the observation that the mean cumulative rainfall for a number of gauges is not very sensitive to changes at individual stations because many errors tend to cancel out. On the other hand, for a single station the cumulative curve is affected by changes at the station.

Adjustments by the double mass curve are relative. Usually, it is assumed that the present trend is more reliable than the past trend. Therefore adjustments are made to the past values relative to the present values. The adjustment factor is m_1/m_2, where m_1 is the slope of the new trend and m_2 is the slope of the past trend. The adjustment is the same for cumulative values as well as for individual values.

There are also times when gross errors are shown in the double mass curve. Such errors could be introduced while typing and/or transcribing the data from the originals.

31.4 MEAN RAINFALL OVER AN AREA

31.4.1 Arithmetic mean

Mean rainfall over an area can be estimated by taking the arithmetic mean of a number of gauges.

31.4.2 Thiessen polygon method (1911)

This method is considered superior to the arithmetic mean method. Linear variation between stations is assumed (maybe a source of error). The method consists of dividing the region into sub-areas by drawing the perpendicular bisectors of the lines joining the gauging stations. The mean rainfall is given by

$$AP = \sum A_i P_i \tag{31.2}$$

where
  A –total area of the region
  P – mean rainfall for the region
  P_i – measured rainfall for the sub region 'i'
  A_i – area of the sub region 'i' of the Thiessen polygon.

For a complex network, such subdivision may not be unique. The method also gives no allowance for topographical effects. More recently, Whitmore et al. (1960) modified the method to take into account the station altitudes. Diskin (1969; 1970) improved the method to automate it by using the Mont–Carlo simulation.

The Thiessen polygon method does not always give representative areal rainfalls. There are situations where the simple arithmetic mean would be more representative.

31.4.3 Isohyetal method

This method is considered most suitable for obtaining the areal mean rainfall. Prior to plotting the isohyetal contours, the data must be checked for consistency, etc. Mean rainfall is obtained as follows:

$$\text{Mean rainfall} = \frac{\sum (\text{Area between isohyets} \times \text{Mean rainfall between isohyets})}{\text{Total area}}$$

31.5 INTENSITY-DURATION-FREQUENCY CURVES

Intensity-duration-frequency curves give the variation of the intensity of rainfall as a function of two independent parameters; the duration of rainfall and the frequency (or the return period) of rainfall. Some definitions related to intensity-duration-frequency curves are given below:

31.5.1 Return period

Return period is the average time interval between events that equal or exceed the considered magnitude. In other words, it is the average interval in years between the occurrence of an event of specified magnitude and an equal or larger event.

31.5.2 Data series

- **Complete series**: Complete series refers to the entire record. Such series are usually not independent.
- **Partial series**: Partial series refers to the series obtained by taking all values above an assumed base value. The advantage in using the partial series is that in some years it may be such that the second or third largest value is greater than the largest values in some other years. In selecting the data for the partial series, it is important that the events be independent. For rainfall data, the conditions that produce significant precipitation tend to become independent after the lapse of about 2 weeks. It is not so for floods however. In a partial series, it is the normal practice to consider the first N largest values where N is the number of years of record.

  The partial series is not a true distribution series because it is defined not in terms of its occurrence but in terms of its magnitude.
- **Annual series**: Annual series are obtained by taking the annual maximum values.
- **Extreme value series**: Mostly applicable to flood analysis.

31.5.3 Plotting positions

Plotting positions are used to estimate the return period or the frequency from the selected data series. There are several methods available:

$$\text{California method:} \; p = \frac{1}{T_r} = \frac{m}{n} \tag{31.3}$$

$$\text{Weibull method:} \; p = \frac{1}{T_r} = \frac{m}{n+1} \tag{31.4}$$

Hazen method: $p = \dfrac{1}{T_r} = \dfrac{2m-1}{2n}$ (31.5)

where m is the rank of the event in order of magnitude, n is the number of events, p is the probability of the magnitude being equalled or exceeded, and, T_r is the return period.

31.5.4 Procedure for obtaining the intensity-duration-frequency curves

- Select a suitable gauging station
- Starting from the current complete year, scan through the records and arrange the data in groups of different durations, e.g. 24, 12, 6 h, etc. For each storm of each duration, obtain the total depth of rainfall from data books and hence obtain the intensity.
- For each duration, arrange the data in descending order so that the maximum depth has the rank 1.
- Determine the return period for each event using a plotting position formula. For data having the same magnitude, use the same ranking, and skip the next value of ranking. For example, 5, 5, 7 or 6, 6, 6, 9, etc.
- Plot the rainfall depth against return period for different durations (A log-log or semi-log will give a better coverage).
- To obtain the intensity–duration curve for different return periods, obtain the data points for a given return period from the graphs plotted.

31.5.5 Theoretical relationships

In general, the greater the intensity of rainfall, the shorter the length of time it lasts. Therefore the intensity can be expressed in a form

$$i = \frac{A}{(t+B)^n}$$ (31.6)

where i is the average rainfall intensity, t is the duration and A, B and n are constants.

Taking logarithms, Eq. 31.6 can be written as

$$\log(i) = \log(A) - n\log(t+B)$$ (31.7)

If B is chosen properly, the plot of $\log(i)$ vs. $\log(t+B)$ would be a straight line with slope n. The values of n and B can be obtained from the straight line.

Given the data set $(t_i, i_i; i = 1, ..., N)$, various values of B are tried such that

$$\sum \{\log i_i - \log i\}^2 = \sum_1^N \{\log i_i - n\log(t_i + B) - \log(A)\}^2$$

is a minimum. The value of B thus obtained is used in the linear regression referred to above.

When the duration is >2 h, the following approximation has been found to be applicable:

$$i = \frac{C}{t^n}$$ (31.8)

C is a constant.

31.6 WORLD'S HIGHEST RECORDED RAINFALL INTENSITIES

Table 26.10 (Chapter 26) gives the highest average annual precipitation in different continents, ranging from 4,648 mm in Europe to 13,299 mm South America. These values cannot be taken as precise since the methods of measurement differ from country to country. The periods of record used to obtain average annual rainfall also differ from country to country. Table 31.1 shown below gives the world's highest recorded rainfall for different durations. The highest recorded annual rainfall of 26,000 mm has been in Cherrapunji in north-east India in the year 1985.

31.7 TYPICAL RETURN PERIODS USED IN DESIGN

- Bridges: 50–100 years
- Culverts on important roads: 25 years
- Culverts on secondary roads: 5–10 years
- Stormwater culverts: 1–2 years.

31.8 DESIGN STORMS

Some design problems require only the total volume of rainfall for a given duration and frequency. However, it is often necessary to give the time distribution of the rainfall volumes as well in the form of a hyetograph. Typical characteristics are the peak value, the time to peak and the distribution. The development of the design storm is based on the empirical analysis of a large number of storms.

31.8.1 Constant intensity design storm

Constant intensity design storms are applicable to very small catchments.

31.8.2 Variable intensity design storm

A variable intensity design storm can have either a symmetrical distribution or a skewed distribution (left or right). In practice, most design storms are made to be reasonably symmetrical. Example 31.2 (for a 6 h duration) illustrates the procedure for deriving a design storm of a given duration and frequency:

Table 31.1 World's highest recorded rainfalls

Duration	Depth (mm)	Intensity (mm/h)	Location
1 min	31.2	1874.5	Unionville, MD; USA
15 min	198.1	792.5	Jamaica
130 min	482.6	222.5	Rockport, WV; USA
15 h	876.3	58.4	Smethport, PA; USA
24 h	1168.1	48.5	Philippines
48 h	1671.0	34.8	Taiwan
7 days	3331.2	19.8	India
1 year	26461.2	3.0	India
2 years	40768.3	2.3	India

Source: R.M. McCuen, *Hydrologic Analysis and Design*, Second Edition, Prentice Hall, Upper Saddle River, NJ, 1989, Table 4.3, p. 186.

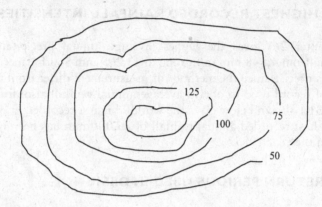

Figure 31.1 Depth–area analysis.

31.9 DEPTH–AREA ANALYSIS

In the depth–area analysis (Figure 31.1), the areal rainfall depths are shown as functions of the area. For small regions, this analysis may not mean much, but for large regions, the areal rainfall expected from a given area can be taken from the depth–area diagram. The following hypothetical example illustrates the procedure.

(1)	(2)	(3)	(4)	(5)	(6)	7 = (6)/(2)
Isohyet (mm)	Total area enclosed (km²)	Area between isohyets	Average rain between isohyets	Volume of rain between isohyets	Cumulative volumes	Areal rainfall (mm)
150	10	10	155.0	1,550	1,550	155.0
125	25	15	137.5	2062.5	3612.5	144.5
100	45	20	112.5	2,250	5867.5	130.4
75	75	30	87.5	2,625	8487.5	113.2
50	110	35	62.5	2187.5	1067.5	97.0

Example 31.1

Construct a double mass curve for the data given in the following table.

	Annual rainfall		Cumulative rainfall	
Year	at Chiangmai	at Base stations	at Chiangmai	at Base stations
1960	0.1339E+04	0.1250E+04	0.1339E+04	0.1250E+04
1959	0.9577E+03	0.1310E+04	0.2297E+04	0.2560E+04
1958	0.1130E+04	0.1126E+04	0.3427E+04	0.3686E+04
1957	0.1151E+04	0.1155E+04	0.4578E+04	0.4840E+04
1956	0.1175E+04	0.1257E+04	0.5753E+04	0.6097E+04
1955	0.1257E+04	0.1296E+04	0.7010E+04	0.7393E+04
	0.1277E+04	0.1256E+04	0.8287E+04	0.8649E+04
	0.2033E+04	0.1522E+04	0.1032E+05	0.1017E+05

(Continued)

Year	Annual rainfall		Cumulative rainfall	
	at Chiangmai	at Base stations	at Chiangmai	at Base stations
	0.1364E+04	0.1413E+04	0.1168E+05	0.1158E+05
	0.1473E+04	0.1314E+04	0.1316E+05	0.1290E+05
1950	0.1618E+04	0.1349E+04	0.1478E+05	0.1425E+05
	0.1390E+04	0.1425E+04	0.1616E+05	0.1567E+05
	0.1544E+04	0.1315E+04	0.1771E+05	0.1699E+05
	0.1415E+04	0.1123E+04	0.1912E+05	0.1811E+05
	0.9380E+03	0.9565E+03	0.2006E+05	0.1906E+05
1945	0.1036E+04	0.1106E+04	0.2110E+05	0.2017E+05
	0.1010E+04	0.1078E+04	0.2211E+05	0.2125E+05
	0.1148E+04	0.1353E+04	0.2325E+05	0.2260E+05
	0.1752E+04	0.1714E+04	0.2501E+05	0.2432E+05
	0.9705E+03	0.1133E+04	0.2598E+05	0.2545E+05
1940	0.1073E+04	0.1174E+04	0.2705E+05	0.2662E+05
	0.1210E+04	0.1358E+04	0.2826E+05	0.2798E+05
	0.1364E+04	0.1494E+04	0.2962E+05	0.2948E+05
	0.1375E+04	0.1438E+04	0.3100E+05	0.3091E+05
	0.9139E+03	0.1060E+04	0.3191E+05	0.3197E+05
1935	0.1255E+04	0.1341E+04	0.3317E+05	0.3331E+05
	0.1189E+04	0.1317E+04	0.3436E+05	0.3463E+05
	0.1198E+04	0.1232E+04	0.3556E+05	0.3586E+05
	0.1281E+04	0.1167E+04	0.3684E+05	0.3703E+05
	0.5005E+03	0.9421E+03	0.3734E+05	0.3797E+05
1930	0.1135E+04	0.1114E+04	0.3847E+05	0.3909E+05
	0.1210E+04	0.1142E+04	0.3968E+05	0.4023E+05
	0.1071E+04	0.1182E+04	0.4075E+05	0.4141E+05
	0.1410E+04	0.1429E+04	0.4216E+05	0.4284E+05
	0.1076E+04	0.1195E+04	0.4324E+05	0.4403E+05
1925	0.1163E+04	0.1167E+04	0.4440E+05	0.4520E+05
	0.1219E+04	0.1246E+04	0.4562E+05	0.4645E+05
	0.1282E+04	0.1234E+04	0.4690E+05	0.4768E+05
	0.1256E+04	0.1297E+04	0.4816E+05	0.4898E+05
	0.1255E+04	0.1320E+04	0.4941E+05	0.5030E+05
1920	0.8937E+03	0.1094E+04	0.5031E+05	0.5139E+05
	0.1166E+04	0.1078E+04	0.5147E+05	0.5247E+05
	0.1395E+04	0.1256E+04	0.5287E+05	0.5373E+05
	0.1187E+04	0.1210E+04	0.5405E+05	0.5494E+05
	0.1429E+04	0.1404E+04	0.5548E+05	0.5634E+05
1915	0.8649E+03	0.1285E+04	0.5635E+05	0.5763E+05
	0.1070E+04	0.1368E+04	0.5742E+05	0.5899E+05
	0.8896E+03	0.1245E+04	0.5831E+05	0.6024E+05
	0.1139E+04	0.1200E+04	0.5945E+05	0.6144E+05
1911	0.1306E+04	0.1221E+04	0.6075E+05	0.6266E+05
	0.1070E+04	0.1368E+04	0.5742E+05	0.5899E+05

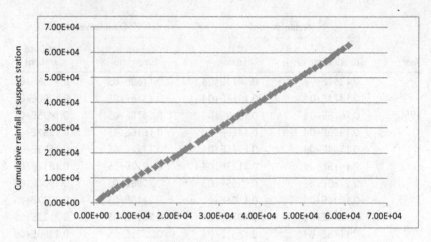

Cumulative rainfall at base stations

Example 31.2

The following example (for a 6 h duration) illustrates the procedure for deriving a design storm of a given duration and frequency:

Duration(h)	Intensity (mm/h)	Depth (mm)	Incremental depth (mm)	Design storm (mm)	Cumulative design storm (mm)
1	30.0	30.0	30.0	2.5	2.5
2	17.5	35.0	5.0	4.0	6.5
3	13.0	39.0	4.0	30.0	36.5
4	10.5	42.0	3.0	5.0	41.5
5	8.9	44.5	2.5	3.0	44.5
6	7.7	46.2	1.7	1.7	46.2

In this example, the data are taken from intensity-duration-frequency curves and are shown in columns 1 and 2. Column 3 shows the products of columns 1 and 2. Column 4 gives the incremental depths which would be the differences between durations. Column 5 gives the design storm. It is the same as column 4 except that the sequence is different. The rainfall depths have been re-arranged to give a reasonably symmetrical distribution while keeping the maximum depth for any duration (within the 6 h period) to be equal to the depths shown in column 3. For example, for a 3 h duration, the intensity from column 3 is 39.0 mm. In the design storm, the sum of the second, third and fourth hours is 39.0.

REFERENCES

Diskin, M. H. (1969): Thiessen coefficients by a Monte Carlo procedure, *Journal of Hydrology*, vol. 8, pp. 323–335.

Diskin, M. H. (1970): On the computer evaluation of Thiessen weights, *Journal of Hydrology*, vol. 11, pp. 69–78.

McCuen, R. M. (1989): *Hydrologic Analysis and Design*, Prentice Hall, Upper Saddle River, NJ, p. 867.

Thiessen, A. H. (1911): Precipitation averages for large areas, *Monthly Weather Review*, vol. 39, no. 7, pp. 1082–1084.

Whitmore, J. S., van Eeden, F. J. and Harvey, K. J. (1960): Assessment of average annual rainfall over large catchments, South African Department of Water Affairs Technical Report No. 14.

Chapter 32

Unit hydrograph methods

32.1 INTRODUCTION

A unit hydrograph is a discharge hydrograph resulting from one unit of direct runoff (1 mm or 1 cm) generated uniformly over the catchment area at a uniform rate during a specified period of time called the unit duration.

The unit hydrograph method is one of the basic tools in hydrological computations. It was originally presented by L.K. Sherman (1932). It is founded upon the assumption that the rainfall excess – direct runoff process is linear which implies that the principles of superposition and proportionality hold. In this context, direct runoff is the difference between the total runoff and base flow. It is that component directly resulting from rainfall. Rainfall excess is that portion of the rainfall that contributes to direct runoff.

The principles of proportionality and superposition lead to very useful practical applications of the unit hydrograph concept. For example, a hydrograph of discharge resulting from a series of rainfall excesses may be constructed by summing up the hydrographs due to each single unit of rainfall excess. They also imply that the time base of direct runoff hydrographs resulting from rainfall excesses of the same unit duration is the same regardless of the intensity.

The unit duration is the rate at which the catchment is filled up to a rainfall excess of unit depth. e.g., 1 mm of rainfall excess could be applied at different rates.

The catchment can be thought of as a tank or a reservoir that receives the rainfall excess and releases it through an outlet. The catchment characteristics are reflected in the discharge coefficient of the tank or reservoir outlet which affects the recession part of the hydrograph. The rising part is affected by the rate of application of the unit rainfall excess.

32.2 DERIVATION OF A UNIT HYDROGRAPH FOR A UNIFORM STORM

32.2.1 Procedure

- Select a storm that is isolated, intense and uniform over the catchment and time by scanning through rainfall and runoff records. The duration of the storm should not be greater than the period of rise (time of concentration).

- Plot the observed discharge hydrograph and separate the base flow.
- Determine the area under the direct runoff hydrograph. The area may be expressed as a volume of runoff (discharge×time = volume) or as a depth of runoff by dividing by the catchment area.
- Divide the direct runoff hydrograph ordinates by the depth of runoff, referred to above, to obtain the unit hydrograph ordinates. The area under the unit hydrograph must be unity.

32.2.2 Determination of the unit duration

A catchment can have an infinite number of unit hydrographs because each duration of rainfall excess produces its own unit hydrograph. The unit duration t_0 is determined by assuming that the volume (or depth) of direct runoff is equal to the volume (or depth) of rainfall excess. The difference between the recorded rainfall and the direct runoff is taken as losses due to infiltration and other causes.

The recorded rainfall is therefore adjusted to give a depth equal to the depth of direct runoff. The effective duration of the adjusted rainfall is the unit duration.

In this adjustment exercise, there is a certain amount of subjectiveness or arbitrariness involved. There are no objective methods of distributing the losses over the duration of the rainfall. The assumptions made in the unit hydrograph concept should, however, be not violated.

32.2.3 Conversion of a unit hydrograph of one unit duration to that of a different unit duration

32.2.3.1 When the required unit duration is an integer multiple of the given unit duration

If the given unit duration is t_0 hours, then superposition of n 't_0' hour unit hydrographs will give a 'nt_0' hour hydrograph (not a unit hydrograph). The 'nt_0' hour unit hydrograph can be obtained by dividing the ordinates of the 'nt_0' hydrograph by 'n'.

32.2.3.2 When the required unit duration is not an integer multiple of the given unit duration (S-curve method)

The S-curve is the hydrograph resulting from an infinite series of runoff increments of unit magnitude with a unit duration of t_0 hours. In other words, it is the hydrograph for a continuous storm with an intensity of $1/t_0$. With such a continuous storm, an equilibrium stage will be attained after some time. Thus, the name S-curve. It is constructed by adding the unit hydrographs of t_0 hour duration, each displaced by t_0 hours.

In practice, this curve will not be smooth due to lack of precision in selecting the unit duration t_0 and due to the fact that the sum of the ordinates × time interval not equal to unity. It is therefore smoothed by averaging.

The difference between two S-curve displaced by t_0' hours where t_0' is the required unit duration gives a hydrograph of duration t_0' hours. It is then made a unit hydrograph by multiplying by t_0/t_0'.

32.3 APPLICATION OF THE UNIT HYDROGRAPH

The unit hydrograph can be used to determine the direct runoff hydrograph resulting from any other storm having the same unit duration by multiplying the ordinates of the unit hydrograph by rainfall excess of the storm. The base flow is then added to obtain the total hydrograph.

32.4 MEAN UNIT HYDROGRAPH

Unit hydrograph derived for one storm for a given catchment may not always be unique for a specified duration. In such situations, it is necessary to derive the mean unit hydrograph from a number of unit hydrographs.

The peak value and the time to peak of the mean unit hydrograph are obtained by averaging the individual unit hydrographs. Other values are not averaged, as that would yield an untypical shape. The mean unit hydrograph is obtained by sketching so that the area under the graph is equal to unity.

32.5 DERIVATION OF THE UNIT HYDROGRAPH FOR A COMPLEX STORM

Using the convolution method which is a linear superposition involving multiplication, translation in time and addition, the discharge corresponding to a multiple unit rainfall excess can be obtained. The convolution integral is given as follows:

$$y_t = \int_0^t x_\tau u_{t-\tau} \, d\tau = \int_0^t x_{t-\tau} u_\tau d\tau \tag{32.1a}$$

In discrete form, it can be written as

$$y_t = \sum_{\tau=0}^t x_\tau u_{t-\tau} \tag{32.1b}$$

For digital computations, the following form is used:

$$y_t = \sum_{j=1}^t x_j u_{t-j+1} \text{ if } t \le r \tag{32.1c}$$

$$y_t = \sum_{j=1}^r x_j u_{t-j+1} \text{ if } r \le t \le m \tag{32.1d}$$

because $u_t = 0$ for $t < 1$ and $t > m$.

Let $x_1, x_2, ..., x_r$ be the rainfall excess values,
$u_1, u_2, ..., u_m$ be the unit hydrograph ordinates, and,
$y_1, y_2, ..., y_n$ be the direct runoff hydrograph ordinates.

Then, by the principle of superposition

$$
\left.\begin{aligned}
y_1 &= u_1 x_1 \\
y_2 &= u_2 x_1 + u_1 x_2 \\
y_3 &= u_3 x_1 + u_2 x_2 + u_1 x_3 \\
y_4 &= u_4 x_1 + u_3 x_2 + u_2 x_3 + u_1 x_4 \\
\vdots\ \ \vdots\ \ &\qquad\qquad \vdots \\
y_r &= u_r x_1 + u_{r-1} x_2 + u_{r-2} x_3 + \cdots + u_2 x_{r-1} + u_1 x_r \\
y_{r+1} &= u_{r+1} x_1 + u_r x_2 + u_{r-1} x_3 + \cdots + u_3 x_{r-1} + u_2 x_r \\
\vdots\ \ \vdots\ \ &\qquad\qquad \vdots \\
y_{r+m-2} &= 0 + 0 + \cdots + x_r u_{m-1} + x_{r-1} u_m \\
y_{r+m-1} &= 0 + 0 + \cdots + x_r u_m
\end{aligned}\right\}
\tag{32.2}
$$

where $n = r + m - 1$.

The same system of equations can be used to derive the unit hydrograph ordinates $u_1, u_2, \ldots, u_m$.

In matrix form, the above system can be written as

$$
\begin{bmatrix}
x_1\ 0 \\
x_2\ x_1\ 0 \\
x_3\ x_2\ x_1\ 0 \\
\vdots \\
\vdots \\
x_r\ x_{r-1}\ x_{r-2} \\
0\ \ x_r\ \ x_{r-1}\qquad x_2\ x_1\ 0 \\
0\ \ 0\ \ x_r \\
0\qquad\qquad\quad x_r\ x_{r-1} \\
\qquad\qquad\qquad\quad x_r
\end{bmatrix}
\begin{bmatrix}
u_1 \\
u_2 \\
u_3 \\
\vdots \\
u_m
\end{bmatrix}
=
\begin{bmatrix}
y_1 \\
y_2 \\
y_3 \\
\vdots \\
y_r \\
y_{r+1} \\
\vdots \\
y_n
\end{bmatrix}
\tag{32.3}
$$

which is of the form

$$
[x][u] = [y]
\tag{32.4}
$$

where $[x]$ is a $(m + r - 1) \times m$ matrix
$[u]$ is a m vector
$[y]$ is a $m + r - 1$ vector.

The solution of Eq. 32.4a is not straightforward because $[x]$ is not a square matrix. There are more equations than the number of unknowns in the system. To get the optimal solution which will minimize the error (least squares method), the following matrix manipulation is performed:

Let $[z] = [x]^T [x]$

It would be a square matrix of size $m \times m$ because, $[x]^T$ is $m \times (m+r-1)$; and $[x]$ is $(m+r-1) \times m$. Then,

$$[z][u] = [x]^T [y] \tag{32.5}$$

Therefore,

$$[u] = [z]^{-1} [x]^T [y] \tag{32.6}$$

Solutions could also be obtained in a sequential manner as follows:

Eq. (1.1) gives $u_1 = y_1/x_2$
Eq. (1.2) gives u_2 with u_1 from Eq. (1.1).

Therefore all the equations can be solved. This method, however, does not lead to accurate solutions because the errors get accumulated with each equation solved and that the solutions will be different depending on whether the procedure is started from the beginning or from the end.

To obtain the normal equations for the least squares fit, multiply each of the original equations by the coefficient of u_1 and add the results as follows:

$$x_1 y_1 = x_1 (u_1 x_1)$$

$$x_2 y_2 = x_2 (u_2 x_1 + u_1 x_2)$$

$$x_3 y_3 = x_3 (u_3 x_1 + u_2 x_2 + u_1 x_3)$$

$$\vdots \quad \vdots \quad \vdots$$

$$x_1 y_1 + x_2 y_2 + x_3 y_3 = u_1 (x_1^2 + x_2^2 + x_3^2) + u_2 (x_1 x_2 + x_2 x_3) + u_3 (x_1 x_3)$$

Similarly, the second normal equation is obtained by multiplying by the coefficient of u_2 and adding

$$x_1 y_2 = x_1 (u_2 x_1 + u_1 x_2)$$

$$x_2 y_3 = x_2 (u_2 x_1 + u_2 x_2 + u_1 x_3)$$

$$x_3 y_4 = x_3 (u_4 x_1 + u_3 x_2 + u_2 x_3 + u_1 x_4)$$

$$\vdots \quad \vdots$$

$$x_1 y_2 + x_2 y_3 + x_3 y_4 = u_1 (x_1 x_2 + x_2 x_3 + x_3 x_4) + u_2 (x_1^2 + x_2^2 + x_3^2 + x_4^2)$$

$$+ u_3 (x_1 x_2 + x_2 x_3 + x_3 x_4) + u_4 (x_1 x_3 + x_1 x_2)$$

and so on.

32.6 DIMENSIONLESS UNIT HYDROGRAPH

32.6.1 SCS Dimensionless unit hydrograph

The US Soil Conservation Service (now known as Natural Resources Conservation Service, NRCS) Dimensionless Unit Hydrograph has been obtained by the following transformation:

$$\text{Dimensionless discharge} = \frac{\text{Discharge}}{\text{Peak discharge}},$$

$$\text{Dimensionless time} = \frac{\text{Time}}{\text{Time to peak}}.$$

A large number of unit hydrographs are computed, and the average of their dimensionless unit hydrographs is taken as representative.

32.7 SYNTHETIC UNIT HYDROGRAPH

A unit hydrograph can be reasonably adequately described by three parameters, namely, the lag time (in the context of the synthetic unit hydrograph, lag time is defined as the time from the centre of the rainfall excess to the time of peak flow), the peak flow and the time base. The simplest synthetic unit hydrograph is a triangle satisfying the relationship

$$\frac{q_p t_b}{2A} = 1 \tag{32.7}$$

where q_p is the peak discharge of the unit hydrograph and t_b is the time base. q_p is estimated using some empirical formula and t_b is related to the time lag of the catchment.

Snyder (1938) proposed a method of deriving a synthetic unit hydrograph for ungauged catchments using the above and some other parameters. The method needs calibration and is applicable to large catchments only. Snyder's unit hydrograph depend on the

- time to peak t_p which depend on the duration of rainfall excess t_d (unit duration) and time lag, t_L,
- time base, t_b,
- peak discharge, and,
- width of the unit hydrograph at 50% and 75% of peak discharge, W_{50} and W_{75}.

For a steady rainfall excess,

$$t_p = \frac{1}{2} t_d + t_L \tag{32.8}$$

He defined,

(i) $t_L = C_t \left(L L_c \right)^{0.3}$ \hspace{2cm} (32.9)

where t_L – time lag in hours
   L – length of the main channel (distance from the outlet to boundary)

L_c – a catchment shape parameter, length measured along the main channel from the outlet to a point on the main channel that is normal to the centre of area of the catchment

C_t – a catchment storage coefficient, determined by calibration. (Depends on slope and storage, steep slopes have lower C_t values; typical values used by Snyder: 1.35–1.65 with a mean of 1.5 when L and L_{ca} are in km; 1.82. 2 with a mean of 2 when L and L_{ca} are in miles.)

(ii)　　$t_d = t_L/5.5$　　　　　　　　　　　　　　　　　　(32.10a)

where t_d is the duration of the rainfall excess in hours.

This equation gives only one value of t_d for a catchment. For other values of duration of rainfall excess (unit duration), Eq. 32.10a can be modified as follows:

$$t_{La} = t_L + 0.25(t_{da} - t_d)$$　　　　　　　　　(32.10b)

where t_{da} is another duration of rainfall excess (unit duration), and t_{La} is the adjusted time lag. The adjusted time lag is used in Eqs. 32.9 and 32.10.

(iii)　$t_b = 72 + 3t_L$　　　　　　　　　　　　　　　　(32.11)

where t_b is the time base of the unit hydrograph in hours.

(iv)　$q_p = 0.278A\,C_p/t_L$　　　　　　　　　　　　　(32.12a)

where q_p is the peak flow of the unit hydrograph in m³/s/mm, A is the catchment area in km² and C_p is a coefficient that needs to be determined by calibration ($C_p = 0.56 - 0.69$ for $t_b/t_L = 3.57 - 2.90$). In US units, with q_p in ft³/s/in, and A in mi², it is of the form

$$q_p = 640A\,C_p/t_L$$　　　　　　　　　　　　　　(32.12b)

(v)　　$W_{50} = \dfrac{0.175}{\left\{q_p\big/A\right\}^{1.08}}$　　　　　　　　　　　(32.13)

(vi)　$W_{75} = \dfrac{0.104}{\left\{q_p\big/A\right\}^{1.08}}$　　　　　　　　　　　(32.14a)

where q_p is the peak discharge in m³/s, A is in km² and W_{50} and W_{75} are in hours. In US units, they are

$$W_{50} = 756\,q_a^{-1.081}$$　　　　　　　　　　　　(32.14b)

$$W_{75} = 450q_a^{-1.081}$$　　　　　　　　　　　　(32.14c)

where q_a is the peak discharge in ft³/s/mi².

In the synthetic unit hydrograph, the co-ordinates for W_{50} and W_{75} are fixed, such that one-third of the time bases W_{50} and W_{75} lie to the left of the peak value and the other two-thirds to the right of the peak value. Seven points together with the condition that the area under the synthetic unit hydrograph must be unity define the unit hydrograph. The seven points are the peak value and the time to peak, the origin, the endpoint defined by the time base and two points each for W_{50} and W_{75}.

Example 32.1

Derive the unit hydrograph for a catchment of area 19.44 km² given the following hourly data:

> Rainfall (mm/h): 5 5 5
> Direct runoff (m³/s): 3 6 9 8 7 6 5 4 3 2 1

Hence, compute the direct runoff hydrograph resulting from a storm of intensity 2 mm/h rainfall excess lasting for 5 h.

$$\text{Area under the direct runoff hydrograph}\left(\sum\text{column 1}\right) = 54\left(m^3/s\right)h$$

$$\text{Depth of runoff} = \frac{54 \times 60 \times 60}{19.44 \times 1,000 \times 1,000} \times 1,000\,mm$$

$$= 10\,mm$$

Column 3 is obtained by dividing Column 2 by 10.

$$\text{Area under the unit hydrograph}\left(\sum\text{column 3}\right)$$

$$= 5.4\left(m^3/s\right)h$$

$$= 1.0\,mm$$

The unit duration is obtained by adjusting the rainfall so that the rainfall excess is equal to the direct runoff. In this hypothetical example, the rainfall is steady for 3 h, with a total rainfall of 15 mm. The direct runoff is only 10 mm. Therefore, the losses would be 5 mm spread over the rainfall duration. Assuming a uniform distribution of 1.667 mm/h, the rainfall excess for the storm would be 3.33 mm/h for 3 h. Hence, the unit hydrograph has a unit duration of 3 h.

If on the other hand, a non-steady rainfall is considered, the adjustments may be done using the ϕ-index as follows:

Time	Rainfall	Losses	Rainfall excess
1	7.50	3.03	0.97
2	7.25	2.98	0.97
3	7.75	2.93	0
4	1.5	1.88	0
5	0.5	1.83	0
6	0.3	1.83	0

Total depth of rainfall, excluding the last 3 h which would not contribute to direct runoff, is 22.5 mm. Therefore,

$$\phi\text{-index} = (22.5 - 10)/3 = 4.167 \text{ mm/h}$$

distributed over the first 3 h. The adjusted rainfalls are

7.50	4.166	3.334
7.25	4.166	3.084
7.75	4.166	3.584
		10.002 mm

Since 10.002 is >10, another refinement is necessary.
The new ϕ-index is

$$(10.002 - 10)/3 = 0.00067 \text{ mm/h distributed over 3 h.}$$

The adjusted rainfalls then are

3.334	0.00067	3.333
3.084	0.00067	3.083
3.584	0.00067	3.583
		0.999 mm

Conversion of the 3 h unit hydrograph into a 6 h unit hydrograph

The principle of superposition is used as illustrated below:

(1)	(2)	(3)	(4)	(5)
Time/(h)	3 h. Unit graph	UG Shifted by 3 h	(2)+(3)	(4)/2
0	0		0	0
1	0.3		0.3	0.15
2	0.6		0.6	0.3
3	0.9	0.0	0.9	0.45
4	0.8	0.3	1.1	0.55
5	0.7	0.6	1.3	0.65
6	0.6	0.9	1.5	0.75
7	0.5	0.8	1.3	0.65
8	0.4	0.7	1.1	0.55
9	0.3	0.6	0.9	0.45
10	0.2	0.5	0.7	0.35
11	0.1	0.4	0.5	0.25
12	0	0.3	0.3	0.15
13		0.2	0.2	0.1
14		0.1	0.1	0.05
15		0.0	0	0

Conversion of the 3 h unit hydrograph into a 5 h unit hydrograph

The S-curve method, which is obtained by summing unit hydrographs each displaced by 3 h, is used as follows:

Time	Direct runoff	3 h Unit graph	S-Curve	Adjusted S-curve	Shifted S-curve	5 h 5/3 Graph	5 h Unit graph	Direct runoff
1	2	3	4	5	6	7	8	9
1	3	0.3	0.3	0.3		0.3	0.18	0.36
2	6	0.6	0.6	0.6		0.6	0.36	0.72
3	9	0.9	0.9	0.9		0.9	0.54	1.08
4	8	0.8	1.1	1.1		1.1	0.66	1.32
5	7	0.7	1.3	1.3		1.3	0.78	1.56
6	6	0.6	1.5	1.5	0.3	1.2	0.72	1.44
7	5	0.5	1.6	1.6	0.6	1.0	0.60	1.20
8	4	0.4	1.7	1.7	0.9	0.8	0.48	0.96
9	3	0.3	1.8	1.8	1.1	0.7	0.42	0.84
10	2	0.2	1.8	1.8	1.3	0.5	0.30	0.60
11	1	0.1	1.8	1.8	1.5	0.3	0.18	0.36
12	0	0	1.8	1.8	1.6	0.2	0.12	0.24
13			1.8	1.8	1.7	0.1	0.06	0.12
14			1.8	1.8	1.8	0.0	0.0	0.0
15			1.8	1.8	1.8	0.0	0.0	0.0

Column 5 is obtained by smoothing the S-curve of column 4. Column 6 is the smoothed S-curve displaced by 5 h, which is the new unit duration for which the direct runoff hydrograph needs to be computed. Column 7 is the difference between columns 5 and 6 and gives the hydrograph corresponding to a unit duration of 5 h. The area under this hydrograph will not be unity. To convert it into a unit hydrograph, the ordinates must be multiplied by 3/5 which is shown in column 8. Column 9 corresponds to the direct runoff hydrograph which is obtained by multiplying column 8 by the given rainfall excess.

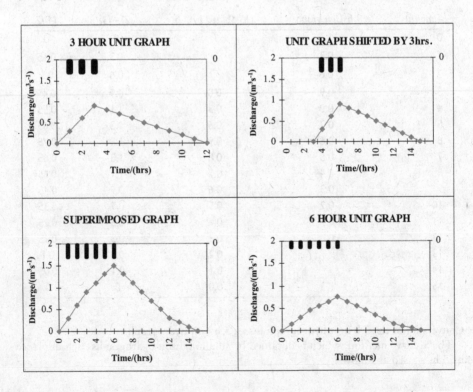

(1)	(2)	(3)	(4)	(5)	(6)	(7)	(A)	(B)	(C)	(D)
0	0					...	0		0	0
1	0.3					...	0.3		0.3	0.2
2	0.6					...	0.6		0.6	0.4
3	0.9	0				...	0.9		0.9	0.5
4	0.8	0.3				...	1.1		1.1	0.7
5	0.7	0.6				...	1.3	0	1.3	0.8
6	0.6	0.9	0			...	1.5	0.3	1.2	0.7
7	0.5	0.8	0.3			...	1.6	0.6	1	0.6
8	0.4	0.7	0.6			...	1.7	0.9	0.8	0.5
9	0.3	0.6	0.9	0		...	1.8	1.1	0.7	0.4
10	0.2	0.5	0.8	0.3		...	1.8	1.3	0.5	0.3
11	0.1	0.4	0.7	0.6		...	1.8	1.5	0.3	0.2
12	0	0.3	0.6	0.9	0	...	1.8	1.6	0.2	0.1
13		0.2	0.5	0.8	0.3	...	1.8	1.7	0.1	-0.1
14		0.1	0.4	0.7	0.6	...	1.8	1.8	0	0
15		0	0.3	0.6	0.9	...	1.8	1.8	0	0

Column (1) $\Rightarrow$ Time in hours
Column (2) $\Rightarrow$ 3 h Unit Hydrograph
Column (3) $\Rightarrow$ 3 h Unit Hydrograph is shifted by 3 h.
Column (4) $\Rightarrow$ 3 h Unit Hydrograph is shifted by 6 h.
Column (5) $\Rightarrow$ 3 h Unit Hydrograph is shifted by 9 h.
Column (6) $\Rightarrow$ 3 h Unit Hydrograph is shifted by 12 h.
Column (7) $\Rightarrow$ Continuation.
Column (A) $\Rightarrow$ S-Curve (summation of columns (1), (2), (3), ...)
Column (B) $\Rightarrow$ S-Curve is shifted by 5 h.
Column (C) $\Rightarrow$ Column (A) $-$ Column (B).
Column (D) $\Rightarrow$ Column(C) $\times 3/5$

Example 32.2

The 1 mm 1 h unit hydrograph of a catchment can be approximated by a triangle of height 0.65 m³/s at a time of 2 h and a time base of 8 h. As a result of development work, the time to peak and the time base of the unit hydrograph are reduced by 50% and 3 h, respectively. A rainstorm over the catchment has intensities of 5, 20 and 10 mm/h for a 3 h period. If the effective rainfall intensities before and after the development work are taken as 0.25 and 0.5, respectively, of the actual rainfall intensities, calculate the resulting increase in the peak runoff.

Effective rainfalls are

Actual (mm/h)	Before development (mm/h)	After development (mm/h)
5	1.25	2.5
20	5	10
10	2.5	5

The total outflow hydrographs are
    Before development

Time (h)	Unit hydrograph (UH) (m³/s/mm)	UH × 1.25 (m³/s)	UH × 5 (m³/s)	UH × 2.5 (m³/s)	Total (m³/s)
0	0	0	0	0	0
1	0.325	0.406	0	0	0.406
2	0.650	0.812	1.625	0	2.437
3	0.542	0.678	3.25	0.812	4.74
4	0.433	0.540	2.71	1.625	4.875
5	0.325	0.406	2.165	1.35	3.921
6	0.217	0.271	1.625		

After development

Time (h)	Unit hydrograph (UH) (m³/s/mm)	UH × 1.25 (m³/s)	UH × 5 (m³/s)	UH × 2.5 (m³/s)	Total (m³/s)
0	0	0	0	0	0
1	1.04	2.6	0	0	2.6
2	0.78	1.95	10.4	0	12.35
3	0.52	1.3	7.8	5.2	14.3
4	0.26	0.65	5.2	3.9	9.25
5	0		2.6	2.6	
6				1.3	

Resulting increase in peak flow = 14.3 − 4.875 = 9.425 m³/s.

REFERENCES

Sherman, L. K. (1932): Streamflow from rainfall by the unit hydrograph method, *Engineering News Record*, vol. 108, pp. 501–505.

Snyder, F. F. (1938): Synthetic unit-graphs, *Transactions of the American Geophysical Union*, vol. 19, pp. 447–454.

Chapter 33

Rainfall-runoff modelling

33.1 INTRODUCTION

Rainfall-runoff models of the deterministic type can be broadly classified into hydrologic models and hydraulic models. Hydrologic models are usually based on the continuity equation only whereas the hydraulic models require an additional equation, which may be a (simplified) form of the dynamic equation. Rainfall-runoff models may also be classified as continuous-time simulation models or event simulation models. The former is usually of the hydrologic type whereas the latter is usually of the hydraulic type. Hydrologic type models widely use the linear reservoir, the linear channel and their combinations as the basic building blocks.

Rainfall-runoff models can also be classified as analog (now outdated), conceptual, physics-based or data driven. They may also be classified as lumped or distributed, linear or non-linear, continuous-time simulation or event simulation, and/or, according to the process description (deterministic, conceptual, stochastic), domain representation (lumped, semi-distributed, distributed), temporal scale (annual, monthly, daily, hourly, etc.), solution technique and model application.

The first rainfall-runoff model was probably the rational method (Mulvany, 1851) which relates the peak runoff to the rainfall intensity. The unit hydrograph method (Sherman, 1932) is a conceptual model that linearly relates the rainfall excess to direct runoff and which has stood the test of time due to its simplicity. Non-linear versions of the unit hydrograph (Amorocho, 1973), as well as many other versions that employ the systems theory approach, have been suggested and used over the years (see Delleur and Rao, 1971; Lattermann, 1991; amongst others).

In addition to the rational method and the unit hydrograph method, other historical highlights of rainfall-runoff models include the Stanford Watershed model (Crawford and Linsley, 1966), Tank model (Sugawara et al., 1984), Xinanjiang model (Zhao, 1977, 1984; Zhao and Liu, 1995; Zhao et al., 1980), HEC-1 Model (USACE, 1998), Topmodel (Beven and Kirkby, 1979), Systeme Hydrologique European (Abbott et al., 1986a,b) and Arno model (Todini, 1996).

Since the advent of the digital computer, there has been a proliferation of models and modelling techniques, giving rise to a plethora of models. Among them, the more widely used ones include HEC-HMS, the standard model used in the private sector in the US, NWS, the standard flood forecasting model in the US, HSPF/BASINS, the standard water quality model in the US, WATFLOOD, a popular model used in Canada, RORB, a popular model used in Australia, HBV, the standard model used in Scandinavia, SHE, the standard distributed model in Europe, Tank, a popular model used in Japan, Xinanjiang, a popular model used in China, ARNO, TOPIKAPI, LCS, popular models used in Italy and VIC, a popular model used in the USA.

There are also other less widely used models such as SWMM – Storm Water Management Model, SWAT – Soil and Water Assessment Tool, QUALHYMO – Storm Water Runoff Model, HSPF – Hydrological Simulation Program Fortran, AGNPS – Agricultural Non-Point Source Pollution Model, PRMS – Precipitation Runoff Modelling System, SSARR – Streamflow Synthesis and Reservoir Regulation Model, HELP – Hydrological Evaluation of Landfill Performance, WATER BUDGET – Cumming Cockburn Limited Water Budget Model and ANSWERS – Areal Non-Point Source Watershed Environment.

Despite the abundance, there is no perfect model that suits all purposes. Each model has its own pros and cons. Moreover, more and more models are being developed continuously. One should take note of the saying that 'all models are wrong, but some are useful', and exercise careful judgment in choosing a model or a modelling approach for a specific purpose.

33.2 CONCEPTUAL TYPE HYDROLOGIC MODELS

33.2.1 Stanford Watershed Model (SWM)

Originated in Stanford University (Crawford and Linsley, 1966), the Stanford Watershed Model is suitable for hourly or daily simulation. It uses the lumped parameter approach.

Inputs consist of hourly or daily precipitation data, evaporation data and catchment parameters. The model can carry out continuous-time simulation and is based on the continuity equation:

$$P = E + R + \Delta S \tag{33.1}$$

where
  P – Precipitation
  E – Evapo-transpiration
  R – Runoff
  ΔS – Total change in storage in each zone.

At each step, all hydrological activities (precipitation, depression storage, infiltration, percolation, surface runoff, groundwater flow, etc.) are simulated and balanced before proceeding to the next time step. The following processes are modelled.

33.2.1.1 Interception

Rainfall is intercepted up to **interception rate** which depends upon the precipitation rate and vegetation cover.

33.2.1.2 Evapo-transpiration

Evapo-transpiration (ET) takes place from

- Interception storage – at potential rate
- Upper zone storage – at potential rate
- Lower zone storage – not at potential rate
- Stream and lake surfaces – at potential rate
- Ground water storage – at potential rate.

When interception storage is depleted, ET will take place from upper zone storage at the potential rate. Once upper zone is depleted, ET will take place from the lower zone storage, but **not** at the potential rate.

33.2.1.3 Infiltration

There are a number of parameters to define infiltration which depends upon soil permeability and the moisture storage capacity of the soil. The following storages are defined:

- Upper zone storage
  - Input is from rainfall
  - Outflow is through percolation, evaporation, surface runoff and interflow
- Lower zone storage
  - Input is from infiltration and percolation
  - Outflow is through base-flow

33.2.1.4 Overland flow

Discharge is a function of detention storage which depends on parameters such as length of flow, slope and roughness

It is empirically related to the above parameters as

$$D_e = ki^\alpha n^\beta L^\gamma S^\theta \tag{33.2}$$

where

i – rainfall intensity
n – Manning's 'n'
L – length of flow
S – slope
k – an empirical constant.

33.2.1.5 Interflow

Channel routing: by linear reservoir

$$0_2 = \overline{I} - KS_1\left(\overline{I} - 0_1\right) \tag{33.3}$$

where

0_2 – outflow at time t_2
$\overline{I}$ – average inflow
0_1 – outflow at time t_1
S_1 – storage at time t_1
K – routing constant.

Calibration of the model is iterative. Several versions of SWM are available. In the early 1970s, SWM expanded and refined to create Hydrocomp Simulation Program (HSP) which includes water quality simulation as well. It led to the development of Hydrological Simulation Program – Fortran (HSPF).

33.2.2 Tank model

The Tank model, proposed and developed by Sugawara and his colleagues in the sixties, is the most widely used rainfall-runoff model in Japan. The publication of the details of this model in the English language came much later (Sugawara et al., 1984) and its application has since then spread to regions outside Japan (e.g., Jayawardena, 1988). It is a deterministic, non-linear, lumped, continuous and time-invariant model. It transforms the measured rainfall into a corresponding runoff without having to estimate losses and base flows separately.

The model is composed of tanks (usually 4) arranged vertically in series. Precipitation is fed into the top tank, and evaporation is subtracted sequentially from the top tank downwards. As each tank is emptied, the evaporation shortfall is taken from the next lower tank until all tanks are empty. The outputs from the side outlets are the calculated runoffs. The output from the top tank is considered as surface runoff, output from the second tank as intermediate runoff, from the third tank as sub-base runoff and output from the fourth tank as base flow.

The basic principle of the Tank model is that discharge is proportional to storage. The catchment is considered as a reservoir, which temporarily stores the rainfall and subsequently releases as runoff. The discharge is considered as consisting of surface runoff, subsurface runoff and groundwater flow, each component having its own response time. The inputs to the Tank model are precipitation and potential evaporation.

In the model (Figure 33.1a), the tanks can have different outlets in the same tank, or different tanks arranged in series. There are threshold values in each tank to account for the initial rainfall which does not produce any runoff. Flooding can be modelled by having two outlets in the top tank. Highly porous surfaces can be modelled by having a bottom outlet in the top tank with a large discharge coefficient. In the latter case, the discharge will be mainly from the second tank. Large catchments can be modelled by having the series arrangement of tanks in parallel.

The soil moisture is modelled by having a primary soil moisture storage in the top tank with a saturation capacity S_1 (Figure 33.1b). There will be no runoff or infiltration until the storage in the top tank attains this value. The difference between the storage in the top tank and the saturation capacity of the primary soil moisture storage $(XA - S_1)$ is the free water available.

After the primary soil moisture storage is saturated, there will be transfer of water from the primary to secondary. When evaporation takes place, primary soil moisture is depleted first, and then water is drawn laterally from the secondary storage. If there is no free water in the second tank, vertical transfer takes place from the free water in the third and fourth tanks.

When $XA \leq S_1$,

$$XP = XA$$
$$XF = 0$$

When $XA > S_1$

$$XP = S_1$$
$$XF = XA - S_1$$

Where
XA–Storage in tank 1
XP – Primary soil moisture storage
XF – Free water available
S_1 – Threshold value in tank 1
The basic model has 17 parameters, some of which have inter-dependence. They are

Five discharge parameters; $A_1, A_2; B_1; C_1; D_1$
Three infiltration parameters; $A_0; B_0; C_0$
Four discharge outlet parameters; $HA_1, HA_2; HB; HC$
Two soil moisture parameters; $S_1; S_2$

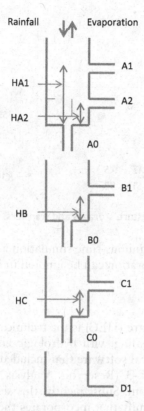

Figure 33.la Structure of the tank model.

Two soil moisture transfer parameters; K_1; K_2
One time lag parameter, TLAG < 1 for daily calibrations

Parameters are grouped into four categories:

- Discharge and infiltration parameters
- Soil moisture parameters
- Discharge threshold parameters
- Time lag parameters.

These categories are assumed to be independent. The first category is determined by feed-back mechanism. The remaining categories are by an iterative procedure using all possible combinations within limits determined from experience.

Calibration can be done either by trial and error or by systematically optimizing an objective function. In the latter case, the objective function used is

$$CRHY = \frac{1}{2}(MSEQ + MSELQ) \tag{33.4}$$

where $MSEQ = \dfrac{\text{Mean square error}}{\text{Mean discharge}}$

$MSEL\,Q = $ Mean square error of log Q

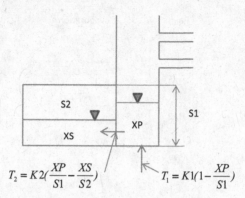

$$T_2 = K2(\frac{XP}{S1} - \frac{XS}{S2}) \qquad T_1 = K1(1 - \frac{XP}{S1})$$

Figure 33.1b Representation of soil moisture storages and transfers.

Tank model can be used for continuous-time simulation as well as for event simulation. In the latter case losses due to evaporation can be ignored in flood analysis.

33.2.3 HEC series

The Hydrologic Engineering Centre (HEC) is the technical arm of the United States Army Corps of Engineers (USACE) dealing with hydrologic engineering. It was established in 1964 and started with hydrological software that included HEC-1 (Watershed Hydrology), HEC-2 (River Hydraulics), HEC-3 (Reservoir Analysis and Conservation) and HEC-4 (Stochastic Streamflow Generation). Subsequently, this series of software has been super-seded with a more up-to-date family that incorporates the same basic concepts as the first series but with improved numerical procedures. The new family of software includes HEC-HMS (Hydrologic Modelling System), HEC-RAS (River Analysis System), HEC-FDA (Flood Damage Reduction Analysis) and HEC-ResSim (Reservoir System Simulation) (USACE, 1998, 2000).

HEC-1

The HEC-1 model has several components that are linked to each other. The catchment is divided into a number of sub-catchments on the basis of the topography and geography. Each sub-catchment is modelled to simulate the following components.

33.2.3.1 Land surface runoff component

To model the land surface runoff, rainfall is the input variable with losses assumed accord-ing to some empirical relationship. In each sub-catchment, rainfall and losses are assumed to be uniform. Infiltration losses are assumed as an initial and uniform loss rate or as an exponential loss rate. Routing the effective rainfall is by the unit hydrograph method or by the kinematic wave method. If the unit hydrograph method is used, synthetic unit hydro-graphs such as Clark's, Snyder's or SCS are used. If kinematic routing is adopted, the sub-catchment is divided into flow planes, collector channels and a main channel. Solution for different sections is by the finite difference method.

33.2.3.2 River routing component

The inflow hydrograph from the land surface area is routed downstream along the river using standard flood routing procedures such as the **Muskingum** method.

33.2.3.3 Reservoir component

The routing procedure is the same as that for rivers, excepting that reservoir provides the storage only.

33.2.3.4 Abstractions and inflows

These if present, are linked to the model nodes.

The parameters of HEC-1 model are obtained mostly by optimization.

33.2.3.4.1 HEC-HMS

HEC-HMS is a generalized modelling system, developed in 1992 to supersede HEC-1, and is the standard model used in the private sector in the US. It is, in some sense, distributed in that the catchment is sub-divided into smaller sub-catchments. Any mass or energy flux in the hydrological cycle can then be represented by a mathematical model. In most cases, several model choices are available for representing each flux. Each mathematical model included in the program is suitable in different environments and under different conditions. Making the correct choice requires knowledge of the catchment, the goals of the hydrologic study and engineering judgment. HEC-HMS provides a number of options for simulating the rainfall-runoff process. These include precipitation modelled using either actual gauged events or hypothetical, frequency-based storms, rainfall losses represented empirically (SCS) or with physically based algorithms (Green and Ampt), runoff generated from unit hydrographs or kinematic wave method, stream routed by Muskingum, Muskingum-Cunge and kinematic wave methods, and reservoir routing, base flow and diversions modelling.

The inputs to HEC-HMS consist of three components:

- A basin component, which is a description of the different elements of the hydrologic system (sub-basins, channels, junctions, sources, sinks, reservoirs and diversions), including their hydrologic parameters and topology;
- A precipitation component, which is a description – in space and time – of the precipitation event to be modelled, and consists of time series of precipitation at specific points or areas and their relation to the hydrologic elements (precipitation distribution can be by arithmetic mean, Thiessen polygon, or, isohyetal method);
- A control component, which defines the time window for the precipitation event and for the calculated flow hydrograph.

Parameters are estimated by optimization. The software can support GIS and is well documented. The source code is not freely available.

Evaporation, as modelled in HEC-HMS, includes evaporation from open water surfaces as well as transpiration from vegetation combined into ET. In this input, monthly-varying ET values are specified, along with an ET coefficient. The potential ET rate for all time periods within the month is computed as the product of the monthly value and the coefficient.

33.2.3.4.2 HEC-RAS

HEC-RAS is a program released in 1995 that can carry out one-dimensional steady flow, unsteady flow, sediment transport/mobile bed computations and water temperature modelling. Steady flow is modelled by solving the one-dimensional energy equation. The momentum equation is used in situations where the flow is rapidly varied such as for example in the case of a hydraulic jump. Unsteady flow is modelled by solving the full, dynamic, Saint-Venant equations using an implicit finite difference method.

HEC-RAS can model a network of channels, a dendritic system or a single river reach as well as subcritical, supercritical and mixed flows with the effects of bridges, culverts and other hydraulic structures. HEC-RAS is well tested, peer-reviewed and can be downloaded freely. On the negative side, numerical instability may occur in unsteady flow computations under steep flow conditions.

33.2.3.4.3 HEC-FDA

HEC-FDA provides the capability to perform an integrated hydrologic engineering and economic analysis during the formulation and evaluation of flood risk management plans. It is designed to assist in using risk analysis procedures for formulating and evaluating flood risk management measures.

33.2.3.4.4 HEC-ResSim

HEC-ResSim is designed to model reservoir operations at one or more reservoirs whose operations are defined by multiple goals and constraints.

33.2.4 Xinanjiang model

The Xinanjiang model has been successfully and widely applied in humid and semi-humid regions in China since its development in the 1970s (Zhao, 1977, 1984; POYB, 1979; Zhao et al., 1980; Zhang, 1990; Zhao and Liu, 1995). The Xinanjiang model provides an integral structure to statistically describe the non-uniform distribution of runoff producing areas, which features it as one of the conceptual, semi-distributed hydrological models developed. The original Xinanjiang model uses a single parabolic curve to represent the spatial distribution of the soil moisture storage capacity over the catchment where the exponent parameter b measures the non-uniformity of this distribution. The same function is also employed in the variable infiltration capacity (VIC) model (Wood et al., 1992) and in the Arno model (Dümenil and Todini, 1992).

The Xinanjiang model consists of a runoff generating component and a runoff routing component. The runoff generation of the Xinanjiang model acts on three vertical layers

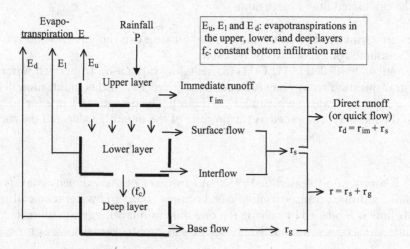

Figure 33.2 Schematic diagram of the Xinanjiang model.

identified as upper, lower and deep (Figure 33.2). The upper layer refers to the vegetation, water surface and the very thin top-soil. The lower layer refers to the soil in which the vegetation roots dominate, and the moisture transportation is mainly driven by the potential gradient. The deep layer refers to the soil beneath the lower layer, where only the deep-rooted vegetation can absorb water and the potential gradient is very small. Replenishment and depletion of soil moisture storage take place, respectively, via rainfall and ET. In the upper layer, ET takes place at the potential rate. On exhaustion of moisture content in the upper layer, ET proceeds to the lower layer at a decreased rate that is proportional to the moisture content in that layer. Only when the total ET in the upper and lower layers is less than a pre-set threshold, represented as a fraction of the potential ET, does it further proceed to the deep layer to keep this pre-set minimum value.

During rainfall, evaporation is first subtracted, and the runoff generation is computed by considering the respective soil moisture states and the storage capacities on the three layers. The runoff is separated into three components: immediate runoff, surface runoff and groundwater runoff, according to their generating levels in the vertical profile (Figure 33.2). Immediate runoff (r_{im}) is the net rainfall directly falling on the impervious and saturated area expressed as a percentage of the total watershed area (IMP). After the storage deficit is satisfied in the upper and lower layers, the remaining excess rainfall infiltrates into the deep layer at a constant rate (f_c) through the bottom of the lower layer to form the groundwater runoff (r_g). Surface flow and interflow are generated to form the surface runoff (r_s) only when the excess rainfall intensity is greater than this constant infiltration. Immediate runoff, surface flow and interflow together contribute to the direct runoff (r_d).

In general, the Xinanjiang model is used with spatially uniform rainfall events. For non-uniform rainfalls, the basin can be divided into several sub-basins. The Xinanjiang model is then applied to each sub-basin with the average rainfall of each sub-basin as the input. The complete Xinanjiang model has a total of 11 parameters, of which eight (IMP, W_{um}, W_{lm}, W_{dm}, α, ϕ, f_c and b) as listed and defined in Table 33.1 are for runoff generation and three (K_r, V_m and β) for runoff routing.

The Xinanjiang model simulates the runoff generation on partial areas by considering the non-uniformity of the spatial distribution of soil moisture storage capacity over the catchment. In the original Xinanjiang model, it is represented by a soil moisture storage capacity curve (Figure 33.3) defined as follows:

Table 33.1 Parameters of the rainfall-runoff transformation component in Xinanjiang model

α	Ratio of potential evapo-transpiration to pan evaporation
ϕ	Pre-set minimum coefficient of evapo-transpiration
IMP	Percentage of the impervious surface
W_{um}	Soil moisture storage capacity of the upper layer
W_{lm}	Soil moisture storage capacity of the lower layer
W_{dm}	Soil moisture storage capacity of the deep layer
W_m	Average soil moisture storage capacity in the lower and deep layers for the whole catchment, $W_m = W_{lm} + W_{dm}$
f_c	Final constant infiltration rate
b	Exponent in the spatial distribution curve of soil moisture storage capacity
c	Weighting factor in the spatial distribution curve of soil moisture storage capacity
Runoff-routing parameters	
K_r	Recession constant of the ground water storage
V_m	Average velocity of pure translation for all cells in the catchment
β	Attenuation coefficient of surface runoff routing

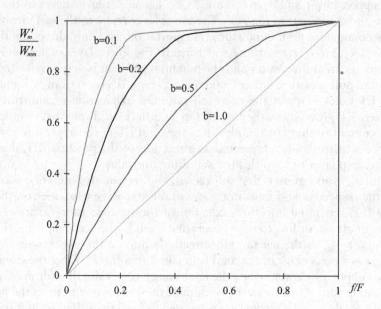

Figure 33.3 Distribution of soil moisture storage capacity represented by a single parabolic curve over the watershed.

$$\frac{f}{F} = 1 - \left(1 - \frac{W'_m}{W'_{mm}}\right)^b; \quad 0 \le \frac{W'_m}{W'_{mm}} \le 1 \tag{33.5}$$

where f is the partial pervious area of the watershed whose soil moisture storage capacity is less than or equal to the ordinate W'_m, which varies from zero to its maximum value W'_{mm} (Figure 33.3), F the total pervious area of the watershed and b the exponent of the curve.

The original Xinanjiang model has subsequently been modified (Jayawardena and Zhou, 2000) by introducing a double parabolic curve in place of the original single parabolic curve. The double parabolic curve (Figure 33.4) takes the form

$$\frac{f}{F} = (0.5 - c)^{1-b} \left(\frac{W'_m}{W'_{mm}}\right)^b; \quad 0 \le \frac{W'_m}{W'_{mm}} \le 0.5 - c \tag{33.6a}$$

and

$$\frac{f}{F} = 1 - (0.5 - c)^{1-b} \left(1 - \frac{W'_m}{W'_{mm}}\right)^b; \quad 0.5 - c \le \frac{W'_m}{W'_{mm}} \le 1 \tag{33.6b}$$

where all notations have the same meanings as in Eq. 33.5 except for the additional parameter c, a weighting factor. It is continuous and smooth at the point where the two branches meet and has its function and first derivative values of $(0.5 - c)$ and (b/W'_{mm}). It can be seen that this double parabolic curve is more flexible and easier to be calibrated than the original single parabolic curve.

The double parabolic curve consists of lower and upper branches, taking weights of $(0.5 - c)$ and $(0.5 + c)$, respectively, as shown in Figure 33.3. In the interval $[0,1]$ of f/F, the lower branch occupies $[0, 0.5 - c]$ part, whereas the upper branch occupies the remaining part, $(0.5 - c, 1]$. The parameter c varies in an interval $(-0.5, 0.5)$. The curve is continuous and smooth and has its function and first derivative values of $(0.5 - c)$ and (b/W'_{mm}) at the point where the two branches meet. When the weighting parameter c takes a value of (-0.5), the double parabolic curve reduces to the single lower parabolic curve:

$$\frac{f}{F} = \left(\frac{W'_m}{W'_{mm}}\right)^b ; \ 0 \le \frac{W'_m}{W'_{mm}} \le 1 \tag{33.7}$$

When the weighting parameter c takes a value of 0.5, the double parabolic curve reduces to the single upper parabolic curve, which is used in the original Xinanjiang model as Eq. 33.5. Thus, the single parabolic curve is a special case of the double parabolic curve which uses an additional weighting parameter c to adjust the shape of the curve, so that it is closer to the spatial distribution of soil moisture storage capacity.

Runoff computation in the Xinanjiang model is carried out as follows:

The rainfall PE reaching the lower layer can be written as

$$PE = (P - \alpha E_m) - r_{im} - (W_{um} - W_u) \tag{33.8}$$

subject to

$$P - \alpha E_m > 0 \text{ and } PE \ge 0$$

where P is the rainfall, E_m is the pan evaporation, α is a pan coefficient, r_{im} is the immediate runoff in the impervious area and W_{um} and W_u respectively represent the moisture capacity and antecedent moisture content in the upper layer. In using Eq. 33.8 and subsequent equations, rainfall P, evaporation E_m and runoff r_{im} are all measured in units of depth per unit time which is usually a day or an hour. Soil moisture storage capacities and actual contents are measured in units of depth.

The immediate runoff r_{im} can be expressed as

$$r_{im} = \text{IMP}(P - \alpha E_m) \tag{33.9}$$

where IMP is the fraction of the catchment area that is impervious as a percentage.

As rainfall reaches the lower and deep layers, the soil moisture increases, and the point x on the curve in Figure 33.4 moves upwards along the curve. The combined runoff in the lower and deep layers, $r(= r_s + r_g)$, is obtained by integrating the distribution curve of soil moisture storage capacity as illustrated in Figure 33.5, in which W_0 is the antecedent soil moisture state in the two layers, ΔW is the increase of the soil moisture content in the two layers due to the rainfall PE and H is the ordinate corresponding to the point x, which is obtained by solving the equations below (Eq. 33.10a or 33.10b) iteratively for an assumed value of W_0. When W_0 is located in the lower branch of the parabolic curve (Figure 33.4), it can be expressed as

$$W_0 = H - \frac{(0.5 - c)^{1-b}}{(1+b)(W'_{mm})^b} H^{1+b} \tag{33.10a}$$

whereas when it is in the upper branch, it is expressed as

$$W_0 = (0.5 - c)W'_{mm} + \frac{2cW'_{mm}}{(1+b)} - \frac{(0.5 + c)^{1-b} W'_{mm}}{(1+b)} \left(1 - \frac{H}{W'_{mm}}\right)^{1+b} \tag{33.10b}$$

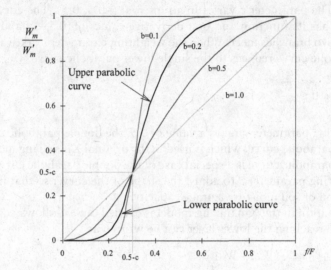

Figure 33.4 Distribution of soil moisture storage capacity represented by a double parabolic curve over the watershed.

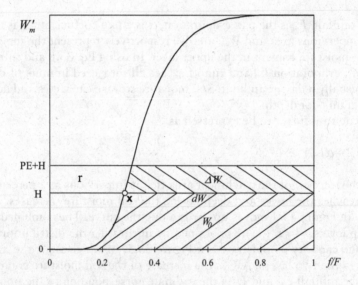

Figure 33.5 Runoff calculation in the Xinanjiang model.

Thus,

$$r = \int_{H}^{PE+H} \left(\frac{f}{F} \right) dW'_m \tag{33.11}$$

Equation 33.11 takes different forms depending on the locations of its integrating limits, H and $PE + H$ in Figure 33.5. The corresponding runoff from the lower and deep layers for different antecedent soil moisture states W_0 and rainfall PE in the two layers can be written as follows:

When the antecedent and new soil moisture contents are both located in the lower branch of the capacity curve, i.e. $H \leq (0.5 - c) W'_{mm}$ and $PE + H \leq (0.5 - c) W'_{mm}$,

$$r = \frac{(0.5-c)^{1-b}W'_{mm}}{(1+b)}\left[\left(\frac{PE+H}{W'_{mm}}\right)^{1+b} - \left(\frac{H}{W'_{mm}}\right)^{1+b}\right]$$ (33.12a)

When the antecedent soil moisture content is located in the lower branch, but the new one in the upper branch, i.e. $H \leq (0.5-c)W'_{mm}$ and $(0.5-c)W'_{mm} < PE+H \leq W'_{mm}$,

$$r = PE+H-(0.5-c)W'_{mm}$$

$$-\frac{W'_{mm}}{(1+b)}\left[2c+(0.5-c)^{1-b}\left(\frac{H}{W'_{mm}}\right)^{1+b} - (0.5+c)^{1-b}\left(1-\frac{PE+H}{W'_{mm}}\right)^{1+b}\right]$$ (33.12b)

When the antecedent soil moisture content is located in the lower branch and the new one is saturated, i.e. $H \leq (0.5-c)W'_{mm}$ and $PE+H > W'_{mm}$,

$$r = PE-(W_m - W_0)$$ (33.12c)

Similarly, when the antecedent and new soil moisture contents are both located in the upper branch (in this case, it is impossible for the new one to be located in the lower branch), i.e. $(0.5-c)W'_{mm} < H \leq W'_{mm}$ and $(0.5-c)W'_{mm} < PE+H \leq W'_{mm}$,

$$r = PE + \frac{(0.5+c)^{1-b}W'_{mm}}{(1+b)}\left[\left(1-\frac{PE+H}{W'_{mm}}\right)^{1+b} - \left(1-\frac{H}{W'_{mm}}\right)^{1+b}\right]$$ (33.12d)

When the antecedent soil moisture content is located in the upper branch and the new one is saturated, i.e. $H > (0.5-c)W'_{mm}$ and $PE+H > W'_{mm}$,

$$r = PE-(W_m - W_0)$$ (33.12e)

Equation 33.12e also includes the situation when the antecedent soil moisture content is saturated because in this case $W_0 = W_m$, and all net rainfall generates runoff.

The Xinanjiang model gives only the runoff generating rainfall (or rainfall excess). It has then got to be routed along the channel network to obtain the required discharge. Several approaches of routing such as the Muskingum method, unit hydrograph method and kinematic wave method can be used.

33.2.5 Variable infiltration capacity (VIC) model

The variable infiltration capacity (VIC) model is a semi-distributed conceptual hydrological model that characterizes a watershed as multi-layered grids and employs distributed information of the watershed. It was developed originally as a Soil-Vegetation-Atmosphere Transfer scheme (SVATs) for the purpose of representing the land surface in GCMs used for climate simulation and numerical weather prediction (Wood et al., 1992). Distinguishing characteristics of the VIC model include the representation of sub-grid variability in land surface vegetation classes, the representation of sub-grid variability in soil moisture storage capacity as a spatial probability distribution and the representation of drainage from the lower soil moisture zone (base flow) as a nonlinear recession (Liang et al., 1994).

Using the infiltration formulation used in the Xinanjiang model (Zhao et al., 1980), the VIC model has been developed for decades to form a family of models that includes model structures from a single layer to three layers sequenced as the VIC-1L, VIC-2L and VIC-3L. The original VIC model with one layer to represent the variation in infiltration capacity within a GCM grid cell (Wood et al., 1992) was progressed to consist of a two-layer characterization of soil column (i.e., VIC-2L) that uses an aerodynamic representation of the latent and sensible heat fluxes at the land surface with different land surface and ET description (Liang et al., 1994). The VIC-2L model was further modified to add one thin surface layer (i.e., VIC-3L) to better capture the dynamic behaviour of soil moisture content. In addition, diffusion process was included in the representation of the VIC soil column (Liang et al., 1996b). Additional progress includes improved snow representation (Storck and Lettenmaier, 1999), improved ground heat flux parameterization (Liang et al., 1999), efficient parameterization of sub-grid precipitation (Liang et al., 1996a), improvements for model performance in cold regions (Cherkauer and Lettenmaier, 1999), efforts to incorporate a double parabolic curve for describing the infiltration capacity (Jayawardena and Mahanama, 2002; Jayawardena and Zhou, 2000) and improved runoff generation strategies and interactions between surface and groundwater (Liang and Xie, 2001).

33.2.5.1 Model description

Among the VIC series models, the VIC-3L is the latest version (as of 2007), and it can combine most model characteristics. The components of the model can be described as follows (Liang et al., 1996b; Matheussen et al., 2000):

- The land surface is described horizontally by $N+1$ land cover (vegetation) types, where $n=1, 2, ..., N$ represents N different types of vegetation, and $n=N+1$ represents bare soil.
- The subsurface is characterized vertically with three soil layers (Figure 33.6; Table 33.2).
- It consists of an upper layer, which is partitioned into a top thin layer (layer 0) and a thicker layer (layer 1, referred to as upper thick layer hereafter) and a lower layer

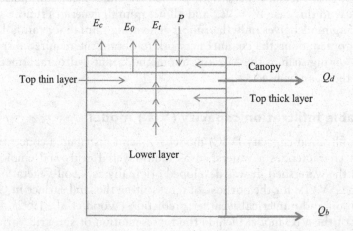

Figure 33.6 Schematic representation of vertical soil moisture movement and grid-based river flow generation in the VIC model.

Table 33.2 Parameter description of the VIC-3L model

Name	Units	Description
P, P_t	mm/day	Precipitation and throughfall precipitation
E_c, E_0	mm/day	Evaporation from canopy layer and bare soil
E_t	mm/day	Transpiration from vegetation
W_0, W_1, W_2	mm	Soil moisture content in layer 0, layer 1, and layer 2
Q_{01}, Q_{12}	mm/day	Drainage from layer 0 to layer 1 and from layer 1 to layer 2
$D_{f,10}, D_{f,21}$	mm/day	Diffusion from layer 1 to layer 0 and from layer 2 to layer 1
D_1, D_2, D_3	m	Soil depths of layer 0, layer 1, and layer 2
Q_d, Q_b	mm/day	Direct flow and base flow

Note: P is the precipitation (mm/day), Q_d is the surface runoff (mm/day), W_u and ΔW_u are the soil moisture storage and the change in the soil moisture storage in the upper layer (mm), A_s is the fraction of an area that is saturated, i_0 is the corresponding point infiltration capacity (mm), and Δt is the time step.

(layer 2). The top thin layer allows a quick response due to changes in surface conditions. The infiltration algorithm of the upper layer represents the dynamic fast response to rainfall. The lower layer characterizes the slowly varying soil moisture behaviour.

- Associated with each land cover class is a single canopy layer and three soil layers. The land cover classes are specified by their leaf area index (LAI), canopy resistance and relative fraction of roots in each of the three soil layers (Figure 33.6) which depend on the vegetation class and the soil type.
- Total ET over a grid cell is computed as the summation of the canopy, vegetation and bare soil components, weighted by the respective surface cover area fractions. The ET from each vegetation type within the grid cell is characterized by its potential ET, aerodynamic resistance to the transfer of water and architectural resistance.
- Surface runoff (direct flow) is generated only from the upper layer. There is vertical soil moisture movement by drainage and diffusion. Drainage from the top thin layer and the upper thick layer to the corresponding next layer is based on the hydraulic conductivity of the current layer. The lower layer produces subsurface flow (base flow) through the ARNO model (Franchini and Pacciani, 1991). In this way, the model separates subsurface runoff from a quick storm response.
- The infiltration, the fluxes of moisture between soil layers and runoff vary with vegetation cover class. Surface runoff and subsurface runoff are computed for each cover type and summed over all land cover types within a grid cell as model output.

As the core of the model, the VIC model assumes that the infiltration capacity (defined as the maximum depth of water that can be stored in the soil column), and therefore, runoff generation and ET, vary within an area (grid cell or catchment) due to variations in topography, soil and vegetation (Wood et al., 1992). The variation in infiltration capacity i over an area is expressed as

$$i = i_m \left[1 - (1-A)^{1/B} \right]$$

(33.13)

where A is the fraction of a grid cell area for which the infiltration capacity is less than i ($0 \le A \le 1$), i_m is the maximum infiltration capacity within the grid cell (mm) and B is the infiltration shape parameter (dimensionless) which is a measure of the spatial variability of the infiltration capacity. By integration of Eq. 33.13,[1] the maximum infiltration capacity i_m can be expressed as

$$i_m = W_u^C \times (1 + B)$$
(33.14)

where W_u^C is the maximum soil moisture content in the upper layer (mm).

Helpful information on the summary of the VIC processes, model operation overview and general guide for inputs and outputs processing, as well as important references, are available from the homepage of the VIC model: http://www.hydro.washington.edu/Lettenmaier/Models/VIC/VIChome.html.

The VIC model assumes that surface runoff (direct flow) is generated by those areas for which precipitation, when added to the soil moisture storage at the end of the previous time step, exceeds the storage capacity of the soil. When the soil is saturated, the drainage term follows the saturated hydraulic conductivity. Otherwise, the drainage term between soil layers is calculated based on the Clapp–Hornberger relationship (Clapp and Hornberger, 1978). Finally, the grid-based runoff is routed through the channel network in the same way as in the Xinanjiang model. The VIC model and its variations have been applied to two major rivers in the Asia Pacific region (Jayawardena, 2001; Jayawardena, 2006; Jayawardena and Tian, 2005).

33.2.6 Système Hydrologique Europèen (SHE) model

Système Hydrologique Europèen (SHE) is a physically-based, distributed, integrated hydrological modelling system produced jointly by the Danish Hydraulic Institute (DHI, http://www.dhisoftware.com), the British Institute of Hydrology (now the Centre for Ecology and Hydrology) and the French consulting company SOGREAH with the financial support of the Commission of the European Communities. Since 1987, SHE has been further developed independently by three respective organizations which are now the University of Newcastle (UK), Laboratoire d'Hydraulique de France and DHI. DHI's version of the SHE, known as the MIKE SHE, represents significant new developments with respect to the user interface, computational efficiency and process descriptions. For the background to the development of SHE, as well as the descriptions of model structures and process equations in SHE, see the review by Abbott et al. (1986a,b), Refsgaard and Knudsen (1996), and Refsgaard (1997). The main modules consist of

- ET component
- Unsaturated zone flow component
  - 1-D Richards' equation
- Saturated zone flow component
  - 3-D Boussinesq equation
- Overland flow (2-D) and channel flow (1-D) components
  - Diffusion wave approximation of the Saint Venant equations.

SHE enables the simulation of water, solutes and sediments in the entire land phase of the hydrological cycle. It is a dynamic, user-friendly modelling tool for the analysis, planning and management of a wide range of water resources and environmental problems related to surface water and groundwater, including

---

[1] $W_u^C = \int_0^1 i(A)\,dA = \int_0^1 i_m \left[1 - (1-A)^{1/B} \right] dA = \dfrac{i_m}{B+1}$

- surface water/groundwater interaction,
- conjunctive use of water,
- water resources management,
- irrigation management,
- changes in land use practices,
- contaminant transport in the subsurface, and,
- farming practices, including fertilizers and agrochemicals.

SHE is an integrated modelling system with a modular structure. Individual components can be used independently and customized to local needs depending on data availability and aims of the given study. Powerful pre-processing and results presentation tools are included in the SHE software package. SHE contains a number of process simulation modules that, in combination, describes the entire land phase of the hydrological cycle.

MIKE SHE solves the partial differential equations governing overland and channel flow, unsaturated flow and saturated subsurface flow. In the complete model, processes such as snowmelt, interception and ET are also included. The flow equations are solved numerically using finite difference methods. In the horizontal plane, the catchment is discretized into a network of grid squares. The river system is assumed to run along the boundaries of these. Within each square, the soil profile is described at a number of nodes, which above the groundwater table may become partly saturated. Lateral subsurface flow is only considered in the saturated part of the profile. The commercially available **MIKE-11** is a comprehensive, one-dimensional modelling system for the simulation of flows, sediment transport and water quality in estuaries, rivers, irrigation systems and other water bodies. Its two-dimensional counterpart **MIKE-21** is a comprehensive modelling system for two-dimensional free surface flows applicable to studies of lakes, reservoirs, estuaries, bays, coastal areas and seas where stratification can be neglected.

Since SHE model is a physically-based one, the parameters should be physically identifiable and measurable, at least in theory. This implies that calibration is not necessary if sufficient data are available. In practice, however, it is difficult if not impossible to make such measurements in the field because of catchment heterogeneities. A compromise is often made in resorting to some calibration technique, thereby diluting the true meaning of a physically-based distributed model. The goal then is to find an optimal set of parameters that are also realistic that will give the expected outcome. Some aspects of calibration and validation of MIKE-SHE model can be found in Sahoo et al. (2006).

33.3 HYDRAULICS OF PHYSICS-BASED MODELS

33.3.1 Governing equations

Surface runoff is considered as overland flow when the precipitated water in excess of the infiltration capacity which, for the purpose of analysis, is imagined to flow down a slope in the form of a uniform thin sheet. In a one-dimensional domain, the governing equations may then be written as

$$\frac{\partial q}{\partial x} + \frac{\partial h}{\partial t} = (i - f) = Q \quad \textbf{Continuity Equation} \tag{33.15}$$

$$\frac{\partial v}{\partial t} + v\frac{\partial v}{\partial x} + g\frac{\partial h}{\partial x} + \frac{Qv}{h} = g\left(S_0 - S_f\right) \quad \textbf{Momentum Equation} \tag{33.16}$$

where

> q – discharge per unit width
> h – depth of flow
> Q – lateral inflow per unit length per unit width
> v – velocity of flow
> S_0 – bed slope of the flow plane
> S_f – friction slope of the flow plane
> i – rainfall rate
> f – infiltration rate
> x, t – distance along the flow plane, and time.

These two equations are also called the **St. Venant equations or the shallow water equations**. In the momentum equation, the first term is the acceleration term, the second term is the convection term, the third term is the diffusion term and the fourth term the lateral momentum term.

33.3.2 Dynamic wave equations

When the complete shallow water equations (Eq. 33.15 and 33.16) are used, the formulation is called the dynamic wave formulation. The problem is unsteady and non-uniform flow.

Then, writing $q = hv$

$$h\frac{\partial v}{\partial x} + v\frac{\partial h}{\partial x} + \frac{\partial h}{\partial t} = Q \tag{33.17}$$

Replacing the derivatives by the finite differences

$$h_{i,j}\frac{v_{i+1,j} - v_{i-1,j}}{2\Delta x} + v_{i,j}\frac{h_{i+1,j} - h_{i-1,j}}{2\Delta x} + \frac{h_{i,j+1} - h_{i,j}}{\Delta t} = Q_{i,j} \tag{33.18}$$

and

$$\frac{v_{i,j+1} - v_{i,j}}{\Delta t} + v_{i,j}\frac{v_{i+1,j} - v_{i-1,j}}{2\Delta x} + g\frac{h_{i+1,j} - h_{i-1,j}}{2\Delta x} + \frac{Q_{i,j}v_{i,j}}{h_{i,j}} = g\left(S_0 - S_f\right) \tag{33.19}$$

The problem has two degrees of freedom, h and v. It is also non-linear. Hence an iterative simultaneous solution should be attempted. The boundary conditions for this problem are usually of the **Dirichlet** type. Prescribed values of the variable must be given.

The frictional slope S_f is usually represented by the Manning or Chezy equations.

33.3.3 Diffusion wave approximation

$$\frac{\partial q}{\partial x} + \frac{\partial h}{\partial t} = (i-f) = Q \quad \textbf{Continuity Equation} \tag{33.20}$$

$$g\frac{\partial h}{\partial x} = g\left(S_0 - S_f\right) \quad \textbf{Momentum Equation} \tag{33.21}$$

The flow is usually considered as quasi-steady non-uniform. Changes in water surface profile are caused only by inflows.

33.3.4 Kinematic wave approximation

The momentum (or dynamic) equation under certain circumstances can be reduced to a more simple form by neglecting the inertia, inflow momentum and free surface slope terms. By an order of magnitude analysis, it can be seen that these terms are small compared with the bed and friction slopes. For example,

If $S_0 = 0.01$, then it is $O(10^{-2})$; $g = O(10)$ m/s^2;

$(i - f) = O(10^{-6})$ m/s or $O(10$ mm/$h)$; $h = O(10^{-2})$ m;

$v = O(10^{-1})$ m/s; $Q = O(10^{-6})$.

Thus all the terms on the LHS of the dynamic equation are small relative to the terms on the RHS. Therefore it is possible to simplify the dynamic equation to

$$g(S_0 - S_f) = 0, \text{ or, }(S_0 - S_f) = 0 \tag{33.22}$$

This condition implies uniform flow in which the weight of the fluid element is balanced by the frictional force due to boundary shear.

$$(A\Delta x)\rho g S_0 = \tau_0(P\Delta x) \Rightarrow \tau_0 = \rho g R S_0 \tag{33.23}$$

where
  A – cross-sectional area
  P – wetted perimeter
  R – hydraulic mean radius $(= A/P)$
  Δx – length of the fluid element along the slope
  τ_0 – boundary shear stress

For overland flow, $R \cong h$, and therefore,

$$\tau_0 = \rho g h S_0 \tag{33.24}$$

Also, τ_0 can be defined using the boundary layer concepts as

$$\tau_0 = c_f \frac{\rho v^2}{2} \tag{33.25}$$

where c_f is a function of **Reynolds number, Re,** and roughness.
  For **laminar flow,** c_f is given as

$$c_f = \frac{4}{Re} = \frac{4v}{vh} \text{ (}v \text{ - kinematic viscosity of water)} \tag{33.26}$$

Therefore, $\tau_0 = \dfrac{4v}{vh} \dfrac{\rho v^2}{2} = \dfrac{2v\rho v}{h}$ \hfill (33.27)

Equating Eqs. 33.24 and 33.27,

$$v = \frac{g S_0 h^2}{2v} \tag{33.28}$$

and, $\quad q = Av = hv = \dfrac{gS_0}{2v} h^3$ \hfill (33.29)

which is of the form

$\quad q = \alpha h^m$ \hfill (33.30)

For **turbulent flow, Manning's** friction factor may be used.

$$v = \frac{1}{n} R^{2/3} S_f^{1/2}$$
$$= \frac{1}{n} h^{2/3} S_0^{1/2} \tag{33.31}$$

(Here, h is the normal depth of flow.)
   and, $q = Av = hv$ gives

$$q = \frac{1}{n} h^{5/3} S_0^{1/2} \tag{33.32}$$

which is again of the form $q = \alpha h^m$.

   This approximation is known as the **kinematic wave approximation,** and has been applied to model overland flow, flood flow and traffic flow. The pair of Eqs. 33.20 and 33.30

$$\frac{\partial q}{\partial x} + \frac{\partial h}{\partial t} = Q$$
$$q = \alpha h^m$$

gives a quasi-steady state solution in which the changes in water surface profile are caused by changes in flow rate q only. The solutions are not valid at the two boundaries where the condition $S_0 = S_f$ is not satisfied, i.e. the bed slope and the free surface slope are not the same. For natural surfaces, $m = 2$ has been used.

   Woolhiser and Liggett (1967) concluded that the kinematic wave solution is a good approximation to the dynamic wave equations provided

$$\frac{S_0 L}{h Fr_L^2} > 10$$

where Fr_L is the Froude number[2] at the end of the flow plane of length, L.

   Lighthill and Whitham (1955) stated the following condition:

$$Fr_L < 2$$

Eagleson (1970), summarizing the findings of the above researchers, stated that the kinematic wave solution is a good approximation to the dynamic wave equations provided

$$Fr < 2 \text{ and } \frac{S_0 L}{h Fr_L^2} > 10$$

---

[2] $Fr = \dfrac{v}{\sqrt{gh}}$; $Fr = 1$ corresponds to critical flow; $Fr < 1$ corresponds to sub-critical flow and $Fr > 1$ corresponds to super-critical flow.

Taylor et al. (1974) claims that the condition $Fr < 2$ is unnecessary provided

$$\frac{S_0 L}{h Fr_L^2} > 10$$

These conditions are generally satisfied for the field conditions in catchment hydrology except where flat land dominates.

33.3.4.1 Boundary and initial conditions

At $t = 0$, $h = 0$ for $0 \le x \le L$ (L is the length of slope)

At $t > 0$, $h = 0$ at $x = 0$

An important property of the kinematic wave approximation is that the solution is always upstream driven. A disturbance made at the downstream has no influence on the upstream. Solutions can, therefore, be obtained with only one boundary condition.

33.3.4.2 Analytical solution

An analytical solution has been provided by Henderson and Wooding (1964), and later critically reviewed by Eagleson (1970) for constant intensity rainfall, i_*.

Case i: $t_r \ge t_c$
   When $0 < t \le t_r$

$$= \left[\frac{x i_*}{\alpha} \right]^{1/m} \quad \text{for } 0 \le x \le x_w \tag{33.33a}$$

$$= i_* t \text{ for } x > x_w \tag{33.33b}$$

where t_r – duration of rainfall excess

$$t_c\text{-time of concentration} = \left[\frac{1}{\alpha} L i_*^{(1-m)} \right]^{1/m}$$

t_p – time up to which the build-up phase remains constant before beginning to recede (applicable only when $t_r < t_c$)
   x_w – distance along the plane up to which the depth continues to grow for any time t
   When $t > t_r$
   The depth h is given by the implicit relationship

$$x = \alpha h^{m-1} \left[h i_*^{-1} + m \left(t - t_r \right) \right] \tag{33.33c}$$

Case ii : $t_r < t_c$
   When $0 < t \le t_r$

the same as Eqs. 33.33a and 33.33b.
When $t_r < t \le t_p$

$$h = i_* t_p \tag{33.33d}$$

When $t > t_p$
the same as Eq. 33.33c.

33.3.4.3 V-shaped gutter flow

$$Q = \alpha A^m \tag{33.34a}$$

Also,

$$Q = AV$$

$$= A \frac{1}{n} R^{\frac{2}{3}} S_0^{\frac{1}{2}}$$

For the V-shaped gutter with side slopes z $(H{:}V = z)$,

$$A = zh^2$$

$$P = 2h\left\{1+z^2\right\}^{\frac{1}{2}}$$

$$R = \frac{zh}{2\left(1+z^2\right)^{\frac{1}{2}}}$$

Therefore,

$$Q = \frac{1}{n} S_0^{\frac{1}{2}} zh^2 \left[\frac{zh}{2\left(1+z^2\right)^{\frac{1}{2}}}\right]^{\frac{2}{3}}$$

$$= \frac{1}{n} S_0^{\frac{1}{2}} \frac{1}{2^{\frac{2}{3}}} \left[\frac{z}{1+z^2}\right]^{\frac{1}{3}} z^{\frac{2}{3}} h^{\frac{8}{3}}$$

$$= \frac{1}{n} S_0^{\frac{1}{2}} \frac{1}{2^{\frac{2}{3}}} \left[\frac{z}{1+z^2}\right]^{\frac{1}{3}} \left\{zh^2\right\}^{\frac{4}{3}}$$

$$= 0.6299 \frac{1}{n} S_0^{\frac{1}{2}} \left[\frac{z}{1+z^2}\right]^{\frac{1}{3}} (A)^{\frac{4}{3}}$$

$$= \alpha A^m$$

where

$$\alpha = 0.6299 \frac{1}{n} S_0^{1/2} \left[\frac{z}{1+z^2} \right]^{1/3} \tag{33.34b}$$

$$m = \frac{4}{3} \tag{33.34c}$$

33.3.4.4 Ground water flow

The governing equations are

$$\frac{\partial q}{\partial x} + \frac{\partial w}{\partial t} = Q' \tag{33.35}$$

where w – water content per unit area ($= \theta h$, where θ is the volumetric moisture content)
  Q' – lateral recharge per unit area
  and

$$q = \text{Area} \times \text{velocity}$$

$$= h \times 1 \times v$$

$$= -hK \frac{\partial \phi}{\partial x}$$

$$= -Kh \left\{ \frac{\partial h}{\partial x} + \frac{\partial z}{\partial x} \right\}$$

$$= Kh \left(S_0 - \frac{\partial h}{\partial x} \right)$$

If $\dfrac{\partial h}{\partial x} \ll S_0$, then $q \cong KhS_0$ which is of the form $q = \alpha h^m$ with $\alpha = KS_0$ and $m = 1$.

Therefore,

$$\frac{d}{dx}(KS_0 h) + S \frac{dh}{dt} = Q' \tag{33.36}$$

33.3.4.5 Finite difference solution scheme

Combining the continuity equation with the kinematic wave approximation, the following equation can be obtained:

$$m\alpha h^{m-1} \frac{\partial h}{\partial x} + \frac{\partial h}{\partial t} = Q \tag{33.37}$$

In explicit form, it is given by

$$m\alpha h_{i,j}^{m-1} \frac{h_{i+1,j} - h_{i-1,j}}{2\Delta x} + \frac{h_{i,j+1} - h_{i,j}}{\Delta t} = Q_{i,j} \tag{33.38}$$

In implicit form, one possible scheme is

$$m \alpha h_{i,j}^{m-1}\left[\beta \frac{h_{i+1,j+1} - h_{i-1,j+1}}{2\Delta x} + (1-\beta)\frac{h_{i+1,j} - h_{i-1,j}}{2\Delta x} \right] + \frac{h_{i,j+1} - h_{i,j}}{\Delta t} = Q_{i,j} \qquad (33.39)$$

It is also possible for $h_{i,j}$ in the product term to be taken as the mean at the two time levels considered.

The quantity $m\,\alpha h^{m-1}$ is called the wave celerity $\left(= \dfrac{\partial q}{\partial h} \right)$.

33.3.4.6 Finite element solution scheme

An alternative approach for solving the kinematic wave approximation is the application of the finite element method instead of the finite difference method. The finite element method is more flexible and has been used to solve several types of field problems although its origin is in structural engineering. The finite elements, in this case, have been obtained by discretizing the catchment into strips normal to the topographical contour lines. Details of the approach, including problem formulation, numerical solution with finite elements in space and finite differences in time and application to real catchments can be found in author's early publications (White and Jayawardena, 1975; Jayawardena and White, 1977, 1979).

33.4 DATA DRIVEN MODELS

The type of hydrological modelling and prediction used for a particular situation depends on the richness of the theory as well as the richness of the data. For theory rich situations, knowledge based models that use the principles and laws of physics can be used. They generally lead to the problem of finding solutions to a set of partial differential equations. Even under the simplest assumptions, theoretical solutions to practical problems cannot be found, and therefore resort is often made to numerical solutions in a simplified spatial domain. They are useful for understanding the system dynamics under often unrealistic assumptions and cannot be applied to real situations. The alternative then is to go for data driven approaches that do not require a prior understanding of the dynamics of the system. Such approaches are particularly suited to data-rich theory weak situations.

The earliest data driven approach may perhaps be the regression analysis in which a simple relationship is sought between dependent and independent variables of a system. A more traditional data driven approach of modelling and predicting complex hydrological systems considers such systems as stochastic, which at least in theory, can have an infinite number of degrees of freedom. Such approaches have been used in the past, but researchers are now turning towards the realization that certain systems that are seemingly stochastic are in fact driven by fully deterministic processes. Several new techniques of analysis and prediction have emerged in the last decade or so, and the frontier in this field is still moving forward. Emphasis has over the years shifted from global modelling in which attempts are made to find a function or a model that will fit into the entire data set to local modelling in which function approximations are done in a piecewise manner. Many of the new techniques have originated in mathematical and statistical domains and subsequently made their way into engineering fields. More details of different types of data driven modelling approaches as well as applications can be found in a separate publication by the author (Jayawardena, 2014).

33.4.1 Guiding principles and criteria for choosing a model

- Should be useful to solve or understand a particular problem under a given set of conditions and constraints.
- A reasonable balance between the costs and benefits; many models and modelling techniques add only a marginal value at an unjustifiable cost.
- Data-driven vs. physics-based. Data-driven models are relatively easier to implement, but not without problems. Physics-based models are more difficult. Their formulation, calibration and implementation are quite resource and expertise demanding. Their problems are also of a higher magnitude.
- Resource-driven or needs-driven?
- Simple models or complex models?
- Whether the model is for a specific purpose to solve a problem or for an academic purpose for better understanding of the system?
- Opinions are divided – whether it is the end result that matters or how it is obtained?

33.4.2 Why data driven models?

- Data driven models are relatively simple to formulate and easy to implement.
- Data contain all the information about the system, and their use is quite logical.
- Can learn from experience.
- Can generalize from examples.
- Less reliance on expertise.
- Particularly suited for data rich theory weak problems.
- Non-linear mapping.
- Physics-based models consider the catchment processes from a physics point of view, but their formulation, calibration and implementation are quite resource and expertise demanding.
- So far, no fully physics-based model has been successfully applied to a catchment without making drastic assumptions and simplifications.
- When other approaches are not feasible.

33.4.3 Types of data driven models

- Regression models
- Stochastic models
- Artificial neural networks
  - Multi-layer perceptron
  - Radial basis functions
  - Recurrent neural networks
  - Wavelet neural networks
  - Product unit neural networks
  - …
  - …
- Fuzzy logic systems
- Support vector machines
- Genetic algorithms
- Genetic programming
- Dynamical systems approach

- ...
- ...

33.4.4 Challenges in catchment modelling

- The main challenges in the data-driven modelling front include choosing between stochastic and deterministic approaches, lumped and semi-distributed approaches, linear and non-linear approaches and stationary and non-stationary assumptions.
- Most data-driven models are lumped. Linear assumption makes subsequent analysis and application simple but, in many instances, it is far from reality. Non-linear assumption makes the problem more realistic, but at a cost and lacks generality. Similarly, stationarity assumption makes analysis and application simpler, but with human influence (such as climate change) in the hydrologic system, the stationarity assumption no longer holds in many situations.
- The next modelling challenge comes from scale issues. For physics-based distributed models, it is necessary to define a set of governing equations. They are defined for a continuum, and whether such equations are valid in the scale of typical distributed models is an unresolved issue.
- Physically identifiable and measurable parameters vs. optimized parameters.
- Spatially and temporally homogeneous or non-homogeneous?
- Global and local optima – Popular global search methods include population-evolution-based search strategies, such as the shuffled complex evolution (SCE) algorithm (Duan et al., 1993) and genetic algorithm (GA) (Wang, 1991).
- Single objective function vs. multi-objective function.
- Based on the original SCE algorithm, recent studies have led to the development of the shuffled complex evolution metropolis (SCEM) and the multi-objective shuffled complex evolution metropolis (MOSCEM) algorithms (Vrugt et al., 2003a,b).
- Direct comparison of these methods would be helpful in selecting the most suitable calibration algorithm from the extensively used SCE family of algorithms.
- For physics-based models that are necessarily of a distributed nature, the use of optimization techniques for calibration defeats the purpose.
- As a result, most models that start with laws of physics end up as data-driven models, thereby defeating the very purpose of adopting such an approach.
- Equi-finality problem in multi-parameter optimization – a concept originated in the general systems model of von Bertalanffy (1968), meaning that the same final result may be arrived from different initial conditions and in different ways. In the context of multi-parameter optimization, what this means is that there is no unique set of parameter values, but rather a feasible parameter space from which a Pareto set of solutions is sought.

REFERENCES

Abbott, M. B., Bathurst, J. C., Cunge, J. A., O'Connell, P. E. and Rasmussen, J. (1986a): An introduction to the European Hydrological System, Système hydrologique Europèen, SHE-1, History and philosophy of a physically based, distributed modelling system, *Journal of Hydrology*, vol. 87, pp. 45–59.

Abbott, M. B., Bathurst, J. C., Cunge, J. A., O'Connell, P. E. and Rasmussen, J. (1986b): An introduction to the European Hydrological System, Système hydrologique Europèen, SHE, 2, Structure of a physically based, distributed modelling system, *Journal of Hydrology*, vol. 87, pp. 61–77.

Amorocho, J. (1973): *Advances in Hydroscience*, Academic Press, New York, vol. 9, pp. 203–251.

Beven, K. J. and Kirkby, M. J. (1979): A physically based, variable source area model of basin hydrology, *Hydrological Sciences Bulletin*, vol. 24, pp. 43–69.

Cherkauer, K. A. and Lettenmaier, D. P. (1999): Hydrologic effects of frozen soils in the upper Mississippi River basin, *Journal of Geophysical Research-Atmospheres*, vol. 104, no. D16, pp. 19599–19610.

Clapp, R. B. and Hornberger, G. M. (1978): Empirical equations for some soil hydraulic properties, *Water Resources Research*, vol. 14, pp. 601–604.

Crawford, N. H. and Linsley, R. K. (1966): Digital simulation in hydrology: Stanford watershed model IV. Technical Report No. 39, Department of Civil Engineering, Stanford University, p. 210.

Delleur, J. W. and Rao, R. A. (1971): Linear systems analysis in hydrology: The transform approach, the kernel oscillations and the effect of noise, *Proceedings of US-Japan Bi-Lateral Seminar in Hydrology*, Honolulu, HI, pp. 116–129.

Duan, Q., Gupta, V. K. and Sorooshian, S. (1993): A shuffled complex evolution approach for effective and efficient global minimization, *Journal of Optimization Theory and Applications*, vol. 76, no. 3, pp. 501–521.

Dümenil, L. and Todini, E. (1992): A rainfall–runoff scheme for use in the Hamburg climate model. In: J. P. O'kane (ed.), *Advances in Theoretical Hydrology: A Tribute to James Dooge*, European Geophysical Society Services Hydrological Science, Elsevier, Amsterdam, vol. 1, pp. 129–157.

Eagleson, P. S. (1970): *Dynamic Hydrology*, McGraw Hill, New York.

Franchini, M. and Pacciani, M. (1991): Comparative analysis of several conceptual rainfall-runoff models, *Journal of Hydrology*, vol. 122, pp. 161–219.

Henderson, F. M. and Wooding, R. A. (1964): Overland flow and groundwater flow from a steady rainfall of finite duration, *Journal of Geophysical Research*, vol. 69, no. 8, pp. 1531–1540.

Jayawardena, A. W. (1988): Stream flow simulation using Tank Model: Application to Hong Kong catchments, Hong Kong Engineer, *Journal of the Hong Kong Institution of Engineers*, vol. 16, no. 7, pp. 33–36.

Jayawardena, A. W. (2001): Coupling of land surface and river runoff models: Application to Mekong River Basin, *Proceedings of the International Symposium on Achievements of IHP-V in Hydrological Research*, held during November 19–22, 2001, Ha Noi, Vietnam, IHP-V Technical Document in Hydrology No. 8, UNESCO Jakarta Office, 2001 pp. 23–31.

Jayawardena, A. W. (2006): Calibration of VIC model for daily discharge prediction of Mekong River using MOSCEM algorithm, *Proceedings of the 3rd APHW Conference*, held in Bangkok, Thailand, October 16–18, 2006 (Abstract in CD ROM, p. 256).

Jayawardena, A. W. (2014): *Environmental and Hydrological Systems Modelling*, CRC Press, Taylor & Francis Group, Boca Baton, FL, 33487, USA, 516 p.

Jayawardena, A. W. and White, J. K. (1977): A finite element distributed catchment model, I: Analytical basis, *Journal of Hydrology*, vol. 34, no. 3–4, pp. 269–286.

Jayawardena, A. W. and White, J. K. (1979): A finite element distributed catchment model, II: Application to real catchments, *Journal of Hydrology*, vol. 42, no. 3–4, pp. 231–249.

Jayawardena, A. W. and Zhou, M. C. (2000): A modified spatial soil moisture storage capacity distribution curve for the Xinanjiang model, *Journal of Hydrology*, vol. 227, no. 1–4, pp. 93–113.

Jayawardena, A. W. and Mahanama, S. P. P. (2002): Meso-scale hydrological modelling: Application to Mekong and Chao Phraya Basins, *Journal of Hydrologic Engineering, ASCE*, vol. 7, no. 1, pp. 12–26.

Jayawardena, A. W. and Tian, Y. (2005): Flow modelling of Mekong River with variance in spatial scale, *Proceedings of the International Symposium on "Role of Water Sciences in Transboundary River Basin Management"*, held in Ubon Ratchathani, Thailand during March 10–12, 2005, S. Herath et al. (Ed.), pp. 147–154.

Lattermann, A. (1991): *System-Theoretical Modelling in Surface Water Hydrology*, Springer-Verlag, Berlin, Heidelberg, p. 200.

Liang, X. and Xie, Z. (2001): A new surface runoff parameterization with subgrid-scale soil heterogeneity for land surface models, *Advances in Water Resources*, vol. 24, no. 9–10, pp. 1173–1193.

Liang, X., Lettenmaier, D. P., Wood, E. F., and Burges, S. J. (1994): A simple hydrologically based model of land surface water and energy fluxes for general circulation models, *Journal of Geophysical Research*, vol. 99, no. D7, 14415–14428.

Liang, X., Lettenmaier, D. P. and Wood, E. F. (1996a): One-dimensional statistical dynamic representation of subgrid spatial variability of precipitation in the two-layer variable infiltration capacity model, *Journal of Geophysical Research*, vol. 101, no. D16, pp. 21403–21422.

Liang, X., Wood, E. F. and Lettenmaier, D. P. (1996b): Surface soil moisture parameterization of the VIC-2L model: Evaluation and modification, *Global and Planetary Change*, vol. 13, no. 1–4, pp. 195–206.

Liang, X., Wood, E. F. and Lettenmaier, D. P. (1999): Modeling ground heat flux in land surface parameterization schemes, *Journal of Geophysical Research*, vol. 104, no. D8, pp. 9581–9600.

Lighthill, M. J. and Whitham, G. B. (1955): On kinematic floods: Flood movement in long rivers, *Proceedings of Royal Society, London A*, vol. 220, pp. 281–316.

Matheussen, B., Kirschbaum, R. L., Goodman, I. A., O'Donnell, G. M., and Lettenmaier, D. P. (2000): Effects of land cover change on streamflow in the interior Columbia River Basin (USA and Canada), *Hydrological Processes*, vol. 14, no. 5, pp. 867–885.

Mulvany, W. T. (1851): On the use of self-registering rain and flood gauges in making observations of the relation of rainfall and flood discharges in given catchment, *Transactions of the Institution of Civil Engineers, Ireland*, vol. 4, pp. 18–33.

POYB (Planning Office of Yangtze Basin) (1979): *Methods of Hydrological Forecasting*, Water Resource and Electric Press, Beijing, pp. 89–112, in Chinese.

Refsgaard, J. C. (1997): Parameterisation, calibration and validation of distributed hydrological models, *Journal of Hydrology*, vol. 198, pp. 69–97.

Refsgaard, J. C. and Knudsen, J. (1996): Operational validation and intercomparison of different types of hydrological models, *Water Resources Research*, vol. 32, no. 7, pp. 2189–2202.

Sahoo, G. B., Ray, C. and De Carlo, E. H. (2006): Calibration and validation of a physically distributed hydrological model, MIKE SHE, to predict streamflow at high frequency in a flashy mountainous Hawaii stream, *Journal of Hydrology*, vol. 327, pp. 94–109.

Sherman, L. K. (1932): Streamflow from rainfall by the unit hydrograph method, *Engineering News Record*, vol. 108, pp. 501–505.

Sugawara, M., Watanabe, I., Ozaki, E. and Katsuyama, Y. (1984): Tank model with snow component, Research Notes No. 65, National Research Center for Disaster Prevention, Japan.

Storck, P. and Lettenmaier, D. P. (1999): Predicting the effect of a forest canopy on ground snow accumulation and ablation in maritime climates. In: C. Troendle (ed.), *Proceedings of the 67th Western Snow Conference*, April 19–22, Lake Tahoe, CA, pp. 1–12.

Taylor, C., Al-Mashidani, G. and Davis, J. M. (1974): A finite element approach to watershed runoff, *Journal of Hydrology*, vol. 21, no. 3, pp. 231–246.

Todini, E. (1996): The ARNO rainfall-runoff model, *Journal of Hydrology*, vol. 175, pp. 339–382.

USACE (1998): *HEC-1 Flood Hydrograph Package User's Manual*, Hydrologic Engineering Center, US Army Corps of Engineers, Davis, CA.

USACE (2000): *HEC-HMS Hydrologic Modeling System User's Manual*, Hydrologic Engineering Center, US Army Corps of Engineers, Davis, CA.

von Bertalanffy, L. (1968): *General Systems Theory, Foundations, Development, Applications*, George Braziller, New York.

Vrugt, J. A., Gupta, H. V., Bastidas, L. A., Bouten, W. and Sorooshian, S. (2003a): Effective and efficient algorithm for multiobjective optimization of hydrologic models, *Water Resources Research*, vol. 39, no. 8, pp. SWC51–SWC519.

Vrugt, J. A., Gupta, H. V., Bouten, W. and Sorooshian, S. (2003b): A Shuffled Complex Evolution Metropolis algorithm for optimization and uncertainty assessment of hydrologic model parameters, *Water Resources Research*, vol. 39, no. 8, pp. SWC11–SWC116.

Wang, Q. J. (1991): The genetic algorithm and its application to calibrating conceptual rainfall-runoff models, *Water Resources Research*, vol. 27, no. 9, pp. 2467–2471.

White, J. K. and Jayawardena, A. W. (1975): Discussion of 'a finite element approach to watershed hydrology' by C. Taylor et al., *Journal of Hydrology*, vol. 27, pp. 357–358.

Wood, E. F., Lettenmaier, D. P., and Zartarian, V. G. (1992): A land-surface hydrology parameterization with subgrid variability for General Circulation Models, *Journal of Geophysical Research*, vol. 97, no. D3, pp. 2717–2728.

Woolhïser, D. A and Liggett, J. A. (1967): Unsteady, one-dimensional flow over a plane - the rising hydrograph, *Water Resources Research*, vol. 3, no. 3, pp. 753–771.

Zhang, W. H. (1990): *Theory and Practices of Flood Forecasting of Storms*, Water Resource and Electric Press, Beijing, pp. 161–176, (in Chinese).

Zhao, R. J. (1984): *Hydrological Simulation of Watersheds*, Water Resource and Electric Press, Beijing, pp. 32–47, in Chinese; see also pp. 71–82 and 106–130.

Zhao, R. J. (1977): Flood Forecasting Method for Humid Regions of China, East China College of Hydraulic Engineering, Nanjing, pp. 19–51, in Chinese; see also pp. 135–170 and 206–224.

Zhao, R. J. and Liu, X. R. (1995): The Xinanjiang model. In: V.P. Singh (ed.), *Computer Models of Watershed Hydrology*, Water Resources Publication, Reprinted in 2012, pp. 215–212.

Zhao, R. J., Zhuang, Y. L., Fang, L. R., Liu, X. R. and Zhang, Q. S. (1980): The Xinanjiang model. *Hydrological Forecasting Proceedings Oxford Symposium*, vol. 129, IAHS Publication, pp. 351–356.

Chapter 34

Flood routing

34.1 INTRODUCTION

Channel storage can be significant in large rivers. When a flood wave propagates downstream, the effect of channel storage should be taken into consideration in the computation of the downstream hydrograph from an upstream hydrograph. This can be done using a hydrologic approach or a hydraulic approach. The hydrologic approach is a kind of system theory approach in which the input is the inflow (upstream) hydrograph $I(t)$, and the output is the outflow (downstream) hydrograph $O(t)$, which is assumed to be related to the inflow hydrograph in a form

$$O(t) = f[c, I(t)] \tag{34.1}$$

where c is a channel characteristic and f is a transfer function.

Flood routing is a hydrological modelling approach in which the continuity equation is used with an outflow–storage relationship. The commonly used methods are the inventory method, the Muskingum method, the modified Puls method and the Muskingum–Cunge method.

34.2 INVENTORY METHOD

The continuity equation is

$$I - O = \frac{dS}{dt} \tag{34.2}$$

where I is the inflow, O is the outflow and S is the storage. For a time interval Δt, it is possible to write

$$S_2 - S_1 = \bar{I}\Delta t - \bar{O}\Delta t \tag{34.3a}$$

or

$$\Delta S = \frac{(I_1 + I_2)}{2}\Delta t - \frac{(O_1 + O_2)}{2}\Delta t \tag{34.3b}$$

where the over-bar indicates the mean value over the time interval. In Eq. 34.3, I_1, I_2, O_1 and S_1 are known; only S_2 and O_2 are unknowns. Therefore, another relationship is needed to solve for these two unknowns. If the outflow–storage relationship is known, Eq. 34.3 can be solved sequentially.

34.3 MUSKINGUM METHOD

The Muskingum method (McCarthy, 1938) considers a prism storage (KO) and a wedge storage $Kc(I - O)$ when added together gives the total storage as $S = KO + Kc(I - O)$. Prism storage refers to the volume between the stream bed and a line parallel to the bed and intersecting the water surface at the end of the reach considered whereas the wedge storage refers to the volume contained between this line and the water surface (Figure 34.1).

It assumes a single stage–discharge relationship which may not always be valid. Under certain flow conditions, the rising part and the recession part of the hydrograph may have different friction slopes, thereby producing a loop (hysteresis) rather than a single relationship. Manning's equation is sometimes used as the stage–discharge relationship. The combined storage can also be obtained as follows:

For an upstream reach, it is assumed that

$$I = a y_u^n \tag{34.4}$$

$$S_I = b y_u^m \tag{34.5}$$

Similarly, for a downstream reach,

$$O = a y_d^n \tag{34.6}$$

$$S_o = b y_d^m \tag{34.7}$$

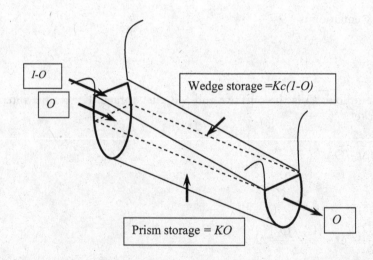

Figure 34.1 Prism and wedge storages.

where

S_I – inflow storage
S_o – outflow storage
y_u – upstream depth of flow
y_d – downstream depth of flow
a, b, n, m – constants for the reach.

Equating upstream and downstream depths, it can be shown that

$$S_I = \frac{b}{a^{m/n}} I^{m/n} = K I^{m/n} \tag{34.8}$$

$$S_o = \frac{b}{a^{m/n}} O^{m/n} = K O^{m/n} \tag{34.9}$$

Assuming that the storage within the reach is a weighted sum of the input and output storages, it is possible to write

$$S = c S_I + (1 - c) S_o \tag{34.10}$$

where c is a weighting factor between 0 and 0.5 (usually about 0.2). Δt is chosen such that it is <20% of the time to peak.

When $c = 0$, $S = S_o$ (maximum attenuation). This condition leads to the **Convex Routing Method,** and is useful for reservoir routing.

When $c = 0.5$, $S = 0.5(S_I + S_o)$ (no attenuation, translation only)

$$\text{If } K = \frac{b}{a^{m/n}} = \frac{b}{a^x}$$

where $x = \dfrac{m}{n}$, then Eq. 34.10 can be written as

$$S = K\left[c I^x + (1 - c) O^x \right] \tag{34.11}$$

where K is a storage constant for the reach which has the unit of time.

For rectangular channels, $n = 5/3$, $m = 1$ and therefore $x = 0.6$. (McCuen, 2004, p. 606).

For natural channels, x is larger, and **Muskingum** method assumes that $x = 1$. Then, Eq. 34.11 becomes

$$S = K\left[c I + (1 - c) O \right] \tag{34.12}$$

Streamflow data generally show that K is approximately equal to the travel time of the reach. Equation 34.12 can be written as

$$S_2 = K\left[c I_2 + (1 - c) O_2 \right] \tag{34.13a}$$

and

$$S_1 = K\left[c I_1 + (1 - c) O_1 \right] \tag{34.13b}$$

or

$$S_2 - S_1 = K\left[c(I_2 - I_1) + (1-c)(O_2 - O_1)\right] \tag{34.13c}$$

Combining with the inventory equation, Eq. 34.3, $S_2 - S_1 = \dfrac{(I_1 + I_2)}{2}\Delta t - \dfrac{(O_1 + O_2)}{2}\Delta t$

$$0.5(I_1 + I_2)\Delta t - 0.5(O_1 + O_2)\Delta t = K\left[c(I_2 - I_1) + (1-c)(O_2 - O_1)\right] \tag{34.14a}$$

or

$$O_2\left[K(1-c) + 0.5\Delta t\right] = I_2(-Kc + 0.5\Delta t) + I_1(Kc + 0.5\Delta t) + O_1\left[K(1-c) - 0.5\Delta t\right] \tag{34.14b}$$

or

$$K = \frac{0.5\Delta t\left[(I_{t+1} + I_t) - (O_{t+1} + O_t)\right]}{\left[c(I_{t+1} - I_t) + (1-c)(O_{t+1} - O_t)\right]} = \frac{Y}{X} \tag{34.14c}$$

which can be written as

$$O_2 = C_0 I_2 + C_1 I_1 + C_2 O_1 \tag{34.15}$$

where

$$C_0 = \frac{-Kc + 0.5\Delta t}{K(1-c) + 0.5\Delta t} \tag{34.16a}$$

$$C_1 = \frac{Kc + 0.5\Delta t}{K(1-c) + 0.5\Delta t} \tag{34.16b}$$

$$C_2 = \frac{K(1-c) - 0.5\Delta t}{K(1-c) + 0.5\Delta t} \tag{34.16c}$$

It can be seen that $C_0 + C_1 + C_2 = 1$.

Therefore, if the input hydrograph is known, the output hydrograph can be determined for the reach. The initial value of the outflow hydrograph, the routing time step Δt and routing parameters K and c must be specified. Implied in this method are the assumptions that the water surface profile within the reach is uniform and continuous and that K and c are constants for the reach.

34.3.1   Estimation of the routing parameters K and c

The parameters K and c can be estimated if the inflow and outflow hydrographs are known for the reach. It can be done graphically or statistically. The statistical method is more precise. They can also be related to the channel characteristics such as the slope, roughness as measured by Manning's 'n', hydraulic radius or other equivalent, etc. This approach requires extensive field studies to establish a relationship between site characteristics and the routing parameters.

The value of K is estimated to be the travel time through the reach. An ambiguity occurs as to whether the travel time should be estimated using the average flow, peak flow or some other flow. It may be estimated as the kinematic travel time or a travel time based on Manning's equation. The second routing parameter c must lie between 0.0 and 0.5. It is a weighting factor for inflow and outflow. A low value of c implies that the inflow has little or no effect on storage. A reservoir, for example, has a c value of zero as the dominant process there is attenuation. Typical values used in natural streams range from about 0.2 to 0.3, but 0.4 to 0.5 may be used for streams with no storage effects. A value of $c = 0.5$ would represent equal weighting between inflow and outflow and would produce translation with little or no attenuation.

34.3.1.1 Graphical method

By assuming different values for K and c, the numerator and denominator of Eq. 34.14c are calculated for different time steps of the inflow and outflow hydrographs and they are summed and plotted against each other (denominator X on the x-axis, and numerator Y on the y-axis). Note that the cumulative values of X and Y will increase first and then decrease to form a loop. The plot which generally has a loop and the one which is closest to a straight line would be the best choice for K and c. The slope of the line gives the value of K. The value of c that produced the closest deviation from a straight line is taken as its representative value.

34.3.1.2 Statistical method

In this approach, K and c are obtained by the least squares method. The error to be minimized is the difference between the actual downstream hydrograph and the computed hydrograph defined in Eq. 34.15:

$$\text{Error, } E = \sum \left(O_2^{\text{computed}} - O_2^{\text{observed}} \right)^2 \tag{34.17}$$

Differentiating E with respect to the three parameters C_0, C_1 and C_2, the normal equations can be obtained. They are

$$C_0 \sum I_{t+1}^2 + C_1 \sum I_{t+1} I_t + C_2 \sum O_t I_{t+1} = \sum I_{t+1} O_{t+1} \tag{34.18a}$$

$$C_0 \sum I_{t+1} I_t + C_1 \sum I_t^2 + C_2 \sum O_t I_t = \sum I_t O_{t+1} \tag{34.18b}$$

$$C_0 \sum O_t I_{t+1} + C_1 \sum O_t I_t + C_2 \sum O_t^2 = \sum O_t O_{t+1} \tag{34.18c}$$

Once C_0, C_1, C_2 have been determined, the Muskingum parameters K and c can be obtained from the relationships given in Eq. 34.16. The additional condition $C_0 + C_1 + C_2 = 1$ may be used as a check. The estimated parameters can also be used to determine the downstream hydrograph which can be compared with the known hydrograph for verification.

To avoid numerical instability (such as negative values for discharges), the time step should be chosen to satisfy the condition

i. $2Kc < \Delta t < K$.
  Other criteria used include

ii. $2Kc < \Delta t < 2K(1-c)$ (Hjelmfelt, 1985)

iii. $\dfrac{K}{3} < \Delta t < K$ (Maidment, 1993)

iv. $\dfrac{1}{2(1-c)} \le \dfrac{K}{N\Delta t} \le \dfrac{1}{2c}$ (USACE, 1990);

N is the number of sub-reaches so that the total travel time through the reach is K. When $c = 0.5$, the conditions (i) and (ii) are equivalent.

34.3.2 Limitations of the Muskingum method

Muskingum method is limited to slow rising hydrographs when routed along mild sloping channels. The method ignores backwater effects such as when there is a dam downstream, constrictions, bridges and tidal influences. It may also give negative flows in the initial portion of the hydrograph.

34.4 MODIFIED PULS METHOD

The modified Puls method of routing is often applied to reservoir routing, but may also be applied to river routing under certain channel situations. It is based on the finite difference form of the continuity equation. The method requires a stage-discharge-storage relationship to be known, assumed or derived.

34.5 MUSKINGUM–CUNGE METHOD

Cunge (1969) developed a routing method that is equivalent to the Muskingum method under certain conditions. It is therefore referred to as the Muskingum–Cunge method. It is based on combining the continuity equation and the diffusion wave approximation form of the momentum equation followed by a linearization, and the formulation is considered an approximate solution of the convective diffusion equation. It uses physical characteristics to estimate the routing parameters. The physical characteristics consist of the reach length L, the slope S_0, the kinematic wave celerity v and a characteristic unit discharge q_0. The wave celerity is defined as the speed at which a flood wave travels downstream and is equal to the slope of the discharge – area curve at a given discharge. In equation form,

$$v = \frac{dQ}{dA} = \frac{1}{B}\frac{dQ}{dh}, \quad \text{which is also equal to } \frac{dx}{dt} \qquad (34.19)$$

where Q is the discharge, h is the depth of flow, A is the cross-sectional area of flow and B is the width of the water surface. The term $\dfrac{dQ}{dh}$ represents the slope of the discharge–stage relationship (rating curve) at the water surface elevation. The actual discharge q (discharge per unit width) or the mean or peak discharge may be taken as the characteristic discharge q_0.

Cunge gave the following expression for the weighting parameter c:

$$c = \frac{1}{2}\left(1 - \frac{Q_0}{BvS_0L}\right) \tag{34.20a}$$

which may also be written as

$$c = 0.5\left(1 - \frac{q_0}{vS_0L}\right) \tag{34.20b}$$

Muskingum–Cunge method uses the expression

$$K = \frac{L}{v} \tag{34.21}$$

for K which uses Eq. 34.19 for v.

Combining the continuity equation

$$\bar{I} - \bar{O} = \frac{dS}{dt} \tag{34.22}$$

and the **Muskingum** equation Eq. 34.12,[1] it is possible to write

$$K\frac{d}{dt}\left[cI + (1-c)O\right] = \bar{I} - \bar{O} \tag{34.23}$$

In finite difference form ($I = O_i$; $O = O_{i+1}$)

$$\frac{K}{\Delta t}\left\{cO_i^{t+1} + (1-c)O_{i+1}^{t+1} - cO_i^t - (1-c)O_{i+1}^t\right\} = 0.5\left\{O_i^{t+1} + O_i^t - O_{i+1}^{t+1} - O_{i+1}^t\right\} \tag{34.24}$$

If $K = \frac{\Delta x}{v}$, where v is the wave celerity, Eq. 34.24 is a finite difference form of the continuity equation with no lateral inflow:

$$\frac{\partial O}{\partial t} + v\frac{\partial O}{\partial x} = 0 \tag{34.25}$$

From Eq. 34.24,

$$O_{i+1}^{t+1} = C_0 O_i^{t+1} + C_1 O_i^t + C_2 O_{i+1}^t \tag{34.26}$$

where the coefficients C_0, C_1 and C_2 have the same definitions as in the Muskingum method (i.e., Eq. 34.16).

$$C_0 = \frac{(\Delta t/K) - 2c}{2(1-c) + (\Delta t/K)}$$

$$C_1 = \frac{(\Delta t/K) + 2c}{2(1-c) + (\Delta t/K)}$$

---

[1] $S = K\left[cI + (1-c)O\right]$.

$$C_2 = \frac{2(1-c)-(\Delta t / K)}{2(1-c)+(\Delta t / K)}.$$

Since $K\left(= \dfrac{\Delta x}{v}\right)$, is the time of travel for each reach of length Δx and velocity v, **Cunge** showed that v is the celerity of the kinematic wave. When K and Δt are constants, the solution given by Eq. 34.26 is an approximation of the solution by the kinematic wave method. It has also been shown that when the wave celerity is defined by Eq. 34.19 (and therefore $K = \dfrac{\Delta x}{v}$) and the weighting factor c is defined by Eq. 34.20a, the solution given by Eq. 34.26 is an approximation to a modified form of diffusion equation.

When $c = 0.5$, $v \dfrac{\Delta t}{\Delta x} = 1.0$ implying only translation with no attenuation.

In the kinematic wave formulation, $v = mV$; $V = \dfrac{O}{A}$; $m = 5/3$ for wide natural channels, and m is given by

$$O = \alpha A^m \ \left(\text{compare with} \ q = \alpha h^m\right)$$

Negative values of C_0 could be avoided if $\dfrac{\Delta t}{K} > 2c$

34.6 HYDRAULIC APPROACH

Hydraulic routing, which is also known as dynamic routing, is carried out by solving the St. Venant's equations. It is a reference routing method for comparison with other routing methods. The solution of the St. Venant's equations is generally accomplished by the method of characteristics or by direct methods (explicit and implicit). The solutions, either way, are resource intensive. The most popular implicit method is to use a four point weighted finite difference scheme. It is also possible to solve the problem using finite element methods, but the formulation is more complex. One-dimensional problems are usually solved by the finite different methods whereas two- and three-dimensional problems often use the finite element method.

Dynamic routing heavily relies on physically measured parameters rather than using calibration techniques that require past flood data. Added advantages include solutions to sediment transport and pollutant transport problems which can be coupled to the water transport problem. On the negative side, computational instability may occur in situations where the flood wave travels through the sub-reach considered in a time less than the time interval used for computations. To get over this possibility, the Courant stability criterion must be satisfied. i.e.,

$$\Delta t \le \frac{\Delta x}{v}$$

34.6.1 Solution of the St. Venant's equation

$$\frac{\partial q}{\partial x} + \frac{\partial h}{\partial t} = (i-f) = Q \quad \text{Continuity Equation} \tag{34.27}$$

$$\frac{\partial v}{\partial t} + v\frac{\partial v}{\partial x} + g\frac{\partial h}{\partial x} + \frac{Qv}{h} = g\left(S_0 - S_f\right) \quad \textbf{Momentum Equation} \tag{34.28}$$

where
q – discharge per unit width
h – depth of flow
Q – lateral inflow per unit length per unit width
v – velocity of flow
S_0 – bed slope of the flow plane
S_f – friction slope of the flow plane
i – rainfall rate
f – infiltration rate
x, t – distance along the flow plane and time.

34.6.1.1 Kinematic wave method

Replacing q ($L^3T^{-1}L^{-1}$) by $Q(L^3T^{-1})$, h (L) by $A(L^2)$ and $Q'(L^3T^{-1}L^{-2})$ by $Q'(L^3T^{-1}L^{-1})$, the governing equations can be written as

$$\frac{\partial Q}{\partial x} + \frac{\partial A}{\partial t} = Q'(x,t) \tag{34.29}$$

$$S_0 = S_f$$

where Q' is the lateral inflow rate and S_f is the frictional slope.
   The momentum equation may also be written as

$$A = \alpha Q^\beta, \tag{34.30}$$

which upon differentiating gives

$$\frac{\partial A}{\partial t} = \alpha\beta Q^{\beta-1}\frac{\partial Q}{\partial t} \tag{34.31}$$

Eliminating A, in the governing equation gives

$$\frac{\partial Q}{\partial x} + \alpha\beta Q^{\beta-1}\frac{\partial Q}{\partial t} = Q' \tag{34.32}$$

Alternatively, by writing the momentum equation as

$$Q = \alpha A^\gamma, \tag{34.33}$$

which when differentiated gives

$$\frac{\partial Q}{\partial A} = \alpha\gamma A^{\gamma-1} \tag{34.34}$$

Using the relationship,

$$\frac{\partial Q}{\partial x} = \frac{\partial Q}{\partial A}\frac{\partial A}{\partial x} \tag{34.35}$$

and substitution in Eq. 34.29, gives

$$\frac{\partial A}{\partial t} + \frac{\partial Q}{\partial A}\frac{\partial A}{\partial x} = Q' \tag{34.36}$$

which simplifies to

$$\frac{\partial A}{\partial t} + \alpha\gamma A^{\gamma-1}\frac{\partial A}{\partial x} = Q' \tag{34.37}$$

Equations 34.32 and 34.37 are the two forms of the governing equation; in Eq. 34.32, Q is the dependent variable, and in Eq. 34.37, A is the dependent variable. Taking logarithms of the kinematic approximation for A (Eq. 34.30),

$$\ln A = \ln\alpha + \beta\ln Q \tag{34.38}$$

which when differentiated gives

$$\frac{dQ}{Q} = \frac{1}{\beta}\frac{dA}{A} \tag{34.39}$$

Similarly, taking logarithms of the kinematic approximation for Q (Eq. 34.33),

$$\ln Q = \ln\alpha + B\ln A \tag{34.40}$$

which when differentiated gives

$$\frac{dA}{A} = \frac{1}{B}\frac{dQ}{Q}, \quad \text{or} \quad \frac{dQ}{Q} = B\frac{dA}{A} \tag{34.41}$$

Equations 34.39 and 34.41 give relative errors of Q and A.

34.7 RESERVOIR ROUTING

The continuity equation can be written as

$$I - O = \frac{dS}{dt} \tag{same as Eq. 34.2}$$

where I represents the inflow, O, the outflow and S, the storage. Over a finite interval of time between t and $t + \Delta t$,

$$\left(\frac{I_1 + I_2}{2}\right) - \left(\frac{O_1 + O_2}{2}\right) = \frac{S_2 - S_1}{\Delta t} \tag{same as Eq. 34.3b}$$

which can be written as

$$(I_1 + I_2) + \left(\frac{2S_1}{\Delta t} - O_1\right) = \left(\frac{2S_2}{\Delta t} + O_2\right) \tag{34.42}$$

In this equation the left-hand side is known, and the right-hand side contains two unknowns, S_2 and O_2. A second equation (usually the stage-storage characteristic) is needed to obtain a solution.

Example 34.1

A stormwater detention basin is estimated to have the following storage characteristics:

Stage (m)	5.0	5.5	6.0	6.5	7.0	7.5	8.0
Storage (m³)	0	694	1,525	2,507	3,652	4,973	6,484

The discharge weir from the detention basin has a crest elevation of 5.5 m, and the weir discharge, Q, is given by

$$Q = 1.83 H^{(3/2)}$$

where H is the height of the water surface above the crest of the weir. The inflow hydrograph is given by

Time (min)	0	30	60	90	120	150	180	210	240	270	300	330	360	390
Runoff (m³/s)	0	2.4	5.6	3.4	2.8	2.4	2.2	1.8	1.5	1.2	1.0	0.56	0.34	0

If the pre-storm stage in the detention basin is 5.0 m, estimate the outflow hydrograph from the detention basin.

This requires runoff routing through a reservoir. Using a time step of 30 min ($\Delta t = 30 \times 60 = 1,800$ s), the calculations can be done in the following tabular form.

In this table, columns 1 and 2 are given. Column 3 is obtained from the weir formula. Column 4 is obtained using columns 2 and 3 (Stage $\rightarrow$ storage $S \rightarrow$ discharge $O \rightarrow 2S/\Delta t + O$). Columns 3 and 4 then gives a relationship between discharge O and $2S/\Delta t + O$, which will be used in subsequent calculations.

At the beginning, $I_1 = 0$, $I_2 = 2.4$; therefore $I_1 + I_2 = 2.4$; $O_1 = 0$; $S_1 = 0$ (because the weir operates only when the head exceeds 5.5 m); therefore $2S_1/\Delta t - O_1 = 0$. For the next time step (30 min), from continuity equation,

$$(I_1 + I_2) + \left(\frac{2S_1}{\Delta t} - O_1\right) = \left(\frac{2S_2}{\Delta t} + O_2\right) \Rightarrow \frac{2S_2}{\Delta t} + O_2 = (I_1 + I_2) + \left(\frac{2S_1}{\Delta t} - O_1\right) = 0 + 2.4 + 0 = 2.4$$

Next, O_2 corresponding to $\frac{2S_2}{\Delta t} + O_2 = 2.4$ is obtained by interpolation using the relationship between O and $\frac{2S}{\Delta t} + O$ given in columns 3 and 4 of Table 34.1 above. This works out to 0.68. Then $\frac{2S_1}{\Delta t} + O_1$ is obtained from the already calculated values of $\frac{2S_2}{\Delta t} + O_2$ and O_2 as

$$\left(\frac{2S_1}{\Delta t} - O_1\right) = \left(\frac{2S_2}{\Delta t} + O_2\right) - 2O_2 = 2.4 - 2 \times 0.68 = 1.04$$

Table 34.1 Stage-storage-discharge characteristics

Stage (m)	S (m³)	O (m³/s)	2S/Δt + O (m³/s)
5.0	0	0	0
5.5	694	0	0.771
6.0	1,525	0.647	2.34
6.5	2,507	1.83	4.62
7.0	3,652	3.36	7.42
7.5	4,973	5.18	10.7
8.0	6,484	7.23	14.4

Table 34.2 Routing through the reservoirs

Time (min)	I_1 (m³/s)	$I_1 + I_2$ (m³/s)	2S/Δt − O (m³/s)	2S/Δt + O (m³/s)	O (m³/s)
0	0	2.4	0	0	0
30	2.4	8.0	1.04	2.4	0.68
60	5.6	9.0	0.52	9.04	4.26
90	3.4	6.2	0.469	9.52	4.52
120	2.8	5.2	0.769	6.669	2.95
...	2.4	...	...	...	...
...	...	...	...	...	...

The rest of the calculations are carried out recursively and given in Table 34.2. The computational sequence may be summarized as follows:

- Stage → Storage S → Discharge $O \to \dfrac{2S}{\Delta t} + O$. Initially S, O, and therefore $\dfrac{2S}{\Delta t} + O$

  are all zero. In the next time step, I_1 and I_2 are known and therefore $\left(\dfrac{2S_2}{\Delta t} + O_2\right)$

  can be calculated.

- Find O corresponding to the calculated value of $\dfrac{2S}{\Delta t} + O$ by interpolation in the
  stage-storage-discharge characteristic.

- Find $\dfrac{2S}{\Delta t} - O$ as $\dfrac{2S}{\Delta t} + O - 2O$, and repeat the procedure.

Example 34.2 (using Muskingum–Cunge method)

The hydrograph at the upstream end of a river is given in the following table. The reach of interest is 18 km long. Using a sub-reach length Δx of 6 km, determine the hydrograph at the end of the reach using the Muskingum-Cunge method. Assume $\nu = 2\,\text{m/s}$, $B = 25.3\,\text{m}$, $S_0 = 0.001$, $Q_0 = 150\,\text{m}^3/\text{s}$ and no lateral flow.

Time (h)	0 km	6 km	12 km	18 km
0	10	10	10	10
2	18	13.89	11.89	10.92
4	50	34.51	24.38	18.19
6	107	81.32	59.63	42.96
8	147	132.44	111.23	88.60
10	146	149.91	145.88	133.35
12	105	125.16	138.82	145.37
14	59	77.93	99.01	117.94
16	33	41.94	55.52	73.45
18	17	23.14	29.63	38.75
20	10	12.17	16.29	21.02
22	10	9.49	9.91	12.09
24	10	10.12	9.70	9.30
26	10	9.97	10.15	10.01
28	10	10.01	9.95	10.08

$$K = \frac{\Delta x}{v} = \frac{6 \times 1,000}{2} = 3,000 \ s; \quad c = \frac{1}{2}\left(1 - \frac{Q_0}{B v S_0 L}\right) = 0.5\left(1 - \frac{150}{25.3 \times 2 \times 0.001 \times 6,000}\right) = 0.253$$

Using $\Delta t = 2\,\text{h}$,

$$C_0 = \frac{(\Delta t/K) - 2c}{2(1-c) + (\Delta t/K)} = \frac{(7,200/3,000) - 2 \times 0.253}{2(1-0.253) + (7,200/3,000)} = \frac{1.894}{3.89} = 0.4869$$

$$C_1 = \frac{(\Delta t/K) + 2c}{2(1-c) + (\Delta t/K)} = \frac{(7,200/3,000) + 2 \times 0.253}{2(1-0.253) + (7,200/3,000)} = \frac{2.906}{3.89} = 0.7466$$

$$C_2 = \frac{2(1-c) - (\Delta t/K)}{2(1-c) + (\Delta t/K)} = \frac{2 \times 0.747 - (7,200/3,000)}{2(1-0.253) + (7,200/3,000)} = -\frac{0.91}{3.89} = -0.2339$$

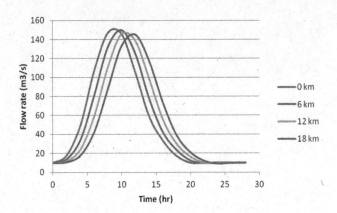

Hence, from Eq. 34.26, all flows can be computed.

$$O_{i+1}^{t+1} = C_0 O_i^{t+1} + C_1 O_i^t + C_2 O_{i+1}^t$$

e.g. $0.4869 \times 18 + 0.747 \times 10 - 0.2339 \times 10 = 13.89$

$0.4869 \times 50 + 0.747 \times 18 - 02339 \times 13.89 = 34.54$

$0.4869 \times 107 + 0.747 \times 50 - 0.2339 \times 34.54 = 81.32$

. . . .

. . . .

REFERENCES

Cunge, J. A. (1969): On the subject of a flood propagation computation method (Muskingum method), *Journal of Hydraulic Research*, vol. 7, no. 2, pp. 205–230.

Hjelmfelt, A. T., Jr. (1985): Negative outflows from Muskingum flood routing, *Journal of Hydraulic Engineering, ASCE*, vol. 111, no. 6, pp. 1010–1014.

Maidment, D. R. (1993): *Handbook of Hydrology*, McGraw Hill Inc., New York.

McCarthy, G. T. (1938): The unit hydrograph and flood routing, *Conference North Atlantic Division*, US Corps of Engineers, New London, CT.

McCuen, R. M.: (2004): *Hydrologic Analysis and Design*, Prentice Hall, Upper Saddle River, NJ, p. 859.

USACE (1990): HEC-1 Flood hydrograph package user's manual, Hydrologic Engineering Centre, US Army Corps of Engineers, Davis, CA.

Chapter 35

Flow through saturated porous media

35.1 INTRODUCTION

Groundwater is stored in aquifers which are groundwater reservoirs in water bearing formations. They may be confined or unconfined. A confined aquifer holds water under pressure between two impermeable zones. An unconfined aquifer holds water at atmospheric pressure with a lower confining impermeable zone and a free surface at the top. The different types of water are

- **Connate water**: water that has not been in contact with the atmosphere for an appreciable part of the geologic time scale
- **Juvenile water**: new water of cosmic or magmatic origin
- **Meteoric water**: water from the atmosphere
- **Metamorphic water**: water associated with rocks.

Groundwater is an important source of water used in municipal and domestic water supply, irrigation and industrial use. On a long-term basis, it is replenished by rainfall and other sources of surface water. The movement of groundwater is slow as it flows through the soil which is a porous medium. The study of flow through saturated porous media is therefore an important area for hydrologists and water resources engineers.

35.2 ZONE OF SATURATION

Zone of saturation refers to the region below the ground surface in which all the voids are filled with water. The upper limit of the zone of saturation is the water table although there may be a shallow zone above the water table where the voids can be saturated as a result of capillary rise.

35.3 SOIL PROPERTIES

35.3.1 Grain size distribution

A typical grain size distribution is shown in Figure 35.1. The ordinate shows the percentage of materials finer than that of a given size on a dry weight basis.

The effective particle size is the 10% finer than value, d_{10}. The distribution of particle sizes is also identified by the uniformity coefficient, u_c, which is defined by Hazen (1904) as

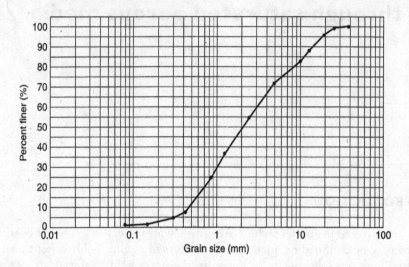

Figure 35.1 A typical grain size distribution curve.

$$u_c = \frac{d_{60}}{d_{10}} \tag{35.1}$$

There are other definitions too. For example, Kramer (1935) defined u_c as

$$u_c = \frac{\sum\limits_{0}^{50} \Delta p_i d_i}{\sum\limits_{50}^{100} \Delta p_i d_i} \tag{35.2}$$

where Δp_i is the fraction of the particle size d_i.

A low uniformity coefficient implies a uniform material whereas a high uniformity coefficient implies a well-graded material.

35.3.2 Porosity, *n*

$$n = \frac{\text{Volume of interstices}}{\text{Total volume}} = \frac{V_i}{V} \tag{35.3}$$

Typical values are 0.35–0.45 for sandy soils and 0.40–0.60 for clays and peats.

35.3.3 Specific retention, S_r

$$S_r = \frac{\text{Volume of water retained in pores against gravity}}{\text{Bulk volume of soil}} = \frac{W_r}{V} \tag{35.4}$$

35.3.4 Specific yield, S_y (or, effective porosity)

$$S_y = \frac{\text{Volume of water that can be drained by gravity}}{\text{Bulk volume of soil}} = \frac{W_y}{V} \qquad (35.5)$$

Specific yield values range from about 0.05 to about 0.40.

It can be seen that

$$W_r + W_y = \alpha = \text{Volume of water in pores}$$

Note: Volume of water in pores is not necessarily equal to the volume of pores (i.e., $\alpha \neq n$).

Table 35.1 shows the variation of porosity, specific yield and permeability for some types of soils.

35.4 AQUIFERS

An aquifer is a water-bearing geological formation. Most aquifers are found in sand and gravel formations, limestone formations and sedimentary rock formations. There are different types of aquifers (Figure 35.2).

35.4.1 Confined aquifers

In a confined aquifer, water is held under pressure between two impermeable zones. The height to which the water level rises in a bore bole through a confined aquifer is called the piezometric head, and if it is above the ground surface, water will flow out of the borehole.

35.4.2 Unconfined aquifers

In an unconfined aquifer, the water is held at hydrostatic pressure between a lower confining impermeable boundary and an upper free surface boundary. The pressure at the free surface is atmospheric.

35.4.3 Perched aquifers

Perched aquifers are unconfined and above the water table. Capacities are relatively small.

35.4.4 Leaky aquifers

When the confining zones are not well defined, confined and unconfined aquifers may exist one above the other. In such situations, transfer of water can take place vertically between the aquifers.

35.4.5 Aquitards

An aquitard is a poorly permeable formation that impedes water movement and, therefore extraction, but allows the transfer of water between aquifers.

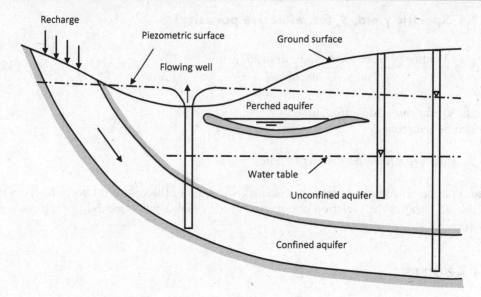

Figure 35.2 Types of aquifers.

Table 35.1 Hydraulic properties of some soils

Type of soil	Porosity	Specific yield	Permeabilty (m/day)
Gravel	0.25–0.35	0.12–0.25	100–1,000
Sand	0.30–0.42	0.10–0.25	1.0–50.0
Silt	0.40–0.45	0.05–0.10	0.0005–0.1
Sand and silts	0.30–0.40	0.02–0.10	0.10–5.0

35.5 IDEALIZATION OF AQUIFERS

Most real aquifers are neither confined nor unconfined in the strict sense. Furthermore, their properties vary from point to point and in all directions. Mathematical treatment of such variations is impractical. Therefore, idealization is done by assuming homogeneity and isotropy. By homogeneity, it is implied that aquifer properties are invariant with position; by isotropy, it is implied that the properties at a point are invariant with direction.

35.6 HYDRAULIC PROPERTIES OF AQUIFERS

35.6.1 Specific storage, S_s

Specific storage is the volume of water that can be released or taken into storage from a unit volume of saturated aquifer by unit reduction in hydraulic head. It has dimension of (vol/vol)/length, i.e. L^{-1}.

35.6.2 Storage coefficient, S

Storage coefficient, which is a measure of the water yielding capacity, is the volume of water that can be released, or taken into storage per unit area per unit change in pressure. It is dimensionless and is given by

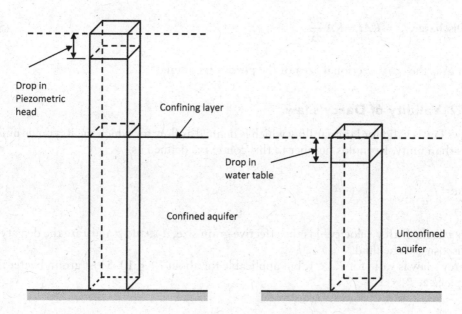

Drop in
Piezometric
head

Confining layer

Drop in
water table

Confined aquifer

Unconfined
aquifer

Figure 35.3 Mechanisms of release of water in confined and unconfined aquifers.

$$S = S_s b \tag{35.6}$$

where b is the thickness of the aquifer.

Typical values of S for confined aquifers vary from about 5×10^{-3} to about 5×10^{-5} whereas for unconfined aquifers values range from about 0.01 to about 0.3. In confined aquifers, the storage coefficient is related to porosity and compressibility factors (reduction in elastic storage). Reduction in pressure causes expansion of the aquifer, thereby releasing water. There is no dewatering taking place. In unconfined aquifers, release of water is by dewatering. Therefore the storage coefficients for unconfined aquifers are very much greater than those for confined aquifers (Figure 35.3).

35.7 GROUND WATER MOVEMENT

35.7.1 Darcy's law

On a macroscopic scale, the velocity of flow is proportional to the hydraulic gradient:

$$v = -Ki = -K\frac{dh}{dx} \tag{35.7}$$

where v is called the Darcy velocity, h is the hydraulic head and K is the hydraulic conductivity.
   Typical values of K are

10^{-3}-10^{-1} m/s for coarse sand and gravel
10^{-9}-10^{-4} m/s for clays and consolidated soils.

$$\text{Average interstitial velocity} = \frac{v}{n} = \frac{\text{Darcy velocity}}{\text{Porosity}} \tag{35.8}$$

Discharge, $Q = KAi = KA\dfrac{dh}{dx}$ $\qquad\qquad\qquad\qquad\qquad\qquad\qquad$ (35.9)

where A is the cross-sectional area of the porous medium.

35.7.2 Validity of Darcy's law

Darcy's law is valid for laminar flow which is defined as flow for which the Reynolds number is less than unity. Reynolds number in this context is defined as

$$Re = \frac{\rho v d}{\mu}$$ $\qquad\qquad\qquad\qquad\qquad\qquad$ (35.10)

where v is the Darcy velocity, d is the effective grain size, d_{10}, and ρ and μ are the density and the viscosity of the fluid.

Darcy's law is valid for $Re < 1$, but applicable for about $Re < 10$. Most groundwater flows take place at $Re < 1$.

35.8 HYDRAULIC PARAMETERS

35.8.1 Hydraulic conductivity

According to Darcy's law, the constant of proportionality K has the dimension of L/T. It therefore represents a velocity and is called the hydraulic conductivity or the coefficient of permeability. It is a function of both the fluid and the porous medium. An equation for K can be obtained by using Buckingham's π theorem in dimensional analysis. It can be expressed as

$$K = c\rho^\alpha g^\beta \mu^\gamma d^\theta$$ $\qquad\qquad\qquad\qquad\qquad$ (35.11)

where the indices can be determined by dimensional analysis giving

$$K = \frac{Cd^2 \rho g}{\mu}$$ $\qquad\qquad\qquad\qquad\qquad\qquad$ (35.12)

Since ρ and μ are properties of the fluid, they together with g may be separated from the rest of the quantities. Then,

$$K = k\frac{\rho g}{\mu}$$ $\qquad\qquad\qquad\qquad\qquad\qquad$ (35.13a)

where

$$k = Cd^2$$ $\qquad\qquad\qquad\qquad\qquad\qquad$ (35.13b)

and is called the intrinsic (specific) permeability. It is a property of the porous medium. The dimensionless constant C depends on porosity and grain shape. Table 35.2 shows typical values of hydraulic conductivities for certain types of porous media.

Table 35.2 Typical values of hydraulic conductivity for certain types of porous media

Porous medium	k (Darcy)	K (m/s)
Shale	$10^{-5}-10^{-7.5}$	$10^{-10}-10^{-13}$
Silt, loam	$1-10^{-3}$	$10^{-5}-10^{-7}$
Silty sand	$10-10^{-2}$	$10^{-4}-10^{-7}$
Gravel	10^4-10^0	$10^{-1}-10^{-4}$

35.8.1.1 Experimental determination of hydraulic conductivity

In the laboratory, the hydraulic conductivity can be determined by using permeameters which may be of the constant head type or the falling head type. In the constant head type,

$$Q = KiA \Rightarrow \frac{\text{Volume of flow}}{\text{Time}} = K\frac{h}{L}A \Rightarrow K = \frac{(\text{Vol})}{A(\text{time})}\frac{L}{h} \tag{35.14}$$

where L is the length of the porous column of horizontal area A, and h is the constant head.
In the falling head type,

$$Q = \pi r_t^2 \frac{dh}{dt} = \pi r_c^2 \frac{Kh}{L} \tag{35.15a}$$

Therefore,

$$\pi r_t^2 \frac{dh}{dt} = \pi r_c^2 \frac{Kh}{L} \Rightarrow K = \frac{r_t^2 L}{r_c^2 t}\ln\left(\frac{h_1}{h_2}\right) \tag{35.15b}$$

where r_t and r_c are the radii of the tube and the cylindrical soil sample respectively, and t is the time interval for the water level in the tube to fall from h_1 to h_2.

35.8.2 Transmissivity, T

$$T = Kb \tag{35.16}$$

35.8.3 Coefficient of compressibility, β

Coefficient of compressibility β is defined by

$$\frac{dh}{d\sigma'} = -\beta h \tag{35.17a}$$

$$\beta = -\frac{dh}{h}\Big/d\sigma' \tag{35.17b}$$

where σ' is the vertical pressure (normal stress) on a soil sample of height h; the sample being confined horizontally so that its cross-sectional area remains constant.
Typical values of β are

$10^{-8}-10^{-7}$ m^2/N for sands, and

$10^{-7}-10^{-6}$ m^2/N for clays.

It is related to other material parameters by the equation

$$\beta = \frac{(1+v)(1-2v)}{E(1-v)} = \frac{1}{K+(4/3)G}$$ (35.18)

where E – Young's modulus
  v – Poisson's ratio
  K – Bulk modulus
  G – Shear modulus.

35.9 GROUND WATER FLOW

35.9.1 Flow rates

Groundwater flows very slowly. Typical velocities are of the order of 1 m/day. Velocities tend to decrease with increasing depth as porosities and hydraulic conductivities also decrease. Flow rates would however be substantial because of the large area through which flow takes place.

35.9.2 Flow nets

A flow net is a map of flow lines and equipotential lines. A flow line is a line such that the macroscopic velocity vector is everywhere tangential to it (similar to a streamline). An equipotential line is a line of constant hydraulic head. The two sets of lines are orthogonal to each other. The flow rates can be computed with knowledge of the geometry of the flow net, the head difference and the hydraulic conductivity.

Referring to Figure 6.3 (Chapter 6) where the hydraulic head is denoted as ϕ and the stream function as ψ and replacing ϕ by h and, ψ by q,

Hydraulic gradient, $i = \dfrac{dh}{ds}$

Flow 'dq' between two adjacent flow lines separated by 'dn' is

$dq = K\dfrac{dh}{ds}dn$ per unit width

For a square net, $ds \cong dn$
  Therefore,

$dq \cong Kdh$

Applying to the whole flow net, with the total head loss divided into 'n' squares,

$dh = \dfrac{h}{n}$

If the flow is divided into 'm' channels,

$Q = mdq = Kh\dfrac{m}{n}$ (35.19)

For anisotropic porous media, the flow lines and equipotential lines are not usually orthogonal. A transformation can rectify this situation.

If $K_x > K_z$, the horizontal dimensions are reduced by the ratio $\sqrt{K_z/K_x}$. The transformed isotropic medium has an equivalent hydraulic conductivity of $\sqrt{K_z\,K_x}$.

35.10 EQUATIONS OF FLOW THROUGH POROUS MEDIA

The governing equations of flow through porous media are obtained from the consideration of continuity and Darcy's law. For one dimensional flow, continuity gives

$$Q_x - \left[Q_x + \frac{\partial Q_x}{\partial x}\,\delta x \right] = S_s \delta x \frac{\partial h}{\partial t} \tag{35.20a}$$

where S_s is the specific storage.

Darcy's law gives

$$Q_x = -(\delta x) K_x \frac{\partial h}{\partial x} \tag{35.21a}$$

Therefore,

$$\frac{\partial}{\partial x}\left(K_x \frac{\partial h}{\partial x} \right) = S_s \frac{\partial h}{\partial t} \tag{35.22a}$$

Extending it to three dimensions,

$$\frac{\partial}{\partial x}\left(K_x \frac{\partial h}{\partial x} \right) + \frac{\partial}{\partial y}\left(K_y \frac{\partial h}{\partial y} \right) + \frac{\partial}{\partial z}\left(K_z \frac{\partial h}{\partial z} \right) = S_s \frac{\partial h}{\partial t} \tag{35.23}$$

In terms of the storage coefficient, S, it is

$$\frac{\partial}{\partial x}\left(K_x \frac{\partial h}{\partial x} \right) + \frac{\partial}{\partial y}\left(K_y \frac{\partial h}{\partial y} \right) + \frac{\partial}{\partial z}\left(K_z \frac{\partial h}{\partial z} \right) = \frac{S}{b} \frac{\partial h}{\partial t} \tag{35.24}$$

where b is the thickness of the aquifer.

For a homogeneous and isotropic medium, $K_x = K_y = K_z = K$. Then,

$$\frac{\partial^2 h}{\partial x^2} + \frac{\partial^2 h}{\partial y^2} + \frac{\partial^2 h}{\partial z^2} = \frac{S}{Kb} \frac{\partial h}{\partial t} = \frac{S}{T} \frac{\partial h}{\partial t} \tag{35.25}$$

For steady-state conditions, $\frac{\partial h}{\partial t} = 0$. Then,

$$\frac{\partial^2 h}{\partial x^2} + \frac{\partial^2 h}{\partial y^2} + \frac{\partial^2 h}{\partial z^2} = 0 \tag{35.26}$$

which is **Laplace equation**.

For unconfined aquifers, the continuity equation is slightly different. The difference arises from the fact that the upper limit of the control volume needed to be considered is not fixed in space. It is the water table which can vary from point to point. Hence, the governing equation becomes

$$Q_x - \left[Q_x + \frac{\partial Q_x}{\partial x} \delta x \right] = S_s \, \delta x \, \delta y \, h \frac{\partial h}{\partial t} \tag{35.20b}$$

$$Q_x = -(\delta y h) K_x \frac{\partial h}{\partial x} \tag{35.21b}$$

$$\frac{\partial}{\partial x} \left(K_x h \frac{\partial h}{\partial x} \right) = S_s h \frac{\partial h}{\partial t} \tag{35.22b}$$

$$\frac{\partial}{\partial x} \left(K_x h \frac{\partial h}{\partial x} \right) = S \frac{\partial h}{\partial t} \tag{35.22c}$$

For steady state,

$$\frac{\partial}{\partial x} \left(K_x h \frac{\partial h}{\partial x} \right) = 0 \tag{35.22d}$$

which is equivalent to (if K_x is constant)

$$\frac{\partial^2 h^2}{\partial x^2} = 0 \tag{35.22e}$$

For three dimensions,

$$\frac{\partial^2 h^2}{\partial x^2} + \frac{\partial^2 h^2}{\partial y^2} + \frac{\partial^2 h^2}{\partial z^2} = 0 \qquad \text{(same as Eq. 35.26)}$$

which is also Laplace equation.

For anisotropic conditions, the unsteady equation can be written as

$$T_x \frac{\partial^2 h}{\partial x^2} + T_y \frac{\partial^2 h}{\partial y^2} + T_z \frac{\partial^2 h}{\partial z^2} = S \frac{\partial h}{\partial t} \tag{35.27}$$

For two-dimensional flow, it is

$$T_x \frac{\partial^2 h}{\partial x^2} + T_y \frac{\partial^2 h}{\partial y^2} = S \frac{\partial h}{\partial t} \tag{35.28}$$

which, written in radial co-ordinates, is (assuming isotropy)

$$\frac{\partial^2 h}{\partial r^2} + \frac{1}{r} \frac{\partial h}{\partial r} = \frac{S}{T} \frac{\partial h}{\partial t} \tag{35.29}$$

For unconfined aquifers, $T = Kh$, and therefore,

$$\frac{\partial}{\partial x}\left(K_x h \frac{\partial h}{\partial x}\right) + \frac{\partial}{\partial y}\left(K_y h \frac{\partial h}{\partial y}\right) + \frac{\partial}{\partial z}\left(K_z h \frac{\partial h}{\partial z}\right) = S \frac{\partial h}{\partial t} \tag{35.30}$$

which is highly non-linear. Linearization is done by assuming K is constant and $\dfrac{\partial h}{\partial x}$ and $\dfrac{\partial h}{\partial y}$ are small. Then,

$$\frac{\partial}{\partial x}\left(h \frac{\partial h}{\partial x}\right) = h \frac{\partial^2 h}{\partial x^2} + \left(\frac{\partial h}{\partial x}\right)^2 \cong h \frac{\partial^2 h}{\partial x^2}$$

Equation 35.30 then becomes

$$K h \frac{\partial^2 h}{\partial x^2} + K h \frac{\partial^2 h}{\partial y^2} + K h \frac{\partial^2 h}{\partial z^2} = S \frac{\partial h}{\partial t} \tag{35.31}$$

Substituting $T = Kh$, the coefficient of transmissivity,

$$\frac{\partial^2 h}{\partial x^2} + \frac{\partial^2 h}{\partial y^2} + \frac{\partial^2 h}{\partial z^2} = \frac{S}{T} \frac{\partial h}{\partial t} \qquad\qquad \text{(same as Eq. 35.25)}$$

Thus, the same equation can be used for both confined and unconfined aquifers although the assumptions involved in arriving at the simplified forms are different.

35.10.1 Steady uni-directional flow

35.10.1.1 Confined aquifers

The steady-state one-dimensional flow equation is (Figure 35.4a)

$$\frac{d^2 h}{dx^2} = 0 \tag{35.32}$$

which has a solution of the form

$$h = Ax + B \tag{35.33}$$

Assuming $h = h_0$ at $x = 0$, $B = h_0$

Using Darcy's law, $\dfrac{dh}{dx} = -\dfrac{v}{K}$, which when substituted in the differentiated form of Eq. 35.33 gives

$$A = -\frac{v}{K}$$

Therefore,

$$h = -\frac{v}{K}x + h_0 \tag{35.34}$$

The head h, therefore, decreases linearly in the direction of flow.

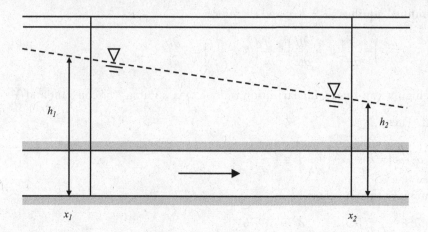

Figure 35.4a Steady uni-directional flow in a confined aquifer.

35.10.1.2 Unconfined aquifers

Direct analytical solution of the Laplace equation is not possible for unconfined flow. It is because the water table in the two-dimensional case represents a flow line, and its shape determines the flow distribution. At the same time, the flow distribution governs the water table shape (Figure 35.4b).

Dupuit assumptions (1863):

To obtain a solution under the above conditions, Dupuit made the following assumptions:

- $v\alpha\dfrac{\partial h}{\partial x}$ instead of $\dfrac{\partial h}{\partial s}$
- The flow is horizontal and uniform everywhere in a vertical section

These assumptions limit the applications but permit a solution to be obtained. The discharge per unit width q in a vertical section can be given as

$$q = -Kh\frac{dh}{dx} \tag{35.35}$$

Integration gives,

$$qx = -\frac{K}{2}h^2 + B \tag{35.36a}$$

At $x = 0$, $h = h_0$; therefore, $B = \dfrac{K}{2}h_0^2$

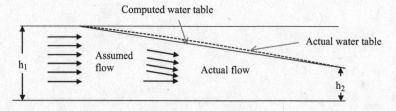

Figure 35.4b Steady uni-directional flow in an unconfined aquifer.

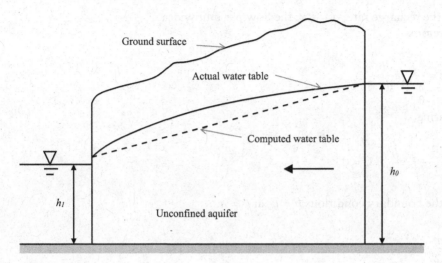

Figure 35.5 Flow between two water bodies.

Substituting,

$$qx = \frac{K}{2}\left(h_0^2 - h^2\right) \tag{35.36b}$$

which is a parabolic variation.

For flow between two fixed bodies of water (Figure 35.5) of constant heads h_0 and h_1, the water table slope at the upstream boundary of the aquifer is (from Eq. 35.35)

$$\frac{dh}{dx} = -\frac{q}{Kh_0} \tag{35.37}$$

But the boundary $h = h_0$ is an equipotential line because the fluid potential in the water body is constant. Therefore, the water table must be horizontal $\left(\frac{dh}{dx} = 0\right)$ at this boundary which is inconsistent with Eq. (35.37).

From Eq. 35.36 or 35.37, it can be seen that h decreases with increasing x, and $\frac{dh}{dx}$ increases with increasing x.[1] By this, the Dupuit assumption becomes increasingly poor approximations to the actual flow in the direction of flow. Therefore, the actual water table deviates more and more from the computed positions in the direction of flow as shown in Figures 35.4b and 35.5.

35.10.1.3 Base flow entering a stream

By Dupuit assumptions,

$$q = -K\,h\,\frac{dh}{dx}$$

---

[1] From Eq. 35.36b, $h^2 = -\frac{2qx}{K} + h_0^2$. Therefore, h decreases with increasing x and $\frac{dh}{dx} = -\frac{q}{Kh}$ increases with increasing x.

If R is the recharge rate, $q = Rx$, the flow per unit width.
 Therefore,

$$Rx = -K\,h\,\frac{dh}{dx}$$

Integrating,

$$R\frac{x^2}{2} + K\,\frac{h^2}{2} + C = 0$$

Using the boundary conditions $h = h_a$ at $x = a$,

$$R\frac{a^2}{2} + K\,\frac{h_a^2}{2} = -C$$

and, substituting for C,

$$R\left(\frac{x^2}{2} - \frac{a^2}{2}\right) + \frac{K}{2}\left(h^2 - h_a^2\right) = 0$$

from which

$$h^2 = h_a^2 + \frac{R}{K}\left(a^2 - x^2\right) \tag{35.38}$$

From symmetry and continuity,

$$Q_b = 2aR \tag{35.39}$$

where Q_b is the base flow entering the stream from the two sides per unit length of stream. The factor 2 accounts for the two sides.

35.10.1.4 Steady recharge due to infiltration

Considering a one-dimensional flow in the x-direction, the continuity relationship gives

$$R\delta x + Q - \left(Q + \frac{dQ}{dx}\,\delta x\right) = 0 \Rightarrow R\delta x - \frac{dQ}{dx}\,\delta x = 0$$

where R is the recharge rate per unit length due to infiltration.

But, $Q = -Kh\dfrac{dh}{dx}\,\delta x$

which leads to

$$R + \frac{d}{dx}\left(Kh\frac{dh}{dx}\right) = 0$$

Integration gives

$$Rx + Kh\frac{\partial h}{\partial x} + C_1 = 0$$

Integration again gives

$$R\frac{x^2}{2} + K\frac{h^2}{2} + C_1 x + C_2 = 0$$

At $x = 0, h = h_1;\ \Rightarrow C_2 = -\frac{1}{2}Kh_1^2$

At $x = L,\ h = h_2;\ \Rightarrow C_1 = -\frac{1}{2L}\left\{RL^2 + Kh_2^2 - Kh_1^2\right\}$

Therefore,

$$h^2 = h_1^2 - \left\{h_1^2 - h_2^2 - \frac{RL^2}{K}\right\}\frac{x}{L} - \frac{Rx^2}{K} \tag{35.40}$$

The hydraulic head here varies non-linearly with the distance x. The hydraulic gradient from Eq. 35.40 is

$$2h\frac{dh}{dx} = -\left\{h_1^2 - h_2^2 - \frac{RL^2}{K}\right\}\frac{1}{L} - \frac{2Rx}{K} \Rightarrow \frac{dh}{dx} = \frac{h_2^2 - h_1^2}{2Lh} + \frac{RL}{2Kh} - \frac{Rx}{Kh}$$

At the water divide, $\frac{dh}{dx} = 0$. This condition gives the position of the water divide as

$$x = \frac{L}{2} - \frac{K}{2LR}\left(h_1^2 - h_2^2\right) \tag{35.41}$$

35.11 WELL HYDRAULICS

35.11.1 Steady radial flow to a well

35.11.1.1 Confined aquifers

Certain assumptions are necessary to describe the steady radial flow to a well:

- Pumping (or recharging) is at a constant rate
- The well fully penetrates the aquifer and is perforated or screened
- The aquifer is homogeneous and isotropic and of infinite horizontal extent
- Water is released from storage as an immediate response to pumping.

Since the flow is horizontal everywhere, the following continuity equation can be written by considering an annular ring of thickness dr (Figure 35.6):

$$Q = 2\pi r Kb\frac{dh}{dr} \tag{35.42}$$

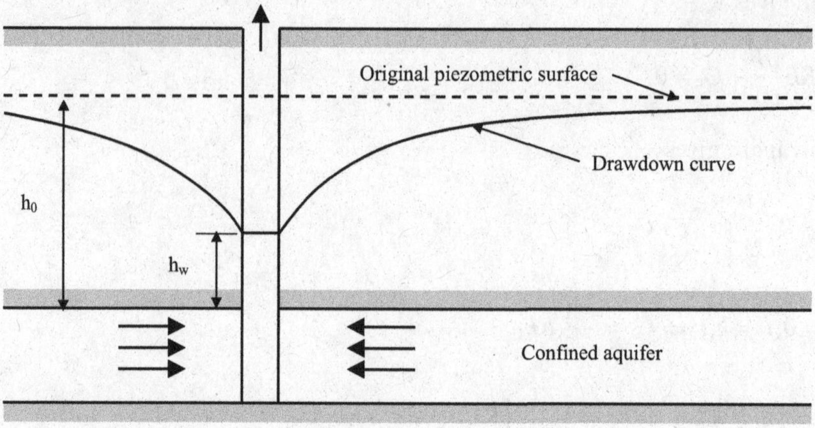

Figure 35.6 Radial flow to a well in a confined aquifer.

The negative sign disappears because r is measured in the opposite direction, i.e. h increases in the direction of increasing r. Therefore,

$$dh = \frac{Q}{2\pi Kb}\frac{dr}{r}$$

Integrating and using the boundary conditions $h = h_w$ at $r = r_w$ and $h = h_0$ at $r = r_0$,

$$h_0 - h_w = \frac{Q}{2\pi Kb}\ln\left(\frac{r_0}{r_w}\right) \tag{35.43}$$

The quantity $(h_0 - h)$ is known as the drawdown.

In a different form,

$$Q = \frac{2\pi Kb\left(h_0 - h_w\right)}{\ln\left(\dfrac{r_0}{r_w}\right)}$$

If the aquifer is extensive, there is no external limit for r. Then,

$$Q = \frac{2\pi Kb\left(h - h_w\right)}{\ln\left(\dfrac{r}{r_w}\right)} \tag{35.44a}$$

or

$$h - h_w = \frac{Q}{2\pi Kb}\ln\left(\frac{r}{r_w}\right) \tag{35.44b}$$

Equation 35.44 indicates that h increases infinitely with increasing r. But the maximum possible value of h is the initial head h_0. Therefore, from a theoretical point of view, steady radial flow in an extensive aquifer does not exist.

However, from a practical point of view, h approaches h_0 with distance from the well.

Equation 35.44 is called the equilibrium or Thiem (1906) equation. It enables the determination of aquifer permeability from the results of a pumping test. Considering any two points distant r_1 and r_2,

$$K = \frac{Q}{2\pi b\left(h_2 - h_1\right)} \ln\left(\frac{r_2}{r_1}\right), \quad \text{and,} \quad T = Kb \tag{35.45}$$

35.11.1.2 Unconfined aquifers

Using Dupuit assumptions, and as before (Figure 35.7),

$$Q = 2\pi r K h \frac{dh}{dr} \tag{35.46}$$

which when integrated between the limits $h = h_w$ at $r = r_w$ and $h = h_0$ at $r = r_0$, gives

$$Q = \frac{\pi K\left(h_0^2 - h_w^2\right)}{\ln\left(\dfrac{r_0}{r_w}\right)} \tag{35.47}$$

Equation 35.47 is not accurate near the well because the large vertical flow contradicts Dupuit assumptions.

The hydraulic conductivity K may be determined by using any pair of values of h:

$$K = \frac{Q}{\pi\left(h_2^2 - h_1^2\right)} \ln\left(\frac{r_2}{r_1}\right) \tag{35.48}$$

For small drawdowns, the transmissivity can be approximated as

$$T \approx K \frac{h_1 + h_2}{2} \tag{35.49}$$

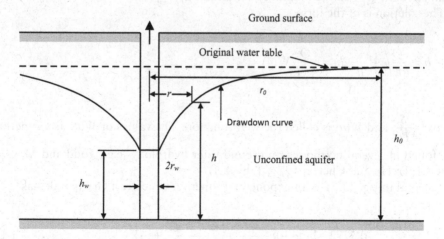

Figure 35.7 Steady radial flow to a well in an unconfined aquifer.

When the drawdowns are appreciable, the heads h_1 and h_2 can be replaced by $h_0 - s_1$ and $h_0 - s_2$ where s represent the depth from the equilibrium water table elevation to the water table level after pumping has started. Then (using Eq. 35.48),

$$T = Kh_0 \frac{Q \ln\left(\frac{r_2}{r_1}\right)}{2\pi\left[\left(s_1 - \frac{s_1^2}{2h_0}\right) - \left(s_2 - \frac{s_2^2}{2h_0}\right)\right]} \tag{35.50}$$

In the case of multiple wells, the drawdown at any point in the well field can be obtained by using the principle of superposition as the sum of the individual drawdown of each well. The individual pumping rates of each well should be known for determining the drawdown caused by each well.

35.11.2 Unsteady radial flow

Because the same non-equilibrium equation is used for confined and unconfined aquifers, the methods of solution may also be the same provided the Dupuit assumptions are satisfied and that the drawdown is small in relation to the saturation thickness. The solutions for unconfined aquifers are only approximate.

The governing equation for unsteady radial flow (confined and unconfined) is

$$\frac{\partial^2 h}{\partial r^2} + \frac{1}{r}\frac{\partial h}{\partial r} = \frac{S}{T}\frac{\partial h}{\partial t} \tag{Eq. 35.29}$$

where t is measured from the time since the beginning of pumping.

35.11.2.1 Theis' solution

Theis (1935), using the analogy between groundwater flow and heat conduction, obtained a solution for the above governing equation. The well is replaced by a mathematical sink of constant strength Q with the boundary conditions $h = h_0$ for $t = 0$ and $h \to h_0$ as $r \to \infty$ for $t \geq 0$. The solution is of the form

$$h_0 - h = \frac{Q}{4\pi T} \int_u^\infty \frac{e^{-u}}{u}\, du = \frac{Q}{4\pi T} W(u) \tag{35.51}$$

where $u = \dfrac{r^2 S}{4Tt}$, and $W(u)$ is called the well function. The values of $W(u)$ as a function of u can be found in several textbooks on groundwater hydrology (e.g., Todd and Mays, 2005, Table 4.4.1; Freeze and Cherry, 1979, Table 8.1).

The integral in Eq. 35.51 is an exponential function which can be expanded as

$$h_0 - h = \frac{Q}{4\pi T}\left\{-0.5772 - \ln(u) + u - \frac{u^2}{2.2!} + \frac{u^3}{3.3!} - \frac{u^4}{4.4!} + \cdots\right\} \tag{35.52}$$

35.11.2.2 Jacob's solution

When u is small (small values of r and large values of t make u small), $W(u)$ can be approximated as

$$W(u) \simeq -0.5772 - \ln(u) \tag{35.53}$$

Therefore,

$$h_0 - h = \frac{Q}{4\pi T}\{-0.5772 - \ln(u)\}$$

$$= \frac{Q}{4\pi T}\left\{\ln\frac{4Tt}{r^2 S} - 0.5772\right\}$$

$$= \frac{Q}{4\pi T}\frac{\ln(2.25Tt)}{r^2 S}$$

$$= \frac{2.3Q}{4\pi T}\log\left\{\frac{2.25Tt}{r^2 S}\right\}$$

$$= \frac{2.3Q}{4\pi T}\{\log(2.25Tt) - 2\log(r) - \log(S)\} \tag{35.54}$$

For a given t, Eq. 35.54 represents a straight-line plot of $h_0 - h$ vs. $\log(r)$.
Alternatively, for a given r, it also represents a straight-line plot of $h_0 - h$ vs. $\log(t)$.
From either of these plots, T can be determined.

35.11.2.2.1 T from the plot of $h_0 - h$ vs. log (r)

$$(h_0 - h_1)_1 = \frac{Q}{4\pi T}\left[\ln\left(\frac{1}{u_1}\right) - 0.5772\right]$$

$$(h_0 - h_1)_2 = \frac{Q}{4\pi T}\left[\ln\left(\frac{1}{u_2}\right) - 0.5772\right]$$

Therefore,

$$(h_0 - h_1)_1 - (h_0 - h_2)_2 = \frac{Q}{4\pi T}\left[\ln\left(\frac{u_2}{u_1}\right)\right]$$

$$= \frac{Q}{4\pi T}\left[\ln\frac{4Ttr_2^2 S}{4Ttr_1^2 S}\right]$$

$$= \frac{2.3Q}{4\pi T}\log(r_2/r_1)^2$$

$$= \frac{2.3Q}{4\pi T}\, 2\log(r_2/r_1) \tag{35.55}$$

$$= \frac{4.6Q}{4\pi T} \log(r_2/r_1) \tag{35.56}$$

For a log cycle, $\log(r_2/r_1) = 1$ (e.g., $\log(1{,}000/100) = \log 10 = 1$
Therefore,

$$(h_0 - h_1)_1 - (h_0 - h_2)_2 = \frac{4.6Q}{4\pi T}$$

and

$$T = \frac{4.6Q}{4\pi \Delta(h_0 - h)} \tag{35.57}$$

where $\Delta(h_0 - h)$ is the drawdown in one log cycle.

35.11.2.2.2 T from the plot of $h_0 - h$ vs. log t: Cooper-Jacob method (Cooper and Jacob, 1946; Jacob, 1950)

As before (Figure 35.8),

$$(h_0 - h_1)_1 - (h_0 - h_2)_2 = \frac{Q}{4\pi T} \left[\ln\left(\frac{u_2}{u_1}\right) \right]$$

$$= \frac{Q}{4\pi T} \left[\ln\left(\frac{t_2}{t_1}\right) \right]$$

$$= \frac{2.3Q}{4\pi T} \left[\log\left(\frac{t_2}{t_1}\right) \right] \tag{35.58}$$

For one log cycle, $\log(t_2/t_1) = 1$. Therefore,

$$T = \frac{2.3Q}{4\pi \Delta(h_0 - h)} \tag{35.59}$$

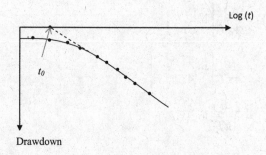

Figure 35.8 T from the plot of $h_0 - h$ vs. log t: Cooper–Jacob method.

35.11.2.2.3 Storage coefficient, S

By extrapolating the plots to intersect the time (or distance) axis, the time (or distance) at which the drawdown becomes zero can be found (Figure 35.8). Then,

$$\ln\left(\frac{1}{u}\right) - 0.5772 = 0$$

If the drawdown is plotted against time,

$$\ln\left(\frac{4Tt_0}{r^2 S}\right) = 0.5772 \tag{35.60}$$

The intercept t_0 is determined from the graph (Figure 35.8). Since r is fixed and known, and T is known, S can be computed from Eq. 35.60.

The same procedure may be used when the drawdown vs. log r is plotted. In this case, the plot gives r_0 as the intercept of the x-axis. Then,

$$\ln\left(\frac{4Tt}{r_0^2 S}\right) = 0.5772 \tag{35.61}$$

This method of solving for S and T is called the Cooper-Jacob method.

35.11.2.3 Theis graphical method of solution

The non-equilibrium equation, Eq. 35.51, is

$$(h_0 - h) = \frac{Q}{4\pi T} \int_u^\infty \frac{e^{-u}}{u}\, du = \left[\frac{Q}{4\pi T}\right] W(u)$$

and u can be written as

$$\frac{r^2}{t} = \left[\frac{4T}{S}\right] u \tag{35.62}$$

Since the terms inside the square brackets are constants, the relationship between $h_0 - h$ vs. r^2/t must be similar to the relationship between $W(u)$ vs. u.

The values of $W(u)$ for various values of u are given in the form of tables, and a plot of $W(u)$ vs. u on a log - log scale is called a type curve (Figure 35.9). The values of $h_0 - h$ vs. r^2/t are also plotted on a log–log scale on log–log paper of the same size. Then, by superimposing the two plots keeping the two axes parallel until the two curves coincide, a matching point is chosen, which gives the values of u, r^2/t, $W(u)$ and $h_0 - h$. Substitutions in Eqs. 35.51 and 35.62 give T and S. The matching point needs not be on the curve.

35.11.2.4 Recovery tests

Transmissivity can also be determined by observing the recovery of the water levels after pumping has stopped. Storage coefficient however cannot be determined.

Residual drawdown = Drawdown below the original static water table

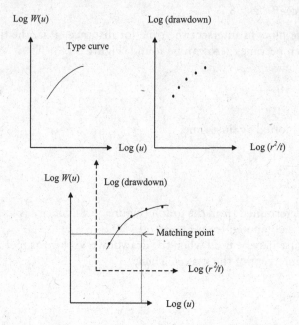

Figure 35.9 Graphical method of determining T and S from pumping test data using Well function.

If pumping is done for some time and then stopped, the drawdown thereafter will be the same as if the discharge had been continued and a hypothetical recharge well with the same flow superimposed on the discharge well at the instant when pumping is stopped. Then, the residual drawdown s' is given by

$$s' = \frac{Q}{4\pi T}\big[W(u) - W(u')\big] \tag{35.63}$$

where $u = \dfrac{r^2 S}{4Tt}$ and $u' = \dfrac{r^2 S}{4Tt'}$

For small r and large t', the well function can be approximated as before. In the above equations, t is measured from the time pumping started, and t' is measured from the time pumping stopped. Then,

$$s' = \frac{2.3Q}{4\pi T} \log\left(\frac{t}{t'}\right) \tag{35.64}$$

By plotting s' vs. $\log(t/t')$, a straight line can be obtained. As before,

$$T = \frac{2.3Q}{4\pi\Delta s'} \tag{35.65}$$

where $\Delta s'$ is the drawdown per log cycle of t/t'. The storage coefficient cannot be determined by the recovery test.

The advantage of using the pumping tests for determining the aquifer parameters is that one test enables the evaluation of K, S and T, and that the values obtained will be field values. The disadvantage is that they are expensive and that the interpretation of pumping test results may not be unique due to the underlying aquifers been leaky, unconfined, etc.

35.11.3 Yield

Yield from a well can be obtained if the aquifer parameters are known. On a long term basis, the yield from aquifers may also be determined from the hydrological equilibrium equation:

Inflow – Outflow = Change in storage

In most aquifers, safe yield is more important. It is the amount that can be drawn without causing undesirable effects to the aquifer. Any abstraction in excess of the safe yield is an overdraft and is called mining. The safe yield depends on

- availability of water supply (safe yield ≤ long term mean annual supply to the basin),
- economy –cost of pumping, installation of machinery, etc.,
- water quality – seawater intrusion can be a problem, and,
- water rights – legal considerations.

The safe yield of unconfined aquifers may be obtained by observing the long term fluctuations of groundwater levels.

35.12 DISPERSION THROUGH POROUS MEDIA

The equation for transport of a solute in a saturated porous medium is derived using the law of conservation of mass under the following assumptions:

- The porous medium is homogeneous and isotropic
- The medium is saturated, and the flow is at steady state
- Darcy's law is applicable.

According to Darcy's law, the solute will be advected by the average linear velocity. If this is the only transport mechanism, the solute will move as a plug. But, due to hydrodynamic dispersion, which is caused by variations in the microscopic velocity within each pore channel and from one channel to another, mixing of the solute will also take place.

Therefore, to describe the transport mechanism using macroscopic parameters yet taking into account microscopic mixing, it is necessary to introduce a second mechanism of transport in addition to advection.

Average linear velocity $\bar{u} = \dfrac{u}{n}$

where u is the Darcy velocity and n is the porosity of the medium

The mass of solute per unit volume of porous medium is nc, where c is the concentration of the solute which is in mass of solute per unit volume of solution (ML^{-3}).

For a homogeneous medium, n is a constant. Therefore,

$$\frac{d}{dx}(nc) = n\frac{dc}{dx} \tag{35.66}$$

Transport by advection = $\bar{u}ncdA$

Transport by dispersion = $nK\dfrac{dc}{dx}dA$

The dispersion coefficient is related to the dispersivity α and the diffusion coefficient D by the relationship

$$K = \alpha \bar{u} + D \tag{35.67}$$

where α is the dispersivity which is a characteristic property of the porous medium (L), and D is the coefficient of molecular diffusion for the solute in the porous medium (L^2T^{-1}).

The form of the dispersive component of the transport by dispersion is analogous to Fick's law. Total mass transport/unit area,

$$q = \bar{u}nc - nK\frac{dc}{dx} \tag{35.68}$$

The negative sign indicates that the transport is towards the zone of lower concentration.

Conservation of mass for a conservative solute gives

$$qdA - \left(q + \frac{dq}{dx}\delta x\right)dA = n\frac{dc}{dt}dAdx$$

which simplifies to

$$-\frac{dq}{dx} = n\frac{dc}{dt}$$

Substituting from Equation 35.68,

$$-\frac{\partial}{\partial x}(\bar{u}nc) + \frac{\partial}{\partial x}\left(nK\frac{\partial c}{\partial x}\right) = n\frac{\partial c}{\partial t}$$

which simplifies to

$$-\bar{u}\frac{\partial c}{\partial x} + \frac{\partial}{\partial x}\left(K\frac{\partial c}{\partial x}\right) = \frac{\partial c}{\partial t} \tag{35.69}$$

In a homogeneous medium, K does not vary in space. Therefore,

$$K\frac{\partial^2 c}{\partial x^2} - \bar{u}\frac{\partial c}{\partial x} = \frac{\partial c}{\partial t} \tag{35.70}$$

By extending the formulation to three dimensions,

$$K_x\frac{\partial^2 c}{\partial x^2} + K_y\frac{\partial^2 c}{\partial y^2} + K_z\frac{\partial^2 c}{\partial z^2} - \left(\bar{u}\frac{\partial c}{\partial x} + \bar{v}\frac{\partial c}{\partial y} + \bar{w}\frac{\partial c}{\partial z}\right) = \frac{\partial c}{\partial t} \tag{35.71}$$

In some applications, the one-dimensional direction is taken as a curvilinear coordinate in the direction of flow along a flow line. The transport equation then is

$$K_1\frac{\partial^2 c}{\partial 1^2} - \left(\bar{u}_1\frac{\partial c}{\partial 1}\right) = \frac{\partial c}{\partial t} \tag{35.72}$$

where l is the coordinate direction along the flow line.

If u and K vary spatially, then

$$\frac{\partial}{\partial x}\left(K\frac{\partial c}{\partial x}\right) - \frac{\partial}{\partial x}(uc) = \frac{\partial c}{\partial t} \tag{35.73}$$

A solution to the dispersion equation with the following boundary and initial conditions obtained by Ogata and Banks (1961) is as follows:

$$c(x,0) = 0 \text{ for } x \geq 0$$

$$c(0,t) = c_0 \text{ for } t \geq 0$$

$$c(\infty,t) = 0 \text{ for } t \geq 0$$

$$c(x,t) = \frac{c_0}{2}\left\{\text{erfc}\left[\frac{x-ut}{2\sqrt{Kt}}\right] + \exp\left[\frac{ux}{K}\right]\text{erfc}\left[\frac{x+ut}{2\sqrt{Kt}}\right]\right\} \tag{35.74}$$

The error function erf and the complementary error function erfc are defined as

$$\text{erf}(x) = \frac{1}{\sqrt{\pi}}\int_{-x}^{x} e^{-t^2}\, dt = \frac{2}{\sqrt{\pi}}\int_{0}^{x} e^{-t^2}\, dt; \quad \text{erfc}(x) = 1 - \text{erf}(x)$$

For low velocities, K in this equation is approximately equal to D (Eq. 35.67).

35.13 GROUNDWATER FLOW MODELLING

35.13.1 Governing equations

35.13.1.1 Classification of partial differential equations

A second-order linear partial differential equation (PDE) in $\phi(x, t)$ can be written as

$$\left[A_1(x,t)\frac{\partial^2 \phi}{\partial x^2} + A_2(x,t)\frac{\partial^2 \phi}{\partial x\, \partial t} + A_3(x,t)\frac{\partial^2 \phi}{\partial t^2}\right.$$

$$\left. + A_4(x,t)\frac{\partial \phi}{\partial x} + A_5(x,t)\frac{\partial \phi}{\partial t} + A_6(x,t)\phi + A_7(x,t)\right] = 0 \tag{35.75a}$$

Eq. 35.75a can also be written as

$$A_1(x,t)\frac{\partial^2 \phi}{\partial x^2} + A_2(x,t)\frac{\partial^2 \phi}{\partial x\, \partial t} + A_3(x,t)\frac{\partial^2 \phi}{\partial t^2} + f\left(x,t,\phi,\frac{\partial \phi}{\partial x},\frac{\partial \phi}{\partial t}\right) = 0 \tag{35.75b}$$

If $f\left(x,t,\phi,\dfrac{\partial \phi}{\partial x},\dfrac{\partial \phi}{\partial t}\right)$ is linear or non-linear, Eq. 35.75 is classified as quasi-linear.

For purposes of classification and study, a second-order quasi-linear PDE is grouped as follows:

Elliptic if $A_2^2 - 4A_1A_3 < 0$

Parabolic if $A_2^2 - 4A_1A_3 = 0$

Hyperbolic if $A_2^2 - 4A_1A_3 > 0$.

In general, elliptic equations are of the steady-state type, meaning applicable in a closed domain. Parabolic equations of the non-steady-state or propagation type have open domains.
Examples include

$$\text{Laplace equation} \frac{\partial^2 \phi}{\partial x^2} + \frac{\partial^2 \phi}{\partial y^2} = 0 \, (\text{Elliptic})$$

$$\text{Poisson's equation} \frac{\partial^2 \phi}{\partial x^2} + \frac{\partial^2 \phi}{\partial y^2} = f(x, y) \, (\text{Elliptic})$$

$$\text{Diffusion equation} \, K \frac{\partial^2 \phi}{\partial x^2} = \frac{\partial \phi}{\partial t} \, (\text{Parabolic})$$

$$\text{Wave equation} \, c \frac{2\partial^2 \phi}{\partial x^2} = \frac{\partial^2 \phi}{\partial t^2} \, (\text{Hyperbolic})$$

The governing equation for flow through porous media is of the elliptic type for steady-state or potential flow problems and of the parabolic type for transient problems. All problems are considered in at least a two-dimensional spatial domain.

35.13.2 Boundary and initial conditions

35.13.2.1 Boundary conditions

- **Dirichlet type:** The magnitude of the variable is specified at certain parts of the boundary. e.g. $\phi = f(x, y, z, t)$ along S_1.

  Imposition of the Dirichlet type boundary condition is quite easy as it requires simple substitution only.
- **Neumann type:** The flux across the remaining part of the boundary is specified, e.g. $\frac{\partial \phi}{\partial n} = f_1(x, y, z, t)$. The most common flux boundary condition is a no flow boundary in which case

$$\frac{\partial \phi}{\partial n} = 0$$

Examples include impervious boundaries, streamlines and water divides.

Imposition of Neumann type boundary conditions can be done by having imaginary nodes which are mirror images of internal nodes, e.g.

$\phi_{i-1,j} \; \phi_{i+1,j}$

This condition is substituted into the general algorithm.

35.13.2.2 Initial conditions

This is applicable to transient problems and can be of the form

$\phi = \phi_0$ at $t = 0$ in the entire domain.

35.13.3 Solution domain

35.13.3.1 Identification

The solution domain consists of the aquifer through which flow takes place.

35.13.3.2 Discretization

The continuous domain is discretized into a finite number of sub-domains by a regular or irregular solution mesh.

35.13.4 Solution algorithms

a. Finite difference method (FDM) – see Chapter 16
b. Finite element method – see Chapter 16.

35.13.5 Problems associated with numerical methods

see Chapter 16.

35.13.6 MODFLOW Software package for groundwater flow modelling

MODFLOW software package was developed by the US Geological Survey in the early 1980s. It has subsequently undergone revisions, and several versions have been released since the launch of the original version. The current version of MODFLOW 6 is version 6.0.4, released on March 11, 2019. The computer code written in FORTRAN solves the transient governing equation for flow through saturated porous media in one, two or three dimensions using the finite difference method.

35.13.7 Examples of flow through porous media

35.13.7.1 Potential flow – steady state, homogeneous and isotropic

Consider the Laplace equation in a rectangular domain

$$\frac{\partial^2 \phi}{\partial x^2} + \frac{\partial^2 \phi}{\partial y^2} = 0 \quad \text{on the interior of the domain} \tag{35.76a}$$

$$\phi = \phi_0 \quad \text{along the upper boundary} \tag{35.76b}$$

$$\frac{\partial \phi}{\partial y} = 0 \quad \text{along the lower boundary} \tag{35.76c}$$

$$\frac{\partial \phi}{\partial x} = 0 \quad \text{along the vertical boundaries} \tag{35.76d}$$

Dividing the domain into 16 rectangles as shown above in Figure 35.10, there will be 9 internal nodes and 11 boundary nodes where the ϕ values are unknown.

Using the central differencing scheme,
$\nabla\phi^2 = 0$ may be written as

$$\frac{\phi_{i+1,j} - 2\phi_{i,j} + \phi_{i-1,j}}{(\Delta x)^2} + \frac{\phi_{i,j+1} - 2\phi_{i,j} + \phi_{i,j-1}}{(\Delta y)^2} = 0 \tag{35.77a}$$

from which (assuming $\Delta x = \Delta y$)

$$4\phi_{i,j} = \phi_{i,j-1} + \phi_{i,j+1} + \phi_{i-1,j} + \phi_{i+1,j} \tag{35.77b}$$

This formulation is applicable in the interior of the domain. If all the boundaries are of the Dirichlet type, then it is quite easy to solve the problem because there will be as many equations as there are unknowns. The above scheme is implicit, and therefore the equations will have to be solved simultaneously.

However, if some boundaries are of the Neumann type, as is the case in this problem, the solution scheme requires the imposition of some imaginary nodes which are the mirror images of the Neumann type boundaries. For the derivative to be zero, the imaginary node on the outside of the Neumann boundary must have a potential equal to that next to the boundary and on the inside. For examples,

$$\phi_{6,1} = \phi_{4,1}; \quad \phi_{0,2} = \phi_{2,2}; \quad \phi_{6,4} = \phi_{4,4}; \text{etc.}$$

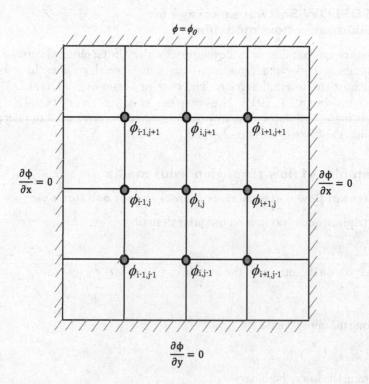

Figure 35.10 Discretization of a domain into 16 rectangles.

At the interior nodes, the following equations can be written:

$$4\phi_{22} = (\phi_{12}) + \phi_{32} + (\phi_{21}) + \phi_{23}$$

$$4\phi_{32} = \phi_{22} + \phi_{42} + (\phi_{31}) + \phi_{33}$$

$$4\phi_{42} = \phi_{32} + (\phi_{52}) + (\phi_{41}) + \phi_{43}$$

$$4\phi_{23} = (\phi_{13}) + \phi_{33} + \phi_{22} + \phi_{24}$$

$$4\phi_{33} = \phi_{23} + \phi_{43} + \phi_{32} + \phi_{34}$$

$$4\phi_{43} = \phi_{33} + (\phi_{53}) + \phi_{42} + \phi_{44}$$

$$4\phi_{24} = (\phi_{14}) + \phi_{34} + \phi_{23} + (\phi_{25})$$

$$4\phi_{34} = \phi_{24} + \phi_{44} + \phi_{33} + (\phi_{35})$$

$$4\phi_{44} = \phi_{34} + (\phi_{54}) + \phi_{43} + (\phi_{45})$$

In this system of equations, the nodes marked in parenthesis are boundary notes. The nodes where Dirichlet boundary conditions apply have their subscripts marked in *italics*. If ϕ values are prescribed at these boundary nodes, then the solution of the nine simultaneous equations enables the evaluation of the ϕ values at the nine internal nodes. It can also be seen that with this scheme, ϕ_{11} and ϕ_{51} are not needed.

At the boundary nodes, the following equations can be written:

$$4\phi_{11} = 2\phi_{12} + 2\phi_{21}$$

$$4\phi_{21} = \phi_{11} + \phi_{31} + 2\phi_{22}$$

$$4\phi_{31} = \phi_{21} + \phi_{41} + 2\phi_{32}$$

$$4\phi_{41} = \phi_{31} + \phi_{51} + 2\phi_{42}$$

$$4\phi_{51} = 2\phi_{41} + 2\phi_{52}$$

$$4\phi_{12} = \phi_{11} + \phi_{13} + 2\phi_{22}$$

$$4\phi_{13} = \phi_{12} + \phi_{14} + 2\phi_{23}$$

$$4\phi_{14} = \phi_{13} + (\phi_{15}) + 2\phi_{24}$$

$$4\phi_{52} = \phi_{51} + \phi_{53} + 2\phi_{42}$$

$$4\phi_{53} = \phi_{52} + \phi_{54} + 2\phi_{43}$$

$$4\phi_{54} = \phi_{53} + (\phi_{55}) + 2\phi_{44}$$

There are 20 equations with 20 unknowns; hence a solution can be obtained.

35.13.7.2 Transient flow – unsteady state, homogeneous and isotropic

Assuming homogeneous and isotropic conditions, the governing equation is

$$S\frac{\partial \phi}{\partial t} = T\left\{\frac{\partial^2 \phi}{\partial x^2} + \frac{\partial^2 \phi}{\partial y^2}\right\} \pm I \tag{35.78}$$

subject to Dirichlet and Neumann type boundary conditions and an initial condition. In Eq. 35.78, I represents a source/sink term.

35.13.7.2.1 Explicit methods

Taking the spatial derivatives at the previous time level, and using forward differencing for the time derivative,

$$\phi_{i,j}^{k+j} = \phi_{i,j}^{k} \pm \frac{I\Delta t}{S} + \alpha\left[\phi_{i-1,j}^{k} - 2\phi_{i,j}^{k} + \phi_{i+1,j}^{k}\right] + \beta\left[\phi_{i,j-1}^{k} - 2\phi_{i,j}^{k} + \phi_{i,j+1}^{k}\right] \tag{35.79}$$

where

$$\alpha = \frac{T\Delta t}{S(\Delta x)^2}; \text{ and } \beta = \frac{T\Delta t}{S(\Delta y)^2} \tag{35.80}$$

In this formulation, the superscripts indicate the time levels.

By this method, the ϕ values at all interior nodes can be determined.

An important limitation in explicit methods is that the stability of the solution depends upon the time step Δt.

35.13.7.2.2 Implicit methods

Alternative formulations could be made by taking the spatial derivatives at

- the present time level
- halfway between the previous and present time levels
- any intermediate point between the previous and present time levels.

In general,

$$\phi_{i,j} = \varepsilon\phi_{i,j}^{k} + (1-\varepsilon)\phi_{i,j}^{k+1} \tag{35.81}$$

where ε $(0 \leq \varepsilon \leq 1)$ is an interpolation parameter.

35.13.7.2.3 Special cases

- When $\varepsilon = 1$, Eq. 35.81 leads to the explicit formulation
- When $\varepsilon = 0$, Eq. 35.81 becomes

$$\phi_{i,j} = \phi_{i,j}^{k+1} \tag{35.82}$$

Then,

$$\phi_{i,j}^{k+1} = \phi_{i,j}^k \pm \frac{I\Delta t}{S} + \alpha\left[\phi_{i-i,j}^{k+1} - 2\phi_{i,j}^{k+1} + \phi_{i+1,j}^{k+1}\right] + \beta\left[\phi_{i,j-1}^{k+1} - 2\phi_{i,j}^{k+1} + \phi_{i,j+1}^{k+1}\right] \tag{35.83}$$

This method is called the fully implicit method. It is unconditionally stable.

The fully implicit method ($\varepsilon = 0$) is biased toward the present state. The bias towards either of the time levels can be avoided by making $\varepsilon = 0.5$.

- When $\varepsilon = 0.5$, Eq. 35.81 leads to the Crank–Nicholson method. However, the Crank–Nicholson method offers no major advantage over the fully implicit method because the latter can be used to solve the steady-state problem as well when $S = 0$, i.e. $S\dfrac{\partial \phi}{\partial t} = 0$.

The formulation for the Crank–Nicholson scheme takes the form

$$\phi_{i,j}^{k+1} = \phi_{i,j}^k \pm \frac{I\Delta t}{S} + \frac{\alpha}{2}\left\{\left[\phi_{i-1,j}^k - 2\phi_{i,j}^k + \phi_{i+1,j}^k\right] + \left[\phi_{i-1,j}^{k+1} - 2\phi_{i,j}^{k+1} + \phi_{i+1,j}^{k+1}\right]\right\}$$

$$+ \frac{\beta}{2}\left\{\left[\phi_{i,j-1}^k - 2\phi_{i,j}^k + \phi_{i,j+1}^k\right] + \left[\phi_{i,j-1}^{k+1} - 2\phi_{i,j}^{k+1} + \phi_{i,j+1}^{k+1}\right]\right\} \tag{35.84}$$

35.13.7.2.4 Alternating direction implicit method (ADI)

In this method, the x and y derivatives are taken at alternating time levels as follows:

At time level $k + 1$

$$\phi_{i,j}^{k+1} = \phi_{i,j}^k \pm \frac{I\Delta t}{S} + \alpha\left[\phi_{i-1,j}^{k+1} - 2\phi_{i,j}^{k+1} + \phi_{i+1,j}^{k+1}\right] + \beta\left[\phi_{i,j-1}^k - 2\phi_{i,j}^k + \phi_{i,j+1}^k\right] \tag{35.85}$$

Equation 35.85 has three unknowns, $\phi_{i-1,j}^{k+1}$, $2\phi_{i,j}^{k+1}$ and $\phi_{i+1,j}^{k+1}$. When similar equations are written for all the internal nodes in the i-direction at time level $k + 1$, there will be enough equations as there are unknowns provided the two boundary values are known. Therefore, the equations can be solved to obtain the ϕ values along one line of nodes at the jth level in the i-direction.

At time level $k + 2$

$$\phi_{i,j}^{k+2} = \phi_{i,j}^{k+1} \pm \frac{I\Delta t}{S} + \alpha\left[\phi_{i-1,j}^{k+1} - 2\phi_{i,j}^{k+1} + \phi_{i+1,j}^{k+1}\right] + \beta\left[\phi_{i,j-1}^{k+2} - 2\phi_{i,j}^{k+2} + \phi_{i,j+1}^{k+2}\right] \tag{35.86}$$

Here too, there are three unknowns, $\phi_{i-1,j}^{k+2}$, $2\phi_{i,j}^{k+2}$ and $\phi_{i+1,j}^{k+2}$. Writing down equations for all the internal nodes in the j-direction at time level $k + 2$, there will be enough equations as there are unknowns provided the two boundary values are known. Therefore, the equations can be solved to obtain the ϕ values along one line of nodes at the ith level in the j-direction.

35.13.7.3 Transient state, unsteady, non-homogeneous, non-isotropic

The equations are approximated as follows:

$$T\frac{\partial \phi}{\partial x} \cong \frac{1}{2\Delta x}\left(T_{i,j} + T_{i+1,j}\right)\left(\phi_{i+1,j} - \phi_{i,j}\right) \tag{35.87a}$$

or

$$T \frac{\partial \phi}{\partial x} \cong \frac{1}{2\Delta x} \left(T_{i-1,j} + T_{i,j}\right) \left(\phi_{i,j} - \phi_{i-1,j}\right) \tag{35.87b}$$

35.14 RECHARGE OF GROUNDWATER

Aquifers can be recharged by a number of methods for a number of reasons. Recharge may be by natural means such as by rainfall or by artificial means.

35.14.1 Basin method

Basins may be natural or artificially constructed using levees to contain the water which may be natural from rainfall or artificially diverted from streams. The soil in the basin needs to be silt free to prevent clogging of pores which will retard and/or prevent infiltration. They include infiltration basins using the soil aquifer treatment (SAT) system in which waste water is allowed to infiltrate into the sub-soil which undergoes natural physical, biological and chemical purification.

35.14.2 Stream-channel method

This method allows water to spread in streams and channels with slow-moving water. They may be natural or artificial by installing check dams to slow down the flow velocities.

35.14.3 Ditch method

Ditch and furrows may be long and winding types as well as a tree-like type. The velocities should be sufficient to carry the sediments which would otherwise clog the pores and retard infiltration.

35.14.4 Irrigation method

This method uses excess irrigation water during non-irrigating seasons.

35.14.5 Recharge pit method

This method uses pits that may occur naturally or artificially made to store water which will infiltrate with time.

35.14.6 Recharge well method

In this method, the aquifer is recharged using recharge wells.

35.14.7 Waste water disposal Soil Aquifer Treatment (SAT) method

In this method, wastewater is allowed to seep into the sub-soil naturally or pumped into an aquifer via a recharge well. The wastewater gets filtered during the process of passage through the porous medium, thereby achieving two objectives – treating the wastewater and recharging the aquifer.

35.15   PROBLEMS ASSOCIATED WITH GROUNDWATER EXTRACTION

35.15.1   Land subsidence

Land subsidence is an irreversible process in which the ground is sinking over large areas. Sometimes, such occurrences can also be seen locally as 'sinkholes'. The primary cause of land subsidence is excessive groundwater pumping which causes compaction of soils in aquifer systems. Many cities in the world that pump excessive quantities of groundwater have been subjected to land subsidence. Examples include the San Joaquin Valley and New Orleans in the USA, some parts of Tokyo metropolitan area in Japan, Bangkok in Thailand, Jakarta in Indonesia, the Hague in the Netherlands, Venice in Italy, Shanghai in China and Mexico city in Mexico. In some cities, the subsidence has been controlled by legal means restricting the quantities of groundwater pumping. For example, in Tokyo Metropolitan area, which has experienced land subsidence since the early 1900s and accelerated since the end of Second World War due to expansion of economic activities have brought the rate of land subsidence from a maximum of about 24 cm/year in some areas to about 2 cm/year in recent years with the introduction of the Industrial Water Law which limits the quantities of groundwater pumping. (Sato et al., 2006). International initiatives to address this problem was the formation of the UNESCO Land Subsidence International Initiative (LASII) in 2018 which was originally called the Working Group, with missions that include improving the scientific and technical knowledge needed to identify and characterize threats related to natural and anthropogenic land subsidence, and to stimulate and enable international exchange of information for sustainable groundwater resources development in areas susceptible to land subsidence. Several International Symposia on Land Subsidence have been held in different countries from Asia, Europe and North America, with the first one in Tokyo in 1969 (Tison, 1969).

35.15.2 Saltwater intrusion

Saltwater intrusion is a common problem in coastal aquifers. The fresh groundwater in coastal aquifers is usually replenished by rainfall. When the rate of abstraction by pumping exceeds the replenishment rate, the freshwater/saltwater interface will move landwards due to the reduction of the hydraulic head in the freshwater aquifer. The saltwater has a higher density and has a higher mineral content resulting in a higher pressure that pushes the interface landward. The aquifer water then becomes contaminated and any pumping then yields freshwater mixed with saltwater which is not suitable for domestic and/or agricultural purposes. Further pumping, if continued, will aggravate the problem as a result of reducing the hydraulic head on the freshwater side. The interface occurs not at sea level but at a depth below the sea level of about 40 times the height of freshwater above sea level. This relationship is known as the Ghyben–Herzberg principle, named after its originators. It can be demonstrated using a U-tube filled with seawater and freshwater that will create an interface. The Ghyben–Herzberg equation then is

$$\rho_s g z = \rho_f g \left(z + h_f \right) \Rightarrow z = \frac{\rho_f}{\rho_s - \rho_f} h_f \tag{35.88}$$

where z is the height of seawater, h_f is the height of freshwater, ρ_f and ρ_s the densities of freshwater and seawater, respectively.

There can also be saltwater intrusion due to tidal effects, storm surges, hurricanes and inland areas where there are seawater conveyancing canals. Under normal conditions, the intrusion of saltwater inland is limited because of the high ground level of aquifer areas which offers a high hydraulic head. Saltwater intrusion is a problem in many regions in the world where the main source of drinking water is groundwater. Examples of places with problems associated with saltwater intrusion include several coastal regions in the USA, including Florida, Louisiana, California, and Washington, and in Pakistan, Cyprus, Morocco and the Mekong Delta.

35.15.3 Water rights

Water rights is a complex issue. In general, water rights give access to water bodies adjacent to a property which means the owner of the property will have access to adjoining waterbodies that include atmospheric, surface and groundwater. Two other closely associated rights are the riparian rights which allow access and usage of flowing bodies of waters like rivers and streams, and littoral rights which allow access to lakes, seas and oceans. Property owners can have the right to use the water, but they do not own the water. Water rights is a legal entitlement that depends upon the legal system in the relevant region. For example, in the USA, each state has its own laws that govern water rights.

In common law, a property owner has the right to use the water beneath his/her property. However, a problem arises when the property owner extracts excessive water from wells drilled in his/her property but draws water not only from the groundwater beneath his/her property but also from adjoining properties as well since there are no property boundaries for groundwater. It then becomes a legal issue.

There is no unified 'water law' in the world. Even in a particular jurisdiction area, water laws have not reached maturity and are still evolving as new environmental and other challenges come in to play. The oldest water law may perhaps be the 'Roman law' for water administration which provided free access to 'aqua publica' but allowed right to request private access to public water with permission from relevant authorities.

Another issue of concern is the contamination of groundwater resulting from the disposal of industrial and other harmful wastes. Once a pollutant enters the groundwater, it gets dispersed by diffusion and dispersion, and the consequences can be far reaching. An incident that attracted attention and publicity was the case of well water contamination in the city of Woburn in Massachusetts in USA during the period from mid to late 1970s which led to the suspicion by the community that the unusually high incidence of leukemia, cancer and a wide variety of other health problems were linked to the possible exposure to volatile organic chemicals in the groundwater pumped from two wells. The incident culminated in a court case that also received wide publicity as well as a book entitled *A Civil Action* (Harr, 1995) written about the case which later in 1998 turned into a movie by the same title.

Example 35.1

A tracer is introduced into an upstream well, fully penetrating an unconfined aquifer with formations of hydraulic conductivities of 100 m/day on the upstream side and 120 m/day on the downstream side. The upstream and downstream parts of the aquifer have lengths of 1,500 and 1,200 m, respectively. The water levels in the upstream and downstream wells are 50 and 25 m, respectively. The presence of the tracer was detected at the downstream well 675 days later. Determine the average porosity of the aquifer formation. Assume that the tracer is well mixed across the depth of the aquifer at the upstream well.

The flow is horizontal everywhere on a vertical plane. Therefore,

$$\frac{\partial h}{\partial s} \approx \frac{\partial h}{\partial x}$$

$$q = \frac{(50+h)}{2} \times 100 \times \frac{(50-h)}{1,500} = \frac{2,500-h^2}{30} \Rightarrow h^2 = 1,375 \Rightarrow h = 37.08$$

h is the height of the phreatic surface at the interface.

$$\text{Also, } q = \frac{h+25}{2} \times 120 \times \frac{h-25}{1,200} = \frac{h^2-625}{20} \Rightarrow q = 37.5 \text{ m}^3/\text{day}$$

$$q = h\bar{K}i = \frac{50+25}{2}\bar{K}\frac{25}{2,700} = 37.5 \text{ m}^3/\text{day} \Rightarrow \bar{K} = 108 \text{ m/day}$$

$$\text{Darcy velocity} = \bar{K}i = 108 \times \frac{25}{2,700} = 1 \text{ m/day}$$

$$\text{Seepage velocity} = \frac{2,700}{675} = 4 \text{ m/day}$$

$$\text{Average porosity} = \frac{1}{4} = 0.25$$

Example 35.2

The following data refer to the drawdown in an observation well located at a distance of 20 m from a production well.

Time (h)	0.5	1.0	1.5	2.0	5.0	10.0	20.0	40.0
Drawdonw (m)	2.0	3.0	3.5	4.0	5.0	6.0	7.0	8.0

The pumping rate is 0.1 m³/s. Determine the storage and transmissivity coefficients of the aquifer.

TRANSMISSIVITY COEFFICIENT, T

$$W(u) = \left\{ -0.5772 - \ln(u) + u - \frac{u^2}{2.2!} + \frac{u^3}{3.3!} + \cdots \right\}$$

$$u = \frac{r^2 S}{4Tt}$$

$$h_0 - h = \frac{Q}{4\pi T} W(u)$$

When u is small (small values of r and large values of t make u small), $W(u)$ can be approximated as

$$W(u) = -0.5772 - \ln(u)$$

Therefore,

$$h_0 - h = \frac{Q}{4\pi T}(-0.5772 - \ln(u)) = \frac{Q}{4\pi T}\left(-0.5772 + \ln\left(\frac{4Tt}{r^2 S}\right)\right) = \frac{Q}{4\pi T}\left(\ln\left(\frac{2.25Tt}{r^2 S}\right)\right)$$

$$= \frac{2.3Q}{4\pi T}\left(\log\left(\frac{2.25Tt}{r^2 S}\right)\right) = \frac{2.3Q}{4\pi T}\left\{\log(2.25Tt) - 2\log(r) - \log(S)\right\}$$

For a given r, this equation represents a straight-line plot of $h_0 - h$ vs. $\log(t)$.

$$\left(h_0 - h\right)_1 - \left(h_0 - h\right)_2 = \frac{Q}{4\pi T}\ln\frac{u_2}{u_1} = \frac{Q}{4\pi T}\ln\frac{t_2}{t_1} = \frac{2.3Q}{4\pi T}\log\frac{t_2}{t_1}$$

For one log cycle, $\log\dfrac{t_2}{t_1} = 1$. Therefore,

$$T = \frac{2.3Q}{4\pi\Delta\left(h_0 - h\right)}$$

From the graph (not shown), $\Delta\left(h_0 - h\right) = 3.2\,\text{m}$. Therefore, $T = 5.72 \times 10^{-3}\ \text{m}^2/\text{s}$.

STORAGE COEFFICIENT, S

When the drawdown is zero,

$$\ln\left(\frac{1}{u}\right) - 0.5772 = 0$$

The time t_0 corresponding to the zero drawdowns from the graph (not shown) is 0.115 h.

$$\ln\left(\frac{4Tt_0}{r^2 S}\right) = 0.5772 \Rightarrow S = 0.0133$$

REFERENCES

Cooper, H. H., Jr. and Jacob, C. E. (1946): A generalized graphical method for evaluating formation constants and summarizing well field history, *Transactions of the American Geophysical Union*, vol. 27, pp. 526–534.

Dupuit, J. (1863): *Estudes Thèoriques et Pratiques sur le mouvement des Eaux dans les canaux dècouverts et à travers les terrains permèables*, (2nd Edition), Dunod, Paris, 304 p.

Freeze, R. A. and Cherry, J. A. (1979): *Groundwater*, Prentice Hall, Englewood Cliffs, NJ, 605 p.

Harr, J. (1995): *A Civil Action*, Random House, Manhattan, NJ, 500 p.

Hazen, A. (1904): On sedimentation, *Transactions of the ASCE*, vol. 53, pp. 45–71.

Jacob, C. E. (1950): Flow of groundwater, Chapter 5, pp. 321–386. In: H. Rouse (ed.), *Engineering Hydraulics*, John Wiley & Sons, New York.

Kramer, H. (1935): Sand mixtures and sand movement in fluvial models, *Transactions of the ASCE*, vol. 100, pp. 798–873.

Ogata, A. and Banks, R. B. (1961): A solution of the differential equation of longitudinal dispersion in Porous Media, *US Geological Survey Professional Paper 411-A*, US Government Printing Office, Washington DC.

Sato, C., Haga, M., and Nishino, J. (2006): Land subsidence and groundwater management in Tokyo, special feature on groundwater management and policy, *International Review for Environmental Strategies*, vol. 6, no. 2, pp. 403–424, Institute for Global Environmental Strategies.

Theis, C. V. (1935): The relation between the lowering of piezometric surface and the rate and duration of discharge of a well using groundwater storage, *Transactions of the American Geophysical Union*, vol. 16, pp. 519–524.

Tison, L. J., ed. (1969): *Land Subsidence - Proceedings of the Tokyo Symposium*, September 1969, vols. 1–2, IAHS Publications, 88–89, pp. 661. http://iahs.info/Publications-News.do.

Thiem, G. (1906): *Hydrologische Methoden*, Gebhardt, Leipzig, 56 p.

Todd, D. K. and Mays, L. W. (2005): *Groundwater Hydrology*, (3rd Edition), John Wiley & Sons, New York, 636 p.

Chapter 36

Statistical methods in hydrology

36.1 INTRODUCTION

The hydrological cycle links all hydrological processes, and, therefore, it is quite logical and natural for a set of hydrological data generated by one hydrological process to have some dependence on similar data generated by another process. For example, river flow or runoff is the outcome of rainfall, and, therefore, runoff is dependent upon rainfall. Likewise, many types of hydrological data will have some dependence on other types of data generated within the broad frame of the hydrological cycle. As a first step in hydrological data analysis, it is therefore quite logical to seek simple relationships between a set of hydrological data that depend upon another set of hydrological data. Statistically, such relationships can be obtained by regression and correlation analysis. For example, runoff is dependent firstly upon rainfall but also upon antecedent conditions reflected by past values of runoff, soil moisture condition, topography, geology, etc. For simplicity, one could seek a simple linear regression between runoff and rainfall but could also look for a multi-linear regression linking several variables. A more sophisticated relationship can be sought via non-linear regression.

Most hydrological processes have some kind of random behaviour, and therefore, statistical concepts are used when it is desired to estimate the likelihood of occurrence or non-occurrence of an event of a certain magnitude. For example, the likelihood of occurrence (probability) of a flood of specified magnitude in a specified period of time which is related to the return period used for the design of any hydraulic structure. Such problems come under frequency analysis in hydrology. The basic procedure involves the following steps:

- Hypothesizing the underlying probability distribution function
- Calculating the sample parameters such as the mean and standard deviation
- Equating the sample parameters to the population parameters
- Constructing a frequency curve that represents the underlying population

The computed frequency curve which represents the population is then used to make probability statements about the likelihood of occurrence or non-occurrence of the random variable such as the magnitude of the flood. For example,

- Finding the probability corresponding to a specific value of the random variable
- Finding the value of the random variable corresponding to a specified probability.

Frequency analysis may be carried out using the probability paper method or the frequency factor method which depends upon the assumed or known probability distribution of the variable involved. Empirical probabilities (or return periods) are estimated from available data using plotting position formulas. These will be highlighted in the section on frequency analysis.

Any statistical method involves some probability distributions. The commonly used discrete probability distributions in hydrology are the Binomial and the Poisson whereas continuous distributions include the Normal (or Gaussian), log-normal (two-parameter and three-parameter), Gamma (three-parameter, two-parameter and one-parameter type), Pearson (Type III), log-Pearson (Type III), Chi-squared, F, and the extreme value distribution. These distributions, their basic properties and how they can be used for frequency analysis will be briefly described in this chapter.

It is also important to note the difference between a population and a sample. A population consists of all possible outcomes of a variable. It may be finite or infinite. A true statistic such as the population mean can be determined only if the population is known. In many instances, however, the population is unknown and information about a sample only is available. Statistical inferences are then made based on the sample information by assuming that the sample information is representative of the population information. This assumption leads to uncertainties in the statistical conclusions made.

Another area in statistics relevant to hydrology is hypothesis testing. Typical examples include testing whether the peak discharge at a gauging station is equal to or greater than a prescribed value and testing whether the concentration of dissolved oxygen in a water body has improved as a result of installing a treatment plant. In such cases, the truth is unknown, and therefore a hypothesis is made and statistical tests are carried out to accept or reject the hypothesis at a chosen confidence level based on the sample data.

In this chapter, the statistical background, including their mathematical formulations, as well as possible errors in analysis and statistical conclusions will be highlighted.

36.2 SOME PRELIMINARIES

Statistics are used in many disciplines, among them Engineering, and Civil Engineering in particular, being a major. For example, the design of a dam to withstand a flood of a certain return period. Statistics are useful for making decisions which otherwise must be made based on experience and subjective judgement. Statistics constitute a class of methods and techniques (theory) applied to a body of information (data) to help make decisions. In any decision which is based on statistics, there is always an element of *uncertainty*.

The main task of modern statistics is to make statements or conclusions about the 'whole' (*population*) from a knowledge of the 'part' (*sample*). Such conclusions are called statistical inferences, which are different from logical deductions. On the other hand, logical deduction of the sample property from knowledge of the population properties (similar to interpolation) is referred to as deductive statistics. Generalization about the population from knowledge of the sample (similar to extrapolation) is referred to as inductive statistics.

36.2.1 Data representation

36.2.1.1 Frequency diagram (distribution)

Raw data refers to the original data without any treatment. For ease of interpretation, they may be arranged in class intervals. The number of observations in each class interval represents the class frequency. Such an arrangement of the raw data represents a frequency distribution.

36.2.1.2 Histogram

A graphical representation of the class frequency against class interval as a bar chart gives a histogram.

36.2.1.3 Frequency polygon

When the class frequency is plotted against the mid-point of the class interval, the resulting diagram is called a frequency polygon. Frequency curve is a smoothed version of either the frequency polygon or the histogram. It is usually represented as a relative frequency. The lower the class interval, the smoother the frequency curves. A summation of the relative frequency distribution gives the cumulative frequency distribution. The latter is sometimes called the ogive.

36.2.1.4 Stem and leaf plots

First digit (s) in the value of data is placed on the 'stem' and the second digit (s) on the 'leaf' (Table 36.1 and Figure 36.1). The main advantage of this representation is that it becomes a sort of histogram when rotated through 90 degrees. They can also be placed back to back when displaying different groups of the same data set, e.g. male and female incomes in a company. The data may be ordered or unordered.

Table 36.1 Annual rainfall (mm) measured at the Hong Kong Observatory for the period 1884–1941 (58 years)

Years	Annual rainfall (mm)	Years	Annual rainfall (mm)	Years	Annual rainfall (mm)
1884	1915.8	1904	2042.5	1924	2503.2
1885	2766.7	1905	1802.1	1925	2224.4
1886	1756.9	1906	1975.9	1926	2559.6
1887	1684.3	1907	2376.0	1927	2740.0
1888	3656.5	1908	2333.7	1928	1807.2
1889	2040.7	1909	1923.6	1929	1773.5
1890	1801.8	1910	1781.0	1930	2440.3
1891	2974.7	1911	2300.0	1931	2041.8
1892	2310.6	1912	1624.1	1932	2323.3
1893	2538.9	1913	2126.8	1933	1583.5
1894	2648.1	1914	2545.3	1934	2480.6
1895	1164.3	1915	1931.1	1935	1811.6
1896	1848.8	1916	2028.3	1936	1772.2
1897	2540.7	1917	2069.7	1937	2095.5
1898	1448.6	1918	2580.8	1938	1406.0
1899	1846.7	1919	1934.0	1939	2202.3
1900	1872.7	1920	2740.0	1940	2989.2
1901	1417.0	1921	2472.2	1941	2433.1
1902	2476.4	1922	1763.6		
1903	2378.7	1923	2711.1		

10	
11	64.3
12	
13	
14	06.0 17.0 48.6
15	83.5
16	24.1 84.3
17	56.9 63.6 72.2 73.5 81.0
18	01.8 02.1 07.2 11.6 46.7 48.8 72.7
19	15.8 31.1 34.0 23.6 75.9
20	28.3 40.7 41.8 42.5 69.7 95.5
21	26.8
22	02.3 24.4
23	00.0 10.6 23.3 33.7 76.0 78.7
24	33.1 40.3 72.2 76.4 80.6
25	03.2 38.9 40.7 45.3 59.6 80.8
26	48.1
27	11.0 40.0 40.0 66.7
28	
29	74.7 89.2
30	
31	
32	
33	
34	
35	
36	56.5

Figure 36.1 Stem and leaf plots.

36.2.1.5 Pie diagrams

Pie diagrams are used mainly to display qualitative data. Bar charts can also be used to represent qualitative data.

36.2.1.6 Scatter diagram

A graphical representation of two (or more) variables, one dependent on the other(s) is called a scatter diagram. It is a multi-dimensional data display.

36.2.1.7 Box plots

Box plots (Figure 36.2) are diagrams containing the minimum, first quartile, second quartile (median), third quartile and the maximum.

36.2.2 Statistical measures of location

Measures of location are characterized by the central tendency or positions other than the centre.

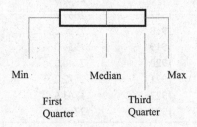

Figure 36.2 Box plots.

36.2.2.I Mean – a measure of the central tendency

Arithmetic sample mean is given as

$$\bar{x} = \frac{1}{n}\sum_{1}^{n} x_i \tag{36.1}$$

Weighted average sample mean is given as

$$\bar{x} = \frac{1}{n}\sum f_k x_k \tag{36.2}$$

where $n = \sum f_k$. It is equivalent to summing the products of each class mid-point and relative frequency $\frac{f_k}{n}$.

There are other 'means' such as the geometric mean which refers to the nth root of the products of n numbers and the harmonic mean which refers to the number of data divided by the summation of the inverse of each data.

36.2.2.I Median – another measure of the central tendency

The median divides the number of observations into two halves – one half above the median value and the other half below the median value. For odd samples, it is the value of the middle size whereas for even samples it is found by averaging the two middle values.

The median is also an average of position. For skewed data, or data with outliers, it may give better meaning than the mean. It is insensitive to extreme values whereas the mean is dependent on all values.

36.2.2.3 Mode

The mode is also a measure of the location, and it refers to the most frequently occurring value. When no single value occurs more than once, the mid-point of the class interval with the highest frequency is taken as the mode.

The relative location of the mean, median and the mode depend upon the frequency distribution forms. In symmetric distributions, the mean, median and the mode coincide. In negatively skewed distributions (skewed to the left), the mean is the lowest, followed by median and mode. In positively skewed distribution (skewed to the right), the mean is the highest preceded by the median and the mode.

Bimodal distributions have two modes. For example, when the heights of population in a class are taken without considering the gender, the distribution will be bimodal if males and females are represented equally. Bimodal distribution implies an underlying non-homogeneity in the raw data.

36.2.3 Percentiles, fractiles and quartiles

These are also location measures concerned with the frequency distribution. Percentile is a point below which a certain percentage of the data lie. For example, 75th percentile refers to the value below which 75% of the data lie. Fractile conveys the same information as the percentile. For example, 75th percentile is the 0.75 fractile; 65th percentile is the 0.65 fractile. The median is the 50th percentile or the 0.5 fractile.

Quartiles are defined by splitting the frequency distribution into four groups of equal frequency. Thus, first quartile is the same as the 25th percentile; second quartile is the same as the 50th percentile and the third quartile is the same as the 75th percentile.

Percentiles, fractiles and quartiles can be easily determined from a graphical plot of the cumulative frequency distribution.

Interquartile range represents the scatter in the middle 50% of the data, e.g. $Q_{0.75} - Q_{0.25}$.

36.2.4 Statistical measures of variability

The measures of variability represent important average descriptors of the data next to the measures of location. Variability or dispersion provides an additional insight to the data which is not contained in the measures of location such as the mean.

36.2.4.1 Range

This is the simplest measure of the variability and is the difference between the largest and the smallest data values. The presence of outliers can distort the effect of the range.

36.2.4.2 Variance and standard deviation

Population variance σ^2 is given as

$$\sigma^2 = \frac{1}{n} \sum (x_i - \mu)^2 \tag{36.3a}$$

Sample variance is given as

$$s^2 = \frac{1}{n-1} \sum (x_i - \bar{x})^2 \tag{36.3b}$$

The square root of the variance is the standard deviation. It has the same units as that of the data.

For grouped data, the variance is approximately given by

$$s^2 = \frac{1}{n-1} \sum f_k x_k^2 - n\bar{x}^2 \tag{36.3c}$$

According to Chebyshev's theorem, the following can be said:

- At least 75% of the population values lie within $\mu \pm 2\sigma$
- At least 89% of the population values lie within $\mu \pm 3\sigma$
- At least 96% of the population values lie within $\mu \pm 5\sigma$.

When the shapes of the distributions are known, these limits can be narrowed.

36.2.4.3 Coefficient of variation, c_v

$$c_v = \frac{s}{\bar{x}} \tag{36.4}$$

36.2.4.4 Coefficient of skewness, s_k

$$s_k = \frac{1}{s} 3(\bar{x} - m) \tag{36.5a}$$

where m is the median. $s_k > 0$ means positive skew; $s_k < 0$ means negative skew.
Or

$$s_k = \frac{\frac{1}{n} \sum (x_i - \bar{x})^3}{s^3} \tag{36.5b}$$

36.2.4.5 Coefficient of kurtosis (measure of peakedness)

$$\kappa = \frac{m_4}{m_2^2} \tag{36.6}$$

where m_4 and m_2 refer to the fourth and second moments (defined below). Kurtosis excess ε is defined as

$$\varepsilon = \kappa - 3 \tag{36.7}$$

- When $\varepsilon > 0$, i.e. high peak, the distribution is called leptokurtic
- When $\varepsilon = 0$, i.e. normal peak, the distribution is called mesokurtic
- When $\varepsilon < 0$, i.e. low peak, the distribution is called platykurtic.

In general, the moments of the distribution are defined as

$$m_r = \frac{1}{n} \sum x_i^r \quad \text{(about zero)} \tag{36.8a}$$

$$m_r = \frac{1}{n} \sum (x_i - \bar{x})^r \quad \text{(about the mean)} \tag{36.8b}$$

In Eq. 36.8, $r = 1$ gives the mean; $r = 2$ gives the variance; $r = 3$ gives a measure of the skewness; $r = 4$ gives a measure of the kurtosis.

36.3 REGRESSION AND CORRELATION

The process of estimation of a dependent variable from the independent variable(s) is called regression. When y is to be estimated from x, it is called a regression of y on x.

Regression analysis involves finding a relationship between two (or more) variables. At least one variable is usually fixed (independent) while the other depends upon the magnitude(s) of the fixed variable(s). Regression relationships are useful

- for summarizing the data in terms of a few parameters,
- to determine any functional relationship between the variables involved,
- for prediction purposes, and,
- for interpolation and extrapolation.

Correlation refers to the joint variation of two variables independently.

A plot of the values of the dependent variable against the values of the independent variable is called a scatter diagram. Each point on the scatter diagram represents one observation, i.e. a pair of values if it involves only two variables.

Visual observation of the scatter diagram may very often indicate a definitive pattern of behaviour. This may easily be approximated by a free hand curve or a straight line. A more accurate representation can be obtained by fitting a regression equation to the set of data, which, if linear, will lead to a straight line.

36.3.1 Linear regression

A simple linear regression equation can be written in the form

$$y_c = a + bx \tag{36.9}$$

where y_c is the value of y corresponding to the value of x as computed by the regression equation, and, a and b are the regression coefficients.

The vertical distance between an individual point and the regression line is considered as the error 'e', i.e. $e = y - y_c$.

More than one line can be fitted to a given set of data. The best fitting curve is obtained by using some criterion. The method of minimizing the error is called the least squares method, and the resulting regression equation is called the least squares regression equation.

Sometimes, other criteria can be used. For example, another least square error can be defined by considering the normal distance between the point and the regression line.

The coefficients 'a' and 'b' are obtained by minimizing the square of the error, i.e.

$$E = \sum (\text{error})^2 = \sum (y - y_c)^2 = \sum (y - a - bx)^2 \tag{36.10}$$

Minimizing the error can be done by setting the partial derivatives of the error with respect to the two constants a and b to zero as given below:

$$\frac{\partial E}{\partial a} = 0 \Rightarrow -2 \sum [y - a - bx] = 0 \tag{36.11a}$$

$$\frac{\partial E}{\partial b} = 0 \Rightarrow -2 \sum [y - a - bx] x = 0 \tag{36.11b}$$

Equations 36.11a,b are called the normal equations, and in simplified form they are

$$\sum a + b \sum x = \sum y \tag{36.12a}$$

$$a \sum x + b \sum x^2 = \sum xy \tag{36.12b}$$

or, using the condition that $\sum a = na$

$$\begin{bmatrix} n & \sum x \\ \sum x & \sum x^2 \end{bmatrix} \begin{bmatrix} a \\ b \end{bmatrix} = \begin{bmatrix} \sum y \\ \sum xy \end{bmatrix}$$

which, when solved give

$$b = \frac{\sum xy - \dfrac{\sum x \sum y}{n}}{\sum x^2 - \left(\sum x\right)^2 / n} = \frac{n\sum xy - \sum x \sum y}{n\sum x^2 - \left(\sum x\right)^2} \tag{36.13a}$$

$$a = \frac{\sum y \sum x^2 - \sum x \sum xy}{n\sum x^2 - \left(\sum x\right)^2} \tag{36.13b}$$

In Eqs. 36.12 and 36.13, n represents the number of data points. The two solutions given in Eq. 36.13 are also equivalent[1] to

$$b = \frac{\sum (x - \bar{x})(y - \bar{y})}{\sum (x - \bar{x})^2} \tag{36.14a}$$

where $\bar{x}$ and $\bar{y}$ are the means of x and y, and,

$$a = \frac{\sum y}{n} - b\frac{\sum x}{n} = \bar{y} - b\bar{x} \tag{36.14b}$$

Equation 36.14b is obtained by dividing Eq. 36.12a by n.

36.3.1.1 Standard error of estimate

The sample standard error of y on x, $s_{y,x}$ is defined[2] as

$$s_{y,x} = \sqrt{\frac{\sum (y - y_c)^2}{n}} \tag{36.15}$$

It is analogous to the standard deviation. For example, if lines parallel to the regression line at vertical distances $s_{y,x}$, $2s_{y,x}$ and $3s_{y,x}$ are constructed, then, if n is large, approximately 68%, 95% and 99.7% respectively of the points would lie between the respective lines.

The definition in Eq. 36.15 is a biased estimate. An unbiased[3] estimate is obtained by multiplying Eq. 36.15 by $\sqrt{\dfrac{n}{(n-2)}}$. The factor 2 corresponds to the number of degrees of

---

[1] $\sum (x - \bar{x})(y - \bar{y}) = \sum xy - \sum x\bar{y} - \sum \bar{x}y + \sum \bar{x}\bar{y} = \sum xy - \bar{x}\sum y - \bar{y}\sum x + \sum \bar{x}\bar{y}$

$$= \sum xy - \bar{x}n\bar{y} - \bar{y}n\bar{x} + n\bar{x}\bar{y} = \sum xy - n\bar{x}\bar{y} = \sum xy - \frac{\sum x \sum y}{n}$$

$$\sum (x - \bar{x})^2 = \sum x^2 - 2\bar{x}\sum x + \sum \bar{x}^2 = \sum x^2 - 2\bar{x}^2 n + n\bar{x}^2 = \sum x^2 - \frac{\left(\sum x\right)^2}{n}$$

[2] $s_{y,x}$ is not the same as s_{yx}, the cross correlation between x, y.
[3] When the expected (mean) value of a statistic is equal to the corresponding population parameter, the statistic is called an unbiased estimate. It is obtained by multiplying the biased estimate by the factor $\dfrac{n}{(n - \text{dof})}$ where dof indicate the degrees of freedom.

freedom which is equal to the number of parameters to be determined (a and b in this case). The corresponding population standard error of estimate is denoted by $\sigma_{y,x}$. The standard error of estimate is also referred to as the sample standard deviation of regression.

36.3.1.2 Sample covariance and correlation coefficient

The sample covariance of x and y is given by

$$s_{xy} = \frac{\sum (x - \bar{x})(y - \bar{y})}{n} \qquad (36.16)$$

The correlation coefficient is defined as

$$r = \frac{s_{xy}}{s_x s_y} \qquad (36.17)$$

where s_x and s_y are the standard deviations of x and y, respectively.

The correlation coefficient gives a measure of the dependence of the two variables. If the covariance is zero, then there is no correlation between the two variables. The two variables are then uncorrelated but may or may not be independent. The correlation coefficient has the sign of b.

It can also be shown that[4]

$$s_{y,x}^2 = s_y^2 \left(1 - r^2\right) \qquad (36.18)$$

---

[4] $\displaystyle \sum (y - y_c)^2 = \sum (y - a - bx)^2 = \sum \left[y - (\bar{y} - b\bar{x}) - bx\right]^2 = \sum \left[(y - \bar{y}) - b(x - \bar{x})\right]^2$

$$= \sum \left\{ (y - \bar{y})^2 - 2b(x - \bar{x})(y - \bar{y}) + [b(x - \bar{x})]^2 \right\}$$

$$= \left\{ \sum (y - \bar{y})^2 - \sum 2b(x - \bar{x})(y - \bar{y}) + \sum bb(x - \bar{x})^2 \right\}$$

But $\displaystyle bb \sum (x - \bar{x})^2 = \frac{b \sum (x - \bar{x})(y - \bar{y})}{\sum (x - \bar{x})^2} \sum (x - \bar{x})^2 = b \sum (x - \bar{x})(y - \bar{y})$

Therefore, $\displaystyle \sum (y - y_c)^2 = \left\{ \sum (y - \bar{y})^2 - b \sum (x - \bar{x})(y - \bar{y}) \right\}$, and

$$s_{y,x}^2 = \frac{\sum (y - \bar{y})^2}{n} - \frac{b \sum (x - \bar{x})(y - \bar{y})}{n} = s_y^2 - b s_{xy}$$

$$= s_y^2 - b r s_x s_y = s_y^2 - \frac{\sum (x - \bar{x})(y - \bar{y})}{\sum (x - \bar{x})^2} \frac{\sum (x - \bar{x})(y - \bar{y})}{n}$$

$$= s_y^2 - \frac{\sum [(x - \bar{x})(y - \bar{y})]^2}{n^2} \frac{n^2}{n \sum (x - \bar{x})^2}$$

$$= s_y^2 - s_{xy}^2 \frac{n}{(x - \bar{x})^2} = s_y^2 \left\{ 1 - r^2 s_x^2 \frac{n}{\sum (x - \bar{x})^2} \right\} = s_y^2 (1 - r^2).$$

from which

$$r^2 = 1 - \frac{s_{y,x}^2}{s_y^2} = 1 - \frac{\sum(y - y_c)^2}{\sum(y - \bar{y})^2} \qquad (36.19)$$

It follows that $r^2 \leq 1$, i.e. $-1 \leq r \leq 1$.

36.3.1.3 Regression error

Any regression equation is an approximation. The least squares method minimizes the error. It is possible to express the total error as a sum of two components as follows:

Total error = Explained error + unexplained error

which is equivalent to

$$(y - \bar{y}) = (y_c - \bar{y}) + (y - y_c) \qquad (36.20)$$

The total error is considered as the difference between the individual y value and it's mean $\bar{y}$ which is the estimator of y when no regression is used. The explained error can be thought of as the amount of error removed as a result of the regression equation. The difference between the total error and the explained error is identified as the unexplained error, i.e. the error still remaining after fitting the regression equation.

By squaring both sides of Eq. 36.20 and taking summation,

$$\sum(y - \bar{y})^2 = \sum(y_c - \bar{y})^2 + \sum(y - y_c)^2 + 2\sum(y - y_c)(y_c - \bar{y}) \qquad (36.21a)$$

The last term of Eq. 36.20 can be written as

$$\sum[y - a - bx][a + bx - \bar{y}] = a\sum(y - a - bx) + b\sum x(y - a - bx) - \bar{y}\sum(y - a - bx) = 0$$

because of the normal equations (Eq. 36.11).

Therefore,

$$\sum(y - \bar{y})^2 = \sum(y_c - \bar{y})^2 + \sum(y - y_c)^2 \qquad (36.21b)$$

Dividing Eq. 36.22 by $\sum(y - \bar{y})^2$ and comparing with Eq. 36.19

$$r^2 = \frac{\sum(y_c - \bar{y})^2}{\sum(y - \bar{y})^2} = \frac{\text{Explained variance}}{\text{Total variance}} = \text{Coefficient of determination, } R^2. \qquad (36.22)$$

It is to be noted that for linear regression, the coefficient of determination R^2 is equal to the square of the correlation coefficient r^2.

36.3.1.4 Interpretation of $\sigma^2_{y,x}$ and the coefficient of determination R^2

From Eq. 36.15, the minimum value of the variance occurs when each individual value of y coincides with the corresponding value of y_c ($y = y_c$). This implies a perfect fit. The coefficient of determination, R^2, then is unity.

When $y_c = \bar{y}$, there is no improvement due to the regression line. The unexplained error is a maximum and is equal to the total error. The explained error is zero. The coefficient of determination, R^2, then is zero.

The coefficient of determination

- gives a measure of the amount of improvement due to the regression line,
- measures the closeness of fit of the regression line to the points, and,
- measures the degree of linearity of the scatter of points.

when R^2 is close to unity, the points are very close to the straight line and vice versa.

36.3.1.5 Variance of the regression coefficients 'a' and 'b'

Each sample of the population gives a regression equation which may be different from the population regression equation. It is, therefore, necessary to determine the variances of the regression coefficients as a means of measuring how close the sample regression lines are in relation to the population regression line. If the variation is large, then the sample lines are less reliable estimates of the population line. The variances of 'a' and 'b' can be defined as

$$\sigma^2_a = \left(\frac{1}{M}\right)\sum_{}^{M}\left[a - E(a)\right]^2 \tag{36.23a}$$

$$\sigma^2_b = \left(\frac{1}{M}\right)\sum_{}^{M}\left[b - E(b)\right]^2 \tag{36.23b}$$

where M is the number of all possible samples of size 'n' selected from a population of size N and $E(a)$, $E(b)$ represent the expected values of 'a' and 'b'.

However, it is not possible to use Eq. 36.23 to estimate σ^2_a and σ^2_b because $E(a)$ and $E(b)$ are not known and that M can be very large. Instead, the following relationships are used (they have theoretical bases). (Wonnacott and Wonnacott, 1990; Scheaffer and McClave, 1990):

$$\sigma^2_a = \frac{\left[\sigma^2_{yx}\right]\left[\sum x^2\right]}{\left[n\sum(x - \bar{x})^2\right]} \tag{36.24a}$$

$$\sigma^2_b = \frac{\sigma^2_{yx}}{\sum(x - \bar{x})^2} \tag{36.24b}$$

where σ^2_{yx} is the variance of the error, and the summation is taken over the sample.

The estimators of σ^2_a and σ^2_b are therefore

$$s^2_a = \frac{\left[s^2_{yx}\right]\left[\sum x^2\right]}{n\sum(x - \bar{x})^2} \tag{36.25a}$$

$$s_b^2 = \frac{\left[s_{yx}^2\right]}{\left[\sum (x - \bar{x})^2\right]}$$

(36.25b)

It can be seen that as the sample size increases, $\Sigma(x - \bar{x})^2$ will become larger and larger and hence s_a^2 and s_b^2 will become smaller and smaller. Also, if s_{yx} is small implying the variability of y being small, s_b will become small. By increasing s_x, i.e. the spread of x values, the reliability of s_b is increased. The sample regression lines will then tend to get closer and closer to the population regression line.

36.3.1.6 Confidence intervals for the expected values of 'a' and 'b'

The confidence intervals for 'α' and 'β' which are the expected values of a and b can be established if some assumptions about the probability distribution of the error 'e' (= $y - y_c$) are made. In linear regression, it is generally assumed[5] that the arithmetic mean of 'e' is zero, the variance of 'e' (Var(e)) is independent of the values of x and that 'e' is normally distributed. i.e. $e = N(o, \sigma_{yx}^2)$. Then, if α and β are the expected values of 'a' and 'b',

$$a \cong N(\alpha, \sigma_a^2)$$

(36.26a)

$$b \cong N(\beta, \sigma_b^2)$$

(36.26b)

and the statistics

$$T = \frac{(a - \alpha)}{\sigma_a}, \text{ and } T = \frac{(b - \beta)}{\sigma_b}$$

(36.27)

follow a Student's 't' distribution with $(n - 2)$ degrees of freedom.

The $(1 - \alpha^*)$ confidence intervals[6] for 'α' and 'β' are given by

$$a \pm \sigma_a \left[t_{\frac{\alpha}{2}}^* (n - 2) \right] \text{ for } \alpha$$

(36.28a)

$$b \pm \sigma_b \left[t_{\frac{\alpha}{2}}^* (n - 2) \right] \text{ for } \beta$$

(36.28b)

α^* is not the same α in Eqs. 36.27 and 36.28.

---

[5] A very simple sample is one whose N observations $x_1, x_2, ..., x_n$ are independent. The distribution of each x is the population distribution $p(x)$, i.e. $p(x) = p(x_1) = p(x_2) = ... = p(x_n)$. Then, each observation has the mean μ and the standard deviation σ of the population.

$$\bar{x} = \frac{1}{n}[x_1 + x_2 + \cdots + x_n]; \quad E(\bar{x}) = \frac{1}{n}[E(x_1) + E(x_2) + \cdots + E(x_n)] = \frac{1}{n}[\mu + \mu + \cdots + \mu] = \mu$$

$$\text{Var}(\bar{x}) = \frac{1}{n^2}[\text{Var}(x_1) + \text{Var}(x_2) + \cdots + \text{Var}(x_n)] = \frac{1}{n^2}\sigma^2 n = \frac{\sigma^2}{n}$$

[6] When $\alpha^* = 0.05$, the 95% confidence interval is considered. For a two-tailed test, this is 2.5% on each side $\left(\frac{\alpha^*}{2} = 0.025\right)$. It means that $Pr(a - 1.86\sigma_a < \alpha < a + 1.86\sigma_a) = 95\%$ where the 1.86 is taken from the 't' distribution at the 5% significance level. It must be noted that α, β are constants and that a, b are random variables. It is the confidence interval that can vary. The more confident 99% level gives a wider range implying a vaguer statement. This is expected because the more certain we want to be about a statement, the more vague we must make it.

These confidence intervals can be used to test the validity of hypotheses about 'α' and 'β'.

For example, the hypothesis that $\alpha = 0$ can be tested by taking the critical values of the 't' distribution at the α^* level of significance with $(n - 2)$ degrees of freedom. If the value of 'a' lies between the interval

$$\pm \sigma_a \left[t^*_{\frac{\alpha}{2}}(n-2) \right]$$

the hypothesis that $\alpha = 0$ can be accepted at α^* level of significance. The same criterion applies for β also. It is also possible to use other values of α and β.

36.3.1.7 Prediction of 'y' at a given level of x

Predicted value of y by sample regression line at $x = x_0$ is

$$y_c = a + bx_0 \tag{36.29}$$

Predicted value of y by population regression line at $x = x_0$ is

$$E(y_c) = \alpha + \beta x_0 \tag{36.30}$$

Different samples give different regression lines and hence different values of y_c at x_0.

$$\text{Var}(y_c) = \sigma^2_{yx} \left[\frac{1}{n} + \frac{(x_0 - \bar{x})^2}{\sum (x - \bar{x})^2} \right] \tag{36.31}$$

The 95% confidence interval for the mean of y_c at $x = x_0$ is

$$(a + bx_0) \pm t_{0.025} \sqrt{\text{Var}(y_c)} \tag{36.32}$$

36.3.2 Multiple linear regression

Multiple linear regression is an extension to simple regression where more than one independent variable is considered. A multiple linear regression equation can be written in the form

$$y_c = a + b_1 x_1 + b_2 x_2 \tag{36.33}$$

where a, b_1 and b_2 are the partial regression coefficients. It represents a plane in a 3-D space, and the coefficients b_1 and b_2 represent the slopes of the planes when one independent variable is kept constant. The error as before is $(y - y_c)$ and the squares of the sum of errors is given by

$$E = \sum (\text{error})^2 = \sum (y - y_c)^2 = \sum (y - a - b_1 x_1 - b_2 x_2)^2 \tag{36.34}$$

The normal equations are obtained by minimizing E with respect to a, b_1 and b_2:

$$\frac{\partial E}{\partial a} = 0 \Rightarrow 2 \sum (y - a - b_1 x_1 - b_2 x_2)(-1) = 0 \tag{36.35a}$$

$$\frac{\partial E}{\partial b_1} = 0 \Rightarrow 2 \sum (y - a - b_1 x_1 - b_2 x_2)(-x_1) = 0 \tag{36.35b}$$

$$\frac{\partial E}{\partial b_2} = 0 \Rightarrow 2\sum (y - a - b_1 x_1 - b_2 x_2)(-x_2) = 0 \tag{36.35c}$$

which simplify to

$$\begin{bmatrix} n & \sum x_1 & \sum x_2 \\ n\sum x_1 & \sum x_1^2 & \sum x_2 x_1 \\ n\sum x_2 & \sum x_1 x_2 & \sum x_2^2 \end{bmatrix} \begin{bmatrix} a \\ b_1 \\ b_2 \end{bmatrix} = \begin{bmatrix} \sum y \\ \sum y x_1 \\ \sum y x_2 \end{bmatrix} \tag{36.36}$$

For computational purposes, it is convenient to write these equations by using the following properties:

$$\sum (x - \bar{x}) = \sum x - n\bar{x} = \sum x - \sum x = 0 \tag{36.37}$$

If the origin of the above equations is shifted from $(0, 0, 0)$ to $\bar{y}, \bar{x}_1, \bar{x}_2$, then, the normal equations can be written as

$$\begin{bmatrix} n & \sum (x_1 - \bar{x}_1) & \sum (x_2 - \bar{x}_2) \\ n\sum (x_1 - \bar{x}_1) & \sum (x_1 - \bar{x}_1)^2 & \sum (x_2 - \bar{x}_2)(x_1 - \bar{x}_1) \\ n\sum (x_2 - \bar{x}_2) & \sum (x_1 - \bar{x}_1)(x_2 - \bar{x}_2) & \sum (x_2 - \bar{x}_2)^2 \end{bmatrix} \begin{bmatrix} a \\ b_1 \\ b_2 \end{bmatrix}$$

$$= \begin{bmatrix} \sum (y - \bar{y}) \\ \sum (y - \bar{y})(x_1 - \bar{x}_1) \\ \sum (y - \bar{y})(x_2 - \bar{x}_2) \end{bmatrix}$$

But from Eq. 36.37,

$$\sum (x_1 - \bar{x}_1) = 0; \quad \sum (x_2 - \bar{x}_2) = 0; \quad \sum (y - \bar{y}) = 0 \tag{36.38}$$

Therefore, the normal equations simplify to

$$\begin{bmatrix} \sum (x_1 - \bar{x}_1)^2 & \sum (x_2 - \bar{x}_2)(x_1 - \bar{x}_1) \\ \sum (x_1 - \bar{x}_1)(x_2 - \bar{x}_2) & \sum (x_2 - \bar{x}_2)^2 \end{bmatrix} \begin{bmatrix} b_1 \\ b_2 \end{bmatrix} = \begin{bmatrix} \sum (y - \bar{y})(x_1 - \bar{x}_1) \\ \sum (y - \bar{y})(x_2 - \bar{x}_2) \end{bmatrix}$$

Then, changing the variables,

$$X_1 = x_1 - \bar{x}_1 \tag{36.39a}$$

$$X_2 = x_2 - \bar{x}_2 \tag{36.39b}$$

$$Y = y - \bar{y}, \tag{36.39c}$$

The normal equations become

$$b_1 \sum X_1^2 + b_2 \sum X_1 X_2 = \sum X_1 Y \tag{36.40a}$$

$$b_1 \sum X_1 X_2 + b_2 \sum X_2^2 = \sum X_2 Y \tag{36.40b}$$

The solutions to Eq. 36.40 are given by

$$b_1 = \frac{\sum X_2 Y \sum X_1 X_2 - \sum X_1 Y \sum X_2^2}{\sum X_1 X_2 \sum X_1 X_2 - \sum X_1^2 \sum X_2^2} \tag{36.41a}$$

$$b_2 = \frac{\sum X_1 Y \sum X_1 X_2 - \sum X_2 Y \sum X_1^2}{\sum X_1 X_2 \sum X_1 X_2 - \sum X_1^2 \sum X_2^2} \tag{36.41b}$$

and, the regression line becomes

$$Y_c = b_1 X_1 + b_2 X_2 \tag{36.42}$$

which, after substituting from Eq. 36.39 is

$$(y_c - \bar{y}) = b_1 (x_1 - \bar{x}_1) + b_2 (x_2 - \bar{x}_2) \tag{36.43}$$

Equation 36.43 may be written as

$$y_c = (\bar{y} - b_1 \bar{x}_1 - b_2 \bar{x}_2) + b_1 x_1 + b_2 x_2 \tag{36.44}$$

Therefore,

$$a = \bar{y} - b_1 \bar{x}_1 - b_2 \bar{x}_2 \tag{36.45}$$

36.3.2.1 Multiple linear regression with a = 0

Sometimes it is necessary to force regression lines or planes pass through the origin thus making the intercept equal to zero. In such situations, the multiple regression equation can be written as

$$y_c = b_1 x_1 + b_2 x_2 \tag{36.46}$$

As before, the normal equations are

$$2 \sum (y - b_1 x_1 - b_2 x_2)(-x_1) = 0 \tag{36.47a}$$

$$2\sum (y - b_1x_1 - b_2x_2)(-x_2) = 0 \qquad (36.47\text{b})$$

which can be written as

$$b_1 \sum x_1^2 + b_2 \sum x_1x_2 = \sum x_1y \qquad (36.48\text{a})$$

$$b_1 \sum x_1x_2 + b_2 \sum x_2^2 = \sum x_2y \qquad (36.48\text{b})$$

which gives

$$b_1 = \frac{\sum x_2y \sum x_1x_2 - \sum x_1y \sum x_2^2}{\sum x_1x_2 \sum x_1x_2 - \sum x_1^2 \sum x_2^2} \qquad (36.49\text{a})$$

$$b_2 = \frac{\sum x_1y \sum x_1x_2 - \sum x_2y \sum x_1^2}{\sum x_1x_2 \sum x_1x_2 - \sum x_1^2 \sum x_2^2} \qquad (36.49\text{b})$$

The only difference between Eqs. 36.49 and 36.41 is the co-ordinate transformation of Eq. 36.39. The x's in Eq. 36.49 are the original values, whereas the X's in Eq. 36.39 are the deviations.

36.3.2.2 Coefficient of multiple determination

The coefficient of multiple determination is a measure of the closeness of fit of the regression plane to the actual points. It also gives an indication of the significance of the partial regression coefficients.

A horizontal plane passing through the points $(\bar{y}, \bar{x}_1, \bar{x}_2)$ corresponds to the mean line $\bar{y}$ in linear regression. Any improvement brought about by the regression is measured with respect to this plane.

$$\text{Total error} = y - \bar{y} = (y_c - \bar{y}) + (y - y_c) = \text{Explained error} + \text{unexplained error} \qquad (36.50\text{a})$$

As before,

$$\sum (y - \bar{y})^2 = \sum (y_c - \bar{y})^2 + \sum (y - y_c)^2 \qquad (36.50\text{b})$$

and

$$R^2 = \frac{\text{Explained error}}{\text{Total error}} = \frac{\sum (y_c - \bar{y})^2}{\sum (y - \bar{y})^2} = 1 - \frac{\sum (y - y_c)^2}{\sum (y - \bar{y})^2} \qquad (36.50\text{c})$$

Coefficient of multiple correlation = R.

36.3.2.3 Adjusted coefficient of multiple determination

Very often, the R^2 value is positively biased and tends to overestimate the true value. Adjusted R_*^2 is given by

$$R_*^2 = 1 - \frac{(n-1)}{(n-k-1)} \frac{\sum (y - y_c)^2}{\sum (y - \bar{y})^2} \tag{36.50d}$$

where k is the number of partial regression coefficients (number of b_i's). The adjusted $R_*^2 < R^2$.

36.3.2.4 Significance of the regression coefficients b_1 and b_2

The null hypothesis is that the population regression coefficients β_1 and β_2 are zero. This may be stated in two ways:

- $H_0 : \beta_1 = 0$ and $H_0 : \beta_2 = 0$
  $H_a : \beta_1 \neq 0$ and $H_a : \beta_2 \neq 0$
  (taking each regressor separately)
  A two-tailed test is carried out using the 't' distribution.
- Find whether $\beta_1 = \beta_2 = \beta_3 = \cdots = 0$
  (for the entire regression equation)
  The F-test is used where

$$F = \frac{\dfrac{\sum (y_c - \bar{y})^2}{k}}{\dfrac{\sum (y - y_c)^2}{(n-k-1)}} \tag{36.51}$$

To accept or reject the null hypothesis, the F value above is compared with the critical value of the F-distribution at a chosen level of significance.

36.3.3 Non-linear regression

It is sometimes possible to find a non-linear relationship between the variables involved. For two variables, the first approximate relationship is of the form

$$y_c = ax^2 + bx + c \tag{36.52}$$

The normal equations are obtained by taking summations:

$$\sum y = a \sum x^2 + b \sum x + c \tag{36.53a}$$

Multiplying by the coefficient of a,

$$\sum x^2 \sum y = a \sum x^4 + b \sum x^3 + c \sum x^2 \tag{36.53b}$$

Multiplying by the coefficient of b,

$$\sum x \sum y = a \sum x^3 + b \sum x^2 + c \sum x \qquad (36.53c)$$

Equation 36.53 can be solved for a, b and c.

36.3.4 Correlation

The mutual relationship between two variables, which can vary independently, is measured by the correlation coefficient which was defined in Eq. 36.17. It is a measure of the linear relationship between the two variables and lies between ± 1. A positive value indicates a tendency for the two variables to increase together whereas a negative value indicates that one variable decreases when the other increases.

Other forms of definitions equivalent to Eq. 36.17 for the correlation coefficient are available. One such form often used for computational purposes is

$$r = \frac{N \sum xy - (\sum x)(\sum y)}{\sqrt{\left[N \sum x^2 - (\sum x)^2\right]\left[N \sum y^2 - (\sum y)^2\right]}} \qquad (36.54)$$

36.3.4.1 Sampling theory of correlation

Like the regression problem, it is often desired that the population correlation coefficient ρ be estimated from the sample correlation coefficient 'r'. The sampling distribution of 'r' must be known for this purpose. If $\rho = 0$, the sampling distribution is symmetric and the Student's 't' can be used for hypothesis testing if the sample size is small (<30). If $\rho \neq 0$, the distribution is skewed. In this case, a transformation called Fisher's z transformation gives a statistic which is approximately normally distributed.

36.3.4.2 Test of hypothesis that $\rho = 0$: (i.e. no association)

The statistic 't' defined below will follow a Student's 't' distribution with $(n - 2)$ degrees of freedom. The sampling distribution is symmetric when $\rho = 0$ and has a variance of $\dfrac{(1 - r^2)}{(n - 2)}$. Then the statistic t is

$$t = \left[\frac{r\sqrt{(n-2)}}{\sqrt{(1-r^2)}}\right]. \qquad (36.55).$$

The critical values of r at various rejection levels and degrees of freedom are given in statistical tables. For example, for $n = 11$, dof = 9, at 5% significance level, the critical value of $r = 0.602$. Therefore, if $|r| < 0.602$, the hypothesis that $\rho = 0$ can be accepted.

36.3.4.3 Test of hypothesis that $\rho \neq 0$

The table of limits of correlation coefficients can be used to test the null hypothesis $\rho = 0$. However, they are not suitable for testing other null hypotheses such as $\rho = \rho'$ (specific value such as, say 0.5) or $\rho_1 = \rho_2$ etc. When $\rho \neq 0$, the shape of the distribution of r changes and becomes skewed.

Fisher (1915, 1921) provided a solution for this problem by transforming r to a quantity z, distributed almost normally with standard error

$$\sigma_Z = \frac{1}{\sqrt{n-3}} \qquad (36.56a)$$

The 'z' transformation is given by

$$z = \frac{1}{2} \ln \frac{(1+r)}{(1-r)} \qquad (36.56b)$$

This has a mean of approximately

$$\frac{1}{2} \ln \frac{(1+\rho)}{(1-\rho)}$$

The values of z are given in statistical tables ($r \to z$ or $z \to r$).

36.3.4.4 *Significance of a difference between correlation coefficients*

When it is necessary to establish whether or not there is a significant difference between two correlation coefficients estimated from two different samples, the following statistic based on the z transformation is used:

$$z = \frac{z_1 - z_2 - \mu_{z_1 - z_2}}{\sigma_{z_1 - z_2}} \qquad (36.57)$$

where $\mu_{z_1 - z_2} = \mu_{z_1} - \mu_{z_2}$; $\sigma_{z_1 - z_2} = \sqrt{\sigma_{z_1}^2 + \sigma_{z_2}^2} = \sqrt{\dfrac{1}{n_1 - 3} + \dfrac{1}{n_2 - 3}}$

The statistic 'z' is approximately normally distributed.

36.4 FREQUENCY ANALYSIS

36.4.1 Data series

Data series can be a complete series that would normally be not independent, an annual series that can be considered as an independent series or a partial duration series. In annual or partial duration series, there is some loss of information when a conversion is done from a complete data series. In an annual series, the annual maximum (or minimum) values are taken. They would normally be independent, but the second (or third, etc.) highest in some years which may be greater than the highest in some other years are ignored. In a partial duration series, all values greater than a certain threshold value are taken. The threshold value may be taken small enough to cover all significant events. In a partial series, it is the common practice to consider the highest N records where N is the number of years of records.

36.4.2 Recurrence interval (return period)

Recurrence interval is the average time interval between events that equal or exceed a given magnitude once. If the probability of occurrence of an event with its magnitude X greater than or equal to a specified value x is denoted by $Pr(X \geq x)$, then the return period T_r is given by (X is a random variable; x is a specific value of X)

$$T_r = \frac{1}{Pr(X \geq x)} \quad \text{or} \quad Pr(X \geq x) = \frac{1}{T_r} \tag{36.58a}$$

Because $Pr(X \geq x) = 1 - Pr(X \leq x)$

$$T_r = \frac{1}{1 - Pr(X \leq x)} \quad \text{or} \quad Pr(X \leq x) = 1 - \frac{1}{T_r} \tag{36.58b}$$

If $Pr(X \leq x)$ represents the probability that x will not be equalled or exceeded in a certain period of time, then $Pr(X \leq x)_n$ represents the probability that x will not be equalled or exceeded in 'n' such periods. For independent events, the multiplication rule gives

$$Pr(X \leq x)_n = [Pr(X \leq x)][Pr(X \leq x)][Pr(X \leq x)]....'n' \text{ times}$$

$$= Pr(X \leq x)^n$$

$$= \left[1 - \frac{1}{T_r} \right]^n \tag{36.59}$$

The probability $\left[Pr(X \geq x) \right]$ at least once in 'n' years $= 1 - \left[Pr(X \leq x) \right]^n$ $\tag{36.60a}$

In terms of the return period, T_r,

$$\left[Pr(X \leq x) \right]^n = 1 - \left[1 - \frac{1}{T_r} \right]^n \tag{36.60b}$$

which is called the risk, an alternative to the concept of return period.

36.4.3 Plotting positions

The probability p of an event being equalled or exceeded in any year can be estimated by using the plotting position, which is defined by a number of formulas as given below:

California (1923) $\dfrac{m}{n}$ $\tag{36.61a}$

Weibull (1939) $\dfrac{m}{n+1}$ $\tag{36.61b}$

Hazen (1930) $\dfrac{2m-1}{2n}$ $\tag{36.61c}$

Beard (1943) $1 - (0.5)^{1/n}$ $\tag{36.61d}$

(only for $m = 1$; other plotting positions are interpolated linearly between this and the value of 0.5 for median event)

Blom (1958) $\dfrac{m - 3/8}{n + 1/4}$ (36.61e)

Tukey (1962) $\dfrac{3m - 1}{3n + 1}$ (36.61f)

Gringorten (1963) $\dfrac{m - 0.44}{n + 0.12}$ (36.61g)

Cunnane (1978) $\dfrac{m - 0.4}{n + 0.2}$ (36.61h)

In these equations, n is the number of values; m is the rank of the value when arranged in descending order. The inverse of p gives the return period T_r.

Different plotting position formulas give practically the same results in the middle but produce different positions near the tails. The California method produces 100% probability at the edges which cannot be plotted on probability paper.

36.4.4 Probability paper method

In probability papers, the scales are modified such that the magnitude of the random variate vs. probability plots as a straight line. The common probability papers used correspond to the Normal and Gumbel distributions although they can be constructed for other distributions as well.

36.4.4.1 Normal probability paper, N(0,1)

z:	−2	−1	0	1	2
F(z):	0.0228	0.1587	0.5	0.8413	0.9772

The magnitude of the random variable is plotted on the linear vertical axis. The horizontal axis is non-linear and gives the probability. {F(z), z} form a pair of co-ordinates.

36.4.4.2 Gumbel probability paper

In Gumbel paper too, the magnitude of the random variate is plotted on the vertical axis to a linear scale. The horizontal axis, which has a non-linear scale, gives the probability. It is constructed as follows by using a reduced variate:

$$y = \alpha (x - \beta)$$ (36.62a)

$$F(x) = Pr(X \leq x)$$ (36.62b)

$$Pr(X > x) = 1 - Pr(X \leq x) \tag{36.62c}$$

$$T_r = \frac{1}{1 - Pr(X \leq x)} \quad \text{or} \quad Pr(X \leq x) = \frac{T_r - 1}{T_r} \tag{36.62d}$$

Therefore,

$$Pr(x) = \frac{T_r - 1}{T_r} = e^{-e^{-y}} \tag{36.62e}$$

36.4.5 Frequency factor method

After selecting the distribution, the frequency analysis can be done according to the following general equation (Chow, 1951):

$$X = \mu + K\sigma \tag{36.63}$$

where
X – event magnitude of return period T_r
μ – population mean (estimated by sample mean $\bar{X}$)
σ – population standard deviation (estimated by sample standard deviation, s)
K – frequency factor which depends upon the return period and distribution parameters.

For a given distribution, a relationship between the return period and the frequency factor can be obtained. The frequency factor gives the number of standard deviations above the mean value to attain the required non-exceedance probability. For a normal distribution, the frequency factors for some selected return periods are

T_r	5	10	20	50	100
K	0.8416	1.2816	1.6449	2.0538	2.3264

For Gumbel distribution, the frequency factors depend on the sample size and the return period as well as the method of estimation. Some typical values are given in Table 36.4.

36.5 PROBABILITY

36.5.1 Some preliminaries

If an event happens a times in n trials, then

$$L\frac{a}{n} = p \tag{36.64}$$

$$n \to \infty$$

is the probability which is also the limit of frequency.

For a random variable X with a probability density function $f(x)$, the following are true:

$$f(x) \geq 0 \text{ for all } x \tag{36.65a}$$

$$\int_{-\infty}^{\infty} f(x)\,dx = 1 \tag{36.65b}$$

$$Pr(a \le X \le b) = \int_{a}^{b} f(x)\,dx \tag{36.65c}$$

For any specific value of X, $Pr(X) = 0$, i.e.

$$Pr(X = a) = \int_{a}^{a} f(x)\,dx = 0 \tag{36.65d}$$

This is only a probability, but by coincidence its occurrence cannot be ruled out.

$$Pr(X \le a) = \int_{-\infty}^{a} f(x)\,dx = F(a) \tag{36.65e}$$

where $F(a)$ is the area under the probability density function curve from $-\infty$ to a.

The cumulative probability distribution function $F(x)$ and the probability density function $f(x)$ are related to each other by

$$F(x) = \int f(x)\,dx; F(x) = \frac{dF(x)}{dx} \tag{36.66}$$

36.5.2 Expected values

The expected value of a random variable X is given by

$$E(X) = \int_{-\infty}^{\infty} x f(x)\,dx \tag{36.67}$$

The variance of a random variable X with expected value μ is given by

$$\begin{aligned}
\text{Var}(X) &= E\left[(X - \mu)^2\right] \\
&= E\left[X^2 - 2X\mu + \mu^2\right] \\
&= E(X^2) - 2\mu E(X) + \mu^2 \\
&= E(X^2) - 2\mu^2 + \mu^2 \\
&= E(X^2) - \mu^2 \\
&= \int_{-\infty}^{\infty} x^2 f(x)\,dx - \mu^2
\end{aligned} \tag{36.68}$$

A probability density function has three types of parameters:

- Location parameter
- Scale parameter
- Shape parameter

The location parameter gives the location relative to some specific point. For some distributions, the location parameter is given by the mean value. Two probability distribution functions can have the same location parameter but different scale parameters. The shape parameter determines the geometrical configuration. Some distributions have more than one shape parameter, while some others have none.

36.6 DISCRETE PROBABILITY DISTRIBUTIONS

36.6.1 Binomial distribution

Binomial distribution comes from Bernoulli processes, which are series of random experiments having only two outcomes, one the complement of the other. For example, success or failure, good or bad, yes or no, head or tail, etc. Bernoulli processes assume that each random experiment is a trial that the trial success probability in a large number of trials is a constant (for example, in the case of tossing a fair coin, the number of occurrences of heads or tails is 50% of the total tosses) and that each trial is independent of the previous trial.

If 'p' is the probability of occurrence of an event in an infinite number of trials and 'q' is the corresponding probability of non-occurrence ($q = 1 - p$), then the probability density function is given as

$$f(x) = {}^{n}C_{x}p^{x}q^{n-x} \tag{36.69}$$

where
$f(x)$ – the probability of the event, i.e. x times in n trials

${}^{n}C_{x}$ – the number of combinations of n things taken x at a time, $= \dfrac{n!}{(n-x)!x!}$

x – the number of occurrences; 0, 1, 2, 3,..., n.

$(n - x)$ – number of occurrences of its complementary event with $n - x = n$, $n - 1$, $n - 2$,..., 3, 2, 1, 0.

The quantity $p^{x}q^{n-x}$ represents the probability of getting x successes and $n - x$ failures in n trials. The function $f(x)$ is given in table form in statistical books as a function of n, x and p for n up to about 20.

36.6.1.1 Properties and parameters

$$E(X) = np \tag{36.70a}$$

$$\text{Var}(X) = npq \tag{36.70b}$$

$$\text{Skewness Coefficient} = \frac{q - p}{\sqrt{pqn}} \tag{36.70c}$$

When $q > p$, the skewness coefficient is positive and the distribution is positively skewed. When $q < p$, the skewness coefficient is negative, and the distribution is negatively skewed. When $q = p$, the distribution is symmetrical. As $n \to \infty$, the Binomial distribution tends to the Normal distribution. Then,

$$Pr(X \le x) = B(x) \cong \phi \left[\frac{x + 0.5 - \mu}{\sigma} \right] \tag{36.71}$$

where the 0.5 is added to approximate the discontinuous function to a continuous function with μ and σ defined as above. Without this correction, the Binomial distribution can be approximated by a normal distribution by the transformation

$$z = \frac{x - \mu}{\sigma} = \frac{x - np}{\sqrt{npq}} \tag{36.72}$$

36.6.2 Poisson distribution

It was noted that the Binomial distribution tends to a normal distribution when $n \to \infty$ and when p and q are near 0.5. But, when $n \to \infty$ and either p or $q \to 0$, the normal distribution cannot be used as an approximation to the Binomial distribution. The calculations are prohibitively difficult when n is large and p is small. However, when n is large and p is small, but np is finite and constant, the Binomial distribution can be approximated by the Poisson distribution.

Although it can be considered as a limiting case of the Binomial distribution, it can also be considered as a distribution on its own right by considering a Poisson process. It is one of the three main distributions widely used in statistics.

Poisson distribution is concerned with a single event. It gives the probability of occurrence of that event. Poisson events will occur at some average rate over a given time or space. A Poisson process has no memory, and the mean process rate is a constant.

Examples of Poisson processes include

- The number of accidents taking place at a certain junction during a certain period of time,
- Number of times the temperature is greater than a specified value,
- Number of times the discharge in a river is greater than a specified value,
- Bacteria count in a sample of water,
- Number of times a machine fails in a specified working period of time,
- Number of defects of an item,
- Number of occurrences of a given extreme value event,
- Number of phone calls arriving at a switch board,
- Number of buses that pass a bus stop, etc.

A Poisson distribution is given by

$$f(x) = \frac{\lambda^x e^{-\lambda}}{x!} \tag{36.73a}$$

where

$$\lambda = np \tag{36.73b}$$

$x = 0, 1, 2,...$ is the number of occurrences of an event of a given small probability p in a large number of trials n. It can be shown that the Binomial function converges to Poisson function:

$$L^nC_x p^x q^{n-x} \Rightarrow \frac{\lambda^x e^{-\lambda}}{x!} \tag{36.74}$$

$$n \to \infty$$

$$p \to 0$$

36.6.2.1 Properties and parameters

$$E(X) = \lambda \tag{36.75a}$$

$$Var(X) = \lambda \tag{36.75b}$$

$$\text{Coefficient of skewness} = \lambda^{-1/2} \tag{36.75c}$$

When $\lambda \to \infty$, the Poisson distribution converges to the normal distribution. The shape of the function is such that $f(x)$ increases up to $(x + 1) < \lambda$ and then decreases towards zero. For $\lambda < 1$, the distribution is J shaped. Other properties include

- For $\lambda = 1$, the coefficient of skewness = 1,
- $f(0) = f(1) = 1/e$
- From $x = 2$, $f(x)$ decreases
- For $\lambda > 1$, the distribution is unimodal.

36.7 CONTINUOUS PROBABILITY DISTRIBUTIONS

36.7.1 Normal Distribution

Normal distribution which is also known as Gaussian distribution is defined as

$$f(x) = \frac{1}{\sqrt{2\pi}\,\sigma} e^{-\frac{(x-\mu)^2}{2\sigma^2}} \quad \infty < x < \infty \tag{36.76a}$$

$$F(x) = Pr(X \le x) = \int_{-\infty}^{x} \frac{1}{\sqrt{2\pi}\,\sigma} e^{-\frac{(x-\mu)^2}{2\sigma^2}}\, dx \tag{36.76b}$$

36.7.1.1 Properties

$$E(X) = \mu \tag{36.77a}$$

$$Var(X) = \sigma^2 \tag{36.77b}$$

$$\text{Skewness coefficient} = 0 \tag{36.77c}$$

$$f(\mu) = \frac{1}{\sigma\sqrt{2\pi}} = \frac{0.4}{\sigma} \tag{36.77d}$$

The points of inflexion are at $x = \mu \pm \sigma$ with

$$f(\mu \pm \sigma) = \frac{1}{\sigma\sqrt{2\pi e}} \cong \frac{0.242}{\sigma} \tag{36.77e}$$

$$F(\mu - \sigma) = 0.1587 \left(\text{corresponds to } Z = -1 \quad Z = \frac{X - \mu}{\sigma} \right) \tag{36.77f}$$

$$F(\mu + \sigma) = 0.8413 \left(\text{corresponds to } Z = +1 \right) \tag{36.77g}$$

$$F(m) = 0.5 \left(\text{corresponds to } Z = 0 \right) \tag{36.77h}$$

Therefore, the area between $\mu + \sigma$ and $\mu - \sigma = 0.8413 - 0.1587 = 0.6826$.

$F(X)$ for the normal distribution cannot be calculated exactly. It is done numerically and compiled into tables of areas under the normal curve for $\mu = 0$ and $\sigma = 1$.

In a normal distribution, approximately 68%, 95% and 99.7% of all probabilities lie within $\mu \pm \sigma$, $\mu \pm 2\sigma$, and $\mu \pm 3\sigma$ respectively. This can be verified using the statistical tables.

36.7.1.2 Standard normal distribution

When $\mu = 0$ and $\sigma^2 = 1$ in a normal distribution, the resulting distribution is called a standard normal distribution. It is given as

$$\phi(z) = \frac{1}{\sqrt{2\pi}} e^{-\frac{z^2}{2}} \tag{36.78a}$$

$$\Phi(z) = Pr(Z \le z) = \int_{-\infty}^{z} \frac{1}{\sqrt{2\pi}} e^{-\frac{z^2}{2}} dz \tag{36.78b}$$

For example,

$$Pr\,(Z \le 1.96) = 0.975\,(97.5\%)$$

$$Pr\,(-1.96 \le Z \le 1.96) = \Phi(1.96) - \Phi(-1.96) = 0.975 - 0.025 = 0.95\,(95\%)$$

Any other normal random variable can be transformed into a standard normal random variable Z, referred to as a normal deviate and defined as

$$Z = \frac{X - \mu}{\sigma} \tag{36.79}$$

Then,

$$F(x) = Pr\,(X \le x) = Pr\left[\frac{X - \mu}{\sigma} \le \frac{x - \mu}{\sigma} \right] = Pr[Z \le z] \tag{36.80}$$

Practical limitation of the normal distribution is that it has infinite limits (or tails) which empirical distributions do not have.

36.7.1.3 Frequency analysis using probability paper

For normal distributions

- Assume normal distribution $N(\mu, \sigma)$
- Compute $\bar{X}$ and s
- Assume $\mu = \bar{X}$ and $\sigma = s$
- Plot a straight line using any two points. Usually, the following two points are used: $(\bar{X} - s, 0.1587)$ and $(\bar{X} + s, 0.8413)$

The values 0.1587 and 0.8413 are the probabilities of exceedance for the magnitudes $(\bar{X} - s)$ and $(\bar{X} + s)$. Probability statements are made using the fitted line.

For these values, the standard deviates are

$$Z_1 = \frac{(\bar{X} - s) - \bar{X}}{s} = -1$$

$$Z_2 = \frac{\bar{X} - \bar{X}}{s} = 0$$

$$Z_3 = \frac{(\bar{X} + s) - \bar{X}}{s} = +1$$

The corresponding probabilities of non-exceedance are

$$Pr(Z \leq -1) = 0.5 - Pr(0 \leq Z \leq 1)$$

$$= 0.5 - 0.3413 \,(\text{from normal tables for } Z = 1)$$

$$= 0.1587$$

$$Pr(Z \leq 0) = 0.5 \,(\text{from Tables})$$

$$Pr(Z \leq 1) = 0.5 + Pr(0 \leq Z \leq 1)$$

$$= 0.5 + 0.3413 \,(\text{form normal tables for } Z = 1)$$

$$= 0.8413$$

For visual comparison of the sample data with the fitted straight line, the data can be arranged in descending order and the plotting positions calculated using any of the plotting position formulas. If the fit is not good, it may be due to either.

- the normal distribution is not suitable, or,
- the sample statistics are not good estimators of population parameters.

If the sample data is concave upwards, it is reasonable to try to fit a log-normal or an extreme value distribution. For log-normal assumption, the following procedure is used:

- Transform the data using $Y = \log X$
- Compute $\bar{Y}$ and s_y of the transformed data
- Using normal probability paper, plot the two points $(\bar{Y} - s_y, 0.1587)$ and $(\bar{Y} + s_y, 0.8413)$ and join by a straight line
- Estimate the return periods of the transformed data and plot on the graph to see if they approximate the straight line.

Note: The mean and the standard deviation of the logarithms are not the same as the logarithms of the mean and the standard deviation.

36.7.1.4 Frequency analysis using frequency factor method

The cumulative probability refers to $Pr\,(X \le X_{T_r})$ which is equal to $1 - (1/T_r)$
For example,

If $T_r = 100$, $Pr\,(X \le X_{100}) = 1 - \dfrac{1}{100} = 0.99$

If $T_r = 50$, $Pr\,(X \le X_{50}) = 1 - \dfrac{1}{50} = 0.98$

If $T_r = 20$, $Pr\,(X \le X_{20}) = 1 - \dfrac{1}{20} = 0.95$

If $T_r = 5$, $Pr\,(X \le X_5) = 1 - \dfrac{1}{5} = 0.8$

Frequency factor $K = f$ (type of distribution, return period)
    In the case of the normal distribution, K is the value of the standard variate corresponding to the cumulative probability and is equal to the z value corresponding to $Pr\,(X \le X_{T_r})$ e.g.

$T_r = 100$, $p = 0.99$, and $z = 2.326$

$T_r = 50$, $p = 0.98$, and $z = 2.054$

$T_r = 20$, $p = 0.95$, and $z = 1.645$

$T_r = 5$, $p = 0.80$, and $z = 0.8416$

$T_r = 2$, $p = 0.50$, and $z = 0$

36.7.1.5 Sampling errors

The magnitude of a random variable in terms of the frequency factor is given as

$$X_{T_r} = \mu + \sigma K_{T_r} \tag{36.81a}$$

where μ and σ refer to the population parameters. They are usually unknown and therefore replaced by sample mean and sample standard deviation:

$$\hat{X}_{T_r} = \bar{X} + s K_{T_r} \tag{36.81b}$$

If a different sample is taken, then there will be a different $\overline{X}$ and s and therefore a different $\hat{X}_{T_r}$. For a Normal population, the sampling distribution of means is also Normal and has the mean μ and variance σ^2/n. The sampling variation of $\hat{X}_{T_r}$ is measured by its standard deviation[7] s_{T_r} which is also called the standard error of estimate. It is given as

$$s_{T_r} = \delta_{T_r} \frac{\sigma}{\sqrt{n}} \qquad (36.82a)$$

where δ_{T_r} depends upon the return period T_r.

Standard error accounts for the incorrect choice of sample parameters. For a Normal population,

$$\delta_{T_r} = \sqrt{1 + \frac{K_{T_r}^2}{2}} \qquad (36.82b)$$

For example,

$$T_r = 100 \quad K_{T_r} = 2.326 \quad \delta_{T_r} = 1.925$$

$$T_r = 2 \quad K_{T_r} = 0 \quad \delta_{T_r} = 1.0$$

These values are given in table form in several texts, e.g. Kite (1977), Table 5.5.

For the computation of s_{T_r}, the population standard deviation σ is replaced by sample standard deviation if the population parameters are not known.

36.7.1.6 Confidence intervals

$\hat{X}_{T_r}$ is an estimate of X_{T_r} and is a random variable. Therefore it is possible to establish a confidence interval for $\hat{X}_{T_r}$.

For a large sample ($n > 30$), the confidence interval is given by

$$\hat{X}_{T_r} \pm z_{\alpha/2} s_{T_r} = \hat{X}_{T_r} \pm z_{\alpha/2} \delta_{T_r} \frac{s}{\sqrt{n}} \qquad (36.83a)$$

where the standard error of estimate s_{T_r} is given in Eq. 36.82a as $\delta_{T_r} \frac{\sigma}{\sqrt{n}}$. For small samples ($n < 30$), it is

$$\hat{X}_{T_r} \pm t_{\alpha/2} s_{T_r} = \hat{X}_{T_r} \pm t_{\alpha/2} \delta_{T_r} \frac{s}{\sqrt{n}} \qquad (36.83b)$$

where $z_{\alpha/2}$ and $t_{\alpha/2}$ refer to the z and t critical values at α level of significance using a two-tailed test of a Normal distribution and a Students' t distribution, respectively. For example, $\alpha = 0.05$ corresponds to 95% confidence and $z_{0.025} = 1.96$ and $t_{0.025} = 2.10$.

The confidence interval depends on δ_{T_r} which increases with increasing return period.

---

[7] Standard error of estimate is given by $e = \left\{ \dfrac{\sum (x_i - \hat{x}_i)^2}{n} \right\}^{1/2}$ where $\hat{x}_i$ is the computed value of the recorded event of magnitude x_i.

36.7.2 Two parameter log-normal distribution

If the data, when plotted on a Normal probability paper, are concave upwards, then it is reasonable to assume a log-normal distribution.

If $Y = \ln(X)$ and Y is normally distributed with mean μ and variance σ^2, then X is log-normally distributed with the same mean μ and variance σ^2. The probability density and cumulative probability functions are given by

$$f(x) = \frac{1}{\sqrt{2\pi}\sigma x} e^{-\frac{[\ln(x)-\mu]^2}{2\sigma^2}} \quad \text{for } x > 0 \tag{36.84a}$$

$$f(x) = 0 \quad \text{otherwise}$$

$$F(x) = Pr(X \le x) = \int_{-\infty}^{X} \frac{1}{\sqrt{2\pi}\sigma x} e^{-\frac{[\ln(x)-\mu]^2}{2\sigma^2}} \, dx \tag{36.84b}$$

36.7.2.1 Properties

The expected and the variance values are given by

$$E(X) = e^{\mu+\frac{\sigma^2}{2}} \tag{36.85a}$$

$$Var(X) = e^{2\mu+2\sigma^2} - e^{2\mu+\sigma^2} \tag{36.85b}$$

Log-normal is useful for transforming positive data to positive and negative data.

36.7.2.2 Frequency analysis using probability paper method

Population line can be obtained by joining the two points:

$$\left(\bar{Y} - s_y, \ 0.1587 \right) \text{ and } \left(\bar{Y} + s_y, 0.8413 \right)$$

In the same way as in Normal distributions, probability statements can be made using the fitted line.

36.7.2.3 Frequency analysis using frequency factor method

$$\ln\left(X_{T_r}\right) = Y_{T_r} = \mu_y + z\sigma_y \Rightarrow \frac{Y - \mu_y}{\sigma_y} = z \tag{36.86}$$

which is the same as for Normal, excepting μ_y, σ_y are used. It can also be computed without using μ_y and σ_y. Instead, the mean and variance of the original data series can be used.

Some extracts of frequency factors for the two-parameter log-normal distribution taken from a Table compiled by Kite (1977) as functions of the coefficient of variation C_v are given in Table 36.2. The coefficient of variation can be determined from the mean and standard deviation of the data. For a two-parameter log-normal distribution, it is sometimes convenient to use the frequency factor method to determine the T_r year event than to transform the original data to logarithms.

Table 36.2 Frequency factors for two parameter Log-normal distribution for some selected coefficients of variation and return periods

Coefficient of variation	Frequency factor					
	Return period					
	2	5	10	20	50	100
0.05	−0.025	0.8334	1.2965	1.6863	2.1341	2.4370
0.10	−0.0496	0.8222	1.3078	1.7247	2.2130	2.5489
0.50	−0.2111	0.6626	1.2778	1.8909	2.7202	3.3673
0.75	−0.2667	0.5387	1.1784	1.8677	2.8735	3.7118
1.0	−0.2929	0.4254	1.0560	1.7815	2.9098	3.9035

36.7.3 Three parameter log-normal distribution

The probability density function for a three-parameter log-normal distribution is given as

$$f(x) = \frac{1}{\sqrt{2\pi}\sigma_y(x-a)} e^{-\frac{[\ln(x-a)-\mu_y]^2}{2\sigma_y^2}} \quad \text{for } x > 0 \tag{36.87}$$

where μ_y is the form (scale) parameter {=mean of $\ln(x - a)$}, and σ_y^2 is the shape parameter (= $\text{Var}(\ln(x - a))$) and a is the location parameter. When $a = 0$, Eq. 36.87 represents a two-parameter log-normal distribution.

If the lower boundary 'a' is known, the reduced variable $(x - a)$ can be used in the two parameter log-normal distribution. If it is unknown, it must be estimated by anyone of the methods of estimation. If the method of moments is used, it is given by

$$a = \mu\left[1 - \frac{(C_v)_1}{(C_v)_2}\right] = \mu - \frac{\sigma}{(C_v)_2} \tag{36.88}$$

where μ and σ are the mean and standard deviation of the original series, and

$$(C_v)_1\text{-Coefficient of variation of the } x \text{ series} = \frac{\sigma}{\mu} \tag{36.89a}$$

$$(C_v)_2\text{-Coefficient of variation of the } (x - a) \text{ series} = \frac{\sigma}{\mu - a} \tag{36.89b}$$

which cannot be calculated without 'a'.

For a log–normal distribution, the skewness coefficient of the x series and the coefficient of variation are related to each other by

$$s_k = 3(C_v)_2 + [(C_v)_2]^3 \tag{36.90a}$$

The solution of this equation gives $(C_v)_2$ as

$$(C_v)_2 = \frac{1 - \omega^{2/3}}{\omega^{1/3}} \tag{36.90b}$$

where

$$\omega = \frac{-s_k + \left(s_k^2 + 4\right)^{1/2}}{2} \tag{36.90c}$$

Hence, by computing μ, σ and s_k of the original series, $(C_v)_1$, $(C_v)_2$ can be computed and then 'a' can be obtained from Eq. 36.88. In terms of these parameters, σ_y and μ_y are given by (without having to take lengths of the x-series).

$$\mu_y = \ln\left(\frac{\sigma}{(C_v)_2}\right) - \frac{1}{2}\ln\left([C_v)_2]^2 + 1\right) \tag{36.91a}$$

$$\sigma_y = \left[\ln\left((C_v)_2]^2 + 1\right)\right]^{1/2} \tag{36.91b}$$

The maximum likelihood method requires an iterative solution. There are also other methods available.

36.7.3.1 Frequency analysis using frequency factor method

$$y_{T_r} = \mu_y + z\sigma_y \tag{36.92a}$$

where μ_y and σ_y are for the series $\ln(x - a)$, and z is the standard normal deviate.
   Therefore, T_r year event X_{T_r} is given by

$$X_{T_r} = a + e^{\mu_y + z\sigma_y} \tag{36.92b}$$

Table 36.3 gives the frequency factors for some selected coefficients of skewness and return periods of the original series extracted from a table given by Kite (1977) which covers a wide range of skewness coefficients using the original series and the relationship $X_{T_r} = \mu + K\sigma$.

36.7.4 Gamma distribution

36.7.4.1 Gamma function

Gamma distribution is defined in terms of the Gamma function (Γ) which is defined as

$$\Gamma(\alpha) = \int_0^\infty x^{\alpha-1} e^{-x}\, dx \quad \text{for } x > 0 \tag{36.93}$$

Table 36.3 Frequency factors for a three-parameter Log-normal distribution for selected skewness coefficients and return periods

| Coefficient of skewness | Frequency factor K | | | | | |
| | Return period (years) | | | | | |
	2	5	10	20	50	100
−2.00	0.2366	−0.6144	−1.2437	−1.8916	−2.7943	−3.5196
−1.00	0.1496	−0.7449	−1.3156	−1.8501	−2.5294	−3.0333
0	0	0	0	0	0	0
2.00	−0.2366	0.6144	1.2437	1.8916	2.7943	3.5196

36.7.4.2 Properties of the Gamma function

$\Gamma(1) = 1 = \Gamma(2)$

$\Gamma(0) \to \infty$

$\Gamma(\alpha + 1) = \alpha\Gamma(\alpha)$

$\Gamma(k+1) = k!$ k is a non-negative integer, $(k \geq 0)$

$\Gamma(1/2) = \sqrt{\pi}$

For any positive integer k, $(k > 0)$

$$\Gamma\left(k + \frac{1}{2}\right) = \frac{1.3.5.7....(2k-1)}{2^k}\sqrt{\pi} \tag{36.94}$$

For $\alpha > 0$, but not an integer, $\Gamma(\alpha)$ is computed by series expansion and numerical integration.

36.7.4.3 Incomplete Gamma function

The Gamma function split into two parts referred to as the upper and lower incomplete Gamma functions are defined as

$$\Gamma(\alpha, x) = \int_x^\infty x^{\alpha-1}e^{-x}dx \tag{36.95a}$$

$$\gamma(\alpha, x) = \int_0^x x^{\alpha-1}e^{-x}dx \tag{36.95b}$$

Depending on the number of parameters in the distribution function, the gamma distribution is defined as one-parameter, two-parameter and three-parameter gamma distribution functions.

36.7.5 Three parameter Gamma distribution

The three-parameter gamma probability density function is given by

$$f(x) = \frac{1}{\Gamma(\alpha)\beta^\alpha}(x - \gamma)^{\alpha-1}e^{-(x-\gamma)/\beta} \quad \text{for} \quad x > 0; \text{ and,}$$

$$\tag{36.96a}$$

$f(x) = 0, \quad \text{otherwise}$

where γ is the location parameter, α is the shape parameter and β is the scale parameter. This form is referred to as the three-parameter gamma distribution with $\gamma \neq 0$ as the lower boundary for x (i.e., $\gamma \leq x \leq \infty$).

This can also be written as

$$f(x) = \frac{1}{\Gamma(\alpha)\beta}\left(\frac{x-\gamma}{\beta}\right)^{\alpha-1} e^{-\frac{(x-\gamma)}{\beta}} \quad \text{for} \quad x > 0; \text{ and,}$$

(36.96b)

$$f(x) = 0, \quad \text{otherwise}$$

Therefore,

$$F(X) = \frac{1}{\Gamma(\alpha)\beta}\int_0^X \left[\frac{(x-\gamma)}{\beta}\right]^{\alpha-1} e^{-\frac{(x-\gamma)}{\beta}} dx$$

(36.96c)

36.7.5.1 Properties

$$E(X) = \gamma + \alpha\beta$$

(36.97a)

$$\text{Var}(X) = \alpha\beta^2$$

(36.97b)

$$\text{Skewness coefficient} = \frac{2}{\sqrt{\alpha}} \text{ (independent of } \beta)$$

(36.97c)

$$\text{Kurtosis coefficient} = \frac{6}{\alpha} + 3 \text{ (independent of. } \beta)$$

(36.97d)

The transformation $y = x - \gamma$, reduce the three parameter distribution to a two parameter one, and $z = \frac{x-\gamma}{\beta}$, with $\beta = 1$ reduces it to a one-parameter distribution. Using the reduced variate z,

$$f(z) = \frac{1}{\Gamma(z)} z^{\alpha-1} e^{-z}$$

(36.98a)

$$F(z) = \frac{1}{\Gamma(\alpha)}\int_0^z z^{\alpha-1} e^{-z} dz$$

(36.98b)

$$= \text{Chi-square distribution with } 2\alpha \text{ degrees of freedom and } \chi^2 = 2z$$

Therefore, the reduced variate event magnitude is

$$Z = \frac{\chi^2}{2}$$

(36.99a)

and, the expected event magnitude is

$$X = \left[\frac{\chi^2}{2}\right]\beta + \gamma$$

(36.99b)

36.7.6 Two parameter Gamma distribution

For a two-parameter Gamma distribution, the probability density function is of the form

$$f(x) = \frac{1}{\Gamma(\alpha)\beta^\alpha} x^{\alpha-1} e^{-\frac{(x)}{\beta}} \quad \text{for} \quad x > 0; \text{ and,}$$

$$f(x) = 0, \quad \text{otherwise}$$

(36.100a)

with the shape parameter $\alpha > 0$ and the scale parameter $\beta > 0$. The function may also be written as

$$f(x) = \frac{1}{\Gamma(\alpha)\beta} \left[\frac{x}{\beta}\right]^{\alpha-1} e^{-\frac{(x)}{\beta}} \quad \text{for} \quad x > 0; \text{ and,}$$

$$f(x) = 0, \quad \text{otherwise}$$

(36.100b)

36.7.6.1 Properties

$$E(X) = \alpha\beta$$

(36.101a)

$$\text{Var}(X) = \alpha\beta^2$$

(36.101b)

$$\text{Skewness coefficient} = \frac{2}{\sqrt{\alpha}}$$

$$\text{Kurtosis coefficient} = \frac{6}{\alpha} + 3$$

(36.101c)

independent of β

When $\alpha = 1$, the two-parameter gamma distribution becomes the Exponential distribution. Examples of this distribution include rainfall data for specific periods of time such as monthly or weekly.

If $X_1, X_2,..., X_n$ are independent Gamma distribution random variables, then

$$Y = X_1 + X_2 + \cdots + X_n$$

is also Gamma distributed with parameters $n\alpha$ and β.

36.7.7 One parameter Gamma distribution

For a one-parameter Gamma distribution, the probability density function is of the form

$$\frac{1}{\Gamma(\alpha)} (x)^{\alpha-1} e^{-x} \quad \text{for} \quad x > 0$$

$$f(x) = 0 \quad \text{otherwise}$$

(36.102)

36.7.7.1 Properties

$$E(X) = \alpha \tag{36.103a}$$

$$\text{Var}(X) = \alpha \tag{36.103b}$$

$$\text{Skewness coefficient} = \frac{2}{\sqrt{\alpha}} \tag{36.103c}$$

$$\text{Kurtosis coefficient} = \frac{6}{\alpha} + 3 \tag{36.103d}$$

If x is normally distributed with $N(\mu, \sigma)$, then

$$y = \frac{1}{2}\left\{\frac{x-\mu}{\sigma}\right\}^2 \tag{36.104}$$

follows the one-parameter Gamma distribution.

The application of one-parameter Gamma distribution is limited because of the relatively small flexibility of all functions with one parameter to fit properly any empirical distribution.

36.7.8 Chi-square distribution

This is a special case of the Gamma distribution with the parameters $\alpha = \frac{r}{2}; \beta = 2$ where r is the degrees of freedom of the Chi-square distribution. The random variable X in the Chi-square is a function of the sample variance of the Normal distribution as

$$\chi^2(n-1) \sim \frac{(n-1)s^2}{\sigma^2}$$

The probability density function of the Chi-square distribution is given as

$$f(x) = \frac{x^{r/2-1}e^{-x/2}}{\Gamma\left(\frac{r}{2}\right)2^{r/2}} \quad x > 0 \tag{36.105}$$

where r is the degrees of freedom which is its single parameter. Its properties include

$$E(\chi^2) = r = n-1 \tag{36.106a}$$

$$\text{Var}(\chi^2) = 2r = 2(n-1) \tag{36.106b}$$

36.7.9 F Distribution

The F distribution named in honour of the statistician R. A. Fisher is used in the analysis of variances. The random variable considered is the ratio of two variances as

$$F = \frac{s_1^2}{s_2^2} \tag{36.107}$$

which follows a F distribution with degrees of freedom a and b. The probability density function takes the form

$$f(x) = \frac{\Gamma\left(\dfrac{a+b}{2}\right) a^{a/2} b^{b/2} x^{(a/2-1)}}{\Gamma\left(\dfrac{a}{2}\right)\Gamma\left(\dfrac{b}{2}\right)(a+bx)^{(a+b)/2}} \quad \text{for} \quad x > 0; a, b > 0 \tag{36.108}$$

Despite the complicated nature of the expression, for particular values of a and b, the density function takes simpler forms. For example, $a = 2$, $b = 2$, $a = b = 2$, $a = b = 1$.

For example, when $a = b = 1$ $f(x) = \dfrac{1}{\pi\sqrt{x}(1+x)}$ for $x > 0$

Expected value and the variance are given by

$$E(X) = \frac{b}{b-2} \quad \text{for } b > 2 \tag{36.109a}$$

$$\text{Var}(X) = 2\left(\frac{b}{b-2}\right)^2 \frac{a+b-2}{a(b-4)} \quad \text{for } b > 4 \tag{36.109b}$$

The F distribution and the Student's t distribution are related. If X follows a t distribution with $(n-1)$ degrees of freedom, then $Y = X^2$ follows a F distribution with $(1,(n-1))$ degrees of freedom.

36.7.10 Pearson type III distribution

Pearson type III is of the form

$$f(x) = \frac{1}{\alpha\Gamma(\beta)}\left(\frac{x-\gamma}{\alpha}\right)^{\beta-1} e^{-\left(\frac{x-\gamma}{\alpha}\right)} \tag{36.110}$$

where α, β and γ are the parameters of the distribution. The parameters are given as

$$\alpha = \frac{\sigma}{\sqrt{\beta}}; \quad \beta = \left(\frac{2}{s_k}\right)^2; \quad \gamma = \mu - \sigma\sqrt{\beta} \tag{36.111}$$

It can be reduced to a Gamma distribution by a suitable substitution.

The frequency factors are given by Kite (1977), and for some selected coefficients of skewness and return periods they are shown in Table 36.4.

36.7.11 Log Pearson type III distribution

Log Pearson type III is quite common in hydrology and has been recommended by the Water Resources Council, Bulletin 17 series. It requires a transformation of data to logarithm of data with the probability density function of Pearson type III.

$$f(x) = \frac{1}{x\,\Gamma(\alpha)\beta}\left[\frac{[\ln(x)]-\gamma}{\beta}\right]^{\alpha-1} e^{-\frac{[\ln(x)]-\gamma}{\beta}} \tag{36.112}$$

Table 36.4 Frequency factors for Pearson Type III distribution for selected skewness coefficients and return periods

Coefficient of skewness	Frequency factor					
	Return period					
	2	5	10	20	50	100
0	0.0000	0.8416	1.2816	1.6448	2.0537	2.3264
0.1	−0.0167	0.8363	1.2917	1.6728	2.1070	2.3997
1.0	−0.1621	0.7537	1.3349	1.8723	2.5428	3.0303
2.0	−0.2977	0.5993	1.2795	1.9684	2.8956	3.6103

where α, β and γ are the shape, scale and location parameters. The computation of these parameters is rather difficult. Hence the data are transformed first, and analysis is carried out using the transformed data.

Parameters in the transformed data are given as

$$\mu_y = \gamma + \alpha\beta; \quad \sigma_y = \beta\sqrt{\alpha}; \quad s_{ky} = \frac{2}{\sqrt{\alpha}} \tag{36.113}$$

These estimates are biased, and therefore, a bias correction should be applied when using them.

The T_r year event can be calculated from

$$Y_{T_r} = \ln(X_{T_r}) = \mu_y + K\sigma_y \tag{36.114}$$

where μ_y and σ_y are the mean and standard deviation of the transformed data. The frequency factors given in Table 36.4 can be used.

36.7.12 Extreme value distribution

The generalized extreme value distribution is given by

$$F(x) = \exp\left[-\{1 - k(x - \beta)/\gamma\}^{1/k}\right] \text{ for } k \neq 0$$

$$= \exp\left[-\exp\{-(x - \beta)/\gamma\}\right] \text{ for } k = 0 \tag{36.115b}$$

($k = 0$ corresponds to type I, or Gumbel distribution, $k < 0$ corresponds to type II distribution and $k > 0$ corresponds to type III distribution). Type I is used for maximum values and type III for minimum values.

For type I distribution, the properties are as follows:

$$E(X) = \mu = \beta + 0.5772\gamma \tag{36.116a}$$

$$\sigma = \frac{\pi\gamma}{\sqrt{6}} = 1.2825\gamma \tag{36.116b}$$

By the probability weighted moment method, the parameters are given by

$$\gamma = \frac{(2b_1 - b_0)k}{\Gamma(1+k)\left(1 - 2^{-k}\right)}; \quad \beta = b_0 + \frac{\gamma\{\Gamma(1+k) - 1\}}{k} \tag{36.116c}$$

where the b_r's are given by

$$b_r = \frac{1}{N}\sum_{j=1}^{N} p_j^r X_j \quad \text{and} \quad p_j = \frac{(j-A)}{N} \tag{36.116d}$$

and k is given by

$$k = 7.8590c + 2.9554c^2 \tag{36.116e}$$

$$c = \frac{2b_1 - b_0}{3b_2 - b_0} - \frac{\ln 2}{\ln 3} \tag{36.116f}$$

X_j are the extreme value data arranged in increasing magnitude; $A = 0.35$ and Γ represents the Gamma function.

When k is expressed as a standard normal variate $\left(z = k\sqrt{\dfrac{N}{0.5635}}\right)$, it lies within the critical range at the 5% significance level.

In the more familiar form in hydrology, the Gumbel distribution is expressed as follows:

$$F(x) = Pr(X \le x) = e^{-e^{-\alpha(x-\beta)}} \tag{36.117a}$$

and

$$f(x) = \frac{dF(x)}{dx} = \alpha\left\{e^{-e^{-\alpha(x-\beta)}}e^{-\alpha(x-\beta)}\right\}$$

$$= \alpha\left\{e^{-e^{-\alpha(x-\beta)}-\alpha(x-\beta)}\right\}$$

$$= \alpha\left\{e^{-\alpha(x-\beta)-e^{-\alpha(x-\beta)}}\right\} \tag{36.117b}$$

This can be obtained by writing $f(x) = \dfrac{dF(x)}{dx} = \dfrac{dF}{dz}\dfrac{dz}{dy}\dfrac{dy}{dx}$ where $F(x) = e^{-z}$ and $z = e^{-y}$.

Substituting $y = \alpha(x - \beta)$, as a reduced variate,

$$F(y) = e^{-e^{-y}} \tag{36.118a}$$

and

$$f(y) = \alpha e^{-y-e^{-y}} \tag{36.118b}$$

Gumbel probability distribution is unbounded with the x-axis as the asymptote.

36.7.12.1 Properties

The parameters are given by

$$E(X) = \mu = \beta + \frac{0.5772}{\alpha} \tag{36.119a}$$

$$\text{Var}(X) = \frac{\pi}{\sqrt{6}\alpha} = \frac{1.2825}{\alpha} \tag{36.119b}$$

$$\text{Median } [F(X \leq x) = 0.5] = \beta + \frac{0.3665}{\alpha} \tag{36.119c}$$

$$\text{Skewness} \cong 1.3 \tag{36.119d}$$

$$\text{Kurtosis} \cong 4.5 \tag{36.119e}$$

The second derivative of $F(x)$, or the first derivative of $f(x)$ is zero when $x = \beta$, or $y = 0$. This means that the mode, which is the maximum of the probability density function, occurs when $x = \beta$, or $y = 0$, i.e.

$$F(m) = F(\beta) = e^{-e^{-0}} = 0.368 \tag{36.119f}$$

$$f(m) = f(\beta) = \alpha\{e^{-0-e^0}\} = \alpha e^{-1} = 0.368\alpha \tag{36.119g}$$

Quartiles $[F(x_1) = 0.25; F(x_2) = 0.75]$

$$F(x) = e^{-e^y}$$

taking logarithms twice,

$$y = -\ln\left\{-\ln\left[F(x)\right]\right\}$$

Therefore,

$$\alpha(x_1 - \beta) = -\ln\left\{-\ln\left[F(x_1)\right]\right\}$$

$$\alpha(x_2 - \beta) = -\ln\left\{-\ln\left[F(x_2)\right]\right\}$$

Subtracting,

$$\alpha(x_2 - x_1) = \ln\left\{-\ln\left[F(x_1)\right]\right\} - \ln\left\{-\ln\left[F(x_2)\right]\right\}$$

$$= \ln\left\{-\ln[0.25]\right\} - \ln\left\{-\ln[0.75]\right\}$$

$$= 1.5724$$

Therefore,

$$\frac{1}{\alpha} = 0.6359(x_2 - x_1) \tag{36.119h}$$

If the x values at the quartile points are known, then α can be estimated. It can be seen that the Gumbel probability paper is in fact rectangular – double logarithmic scale.

Since $Pr(X \leq x) = 1 - \dfrac{1}{T_r}$

$$1 - \frac{1}{T_r} = e^{-e^{-\alpha(x-\beta)}}$$

Substituting $y = \alpha(x - \beta)$,

$$1 - \frac{1}{T_r} = e^{-e^{-y}}$$

Taking logarithms,

$$\ln\left[1 - \frac{1}{T_r}\right] = -e^{-y}$$

Taking logarithms again,

$$\ln\left\{-\ln\left[1 - \frac{1}{T_r}\right]\right\} = -y$$

Therefore,

$$y = -\ln\left\{-\ln\left[1 - \frac{1}{T_r}\right]\right\} \qquad (36.119i)$$

and

$$x = \beta + \frac{1}{\alpha} y \qquad (36.119j)$$

By plotting x vs. y on rectangular graph paper, α and β can also be estimated.

The frequency factors for Gumbel distribution depend on the sample size and the return period. Some typical values taken from a table given by Kite (1977) are shown in Table 36.5.

- **Mean Annual Flood:** In the context of Gumbel distribution, the mean annual flood is defined as follows:

$$\mu = \beta + \frac{0.5772}{\alpha} \qquad (36.120a)$$

$$x = \beta + \frac{1}{\alpha} y \qquad (36.120b)$$

Table 36.5 Frequency factors for Gumbel distribution for some selected sample sizes and return periods

| Sample size | Frequency factor | | | | | |
| | Return period | | | | | |
	2	5	10	20	50	100
10	−0.1355	1.0580	1.8483	2.6063	3.5874	4.3227
20	−0.1478	0.9187	1.6248	2.3020	3.1787	3.8356
30	−0.1526	0.8664	1.5410	2.1881	3.0257	3.6534
50	−0.1568	0.8197	1.4663	2.0865	2.8892	3.4908
100	−0.1604	0.7791	1.4011	1.9977	2.7700	3.3487

If $x = \mu$, then,

$$\beta + \frac{1}{\alpha} \quad y = \beta + \frac{0.5772}{\alpha} \tag{36.120c}$$

which gives

$$y = 0.5772 \, (\text{Euler constant}) \tag{36.120d}$$

The corresponding probability $Pr(Y \le y) = 0.5703$, and the return period is

$$T_r = \frac{1}{1 - Pr(Y \le y)} = 2.33 \tag{36.120e}$$

- **Median Flood:** Corresponds to $F(x) = 0.5$ and has a return period of 2 years
- **Most Probable Flood:** Corresponds to $x = $ mode, i.e. $F(x) = 0.367$, and has a return period of 1.58 years

36.8 PARAMETER ESTIMATION

36.8.1 Method of moments

This method is simple but may not be statistically efficient. The kth moment of a random sample is given by

$$m_k = E\left(X^k\right) \tag{36.121}$$

It involves equating the sample moments with the moments for the underlying population distribution and then algebraically solving the resulting equations. Two frequently used moments are the first and second moments:

$$m_1 = E(X) \tag{36.122a}$$

$$m_2 = E(X^2) \tag{36.122b}$$

36.8.2 Maximum likelihood method

This method maximizes the probability of obtaining the particular sample. It is the most common method used in hydrology.

The likelihood function is the joint probability density function evaluated at the actual sample observations. If X_1, X_2, ..., X_n represent a random sample, and $x_1, x_2, ..., x_n$ represent the observations of this sample, and θ is the estimator, then the likelihood function $L(\theta)$ is defined as follows:

$$L(\theta) = f(x_1; \theta) \, f(x_2; \theta) \, f(x_3; \theta), ..., f(x_n; \theta) \tag{36.123}$$

$L(\theta)$ is maximized as a function of θ to find the maximum likelihood estimator of θ. $L(\theta)$ is the total probability of getting each item of the sample.

If the probability density function is not too complicated, the functional form of the maximum likelihood estimator can be obtained by using calculus.

Since it is difficult to differentiate the product, the fact that the logarithm of a variable must have its maximum at the same place as the place where the variable has its maximum is used.

36.8.3 Least squares method

Least squares method involves fitting a theoretical distribution function to the empirical distribution and then minimizing the sum of squares of the deviations.

$$E = \Sigma \left(y_i - \hat{y}_i \right)^2 \tag{36.124}$$

where

y_i – observed frequency for x_i

$\hat{y}_i$ – expected frequency of the assumed distribution for x_i

The parameters can be obtained by differentiating E with respect to the parameters.
Conditions for an efficient estimator by this method are

- The deviations must be Normal, or at least symmetrical
- Variance of the population deviations be independent of $\hat{y}_i$
- Population variance of the deviations along the least squares curve be constant

36.8.4 Graphical method

Graphical method involves a visual fitting of a function to the plotted data points.

36.9 ERRORS IN FREQUENCY ANALYSIS

The following assumptions are implied or made in frequency analysis:

- The data are random
- The data series is stationary
- The population parameters can be estimated from the sample.

Any violation of these assumptions will cause errors in the analysis.

Errors may also be due to the lack of knowledge of the true distribution. The sample data usually correspond to high probability values, which fill the central part of the distribution with not enough points to define the tails, which have low probability.

The last source of error comes from the parameters of the chosen distribution. The parameters, which are estimated from limited samples may not represent the population values. The method used to estimate them may also give rise to some errors.

36.10 GOODNESS OF FIT ANALYSIS

Goodness-of-fit tests are used to make inferences concerning the unknown population frequency distribution. The sample size should be large enough and at least greater than 5 per class interval.

36.10.1 Chi-square statistic for goodness-of-fit test

- **Null hypothesis, H_0:** The data comes from a Normal distribution
- **Alternative hypothesis, H_a:** The data do not come from a Normal distribution
- **Test statistic:**

$$\chi^2 = \sum \frac{\left(f_i - \hat{f}_i\right)^2}{f_i} \qquad (36.125)$$

where f_i – observed frequency

$\hat{f}_i$ – expected frequency under the null hypothesis
- Decision rule: Accept H_0 if $\chi^2 < \chi_\alpha^2(v)$, where v is the degree of freedom (= number of classes – 2 – 1), and α is the level of significance. Use the upper tail for the test. $\chi_\alpha^2(v)$ values for a given number of degrees of freedom can be found in statistical tables.

36.10.2 Kolmogorov–Smirnov statistic for goodness-of-fit test

This is an alternative to Chi-square test in which the cumulative frequencies are considered. The test statistic is the maximum deviation between matching cumulative frequencies of the respective distribution.

$$D = \max \left| F(x) - \hat{F}(x) \right| \qquad (36.126)$$

The Kolmogorov–Smirnov (K–S) statistic is preferred for small samples. All population parameters must be specified before using this test.

Reject H_0 if D is too large. Cutoff values can be found in statistical tables.

36.11 RISK ASSESSMENT

The lack of knowledge of the true distribution of the random variable leads to uncertainties in the computed flood frequencies. Random sampling is another source of uncertainty. Even when the probability density function is known, there are chance occurrences. For example, a 10-year flood may occur successively in a 10-year period or may not occur even once, both with low probability. Such occurrences (or non-occurrences) are independent of the lack of knowledge of probability density function. They are called design uncertainties or risk of failure and can be estimated by using the concept of binomial risk assessment.

36.12 HYPOTHESIS TESTING

36.12.1 Basic concepts of hypothesis testing

Inferences about populations are usually made in terms of the confidence intervals. However, sometimes it becomes necessary to check the validity of a claim or hypothesis concerning the magnitude of a population parameter. Such checking or testing is based on two complementary assumptions neither of which are known to be true, or hypothesis, regarding the unknown population. The procedure for hypothesis testing involves

- Formulation of the null hypothesis
- Formulation of an alternate hypothesis
- Definition of a test statistic
- Determination of the distribution of the test statistic
- Definition of critical (or acceptance) region of the test statistic
- Testing whether the calculated value of the test statistic falls within the acceptance region.

36.12.1.1 Formulation of the null hypothesis, H_0

The null hypothesis assumes a certain specific value for the unknown population parameter. It is also possible to define the null hypothesis as an inequality – greater than or less than. For example, if the mean of a population is considered, then

$$H_0 : \mu = \mu_0 \tag{36.127a}$$

$$H_0 : \mu \leq \mu_0 \tag{36.127b}$$

$$H_0 : \mu \geq \mu_0 \tag{36.127c}$$

36.12.1.2 Formulation of the alternate hypothesis, H_a

The alternate hypothesis assign the value(s) to the population parameter that are not contained in the null hypothesis. For example,

$$H_a : \mu \neq \mu_0 \tag{36.128a}$$

$$H_a : \mu > \mu_0 \tag{36.128b}$$

$$H_a : \mu < \mu_0 \tag{36.128c}$$

The null hypothesis is accepted or rejected based on the information provided by the sample.

36.12.1.3 Definition of a test statistic

A test statistic must be defined to test the validity of the hypothesis. The test statistic is computed from sample information. For example, if the population mean is to be tested, the test statistic would be

$$Z = \frac{\bar{X} - \mu_0}{\frac{\sigma}{\sqrt{n}}} \tag{36.129a}$$

if large samples are considered, or, if population standard deviation σ is known for small samples. If small samples are considered and only the sample standard deviation s is known,

$$T = \frac{\bar{X} - \mu_0}{\frac{s}{\sqrt{n}}} \tag{36.129b}$$

In Eq. 36.129b, $\bar{X}$ and s are the sample mean and sample standard deviation, respectively.

36.12.1.4 Determination of the distribution of the test statistic

The probability distribution of the test statistic depends on the null hypothesis assumed, the parameter to be tested, and the sample size. Commonly used ones are the Normal, Student's "t", Chi-square and F-distributions.

36.12.1.5 Definition of the critical (or acceptance) region for the test statistic

The set of values of the test statistic that leads to the rejection of H_0 in favour of H_a is called the rejection region or critical region. The critical values depend upon whether the testing is one-sided or two-sided. For example, for the mean, the critical regions are given by

Lower-tailed test ($H_0: \mu \geq \mu_0$)

$$\bar{X}^* = \mu_0 - z_\alpha \sigma_{\bar{X}} \tag{36.130a}$$

Upper-tailed test ($H_0: \mu < \mu_0$)

$$\bar{X}^* = \mu_0 + z_\alpha \sigma_{\bar{X}} \tag{36.130b}$$

Two-sided test ($H_0: \mu = \mu_0$)

$$\bar{X}^* = \mu_0 \pm z_{\frac{\alpha}{2}} \sigma_{\bar{X}} \tag{36.130c}$$

where $\bar{X}^*$ is the critical value of the test statistic, $\sigma_{\bar{X}}$ is the standard deviation of $\bar{X}\left(= \dfrac{\sigma}{\sqrt{n}}\right)$ and z_α is the α-level of significance of the relevant probability distribution.

36.12.1.6 Decision rule

A decision rule is used to accept or reject the null hypothesis. For example, for an upper tailed test, $H_0: \mu \leq \mu_0$

Accept H_0 if $\bar{X} \leq \bar{X}^*$

Reject H_0 if $\bar{X} > \bar{X}^*$

In this context, $\bar{X} \leq \bar{X}^*$ is the acceptance region and $\bar{X} > \bar{X}^*$ is the rejection region.
  Similarly, for a lower tailed test, $H_0: \mu \geq \mu_0$

Accept H_0 if $\bar{X} \geq \bar{X}^*$

Reject H_0 if $\bar{X} < \bar{X}^*$

For a two-tailed test, $\mu = \mu_0$

Accept H_0 if $\bar{X}_1^* \leq \bar{X} \leq \bar{X}_2^*$

Reject H_0 otherwise.

In this case, $\bar{X}_1^* \leq \bar{X} \leq \bar{X}_2^*$ is the acceptance region. Eq. 36.130c provides the two values for $\bar{X}^*$.

36.12.1.7 Outcome

The acceptance or rejection of the hypothesis will lead to the following possible outcomes:

- Accept H_0 when H_0 is true Correct decision
- Reject H_0 when H_0 is false Correct decision
- Reject H_0 when H_0 is true Type I error
- Accept H_0 when H_0 is false Type II error.

36.12.1.8 Error probabilities

The inferences made on the basis of the sample information would always have some degree of error. The probabilities of such errors are defined as follows:

$$\alpha = Pr\left[\text{Type I error}\right] = Pr\left[\text{Reject } H_0 \text{ when } H_0 \text{ is true}\right] \tag{36.131a}$$

$$\beta = Pr\left[\text{Type II error}\right] = Pr\left[\text{Accept } H_0 \text{ when } H_0 \text{ is false}\right] \tag{36.131b}$$

Rejecting a null hypothesis (type I error) is more serious than accepting it when it is false. Therefore, the error probability α is referred to as the significance level. Larger sample sizes tend to reduce both type I and II errors probabilities.

For testing $H_0: \mu \geq \mu_0$ vs. $\mu < \mu_0$, H_0 is rejected in favour of H_a when

$$\frac{\bar{X} - \mu_0}{\frac{\sigma}{\sqrt{n}}} < -z_\alpha$$

For testing $H_0: \mu \leq \mu_0$ vs. $\mu > \mu_0$, H_0 is rejected in favour of H_a when

$$\frac{\bar{X} - \mu_0}{\frac{\sigma}{\sqrt{n}}} > z_\alpha$$

The probability that H_0 is rejected when $\mu = \mu_0$ is given by

$$Pr\{\text{Reject } H_0 \text{ when } \mu = \mu_0\} = Pr\left\{\frac{\bar{X} - \mu_0}{\frac{\sigma}{\sqrt{n}}} < -z_\alpha \text{ when } \mu = \mu_0\right\} = \alpha \tag{36.131c}$$

Therefore,

$$\alpha = Pr\{\text{Type I error when } \mu = \mu_0\} \tag{36.131d}$$

and

$$\beta = Pr\{\text{Type II error when } \mu = \mu_a\} = Pr\{\text{Accept } H_0 \text{ when } \mu = \mu_a\} \tag{36.131e}$$

Figure 36.3 shows two Normal curves with mean values μ_0 and μ_a, with α as the area to the right of K under the Normal curve with centre μ_0 and β as the area to the left of K under the Normal curve centered at μ_a. Note that α is the probability that the null hypothesized value

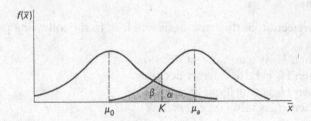

Figure 36.3 Error probabilities.

μ_0 is rejected when in fact $\mu = \mu_0$, and β is the probability that the null hypothesis is accepted when in fact $\mu = \mu_a$. The point K is determined as follows:

For testing $H_0: \mu = \mu_0$ vs. $H_a: \mu > \mu_0$, H_0 is rejected if

$$\frac{\bar{X} - \mu_0}{\dfrac{\sigma}{\sqrt{n}}} > z_\alpha$$

Therefore,

$$\alpha = Pr\left\{ \frac{\bar{X} - \mu_0}{\dfrac{\sigma}{\sqrt{n}}} > z_\alpha \right\} \tag{36.131f}$$

$$\alpha = Pr\left\{ \bar{X} > \mu_0 + z_\alpha \frac{\sigma}{\sqrt{n}} \right\} \tag{36.131g}$$

Therefore,

$$K = \mu_0 + z_\alpha \frac{\sigma}{\sqrt{n}} \tag{36.131h}$$

i.e. $\alpha = Pr\{\bar{X} > K\}$ \hfill (36.131i)

From the figure, it is also clear that

$$\beta = Pr\left\{ \bar{X} < \mu_a - z_\beta \frac{\sigma}{\sqrt{n}} \right\} = Pr\{\bar{X} < K\} \tag{36.131j}$$

Therefore,

$$K = \mu_a - z_\beta \frac{\sigma}{\sqrt{n}} \tag{36.131k}$$

Equating Eqs. 36.131h and 36.131k,

$$\mu_0 + z_\alpha \frac{\sigma}{\sqrt{n}} = \mu_a - z_\beta \frac{\sigma}{\sqrt{n}} \tag{36.131l}$$

$$\frac{\sigma}{\sqrt{n}}\left(z_\alpha + z_\beta \right) = \mu_a - \mu_0$$

or

$$n = \frac{(z_\alpha + z_\beta)\sigma^2}{(\mu_a - \mu_0)^2} \tag{36.131m}$$

This is the sample size for specified values of α and β when testing $H_0: \mu = \mu_0$ vs. $H_a: \mu = \mu_a$.

36.12.2 Hypothesis testing using a single sample

36.12.2.1 Testing the mean: general distribution (large samples)

36.12.2.1.1 Two-sided test

$$H_0 : \mu = \mu_0$$

$$H_a : \mu \neq \mu_0$$

The test statistic is

$$Z = \frac{\bar{X} - \mu_0}{\dfrac{\sigma}{\sqrt{n}}} \tag{36.132a}$$

The null hypothesis is rejected if

$$|Z| = \left| \frac{\bar{X} - \mu_0}{\dfrac{\sigma}{\sqrt{n}}} \right| > z_{\frac{\alpha}{2}} \tag{36.132b}$$

for some small prescribed value of α.

36.12.2.1.2 One-sided test

Lower tailed

$$H_0 : \mu \geq \mu_0$$

$$H_a : \mu < \mu_0$$

The test statistic is the same and H_0 is rejected if $Z < -z_\alpha$

Upper tailed

$$H_0 : \mu \leq \mu_0$$

$$H_a : \mu > \mu_0$$

The test statistic is the same, and H_0 is rejected if $Z > z_\alpha$

36.12.2.2 Testing the mean: Normal distribution (small samples)

For small samples (small for the central limit theorem to provide a good approximation to the distribution of $\bar{X}$), the population is assumed to be Normal. Then $\bar{X}$ will also be Normal, and,

$$T = \frac{\bar{X} - \mu_0}{\dfrac{s}{\sqrt{n}}} \tag{36.133}$$

will have a "t" distribution with $(n - 1)$ degrees of freedom under $H_0\colon \mu = \mu_0$. H_0 is rejected if $|T| \geq t_{\frac{\alpha}{2}}$ where $t_{\frac{\alpha}{2}}$ corresponds to the upper tail of the "t" distribution with $(n - 1)$ degrees of freedom.

36.12.2.3 Testing the probability of success: Binomial distribution

If Y is Binomial with mean "np", then Y/n is approximately Normal for large n with mean p and variance $\dfrac{p(1-p)}{n}$.

If $H_0\colon \hat{p} = \dfrac{Y}{n}$

$$Z = \frac{\hat{p} - p_0}{\sqrt{\dfrac{p_0(1-p_0)}{n}}} \tag{36.134}$$

has a standard Normal distribution.

36.12.2.4 Testing the variance: Normal distribution

$$H_0 : \sigma^2 = \sigma_0^2$$

$$H_a : \sigma^2 \neq \sigma_0^2$$

Test statistic

$$\chi^2 = \frac{(n-1)s^2}{\sigma_0^2} \tag{36.135}$$

Reject H_0 if the test statistic is

$$> \chi^2_{\frac{\alpha}{2}}(n-1) \quad \text{or} \quad < \chi^2_{1-\frac{\alpha}{2}}(n-1)$$

36.12.3 Hypothesis testing using multiple samples

36.12.3.1 Testing the difference between two means: General distribution

$$H_0 : \mu_1 - \mu_2 = D_0$$

$$H_a : \mu_1 - \mu_2 \neq D_0$$

Test statistic

$$Z = \frac{\bar{X}_1 - \bar{X}_2 - D_0}{\sqrt{\dfrac{\sigma_1^2}{n_1} + \dfrac{\sigma_2^2}{n_2}}} \tag{36.136}$$

Reject H_0 if $|Z| > z_{\alpha/2}$

36.12.3.2 Testing the difference between two means: Normal distribution with equal variance (small samples)

$$H_0 : \mu_1 - \mu_2 = D_0$$

$$H_a : \mu_1 - \mu_2 \neq D_0$$

Test statistic

$$T = \frac{\bar{X}_1 - \bar{X}_2 - D_0}{s_p \sqrt{\dfrac{1}{n_1} + \dfrac{1}{n_2}}} \tag{36.137}$$

where

$$s_p^2 = \frac{(n_1 - 1) s_1^2 + (n_2 - 1) s_2^2}{n_1 + n_2 - 2} \tag{36.138}$$

has a "t" distribution with $n_1 + n_2 - 2$ degrees of freedom.
   Reject H_0 if $|T| > t_{\alpha/2}$

36.12.3.3 Testing the difference between two means: Normal distribution with unequal variances (small samples)

This method is only approximate.
   Test statistic

$$T = \frac{\bar{X}_1 - \bar{X}_2 - D_0}{\sqrt{\dfrac{s_1^2}{n_1} + \dfrac{s_2^2}{n_2}}} \tag{36.139}$$

has a "t" distribution with the degree of freedom given by the integral part of

$$\frac{\left(s_1^2 + s_2^2\right)^2}{\left(\dfrac{s_1^2}{n_1 - 1}\right) + \left(\dfrac{s_2^2}{n_2 - 1}\right)}$$

36.12.3.4 Testing the difference between means for paired samples: Normal distribution (small samples)

The two sample "t" test above is applicable to independent samples. If (X_1, Y_1), (X_2, Y_2), (X_3, Y_3), ..., (X_n, Y_n) are random samples of paired observations, and it is required to test the hypothesis concerning the difference between $E(X_i)$ and $E(Y_i)$, the earlier tests cannot be used because of the dependence between X_i and Y_i. Since

$$E(X_i) - E(Y_i) = E(X_i - Y_i)$$

$$H_0 : E(X_i) - E(Y_i) = 0$$

is equivalent to

$$H_0 : E(D_i) = 0$$

where

$$D_i = X_i - Y_i$$

Assume $D_1, D_2, ..., D_n$ is a random sample of differences, each normally distributed with mean μ_D and variance σ_D^2

$$H_0 : \mu_D = 0$$

$$H_a : \mu_D \neq 0$$

Test statistic

$$T = \frac{\bar{D} - \mu_D}{\frac{s_D}{\sqrt{n}}} \qquad (36.140a)$$

where

$$\bar{D} = \frac{1}{n} \sum D_i$$

and

$$s_D^2 = \frac{1}{n-1} \sum \left(D_i - \bar{D} \right)^2 \qquad (36.140b)$$

has a "t" distribution with $(n - 1)$ degrees of freedom.

36.12.3.5 Testing the ratio of variances: Normal distribution

$$H_0 : \quad \sigma_1^2 = \sigma_2^2 \quad \left(\text{or} \quad \frac{\sigma_1^2}{\sigma_2^2} = 1 \right)$$

$$H_a: \sigma_1^2 \neq \sigma_2^2 \quad \left(\text{or} \quad \frac{\sigma_1^2}{\sigma_2^2} \neq 1 \right)$$

Test statistic

$$F = \frac{s_1^2}{s_2^2} \tag{36.141}$$

has a F-distribution under the null hypothesis H_0.

Reject H_0 for $F > F_{\frac{\alpha}{2}}(v_1, v_2)$ or $F < F_{1-\frac{\alpha}{2}}(v_1, v_2)$

Note: $F_{1-\frac{\alpha}{2}}(v_1, v_2) = \dfrac{1}{F_{\frac{\alpha}{2}}(v_2, v_1)}$

One-sided tests can be formulated for testing $\sigma_1^2 \leq \sigma_2^2$ or $\sigma_1^2 \geq \sigma_2^2$. In these cases, reject H_0 if $F > F_\alpha(v_1, v_2)$, or, $F < F_{1-\alpha}(v_1, v_2)$ accordingly.

36.12.4 Tests on frequency data

36.12.4.1 Testing parameters of the Multinomial distribution

Let X_i, $i = 1, 2, ..., k$ denote the number of trials resulting in outcome i with p_i denoting the probability that any open trial will result in outcome i.

$$E(X_i) = np_i$$

$$H_0: p_1 = p_{1_0}, p_2 = p_{2_0}, p_{3_0} = p_{3_0}, ..., p_k = p_{k_0}$$

H_a : At least one equality fails (in the most general case)

Test statistic

$$X^2 = \sum_1^k \frac{\left[X_i - E(X_i) \right]^2}{E(X_i)} \tag{36.142}$$

will be approximately χ^2 – distributed with $(k - 1)$ degrees of freedom. Division by $E(X_i)$ is to standardize the difference of squares.

Reject H_0 if $X^2 > \chi_\alpha^2(k-1)$

36.12.4.2 Testing equality among Binomial parameters

Consider k independent random samples resulting in k Binomially distributed random variables $Y_1, Y_2, ..., Y_k$ where Y_i is based on n_i trials with success probability p_i on each trial.

$$H_0: p_1 = p_2 = \cdots = p_k$$

H_a : atleast one inequality

Observation	1	2	...	k	Total
Successes	y_1	y_2	...	y_k	Y
Failures	$n_1 - y_1$	$n_2 - y_2$	...	$n_k - y_k$	$n - y$
Total	n_1	n_2	...	n_k	N

$$n = \sum_{1}^{k} n_i; \quad y = \sum_{1}^{k} y_i$$

Since Y_i is Binomial,

$$E(Y_i) = n_i p_i$$

$$E(n_i - Y_i) = n_i - n_i p_i = n_i (1 - p_i)$$

Under the null hypothesis that $p_1 = p_2, ..., = p_k = p$, the minimum variance unbiased estimator of p

$$= \frac{1}{n} \sum_{1}^{k} Y_i = \frac{Y}{n} = \frac{\text{Total number of successes}}{\text{Total sample size}}$$

The estimators of the expected cell frequencies are

$$\widehat{E}(Y_i) = n_i \left(\frac{Y}{n} \right) \qquad \text{(successes)}$$

$$\widehat{E}(n_i - Y_i) = n_i \left(1 - \frac{Y}{n} \right) = n_i \left(\frac{n - Y}{n} \right) \quad \text{(failures)}$$

[The expected cell frequency is found from

$$\text{Estimated expected cell frequency} = \frac{(\text{Column total})(\text{row total})}{\text{Overall total}}]$$

Test statistic

$$X^2 = \sum_{1}^{k} \left[\frac{\left[Y_i - \widehat{E}(Y_i) \right]^2}{\widehat{E}(Y_i)} + \frac{(n_i - Y_i) - \widehat{E}(n_i - Y_i)^2}{\widehat{E}(n_i - Y_i)} \right] \tag{36.143}$$

has approximately χ^2 distribution with $(k - 1)$ degrees of freedom.

36.12.5 Sample size for testing the mean

A larger sample improves the reliability of the result. The required (minimum) sample size depends on α and β.

When $\bar{X}^*$ shifts in one direction, α will increase and β will decrease. When it shifts in the opposite direction, α will decrease and β will increase. By choosing the correct sample size, a balance can be achieved.

For upper-tailed test ($H_0: \mu \le \mu_0$),

$$\bar{X}^* = \mu_0 + z_\alpha \alpha_{\bar{X}} = \mu_0 + z_\alpha \frac{\sigma}{\sqrt{n}}$$

Since μ_0, σ and z_α are fixed, $\bar{X}^*$ can shift only by increasing or decreasing n. If z_β is the critical Normal deviate providing a lower tail area equal to β when the Normal curve for $\bar{X}$ is centered at μ_1, n can be adjusted so that the resulting $\bar{X}^*$ has the following property:

$$\bar{X}^* = \mu_1 - z_\beta \frac{\sigma}{\sqrt{n}}$$

α, β must be assumed. Also, the assumed population for β test is μ_1, $\mu_1 > \mu_0$ for upper tailed test. Therefore,

$$n = \left[\frac{\sigma(z_\alpha + z_\beta)}{\mu_1 - \mu_0} \right]^2 \tag{36.144}$$

Example 36.1

A rare daily event (rainfall > 100 mm) occurred five times during a period of 10 years.

   i. What is the probability of this event occurring 6, 7, 8, 9, 10 times in 20 years?
  ii. What is the probability of this event occurring
     a. less than 6 times
     b. more than 10 times
     c. between 7 and 9 times
     d. exactly 7 times.

Using Poisson distribution,

$$p = \frac{5}{3{,}650} = \frac{1}{730}$$

$$n = 20 \times 365 = 7{,}300$$

$$\lambda = np = 10$$

Therefore,

$$f(6) = \frac{10^6 e^{-10}}{6!} = 0.063$$

$$f(7) = \frac{10^7 e^{-10}}{7!} = 0.090$$

$$f(8) = \frac{10^8 e^{-10}}{8!} = 0.113$$

$$f(9) = \frac{10^9 e^{-10}}{9!} = 0.125$$

$$f(10) = \frac{10^{10} e^{-10}}{10!} = 0.125$$

Using statistical tables, which give the cumulative probabilities, it is possible to find the probability of this event occurring

a. less than 6 times

$$F(X \le 6) = \sum_{x=0}^{6} f(x) = 0.130$$

b. more than 10 times

$$F(X \ge 10) = 1 - F(X \le 9) = 0.542$$

c. between 7 and 9 times

$$F(7 \le X \le 9) = F(X \le 9) - F(X \le 7) = 0.458 - 0.220 = 0.238$$

d. exactly 7 times

$$F(X = 7) = F(X \le 7) - F(X \le 6) = 0.220 - 0.130 = 0.090$$

Example 36.2

What is the probability of a 20-year return period flood occurring in a particular 3 year period?

$$Pr(X \ge 20 \text{ year flood})_3 = 1 - \left(1 - \frac{1}{20}\right)^3 = 0.143 = 14.3\%$$

Note: As the length of the particular period increases, the probability increases. For example, if $n = 100$ years,

$$Pr(X \ge 20 \text{ years flood})_{100} = 1 - \left[1 - \frac{1}{20}\right]^{100} \approx 100\%$$

If the probability $Pr(X \ge x)_n$ is pre-defined, the value of 'n', the design period, can be determined as follows:

$$Pr(X \ge x)_n = 1 - \left(1 - \frac{1}{T_r}\right)^n$$

Therefore,

$$n = \frac{\ln\left[1 - Pr(X \ge x)_n\right]}{\ln\left(\frac{T_r - 1}{T_r}\right)}$$

REFERENCES

Beard, L. R. (1943): Statistical analysis in hydrology, *Transactions, American Society of Civil Engineers*, vol. 108, pp. 1110–1160.

Blom, G. (1958): *Statistical Estimates and Transformed Beta Variables*, Wiley, New York, p. 73.

California Department of Public Works (1923): Flow of California streams: Div. Engineering and Irrigation, Bull. 5, chapter 5.

Chow, V. T. (1951): A general formula for hydrologic frequency analysis, EOS, American Geophysical Union, April 1951. doi: 10.1029/TR032i002p00231.

Cunnane, C. (1978): Unbiased plotting positions. A review, *Journal of Hydrology*, vol. 37, pp. 205–222.

Fisher, R. A. (1915): Frequency distribution of the values of the correlation coefficient in samples of an indefinitely large population, *Biometrika*, vol. 10, no. 4, pp. 507–521.

Fisher, R. A. (1921): On the 'probable error' of a coefficient of correlation deduced from a small sample (PDF), *Metron*, vol. 1, pp. 3–32.

Gringorten, I. I. (1963): A plotting rule for extreme probability paper, *Journal of Geophysical Research*, vol. 68, pp. 813–814.

Hazen, A. (1930): *Flood Flows, Study of Frequency and Magnitudes*, Wiley, New York.

Kite, G. W. (1977): *Frequency and Risk Analysis in Hydrology*, Water Resources Publications, Fort Collins, CO.

McCuen, R. H. (1986): *Hydrological Modelling: Statistical Methods and Applications*, Prentice Hall, Upper Saddle River.

Scheaffer, R. L. and McClave, J. T. (1990): *Probability and Statistics for Engineers*, PWS-Kent, Boston, MA, p. 380.

Tukey, W. (1962): The future of data analysis, *Annals of Mathematical Statistics*, vol. 33, pp. 21–24.

Weibull, W. (1939): A statistical theory of strength of materials, *Ingeniors Vetenskaps Akademien, Handlingar*, vol. 151, pp. 1–45.

Wonnacott, T. W. and Wonnacott, R. J. (1990): *Introductory Statistics for Business and Economics*, John Wiley & Sons, Hoboken, NJ, pp. 378–380.

Chapter 37

Systems theory approach to hydrological modelling

37.1 INTRODUCTION

Systems science is an interdisciplinary field of studies that covers a broad range of areas including nature, society, science, engineering and medicine. It covers well-defined fields such as complex systems, cybernetics, dynamical systems and chaos, control theory, operational research, ecology and many others. In view of the diversity of fields covered, systems theory has developed from different fields and its applications extend from microscopic scale to very large scale.

The general systems theory (GST) was originally proposed by biologist Ludwig von Bertalanffy in the 1930s as a modelling tool that accommodates the interrelationships and overlap between separate disciplines. More details of the background of systems theory together with applications in the environmental and hydrological fields can be found in a separate publication by the author (Jayawardena, 2014).

A deterministic model defines a cause–effect relationship and therefore predictions are possible, at least in theory. In a deterministic system, cause is the driving function (input) and effect is the response function (output). Deterministic systems can be considered as lumped parameter systems or as distributed parameter systems. In general, from a mathematical point of view, lumped parameter systems are represented by ordinary differential equations, and distributed parameter systems are represented by partial differential equations. In a linear system, the transformation from input to output takes place via a linear operator.

37.2 BASIC CONCEPTS OF LINEAR SYSTEMS

A linear system can be represented by an ordinary differential equation of the form

$$a_1 y(t) + a_2 y'(t) + a_3 y''(t) + \ldots = b_1 x(t) + b_2 x'(t) + b_3 x''(t) + \ldots \tag{37.1}$$

where $x(t)$ represents an input function, $y(t)$ represents an output function, a_i's and b_i's are system parameters, and the superscripts in x and y indicate the corresponding derivatives.

All linear systems satisfy the principles of proportionality and superposition, i.e. if $y_1(t)$ and $y_2(t)$ are output functions corresponding to the input functions $x_1(t)$ and $x_2(t)$, then,

- $y_1(t) + y_2(t)$ will be the output function corresponding to the input function $x_1(t) + x_2(t)$, and,
- $\alpha y_1(t)$ will be the output corresponding to the input function $\alpha x_1(t)$ (α is a constant).

701

37.3 TIME DOMAIN ANALYSIS

37.3.1 Types of input functions

37.3.1.1 Impulse (or Dirac Delta) function

The unit impulse is known as the Dirac Delta function (Figure 37.1) which has the following properties:

$$\delta(t-a) = 0 \text{ for } t \neq a \tag{37.2a}$$

$$\int_0^\infty \delta(t-a)\,dt = 1 \tag{37.2b}$$

If it is applied at $t = 0$,

$$\delta(t) = 0 \text{ for } t \neq 0 \tag{37.3a}$$

$$\int_0^\infty \delta(t)\,dt = 1 \tag{37.3b}$$

37.3.1.2 Step function

The unit step function $u(t)$ has the following properties (Figure 37.2):

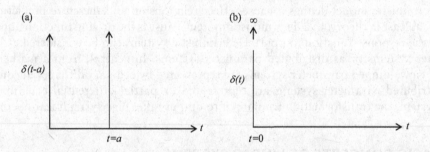

Figure 37.1   (a) Delta function applied at $t = a$. (b) Delta function applied at $t = 0$.

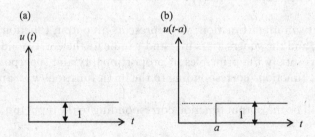

Figure 37.2   (a) Step function applied at $t = 0$. (b) Step function applied at $t = a$.

$$u(t) = 0 \text{ for } t < 0 \tag{37.4a}$$

$$u(t) = 1 \text{ for } t \geq 0 \tag{37.4b}$$

If it is applied at $t = a$, then

$$u(t - a) = 0 \text{ for } t < a \tag{37.5a}$$

$$u(t - a) = 1 \text{ for } t \geq a \tag{37.5b}$$

The delta function and the step function are related to each other by

$$\int_{-\infty}^{t} \delta(\alpha - \tau) \, d\alpha = u(t - \tau); \text{ or, } \frac{du(t)}{dt} = \delta(t) \tag{37.6}$$

37.3.1.3 Arbitrary function

Any real input function can be expressed as a sum of functions of either the delta function or the step function. Decomposing into delta functions, an input function $x(t)$ can be written as

$$x(t) = \int_{-\infty}^{\infty} x(\tau) \, \delta(t - \tau) \, d\tau \tag{37.7a}$$

Similarly, decomposing into step functions,

$$x(t) = \int_{-\infty}^{\infty} \dot{x}(\tau) \, u(t - \tau) \, d\tau \tag{37.7b}$$

In both these equations, the limits of integration can be changed from zero to infinity instead of from minus infinity to infinity.

In discrete form Eqs. 37.7a and 37.7b can be written as

$$x(j) = \sum_{k=0}^{k=\infty} x(k) \, \delta(j - k) \tag{37.8a}$$

and

$$x(j) = \sum_{k=1}^{k=j} \dot{x}(k) \, u(j - k) \tag{37.8b}$$

37.3.2 System response function – convolution integral

The input and output functions of linear systems are related to each other by the convolution integral, which takes the form

$$y(t) = \int_0^t h(\tau) x(t - \tau) d\tau \tag{37.9a}$$

where $h(t)$ is the system response function. For real systems, the variables in the convolution integral are interchangeable which means

$$y(t) = \int_0^t x(\tau) h(t - \tau) d\tau \tag{37.9b}$$

The system response function $h(t)$ corresponds to the output function when the input function is of the impulse (delta) function type. It is, therefore, called the impulse response function (IRF). A linear system is completely known if the IRF is known.

If the system is linear time variant, then the corresponding equation is of the form:

$$y(t) = \int_0^t h(\tau, t) x(t - \tau) d\tau \tag{37.10}$$

implying $h(\tau, t)$ is a function of t.

For discrete systems, the convolution summation can be written as

$$y(j) = \sum_{k=0}^{k=j} x(j - k) h(k) \tag{37.11}$$

37.4 FREQUENCY DOMAIN ANALYSIS

37.4.1 Fourier transform – frequency response function (FRF)

If a linear time invariant system is physically realizable and stable,[1] then it can be described by a frequency response function in the frequency domain. The frequency response function $H(f)$ is the Fourier transform of the IRF.

$$H(f) = \int_0^\infty h(\tau) e^{-2\pi f \tau} d\tau \tag{37.12}$$

$$\left(\int_{-\infty}^\infty = \int_0^\infty \text{ because } h(\tau) = 0, \text{ for } \tau < 0 \right)$$

$H(f)$, the Fourier transform of the IRF $h(t)$ is given by

---

[1] If the input function is bounded, then the output function is also bounded, i.e. if $|x(t)| < N$, then $|y(t)| < M$; N, M constants.

$$H(f) = \int\limits_{-\infty}^{\infty} h(t) e^{-2\pi j f t} \, dt$$

$$= \int\limits_{-\infty}^{\infty} h(t) \left\{ \cos(2\pi f t) - j \sin(2\pi f t) \right\} dt \qquad (37.13)^2$$

$$= R_h(f) + j I_h(f)$$

Real part + Imaginary part

(even)+(odd)[3]

Inverse Fourier transform of $H(f)$ gives the IRF $h(t)$:

$$h(t) = \frac{1}{2\pi} \int\limits_{-\infty}^{\infty} H(f) e^{2\pi j f t} \, df$$

$$= \frac{1}{2\pi} \int\limits_{-\infty}^{\infty} H(f) \left\{ \cos(2\pi f t) + j \sin(2\pi f t) \right\} df \qquad (37.14)$$

Similarly, $X(f)$ and $Y(f)$, the Fourier transforms of $x(t)$ and $y(t)$ are given by

$$X(f) = \int\limits_{-\infty}^{\infty} x(t) \left\{ \cos(2\pi f t) - j \sin(2\pi f t) \right\} dt \qquad (37.15)$$

$$Y(f) = \int\limits_{-\infty}^{\infty} y(t) \left\{ \cos(2\pi f t) - j \sin(2\pi f t) \right\} dt \qquad (37.16)$$

Through Fourier transformation, the convolution integral which relates the input to the output in the time domain can be represented by an algebraic equation in the frequency domain as follows:

In the time domain,

$$y(t) = \int\limits_{-\infty}^{\infty} x(t) h(t - \tau) \, dt \qquad (37.17)$$

In the frequency domain,

$$Y(f) = X(f) H(f) \qquad (37.18)$$

---

[2] In this equation, t can be replaced by τ; hence $h(\tau)$ by $h(t)$.

[3] For even functions $\int\limits_{-\infty}^{\infty} = 2 \int\limits_{0}^{\infty}$; and for odd functions $\int\limits_{-\infty}^{\infty} = 0$.

Therefore,

$$H(f) = \frac{Y(f)}{X(f)} \tag{37.19}$$

37.4.2 Laplace transform

The Laplace transform $H(s)$ of $h(t)$ is given by

$$H(s) = \int_0^\infty h(t) e^{-st} \, dt \tag{37.20}$$

where s is the variable of the Laplace transform which has the dimension $[T]^{-1}$. $H(s)$ is dimensionless.

Similar to Fourier transform, the convolution integral transforms to an algebraic equation by Laplace transformation as well:

$$Y(s) = X(s) H(s) \tag{37.21}$$

where $Y(s)$ and $X(s)$ are the Laplace transforms of $y(t)$ and $x(t)$. Therefore, as before,

$$H(s) = \frac{Y(s)}{X(s)}$$

37.4.2.1 Properties of H(s)

Because $h(t)$ is a bounded function of exponential order, $H(s)$ is finite for all positive values of s. Since $h(t)$ and e^{-st} are both positive functions in the range of integration, $H(s)$ is non-negative. Since $h(t)$ is of exponential order, $H(s)$ must approach zero as s increases.

For $s = 0$, $H(0) = \int_0^\infty h(t) \, dt = 1$

For $s = 1$, $\dfrac{dH(s)}{ds} = -t_L$

$$\left[\text{because} \int_0^\infty h(t) t \, dt = t_L \, (\text{lag time}) \right]$$

Also, $\dfrac{dH(s)}{ds} \left(= \dfrac{dH(s)}{dt} \right)$ is negative

The problem with Laplace transforms is that it is not easy to find the inverse transform to obtain $h(t)$. Usually, $H(s)$ is obtained in a discrete form, and an empirical equation is fitted to the discrete form before inverse transformation is carried out.

37.4.3 z-transform

The z-transform is usually used for discrete time functions.

$$X(z) = \sum_{i=0}^{\infty} x_i z^{-i} \tag{37.22}$$

where $x \equiv x_1, x_2, \ldots x_i$

and $x_i = \dfrac{1}{2\pi j} \oint\limits_{\text{unit circle}} X(z) z^{i-1} \, dz$

where z is a complex variable.
Similar to Laplace transform,

$$Y(z) = H(z)X(z)$$

and $H(z)$ is called the pulsed transfer function.

37.5 SPECIAL CASES OF LINEAR SYSTEMS

37.5.1 Linear reservoir

The concept of linear reservoir has been used in many lumped parameter hydrological models. In a linear reservoir, the output (discharge) is assumed to be proportional to the storage S as follows:

$$S(t) = Ky(t) \tag{37.23}$$

From continuity considerations,

$$x(t) - y(t) = \frac{dS(t)}{dt} \tag{37.24}$$

Combining,

$$x(t) - y(t) = K \frac{dy(t)}{dt} \tag{37.25}$$

or

$$\frac{dy(t)}{dt} + \frac{1}{K} y(t) = \frac{1}{K} x(t)$$

which is the same as Eq. 37.1 with $a_1 = \dfrac{1}{K}$, $a_2 = 1$, $b_1 = \dfrac{1}{K}$, and all other a_i's and b_i's zero.

This relationship can also be obtained as follows:
The response function of a linear reservoir is of the form

$$h(t) = \alpha \, e^{-\alpha t}$$

The output function therefore is

$$y(t) = \int_0^t \alpha\, e^{-\alpha(t-\tau)}\, x(\tau)\, d\tau$$

The differential equation that describes the linear reservoir can now be obtained by differentiating the integral equation:

$$\frac{d}{dt}\{y(t)\} = \frac{d}{dt} \int_0^t \alpha\, e^{-\alpha(t-\tau)} x(\tau)\, d\tau$$

The integration is done using Leibniz's rule, i.e.

If $F(t) = \displaystyle\int_{v(t)}^{u(t)} f(t,\tau)\, d\tau$, then

$$F'(t) = \int_{v(t)}^{u(t)} \frac{df}{dt}\, d\tau + u'(t) f(t, u(t)) - v'(t) f(t, v(t))$$

In this case,

$$f(t,\tau) = \alpha\, e^{-\alpha(t-\tau)} x(\tau); \text{ and therefore, } f'(t) = -\alpha^2\, e^{-\alpha(t-\tau)}$$

$$u(t) = t; \text{ and therefore, } u'(t) = 1$$

$$v(t) = 0; \text{ and therefore, } v'(t) = 0.$$

Substituting,

$$y'(t) = \int_0^t -\alpha^2\, e^{-\alpha(t-\tau)} x(\tau)\, d\tau + \alpha\, e^{-\alpha(t-\tau)} x(t) + 0$$

$$= -\alpha y(t) + \alpha x(t)$$

or

$$y'(t) + \alpha y(t) = \alpha x(t)$$

which is the same as Eq. 37.1 with $a_1 = \alpha$, $a_2 = 1$, $b_1 = \alpha$, and all other a_i's and b_i's zero; $\alpha = \dfrac{1}{K}$]

The solution to this ODE can be obtained as follows:

$$x(t) - y(t) = K \frac{dy(t)}{dt} \Rightarrow \frac{1}{K}\, dt = \frac{dy(t)}{x(t) - y(t)}$$

Integration yields,

$$\frac{t}{K} = -\ln\{x(t) - y(t)\} + \text{const}$$

Initial condition is that at $t = 0$, $y(t) = 0$. Therefore, constant $= \ln[x(t)]$.
  Substituting back,

$$\frac{t}{K} = -\ln\{x(t) - y(t)\} + \ln[x(t)]$$

$$\frac{t}{K} = -\ln\left\{\frac{x(t) - y(t)}{x(t)}\right\}$$

or

$$e^{-t/K} = \left\{\frac{x(t) - y(t)}{x(t)}\right\}$$

or

$$y(t) = x(t)\{1 - e^{-t/K}\} \tag{37.26}$$

When $t \to \infty$, $y(t) \to x(t)$ in the above equation (steady state condition).
  If the inflow stops at time t_0 since outflow began, the outflow at time t in terms of the outflow $y(t_0)$ at time t_0 is obtained as follows:

$$x(t) - y(t) = K\frac{dy(t)}{dt} \Rightarrow -y(t) = K\frac{dy(t)}{dt} \text{ (for } t > t_0)$$

which when integrated gives

$$-\frac{1}{K}t = \ln[y(t)] + \text{constant}$$

The initial condition is that at $t = t_0$, $y(t) = y(t_0)$. Therefore, $\text{constant} = -\dfrac{t_0}{K} - \ln[y(t_0)]$
  Substituting back,

$$-\frac{t}{K} = \ln[y(t)] - \frac{t_0}{K} - \ln[y(t_0)]$$

$$-\frac{t - t_0}{K} = \ln\frac{y(t)}{y(t_0)} \Rightarrow \frac{y(t)}{y(t_0)} = e^{-\frac{t-t_0}{K}}$$

which can be written as

$$y(t) = y(t_0)\,e^{-(t-t_0)/K} \tag{37.27}$$

Since $S(t_0) = Ky(t_0)$, $y(t_0) = \dfrac{S(t_0)}{K}$

When the inflow is instantaneous, $(t_0 = 0)$, then,

$$y(t) = \frac{S(t_0)}{K} e^{-t/K} \tag{37.28}$$

For unit input, $S(t_0) = 1$, and therefore the IRF of a linear reservoir is

$$h(t) = \frac{1}{K} e^{-t/K} \tag{37.29}$$

The storage coefficient K is approximated by the lag time t_L.

One disadvantage of the linear reservoir is that the discharge takes place without any translation.

37.5.2 Linear cascade

In this model, a series of linear reservoirs are arranged in series, and the output from the upstream reservoir is used as the input to the downstream reservoir in the form of a cascade. The resulting response function takes the form

$$y_1(t) = h_1(t) = \frac{1}{K} e^{-t/K} \tag{37.30a}$$

$$y_2(t) = h_2(t) = \int_0^t h(\tau) x(t-\tau)\, d\tau = \int_0^t \frac{1}{K} e^{-\tau/K} \frac{1}{K} e^{-(t-\tau)/K}\, d\tau$$

$$= \frac{1}{K^2} \int_0^t e^{-\tau/K} e^{-(t-\tau)/K}\, d\tau = \frac{1}{K^2} \int_0^t e^{-\tau/K - (t-\tau)/K}\, d\tau \tag{37.30b}$$

$$= \frac{1}{K^2} e^{-t/K} \int_0^t d\tau = \frac{1}{K^2} t e^{-(t-\tau)/K}$$

$$y_3(t) = h_3(t) \int_0^t h(\tau) x(t-\tau)\, d\tau = \int_0^t \frac{1}{K} e^{-\tau/K} \frac{1}{K^2} e^{-(t-\tau)/K} (t-\tau)\, d\tau$$

$$\tag{37.30c}$$

$$= \frac{1}{2} \frac{1}{K^3} t^2 e^{-(t-\tau)/K}$$

$$\cdots\cdots\cdots\cdots\cdots\cdots\cdots\cdots\cdots\cdots\cdots\cdots\cdots\cdots$$

$$\cdots\cdots\cdots\cdots\cdots\cdots\cdots\cdots\cdots\cdots\cdots\cdots\cdots\cdots$$

$$y_n(t) = h_n(t) = \frac{1}{(n-1)! K^n} t^{(n-1)} e^{-t/K} \tag{37.30d}$$

$$= \frac{1}{\Gamma(n)K^n} t^{(n-1)} e^{-t/K} \tag{37.30e}$$

which mathematically, is representing a two-parameter Gamma distribution. The linear cascade hydrological model was proposed by Nash (1957) and is widely known as the Nash model.

37.5.2.1 Determination of K and n

37.5.2.1.1 Moment method[4]

$m_1 = nK$ – first moment of IUH about $t = 0$, the lag time

$m_2 = n(n+1)K^2$ – second moment (about the origin $t = 0$)

n and K are assumed and the IUH determined. Then verify n and K by the method of moments – trial and error.

In the context of the IUH, the lag time is defined as follows:

$$m_{drh1} - m_{reh1} = nK$$

$$\tag{37.31}$$

$$m_{drh2} - m_{reh2} = n(n+1)K^2 + 2nKm_{reh1}$$

where drh – direct runoff hydrograph,

reh – rainfall excess hydrograph,

suffixes 1 and 2 refer to the first and second moments.

Therefore the values of n and K can be determined from the known moments of rainfall excess and direct runoff hydrographs. If K is variable, the cascade becomes non-linear.

37.5.2.2 Variations of the cascade model

A variation of the linear cascade model with two linear cascades of different n and K has been proposed by Diskin (1964). The input is split into two parts by weighting parameters α and β where $\beta = 1 - \alpha$.

$$h(t) = \frac{\alpha}{\Gamma(n_1)K_1^{m_1}} t^{(m_1-1)} e^{-t/K_1} + \frac{\beta}{\Gamma(n_2)K_2^{n_2}} t^{(m_2-1)} e^{-t/K_2} \tag{37.32}$$

37.5.3 Linear channel

The response to a linear system consists of storage effects as well as translation effects. Storage effects are considered by the linear reservoirs and cascades. When an input function is routed through a linear channel, its shape is unchanged but translated downstream. The time lag between the input and output functions is called the translation time (also, time of concentration). This concept was first put forward by Dooge (1959).

---

[4] Zeroth moment, $m_0 = \int h(t)\,dt$; First moment, $m_1 = \int h(t)\,t\,dt$; Second moment, $m_2 = \int h(t)\,t^2\,dt$.

37.5.3.1 Time–area diagram

If only translation effects are considered, the system response function is given by the time–area diagram which is a function which shows the percentage of the total catchment area contributing to runoff at different times.

$$h(t) = A(t)$$

The model proposed by Clark (1945) is an example of a combination of linear reservoir with a time–area model (sometimes referred to as an isochrones model).

Another variation in which the catchment response is considered to be consisting of the storage effects and translation effects has been proposed by Singh (1964). It consists of two linear reservoirs with storage coefficients K_1 and K_2, and a linear channel with translation coefficient τ with two kernel functions, $h_s(t)$ and $w(t)$ and is of the form

$$h(t) = \frac{1}{K_1 - K_2} \int \left\{ e^{-(t-\tau)/K_1} - e^{-(t-\tau)/K2} \right\} w(\tau) \, d\tau \qquad (37.33)$$

where

$$w(\tau) = \frac{dA}{d\tau}$$

37.6 RANDOM PROCESSES AND LINEAR SYSTEMS

In the systems theory approach, the usual practice is to have deterministic inputs and outputs. However, it is also possible to have stochastic inputs and outputs in a linear system and still estimate the system response function using their statistical properties. In such situations,

- Stochasticity can only be described statistically.
- The complete phenomenon is never known. Information about sample(s) only is available. Inferences about the population are made on the basis of the information gathered from the sample(s).
- This places some restrictions. For example, it is assumed that the phenomenon is stationary and ergodic. Normality is also often implied.
- Statistical properties of the function include the moments, correlation, probability density, etc.

For the rainfall-runoff process with random variables $x(t)$ and $y(t)$,

$$y(t) = \int_0^t h(t) x(t - \tau) \, d\tau$$

where the system function $h(t)$ is deterministic. In linear systems (convolution integral assumes linearity), two of the commonly used stochastic functions in hydrology are the white noise and Markov processes (AR(1) is a Markov process).

For using stochastic inputs and outputs, the following deterministic statistical parameters of the input and output functions are used:

- Mean value

$$E[y(t)] = E\left[\int_{-\infty}^{\infty} h(\tau)x(t-\tau)\,d\tau\right] = \left[h(\tau)E\int_{-\infty}^{\infty} x(t-\tau)\,d\tau\right] \tag{37.34}$$

- Auto-correlation

$$R_{yy}(u) = \int_{-\infty}^{\infty} h(\tau)\,R_{xy}(u-\tau)\,d\tau \tag{37.35a}$$

- Cross-correlation

$$R_{yx}(u) = \int_{-\infty}^{\infty} h(\tau)\,R_{xx}(u-\tau)\,d\tau \tag{37.35b}$$

where $R_{yx}(u)$ is the cross-correlation between y and x; R_{xx} is the autocorrelation of x and $u = t - t_i$. Equation 37.35a and 37.35b contain deterministic functions only which can be derived from the stochastic functions.

The cross-correlation functions can be obtained by minimizing the mean square error between the measured and calculated output functions:

$$\text{Min } E\left[\left|y(t) - \int_{-\infty}^{\infty} h(\tau)\,x(t-\tau)\,d\tau\right|^2\right]$$

The minimization is obtained by differentiating the error with respect to the unknown $h(\tau)$ and setting it to zero. The result of the above minimization (for derivation, see Lattermann, 1991: pp. 105–106) leads to Eq. 37.35b.

Equation 37.35 is known as the Wiener filter for non-causal systems. For causal stationary systems, the limits of integration are 0 to ∞, and the resulting form of Eq. 37.35 is known as Wiener–Hopf integral equation. In discrete form, it is

$$R_{yx(m)} = \sum_{k=0}^{m} h(k)R_{xx}(m-k) \tag{37.36}$$

which can be written as

$$R_{yx}(0) = h(0)R_{xx}(0) + h(1)R_{xx}(-1) + \ldots + h(m)R_{xx}(0-m)$$

$$R_{yx}(1) = h(0)R_{xx}(1) + h(1)R_{xx}(0) + \ldots + h(m)R_{xx}(1-m)$$

$$R_{yx}(2) = h(0)R_{xx}(2) + h(1)R_{xx}(1) + \ldots + h(m)R_{xx}(2-m)$$

$$\ldots\ldots\ldots\ldots\ldots\ldots\ldots\ldots\ldots\ldots\ldots\ldots\ldots\ldots\ldots\ldots\ldots\ldots$$

$$\ldots\ldots\ldots\ldots\ldots\ldots\ldots\ldots\ldots\ldots\ldots\ldots\ldots\ldots\ldots\ldots\ldots\ldots$$

$$R_{yx}(m) = h(0)R_{xx}(m) + h(1)R_{xx}(m-1) + \ldots + h(m)R_{xx}(m-m)$$

This set of equations is also called the Yule–Walker equation. From Eq. 37.36, $h(t)$ can be determined.

37.7 NON-LINEAR SYSTEMS

When the system considered is non-linear, the representation becomes very complex. For example, if the general linear representation is to be considered in a non-linear model, the representation will be

$$y(t) = \int_0^\infty h_1(\tau)\,x(t-\tau)\,d\tau + \int_0^\infty \int_0^\infty h_2(\tau_1,\tau_2)\,x(t-\tau_1)\,x(t-\tau_2)\,d\tau_1\,d\tau_2$$

$$+ \int_0^\infty \int_0^\infty \int_0^\infty h_3(\tau_1,\tau_2,\tau_3)\,x(t-\tau_1)\,x(t-\tau_2)\,x(t-\tau_3)\,d\tau_1\,d\tau_2\,d\tau_3 \qquad (37.37)$$

$$+\ldots$$

This is similar to the Volterra series and has been used to solve many non-linear systems, including the Unit hydrograph (Amorocho, 1973).

37.7.1 Determination of the kernel functions

For a second-order system, with an input function of the form

$$x(t) = a\,\delta(t)$$

the output function is

$$y(t) = \int_0^t h_1(\tau)\,a\delta(t-\tau)\,d\tau + \int_0^t \int_0^t h_2(\tau_1,\tau_2)\,a\delta(t-\tau_1)\,a\delta(t-\tau_2)\,d\tau_1\,d\tau_2$$

$$= ah_1(t) + a^2 h_2(t,t)$$

$$\left(\text{because } h(t) = y(t)\int_0^t h(\tau)\,\delta(t-\tau)\,d\tau \right)$$

Therefore,

$$\frac{y(t)}{a} = h_1(t) + ah_2(t,t) = HU(t)$$

where $h_1(t)$ is the first-order unit response function, and $h_2(t,t)$ is the second-order unit response function. The relationship

$$\frac{y(t)}{a} = HU(t)$$

represents a linear system.

The system functions should satisfy

$$\int_0^\infty h_1(t)\,dt = 1$$

and

$$\int_0^\infty h_2(t, t)\,dt = 0$$

Considering two independent input functions a_1 and a_2 (e.g., effective rainfall), the first and second-order kernel functions of the non-linear Volterra series can be obtained as follows:

$$HU_1(t) = h_1(t) + a_1 h_2(t, t) \tag{37.38a}$$

$$HU_2(t) = h_1(t) + a_2 h_2(t, t) \tag{37.38b}$$

Subtracting,

$$HU_1(t) - HU_2(t) = (a_1 - a_2)\,h_2(t, t)$$

$$h_2(t, t) = \{HU_1(t) - HU_2(t)\}/(a_1 - a_2) \tag{37.39a}$$

Eliminating $h_2(t, t)$ from Eqs. 37.38a and 37.38b,

$$h_1(t) = \{a_2 HU_1(t) - a_1 HU_2(t)\}/(a_2 - a_1) \tag{37.39b}$$

It is possible to have three input functions a_1, a_2, a_3. Then h_1, h_2 and h_3 can be simultaneously determined. If there are three input functions and two kernel functions, then some optimization will be needed.

37.8 MULTI-LINEAR OR PARALLEL SYSTEMS

Linear systems can be combined to form multi-linear models. Depending on the arrangement, non-linear effects of the system can be incorporated. In general,

$$y(t) = \sum_{i=1}^{n} y_i(t) = \sum_{i=1}^{n} \int_0^t h_i(\tau) x_i(t - \tau)\,d\tau \tag{37.40}$$

with

$$\sum_{i=1}^{n} x_i(t) = x(t) \tag{37.41}$$

The input and output functions are divided into parts that are then used to determine their system parameters. The subdivision to a considerable extent influences the effect on non-linear modelling and, therefore, cannot be done arbitrarily. Once the individual response functions are determined, the output function for any given input function can be obtained by superposition.

Example 37.1

Given the following input and time-area functions, determine the output function.

Time (h)	1	2	3	4
A(t)	0.25	0.5	0.25	0
x(t)	1	3	2	0

$$y(t) = \sum_0^t x(k)\, h(t-k)$$

$y(1)$	$= 1 \times 0.25$			$= 0.25$
$y(2)$	$= 1 \times 0.50$	$+ 3 \times 0.25$		$= 1.25$
$y(3)$	$= 1 \times 0.25$	$+ 3 \times 0.25$	$+ 2 \times 0.25$	$= 2.25$
$y(4)$	$=$	3×0.25	$+ 2 \times 0.50$	$= 1.75$
$y(5)$	$=$		$+ 2 \times 0.25$	$= 0.50$

Example 37.2

Given the following system response function and the input function determine the output function.

$$h(t) = 0.5\, e^{-0.5t}$$

$$x(t) = 1 \text{ for } 0 < t < 1$$

$$x(t) = 3 \text{ for } 1 < t < 2$$

$$x(t) = 2 \text{ for } 2 < t < 3$$

$$x(t) = 0 \text{ for } t > 3.$$

This is a step function.

$$y(t) = \int_0^t 0.5 e^{-0.5(t-\tau)} x(\tau)\, d\tau$$

$$= 0.5\left\{ 1 \cdot \int_0^1 0.5 e^{-0.5(t-\tau)}\, d\tau + 3 \cdot \int_1^2 0.5 e^{-0.5(t-\tau)}\, d\tau + 2 \cdot \int_2^3 0.5 e^{-0.5(t-\tau)}\, d\tau \right\}$$

$$= \left\{ e^{-0.5(t-1)} - e^{-0.5t} \right\} + 3\left\{ e^{-0.5(t-2)} - e^{-0.5(t-1)} \right\} + 2\left\{ e^{-0.5(t-3)} - e^{-0.5(t-2)} \right\}$$

The first term within {} is applicable for $0 < t < 1$; the first two terms within {} are applicable for $1 < t < 2$; and all terms are applicable for $t > 2$.

Some typical calculations yield the following results:

$y(0.5) = 0.505$

$y(1.0) = 0.394$

$y(1.5) = 1.83$

$y(2.0) = 1.419$

$y(2.5) = 2.115$

$y(3.0) = 1.696$

....

....

REFERENCES

Amorocho, J. (1973): Nonlinear hydrologic analysis, *Advances in Hydroscience*, vol. 9, pp. 203–251.

Clark, C. O. (1945): Storage and the unit hydrograph, *Transactions of the American Society of Civil Engineers*, vol. 110, pp. 1419–1488.

Diskin, M. H. (1964): A basic study of the linearity of rainfall-runoff process in watersheds, Ph.D. thesis, University of Illinois, Urbana.

Dooge, J. C. I. (1959): A general theory of the unit hydrograph theory, *Journal of Geophysical Research*, vol. 64, pp. 241–256.

Jayawardena, A. W. (2014): *Environmental and Hydrological Systems Modelling*, CRC Press, Taylor and Francis Group, Boca Raton, FL, 516 pp.

Lattermann, A. (1991): *System-Theoretical Modelling in Surface Water Hydrology*, Springer-Verlag, Berlin, Heidelberg, 200 pp.

Nash, J. E. (1957): *The Form of the Instantaneous Unit Hydrograph*, International Association of Scientific Hydrology, Wallingford, Publication No. 45, pp. 114–121.

Singh, K. P. (1964): Nonlinear instantaneous unit-hydrograph theory, *Journal of the Hydraulics Division*, vol. 90 no.:HY2, pp. 313–347.

ADDITIONAL READING

Delleur, J. W. and Rao, R. A. (1971): Linear systems analysis in hydrology – the transform approach, the Kernel oscillations and the effect of noise, Proceedings of the US-Japan Bi-Lateral Seminar in Hydrology, Honolulu, Jan, pp. 116–129.

Dooge, J. C. I. (1973): *Linear Theory of Hydrologic Systems*, Technical Bulletin No. 1468, Agricultural Research Services, US Department of Agriculture, Washington, DC.

Dooge, J. C. I. and Harley, B. M. (1967): Linear routing in uniform open channel. Proceedings of the International Hydrology Symposium, Fort Collins, CO, Sept 6–8, 1967, vol. 1, pp. 57–63.

Volterra, V. (1887): Sopra le funzioni che dipendono de altre funzioni. *Rendiconti R. Academia dei Lincei*, vol. 2° Sem.: 97–105, pp. 141–146, 153–158.

Volterra, V. (1959): *Theory of Functionals and of Integral and Integro-Differential Equations*, Dover, New York.

Wiener, N. (1958): *Nonlinear Problems in Random Theory*, Wiley, New York.

Chapter 38

Time series analysis and forecasting

38.1 INTRODUCTION

Most hydro-meteorological and environmental data are measured at regular intervals of time and can therefore be represented as functions of time or time series. These, when observed over a long period of time, exhibit certain patterns which if identifiable can be used for forecasting purposes. Time series analysis involves the identification of such patterns or properties by a process of decomposition and subsequent extrapolation by synthesizing the decomposed components such that the statistical character of the generated series remains the same as that of the historical series.

The objective of this chapter is to introduce basic properties of time series, homogeneity tests, decomposition of a time series, tests for identification of trends and periodicities, time and frequency domain analyses, representation of the dependent stochastic component by various stochastic models, generation of synthetic data using the probability distribution of the independent residuals, forecasting and basic concepts of Kalman filtering. Illustrative examples will be given where possible.

Some parts of this chapter are extracted from an earlier publication by the author (Jayawardena, 2014).

38.2 BASIC PROPERTIES OF A TIME SERIES

38.2.1 Stationarity

In a stationary time series, the statistical properties of the series when computed from different samples do not change except due to sampling variations. This means that the probability density functions estimated from any sample of the time series is the same. To make the definition of stationarity more practical, this condition is restricted to the mean (expected value) and the variance only. A series which is stationary in the mean is called a first order stationary series. A series which is stationary in the covariance and the mean is called second order stationary or 'weakly stationary' (stationarity in the covariance[1] implies stationarity in the variance). If other statistical properties also are time-invariant, the series is said to be 'strongly stationary'. A process can be stationary in one property and not in others.

A series whose statistical properties vary with time is non-stationary. A non-stationary series contains deterministic components such as trends and periodicities.

---

[1] $\mathrm{cov}\left[x(t), x(t+\tau)\right] = \underset{T \to \infty}{\mathrm{Limit}} \frac{1}{T} \int_0^T \left[x(t) - \mu\right]\left[x(t+\tau) - \mu\right] dt = \mathrm{cov}(\tau)$

$\mathrm{Autocorrelation} = \dfrac{\mathrm{cov}(\tau)}{\mathrm{cov}(0)}$

38.2.2 Ergodicity

A stationary time series is said to be ergodic if the time averaged statistical properties are the same as the ensemble or spatial averages over the entire population. An ergodic series is necessarily stationary, but a stationary series may not be ergodic. For practical purposes, all stationary series are assumed to be ergodic.

38.2.3 Homogeneity

A series is said to be non-homogeneous when there is a jump (or drop) in the mean value of magnitude Δ after a certain number of observations. Homogeneity of a data series in statistical terms implies that the data belong to one population and therefore has a mean value which is time invariant.

Homogeneity of data is affected by several factors such as changes in the circumstances under which the data have been collected, changes in the data collection site and procedure, etc. This is particularly applicable to rainfall data. For example, non-homogeneity can occur due to changes in the height of exposure of the rain gauge, changes in site and environment, change of personnel and change of instrument. One way of testing the homogeneity is to compare the cumulative values of the suspect data with the cumulative values of the mean of a number of nearby gauging stations, which are affected by similar meteorological conditions. The method, which leads to the double mass curve, is simple and provides an indication of the relative shift in the mean value if any. It does not distinguish real changes in mean level from changes due to random fluctuations. Therefore it is necessary to examine the significance of the non-homogeneity by using statistical tests.

38.3 STATISTICAL PARAMETERS OF A TIME SERIES

38.3.1 Sample moments

Sample moments around zero are defined as

$$m_r = \frac{1}{N} \sum x_t^r \tag{38.1}$$

where m_r is the rth moment.

Sample moments around the mean $\bar{x}$ are defined as

$$m_r = \frac{1}{N} \sum (x_t - \bar{x})^r \tag{38.2}$$

For example, the first moment ($r = 1$) gives the mean; the second moment ($r = 2$) gives the variance; the third moment ($r = 3$) gives a measure of the skewness and the fourth moment ($r = 4$) gives a measure of the kurtosis. Variance when multiplied by $\frac{N}{N-1}$ gives an unbiased[2] estimate.

When the third moment is calculated, positive and negative values can cancel out if the distribution is symmetrical. The degree to which the third moment deviates from zero is

---

[2] When the expected value of a statistic is equal to the corresponding population parameter, the statistic is called an unbiased estimator. For example, if the mean of the sample means is equal to the population mean, then the mean of samples is an unbiased estimator (i.e., mean of $\bar{x} = \mu$). The unbiased estimate is obtained by multiplying the biased estimate by $\frac{N}{N-\text{DOF}}$ where DOF refers to the number of degrees of freedom. For variance, DOF = 1.

a measure of the skewness of the distribution and the sign of the third moment gives its direction. If the skewness coefficient is negative, the distribution is skewed to the right (mode > median and mean), and if it is positive it is skewed to the left (mode < median and mean). Skewness coefficient is defined as

$$\frac{1}{N} \sum \frac{(x_t - \bar{x})^3}{s^3} \tag{38.3}$$

where s is the sample standard deviation.

The coefficient of kurtosis, which is a measure of peakedness, is defined as $\frac{m_4}{s^4}$, and kurtosis excess ε is defined as

$$\varepsilon = \kappa - 3 \tag{38.4}$$

A low value of the kurtosis indicates relatively few extreme values. It is an indicator of the peakedness or flatness. The coefficient of kurtosis for a normal distribution is 3. If it is greater than 3, then there are more high values concentrated near the mean. If it is less than 3, then there are more low values concentrated near the mean. When $\varepsilon > 0$, i.e. high peak, the distribution is called leptokurtic; when $\varepsilon = 0$, i.e. normal peak, the distribution is called mesokurtic; when $\varepsilon < 0$, i.e. low peak, the distribution is called platykurtic.

38.3.2 Moving averages – low-pass filtering

Moving average, also referred to as running average or rolling average, which has the effect of a low-pass filter[3] will reduce high-frequency oscillations and is useful for eliminating short term fluctuations. For example, a five-term moving average model of a series $x_1, x_2, x_3, x_4, x_5, x_6, x_7, x_8, x_9, x_{10}, \ldots$ will be of the form

$$y_1 = \frac{1}{5}(x_1 + x_2 + x_3 + x_4 + x_5)$$

$$y_2 = \frac{1}{5}(x_2 + x_3 + x_4 + x_5 + x_6)$$

$$y_3 = \frac{1}{5}(x_3 + x_4 + x_5 + x_6 + x_7) \tag{38.5}$$

....

....

As the number of terms in the moving average model increases, the number of terms of the new series is reduced.

38.3.3 Differencing – high-pass filtering

Low-frequency oscillations, on the other hand, can be removed by using high-pass filtering methods such as

---

[3] A low pass filter allows low-frequency (long-period) signals to pass and smoothen high frequencies (short period). A high-pass filter does the opposite.

$$y'_t = x_t - x_{t-1} \text{ for } t = 2, 3, \ldots \text{ (1st order)} \tag{38.6a}$$

$$y_t = y'_t - y'_{t-1}$$
$$= x_t - 2x_{t-1} + x_{t-2} \text{ for } t = 3, 4, \ldots \text{ (2nd order)} \tag{38.6b}$$

In practice, however, most of these filtering methods may not be necessary. Visual observation of the data plot will often indicate whether or not a trend exists. If one exists, it may be described by a polynomial of the form

$$x_t = x_0 + a_1 t + a_2 t^2 + \ldots + a_2 t^n + \zeta_t \text{ (error)} \tag{38.7}$$

which is a linear trend if only the first two terms are applicable. Differencing is a 'whitening filter' and has the effect of transforming the distribution of the original series closer to a normal distribution.

38.3.4 Recursive means and variances

Recursive mean $\bar{x}_k$ is defined as

$$\bar{x}_k = \frac{1}{k} \sum_{i=1}^{k} x_i = \frac{1}{k} \left\{ \sum_{i=1}^{k-i} x_i + x_k \right\} \tag{38.8a}$$

$$= \frac{1}{k} \left\{ (k-1)\bar{x}_{k-1} + x_k \right\} = \frac{(k-1)}{k} \bar{x}_{k-1} + \frac{1}{k} x_k \tag{38.8b}$$

$$= \bar{x}_{k-1} + \frac{1}{k} \left\{ x_k - \bar{x}_{k-1} \right\} = \bar{x}_{k-1} - \frac{1}{k} \left\{ \bar{x}_{k-1} - x_k \right\} \tag{38.8c}$$

Similarly, recursive variances σ_k^2 can be defined as

$$\sigma_k^2 = \frac{1}{k} \sum_{i=1}^{k} \left(x_i - \bar{x}_k \right)^2 \tag{38.9a}$$

which can be shown to be equivalent to

$$\sigma_k^2 = \frac{k-1}{k} \sigma_{k-1}^2 - \frac{1}{k-1} \left\{ x_k - \bar{x}_k \right\}^2 \tag{38.9b}$$

Equations 38.8c and 38.9b are more efficient computationally.

38.4 TESTS FOR STATIONARITY

A test for stationarity can be done by dividing the time series into two or more sub-series and checking whether the statistical character of each sub-series is significantly different from one another. The standard normal variate z or the statistic Student's 't' can be used in this test. Student's 't' instead of z is used when the sample size is small (<30), or when the population standard deviation is replaced by the sample standard deviation. Example 38.1 illustrates the test for stationarity.

Example 38.1: Stationarity test

Stationarity test is illustrated using the annual rainfall data measured at the Observatory of Hong Kong (Hong Kong Observatory, Meteorological Results Part I) for the period 1884–1941 ($N = 58$ years), listed below and shown in Figure 38.1.

1,915.8	2,766.7	1,756.9	1,684.3	3,656.5	2,040.7	1,801.8	2,974.7
2,310.6	2,538.9	2,648.1	1,164.3	1,848.8	2,540.7	1,448.6	1,846.7
1,872.7	1,417.0	2,476.4	2,378.7	2,042.5	1,802.1	1,975.9	2,376.0
2,333.7	1,923.6	1,781.0	2,300.0	1,624.1	2,126.8	2,545.3	1,931.1
2,028.3	2,069.7	2,580.8	1,934.0	2,740.0	2,472.2	1,763.6	2,711.1
2,503.2	2,224.4	2,559.6	2,740.0	1,807.2	1,773.5	2,440.3	2,041.8
2,323.3	1,583.5	2,480.6	1,811.6	1,772.2	2,095.5	1,406.0	2,202.3
2,989.2	2,433.1						

The series is split into three sub-series of approximately equal length. The statistical parameters for the entire series, together with those for the three sub-series, are computed. The mean value is used as the parameter for statistical tests. The results are summarized in Table 38.1.

In this example, the complete data series constitutes the population, and the sub-series constitute the samples. Since the population parameters (mean μ; variance, σ^2) are known, the sample mean $\bar{x}$ is approximately normally distributed with mean μ and standard deviation $\dfrac{\sigma}{\sqrt{N}}$ where N is the sample size. The approximation improves as the sample size N increases. For the above statement to be valid, there is no need to have any assumptions about the distribution of the population.

If the random sample comes from a Normal population, then the above statement is exact regardless of the sample size N.

Therefore,

$$z = \frac{\bar{x} - \mu}{\left(\dfrac{\sigma}{\sqrt{N}} \right)}$$

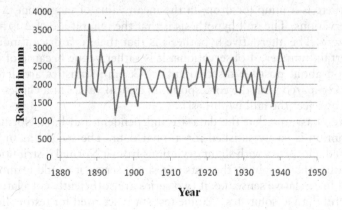

Figure 38.1 Annual rainfall in Hong Kong.

Table 38.1 Statistical parameters of the sub-series

	Whole series $N = 58$	First one-third $N = 19$	Second one-third $N = 19$	Last one-third $N = 20$
Mean (mm)	2,161.0	2,143.0	2,156.0	2,183.0
Standard deviation (mm)	459.8	616.1	307.0	429.5
Skewness coefficient	0.4817	0.7048	0.2228	−0.00956
Student's 't'		−0.124	−0.069	0.223

follows a standard Normal distribution. (This is approximate if the population is not Normal).

The null hypothesis H_0 and the alternative hypothesis H_a, respectively, are

H_0: $\bar{x} = 2161.0$, and the series is stationary,
H_a: $\bar{x} \neq 2161.0$, and the series is not stationary.

A two-tailed test is required. Under the null hypothesis H_0, for the first one-third

$$z = \frac{\bar{x} - \mu}{\left(\frac{\sigma}{\sqrt{19}} \right)} = \frac{2,143 - 2,161}{459.8} \sqrt{19} = -0.171$$

For a two-tailed test at 5% level of significance, adopt the decision rule:

- Accept H_0 if 'z' lies within the interval $-z_{0.975}$ to $z_{0.975}$ which is the interval ± 1.96.
- Reject H_0 otherwise.

Since $z = -0.171$, H_0 is accepted at 5% significance level.

For the second and last one-thirds, the corresponding z values are -0.0474 and 0.214, respectively, which lie well within the interval ± 1.96. Therefore, the hypothesis H_0 (that the series is stationary) is accepted.

38.5 TESTS FOR HOMOGENEITY

Homogeneity of a data series in statistical terms implies that the data belong to one population and therefore has a mean value which is time-invariant. The tests are carried out by assuming that there is a jump (or drop) in the mean value of magnitude Δ after a certain number of observations. The null hypothesis is that the magnitude of Δ is not significantly different from zero. The alternative hypothesis is that there is a discontinuity in the mean value after a certain number of observations. It is rather vague because of the lack of any prior information about possible discontinuities. Certain statistics are defined, and their magnitudes are compared with the corresponding critical values at a chosen level of significance to accept or reject the null hypothesis.

Usually, the tests assume that the data are independent and have Normal distribution. The first assumption is easily satisfied for annual data. The second assumption may not always be satisfied. However, with slight deviations from a Normal distribution the tests can still be used. Rather than applying the tests to a single series, it would be more conclusive if they are applied in a relative sense, i.e. if two series are sufficiently correlated, relative tests are more powerful than absolute tests. Some test statistics used for testing homogeneity are defined below.

38.5.1 von Neumann ratio

The null hypothesis in this test is that the time series variable is independently and identically distributed (random). The alternate hypothesis is that the series is not random. The von Neumann Ration NR (von Neumann, 1941) is defined as

$$NR = \frac{\sum_{t=1}^{N-1}(x_t - x_{t+1})^2}{\sum_{t=1}^{N}(x_t - \bar{x})^2} \tag{38.10}$$

where x_t is the series of sample size N and mean value $\bar{x}$. Under the null hypothesis, $E(NR) = 2$. The mean of NR tends to be smaller than 2 for a non-homogeneous series. Critical values of NR for Normally distributed samples, as given by Owen (1962) and Wijngaard et al. (2003), are shown in Table 38.2.

In this test, if a break occurs, implying non-homogeneity, the value of NR tends to be lower than 2, but it does not give any indication about the location of the break.

38.5.2 Cumulative deviations

In this test too, the null hypothesis is that the time series variable is independently and identically distributed. The alternate hypothesis is that there is a shift in the mean value after a certain time.

In the cumulative deviations test (Buishand, 1982), the departure from homogeneity is tested using the statistics Q and R which are defined as

$$Q = \max_{0 \le k \le N}\left|S_k^{**}\right| \tag{38.11}$$

and,

$$R = \max_{0 \le k \le N} S_k^{**} - \min_{0 \le k \le N} S_k^{**} \tag{38.12}$$

in which the re-scaled adjusted partial sums S_k^{**} are given by

$$S_k^{**} = \frac{S_k^*}{s_x} \quad k = 0, 1, 2, ..., N \tag{38.13}$$

where s_x is the sample standard deviation.

The adjusted partial sums (or cumulative deviations from the mean), S_k^* are given by

$$S_0^* = 0 \tag{38.14a}$$

Table 38.2 Critical values of NR (1% and 5% significance levels) for von Neumann ratio test

N	20	30	40	50	70	100
1%	1.04	1.20	1.29	1.36	1.45	1.54
5%	1.30	1.42	1.49	1.54	1.61	1.67

Source: Owen, D.B., *Handbook of Statistical Tables*, Addison Wesley, Reading, MA, 1962. With permission, Wijngaard, J.B., Klein Tank, A.M.G., Können, G.P., *International Journal of Climatology*, 23, 679–692, 2003. With permission.

$$S_k^* = \sum_{t=1}^{k}(x_t - \bar{x}), \quad k = 1, 2, ..., N \tag{38.14b}$$

where x_t and $\bar{x}$ have the same meanings as before. High values of Q and R indicate departure from homogeneity. Critical values of $\dfrac{Q}{\sqrt{N}}$ and $\dfrac{R}{\sqrt{N}}$ based on 19,999 synthetic Gaussian random numbers, obtained by Buishand (1982), are given in Table 38.3.

In this test, if a break occurs in year k^*, then the statistic S_k^* attains its maximum (negative shift) or minimum (positive shift) near the year $k = k^*$.

Example 38.2: Homogeneity tests

Homogeneity and trend tests are illustrated using the annual maximum temperature (°C) in the city of Dhaka, Bangladesh[4] for the period 1980–2009 (30 years), which is listed below and shown in Figure 38.2.

30.29	30.01	30.34	30.13	30.48	31.00	31.29	31.48	31.49	31.38
30.03	30.11	30.68	30.20	30.93	31.48	31.68	30.62	30.87	31.53
30.13	30.47	30.30	30.23	30.52	30.83	31.28	30.54	30.52	31.61

The mean value of the data set is 30.75°C and the standard deviation is 0.55°C.

For the data given in Example 38.2, the minimum value of the test statistic S_k^* is −2.49 (see also Figure 38.3) and occurs at around the year 1984. It corresponds to a positive shift of the mean value whereas a smaller negative shift occurs at around the year 1999.

The $\dfrac{R}{\sqrt{N}}$ value is 1.18 which is less than the critical value of 1.50 for $N = 30$ at the 5% significance level. Based on this test the data series can be considered as homogeneous.

38.5.3 Bayesian statistics

In the Bayesian procedure, the shift in mean level is tested using the statistics U and A, which are defined as

$$U = \frac{1}{N(N+1)} \sum_{k=1}^{N-1}\left(S_k^{**}\right)^2 \quad \left(S_k^{**} \text{ is defined in Eq. 38.13}\right) \tag{38.15}$$

Table 38.3 Critical values of $\dfrac{Q}{\sqrt{N}}$ and $\dfrac{R}{\sqrt{N}}$

N	10	20	30	40	50	100	∞
1%	1.29	1.42	1.46	1.50	1.52	1.55	1.63
	(1.38)	(1.60)	(1.70)	(1.74)	(1.78)	(1.86)	(2.00)
5%	1.14	1..22	1.24	1.26	1.27	1.29	1.36
	(1.28)	(1.43)	(1.50)	(1.53)	(1.55)	(1.62)	(1.75)

Source: Buishand, T.A., Journal of Hydrology, 58, 11–27, 1982. With permission.

(1% and 5% significance levels) for cumulative deviations test; $\dfrac{R}{\sqrt{N}}$ is in bold face inside parenthesis.

---

[4] The example illustrated using this data set is taken from the thesis submitted by A.K.M Saifuddin (2010) as partial fulfillment for the degree of Master of Disaster Management of the National Graduate Institute for Policy Studies in Japan, under the supervision of the author.

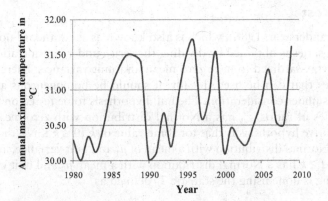

Figure 38.2 Annual maximum temperature in the city of Dhaka, Bangladesh.

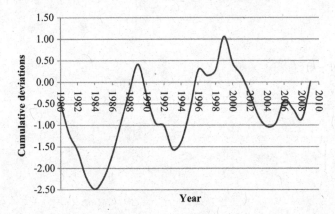

Figure 38.3 Cumulative deviations test (Eq. 38.14b) for the city of Dhaka's annual maximum temperature
data.

and

$$A = \sum_{k=1}^{N-1} \frac{\left(S_k^{**}\right)^2}{k(N-k)} \tag{38.16}$$

Large values of U and A are indications of departure from homogeneity. Their critical
values, based on 19,999 synthetic Gaussian random numbers obtained by Buishand (1982)
are shown in Table 38.4.

Table 38.4 Critical values of U and A

N	10	20	30	40	50	100	∞
1%	0.575	0.662	0.691	0.698	0.718	0.712	0.743
	(3.14)	**(3.50)**	**(3.70)**	**(3.66)**	**(3.78)**	**(3.82)**	**(3.86)**
5%	0.414	0.447	0.444	0.448	0.452	0.457	0.461
	(2.31)	**(2.44)**	**(2.42)**	**(2.44)**	**(2.48)**	**(2.48)**	**(2.49)**

Source: Buishand, T.A., *Journal of Hydrology*, 58, 11–27, 1982. With permission.

(1% and 5% significance levels); A is in bold face inside parenthesis.

38.5.4 Ratio test

In this test (Alexandersson, 1986), which is also known as the standard normal homogeneity test, the ratio series is obtained by dividing the series under consideration by a reference series. It is then standardized to have zero mean and unit variance. There are many forms of reference series that can be used. Ideally, it should be homogeneous and highly correlated to the series under consideration. The null hypothesis to be tested on the standardized (normalized) series of ratios, z_t, has a Normal distribution with zero mean and unit variance. The alternative hypothesis is that for some value of k ($0 \leq k \leq N$), the partial series of z_t for $t \leq k$ has a Normal distribution with a mean of μ_1 and unit variance, and the remaining partial series for $t \geq k$ has a Normal distribution with a mean μ_2 and unit variance ($\mu_1 \neq \mu_2$). Hypothesis testing is done using the statistic T_0 defined as

$$T_0 = \max_{1 \leq k \leq N} (T_k) \tag{38.17}$$

where

$$T_k = k\bar{z}_1^2 + (N-k)\bar{z}_2^2 \tag{38.18}$$

and

$$\bar{z}_1 = \frac{1}{k}\left(\sum_{t=1}^{k} z_t\right) \tag{38.19}$$

$$\bar{z}_2 = \frac{1}{(N-k)}\left(\sum_{t=k+1}^{N} z_t\right) \tag{38.20}$$

High values of T_0 indicate non-homogeneous character of the data series. Critical values of T_0 have been given by Alexanderson (1986), Jaruškova (1994) and Alexandersson and Moberg (1997). Table 38.5 shows these values for different sample sizes at two significant levels. This test also permits the evaluation of the position of possible break in the series which correspond to the maximum value of T_k.

For the data set of Example 38.2, the test statistic T_0 has a value of 5 at around the year 1984 which is less than the corresponding critical value of 7.65 for $N = 30$ at 5% significance level. By this test too, the data set can be considered as homogeneous, and the break seems to occur in or around the year 1984 (Figure 38.4).

Table 38.5 Critical values of T_0 for the ratio test

N	20	30	40	50	70	100
1%	9.56	10.45	11.01	11.38	11.89	12.32
5%	6.95	7.65	8.10	8.45	8.80	9.15

Source: 1% significance level based on simulations carried out by Jaruškova, D., *Monthly Weather Review*, 124, 1535–1543, 1994. With permission.

Source: 5% significance level based on simulations carried out by Alexandersson, H., Moberg, A., *International Journal of Climatology*, 17, 25–34, 1997. With permission.

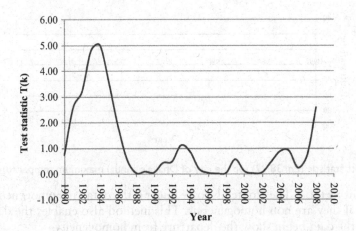

Figure 38.4 Ratio test (Eq. 38.17) for the city of Dhaka's annual maximum temperature data.

38.5.5 Pettitt test

In Pettitt test (Pettitt, 1979) which is non-parametric and based on the Wilcoxon test, the statistics S_k are calculated from ranks $r_1, r_2, ..., r_N$ of the series $x_1, x_2, ..., x_N$:

$$S_k = 2 \sum_{i=1}^{k} r_i - k(N+1) \quad k = 1, 2, ..., N \tag{38.21}$$

Test statistic S_{k*} is defined as

$$S_{k*} = \max_{1 \le k \le N} |S_k| \tag{38.22}$$

Critical values for S_{k*} are given in Table 38.6.

In this test, if a break occurs in year $k*$, then the statistic attains its maximum or minimum near the year $k = k*$.

For the data set of Example 38.2, the test statistic S_{k*} (Eq. 38.22) has a minimum value of -84 at around the year 1984 which is less than the corresponding critical value of -107 for $N = 30$ at 5% significance level (Figure 38.5). It also shows a slightly smaller maximum value of 25 at around the year 1999. By this test too, the data set can be considered as homogeneous.

Homogeneity can also be tested graphically by the double mass curve when applied to hydrological data. It is done by plotting the cumulative sums of the variable at the station under consideration against the cumulative sums of a number of nearby stations. A high degree of correlation of the data set of the suspect station with those of the nearby stations

Table 38.6 Critical values of S_{k*}

N	20	30	40	50	70	100
1%	71	133	208	293	488	841
5%	57	107	167	235	393	677

Source: 1% and 5% significance levels based on simulations carried out by Pettitt, A.N., *Journal of the Royal Statistical Society, Series C (Applied Statistics)*, 28, 126–135, 1979. With permission, and reported by Wijngaard, J.B, Klein Tank, A.M.G., Können, G.P., *International Journal of Climatology*, 23, 679–692, 2003. With permission.

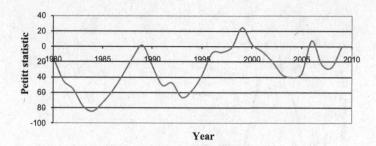

Figure 38.5 Pettitt statistic (Eq. 38.22) for the city of Dhaka's annual maximum temperature data.

is implied in this method. The plot tends to be linear if the data are homogeneous and depart from linearity if they are non-homogeneous. This method also enables the detection of the time at which the data begin show the departure from homogeneity.

In the context of climatology and hydrology, several researchers have carried out homogeneity tests on various climatic and hydrological variables. Examples include the studies on European daily temperature and precipitation data (Wijngaard et al., 2003), rainfall data (Alexandersson, 1986; Buishand, 1982; Jayawardena and Lau, 1990b), Bulgarian temperature data (Syrakova and Stefanova, 2009), Turkish meteorological data (Sahin and Cigizoglu, 2010), annual air temperature data in Croatia (Pandžíc and Likso, 2009), monthly Spanish temperature data (Staudt et al., 2007), and Swedish temperature data (Alexandersson and Moberg, 1997), among others.

38.6 COMPONENTS OF A TIME SERIES

A time series normally has a deterministic component the outcome of which can be predicted with certainty (at least in theory) and a stochastic component, the outcome of which is attributed to chance and cannot be explained physically and therefore cannot be predicted with certainty. The deterministic component may be in the form of a long-term trend, or of a periodic nature or a combination. Therefore, in general, a time series may be considered as a linear combination of a trend, a periodic component, a dependent stochastic component and an independent residual component.

It is also possible for the three effects to be expressed in a multiplicative form but such representations can be easily transformed into linear form by taking the logarithms.

38.7 TREND ANALYSIS

If the mean value of a time series is not time-invariant, then the time series may have a trend. Trends may be either increasing or decreasing. They can be removed if detected. It is best to test for trends after carrying out homogeneity tests. Whether a given time series is credible for carrying out trend analysis is determined based on the statistical results of the homogeneity tests discussed in Section 38.5. Wijngaard et al. (2003) classify data sets as 'useful' if one or zero tests reject the null hypothesis at 1% significance level, 'doubtful' if two tests reject the null hypothesis at 1% significance level, and 'suspect' if three or four tests reject the null hypothesis at 1% significance level. 'Useful' implies no clear signal of non-homogeneity and the data set can be used for trend analysis. 'Doubtful' implies that the trend analysis should be carried out and interpreted with caution, and 'suspect' implies that the data set lacks credibility for trend analysis.

38.7.1 Tests for randomness and trend

When there is no clear visual indication of the presence of a trend, it is necessary to carry out statistical tests to determine whether an apparent trend is significant or not. There are several tests that can be carried out.

38.7.1.1 Turning point test for randomness

In this test, the presence of high and low values is examined by determining the number of turning points in the series. A turning point in a time series x_t, $(t = 1, N)$, occurs at i if

$$x_i > x_{i-1} \text{ and } x_{i+1} \quad \text{or} \quad x_i < x_{i-1} \text{ and } x_{i+1}$$

For three unequal observations, there are six possible orders of magnitudes:

$$x_{i-1} > x_i > x_{i+1}; \quad x_{i-1} > x_{i+1} > x_i; \quad x_i > x_{i-1} > x_{i+1};$$

$$x_i > x_{i+1} > x_{i-1}; \quad x_{i+1} > x_{i-1} > x_i; \quad x_{i+1} > x_i > x_{i-1}$$

They have equal probabilities of occurrence in the case of a random series. Turning points occur in all except the first and the last. Therefore, the probability of having turning points is 4/6 (or 2/3). In a series, the first and the last terms are not turning points. Therefore, the expected number of turning points, p, in a random series is given by

$$E(p) = \frac{2}{3}(N - 2) \tag{38.23}$$

To establish whether the difference between the observed and the expected number of turning points is significant, it is necessary to calculate the variance of the expected number of turning points. This has been shown to be (Kendall, 1975)

$$\text{Var}(p) = \frac{(16N - 29)}{90} \tag{38.24}$$

A two-tailed test of significance is carried out on p which is expressed as a standard Normal deviate z where

$$z = \frac{p - E(p)}{\sqrt{\text{Var}(p)}} \tag{38.25}$$

which should lie within ±1.96 at 5% significance level if the series is random. Too many or too few turning points indicate non-randomness.

> Example 38.3: Turning point test for randomness
>
> The turning point test is illustrated using the data set given in Example 38.1. The results are as follows:
>
> Number of turning points in the series, $p = 37$
> Expected number of turning points if the series is random = 37.33
> Variance of the expected number of turning points = 9.989
> Number of turning points expressed as a standard Normal variate, $z = -0.106$.

The hypotheses are:

H_0: $p = 37.33$, and the series is random,
H_1: $p \neq 37.33$, and the series is not random.

A two-tailed test is carried out. The hypothesis H_0 (that the series is random) is accepted since z lies between ± 1.96 at 5% significance level.

38.7.1.2 Kendall's rank correlation[5] test (τ test)

Rank correlation (Kendall, 1948; Abdi, 2007) can be used to establish whether an apparent trend in a series is significant or not. The number of times p in all pairs of observations x_i, x_j; $j > i$ that $x_j > x_i$ is determined (i.e. For $i = 1, N - 1$ how many times $x_j > x_i$ for $j = i + 1$, $i + 2, ..., N$).

The ordered sub-sets in such a series are:

$x_1: x_2, x_3, x_4, ..., x_N$

$x_2: x_3, x_4, x_5, ..., x_N$

$x_3: x_4, x_5, ..., x_N$

...

$x_{N-1}: x_N$

The maximum number of such pairs occurs in a continuously rising curve when all $x_j > x_i$ for $j = i + 1, N; i = 1, N - 1$. Then,

$$p = (N-1) + (N-2) + (N-3) + ... + 1 = \frac{1}{2} N(N-1) \tag{38.26}$$

In a falling trend, all $x_j < x_i$, and therefore $p = 0$. Therefore for a trend free series,

$$E(p) = \frac{1}{2}\left(0 + \frac{1}{2} N(N-1)\right) = \frac{1}{4} N(N-1) \tag{38.27}$$

The test is carried out using the statistic 'τ' (known as Kendall's τ and which varies between ± 1) defined as

$$\tau = \frac{4p}{N(N-1)} - 1 \tag{38.28}$$

For a random series, $E(\tau) = 0$, and its variance is given as

$$\mathrm{Var}(\tau) = \frac{2(2N+5)}{9N(N-1)} \tag{38.29}$$

---

[5] Kendall rank correlation coefficient, commonly referred to as Kendall's tau (τ) is also used to measure the association between two data sets.

As N increases, $\dfrac{\tau - E(\tau)}{\sqrt{\mathrm{Var}\,(\tau)}}$ converges to a standard Normal distribution. It may also be possible to carry out a test using $E(\tau)$ which takes values of -1 and 1 leading to the inference that there is a rising or falling trend.

Example 38.4: Kendall's rank correlation test for randomness

Kendall's rank correlation test is illustrated using the data set given in Example 38.1. The results are as follows:

Number of times $x_{i+1} > x_i = 857$
Statistic $\tau = 0.0369$
Var $(\tau) = 0{:}008133$
$z = 0.4092$

which again is within the interval ± 1.96 at the 5% significance level. Hence the hypothesis H_0 (that the series is random) is accepted.

38.7.1.3 Regression test for linear trend

A linear trend may be approximated as

$$y_c = a + bx_i + \zeta_i \text{ (error)} \tag{38.30}$$

The hypothesis to be tested is that $b \neq 0$. A linear regression of the form $y_c = a + bx_i$ is first fitted to the data. The regression coefficients, a and b are obtained by minimizing the squared error. For pairs of dependent and independent variables y_i and x_i, it can be shown that

$$a = \bar{y} - b\bar{x} \tag{38.31a}$$

$$b = \frac{\displaystyle\sum_{i=1}^{N}(x_i - \bar{x})(y_i - \bar{y})}{\displaystyle\sum_{i=1}^{N}(x_i - \bar{x})^2} \tag{38.31b}$$

where the over-bar indicates the mean.

For computational purposes, it is convenient to write b as

$$b = \frac{\displaystyle\sum_{i=1}^{N}x_i y_i - \frac{\displaystyle\sum_{i=1}^{N}x_i \sum_{i=1}^{N}y_i}{N}}{\displaystyle\sum_{i=1}^{N}x_i^2 - \frac{1}{N}\sum_{i=1}^{N}x_i} \tag{38.31c}$$

Estimate of the variance of b is given by

$$s_b^2 = \frac{s_{yx}^2}{\displaystyle\sum_{i=1}^{N}(x_i - \bar{x})^2} \tag{38.32}$$

where s_{yx} is the standard error of regression (or estimate), which is given by

$$s_{yx}^2 = \frac{1}{(N-2)} \sum_{i=1}^{N} \left(y_c - (a + bx_i)^2 \right) = \frac{\sum_{i=1}^{N} (\text{error}_i)^2}{N-2} \quad (38.33)$$

Student's t is given by the ratio $\dfrac{(b - E(b))}{s_b}$. (The statistic Student's 't' is used here because σ_b is unknown and s_b is an estimate of σ_b).

Example 38.5: Regression test for linear trend

Regression test for linear trend is illustrated using the data set given in Example 38.1. The results are as follows:
   The student 't' statistic from the calculations is 0.06063. The hypotheses to be tested are

   H_0: $b = 0$, and there is no linear regression,
   H_1: $b \neq 0$, there is a linear regression.

At 5% level of significance, accept H_0 if 't' lies in the interval $-t_{0.975}$ to $t_{0.975}$ which for 57 (= 58 – 1) degrees of freedom is the interval –2.003 to 2.003. Hence H_0 is accepted at 5% significance level.

38.7.1.4 Mann–Kendall test

The Mann–Kendall test (Mann, 1945; Kendall, 1975) is a non-parametric test to detect trends in time series data. For a time series $x_1, x_2,..., x_N$, the Mann–Kendall statistic S is defined as follows:

$$S = \sum_{i=1}^{N-1} \sum_{j=i+1}^{N} \text{Sgn}(x_j - x_i) \quad (38.34)$$

where

$$\text{Sgn}(x_j - x_i) = \begin{cases} 1 & \text{if } x_j > x_i \\ 0 & \text{if } x_j = x_i \\ -1 & \text{if } x_j < x_i \end{cases} \quad (38.35)$$

If a data set displays a consistently increasing trend, the statistic S will be positive, whereas, if it displays a decreasing trend, S will be negative. A larger magnitude S indicates that the trend is more consistent in its direction. The Sgn function enables the detection of trends featuring either large or small increase steps from year to year equally.
   Under the null hypothesis that there is no trend displayed by the time series, the distribution of S is expected to have a zero-mean and variance:

$$\text{Var}(S) = \frac{1}{18} \left[N(N-1)(2N+5) - \sum_{p=1}^{g} t_p (t_p - 1)(2t_p + 5) \right] \quad (38.36a)$$

where N is the number of data under consideration, g is the number of tied[6] groups and t_p is number of data in the pth tied group. If there are no tied groups in the series, Eq. 38.36a simplifies to

$$\text{Var}(S) = \frac{N(N-1)(2N+5)}{18} \tag{38.36b}$$

The test statistic z is calculated as

$$z = \begin{cases} \dfrac{S-1}{\sqrt{\text{Var}(S)}} & \text{if } S > 0 \\[2mm] 0 & \text{if } S = 0 \\[2mm] \dfrac{S+1}{\sqrt{\text{Var}(S)}} & \text{if } S < 0 \end{cases} \tag{38.37}$$

and tested against the critical value of the Normal distribution at a chosen level of significance. With no tied groups, the test is reported to be valid for $N > 10$, and normally valid for $N > 40$ (Gilbert, 1987).

Example 38.6: Mann–Kendall test for trend

Mann–Kendall test is illustrated using the data set given in Example 38.2. This data set has three tied groups; tied group 1 has two data points, tied group 2 has two data points and tied group 3 has two data points. The statistic S (Eq. 38.34) works out to be 70 with a variance of 3,139, giving a standard Normal value (Eq. 38.37) of 1.23 which is less than 1.96 at 5% significance level. Hence the series in Example 38.2 can be considered as having an insignificant positive trend.

Trend analyses of climatological data have been carried out using one or more of the above methods by several researchers from several geographical regions. Examples of data used include daily maximum and minimum temperatures in Catalonia, Spain (Martínez et al., 2010), summer temperature in Tuscany, Italy (Bartolini et al., 2008), precipitation in the Kalu Ganga basin in Sri Lanka (Ampitiyawatta and Guo, 2009), Turkish precipitation (Partal and Kahya, 2006), low flows in the United States (Douglas et al., 2000), monthly water quality data in the United States (Hirsch et al., 1981, 1991), temperature and precipitation in the Yangtze River basin in China (Su et al., 2006), annual and seasonal rainfall data in the Mediterranean area (Longobardi and Villani, 2009), rainfall in Amazonia, (Paiva and Clarke, 1995), among others.

38.7.2 Trend removal

A trend can be linear or non-linear. Removal in the former case is quite simple by fitting a regression line to the data set. The slope and the intercept of the straight line can be estimated by the least-squares method. In the latter case of a non-linear trend, it can be removed by higher order polynomial regression. The parameters, in this case, can also be estimated by the method of least squares. Trends may also be removed by differencing, i.e. by transforming a non-stationary series to a stationary one.

---

[6] If the time series has values 10, 15, 10, 20, 15, 18, 20, 30, 20, 5, then the number of tied groups is 3 and the number of data in tied group 1 (corresponding to 10) is 2, group 2 (corresponding to 15) is 2, and group 3 (corresponding to 20) is 3.

However, higher order polynomial interpolation does not necessarily lead to better approximations. It has been shown by Runge (as cited by Kreyszig, 1999: p. 861) that the maximum error even approaches infinity as the order of the polynomial tends to infinity. An alternative to global polynomial regression is piecewise regression fitting. This is called spline regression that has the ability to avoid such numerical oscillations. Instead of a single higher order polynomial, several lower order polynomials can be used in a piecewise manner.

The piecewise interpolation function $g(x)$ of $f(x)$ should be such that at each end point of the interval

$$g(x_0) = f(x_0); \; g(x_1) = f(x_1); \; ...; \; g(x_N) = f(x_N) \tag{38.38}$$

In each interval, the function $g(x)$ should be several times differentiable (e.g., for a cubic spline, it should be twice differentiable; for a fourth order spline, it should be three times differentiable). A linear spline is the simplest, but it has sharp corners and the derivatives of adjoining splines cannot be matched easily. A necessary condition for the spline to be continuous at the end points of each segment is that their derivatives computed from either side of the intersection should match.

38.7.2.1 Splines

Spline, in the ordinary sense, refers to a flexible strip used in drafting to draw a smooth curve through a set of points. In the mathematical sense too it has the same meaning. Spline methods can be used for interpolation as well as for regression. The basic objective in spline interpolation is to connect a set of data points by a smooth curve via a combination of several piecewise low order curves. The interpolation (or regression) functions can be polynomial, sinusoidal, exponential or their combinations. The cubic spline which consists of N polynomial functions each of which has order not greater than three is by far the most widely used. It is of the form

$$y = a_i + b_i x + c_i x^2 + d_i x^3 \tag{38.39}$$

where x, y, respectively are the independent and dependent variables, a_i, b_i, c_i, d_i are parameters for the ith interval. Thus, in a cubic spline, there are four parameters to be estimated for each interval which require four equations, i.e. $4(N - 1)$ equations in total. These consist of $2N - 2$ equations obtained by conditions of satisfying Eq. 38.38 at all data points (twice at interior points and once at end points), $2(N - 2)$ equations by equating the first and second derivatives (i.e., continuity at each interior point) of the interpolation function at each interior point and two more equations by setting the second derivatives of the function at the two end points to zero, adding up to $4(N - 1)$ equations.

Cubic spline regression is similar to cubic spline interpolation in all aspects excepting that a fewer number of 'knots' ('knot' refers to the intersection of two adjacent splines; k knots link $k + 1$ splines) is used based on the curvature changes of the data trajectory. The cubic regression spline equation has the form

$$y = a_i + b_i x + c_i x^2 + d_i x^3 + \sum_{j=1}^{k} D_j e_j (x_i - x_j)^3 \tag{38.40}$$

where x_j is the location of the ith knot, k is the number of knots, D_j is a dummy variable (zero or one), e_j is an additional parameter. For example,

$$y = a_i + b_i x + c_i x^2 + d_i x^3 \quad \text{(for the first interval; all } D_j\text{'s} = 0\text{)} \tag{38.41a}$$

$$y = a_i + b_i x + c_i x^2 + d_i x^3 + e_1(x - x_1)^3 \quad \text{(for the second interval; } D_1 = 1\text{)} \tag{38.41b}$$

$$y = a_i + b_i x + c_i x^2 + d_i x^3 + e_1(x - x_1)^3 + e_2(x - x_2)^3 \quad \text{(for the third interval; } D_1 = D_2 = 1\text{)},$$
$$\tag{38.41c}$$

and so on (Luo and White, 2005).

In these expressions, the continuity of $g(x)$ and its derivatives are automatically satisfied. If the number of knots and their locations are decided, the parameters a_i, b_i, c_i, d_i, e_j can be obtained by the linear least squares method. The locations can be equally spaced between the end points, in which case it may not be optimum. If the locations of the knots are also to be optimized from a random allocation, then the parameters can only be determined by the non-linear least squares method. More details of the method, including an application is given by Luo and White (2005).

The smoothing spline function has been shown to be equivalent to minimizing the functional (Reinsch, 1967)

$$S(\lambda) = \sum_{i=1}^{N} \left\{ (y_i - g(x_i)) \right\}^2 + \lambda \int_a^b \left\{ g''(x) \right\}^2 dx \tag{38.42}$$

where λ is a positive pre-specified parameter and a and b are the beginning and end values of x. The first term in Eq. 38.42 represents the residual sum of squares, and it penalizes the lack of fit of the function $g(x)$. The second term which is weighted by λ represents a roughness penalty that accounts for the curvature of the function $g(x)$. The λ in Eq. 38.42 is a smoothing parameter which may vary from zero to infinity. If λ is close to zero, $g(x)$ will not be smooth but will fit the data more closely. If λ is large, $g(x)$ will be smooth, but will not fit the data closely. The parameter λ, therefore, controls the trade-off between the closeness of fit to the data as measured by the residual sum of squares $\sum_{i=1}^{N} \left\{ (y_i - g(x_i)) \right\}^2$ and the smoothness as measured by the integral part $\int_a^b \left\{ g''(x) \right\}^2 dx$.

For a given λ, the function $g(x)$ which minimizes $S(\lambda)$ is a cubic spline (Reinsch, 1967). It has the following properties:

$g(x)$ has a continuous derivative everywhere
$g(x)$ is linear for $x < x_1$ and $x > x_N$
$g(x)$ is cubic between each successive values of x_i.

The minimization problem can be solved using variational calculus.

In the case of spline regression in a non-parametric sense the function can be defined as

$$g(x_i) = \sum_{j=1}^{N} w_{ij} y_j \tag{38.43}$$

where the weights w_{ij} are functions of the distance in the x-space and are given in the form of kernel functions which can take many forms. For example, Nadaraya (1964) and Watson (1964), commonly referred to as the Nadaraya–Watson kernel, used the form

$$w_{ij} = \frac{K(u)}{\sum\limits_{j=1}^{N} K(u)}; \quad u = \frac{x_i - x_j}{b} \tag{38.44}$$

The kernel function $K(u)$ can be a probability distribution function such as the Gaussian as well as several other functions. The following sinusoidal function with $b = \lambda^{0.25}$ and $W_{ij} \cong \dfrac{1}{b} K\left(\dfrac{x_i - x_j}{b}\right)$ has also been used:

$$K(u) = 0.5 e^{-\frac{|u|}{\sqrt{2}}} \sin\left(0.25\pi + \frac{|u|}{\sqrt{2}}\right). \tag{38.45}$$

38.8 PERIODICITY

Periodicity is a common feature in most natural phenomena where the effects of the revolution of the earth around the sun, rotation of the earth about its own axis and the rotation of the moon around the earth produce well-defined cyclic patterns. If the period of the periodicity is known, then, it can be removed by harmonic analysis. If it is not well defined, the periodicities could be identified by autocorrelation and spectral analysis.

38.8.1 Harmonic analysis – cumulative periodogram

If, for the annual cycle, $\tau = 1, 2,..., p$ are the periods of the harmonics (for example, in the case of monthly data, $p = 12$; weekly data, $p = 52$; daily data, $p = 365$), the harmonically fitted means, m_τ, for each period τ, can be represented by a series of sine and cosine terms (At this stage the trend component is assumed to have been removed) as follows:

$$m_\tau = \mu + \sum_{i=1}^{b} \left\{ A_i \cos\left(\frac{2\pi i\tau}{p}\right) + B_i \sin\left(\frac{2\pi i\tau}{p}\right) \right\} \tag{38.46}$$

where μ is the population mean, A_i and B_i are the Fourier coefficients, and b, the total number of harmonics. Sample estimates of m_τ, from the series are given by

$$\bar{x}_\tau = \frac{p}{N} \sum_{i=1}^{N/p} x_{\tau + p(i-1)} \tag{38.47}$$

The estimates of the population mean and the Fourier coefficients A_i and B_i are obtained by minimising the sum of squares of the differences, $\sum_{\tau=1}^{p} (\bar{x}_\tau - m_\tau)^2$ i.e.

$$E = \sum_{\tau=1}^{p} (\bar{x}_\tau - m_\tau)^2$$

$$= \sum_{\tau=1}^{p} \left[\bar{x}_\tau - \mu - \sum_{i=1}^{b} \left\{ A_i \cos\left(\frac{2\pi i\tau}{p}\right) + B_i \sin\left(\frac{2\pi i\tau}{p}\right) \right\} \right]^2 \tag{38.48}$$

which is done by setting $\dfrac{\partial E}{\partial \mu} = 0$; $\dfrac{\partial E}{\partial A_i} = 0$, and $\dfrac{\partial E}{\partial B_i} = 0$.

The first condition gives the estimate of the population mean μ as

$$\hat{\mu} = \frac{1}{p} \sum_{\tau=1}^{p} \bar{x}_\tau \tag{38.49}$$

and setting $\dfrac{\partial E}{\partial A_i} = 0$ and $\dfrac{\partial E}{\partial B_i} = 0$, and using the orthogonal properties (See Appendix 38.1A) of the sine and cosine functions, it can be shown (Kottegoda, 1980: pp. 37–38) that

$$A_i = \frac{2}{p} \sum_{\tau=1}^{p} \bar{x}_\tau \cos\left(\frac{2\pi i \tau}{p}\right) \quad \text{for } i = 1, 2, ..., h \tag{38.50a}$$

$$B_i = \frac{2}{p} \sum_{\tau=1}^{p} \bar{x}_\tau \sin\left(\frac{2\pi i \tau}{p}\right) \quad \text{for } i = 1, 2, ..., h \tag{38.50b}$$

When p is even, the last two coefficients are given by

$$A_h = \frac{1}{p} \sum_{\tau=1}^{p} \bar{x}_\tau \cos\left(\frac{2\pi h \tau}{p}\right) \tag{38.51a}$$

$$B_h = 0 \tag{38.51b}$$

In the above equations, i refers to the number of harmonics, and h, the total number of harmonics[7] (see Appendix 38.1A) which is equal to $\dfrac{p}{2}$ (or $\dfrac{p-1}{2}$ if p is odd).

For example, for monthly data, $p = 12$, therefore $h = 6$
for weekly data $p = 52$, therefore $h = 26$
for daily data $p = 365$, therefore $h = 182$.

When m_τ is determined by considering all the harmonics, then it will be the same as the actual periodic means $\bar{x}_\tau$ for all the values of τ. However, in practice only a few harmonics ($h^* < h$) are sufficient. These are called the significant harmonics – those which contribute significantly to the variability of x.

Experience in using the Fourier series representation indicates that for small time interval (daily, weekly, etc.) series, only the first few harmonics are sufficient (about 4–5 out of a total of 182 for daily series). For monthly series, only about one or two harmonics would be sufficient. A more accurate estimate of the number of significant harmonics could be obtained by plotting the cumulative periodogram (Salas et al., 1980: p. 80) which is defined as

---

[7] The harmonics have wave lengths (or periods) of $\dfrac{p}{1}, \dfrac{p}{2}, \dfrac{p}{3}, ..., \dfrac{p}{h}$, where $\dfrac{p}{h} \geq 2\Delta t$ and Δt is the time interval between data. The shortest wave length is $2\Delta t$, because, at least three points are needed to define a curve. If $\Delta t = 1$, $\dfrac{p}{h} = 2$, and, hence $h = \dfrac{p}{2}$ $\left(\text{or } \left(\dfrac{p}{2} - 0.5\right) \text{ if } p \text{ is odd}\right)$.

$$P_j = \sum_{i=1}^{j} \frac{\text{Var}(h_i)}{\text{Var}(x)} \tag{38.52}$$

where $\text{Var}(h_i)$ is the mean square deviation of m_τ around $\hat{\mu}$ for each harmonic $i = 1, 2, ..., h$, and $\text{Var}(x)$ is the mean square deviation of $\bar{x}_\tau$ around $\hat{\mu}$. They are given as[8]

$$\text{Var}(h_i) = \frac{1}{2}\left(A_i^2 + B_i^2\right); \quad i = 1, 2, ..., h \tag{38.53}$$

$$\text{Var}(x) = \frac{1}{p} \sum_{\tau=1}^{p} (\bar{x}_\tau - \hat{\mu})^2 \tag{38.54}$$

It can also be shown that

$$\sum_{i=1}^{h} \text{Var}(h_i) = \text{Var}(x) \tag{38.55}$$

i.e., $P_j = 1$ when $j = h$, in which case the explained variance is equal to the total variance. It should be noted that in the computation of the cumulative periodogram, the $\text{Var}(h_i)$ must be arranged in decreasing order of magnitude.

The criterion for testing the significance is based on the concept that the variation of P_j versus j is composed of a fast increasing periodic part and a slow increasing sampling part. The point of intersection of these two parts corresponds to the critical harmonic h^*. For an independent series, the sampling part is a straight line whereas for a linearly dependent series, it is a curve. The periodogram is also identified as the sample spectrum, or, the line spectrum, or, the discrete spectrum. In the plot, j represents the frequency.

Example 38.7: Periodicity analysis

Monthly rainfall data measured at the Observatory of Hong Kong (Hong Kong Observatory, Meteorological Results Part I) for the period 1884–1939 ($N = 672$) are used to illustrate various tests and analysis of periodic data. The data series, shown in Figure 38.6, has a mean value of 178.4 mm, a standard deviation of 193.4 mm, a skewness coefficient of 1.496, and kurtosis of 5.216. Typical periodogram results for this data set are shown in Table 38.7 and Figure 38.7.

As shown in Table 38.7 and Figure 38.7, the cumulative periodogram of the monthly means shows that the first harmonic is very significant explaining 94.85% of the variance followed by second, third, fourth, fifth, and sixth, explaining 4.292%, 0.4076%, 0.4062%, 0.04498% and 0.001476% respectively of the variance. Hence, the first, second, third and fourth would be sufficient (they together explain 99.9% of the variance). However, all six harmonics have been retained for Fourier series fitting.

As shown in Table 38.8 and Figure 38.7, the cumulative periodogram of the monthly standard deviations shows that the first harmonic explains 89.73% of the variance followed by third, fifth, fourth, second and sixth explaining 5.277%, 2.484%, 1.348%, 1.024% and 0.06896% respectively of the variance. Based on the periodogram, five harmonics (they together explain 99.9% of the variance) would be sufficient to describe the monthly standard deviations, but all six harmonics have been retained.

---

[8] The variance of m_τ around $\hat{\mu}$ is composed of variances of each harmonic, $\text{Var}(h_i)$. It is considered as the part of the variance of m_τ which is contributed by the harmonic i. Therefor P_j would represent the ratio of the explained variance by the first j harmonics to the total variance.

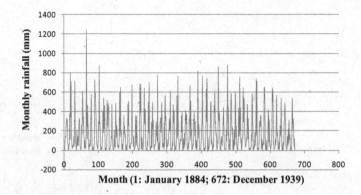

Figure 38.6 Monthly rainfall data in Hong Kong (1884–1939).

Table 38.7 Cumulative periodogram for monthly means of Hong Kong rainfall data: 1884–1939 (56 years)

J	A_i	B_i	$\frac{1}{2}(A_i^2 + B_i^2)$	$\sum_{i=1}^{j} \frac{1}{2}(A_i^2 + B_i^2)$	$\frac{\frac{1}{2}(A_i^2 + B_i^2)}{\text{Variance}}$	$\frac{\sum_{i=1}^{h} \frac{1}{2}(A_i^2 + B_i^2)}{\text{Variance}}$
1	−174.5	−80.33	18450	18,450	0.9485	0.9485
2	21.46	34.78	835.2	19,290	0.04292	0.9914
3	−2.407	12.36	79.30	19,370	0.004076	0.9955
4	6.421	−10.81	79.04	19,450	0.004062	0.9995
5	−1.378	3.950	8.752	19,460	0.0004498	1.0000
6	−0.7577	0.000	0.2871	19,460	0.00001476	1.0000

The comparisons of the sample monthly means and standard deviations with the Fourier fitted means, and standard deviations for different numbers of harmonics up to the maximum are illustrated in Tables 38.9 and 38.10, and Figures 38.8 and 38.9. It is seen that with the maximum numbers of harmonics, the sample monthly means and the standard deviations and the harmonically fitted monthly means and the standard deviations are exactly the same.

The significance of each harmonic can also be tested by the F-test. Analysis of variance is then required.

38.8.2 Auto-correlation analysis

Many time series have some form of dependence between events separated by small time steps. A measure of the degree of linear dependence is the auto-covariance function c_k which represents the covariance between x_t and x_{t+k} where k is the time lag. For a stationary continuous series it is defined as

$$c_k = \lim_{T \to \infty} \frac{1}{2T} \int_{-T}^{T} x_t\, x_{t+k}\, dt \tag{38.56}$$

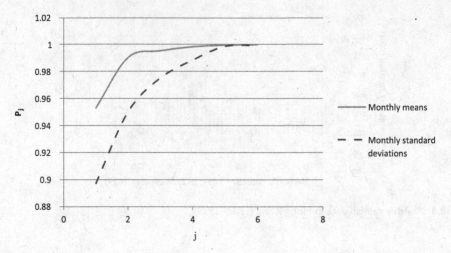

Figure 38.7 Cumulative Periodogram for monthly means and standard deviations of Hong Kong rainfall data (See Eq. 38.52).

Table 38.8 Cumulative periodogram for monthly standard deviations of Hong Kong rainfall data: 1884–1939 (56 years)

J	A_i	B_i	$\frac{1}{2}(A_i^2 + B_i^2)$	$\sum_{i=1}^{j}\frac{1}{2}(A_i^2 + B_i^2)$	$\dfrac{\frac{1}{2}(A_i^2 + B_i^2)}{\text{Variance}}$	$\dfrac{\sum_{i=1}^{h}\frac{1}{2}(A_i^2 + B_i^2)}{\text{Variance}}$
1	−82.37	−38.98	4152	4,152	0.8973	0.8973
3	−7.407	20.82	244.2	4,396	0.05277	0.9501
5	8.254	12.72	114.9	4,511	0.02484	0.9749
4	−5.107	−9.935	62.39	4,574	0.01348	0.9884
2	3.979	−8.885	47.39	4,621	0.01024	0.9986
6	−2.526	0.000	3.191	4,624	0.0006896	0.9993

Table 38.9 Sample and Fourier series fitted means

Month	Sample means	Fourier series fitted means for different numbers of harmonics					
		1 Harmonic	2 Harmonics	3 Harmonics	4 Harmonics	5 Harmonics	6 Harmonics
January	31.72	−12.85	28	40.36	27.79	30.96	31.72
February	44.7	21.62	41.01	43.42	49.57	45.46	44.7
March	75.42	98.12	76.66	64.29	70.72	74.67	75.42
April	136.8	196.1	155.3	152.9	140.3	137.6	136.8
May	290.1	289.4	270	282.4	288.5	289.3	290.1
June	383.9	353	374.4	376.8	383.3	384.6	383.9
July	383.3	369.7	410.6	398.2	385.7	382.5	383.3
August	361.8	335.3	354.7	352.2	358.4	362.5	361.8
September	252.9	258.8	237.3	249.7	256.1	252.1	252.9
October	111.7	160.8	119.9	122.3	109.7	112.5	111.7
November	41.85	67.48	48.09	35.73	41.88	41.1	41.85
December	27.27	39.32	25.4	22.99	29.41	28.03	27.27

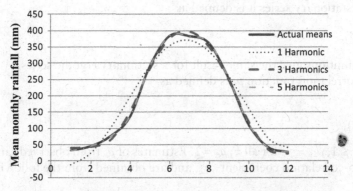

Month of the year (1: January; 12: December)

Figure 38.8 Actual and harmonically fitted monthly means of rainfall in Hong Kong.

Table 38.10 Sample and Fourier series fitted standard deviations

Month	Sample standard deviations	Fourier series fitted standard deviations for different numbers of harmonics					
		1 Harmonic	2 Harmonics	3 Harmonics	4 Harmonics	5 Harmonics	6 Harmonics
January	36.55	25.75	46.57	45.78	39.73	34.02	36.55
February	41.10	41.63	49.04	42.15	53.31	43.63	41.10
March	62.93	77.59	56.77	69.49	64.38	60.40	62.93
April	98.58	124.0	116.6	101.5	95.40	101.1	98.58
May	226.1	168.4	189.2	202.7	213.9	223.6	226.1
June	194.4	198.9	206.3	198.1	193.0	197.0	194.4
July	178.1	207.4	186.6	187.4	181.3	175.6	178.1
August	189.9	191.5	184.1	191.0	202.1	192.5	189.9
September	157.1	155.6	176.4	163.7	158.5	154.6	157.1
October	128.8	109.1	116.6	131.7	125.6	131.3	128.8
November	53.77	64.73	26.80	30.40	41.56	51.24	53.77
December	31.40	34.20	26.80	35.05	29.94	33.92	31.40

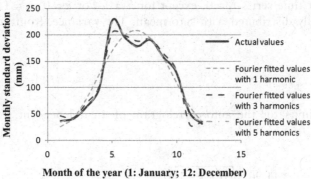

Month of the year (1: January; 12: December)

Figure 38.9 Actual and harmonically fitted monthly standard deviations of rainfall in Hong Kong.

For a weakly stationary series it is defined as

$$c_k = E(x_t\, x_{t+k}) \tag{38.57}$$

Both these definitions give the same result for a stationary ergodic series.

The auto-correlation function ρ_k is defined as

$$\rho_k = \frac{c_k}{E(x_t^2)} \tag{38.58}$$

where $-1 \le \rho_k \le 1$; $\rho_{-k} = \rho_k$ for all k; $\rho_0 = 1$. Estimates of ρ_k from observed samples are called the serial auto-correlation coefficients, r_k, and are obtained from the following equation:

$$r_k = \frac{\dfrac{1}{N-k}\displaystyle\sum_{t=1}^{N-k}(x_t - \bar{x})(x_{t+k} - \bar{x})}{\dfrac{1}{N-k}\displaystyle\sum_{t=1}^{N}(x_t - \bar{x})^2} \tag{38.59}$$

where N is the total number of observations, and $\bar{x}$ is the mean of x_t.

A plot of r_k versus k, called the auto-correlogram, is a useful tool for investigating the existence of periodicities in time series. In a stationary series, the auto-correlogram dies down to zero as k increases. In the computation of r_k, it is assumed that the trend component has already been removed.

For periodic time series, the auto-correlogram will also be periodic. For non-periodic dependent series the strength of correlation depends on the dependence of the series. The serial correlation coefficient usually increases with closeness of data. For example, daily data will show a stronger correlation compared with monthly or annual data.

The variance of the serial auto-correlation coefficient is given approximately by (Bartlett, 1946)

$$\mathrm{Var}\,(r_k) \approx \left(\frac{1}{N}\right)\left(1 + 2\sum_{i=1}^{q} r_i^2\right) \quad \text{for } k > q \tag{38.60a}$$

This is based on the assumption that $r_k = 0$ for $k > q$ and $r_k \ne 0$ for $k \le q$. For $k > q$ the r_k values tend to a normal distribution with zero mean. For a dependent series, it is complicated even for large samples from a Normal population.

For independent time series $r_k = 0$, except for $k = 0$. For $k = 0$, $r_k = 1$. For $k \ne 0$, the r_k values are Normally distributed with zero mean. The variance, from Eq. 38.60a, is then given by

$$\mathrm{Var}\,(r_k) \approx \frac{1}{N} \tag{38.60b}$$

The probability limits for the auto-correlogram of an independent series are given by (Anderson, 1941)

$$r_k\,(95\%) = \frac{\left\{-1 \pm \left(1.96\ \sqrt{(N-k-1)}\right)\right\}}{N-k} \tag{38.61a}$$

$$r_k \, (99\%) = \frac{\left\{ -1 \pm \left(2.326 \; \sqrt{(N-k-1)} \right) \right\}}{N-k} \tag{38.61b}$$

Figures 38.10 and 38.11 respectively show the auto-correlograms for annual and monthly rainfall data measured at the Observatory of Hong Kong. For the annual data, the auto-correlation oscillates within the 95% confidence limits indicating the absence of any clear periodicity. On the other hand, the monthly data clearly shows the annual periodicity.

38.8.3 Spectral analysis

An alternative to the time domain analysis of the dependence of a time series by the auto-correlation method is the spectral analysis which is carried out in the frequency domain. Like the decomposition of a periodic time series into its harmonics of different periods, the objective of spectral analysis is the decomposition on a frequency basis and then the estimation of the frequencies and their relative amplitudes.

By analogy with the spectral analysis in optics, a time series can be considered as a combination of basic frequencies of occurrence of random variables. The white light when passed through a prism is decomposed into fundamental colours violet, indigo, blue, green, yellow, orange and red which have specific wavelengths (or frequencies), with the violet having the highest frequency and the red having the lowest frequency. Similarly, the spectrum of the time series decomposes it into components on a frequency basis.

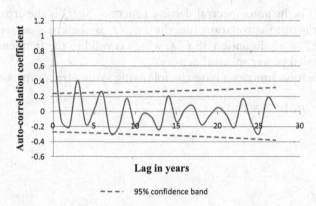

Figure 38.10 Auto-correlation of annual rainfall data in Hong Kong.

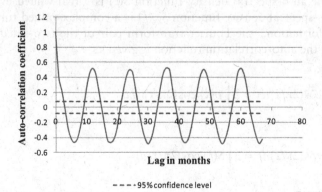

Figure 38.11 Auto-correlation of monthly rainfall data in Hong Kong.

The general definition of spectral density function which is useful for mathematical calculations has been shown to be the Fourier Transform of the correlation function, and the relationship is known as the Wiener Khintchine equations named in honour of the two mathematicians N. Wiener of USA and A.I. Khintchine of USSR in the early 1930s. The equations form a Fourier Transform pair and are given as

$$S(f) = \int_{-\infty}^{\infty} \rho(\tau) e^{-j2\pi f\tau} \, d\tau \qquad\qquad (38.62a)$$

$$\rho(\tau) = \int_{-\infty}^{\infty} S(f) e^{j2\pi f\tau} \, df \qquad\qquad (38.62b)$$

$$S_{xy}(f) = \int_{-\infty}^{\infty} \rho_{xy}(\tau) e^{-j2\pi f\tau} \, d\tau \qquad\qquad (38.62c)$$

$$\rho_{xy}(\tau) = \int_{-\infty}^{\infty} S_{xy}(f) e^{j2\pi f\tau} \, df \qquad\qquad (38.62d)$$

In Eq. 38.62a, $S(f)$ is the auto-spectral density function, $S_{xy}(f)$ is the cross spectral density function, $\rho(\tau)$ is the auto-correlation function, $\rho_{xy}(\tau)$ is the cross-correlation function, f, the frequency and $j^2 = -1$. Equation 38.62a, which is valid in the frequency range $(-\infty, \infty)$, refers to the two-sided power spectral density.

Stationary correlation functions have the following symmetric properties:

$$\rho(-\tau) = \rho(\tau); \quad \rho_{xy}(-\tau) = \rho_{yx}(\tau) \qquad\qquad (38.63)$$

Therefore,

$$S(-f) = S(f) \qquad\qquad (38.64a)$$

$$S_{xy}(-f) = S_{yx}(f) \qquad\qquad (38.64b)$$

indicating that the auto-spectral density function $S(f)$ is a real-valued even function of f, whereas the cross-spectral density function $S_{xy}(f)$ is a complex valued function of f. Using the Cosine transformation,[9] the Fourier transform pair of the two-sided spectral density function $S(f)$ and the auto-correlation function is given as

$$S(f) = \int_{-\infty}^{\infty} \rho(\tau) \cos(2\pi f\tau) \, d\tau = 2\int_{0}^{\infty} \rho(\tau) \cos(2\pi f\tau) \, d\tau \qquad\qquad (38.65a)$$

$$\rho(\tau) = \int_{-\infty}^{\infty} S(f) \cos(2\pi f\tau) \, df = 2\int_{0}^{\infty} S(f) \cos(2\pi f\tau) \, df \qquad\qquad (38.65b)$$

---

[9] The cosine transformation of the complex term is possible because $\rho(\tau)$ is an even function of $\tau \left[\rho(-\tau) = \rho(\tau) \right]$. Therefore $G(f)$ is a real function of f and is also even.

This definition is for physically realizable processes in the frequency range $(0, \infty)$. The one-sided spectral density functions, $G(f)$, defined in the frequency range $(0, \infty)$, are

$$G(f) = 2S(f) \tag{38.66a}$$

$$G_{xy}(f) = 2S_{xy}(f) \tag{38.66b}$$

Therefore, in terms of the one-sided spectral density functions, the relationship between the stationary correlation function and the spectral density function can be written as

$$G(f) = 4 \int_0^\infty \rho(\tau) \cos(2\pi f \tau) \, d\tau \tag{38.66c}$$

$$\rho(\tau) = \int_0^\infty G(f) \cos(2\pi f \tau) \, df \tag{38.66d}$$

$$G_{xy}(f) = 2 \int_{-\infty}^\infty \rho_{xy}(\tau) e^{-j2\pi f \tau} \, d\tau = P_{xy}(f) - jQ_{xy}(f) \tag{38.66e}$$

$$\rho_{xy}(\tau) = \int_0^\infty \left\{ P_{xy}(f) \cos(2\pi f \tau) + Q_{xy}(f) \sin(2\pi f \tau) \right\} df \tag{38.66f}$$

where $P_{xy}(f)$ is the real part (called the co-spectrum) and $Q_{xy}(f)$ is the imaginary part of the complex function but real-valued odd function of f (called the quadrature spectrum). The normalized two sided power spectral density function, $S^*(f)$, is given by

$$S^*(f) = \frac{S(f)}{\sigma_x^2} \tag{38.67}$$

The power spectral density function may also be calculated via the fast Fourier transform (FFT) method in which a direct Fourier transform of the original data is carried out. This method is more efficient computationally for series involving large amounts of data. However, when the volume of data is not excessive, and when the auto-correlation function is anyway determined, the FFT method does not offer a great advantage.

The approximate discrete representation of the power spectral density function given in Eq. 38.66c may be evaluated from the following equation[10]:

$$G(f) = 2\Delta t \left\{ r_0 + 2 \sum_{k=1}^{M-1} r_k \cos(2\pi f k) + r_M \cos 2\pi f M \right\} \tag{38.68}$$

---

[10] By the trapezoidal rule of numerical integration,

$$\int r(\tau) \cos(2\pi f \tau) d\tau = \sum_{k=0}^{M} r(k) \cos(2\pi f k) \Delta t$$

$$= \Delta t \left\{ 0.5 r_0 + \sum_{k=1}^{M-1} r(k) \cos(2\pi f k) + 0.5 r_M \cos(2\pi f M) \right\}$$

where Δt is the time interval between two events, r_k is the sample autocorrelation coefficient, and M is the maximum lag considered in the correlogram. For an independent series, the expected summation in the above equation is zero.

Similar to harmonic analysis in the time domain where the shortest period a harmonic could have is $2\Delta t$, where Δt is the sampling interval (because at least three points are needed to define a curve), the highest frequency in cycles/unit time for which information is sought is $\dfrac{1}{2\Delta t}$ in cycles/unit time. This frequency is called the Nyquist frequency (sometimes called the cut-off frequency, f_c). In the computation of the serial auto-correlation coefficient, the maximum lag corresponds to this limiting frequency. The maximum lag is usually chosen to be around $\dfrac{N}{5}$ to $\dfrac{N}{10}$ (subjectively). For example, if $\Delta t = 1$ month, $f_c = 0.5$ cycles/month.

Therefore the frequency range is $(0, 0.5)$. This range may be divided into frequency intervals as follows:

$$\left(0, \frac{1}{M}, \frac{2}{M}, \frac{3}{M}, \ldots, \frac{M}{M}\right) f_c$$

where M is usually taken as the maximum lag (not necessarily). In terms of the frequencies, $G(f)$ can be written as

$$G(f) = 2\Delta t \left\{ r_o + 2\sum_{k=1}^{M-1} r_k \cos\frac{\pi k f}{f_c} + r_m \cos\frac{\pi M f}{f_c} \right\} \qquad (38.69)$$

In the normalized form (making the area under the power spectral density function equal to unity) $G(f)$ is given by

$$G(f) = \frac{2\Delta t}{2M} \left\{ r_o + 2\sum_{k=1}^{M-1} r_k \cos\frac{\pi k f}{f_c} + r_M \cos\frac{\pi M f}{f_c} \right\} \qquad (38.70a)$$

$$= \frac{1}{M} \left\{ r_0 + 2\sum_{k=1}^{M-1} r_k \cos\frac{\pi k f}{f_c} + r_M \cos\frac{\pi M f}{f_c} \right\} \qquad \text{if } \Delta t = 1 \qquad (38.70b)$$

In this equation, k is an index for the frequency, and varies from $0, 1, 2, \ldots, M$. Frequency, f, is given by

$$f = \frac{k}{2\Delta t M} \qquad (38.71)$$

Therefore, $G(f)$ can be expressed as $G(k)$.

If frequency is measured in radians/unit time, M in the above equation should be replaced by π.

Low values of M in the above equation will result in loss of information in the correlogram leading to excessive smoothing of power spectral density function (fewer points in the power spectral density function). High values of M will lead to a spectrum which is not clear. The sensitivity of the power spectral density function to variations of M may be carried out using different values of M.

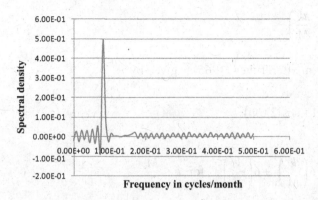

Figure 38.12 Raw spectral density function for monthly rainfall data.

The power spectral density function so obtained is called the raw estimate of the power spectral density. It is the convolution of the true power spectral density function with a window function of the box-car type. It has the net effect of causing leakage by spreading the main lobes of the power spectral density function and by adding an infinite number of smaller side lobes. This leads to the possibility of obtaining negative values for the power spectral density function.

An example of the raw spectral density function for the rainfall data in Example 38.3 is shown in Figure 38.12. Leakage can be reduced by the following procedures.

(a) **Hanning method: (After J. von Hann)**

The Hanning method was named after J. Von Hann (Blackman and Tukey, 1959). This window function (see Eqs. 38.74–38.76), which is of the form

$$\gamma_k = \frac{1}{2}\left(1 + \cos\frac{\pi k}{M}\right),$$

is equivalent to a weighted moving average of three terms as shown in Eq. 38.72 in which GS is the smoothed function.

$GS(1) = 0.5G(1) + 0.5G(2)$ for $k = 1$

$GS(k) = 0.25G(k-1) + 0.5G(k) + 0.25G(k+1)$ for $k = 2$ to $M - 1$ (38.72)

$GS(M) = 0.5G(M-1) + 0.5G(M)$ for $k = M$

(b) **Hamming method: (After Hamming, 1983)**

$GS(1) = 0.54G(1) + 0.46G(2)$ for $k = 1$

$GS(k) = 0.23G(k-1) + 0.54G(k) + 0.23G(k+1)$ for $k = 2$ to $M - 1$ (38.73)

$GS(M) = 0.54G(M-1) + 0.46G(M)$ for $k = M$

(c) **Lag window method: (After Tukey, 1965)**

As the lag increase, r_k becomes less reliable because there are fewer observations. Therefore, it is multiplied by unequal weights called lag windows γ_k of the form

$$\gamma_k = \frac{1}{2}\left(1 + \cos\frac{\pi k}{M}\right) \tag{38.74}$$

One such form after Tukey (1965) is given by

$$(a')\ D(k) = \frac{1}{2}\left(1 + \cos(2\pi f k)\right) \tag{38.75}$$

$$D(k) = 1 - 6\left(\frac{k}{M}\right)^2 + 6\left(\frac{k}{M}\right)^3 \quad \text{for } k \leq \frac{M}{2}$$

$$(b')\ D(k) = 2 - \left(\frac{k}{M}\right)^3 \quad \text{for } \frac{M}{2} < k < M \tag{38.76}$$

$$D(k) = 0 \quad \text{for } k > M$$

An illustration of the smoothed spectral density function by the Hanning method for the rainfall data in Example 38.3 is shown in Figure 38.13. The annual periodicity is shown as a peak in the spectral density function at a frequency of 0.0827 cycles/month $\left(\approx \frac{1}{12}\right)$.

The spectrum usually does not give any more information than the auto-correlogram. It, however, indicates where the high and low frequencies are concentrated. Peaks in the power spectral density function (of series where trends are already removed) indicate periodicities. Sharp peaks indicate regular periodicities. A perfect periodicity such as a sinusoidal curve will produce a vertical line in the spectrum. The spectrum in addition to revealing the above characteristics of the series also helps to guide the choice of the mathematical model that generates the series. The auto-correlation and spectral density are complementary to each other and should be interpreted together.

Confidence limits can be set for spectral density functions based on a known population.

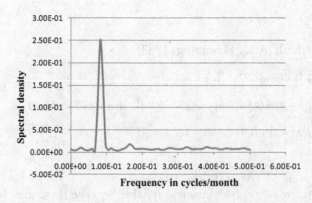

Figure 38.13 Smoothed spectral density function for monthly rainfall data.

38.8.4 Cross correlation

Cross correlation is used to determine the dependence between two series x_t and y_t, not only at the same time level but also for one series leading or lagging the other. For x_t and y_t with zero mean, the cross-covariance $c_{xy}(\tau)$ is defined as

$$c_{xy}(\tau) = E[x_t y_{t+\tau}] \tag{38.77}$$

$$c_{yx}(\tau) = c_{xy}(-\tau) = E[x_{t+\tau} y_t] \tag{38.78}$$

(Note that $c_{xy}(\tau) \neq c_{xy}(-\tau)$, because the cross-covariance is not an even function)
    Cross-correlation $\rho_{xy}(\tau)$ is defined as

$$\rho_{xy}(\tau) = E\left\{ \frac{[x_t - E(x_t)][y_{t+\tau} - E(y_t)]}{\sqrt{\text{Var}(x_t)\,\text{Var}(y_t)}} \right\} \tag{38.79}$$

Sample cross-correlation coefficient $r_{xy}(k)$ is defined as

$$r_{xy}(k) = \frac{\sum\limits_{t=1}^{N-k}(x_t - \bar{x})(y_{t+k} - \bar{y})}{\left\{ \left[\sum\limits_{t=1}^{N}(x_t - \bar{x})^2 \right] \left[\sum\limits_{t=1}^{N}(y_t - \bar{y})^2 \right] \right\}^{1/2}} \tag{38.80}$$

38.8.5 Cross spectral density function

The cross-spectral density function and the cross-correlation were defined in Eqs. 38.66e and 38.66f.
    An estimate of the sample cross-spectral density functions $s_{xy}(\omega)$ can be obtained as

$$s_{xy}(\omega) = \frac{1}{\pi}\left\{ \sum\limits_{\tau=-M}^{M} \gamma(\tau) r_{xy}(\tau)\cos(\omega\tau) \right\} \tag{38.81}$$

where $\gamma(\tau)$ is a lag window and $\omega = 2\pi f$.
    It should be noted that the cross-spectral density function has a real part that is in phase and an imaginary part which is out of phase. The real (in phase) part, known as the normalized sample co-spectrum, is given by

$$P_{xy}(\omega) = \frac{1}{\pi}\left\{ \gamma(0)r_{xy}(0) + \sum\limits_{\tau=1}^{M} \gamma(\tau)\left[r_{xy}(\tau) + r_{xy}(-\tau) \right]\cos(\omega\tau) \right\} \tag{38.82}$$

and the imaginary (quadrature spectrum or out of phase part) is given by

$$Q_{xy}(\omega) = \frac{1}{\pi}\left\{ \sum\limits_{\tau=1}^{M} \gamma(\tau)\left[r_{xy}(\tau) - r_{xy}(-\tau) \right]\sin(\omega\tau) \right\} \tag{38.83}$$

where $s_{xy}(\omega) = P_{xy}(\omega) - jQ_{xy}(\omega)$.

Normalized cross amplitude spectrum which measures the association between the amplitudes in the two series at the same frequency is given by

$$\beta_{xy}(\omega) = \left[P_{xy}^2(\omega) + Q_{xy}^2(\omega) \right]^{1/2} \tag{38.84}$$

Phase spectrum is defined as

$$\phi_{xy}(\omega) = \tan^{-1}\left\{ -\frac{Q_{xy}}{P_{xy}} \right\} \tag{38.85}$$

Coherence spectrum, h_{xy}, is defined as

$$h_{xy}(\omega) = \frac{\beta_{xy}^2}{s_x(\omega)s_y(\omega)} \tag{38.86}$$

The coherence which lies between 0 and 1 measures the interdependence between frequency components in the two series at frequency ω. Phase and coherence are not considered in the auto-correlation or spectral analysis. Coherence is analogous to ρ^2.

A plot of functions such as coherence and phase against frequency enables an assessment of the non-linear nature of the time series which would help in the model formulation. Strong coherence occurs at low frequencies (e.g., annual cycle) whereas at high frequencies, the coherence is small. This means that the periodicities with higher frequencies are spurious.

38.9 STOCHASTIC COMPONENT

After the trend and the periodic component have been removed, the remaining stochastic component which is assumed to be covariance stationary may consist of a dependent (correlated) part and an independent (uncorrelated) random part. Four different types of stochastic models can be used to describe the dependent part, namely,

- Auto-Regressive (AR)
- Moving Average (MA)
- Auto-Regressive Moving Average (ARMA)
- Auto-Regressive Integrated Moving Average (ARIMA).

In all the above four types of models, the present value of the stochastic variable is linearly related to the past values in some form. They are therefore identified as linear stochastic models. Stationarity is also implied in the AR, MA and ARMA type models.

38.9.1 Auto-regressive (AR) models

In auto-regressive models, the current value of the variable is linearly related to the weighted sum of a number of past values and an independent random value.

The general p'th order AR model has the form:

$$z_t = \phi_{p,1} z_{t-1} + \phi_{p,2} z_{t-2} + \ldots + \phi_{p,p} z_{t-p} + \eta_t = \sum_{i=1}^{p} \phi_{p,i} z_{t-i} + \eta_t \tag{38.87}$$

where the $\phi_{p,i}$'s are called the autoregressive coefficients, η_t is an independent (uncorrelated) random number (sometimes referred to as 'white noise') and z_t is the stochastic component which is obtained by subtracting any trends and periodicities from the original time series. For convenience, z_t is reduced to zero mean and unit variance (normalized). An AR(p) model has $p + 2$ parameters: the mean of the stochastic series, the p auto-regressive coefficients and the variance of the random component.

38.9.1.1 Properties of auto-regressive models

The following properties form the basis of model development:

$$
\left.
\begin{aligned}
&E(z_t) = E(\eta_t) = 0 \\
&\mathrm{Var}\left(z_t\right) = E\left(z_t^2\right) = \sigma_z^2 \\
&\mathrm{Var}\left(\eta_t\right) = E\left(\eta_t^2\right) = \sigma_\eta^2 \\
&\rho_k = \frac{E\left(z_t\, z_{t-k}\right)}{\sigma_z^2} \\
&E\left(\eta_t\, \eta_{t-k}\right) = E\left(\eta_t\, z_{t-k}\right) = 0 \quad \text{for } k = 1, 2, 3 \dots \\
&\sigma_\eta^2 = \sigma_z^2 \left(1 - \sum_{i=1}^{p} \phi_{p,i}\, \rho_i \right)
\end{aligned}
\right\}
\tag{38.88}
$$

$$
\sigma_z^2 = \frac{\sigma_\eta^2}{1 - \rho_1 \phi_{1,1} - \rho_2 \phi_{2,2} - \dots \rho_p\, \phi_{p,p}}
$$

In the above equations for σ_η^2, the summation (RHS) is called the coefficient of determination, R^2. It is the square of the multiple correlation coefficient. An unbiased estimate of σ_η^2 may be obtained by multiplying by $\dfrac{N}{N-p}$. An AR process is completely known if $\phi_{p,i}$'s and σ_η^2 are known.

38.9.1.2 Estimation of parameters

Multiplying the general auto-regressive equation by $z_{t-1}, z_{t-2}, \dots, z_{t-p}$ in turn and taking expectations,[11] the following p equations can be obtained. They are called the Yule–Walker equations (After Yule (1927) and Walker (1931)).

$$
\begin{bmatrix} \rho_1 \\ \rho_2 \\ \rho_3 \\ \vdots \\ \rho_p \end{bmatrix}
=
\begin{bmatrix}
1 & \rho_1 & \rho_2 & \cdots & \rho_{p-1} \\
\rho_1 & 1 & \rho_1 & \cdots & \rho_{p-2} \\
\rho_2 & \rho_1 & 1 & \cdots & \rho_{p-3} \\
\vdots & \vdots & \vdots & \cdots & \vdots \\
\rho_{p-1} & \rho_{p-2} & \rho_{p-3} & \cdots & 1
\end{bmatrix}
\begin{bmatrix} \phi_{p,1} \\ \phi_{p,2} \\ \phi_{p,3} \\ \vdots \\ \phi_{p,p} \end{bmatrix}
\tag{38.89}
$$

---

[11]Multiplying Eq. 38.87 by z_{t-k} gives $z_t z_{t-k} = \phi_{p,1} z_{t-1} z_{t-k} + \phi_{p,2} z_{t-2} z_{t-k} + \dots + \phi_{p,p} z_{t-p} z_{t-k} + \eta_t z_{t-k}$. Taking expectations term by term (using the definition that $\rho_k = \dfrac{E\left[z_t z_{t-k}\right]}{\sigma_z^2}$ and the assumption that $\sigma_z^2 = 1$), we have $\rho_k = \phi_{p,1} \rho_{k-1} + \phi_{p,2} \rho_{k-2} + \dots + \phi_{p,p} \rho_{k-p} = \sum_{i=1}^{p} \phi_{p,i} \rho_{k-i}$ for $k > 0$, and also using $\rho(-k) = \rho(k)$.

The diagonal term in the above matrix corresponds to ρ_0. In a different form, these equations can be written as

$$\rho_k = \sum_{i=1}^{p} \phi_{p,i} \rho_{k-i} \quad \text{for } k > 0 \tag{38.90}$$

A necessary condition for stationarity of the AR model is that the auto-correlation matrix $[\rho]$ is positive definite.[12] This means that the determinant and all its minors are positive, i.e.

$$\rho_0 > 0; \quad \begin{bmatrix} \rho_0 & \rho_1 \\ \rho_1 & \rho_0 \end{bmatrix} > 0; \quad \begin{bmatrix} \rho_0 & \rho_1 & \rho_2 \\ \rho_1 & \rho_0 & \rho_1 \\ \rho_2 & \rho_1 & \rho_0 \end{bmatrix} > 0 \text{ etc.}$$

The positive definiteness of the auto-correlation matrix also implies that for AR(1) models, $-1 < \rho_1 < 1$, and for AR(2) models, $-1 < \rho_1 < 1$, $-1 < \rho_2 < 1$, and $-1 < \dfrac{\rho_2 - \rho_1^2}{1 - \rho_1^2} < 1$.

The stationarity condition is also satisfied if the roots of the characteristic equation given below lie within the unit circle given by $u^2 = 1$ (u is a dummy variable).

$$u^p - \phi_{p,1} u^{p-1} - \phi_{p,2} u^{p-2} - \dots \phi_{p,p} = 0 \tag{38.91}$$

The AR coefficients can be determined from the estimates of ρ, or if ϕp, i's are known then ρ's may be determined.

38.9.1.3 First-order model (lag-one Markov model)

First order auto-regressive model (AR(1)) is given by

$$z_t = \phi_{1,1} z_{t-1} + \eta_t \tag{38.92}$$

From the Yule–Walker equations it can be seen that

$$\left\{ \begin{array}{l} \rho_1 = \phi_{1,1} \\ \rho_2 = \phi_{1,1} \rho_1 \\ \rho_3 = \phi_{1,1} \rho_2 \\ \vdots \\ \rho_p = \phi_{1,1} \rho_{p-1} \end{array} \right\} \tag{38.93}$$

The $\rho - \phi$ relationship can be expressed either as

$$\rho_0 = 1; \quad \rho_k = \phi_{1,1}^k; \quad k \geq 0 \tag{38.94}$$

---

[12] Positive semi-definite means ≥ 0.

or as

$$\rho_k = \rho_1^k \tag{38.95}$$

This equation decays exponentially to zero for positive values of $\phi_{1,1}$. For negative $\phi_{1,1}$, the auto-correlation function is damped and oscillates around zero.

The variance of the independent component η_t is given by

$$\sigma_\eta^2 = \left(1 - \rho_1^2\right)\left(R^2 = \rho_1^2\right) \text{ because } \phi_{1,2}, \phi_{1,3} \text{ etc. are zero.} \tag{38.96}$$

$$\sigma_z^2 = \frac{\sigma_\eta^2}{1 - \rho_1\phi_{1,1}} = \frac{\sigma_\eta^2}{1 - \phi_{1,1}^2} \tag{38.97}$$

The stationarity condition[13] for AR(1) models can be shown to be $-1 < \rho_1 < 1$.

38.9.1.3.1 Special properties of AR(1) model

(a')   An AR(1) model can be written as an infinite series as follows:

$$z_t = \phi_{1,1}z_{t-1} + \eta_t$$

$$= \phi_{1,1}(\phi_{1,1}z_{t-2} + \eta_{t-1}) + \eta_t$$

$$= \phi_{1,1}(\phi_{1,1}(\phi_{1,1}z_{t-3} + \eta_{t-2}) + \phi_{1,1}\eta_{t-1}) + \eta_t$$

$$= \eta_t + \phi_{1,1}\eta_{t-1} + \phi_{1,1}^2\eta_{t-2} + \phi_{1,1}^3\eta_{t-3} + \ldots \tag{38.98}$$

$$= \sum_{j=0}^{\infty} \phi_{1,1}^j \eta_{t-j}$$

This is a MA(∞) process (see also MA(1) model).

(b')   An AR(1) model can be used to transform a dependent series into an independent series

$$z_t = \phi_{1,1}z_{t-1} + \rho_t$$

which gives

$$\rho_t = z_t - \phi_{1,1}z_{t-1} \tag{38.99}$$

which is an independent series.

For example, if z_t vs. z_{t-1} from a AR(1) series is plotted, a trend can be seen. On the other hand, no trend or correlation will be seen if z_t vs. z_{t-1} is plotted for the process $z_t = \eta_t$ (i.e. $\phi_{1,1} = 0$), which is white noise.[14]

---

[13]The characteristic equation for an AR(1) model is

$u - \phi_{1,1} = 0$ which gives $u = \phi_{1,1} = \rho_1$ and hence $|u| < 1$.

[14]White noise has $\rho_k = \begin{cases} 1 & \text{for } k = 0 \\ 0 & \text{for } k \neq 0 \end{cases}$, $\phi_{k,k} = \begin{cases} 1 & \text{for } k = 0 \\ 0 & \text{for } k \neq 0 \end{cases}$, and the one sided spectral density function has a constant value for all frequencies.

(c')   For z_t to be finite, $|\phi_{1,1}| < 1$. Otherwise,

$$z_t = \phi_{1,1} z_{t-1} + \eta_t$$

will increase infinitely as t increases thereby making z_t non-stationary.

(d')   Random walk as a limiting case of AR(1)

If $\phi_{1,1} \to 1$, the AR(1) model becomes

$$z_t = z_{t-1} + \rho_t$$

or

$$\eta_t = z_t - z_{t-1} = \nabla z_t$$

z_t can also be written as

$$z_t = z_{t-1} + \eta_t$$
$$= \eta_t + z_{t-1}$$
$$= \eta_t + z_{t-2} + \eta_{t-1}$$
$$= \eta_t + \eta_{t-1} + z_{t-3} + \eta_{t-2} \tag{38.100}$$

$$\cdots$$

$$= \eta_t + \eta_{t-1} + \eta_{t-2} + \cdots$$

$$= \sum_{j=0}^{\infty} \eta_{t-j}$$

which is a random walk process.

In a random walk process, the best forecast at the present time level is the outcome of the process at the previous time level.

38.9.1.4 Second-order model (lag-two model)

The second order auto-regressive model (AR(2)) has the form

$$z_t = \phi_{2,1} z_{t-1} + \phi_{2,2} z_{t-2} + \eta_t \tag{38.101}$$

From the Yule–Walker equations,

$$
\begin{cases}
\rho_1 = \phi_{2,1} + \phi_{2,2}\, \rho_1 \\
\rho_2 = \phi_{2,1}\rho_1 + \phi_{2,2} \\
\rho_3 = \phi_{2,1}\rho_2 + \phi_{2,2}\rho_1 \\
\quad \vdots \\
\rho_p = \phi_{2,1}\, \rho_{p-1} + \phi_{2,2}\, \rho_{p-2}
\end{cases}
\tag{38.102}
$$

By solving the first two equations simultaneously,

$$\phi_{2,1} = \frac{\rho_1(1-\rho_2)}{(1-\rho_1^2)} \tag{38.103a}$$

$$\phi_{2,2} = \frac{(\rho_2 - \rho_1^2)}{(1 - \rho_1^2)} \tag{38.103b}$$

or alternatively,

$$\rho_0 = 1; \quad \rho_1 = \frac{\phi_{2,1}}{1 - \phi_{2,2}}; \quad \rho_2 = \phi_{2,2} + \frac{\phi_{2,1}^2}{1 - \phi_{2,2}} \tag{38.104}$$

For $k \geq p$, the ρ_k's can be obtained recursively by the Yule–Walker equations:

$$\rho_k = \phi_{2,1}\rho_{k-1} + \phi_{2,2}\rho_{k-2} \tag{38.105}$$

The variance of the independent component η_t is given by

$$\sigma_\eta^2 = \frac{1 + \phi_{2,2}}{1 - \phi_{2,2}}\left\{(1 - \phi_{2,2})^2 - \phi_{2,1}^2\right\} \tag{38.106}$$

$$\sigma_z^2 = \frac{\sigma_\eta^2}{1 - \rho_1\phi_{2,1} - \rho_2\phi_{2,2}} \tag{38.107}$$

The stationarity condition[15] for AR(2) models can be shown to be

$$-1 < \rho_1 < 1; \; -1 < \rho_2 < 1; \text{ and } \rho_1^2 < \frac{1}{2}(1 + \rho_2)$$

or,

$$\phi_{2,1} + \phi_{2,2} < 1; \quad \phi_{2,2} - \phi_{2,1} < 1; \quad -1 < \phi_{2,2} < 1$$

Depending on the values of $\phi_{2,1}$ and $\phi_{2,2}$, the auto-correlation function ρ_k of an AR(2) model may have different forms.

For example, if $\phi_{2,1}^2 + 4\phi_{2,2} \geq 0$, ρ_k decays exponentially to zero when $\phi_{2,1} > 0$, but oscillates around zero if $\phi_{2,1} < 0$.

If $\phi_{2,1}^2 + 4\phi_{2,2} < 0$, the auto-correlation function is pseudo periodic or damped.

(e') Partial auto-correlation function (PAF)

The set of parameters $\phi_{1,1}, \phi_{2,2}, \phi_{3,3}, \ldots, \phi_{k,k}$ of the AR models of order 1, 2, 3,..., p constitute the partial auto-correlation function. They may be estimated by the method of least squares or by the Yule–Walker equations. The resulting relationship is of the form

$$\rho_j = \phi_{k,1}\rho_{j-1} + \phi_{k,2}\rho_{j-2} + \ldots \phi_{k,k}\rho_{j-k}; \; j = 1, 2, \ldots, k \tag{38.108}$$

The set of parameters $\phi_{1,1}, \phi_{2,2}, \phi_{3,3}, \ldots, \phi_{k,k}$ can then be determined successively for $k = 1, 2, \ldots$ from Eq. 38.108 as

---

[15] The characteristic equation for an AR(2) model is

$$u^2 - \phi_{2,1}u - \phi_{2,2} = 0$$

the roots of which depend on the values of $\phi_{2,1}$ and $\phi_{2,2}$, and should lie outside the unit circle for stationarity.

$$\phi_{1,1} = \rho_1 \tag{38.109a}$$

$$\phi_{2,2} = \frac{\begin{vmatrix} 1 & \rho_1 \\ \rho_1 & \rho_2 \end{vmatrix}}{\begin{vmatrix} 1 & \rho_1 \\ \rho_1 & 1 \end{vmatrix}} \tag{38.109b}$$

$$\phi_{3,3} = \frac{\begin{vmatrix} 1 & \rho_1 & \rho_1 \\ \rho_1 & 1 & \rho_2 \\ \rho_2 & \rho_1 & \rho_3 \end{vmatrix}}{\begin{vmatrix} 1 & \rho_1 & \rho_2 \\ \rho_1 & 1 & \rho_1 \\ \rho_2 & \rho_1 & 1 \end{vmatrix}} \tag{38.109c}$$

$$\phi_{k,k} = \frac{\begin{vmatrix} 1 & \rho_1 & \rho_2 & \cdots & \rho_{k-2} & \rho_1 \\ \rho_1 & 1 & \rho_1 & \cdots & \rho_{k-3} & \rho_2 \\ \cdots & \cdots & & \cdots & \cdots & \\ \cdots & \cdots & & \cdots & \cdots & \\ \rho_{k-1} & \rho_{k-2} & \rho_{k-3}\cdot & \cdots & \rho_1 & \rho_k \end{vmatrix}}{\begin{vmatrix} 1 & \rho_1 & \rho_2 & \cdots & \rho_{k-1} \\ \rho_1 & 1 & \rho_1 & \cdots & \rho_{k-2} \\ \cdots & \cdots & \cdots & \cdots & \cdots \\ \cdots & \cdots & \cdots & \cdots & \cdots \\ \rho_{k-1} & \rho_{k-2} & \rho_{k-3}\cdots & \cdots & 1 \end{vmatrix}} \tag{38.109k}$$

It is a useful tool for determining the order of a model because of the fact that $\phi_{k,k}$ are theoretically zero for $k > p$ [$\phi_{k,k}$, etc. are the last coefficients in an AR model]. It gives the correlation between x_i and x_{k+i}, adjusted for the intervening observations $x_{t+1}, \ldots, x_k$. For AR(2) model, the PAF from Eq. 38.108 would be

$$\phi_{1,1} = \rho_1; \ \phi_{2,2} = \frac{\rho_2 - \rho_1^2}{1 - \rho_1^2}; \ \phi_{k,k} = 0 \quad \text{for } k > 2 \tag{38.110}$$

In general, $\phi_{k,k} \neq 0$ for $k \leq p$ and $\phi_{k,k} = 0$ for $k > p$.
Variance of the estimated PAF for an AR(p) model is given by

$$\text{Var}\,(\phi_{k,k}) \approx \frac{1}{N} \quad \text{for } k > p \qquad\qquad (38.111)$$

where N is the sample length.

The confidence limits can.be obtained by assuming that the distribution of the PAF with zero expected value is approximately Normal. However, the distribution may be significantly different from Normal if $N < 30$ (Anderson, 1941; Kottegoda, 1980: p. 120). Then, the small sampling distribution theory may be used.

38.9.2 Moving average (MA) models

In a moving average model of order q the current value of the variable is considered as the weighted sum of $q + 1$ independent residuals. It takes the form:

$$z_t = \eta_t - \theta_{q,1}\eta_{t-1} - \theta_{q,2}\eta_{t-2} - \ldots \theta_{q,q}\eta_{t-q} = \eta_t - \sum_{i=1}^{q} \theta_{q,i}\eta_{t-i} \qquad\qquad (38.112)$$

in which $\theta_{q,i}$ are the moving average coefficients. The weight assumed for the current value of the residual is unity. As before, the stochastic component is assumed to be normalised. A MA(q) model has $q + 2$ parameters: the mean of the stochastic series, the q moving average coefficients and the variance of the random component.

38.9.2.1 Properties of MA models

$$E(z_t) = E(\eta_t) = 0 \qquad\qquad (38.113a)$$

$$E(z_t^2) = \sigma_z^2 = 1 \qquad\qquad (38.113b)$$

$$E(\eta_t\,\eta_{t-k}) = 0 \quad \text{for } k \neq 0 \qquad\qquad (38.113c)$$

By squaring Eq. 38.112, taking expectations, and substituting the properties given by Eq. 38.113, the variance of the residuals can be shown to be equal to

$$\sigma_\eta^2 = \frac{1}{\left(1 + \theta_{q,1}^2 + \theta_{q,2}^2 + \ldots + \theta_{q,q}^2\right)}. \qquad\qquad (38.114)$$

A finite MA process is always stationary.

38.9.2.2 Parameters of MA models

Multiplying the LHS of Eq. 38.112 by z_{t-k} and the RHS by $\eta_{t-k} - \theta_{q,1}\eta_{t-1-k} - \theta_{q,2}\eta_{t-2-k} - \ldots \theta_{q,q}\eta_{t-q-k}$, (this is the same as z_{t-k}), taking expectations, and substituting the properties given by Eq. 38.113, gives

$$\rho_k = \sigma_\eta^2\left(-\theta_{q,k} + \theta_{q,1}\theta_{\theta_q,k+1} + \theta_{q,2}\theta_{\theta_q,k+2} + \ldots + \theta_{q,q-k}\theta_{\theta_q,q}\right) \quad \text{for } k = 1, 2, \ldots, q. \qquad (38.115)$$

Substituting for σ_η^2 from Eq. 38.114 gives

$$\rho_k = \frac{\left[-\theta_{q,k} + \displaystyle\sum_{i=1}^{q-k} \theta_{q,i}\theta_{q,i+k}\right]}{\left[1 + \displaystyle\sum_{i=1}^{q} \theta_{q,i}^2\right]} \quad \text{for } k = 1, 2, \ldots, q \tag{38.116}$$

and

$$\rho_k = 0 \quad \text{for } k > q.$$

Equation 38.116 is non-linear and needs to be solved numerically, except for MA(1).

38.9.2.3 MA(1) model

MA(1) model is given by

$$z_t = \eta_t - \theta_{1,1}\,\eta_{t-1} \tag{38.117}$$

From Eq. 38.116,

$$\rho_1 = \frac{\theta_{1,1}}{\left(1 + \theta_{1,1}^2\right)}; \; \rho_k = 0 \quad \text{for } k \geq 2 \tag{38.118}$$

Hence, $\rho_{1,1}$ can be estimated.

It can be seen that one root of this equation is the reciprocal of the other. The root which satisfies the invertibility condition $|\rho_{1,1}| > 1$ is the initial estimate which is not necessarily the best. A better estimate, which will make the generated data look more like the historical data can be obtained by minimising the sum of squares of the residuals ε_i where

$$\varepsilon_1 = z_1^{\text{obs}}$$

$$\varepsilon_2 = z_2^{\text{obs}} + \theta_{1,1}\varepsilon_1 \tag{38.119}$$

$$\varepsilon_3 = z_3^{\text{obs}} + \theta_{1,1}\varepsilon_2, \text{etc.}$$

(z_t^{obs} for $t = 1, 2, \ldots, N$, is an observed series). They can also be determined by the more difficult maximum likelihood estimate method.

38.9.2.3.1 Invertibility

By writing the equation for MA(1) recursively, it can be seen that

$$\eta_t = z_t + \theta_{1,1}\eta_{t-1}$$

$$= z_t + \theta_{1,1}(z_{t-1} + \theta_{1,1}\eta_{t-2})$$

$$= z_t + \theta_{1,1}z_{t-1} + \theta_{1,1}^2(z_{t-2} + \theta_{1,1}\eta_{t-3}) \tag{38.120}$$

$$\cdots$$

$$= z_t + \rho_{1,1}z_{t-1} + \rho_{1,1}^2 z_{t-2} + \cdots$$

which converges for $|\theta_{1,1}| < 1$, or

$$z_t = -\theta_{1,1}z_{t-1} - \theta_{1,1}^2 z_{t-2} + \ldots + \eta_t \tag{38.121}$$

This can be considered as an AR(∞) process provided that $|\rho_{1,1}| < 1$. Thus, MA(1) process is an inverted AR(∞) process.

Similarly, an AR(1) process can be represented as a MA(∞) process as shown in Eq. 38.98.

Because of the invertibility condition between MA and AR models, the auto-correlation and the PAF of an MA process behave like the PAF and the auto-correlation of an AR process and vice versa. The auto-correlation function of a MA(1) has a cut-off at lag 1, whereas that of an AR(1) has an exponential decay. The PAF of an MA(1) has an exponential decay whereas that of an AR(1) has a cut-off after lag 1. A finite order stationary AR(p) process corresponds to an MA(∞) process and a finite order invertible MA(q) process corresponds to an AR(∞) process. This property is known as duality. For the MA(1) model, the PAF is given by

$$\phi_{k,k} = -\frac{\rho_{1,1}^k(1-\rho_{1,1}^2)}{(1-\rho_{1,1}^{2(k+1)})} \tag{38.122}$$

and alternates in sign if $\rho_1 > 0$ (i.e. $\phi_{1,1} < 0$), and negative if $\rho_1 < 0$ (i.e. $\phi_{1,1} > 0$).

38.9.2.4 MA(2) model

MA(2) model is given by

$$z_t = \rho_t - \rho_{2,1}\rho_{t-1} - \rho_{2,2}\rho_{t-2} \tag{38.123}$$

For stationarity, the parameters should satisfy

$$\theta_{2,2} + \theta_{2,1} < 1$$

$$\theta_{2,2} - \theta_{2,1} < 1$$

$$-1 < \theta_{2,2} < 1$$

which is similar to the stationarity condition of AR(2) model. The auto-correlation function from Eq. 38.116 is

$$\rho_1 = \frac{-\theta_{2,1}(1-\theta_{2,2})}{1+\theta_{2,1}^2 + \theta_{2,2}^2}; \tag{38.124a}$$

$$\rho_2 = \frac{-\theta_{2,2}}{1+\theta_{2,1}^2 + \theta_{2,2}^2}; \tag{38.124b}$$

$$\rho_k = 0 \quad \text{for } k > 2 \tag{38.124c}$$

The PAF for MA(2) decays exponentially or is damped with oscillations depending upon the sign and magnitude of the moving average parameters. It behaves like the auto-correlation function of an AR(2) process. Duality exists between the auto-correlation function and the PAF of AR(2) and MA(2) processes.

The PAF from Eqs. 38.109a–38.109k is

$$\phi_{1,1} = \rho; \quad \phi_{2,2} = \frac{\rho_2 - \rho_1^2}{1 - \rho_1^2}; \quad \phi_{3,3} = \frac{\rho_1^3 - \rho_1\rho_2(2 - \rho_2)}{1 - \rho_2^2 - 2\rho_1^2(1 - \rho_2)}; \dots \tag{38.125}$$

38.9.3 Auto-Regressive Moving Average (ARMA) models

Auto-regressive and moving average models constitute a combination of AR and MA models. The ARMA(p, q) model is a combination of an AR(p) model and an MA(q) model. It is of the form

$$z_t = \sum_{i=1}^{p} \phi_{p,i} z_{t-1} + \eta_t - \sum_{i=1}^{q} \theta_{q,i} \eta_{t-i} \tag{38.126}$$

38.9.3.1 Properties of ARMA(p, q) models

The properties of ARMA(p, q) are given in terms of the cross covariance between z and η:

$$c_k^{(z\eta)} = \text{Cov}\left[z_{t-k} \, \eta_t\right]_{=0 \text{ for } k>0}^{\neq 0 \text{ for } k \leq 0} \tag{38.127}$$

because z depends only on the present and past values of η. An ARMA(p, q) model has $p + q + 2$ parameters which include the mean of the stochastic series, p auto-regressive coefficients, q moving average coefficients and the variance of the random component.

Multiplying Eq. 38.126 by z_{t-k} and taking expectations of products of the form $z_t z_{t-k}$, (assuming $\theta_{q,0} = -1$) as before,

$$c_k = \sum_{i=1}^{p} \phi_{p,i} c_{k-i} - \sum_{i=0}^{q} \theta_{q,i} c_{k-i}^{(z\eta)} \quad \text{for } k < q+1$$

$$= \sum_{i=1}^{p} \phi_{p,i} c_{k-i} \quad \text{for } k \geq q+1 \tag{38.128}$$

$$\text{Var}(z_t) = \sigma_\eta^2 + \sum_{i=1}^{p} \phi_{p,i} c_i - \sum_{i=1}^{q} \theta_i c_{-i}^{(z\eta)} \tag{38.129}$$

$$\rho(k) = \sum_{i=1}^{p} \phi_{p,i} \rho_{k-i} \quad \text{for } k \geq q+1 \tag{38.130}$$

For stationarity of the AR part, the roots of the characteristic equation $\phi(B) = 0$ should lie outside the unit circle, and for invertibility of the MA part, the roots of the characteristic equation $\theta(B) = 0$ should lie outside the unit circle.

38.9.3.2 ARMA(1,1) model

ARMA(1,1) model is given by

$$z_t = \phi_{1,1}z_{t-1} + \eta_t - \theta_{1,1}\eta_{t-1} \tag{38.131}$$

which, by repeated substitution may be written as

$$z_t = \eta_t + b_{1,1}\eta_{t-1} + b_{1,2}\eta_{t-2} + \ldots \tag{38.132}$$

where

$$b_{1,i} = (\phi_{1,1} - \theta_{1,1})\phi_{i,1}^{i-1}$$

This represents a MA process of infinite order. Similarly, it may also be written as an AR process of infinite order as follows:

$$z_t = d_{1,1}z_{t-1} + d_{1,2}z_{t-2} + \ldots + \eta_t \tag{7.133}$$

where

$$d_{1,i} = (\phi_{1,i} - \theta_{1,1})\theta_{1,1}^{i-1}$$

The convergence criteria for the two cases are $|\phi_{1,1}| < 1$ and $|\rho_{1,1}| < 1$ respectively.
The auto-correlation function is given as

$$\rho_k = 1 \quad \text{for } k = 0; \quad \rho_k = \frac{(\phi_{1,1} - \theta_{1,1})(1 - \phi_{1,1}\theta_{1,1})}{1 + \theta_{1,1}^2 - 2\phi_{1,1}\theta_{1,1}} \quad \text{for } k = 1; \quad \rho_k = \phi_{1,1}\rho_{k-1} \quad \text{for } k \geq 2 \tag{38.134}$$

The PAF for ARMA(1, 1) is complicated and its shape depends on the magnitudes of the AR and MA coefficients, and therefore cannot be generalized. It behaves like the PAF of a MA(1) process and decays off exponentially like the auto-correlation function. It is dominated by smooth damped exponential decay when $\theta_{1,1}$ is positive, and dominated by oscillating exponential decay when $\theta_{1,1}$ is negative. The Stationarity and invertibility conditions are

$$|\rho_2| < |\rho_1|$$

$$\rho_2 > \rho_1(2\rho_1 + 1) \quad \text{for } \rho_2 < 0$$

$$\rho_2 > \rho_1(2\rho_1 - 1) \quad \text{for } \rho_1 > 0$$

38.9.4 Backshift operator

ARMA type models are usually represented using the backshift operator B. The notation is as follows:

$$Bz_t = z_{t-1}$$

$$B\eta_t = \eta_{t-1}$$

$$B^2 z_t = z_{t-2}$$

$$B^2 \eta_t = \eta_{t-2}$$

$$\dots$$

$$B^p z_t = z_{t-p} \tag{38.135}$$

$$B^q \eta_t = \eta_{t-q}$$

Using this notation, Eq. 38.126 may be written as

$$\phi_p(B)z_t = \theta_q(B)\eta_t \tag{38.136}$$

where

$$\phi_p(B) = 1 - \phi_{p,1}B - \phi_{p,2}B^2 - \dots - \phi_{p,p}B^p \tag{38.137}$$

and,

$$\theta_q(B) = 1 - \theta_{q,1}B - \theta_{q,2}B^2 - \dots - \theta_{q,q}B^q \tag{38.138}$$

38.9.5 Difference operator

The difference operator ∇ has the following meaning:

$$\nabla z_t = (1-B)z_t = z_t - z_{t-1}$$

$$\nabla^2 z_t = (1-B)^2 z_t = (1-2B+B^2)z_t = z_t - 2z_{t-1} + z_{t-2}$$

$$\tag{38.139}$$

$$\dots$$

$$\nabla^d z_t = (1-B)^d z_t$$

38.9.6 Autoregressive Integrated Moving Average (ARIMA) models

ARIMA models are used for non-stationary time series where the periodicities are represented by several parameters. Annual series which are stationary can usually be represented by ARMA models. Representation of non-stationary series such as monthly, weekly and daily by ARMA type models requires a large number of parameters. For example, ARMA(1,1) for monthly series requires 27 parameters (12 monthly means, 12 monthly variances, $\phi_{1,1}, \theta_{1,1}$ and σ_η^2). Non periodic ARIMA(p, d, q) is an alternate way of transforming a time series into a stationary series requiring fewer parameters. The transformation is done by differencing. For example, the d'th order differenced ARMA model is given by

$$\phi_p(B)\nabla^d z_t = \theta_q(B)\eta_t \tag{38.140a}$$

which can also be written as

$$\phi_p(B)(1-B)^d z_t = \theta_q(B)\eta_t \tag{38.140b}$$

where $\phi_p(B)$, $\theta_q(B)$, are polynomials of degree p, q with all roots of the polynomial equations

$$\phi_p(B) = 0 \tag{38.141a}$$

$$\theta_q(B) = 0 \tag{38.141b}$$

outside the unit circle. This is necessary for stationarity and invertibility. For example, the random walk model $(1-B)z_t = \eta_t$ is an ARIMA(0,1,0) model. An ARIMA(p, d, q) model has $p + d + q + 2$ parameters which include the mean of the stochastic series, p autoregressive coefficients, orders of differencing d, q moving average coefficients and the variance of the random component.

Identification of ARIMA models include

- Checking for stationarity
- Taking differences if not stationary, once, twice etc.
- Checking to see if the auto-correlogram decays to zero
- Finding a cut-off value of lag q, for which $\rho_k = 0$ for $k > q$
- Suggesting a MA(q) model
- Determining the PAF, and finding a cut-off value p for which $\phi_{p,p} = 0$ for $k > p$
- Suggesting an AR(p) model.

38.10 RESIDUAL SERIES

Residual series refer to the series represented by η_t in equations that define the stochastic component. It can be obtained recursively by back substitution once the parameters of the model are determined. For example, if an AR(p) model is considered η_t is given by Eq. 38.87. This however requires the past values of z_t which are not known. For lower order AR models, neglecting the first p terms of the residual series would lose very little information. Therefore, η_t is determined from $t = p + 1$ [for AR(1) models, η_2, η_3, η_4,...; for AR(2) models, η_3, η_4, η_5,...; For MA models, assume $\eta_t = 0$, for $t < 0$, then $z_1 = \eta_1 \Rightarrow \eta_1 = z_1$; $z_2 = \eta_2 - \theta_{q,1}\eta_1 \Rightarrow \eta_2 = z_2 + \theta_{q,1}\eta_1$ etc.].

38.10.1 Test of independence

The dependence of the residual series can be tested either by the residual correlogram or by the Porte Manteau lack of fit test (Box and Pierce, 1970). The statistic 'Q' for the Porte Manteau lack of fit test is defined as

$$Q = (N - d)\sum_{k=1}^{M} r_k^2(\eta) \tag{38.142}$$

where d is the number of differences in a general ARIMA (p, d, q) model and M is the maximum lag which is taken to be about $\dfrac{N}{5}$. In the case of AR models, $d = 0$. If η_t is independent, then Q, estimated from Eq. 38.142, is approximately Chi-squared distributed with $(M\text{-}p\text{-}q)$

degrees of freedom (Ljung and Box, 1978). Therefore, if $Q < \chi^2 (M\text{-}p\text{-}q)$ at a given level of significance, then the residual series η_t can be considered as independent implying that the assumed model is adequate.

38.10.2 Test of normality

If the residual series is Normaly distributed the subsequent analysis and synthesis becomes relatively simple. Normality of the residual series can be tested by a number of methods. For instance, a plot of the empirical distribution on Normal probability paper should give a straight line if the residual series is Normaly distributed. Alternatively, the empirical cumulative frequency distribution of the series could be compared with the theoretical cumulative frequency distribution for a Normal series. A third method uses a skewness test based on the criterion that the skewness coefficient for a Normal distribution is zero. Snedecor and Cochran (1967) suggested a method using the skewness coefficient based on the condition that if the series is Normal, the skewness coefficient is asymptotically Normal with zero mean and variance $\dfrac{6}{N}$. For large values of N (> 150) the hypothesis of Normality can be accepted at 5% level of significance if the standard Normal variate z, defined as follows lies within ± 1.96:

$$z = \frac{\gamma - E(\gamma)}{\sigma_\gamma} \tag{38.143}$$

In Eq. 38.143, γ is the skewness coefficient of the residual series, $E(\gamma)$ is the expected value of the skewness coefficient if the series is Normal (= 0), and σ_γ is $\sqrt{\dfrac{6}{N}}$.

The above test is sufficiently accurate for $N > 150$. For smaller values of N, the limiting values of the skewness coefficients for different probability levels are given in Table 38.11 (Snedecor and Cochran: p. 552; or Salas et al., 1980: p. 93). If the computed skewness coefficients fall within these limits, the hypothesis of Normality can be accepted.

If the residual series is not Normal the original series could be Normalized by a transformation (for example a logarithmic transformation). Alternatively, a non-Normal distribution may be fitted to the residual series.

Table 38.11 Critical values for the skewness for Normality test

N	Significance level, α 0.02	0.01	N	Significance level, α 0.02	0.01
25	1.061	0.711	70	0.673	0.459
30	0.986	0.662	80	0.631	0.432
35	0.923	0.621	90	0.596	0.409
40	0.870	0.587	100	0.567	0.389
45	0.825	0.558	125	0.508	0.350
50	0.787	0.534	150	0.464	0.321
60	0.723	0.492	175	0.430	0.298

38.10.3 Other distributions

For other distributions such as Gamma and log-Normal, a test of goodness of fit can be carried out using the Chi-squared test. The test statistic $\hat{\chi}^2$ is defined as

$$\hat{\chi}^2 = \sum_{i=1}^{k} \frac{(O_i - E_i)^2}{E_i} \tag{38.144}$$

where k is the number of classes in the sample, O_i is the observed frequency of the class i and E_i is the expected[16] frequency under the null hypothesis. The χ^2 goodness of fit test is carried out by comparing the statistic $\hat{X}^2$ with the critical values of the χ^2-distribution (with $k - 1 - a$ degrees of freedom) under the null hypothesis at a chosen significance level. For two parameter distributions, $a = 2$, and therefore, the degree of freedom is $k - 3$. Before comparison, the variates of the theoretical distributions under the null hypothesis and those of the residual series must be standardized.

38.10.4 Test for parsimony

Finding a model with the minimum number of parameters, which will adequately reproduce the statistics that will have the smallest variances (or those statistics, which need to be preserved), is called the principle of parsimony. This has its origin in the principle referred to as the Occam's (or, Ockham) razor which states that 'entities must not be multiplied beyond necessity', and which is attributed to the 14th-century English logician, theologian and Franciscan friar William of Ockham.

38.10.4.1 Akaike Information Criterion (AIC) and Bayesian Information Criterion (BIC)

Parsimony of the assumed model is established by using the Akaike Information Criterion (AIC) which for ARMA(p, q) models is defined as (Akaike, 1970, 1973, 1974)

$$AIC(p, q) = N \ln(\sigma_\eta^2) + 2(p + q) \tag{38.145}$$

where σ_η^2 is the maximum likelihood estimate of the residual variance. The model, which gives the minimum AIC value, is the one to be chosen.

A more recent criterion is the Bayesian Information Criterion, BIC (Akaike, 1978, 1979) which takes the form

$$BIC(p, q) = N \ln(\sigma_\eta^2) - (N - p - q) \ln\left\{1 - \frac{(p+q)}{N}\right\} + (p + q) \ln(N) + (p + q) \ln\left\{\frac{\left(\frac{\sigma_z^2}{\sigma_\eta^2} - 1\right)}{(p + q)}\right\} \tag{38.146}$$

---

[16]Expected frequency depends on the assumed distribution.

38.10.4.2 *Schwartz Bayesian Criterion (SBC)*

Schwartz Bayesian criterion (SBC) (Schwartz, 1978) is defined as

$$SBC(p, q) = N \ln(\sigma_\eta^2) + (p + q) \ln(N) \tag{38.147}$$

There are other criteria that can be used for model selection (e.g., Parzen's criterion for autoregressive transfer functions, or CAT (Parzen, 1977), among several others). In all these criteria, $p + q$ should be replaced by $p + q + d$ if an ARIMA (p, d, q) model is considered.

Fitting ARMA models can be accomplished using several standard software such as IMSL, SPSS, SAS, MINITAB among others. Each of such software may have different methods of parameter estimation and therefore the results may differ appreciably from one another.

38.11 FORECASTING

Forecasting is the process of predicting the outcome of a future event before its actual occurrence. Weather forecasting is a typical example in every-day life. Flood forecasting is another example in which the magnitude and time of occurrence of impending floods are used as the bases of early warning systems for disaster mitigation. The methods and tools used for forecasting can also be used for hindcasting which is a process of reproducing events that have already occurred. The time of occurrence of the event measured from the time the forecast is made is known as the lead-time. It is a fact that the reliability of any forecast decreases with increasing lead-time, but its usefulness from the point of view of taking necessary mitigative measures to avert a disaster such as for example in the case of a flood increases with increasing lead-time. Very often, a compromise is needed in satisfying these two conflicting requirements. Forecasting can be made for any length of lead-time from a fixed time origin or, one step at a time from a moving time origin. In the former case, the information that goes into the forecasting system is static. The system only uses the information available up to the time of the fixed origin. In the latter case, new information as and when they become available can be made use of when the forecasts are updated at the new origin. Real-time forecasting systems operate in the latter mode.

Forecasts are made using mathematical models which can be empirical, conceptual, stochastic, deterministic or their combinations. In all these types, some assumptions are necessary, and the forecasts should be interpreted giving due consideration to such assumptions as well as to the uncertainties associated with model formulation, implementation and accuracy of data. In the context of time series analysis, stochastic models are widely used after removing the deterministic components. Of the different approaches, the minimum mean square error type difference equation is by far the simplest.

38.11.1 Minimum mean square error type difference equation

The minimum mean square error forecast (Box and Jenkins, 1976) type difference equation takes the form:

$$\hat{z}_{t+l} = \phi_1 \left[z_{t+l-1} \right] + \ldots + \phi_p \left[z_{t+l-p} \right] - \theta_1 \left[\eta_{t+l-1} \right] - \ldots - \theta_q \left[\eta_{t+l-q} \right] + \left[\eta_{t+l} \right] \tag{38.148}$$

where $\hat{z}_{t+l}$ is the forecast at origin t for lead-time l, z_{t+l-i} is the observed sequence, η_{t+l} is the residual sequence and ϕ_i, θ_j are the auto regressive and moving average coefficients. Square brackets [ ] indicate the conditional expectations (conditional on the past values) which can be calculated as follows:

$$[z_{t-j}] = \underset{t}{E}[[z_{t-j}]] = [z_{t-j}] \quad j = 0, 1, 2,\ldots \tag{38.149a}$$

$$[z_{t+j}] = \underset{t}{E}[z_{t+j}] = \hat{z}_t(j) \quad j = 1, 2,\ldots \tag{38.149b}$$

$$[\eta_{t-j}] = \underset{t}{E}[\eta_{t-j}] = \eta_{t-j} = z_{t-j} - \hat{z}_{t-j-1}(1) \quad j = 0, 1, 2,\ldots \tag{38.149c}$$

$$[\eta_{t+j}] = \underset{t}{E}[\eta_{t+j}] = 0 \quad j = 1, 2,\ldots \tag{38.149d}$$

where $E[\ ]$ denotes the expectation operator.

Note: z_{t-j} $j = 0, 1, 2,\ldots$ which have already happened are left unchanged.

z_{t+j} $j = 1, 2,\ldots$ which have not yet happened are replaced by their forecasts $\hat{z}_t(j)$ at origin t.

ρ_{t-j} $j = 0, 1, 2,\ldots$ which have happened are available from $z_{t-j} - \hat{z}_{t-j-1}(1)$

ρ_{t+j} $j = 0, 1, 2,\ldots$ which have not yet happened are replaced by zeros.

Forecast values are updated when new values are available using the following relationship (as a linear function of present and past shocks):

$$\hat{z}_{t+1}(l) = \psi_l \eta_{t+1} + \psi_{l+1} \eta_t + \psi_{l+2} \eta_{t-1} + \ldots \tag{38.150}$$

$$\hat{z}_t(l+1) = \psi_{l+1} \eta_t + \psi_{l+2} \eta_{t-1} + \ldots \tag{38.151}$$

Subtracting Eq. 38.151 from Eq. 38.150,

$$\hat{z}_t(l) = \hat{z}_t(l+1) + \psi_l \eta_{t+1} \tag{38.152}$$

where $\hat{z}_{t+1}(l)$ – the forecast for lead-time l made at time $t + 1$;

$\hat{z}_t(l+1)$ – the forecast for lead-time $l + 1$ made at time t;

$\eta_{t+1} = z_{t+1} - \hat{z}_t(1)$ is the one step ahead forecast error, and the ψ_l weights used for updating the forecasts are given by the recursive equation

$$\psi_l = \phi_1 \psi_{l-1} + \ldots + \phi_p \psi_{l-p} - \theta_l \tag{38.153}$$

in which $\psi_0 = 1$, $\psi_l = 0$ for $l < 0$ and $\theta_l = 0$ for $l > q$.

i.e. that the forecasts for lead-time l at origin $t + 1$ is an update of forecast at lead-time $l + 1$ at origin t.

$$\psi_1 = \phi_1 - \theta_1 \tag{38.154a}$$

$$\psi_2 = \phi_1 \psi_1 + \phi_2 \psi_0 - \theta_2 \tag{38.154b}$$

$$\psi_3 = \phi_1\psi_2 + \phi_2\psi_1 + \phi_3\psi_0 - \theta_3 \text{ etc.} \tag{38.154c}$$

The forecast at time t with lead-time 1, i.e. $\hat{z}_t(l)$ is obtained by writing down the Equation for z_{t+l}.

38.11.2 Confidence limits

The 95% confidence limits for lead-time forecasts are given by

$$z_{t+l}(\pm) = \hat{z}_t(l) \pm z_{\alpha/2}\left\{1 + \sum_{j=1}^{l-1} \psi_j^2\right\}^{1/2} \sigma_\eta \tag{38.155}$$

This means that given the information up to time t, the actual observation will lie within this confidence band with $(1 - \alpha)$ probability. The quantity $z_{\alpha/2}$ in Eq. 38.155 is the z value of a standard Normal distribution and should not be confused with the stochastic variable z_t.

38.11.3 Forecast errors

Forecast errors can be estimated by

$$\text{MSE} = \frac{1}{N} \sum_{l=1}^{N} \left[z_{t+l} - \hat{z}_t(l) \right]^2 \tag{38.156a}$$

or by

$$\text{MAPE} = \frac{1}{N} \sum_{l=1}^{N} \left| \frac{z_{t+l} - \hat{z}_t(l)}{z_{t+l}} \right| \times 100 \tag{38.156b}$$

or by

$$\text{MAE} = \frac{1}{N} \sum_{l=1}^{N} \left| z_{t+l} - \hat{z}_t(l) \right| \tag{38.156c}$$

where MSE – Mean Square Error
  MAPE – Mean Absolute Percentage Error
  MAE – Mean Absolute Error.

Example 38.8: AR(2) model forecasting

Calculate the lead-time and one-step ahead forecasts for the AR(2) model defined by $(1 - 1.5B + 0.5B^2)z_{t+1} = \eta_{t=1}$ for the time series z_t 23.7, 23.4, 23.1, 22.9, and 22.8 for $t = 1$, 2, 3, 4, 5.

The AR(2) model can be written as

$$z_{t+l} = 1.5z_{t+l-1} - 0.5z_{t+l-2} + \eta_{t+l}$$

Forecasts at origin t are given by

$$\hat{z}_t(1) = 1.5z_t - 0.5z_{t-1} + 0$$

$$\hat{z}_t(2) = 1.5z_{t+1} - 0.5z_t = 1.5\hat{z}_t(1) - 0.5z_t$$

$$\hat{z}_t(3) = 1.5z_{t+2} - 0.5z_{t+1} = 1.5\hat{z}_t(2) - 0.5\hat{z}_t(1)$$

$$\hat{z}_t(4) = 1.5z_{t+3} - 0.5z_{t+2} = 1.5\hat{z}_t(3) - 0.5\hat{z}_t(2)$$

....
....

$$\hat{z}_t(l) = 1.5\,z_{t+l-1} - 0.5\,z_{t+l-2} = 1.5\,\hat{z}_t(l-1) - 0.5\,\hat{z}_t(l-2)$$

Taking $t = 2$ as the origin (present time level),

$$\hat{z}_2(1) = 1.5z_2 - 0.5z_1 = 1.5 \times 23.4 - 0.5 \times 23.7 = 23.25$$

$$\hat{z}_2(2) = 1.5\hat{z}_2(1) - 0.5z_2 = 1.5 \times 23.25 - 0.5 \times 23.4 = 23.175$$

$$\hat{z}_2(3) = 1.5\hat{z}_2(2) - 0.5\hat{z}_2(1) = 1.5 \times 23.175 - 0.5 \times 23.25 = 23.178$$

$$\hat{z}_2(4) = 1.5\hat{z}_2(3) - 0.5\hat{z}_2(2) = 1.5 \times 23.178 - 0.5 \times 23.175 = 23.119$$

After z_3 is observed,

$$e_2(1) = \eta_3 = z_3 - \hat{z}_2(1) = 23.1 - 23.25 = -0.15$$

Now the origin can be shifted to $t = 3$; then, the updated forecasts are (Eq. 38.152)

$$\hat{z}_{2+1}(1) = \hat{z}_2(2) + \psi_1\eta_3$$

$$\hat{z}_{2+1}(2) = \hat{z}_2(3) + \psi_2\eta_3$$

$$\hat{z}_{2+1}(3) = \hat{z}_2(4) + \psi_3\eta_3$$

....
....

The ψ values for updating are given by

$$\psi_0 = 1$$

$$\psi_1 = 1.5 - 0 = 1.5$$

$$\psi_2 = 1.5 \times 1.5 - 0.5 \times 1 - 0 = 1.75$$

$$\psi_3 = 1.5 \times 1.75 - 0.5 \times 1.5 - 0 - 0 = 1.875$$

Updated forecasts at $t = 3$ are given by

$$\hat{z}_{2+1}(1) = 23.175 + 1.50(-0.15) = 22.95$$

$$\hat{z}_{2+1}(2) = 23.178 + 1.75(-0.15) = 22.92$$

$$\hat{z}_{2+1}(3) = 23.119 + 1.875(-0.15) = 22.84$$

Therefore, a new series of lead-time forecasts are obtained.

If on the other hand, one-step-ahead forecasts are required, the procedure is as follows:
Again from Eq. 38.152

$$\hat{z}_{t+1}(1) = \hat{z}_t(2) + \psi_1\eta_{t+1} \quad \text{and} \quad \eta_{t+1} = z_{t+1} - \hat{z}_t(1)$$

$$\hat{z}_{t+2}(1) = \hat{z}_{t+1}(2) + \psi_1\eta_{t+2} \quad \text{and} \quad \eta_{t+2} = z_{t+2} - \hat{z}_t(2)$$

where z_t are the observed values and $\hat{z}_t(1)$ are the forecast values. In this example,

$$\eta_{2+1} = z_{2+1} - \hat{z}_2(1) = 23.1 - 23.25 = -0.15$$

$$\hat{z}_{2+1}(1) = \hat{z}_2(2) + \psi_1\eta_{2+1} = 23.175 + 1.5(-0.15) = 22.95$$

$$\eta_{2+2} = z_{2+2} - \hat{z}_2(2) = 22.9 - 23.175 = -0.275$$

$$\hat{z}_{2+2}(1) = \hat{z}_3(2) + \psi_1\eta_{2+2} = 22.92 + 1.5(-0.275) = 22.51$$

Therefore, one-step-ahead forecasts can be obtained as and when new data become available. Confidence limits for lead time forecasts at time $t = 2$ are

For $\hat{z}_2(2)$: $\hat{z}_2(2) \pm 1.96\left\{1 + 1.5^2\right\}^{1/2}\sigma_\eta = 23.175 \pm 1.803\sigma_\eta$

For $\hat{z}_2(3)$: $\hat{z}_2(3) \pm 1.96\left\{1 + 1.5^2 + 1 + 1.75^2\right\}^{1/2}\sigma_\eta = 23.178 \pm 2.512\sigma_\eta$

Example 38.9: ARMA(2,2) model forecasting

Consider the ARMA(2,2) model defined by $\nabla^2 z_{t+1} = (1 - 0.5B + 0.5B^2)\eta_{t+1}$ which can be written as

$$z_{t+l} - 2z_{t+l-1} + z_{t+l-2} = (1 - 0.5B + 0.5B^2)\eta_{t+1}$$

$$z_{t+l} = 2z_{t+l-1} - z_{t+l-2} + \eta_{t+1} - 0.5\eta_{t+1-1} + 0.5\eta_{t+1-2}$$

Forecasts are given by

$$\hat{z}_t(1) = 2z_t - z_{t-1} - 0.5\eta_t + 0.5\eta_{t+1} + \eta_{t+1}$$

$$\hat{z}_t(2) = 2z_{t+1} - z_t - 0.5\eta_{t+1} + 0.5\eta_t + \eta_{t+2} = 2\hat{z}_t(1) - z_t + 0.5\eta_t$$

$$\hat{z}_t(3) = 2z_{t+2} - z_{t+1} - 0.5\eta_{t+2} + 0.5\eta_{t+1} + \eta_{t+3} = 2\hat{z}_t(2) - \hat{z}_t(1)$$

For shorter lead times, the η's may be set to their unconditional expected value of zero, i.e. up to $\hat{z}_t(q)$ which involve η_t and η_{t-1}. Similarly, for all lead times greater than 2, there will be no η component, i.e.

$$\hat{z}_t(4) = 2\hat{z}_t(3) - \hat{z}_t(2)$$

$$\hat{z}_t(5) = 2\hat{z}_t(4) - \hat{z}_t(3)$$

....
....

For the MA(q) part, the forecasts will depend on η's up to $\hat{z}_t(q)$; for longer lead times, the forecasts will not depend on the η's. In example (a), there is no MA part. Therefore, the forecasts are independent of the η's. In example (b), MA part is of order 2. Therefore, the forecasts up to lead $l = 2$, depend on the η's. Thereafter, the forecasts are independent of the η's.

Using the same data as in Example (a)

$$\hat{z}_t(1) = 2 \times 23.4 - 23.7 = 23.1$$

$$\hat{z}_t(2) = 2 \times 23.1 - 23.4 = 22.8$$

$$\hat{z}_t(3) = 2 \times 22.8 - 23.1 = 22.5$$

$$\hat{z}_t(4) = 2 \times 22.5 - 22.9 = 22.1.$$

38.12 SYNTHETIC DATA GENERATION

The 'end product' of time series analysis is the independent random component η_t. It is ideal for this to be white noise. If so, independent standard Normal variates can be generated using a number of procedures. The dependent structure and the deterministic components are then added to obtain the synthetically generated series.

Random sequences can be generated using the Monte Carlo technique, or tossing a coin. Box and Muller (1958) proposed a method of generating Standard Normal random numbers ε_1 and ε_2:

$$\varepsilon_1 = \left(\ln\left(\frac{1}{u_1} \right) \right)^{1/2} \cos(2\pi u_2) \tag{38.157a}$$

$$\varepsilon_2 = \left(\ln\left(\frac{1}{u_1} \right) \right)^{1/2} \sin(2\pi u_2) \tag{38.157a}$$

where u_1 and u_2 are random numbers of the uniform distribution $(0, 1)$. Normality and independence tests must be carried out for the generated numbers because they are often pseudo Normal or pseudo independent.

The stochastic component of an AR process is obtained as follows:

$$\eta_t = \hat{\sigma}_\eta \varepsilon_t \tag{38.158a}$$

$$\hat{z}_t = \phi_{p,i} \hat{z}_{t-i} + \ldots + \phi_{p,p} \hat{z}_{t-p} + \eta_t \tag{38.158b}$$

where $\hat{\sigma}_\eta$, $\phi_{p,i}$'s and $\hat{z}_t$ respectively are the standard deviation of the residual series, the AR parameters and the generated stochastic series, respectively. The synthetic series is obtained by adding the deterministic components to $\hat{z}_t$.

However, the estimation of $\hat{z}_t$ requires $\hat{z}_{t-p}$ which is unknown. This problem is solved by assuming 'zero initial condition' and having a sufficient 'warm-up length'. For example, if N_g values of generated data are required, the procedure is as follows:

- Generate the standard normal random variates ε_1 and calculate $\hat{z}_1$ assuming that $\hat{z}_0$, $\hat{z}_{-1}, \ldots$ are zeros.
- In the same manner, generate a second standard normal variate ε_2 and calculate $\hat{z}_2$ based on $\hat{z}_1$ already generated and assuming $\hat{z}_0$, $\hat{z}_{-1}, \ldots$ are zeros.
- Repeat the procedure until the series $\hat{z}_1, \hat{z}_2, \hat{z}_3, \ldots \hat{z}_{N'}$ is generated where
  $N' = N_g + N_W$, and, N_W is the warm-up length.
  (The length N_W is necessary to remove the effect of the starting condition)
- Delete the first N_W values and re-initialize the last N_g values.

38.13 ARMAX MODELLING

ARMAX models take the form

$$Q_t = \sum_1^p A_i Q_{t-i} + \sum_0^q B_j I_{t-j} \tag{38.159}$$

where I_t and Q_t are the input and output variables at time t; A_i and B_j are the weighting function coefficients for Q_{t-i} and I_{t-j}; p and q are the orders of the models for Q and I. The coefficients A_i and B_j can be estimated by least squares method. The normal equations are obtained by minimizing the squared errors as follows:

$$E = \left(Q_t - \sum_1^p A_i Q_{t-i} - \sum_0^q B_j I_{t-j} \right)^2 \tag{38.160}$$

Differentiating with respect to the coefficients,

$$\frac{\partial E}{\partial A_1} = -2 \left(Q_t - \sum_1^p A_i Q_{t-i} - \sum_0^q B_j I_{t-j} \right)^2 Q_{t-1} \tag{38.161a}$$

$$\frac{\partial E}{\partial A_2} = -2\left(Q_t - \sum_1^p A_i Q_{t-i} - \sum_0^q B_j I_{t-j}\right)^2 Q_{t-2} \qquad (38.161b)$$

....

....

$$\frac{\partial E}{\partial B_0} = -2\left(Q_t - \sum_1^p A_i Q_{t-i} - \sum_0^q B_j I_{t-j}\right)^2 I_t \qquad (38.161c)$$

....

Setting Eq. 38.161 to zero,

$$
\begin{bmatrix}
Q_{t-1}Q_{t-1} & Q_{t-1}Q_{t-2} & \dots Q_{t-1}I_t & Q_{t-1}I_{t-1} \dots \\
Q_{t-2}Q_{t-1} & Q_{t-2}Q_{t-2} & \dots Q_{t-2}I_t & Q_{t-2}I_{t-1} \dots \\
Q_{t-3}Q_{t-1} & Q_{t-3}Q_{t-2} & \dots Q_{t-3}I_t & Q_{t-3}I_{t-1} \dots \\
\dots \\
I_t Q_{t-1} & I_t Q_{t-2} & \dots I_t I_t & I_t I_{t-1} \dots \\
I_{t-1}Q_{t-1} & I_{t-1}Q_{t-2} & \dots I_{t-1}I_t & I_{t-1}I_{t-1} \dots \\
I_{t-2}Q_{t-1} & I_{t-2}Q_{t-2} & \dots I_{t-2}I_t & I_{t-2}I_{t-1} \dots \\
\dots
\end{bmatrix}
\begin{bmatrix}
A_1 \\ A_2 \\ A_3 \\ \dots \\ B_0 \\ B_1 \\ B_3 \\ \dots
\end{bmatrix}
=
\begin{bmatrix}
Q_{t-1}Q_t \\ Q_{t-2}Q_t \\ Q_{t-3}Q_t \\ \dots \\ I_t Q_t \\ I_{t-1}Q_t \\ I_{t-2}Q_t \\ \dots
\end{bmatrix}
$$

which is of the form

$$[A]\phi = [B] \qquad (38.162)$$

which can be solved for the coefficients A_i and B_j.

38.14 KALMAN FILTERING

Kalman filter (Kalman, 1960; Kalman and Bucy, 1961) recursively estimates the state of a linear dynamic system in a way that minimizes the mean of the squared error. Only the estimate of the state at the previous time step and the current measurement are needed to compute the estimate of the present state. In comparison, batch estimates require the entire history of estimates and measurements which lead to solution of sets of simultaneous equations.

Linear Kalman filter algorithm starts with two equations:

$$x_{t+1} = Ax_t + Bi_t + w_t \quad \text{(State equation)} \qquad (38.163)$$

$$y_t = Cx_t + v_t \quad \text{(Measurement equation)} \qquad (38.164)$$

where x_t is the state which may include more than one variable. In general, it is a vector of the form

$$x_t = \begin{bmatrix} x_t \\ x_{t-1} \\ x_{t-2} \\ \cdots \\ \cdots \end{bmatrix} \qquad (38.165)$$

In a one dimensional system, x_t is a scalar quantity. The term i_t in Eq. 38.163 is an input variable (controllable), which may or may not exist; w_t is an independently distributed random variable which is assumed to be Normaly distributed with zero mean and variance R_w. It accounts for the error in describing the process (or state) by Eq. 38.163; y_t is the measurement which is related to the state x_t with an error v_t, which is also assumed to be Normally distributed with zero mean and variance R_V, i.e.

$$w_t \sim N(0, R_w)$$

$$v_t \sim N(0, R_v)$$

It is further assumed that the two error terms are independent of each other and independent of the state variable x_t and the measurement y_t, i.e.

$$E(w_t v_t) = 0$$
$$E(w_t x_t) = E(w_t y_t) = E(v_t x_t) = E(v_t y_t) = 0 \qquad (38.166a)$$

$$E(w_t w_t^T) = R_w$$
$$E(v_t v_t^T) = R_v \qquad (38.166b)$$

A, B and C are matrices which are generally assumed to be time-invariant, i.e. constants. However, they can be time-dependent too. In linear Kalman filter, they are assumed to be constants.

Kalman Filter is a predictor–corrector type of algorithm which minimizes the error covariance, i.e. Min $(x_t - \hat{x}_t)^2$. Kalman filter does it optimally compared to other error minimization techniques. The equations can be written in two stages:

• **At prediction:**

$$x_{t+1|t} = Ax_{t|t} + Bi_t$$
$$P_{t+1|t} = AP_{t|t}A^T + R_w \qquad (38.167)$$

Since Kalman filter is a recursive algorithm, it is necessary to have initial conditions; in this case $x_{t|t}$, $P_{t|t}$ and R_w.

$x_{t|t}$: State value using the information up to and including the time level t

$P_{t|t}$: State (or process) error covariance using the information up to and including the time level t

R_w: Variance of the random error component w_t.

To start the algorithm, these initial conditions must be provided. Estimates of these can be made in an off-line mode using past information.

- **At correction:**

  Once a new measurement is made, there is new information brought in, which allows a correction to be made to the previous prediction. It is done as follows:

$$P_{t+1|t} = AP_{t|t}A^T + R_w \tag{38.168}$$

$$K_{t+1} = P_{t+1|t}C^T\{CP_{t+1|t}C^T + R_v\}^{-1} \tag{38.169}$$

$$\hat{x}_{t+1|t+1} = \hat{x}_{t+1|t} + K_{t+1}\{y_{t+1} - C\hat{x}_{t+1|t}\} \tag{38.170}$$

and the updated error covariance

$$P_{t+1|t+1} = \{I - K_{t+1}C\}P_{t+1|t} \tag{38.171}$$

The term K is known as the Kalman gain, and the quantity $\{y_{t+1} - C\hat{x}_{t+1|t}\}$ is known as the innovation (or, sometimes the residual).

If R_v (variance of measurement error) is very large, then, from Eq. 38.169,

$$K_{t+1} \to 0 \tag{38.172}$$

Therefore, the state estimate update (Eq. 38.170) becomes

$$\hat{x}_{t+1|t+1} = x_{t+1|t} \tag{38.173}$$

I.e. the state estimate is equal to the predicted value from the mathematical model.

The updated error covariance (Eq. 38.171) then becomes

$$P_{t+1|t+1} = P_{t+1|t} = AP_{t|t}A^T + R_W \tag{38.174}$$

This depends only on the initial error covariance $P_{0|0}$ and the variance of the model error. The effect of $P_{0|0}$ is insignificant.

On the other hand, if R_V is small, then Eq. 38.169 becomes

$$K_{t+1} = P_{t+1|t}C^T\left\{CP_{t+1|t}C^T + 0\right\}^{-1} = C^{-1} \tag{38.175}$$

Therefore, the state estimate update (Eq. 38.170) becomes

$$\hat{x}_{t+1|t+1} = \hat{x}_{t+1|t} + K_{t+1}\left\{y_{t+1} - C\hat{x}_{t+1|t}\right\} = C^{-1}y_{t+1} \tag{38.176}$$

I.e. the updated state estimate at $t + 1$ is equal to the observed state at $t + 1$.

If C is the identity matrix, i.e. $y_t = x_t + v_t$, then

$$x_{t+1|t+1} = y_{t+1} \tag{38.177}$$

Therefore,

$$x_{t+2|t+1} = Ay_{t+1} + Bi_{t+1} \tag{38.178}$$

When R_V is large, $K_{t+1} \to 0$; and, when R_V is small, $K_{t+1} \to 1$. The model error and the measurement error covariance matrices take the form

$$
R_W = \begin{bmatrix}
\rho_w^2 & 0 & 0 & \cdots \\
0 & \rho_w^2 & 0 & \cdots \\
0 & 0 & \rho_w^2 & \\
\cdots & & & \\
0 & 0 & 0 & \rho_w^2
\end{bmatrix}
\tag{38.179}
$$

$$
R_V = \begin{bmatrix}
\rho_v^2 & 0 & 0 & \cdots \\
0 & \rho_v^2 & 0 & \cdots \\
0 & 0 & \rho_v^2 & \\
\cdots & & & \\
0 & 0 & 0 & \rho_v^2
\end{bmatrix}
\tag{38.180}
$$

and the model error variance is given by

$$
\sigma_w^2 = \frac{1}{N} \sum_1^N (x_t - \hat{x}_t)^2
\tag{38.181}
$$

With a knowledge of the initial estimates of $x_{0|0}$, $P_{0|0}$, R_V and R_W,

$$x_{1|0} \to P_{1|0} \to K_1 \to x_{1|1} \to P_{1|1};$$

$$x_{2|1} \to P_{2|1} \to K_2 \to x_{2|2} \to P_{2|2};$$

$$x_{3|2} \to P_{3|2} \to K_3 \to x_{3|3} \to P_{3|3};$$

....

....

can be recursively estimated in that sequence.

The Kalman filtering algorithm has many variables that need to be determined. The model parameters A, B and C can be estimated using any parameter estimation technique. The error estimates, R_W and R_V however are unknown that cannot be estimated from the available input and output data. The state of the process x_t is unknown; only the observation y_t is known. Therefore it is not possible to separate the two error terms w_t and v_t. They are therefore assigned arbitrary values. In particular, $R_V = 0$ make $K = 1$ which gives the best correction of the previous forecast which is the observed value itself.

The performance of the predictor is judged by comparing $\hat{x}_{t+1|t}$ with y_{t+1} and, not with x_{t+1}. The latter is not available.

If, on the other hand, information about the statistics of the measurement error is available *a priori*, then R_V should be assigned a value. Otherwise, it is not unreasonable to set the measurement error equal to zero.

Under-estimation of the model error R_W may reduce the Kalman gain, and hence the filtering of the previous forecast $\hat{x}_{t+1|t}$ using the new observation y_{t+1}. Correction strongly depends upon Kalman gain and hence upon R_W. The propagation of the forecast error to the prediction of $\hat{x}_{t+1|t+1}$ can be prevented by choosing the right value of R_W. It is, therefore, best to have a scheme whereby R_W is re-estimated at each time level (adaptive Kalman filtering) which can be made on the basis of the comparison of y_{t+1} with $C\hat{x}_{t+1|t}$, y_{t+2} with $Cx_{t+2|t+1}$, etc.

The quantity $\left[y_{t+1} - C\hat{x}_{t+1|t} \right]$ is called the innovation – the new information brought about by the current observation. The new observation can be used for

- Filtering the state vector x_t,
- Correcting the error covariance R_W, and,
- For updating the model parameters A, B and C.

If the measurement noise R_V is large, K_t will be small and therefore the measurement will have little or no impact on future state estimate $\hat{x}_{t+1}$. If, on the other hand, R_V is small, then, K_t will be large, and the measurement will have an impact on the future state estimate $\hat{x}_{t+1}$. The state vector update, Kalman gain and the error covariance equations represent an asymptotically stable system and therefore the estimates of state x_t become independent of the initial estimates of $x_{t|t}$ and $P_{t|t}$ as t increases.

Example 38.10: Kalman filtering

Use the following process and measurement equations to forecast the series starting from the origin $t = 3$:

Process equation: $x_t = 1.8x_t - 0.8x_{t-1} + w_{t+1}$

Measurement equation: $y_t = x_t + v_t$; $(y_t = 23.7, 23.4, 23.1, 22.9, 22.8$ for $t = 1, 2, ..., 5)$

- **Initial estimates:**
  Since the state variable x_t and x_{t-1} are not known, the best estimate of these past values will be the measurement. Therefore,

$\hat{x}_{2|2} = 23.4$

$\hat{x}_{1|1} = 23.7$

Initial error covariance $P_{2|2}$ is unknown too. It measures the error between the actual value of the state variable and its estimate, i.e.

A priori: $e_{t|t-1} = x_t - \hat{x}_{t|t-1}$; $P_{t|t-1} = E\left(e_{t|t-1} e_{t|t-1}^T\right)$

Posteriori: $e_{t|t} = x_t - \hat{x}_{t|t}$; $P_{t|t} = E\left(e_{t|t} e_{t|t}^T\right)$

The initial value of $P_{t|t}$ assumed is not crucial as long as it is non-zero, because its value quickly converges as the forecasting progresses in time. Therefore, let $P_{2|2} = 100$.

Values of R_w and R_v are also needed to start the Kalman filtering algorithm. If they can be estimated from prior information, then they could be assigned those values. Since there is not much prior information available, let them be arbitrarily chosen as $R_w = 10$ and $R_v = 5$.

- **Kalman filter estimates:**
  The Kalman filter equations (for this example) are

$$\hat{x}_{t+1|t} = A\hat{x}_{t|t}$$

$$P_{t+1|t} = AP_{t|t}A^T + R_w$$

$$K_{t+1} = P_{t+1|t}C^T\{CP_{t+1|t}C^T + R_v\}^{-1}$$

$$\hat{x}_{t+1|t+1} = \hat{x}_{t+1|t} + K_{t+1}\{y_{t+1} - C\hat{x}_{t+1|t}\}$$

$$P_{t+1|t+1} = \{I - K_{t+1}C\}P_{t+1|t}$$

At time $t = 3$,

$$A\,priori\ \text{estimate:}\ \hat{x}_{3|2} = \begin{pmatrix} 1.8 & -0.8 \end{pmatrix} \begin{bmatrix} \hat{x}_{2|2} \\ \hat{x}_{1|1} \end{bmatrix} = 1.8 \times 23.4 - 0.8 \times 23.7 = 23.16$$

$$P_{3|2} = \begin{pmatrix} 1.8 & -0.8 \end{pmatrix} \begin{bmatrix} 1.8 \\ -0.8 \end{bmatrix} P_{2|2} + R_w = \left(1.8^2 + 0.8^2\right)100 + 10 = 398$$

$$K_3 = 398\left(398 + R_v\right)^{-1} = 398/(398 + 5) = 0.987$$

$$A\,posteriori\ \text{estimate:}\ \hat{x}_{3|3} = \hat{x}_{3|2} + K_3\left(y_3 - \hat{x}_{3|2}\right) = 23.16 + 0.987(23.1 - 23.16) = 23.10$$

$$P_{3|3} = \left(I - K_3\right)P_{3|2} = 398(1 - 0.987) = 5.174$$

At time $t = 4$,

$$A\,priori\ \text{estimate:}\ \hat{x}_{4|3} = \begin{pmatrix} 1.8 & -0.8 \end{pmatrix} \begin{bmatrix} \hat{x}_{3|3} \\ \hat{x}_{2|2} \end{bmatrix} = 1.8 \times 23.10 - 0.8 \times 23.4 = 22.86$$

$$P_{4|3} = \begin{pmatrix} 1.8 & -0.8 \end{pmatrix} \begin{bmatrix} 1.8 \\ -0.8 \end{bmatrix} P_{3|3} + R_w = \left(1.8^2 + 0.8^2\right)5.174 + 10 = \therefore$$

$$K_4 = 30.07\left(30.07 + R_v\right)^{-1} = 30.07/(30.07 + 5) = 0.857$$

$$A\,posteriori\ \text{estimate:}\ \hat{x}_{4|4} = \hat{x}_{4|3} + K_4\left(y_4 - \hat{x}_{4|3}\right) = 22.86 + 0.857(22.9 - 22.86) = 22.89$$

$$P_{4|4} = \left(I - K_4\right)P_{4|3} = 30.07(1 - 0.857) = 4.30$$

At time $t = 5$,

$$A\ priori\ \text{estimate:}\ \hat{x}_{5|4} = \begin{pmatrix} 1.8 & -0.8 \end{pmatrix} \begin{bmatrix} \hat{x}_{4|4} \\ \hat{x}_{3|3} \end{bmatrix} = 1.8 \times 22.89 - 0.8 \times 23.1 = 22.72$$

$$P_{5|4} = \begin{pmatrix} 1.8 & -0.8 \end{pmatrix} \begin{bmatrix} 1.8 \\ -0.8 \end{bmatrix} P_{4|4} + R_w = \left(1.8^2 + 0.8^2\right)4.30 + 10 = 2$$

$$K_5 = 26.68\left(26.68 + R_v\right)^{-1} = 26.68/\left(26.68 + 5\right) = 0.842$$

$$A\ posteriori\ \text{estimate:}\ \hat{x}_{5|5} = \hat{x}_{5|4} + K_5\left(y_5 - \hat{x}_{5|4}\right) = 22.72 + 0.842\left(22.8 - 22.72\right) = 22.79$$

$$P_{5|5} = \left(I - K_5\right)P_{5|4} = 26.68\left(1 - 0.842\right) = 4.21$$

38.15 PARAMETER ESTIMATION

All the models described above have parameters that need to be estimated by some technique. The poor fitting of the model results with the actual observations that may result in some situations may not necessarily be due to inadequacies in the model, but more likely on the inaccuracies in the parameter estimation. Several methods for model parameter estimation are available but their estimates could sometimes differ appreciably. Initial estimates of parameters are usually obtained by the method of moments by solving the Yule–Walker equations. For MA and ARMA models, this procedure is complicated and the moment estimators are also sensitive to round-off errors. They should not be used as final estimates of parameters. MA and ARMA model parameters are estimated iteratively.

A better but more complicated alternative is the maximum likelihood method in which the log likelihood function of the parameters is maximized. With an assumed initial values of z_t and η_t, the method leads to the conditional maximum likelihood estimates. It is also possible to define an unconditional log-likelihood function and an exact likelihood function. More details of these procedures can be found in several textbooks (for example, Box and Jenkins, 1976, Chapter 7; Wei, 1990, Chapter 7).

38.16 APPLICATIONS

Time series analysis and modelling have diverse applications in many disciplines. It is used in statistics, actuarial sciences, banking and insurance, signal processing, control engineering, economics, life sciences and in many areas of science and engineering. In the context of this chapter, the interest is restricted to applications in the hydro-environment. A comprehensive list of relevant literature is beyond the scope of this chapter, but the studies carried out by Thomann (1967), Lohani and Wang (1987), Jayawardena and Lai (1989) and Huck and Farquhar (1974) in water quality modelling, Srikanthan and McMahon (1982) in annual and monthly rainfall simulation, McMicheal and Hunter (1972) in modelling temperature and flows in rivers, Jayawardena and Lau (1990a) on stochastic modelling of evaporation data, Lawrance and Kottegoda (1977) in river flow time series modelling, Gupta and Chanhan (1986) in irrigation requirement modelling and Delleur and Kavvas (1978) in the synthetic generation of monthly rainfall data, among many others, may lead the readers to more applications in the field.

In this chapter, an attempt has been made to describe the various techniques of time series analysis and forecasting in the context of hydrological and environmental time series. It should be noted that time series analysis and modelling is an iterative process which involve diagnostic checking and model selection. The content of this chapter is restricted to techniques and analysis under the assumptions of linearity and stationarity. Non-stationary series can be analysed by transforming them into stationary ones. Also, seasonal models have not been dealt with explicitly, but they can be taken care of by the various techniques of periodicity analysis. Examples to illustrate the techniques have been given where possible. The reader is however encouraged to refer to relevant literature given for further understanding and consolidation.

38.1A APPENDIX

38.1.1A Fourier series representation of a periodic function

Any periodic function $f(x)$ can be represented as a combination of sine and cosine function as follows:

$$f(x) = a_0 + \sum_{n=1}^{\infty} \left\{ a_n \cos(nx) + b_n \sin(nx) \right\} \tag{A1.1}$$

The coefficients a_n and b_n are determined by integration Eq. A1.1 as

$$\int_{-\pi}^{\pi} f(x) \, dx = \int_{-\pi}^{\pi} a_0 \, dx + \sum_{n=1}^{\infty} \left\{ a_n \int_{-\pi}^{\pi} \cos(nx) \, dx + b_n \int_{-\pi}^{\pi} \sin(nx) \, dx \right\} \tag{A1.2}$$

which leads to (because all sine and cosine integrals are zero)

$$a_0 = \frac{1}{2\pi} \int_{-\pi}^{\pi} f(x) \, dx \tag{A1.3a}$$

If in Eq. A1.1, $\dfrac{a_0}{2}$ is used instead of a_0, then, the corresponding equation will be

$$a_0 = \frac{1}{\pi} \int_{-\pi}^{\pi} f(x) \, dx \tag{A1.3b}$$

The coefficients a_n's can be obtained by multiplying Eq. A1.1 by $\cos(mx)$ and integrating from $-\pi$ to π as

$$a_n = \frac{1}{\pi} \int_{-\pi}^{\pi} f(x) \cos(mx) \, dx \tag{A1.4}$$

Similarly, the coefficients b_n's can be obtained by multiplying Eq. A1.1 by $\sin(mx)$ and integrating from $-\pi$ to π as

$$b_n = \frac{1}{\pi} \int_{-\pi}^{\pi} f(x) \sin(mx) \, dx \tag{A1.5}$$

Equations A1.3, A1.4 and A1.5 are referred to as Euler equations for determining the Fourier coefficients (Kreszig, 1999: pp. 530–532).

The above formulation is valid for a periodic function with period 2π. For any other function with period $p = 2L$, Eq. A1.1 can be written as (Kreszig, 1999: p. 537)

$$f(x) = a_0 + \sum_{n=1}^{\infty} \left\{ a_n \cos\left(\frac{n\pi x}{L}\right) + b_n \sin\left(\frac{n\pi x}{L}\right) \right\} \tag{A1.6}$$

and the corresponding Fourier coefficients are

$$a_0 = \frac{1}{2L} \int_{-L}^{L} f(x)\, dx \tag{A1.7}$$

$$a_n = \frac{1}{L} \int_{-L}^{L} f(x) \cos\left(\frac{n\pi x}{L}\right) dx \quad \text{for } n = 1, 2, 3, \ldots \tag{A1.8}$$

$$b_n = \frac{1}{L} \int_{-L}^{L} f(x) \sin\left(\frac{n\pi x}{L}\right) dx \quad \text{for } n = 1, 2, 3, \ldots \tag{A1.9}$$

In the integration of Eq. A1.1, a key property used in deriving Euler formulas is the orthogonality conditions of trigonometric functions (Kreszig, 1999: p. 534). Orthogonality implies that the integral of the products of any two different combinations of these functions is zero over the interval $-\pi$ to π. The period 2π can be easily replaced by any other period $2L$. The orthogonality conditions are (Kreszig, 1999: p. 537) given as follows:

For integers m and n,

$$\int_{-\pi}^{\pi} f(x) \cos(mx) \cos(nx)\, dx = \begin{cases} 0 & \text{if} \quad m \neq n \\ \pi & \text{if} \quad m = n \neq 0 \end{cases} \tag{A1.10}$$

$$\int_{-\pi}^{\pi} f(x) \sin(mx) \sin(nx)\, dx = \begin{cases} 0 & \text{if} \quad m \neq n \\ \pi & \text{if} \quad m = n \neq 0 \end{cases} \tag{A1.11}$$

and for any integers m and n,

$$\int_{-\pi}^{\pi} f(x) \cos(mx) \sin(nx)\, dx = 0 \tag{A1.12}$$

It is also important to note that the Fourier series of an even function of period $2L$ is a *cosine* function whereas that of an odd function is a *sine* function i.e.

$$f(x) = a_0 + \sum_{n=1}^{\infty} \left\{ a_n \cos\left(\frac{n\pi x}{L}\right) \right\} \tag{A1.13}$$

$$f(x) = \sum_{n=1}^{\infty} \left\{ b_n \sin\left(\frac{n\pi x}{L}\right) \right\} \tag{A1.14}$$

REFERENCES

Abdi, H. (2007): Kendall rank correlation, In: Salkind, N. J. (Ed.), *Encyclopaedia of Measurement and Statistics*, Sage, Thousand Oaks, CA.

Akaike, H. (1970): Statistical predictor identification, *Annals of the Institute of Statistical Mathematics*, vol 22, pp. 203–217.

Akaike, H. (1973): Information theory and an extension of the maximum likelihood principle, In: Petrov, B. N. and Csaki, F. (Eds.), Proceedings of the 2nd International Symposium on Information Theory, Akademiai Kaido, Budapest, pp. 267–281.

Akaike, H. (1974): A new look at the statistical model identification, *IEEE Transactions on Automatic Control*, vol 19, pp. 716–723.

Akaike, H. (1978): A Bayesian analysis of the minimum AIC procedure, *Annals, Institute of Statistical Mathematics*, vol 30A, pp. 9–14.

Akaike, H. (1979): A Bayesian analysis of the minimum AIC procedure of auto-regressive model fitting, *Biometrica*, vol 66, pp. 237–242.

Alexandersson, H. (1986): A homogeneity test applied to precipitation data, *Journal of Climatology*, vol 6, pp. 661–675.

Alexandersson, H. and Moberg, A. (1997): Homogenization of Swedish temperature data, Part 1: Homogeneity test for linear trends, *International Journal of Climatology*, vol 17, pp. 25–34.

Ampitiyawatta, A. D., and Guo, S. (2009): Precipitation trends in the Kalu Ganga Basin in Sri Lanka, *The Journal of Agricultural Sciences*, vol 4 no.:1, pp. 10–18.

Anderson, R. L. (1941): Distribution of serial correlation coefficients, *Annals of Mathematical Statistics*, vol 8 no.:1, pp. 1–13.

Bartlett, M. S. (1946): On the theoretical specification of sampling properties of auto-correlated time series, *Journal of the Royal Statistical Society*, vol B8, p. 27.

Bartolini, G., Marco, M., Alfonso, C., Daniele, G., Tommaso, T., Martina, P., Giampiero M. and Simone, O. (2008): Recent trends in Tuscany (Italy) summer temperature and indices of extremes, *International Journal of Climatology*, vol 28, pp. 1751–1760.

Blackman, R. B. and Tukey, J. (1959): Particular pairs of windows, In: Blackman, R. B. and Tukey, J. (Eds.), *The Measurement of Power Spectra, From the Point of View of Communications Engineering*, Dover, New York, pp. 98–99.

Box, G. E. P. and Jenkins, G. M. (1976): *Time Series Analysis: Forecasting and Control*, (Revised Edition), Holden-Day, Oakland, CA, 575 pp.

Box, G.E.P. and Muller, M. E. (1958): A note on generation of random normal deviates, *Annals of Mathematical Statistics*, vol 29, pp. 610–611.

Box, G. E. P. and Pierce, D. A. (1970): Distribution of residual autocorrelations in auto-regressive integrated moving average time series models, *Journal of the American Statistical Association*, vol 65, pp. 1509–1526.

Buishand, T. A. (1982): Some methods for testing the homogeneity of rainfall records, *Journal of Hydrology*, vol 58, pp. 11–27.

Delleur, F. W. and Kavvas, M. L. (1978): Stochastic models for monthly rainfall forecasting and synthetic generation, *Journal of Applied Meteorology*, vol 17, pp. 1528–1536.

Douglas, E. M., Vogel, R. M. and Kroll, C. N. (2000): Trends in floods and low flows in the United States: impact of spatial correlation, *Journal of Hydrology*, vol 240, no.:1–2, pp. 90–105.

Gilbert, R. O. (1987): *Statistical Methods for Environmental Pollution Monitoring*, John Wiley & Sons Inc., New York.

Gupta, R. K. and Chanhan, H. S. (1986): Stochastic model of irrigation requirements, *Journal of Irrigation and Drainage Engineering*, vol 112 no.:1, pp. 65–76.

Hamming, R. W. (1983): *Digital Filters*, (2nd Edition), Prentice Hall, Englewood Cliffs, NJ.

Hirsch, R. M., Alexander, R. B. and Smith, R. A. (1991): Election of methods for the detection and estimation of trends in water quality, *Water Resources Research*, vol 27 no.:5, pp. 803–813.

Hirsch, R. M., Slack, J. R. and Smith, R. A. (1981): Techniques of trend analysis for monthly water-quality data, U.S. Geological Survey Open-File Report No. 81-488.

Hong Kong Observatory. (1884–1939): *Meteorological Results Part I (1884–1939)*, Royal Observatory, Hong Kong, Tsim Sha Tsui.

Huck, P. M. and Farquhar, G. J. (1974): Water quality models using Box-Jenkins method, *Journal of Environmental Engineering*, vol 100 no.:3, pp. 733–753.

Jaruškova, D. (1994): Change-point detection in meteorological measurement, *Monthly Weather Review*, vol 124, pp. 1535–1543.

Jayawardena, A. W. (2014): *Environmental and Hydrological Systems Modelling*, Taylor and Francis Group and CRC Press, Boca Baton, FL, 516 pp.

Jayawardena, A. W. and Lai, F. (1989): Time series analysis of water quality data in Pearl River, China, *Journal Environmental Engineering*,vol 115 no.:3, pp. 590–607.

Jayawardena, A. W. and Lau, W. H. (1990a): Stochastic analysis and generation of monthly and 14-day evaporation data, *Journal of the Japan Society of Hydrology and Water Resources*, vol 3 no.:3, pp. 56–67.

Jayawardena, A. W. and Lau, W. H. (1990b): Homogeneity tests for rainfall data, *Journal of the Hong Kong Institution of Engineers*, vol 18, pp. 22–25.

Kalman, R. E. (1960): A new approach to linear filtering and prediction problems, *ASME Journal of Basic Engineering Series D*, vol 82, pp. 35–45.

Kalman, R. E. and Bucy, R. S. (1961): New results in linear filtering and prediction theory, *ASME Journal of Basic Engineering Series D*, vol 83, pp. 95–108.

Kendall, M. G. (1948): *Rank Correlation Methods*, Charles Griffin & Company Limited, London.

Kendall, M. G. (1975): *Rank Correlation Methods*, Charles Griffin & Company Limited, London.

Kottegoda, N. T. (1980): *Stochastic Water Resources Technology*, John Wiley, New York.

Kreyszig, E. (1999): *Advanced Engineering Mathematics*, (8th Edition), John Wiley & Sons, New York.

Lawrance, A. J. and Kottegoda, N. T. (1977): Stochastic modelling of river flow time series, *Journal of Royal Statistical Society Series A*, vol 140, pp. 1–47.

Ljung, G. M. and Box, G. E. P. (1978): On a measure of lack of fit in time series models, *Biometrika*, vol 65, pp. 297–303.

Lohani, B. N. and Wang, M. M. (1987): Water quality data analysis in Chung Kang River, *Journal of Environmental Engineering*, vol 113 no.:1, pp. 186–195.

Longobardi, A. and Villani, P. (2009): Trend analysis of annual and seasonal rainfall time series in the Mediterranean area, International Journal of Climatology, 30 (10), 1538 – 1546.

Luo, Q. and White, R. E. (2005): Cubic spline regression for the open-circuit potential curves of a lithium-ion battery, *Journal of the Electrochemical Society*, vol 152 no.:2, pp. A343–A350.

Mann, H. B. (1945): Non parametric test against trend, *Econometrica*, vol 13, no.:3, pp. 245–259.

Martínez, M. D., Serra, C, Burgueño, A. and Lana, X. (2010): Time trends of daily maximum and minimum temperatures in Catatonia (NE Spain) for the period 1975-2004, *International Journal of Climatology*, vol 30, pp. 267–290.

McMicheal, F. C. and Hunter, J. S. (1972): Stochastic modelling of temperature and flow in rivers, *Water Resources Research*, vol 8 no.:1, pp. 87–98.

Nadaraya, E. A. (1964): On estimating regression, *Theory of Probability and Its Applications*, vol 9 no.:1: pp. 141–142.

von Neumann, J. (1941): Distribution of the ratio of the mean square successive difference to the variance, *The Annals of Mathematical Statistics*, vol 12 no.:4, 13, pp. 367–395.

Owen, D. B. (1962): *Handbook of Statistical Tables*, Addison Wesley, Reading, MA.

Paiva, E. M. C. D. de and Clarke, R. T. (1995): Time trends in rainfall records in Amazonia, *Bulletin of the American Meteorological Society*, vol 76, pp. 2203–2209.

Pandžíc, K. and Likso, T. (2009): Homogeneity of average annual air temperature time series for Croatia, *International Journal of Climatology*, vol 30 no.:8, pp. 1215–1225.

Partal, T. and Kahya, E. (2006): Trend analysis in Turkish precipitation data, *Hydrological Processes*, vol 20, no.:9, pp. 2011–2026.

Parzen, E. (1977): Multiple time series modelling: determining the order of approximating autoregressive schemes, In: Krishnaiah, P. (Ed.), *Multivariate Analysis IV*, North Holland, Amsterdam, pp. 283–295.

Pettitt, A. (1979): A non-parametric approach to the change-point detection, *Journal of the Royal Statistical Society, Series C (Applied Statistics)*, vol 28 no.:2, pp. 126–135.

Reinsch, C. (1967): Smoothing by spline functions, *Numerische Mathematic*, vol 10, pp. 177–183.

Sahin, S. and Cigizoglu, H. K. (2010): Homogeneity analysis of Turkish meteorological data set, *Hydrological Processes*, vol. 24, pp. 981–992.

Saifuddin, A. K. M. (2010): Homogeneity and trend analysis of temperature for urban and rural areas, Thesis submitted in partial fulfilment of the requirement for the Master's Degree in Disaster Management, National Graduate Institute for Policy Studies, Tokyo, Japan.

Salas, J. D., Delleur, J. W., Yevjevich, V. and Lane, W. L. (1980): *Applied modelling of Hydrologic Time Series*, Water Resources Publications, Littleton, CO.

Schwartz, G. (1978): Estimating the dimension of a model, *Annals of Statistics*, vol 6, pp. 461–464.

Snedecor, G. W. and Cochran, W. G. (1967): *Statistical Methods*, The Iowa State University Press, Ames.

Srikanthan, R. and McMahon, T. A. (1982): Simulation of annual and monthly rainfalls – a preliminary study at five Australian stations, *Journal of Applied Meteorology*, vol 21, pp. 1472–1479.

Staudt, M., Esteban-Parra, M. J. and Castro-Díez, Y. (2007): Homogenization of long-term monthly Spanish temperature data, *International Journal of Climatology*, vol 27, pp. 1809–1823.

Su, B. D., Jiang, T. and Jin, W. B. (2006): Recent trends in observed temperature and precipitation extremes in the Yangtze River basin, China, *Theoretical and Applied Climatology*, vol 83, pp. 139–151.

Syrakova, M. and Stefanova, M. (2009): Homogenization of Bulgarian temperature series, *International Journal of Climatology*, vol 29, pp. 1835–1849.

Thomann, R. V. (1967): Time series analysis of water quality data, *Journal of Sanitary Engineering Division*, vol 93 no.:1, pp. 1–23.

Tukey, J. W. (1965): Data analysis and frontiers of geophysics, *Science*, vol 148, pp. 1283–1289.

Walker, G. (1931): On periodicity in series of related terms, *Proceedings of the Royal Society of London, Series A*, vol 131, pp. 518–532.

Watson, G. S. (1964): Smooth regression analysis, *Sankhya, Series A*, vol 26, pp. 359–372.

Wei, W. W. S. (1990): *Time Series Analysis, Univariate and Multivariate Methods*, reprinted with corrections, 1994, Addison-Wesley, Reading, MA.

Wijngaard, J. B, Klein Tank, A. M. G. and Können, G. P. (2003): Homogeneity of 20th century European daily temperature and precipitation series, *International Journal of Climatology*, vol 23, pp. 679–692.

Yule, G. U. (1927): On a method of investigating periodicities in disturbed series, with special reference to Wolfer's sunspot numbers, *Philosophical Transactions of the Royal Society of London, Series A*, vol 226, pp. 267–298.

Chapter 39

Water – state of the resource

39.1 WATER – THE RESOURCE

Water is a precious resource that is essential for all forms of life. It is abundant in nature but has significant temporal and spatial variability. With increasing population, the per capita share of water on earth is decreasing, and in some regions, it has reached levels where communities face water stress and water scarcity. Whereas lack of safe drinking water is a major problem for over a billion inhabitants of the earth, too much water also brings about misery, agony and destruction to many people, places and infrastructure. The former may be attributed to the physical lack of water, pollution or unaffordability, and the latter is attributed mainly to urbanization and livelihood issues.

39.1.1 Unique properties of water

- Has a density more than that of its solid form.
- All substances which can exist in different states (solid, liquid and gas) have the following general relationship with respect to density

$$\rho_{solid} > \rho_{liquid} > \rho_{gas}$$

In general, ρ_{soild} is of the same order of magnitude as ρ_{liquid} and ρ_{liquid} is about 1,000 times ρ_{gas}. Water is the only exception.
- A fourth state for water when subjected to extreme pressure in a small space called 'tunneling' has been recently discovered as recent as 1998.
- The 'wetness' of water is due to the hydrogen bond in the water molecule. Water has the highest surface tension of all liquids except Mercury.
- Water is a universal solvent
- In Chemistry, water is the least known substance from the point of view of predicting its properties.

39.1.2 Classification of water

- **Blue water**: Liquid water moving above and below the ground surface. As it moves through the land phase of the earth, it can be re-used until it reaches the sea.
- **Green water**: Soil water (moisture) replenished by rainfall (or irrigation) and is used up by plants and returned to the atmosphere by evapo-transpiration. Green water becomes unproductive if evaporated from open water and bare soil.
- **White water**: That part of green water which is non-productive.

- **Grey water:** Wastewater of poor quality which can be re-used for some purposes.
- **Black water:** Heavily polluted water (usually with microbes) and is harmful for human use. It can be treated to acceptable quality at a high cost.

39.2 EARTH'S WATER RESOURCES

The total amount of freshwater, easily accessible which consists of surface waters and groundwater, is about 14,000 km^3. Of this, only about 5,000 km^3 are being used by humans. Thus, there is plenty of freshwater available to meet the demands of the present population of over 7 billion and even to meet the needs of future populations up to about 9 billion. However, there is significant spatial and temporal variation in the distribution of this global resource which can and will lead to water stresses and water scarcities in some places around the globe during certain times. To ensure that all living things in the world share this precious resource in an equitable manner requires effective water resources management, good water governance and sometimes changing the way water is used.

39.2.1 Distribution of water on earth

Water is abundant in nature but not often found in places when and where needed. Table 39.1 gives the approximate distribution of earth's water resources from which it can be seen that the temporary storage of water within the hydrological cycle occurs at four main places. They are:

- Oceans 1,350,400 (10^3 km^3)
- Ice caps and glaciers 26,000 (")
- Groundwater 7,000 (")
- Lakes and inland seas 230 (").

The four main processes of water transfer between these storages are:

- Precipitation 516 (10^3 km^3/year)
- Evaporation 516 (")
- Surface runoff 29.5 (")
- Groundwater flow 1.5 (").

The estimates of annual renewable water resources and access to renewable water resources globally have been reported to be 39.6 and 29.7 km^3, respectively (WWDR3, 2009; Table 10.1). The latter works out to be about 75% of the total renewable water resources. Regionally they are 9.8, 4.0, 13.2, 0.25, 4.4 and 8.1 km^3, respectively, for Asia, Eastern Europe and Central Asia, Latin America, Middle East and North Africa, sub-Saharan Africa and OECD for total renewable resources and 9.3, 1.8, 8.7, 0.24, 4.1 and 5.6 km^3, respectively, for Asia, Eastern Europe and Central Asia, Latin America, Middle East and North Africa, sub-Saharan Africa and OECD for access to renewable water resources (WWDR3, 2009; Table 10.1).

The total water in the land areas works out to about 2.459% of the earth's entire water resources. The extractable percentage of this is of the order of about 0.5% (from lakes, rivers and groundwater). The largest volume of freshwater is found in ice caps and glaciers (26,000×10^3 km^3, or, 1.925%) which is sufficient to keep the world's rivers flowing for nearly 1,000 years (annual surface runoff from rivers is about 29.5×10^3 km^3). The rest of the

Table 39.1 Approximate distribution of earth's water resources

Item	Area (km² × 10³)	Volume (km³ × 10³)	% of Total water
Atmospheric vapour (water equivalent)	510,000 (at sea-level)	13	0.0001
World ocean	362,033	1,350,400	97.6
Water in inland areas:	148,067	124,000	-
Rivers (average channel storage)		1.7	0.0001
Freshwater lakes	825	125	0.0094
Saline lakes; inland seas	700	105	0.0076
Soil moisture; Vadose water	131,000	150	0.0108
Biological water	131,000	Negligible	-
Groundwater	131,000	7,000	0.5060
Ice-caps and glaciers	17,000	26,000	0.9250
Total in land areas (rounded)		33,900	2.4590
Total water, all realms (rounded)		1,384,000	100
Cyclic water:			
Annual evaporation			
From world ocean		445	0.0320
From land areas		71	0.0050
Total		516	0.0370
Annual precipitation			
On world ocean		412	0.0291
On land areas		104	0.0075
Total		516	0.0370
Annual outflow from land to sea			
River outflow		29.5	0.0021
Calving, melting, and deflation from ice-caps		2.5	0.0002
Groundwater outflow		1.5	0.0001
Total		33.5	0.0024

freshwater is stored in inland lakes, soil water, groundwater, atmospheric water and rivers and streams. Of this fresh water, the extractable part is only of the order of about 1%. On a long-term basis, the total water resources on earth are in a stable equilibrium state through the processes of the hydrological cycle. The temporal and spatial variability and changing lifestyles lead to water stress and water scarcity.

The human population in the world has always been increasing except for short term falls in the 14th and 17th centuries. These falls were mainly due to pandemics caused by 'black death'[1] and plague. It is also projected that the population will continue to increase until about 2050. The positive trends can be attributed to a number of factors such as improved medical facilities, low infant mortality rates, increased life expectancy and increased food production. The population which in 1750 was 791 million has exploded to 7.8 billion (according to United Nations and the United States Census Bureau) as of March 2020 (Figure 39.1) (en.wikipedia.org/wiki/World_population), with the highest rate of growth of 2.2% per year recorded in 1963. It has taken over 200,000 years for the world's population to reach 1 billion and only 200 years more to reach 7 billion.

---

[1] Black death was the most devastating pandemic in human history which is reported to have killed about 30%–60% of European population during 1348–1350. It is said to be caused by a bacterium carried by rat fleas living on 'black rats' travelling in merchant ships.

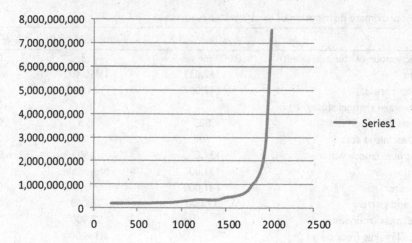

Figure 39.1 World population.

The world population surpassed the 7 billion mark at the end of October 2011. Regionally, Asia is the home to 60% of the world population followed by Africa with 16%, Europe with 10%, Latin America and the Caribbean with 9%, North America with 5% and Oceania with 0.5%. China, India, USA, Indonesia, Pakistan, Brazil, Nigeria, Bangladesh, Russia and Mexico rank as the ten most populous countries whereas Singapore (with over 7,940 persons/km^2), Bangladesh, Taiwan, Lebanon, South Korea, Rwanda, Netherlands, Haiti, India and Israel rank as the ten most densely populated countries in the world in that order (en.wikipedia.org/wiki/World_population). Future projections of world population vary significantly from source to source depending upon the scenario assumed and range from a low of about 3.2 billion to a high of about 24.8 billion by the year 2050.

Despite the past trends, it is also well known that continued increase is not sustainable as the competition for resources will begin to dominate the growth rate sooner or later. An important resource that could dominate the growth rate is water.

The total global runoff to the oceans (exorheic) is estimated to be about 37,200 km^3 annually, and the corresponding runoff to inland receiving waters (endorheic) is about 940 km^3 (WWDR3, 2009). In terms of the runoff discharging into the oceans, the Amazon River takes the first place with its discharge more than the combined discharge of the next few large rivers.

Water needs is another parameter which is constantly on the increase. In Asia and Africa, the needs are likely to exceed the supplies in the foreseeable future, and the alternative then is the re-use of water. In the not so distant future, it is conceivable that water will play the same or even a more important role than oil in geopolitical conflicts.

A country or region may be considered as "water-rich" or "water-poor" by the per capita amount of water available. Water-rich countries include Iceland (highest), Gabon, Papua New Guinea, Canada and New Zealand with respective annual per capita water availabilities of 294,340, 176,370, 154,610, 84,510, and 79,810 m^3, whereas water-poor countries include Botswana (lowest), Chad, Namibia and United Arab Emirates with respective annual per capita water availabilities of –7,460, –3,280, –1,940, and –910 m^3 (https://www.nationmaster.com/country-info/stats/Environment/Water/Availability). The world average is about 14,000 m^3. The negative availability implies that water has to be imported from trans-boundary sources. The numbers quoted above are the average values for the period 1961–1990 and would be much less at the present time due to increase in population.

Table 39.2 Potential water resources in different parts of the world

Continent	Total runoff (km³/annum)	Stable runoff (km³/annum)	Stable runoff (%)
Africa	4,225	1,903	45
Asia (except USSR)	9,544	2,900	30
Australia	1,965	495	25
Europe (except USSR)	2,362	1,020	43
North America	5,960	2,380	40
South America	10,380	3,900	38
USSR	3,484	1,410	32
All countries	38,820	14,010	36

Source: After Lvovich (1979).

39.2.2 Residence time

From the rates of transfer (Table 39.1), it can be seen that the average length of time water resides in any one storage (residence time) varies considerably. For example, the residence time in the oceans is given by

$$\text{Volume in storage/rate of transfer} = \frac{1,350,400}{445} \approx 3,000 \text{ years}$$

Similarly, the residence times in the ice caps and glaciers is about 10,000 years, in groundwater is about 4,700 years and in rivers, lakes and inland seas is about 2.34 years.

Thus, it can be seen that the most dynamic part of the hydrological cycle involves surface water on the continents which has a residence time of some 2.3 years. It should however be noted that these figures are not exact. More recent figures compiled in the WWDR3 show slightly different residence times.

Table 39.2 gives the distribution of water resources in different parts of the world. Africa has the highest percentage (45%) of stable runoff whereas Australia has the lowest (25%). The world average is about 36%.

39.2.3 Water withdrawals

Water withdrawals may be considered as an indicator of the prosperity of a nation. As can be seen in Table 39.3 the industrial use is greater than the domestic use in Asia (mainly China and Japan), North America and Europe whereas the domestic use is greater than the industrial use in Africa, Caribbean and Oceania. Of all the regions, Asia has the highest withdrawals as a percentage of the renewable resources. The two major users of water are agriculture and industry. In terms of quantities, the requirements for domestic and municipal supplies (9% of the total water resources of earth) are far less than those for growing or producing food. Table 39.3 (WWDR3, 2009) shows the water withdrawals by region and sector. Forecasts for the future are likely to increase partly due to an unabated increase in population and partly due to increase in per capita consumption in developing countries.

39.2.4 Virtual water and water trade

The concept of virtual water was introduced in 1993 by the 2008 Stockholm Water Prize Laureate Professor John Allen (Allen, 1993). Virtual water (sometimes known as embedded water)

Table 39.3 Water resources and withdrawals in 2000

Region	Renewable water resources (km³/year)	Withdrawals for agriculture (km³/year)	Withdrawals for industry (km³/year)	Withdrawals for domestic use (km³/year)	Withdrawals as a percentage of renewable resources
Africa	3,936	186	9	22	5.5
Asia	11,594	1,936	270	172	20.5
South America	13,477	178	26	47	1.9
Caribbean	93	9	1	3	14
North America	6,253	203	252	70	8.4
Oceania	1,703	18	3	5	1.5
Europe	6,603	132	223	63	6.3
World	43,659	2,663	784	382	8.8

Source: WWDR3 (2009).

refers to the water needed to produce goods and services imported or exported. For example, when a country imports goods (including food) produced in another country, the imported goods carry a hidden water cost. To produce 1 kg of wheat requires about 1.34 m³ of water. Therefore, when a country (or region) imports wheat, it is also implicitly importing a proportionate quantity of water. This applies to any type of goods and services that require water for their production and transportation. It also means that the importing country is saving the amount of water that has been used for the production and transportation of the goods and services it imports. Virtual water trade is a relatively new concept, but in recent years it has received attention from scientists, businesses and politicians.

Table 39.4 gives the global average water footprints for some typical foods. However, the actual water footprints vary from country to country, depending upon the climate and the production process. For instance, growing wheat in France needs 1/10th of the water required to grow the same crop in Morocco (Mekonnen and Hoekstra 2010). The water used during the growing season of a crop is based on crop evapotranspiration, which is usually obtained from computer models. All computer models have some assumptions, and evapotranspiration values vary widely from place to place. Water required also depends upon whether the crop is rain-fed or irrigated. Hotter and drier countries have larger footprints compared with countries with a temperate climate. For example, the water footprint of wheat grown in Slovakia is reported to be 465 l/kg whereas for the same crop grown in Somalia it is 18,070 l/kg (Chapagain and Hoekstra, 2004).

Table 39.4 The water footprint of some food items

Food/drink product	Global average water footprint (L/kg)
Apple or Pear	700
Banana	860
Beef	15,500
Beer from Barley	75/250 mL
Cheese	5,000
Chicken	3,900
Chocolate	24,000
Milk	250/250 mL
Rice	3,400

Source: https://waterfootprint.org/media/downloads/Hoekstra-2008-WaterfootprintFood.pdf.

Virtual water has major impacts on global trade. Water-rich countries can export virtual water in the form of goods and services to water poor countries, thereby increasing the water trade. Food security can be enhanced by virtual water which can be considered as another water 'source'. For the period 1995–1999, the countries with the largest net virtual water export were United States, Canada, Thailand, Argentina and India. The countries with the largest net virtual water import in the same period were Sri Lanka, Japan, the Netherlands, Republic of Korea and China (Hoekstra and Hung, 2002).

It is also a fact that developing countries produce basic food items with not much value added whereas developed countries produce manufactured goods which have a high water cost. In developing countries, the water cost of producing food items is normally not added to the cost of the primary products because the water needed comes mainly from rainfall and sometimes from irrigation which is not charged. In developed countries, the water cost is added to the cost of production. As a result, there appears to be a disparity in the virtual water trade between developed and developing countries.

There are also other shortcomings in the way virtual water is calculated. For example, it assumes that all sources of water, whether from rainfall or from irrigation, have the same value. Also, whether the water that is saved as a result of importing virtual water is used economically in other less water-intensive activities is questionable.

39.2.5 Bottled water

Bottled water is a lucrative business in the world today. Globally, it is estimated that about 55 million bottles are sold every hour. It can also be considered as a type of virtual water. Water from various sources packed into PET (polyethylene terephthalate) bottles has become a convenient source of drinking water despite their cost in comparison with the cost of tap water. Per unit volume, bottled water can be several thousand times the cost of tap water depending on the domestic water tariff structure. Still, people have got used to bottled water for convenience, perhaps quality and as a substitute for sugary drinks.

Different countries have different specifications for bottled water. The US Food and Drug Administration (FDA) considers bottled water as a packaged food product, and by law the FDA's quality requirements must be at least as stringent as the Environmental Protection Agency standards for tap water. Bottled water cannot contain sweeteners or chemical additives (other than flavours, extracts or essences) and must be calorie-free and sugar-free. If flavours, extracts and essences – derived from spice or fruit – are added to the water, these additions must comprise <1% by weight of the final product.

The origin of most bottled water is groundwater and/or spring water. Some companies use purified tap water. One positive factor in bottled water is hardly any metal (lead, copper, etc.) contamination because the water is not conveyed through pipelines. Bottled water also does not contain Chlorine which is added to tap water during the treatment process. Excessive chlorine can lead to health complications. Despite some positive factors, there are also strong arguments to convince consumers that bottled water is no better than tap water from a well-managed domestic water provider.

An important factor to discourage consumers from using bottled water is the problem of disposal of the used PET bottles. Plastic waste is not biodegradable, and they tend to accumulate causing serious environmental problems. The Great Pacific Garbage Patch could become the final resting place for used PET bottles too.

39.2.6 Major rivers in the world

A river can be classified as large or small using several criteria such as the length, the catchment area, the discharge and the population served. In terms of the discharge, the Amazon

River takes the top place with a discharge larger than the combined discharge of the next few large rivers, followed by Ganges, Congo, Orinoco, Yangtze, La Plata, Yenisei, Lena, Mississippi, Mekong, Irrawaddy and Ob, in that order. In terms of the length, the Nile River takes the top place with a length of 6,650 km. Some statistics of the 15 longest rivers in the world are shown in Table 39.5. As can be seen, 10 of the 15 longest rivers cross national boundaries. Most of them run through countries in Africa, South America and Asia. More details of some of the major rivers in the world can be found in the *Handbook of Applied Hydrology* (Singh, 2016).

Of particular interest in the Asia Pacific region is the Mekong River which runs through six countries. It has several rapids, waterfalls and uneven gradients, thus making it difficult for navigation in the entire river. However, it is an important trading route between the Yunnan Province of China and neighboring Myanmar, Lao PDR and Thailand. The French

Table 39.5 Fifteen longest rivers in the world

Ranking	Name of river	Length (km)	Catchment area (10^6 km^2)	Average Discharge (m^3/s)	Outflow	Riparian countries
1	Nile	6,650	3.35	2,800	Mediterranean sea	Ethiopia, Eritrea, Sudan, Uganda, Tanzania, Kenya, Rwanda, Burundi, Egypt, Democratic Republic of the Congo, South Sudan
2	Amazon	6,436	7.05	209,000	Atlantic Ocean	Brazil, Peru, Bolivia, Colombia, Ecuador, Venezuela, Guyana
3	Yangtze River	6,378	1.80	31,900	East China Sea	China
4	Mississippi/ Missouri	5,970	3.22	16,200	Gulf of Mexico	USA
5	Yenisei/Angara	5,539	2.60	19,600	Kara Sea	Russia, Mongolia
6	Ob-Itysh	5,410	2.98	12,800	Gulf of Ob	Russia, Kazakhstan, China, Mongolia
7	Yellow River	5,400	0.75	2,110	Bohai Sea	China
8	Rio de la Plata	4,880	2.58	18,000	Rio de la Plata	Brazil, Argentina, Paraguay, Bolivia, Uruguay
9	Congo	4,700	3.48	41,800	Atlantic Ocean	Democratic Republic of the Congo, Central African Republic, Angola, Republic of the Congo, Tanzania, Cameroon, Zambia, Burundi, Rwanda
10	Amur (Heilong Jiang)	4,440	1.86	11,400	Sea of Okhotsk	Russia, China, Mongolia
11	Lena	4,400	2.49	17,100	Laptev Sea	Russia
12	Mekong	4,350	0.81	16,000	South China Sea	China, Myanmar, Laos, Thailand, Cambodia, Vietnam
13	Mackenzie	4,241	1.79	10,300	Beaufort Sea	Canada
14	Niger	4,200	2.09	9,570	Gulf of Guinea	Nigeria, Mali, Niger, Algeria, Guinea, Cameroon, Burkina Faso, Côte'Ivoire, Benin, Chad
15	Brahmaputra	3,848	0.71	19,800	Ganges	India, China, Nepal, Bangladesh, Disputed India/ China, Bhutan

colonialists in the late 19th century attempted to make the river navigable up to China, but were not successful in taming the Khone Phapheng falls, which are the largest falls of its kind in Southeast Asia. Khone Phapheng falls are located in Champasak Province in southern Lao PDR near the border with Cambodia (13°56′53″N; 105°56′26″E). It is a succession of rapids extending to about 9.7 km along the river with the highest fall reaching 21 m. The Falls are in the area called Si Phan Don, which means '4000 islands'. This section of the river is about 14 km wide during the monsoon season, making the Falls the widest in the world. The main river branches off into many streams, thereby forming many islands some of which are perennial and inhabited by fishermen and the majority appearing only during the dry season, and hence the name '4000 islands'. The river reach containing the '4000 islands' is about 50 km long. The Mekong River is also connected to Tonle-sap (meaning great lake), a shallow lake in Cambodia via the Tonle-sap River which has a unique feature of flow reversal. It is the largest freshwater lake in Southeast Asia and was designated as a UNESCO ecological hotspot (known as Biosphere) in 1997. During the dry season (November–May), flow takes place from the Tonle-sap Lake to the Mekong River and vice versa during the wet season (monsoon season from June to October).

The Ganges River, which flows through India and Bangladesh, originates in the Indian state of Uttarakhand at the confluence of Bhagirathi and Alaknanda rivers at Devprayaghas in western Himalayas, has a length of 2,525 km and empties into the Bay of Bengal. The Ganges River system considered as the holy river in India, joins with Yamuna River, which has a flow larger than that of Ganges, and after joining with several rivers in India enters Bangladesh. After entering Bangladesh, the main branch of the Ganges is known as the Padma. The Padma is joined by the Jamuna River, the largest distributary of the Brahmaputra River which has its headwaters in Nepal. Further downstream, the Padma joins the Meghna River, the second largest distributary of the Brahmaputra. Thereafter it is called Meghna River as it enters the Meghna Estuary, which empties into the Bay of Bengal.

There also exists a large number of relatively small rivers in the Asia Pacific Region. Details of some of these rivers are given in the UNESCO-IHP publication 'Catalogue of Rivers for Southeast Asia and the Pacific – vols 1–6' (Takeuchi et al., 1995; Jayawardena et al., 1997; Pawitan et al., 2000; Ibbitt et al., 2002; Tachikawa et al., 2004; Chikamori et al., 2012). In these publications, details of 121 rivers from Australia, Cambodia, China, Indonesia, Japan, Laos, Malaysia, Mongolia, Myanmar, New Zealand, Papua New Guinea, Philippines, Peoples Republic of Korea, Republic of Korea, Thailand and Vietnam can be found.

39.2.7 Water stress, water scarcity and water risk

Three indicators used to define the availability of water, or lack of it, are water stress, water scarcity and water risk. Water stress refers to the inability to meet human and ecological demand for freshwater. It is related to water availability, water quality and accessibility of water, including affordability. Due to insufficiency of infrastructure and unaffordability, water stress can occur even when sufficient resources are physically available.

Water scarcity refers to the lack of freshwater resources. It is human-driven and depends on the volume of water consumption relative to the volume of water resources available in a given area. Therefore, an arid region with very little available water with no human consumption would not be considered as water-scarce region.

Water risk refers to the possibility of experiencing a water-related challenge such as water scarcity, water stress, flooding, infrastructure decay, drought, etc. The magnitude of the risk is a function of the probability of occurrence of a specific challenge and is severity.

The severity depends on the intensity of the challenge as well as the vulnerability and coping capacity of those affected.

All these definitions are subjective and may have different interpretations in different regions, cultures and lifestyles. For example, thresholds for clean drinking water and requirements for environmental freshwater ecosystems may have different values in different regions, cultures and lifestyles. A widely used quantitative definition of water stress (or water stress index) is that proposed by Falkenmark et al. (1989) which states that a country experiences water stress if the amount of renewable water resources is below $1,700\,\mathrm{m^3}$ per person per year, and water scarcity if it falls below $1,000\,\mathrm{m^3}$ per person per year and absolute water scarcity if it falls below $500\,\mathrm{m^3}$ per person per year. Other similar indicators include defining water scarcity in terms of each country's water demand compared to the amount of water available, and an indicator that takes into account the water infrastructure such as desalination and recycling facilities available in the country and limiting water demand by consumptive use rather than total withdrawals developed at the International Water Management Institute (IWMI) (Seckler et al., 1998).

According to the latest World Water Development Report, over 2 billion people live in countries experiencing high water stress, and about 4 billion people experience severe water scarcity during at least one month of the year (WWDR, 2019). The report also projects an increase in stress levels because of the increasing demand for water in the future. It is also highlighted that although the global average water stress is only 11%, 31 countries experience water stresses between 25% (which is defined as the minimum threshold of water stress) and 70%, and 22 countries are above 70% and are therefore under serious water stress (United Nations, 2018). About 4 billion people, representing nearly two-thirds of the world population, experience severe water scarcity during at least one month of the year (Mekonnen and Hoekstra, 2016).

39.3 WORLD WATER ASSESSMENT PROGRAMME (WWAP) AND UN-WATER

The World Water Assessment Programme (WWAP), founded in 2000, is the flagship programme of UN-Water, an initiative consisting of 28 members drawn from UN organizations. Organizations outside the UN system are partners in UN-Water. It was initially located in UNESCO headquarters in Paris but later relocated to Perugia, Italy. WWAP monitors freshwater issues to provide recommendations, develop case studies, enhance assessment capacity at a national level and inform the decision-making process.

Its primary product, the World Water Development Report (WWDR) is a periodic comprehensive review providing information on the state of the world's freshwater resources. The basis for the WWDR springs from the Rio Earth Summit of 1992 and the UN Millennium Declaration of 2000. The first four issues of WWDR have been published to coincide with the four World Water Forums held respectively in Kyoto, Japan (2003), Mexico city, Mexico (2006), Istanbul, Turkey (2009), and Marseille, France (2012). The first World Water Forum was held in Marrakesh, Morocco in 1997, followed by the second, third, fourth, fifth, sixth, seventh and eighth held respectively in the Hague, Netherlands (2000), Kyoto, Japan (2003), Mexico City, Mexico (2006), Istanbul, Turkey (2009), Marseille, France(2012), Deigu, Korea (2015), and Brasilia, Brazil (2018). The next World Water Forum is expected to be held in Dakar, Senegal in 2021. The titles of the World Water Development Reports so far have been as follows:

- Water for People, Water for Life (WWDR1, 2003)
- Water: A Shared Responsibility (WWDR2, 2006)

- Water in a Changing World (WWDR3, 2009)
- Managing Water under Uncertainty and Risk (WWDR4, 2012)
- Water and Energy (WWDR, 2014)
- Water for a Sustainable World (WWDR, 2015)
- Water and Jobs (WWDR, 2016)
- Wastewater: An untapped resource (WWDR, 2017)
- Nature based solutions for water (WWDR, 2018)
- Leaving no one behind (WWDR, 2019).

Since 2014, the theme of the World Water Development Report and that of World Water Day have been harmonized to provide a deeper focus and in-depth analysis of a specific water-related issue every year.

These documents can be accessed at http://www.unesco.org/water/wwap/.

Each WWDR report gives some key messages, and the ones from the last report (WWDR, 2019) are as follows.

39.3.1 Key messages from WWDR 2019

- Access to safe, affordable and reliable drinking water and sanitation services are basic human rights.
- The wealthy generally receive high levels of service at very low price, while the poor often pay a much higher price for services of similar or lesser quality.
- Equitable access to water for agricultural production, even if only for supplemental watering of crops, can make the difference between farming as a mere means of survival and farming as a reliable source of livelihoods.
- Refugees and internally displaced people often face barriers in accessing water supply and sanitation services.

Overcoming exclusion and inequality:

- International human rights law obliges states to work towards achieving universal access to water and sanitation for all, without discrimination, while prioritizing those most in need.
- Investing in water supply and sanitation in general, and for the vulnerable and disadvantaged in particular, makes good economic sense.
- Accountability, integrity, transparency, legitimacy, public participation, justice and efficiency are all essential features of 'good governance'.
- Responses that are tailored to specific target groups help ensure that affordable water supply and sanitation services are available to all.

39.4 WATER FOOTPRINTS

Water footprint is defined as the total volume of water used in the production of goods and services consumed by an individual or community or produced by a business. In the context of a country, it is the volume of water used in the production of all goods and services consumed by all inhabitants of the country. USA has a water footprint of 2,480 m³ per capita per year; China has a footprint of about 700 m³ per capita per year while the global average is 1,240 m³ per capita per year (WWDR3, 2009). A country's internal water footprint is the volume of water used from domestic sources whereas the external footprint is the volume of water used in other countries to produce the goods it imports (virtual water).

Water footprints are classified as green, blue and grey. Green water footprint refers to water from precipitation that is stored in the root zone of the soil and evaporated, transpired or absorbed by plants. It is particularly relevant for agricultural, horticultural and forestry products. Blue water footprint refers to water from surface or groundwater and is either evaporated, absorbed into a product or taken from one body of water and returned to another, or returned at a different time. Irrigated agriculture, industry and domestic water use can each have a blue water footprint. Grey water footprint refers to the amount of freshwater required to assimilate pollutants to meet specific water quality standards. It considers point-source pollution as well as through runoff or leaching from the soil, impervious surfaces or other diffused sources.

It is also possible for the water footprint of a country to fall outside the borders of that country. For example, about 10%, 20% and 77% of the water footprints in China, USA and Japan respectively fall outside their national boundaries. It is also reported that the largest external water footprint of US consumption lies in the Yangtze River Basin, China (https://waterfootprint.org/en/water-footprint/what-is-water-footprint/).

A personal water footprint is the amount of water needed by a person for his food, clothing, shelter and the lifestyle. It varies from country to country and within the country too. Globally, on average, a person consumes about 5,000 l/day of water for his/her food and lifestyle. Obviously, a person living in a rich country consumes more water than a person living in a poor country. This amount of water may not necessarily be from sources within the country but may have hidden contributions from other countries too.

Water footprint statistics including national water footprints, international water footprints, product water footprints, etc. can be found in WaterStat, one of the world's most comprehensive water footprint data bases (https://waterfootprint.org/en/resources/waterstat/) the computations of which are based on the Global Water Footprint Assessment Standard.

39.5 WATER USE

Globally, surface water by far is the major source (about 71%) followed by groundwater (about 18%) for all types of uses (drinking, agriculture, energy production and industry). Water re-use is very limited, and desalination is almost insignificant (about 0.34%) when all users are taken into account. For drinking water, the major source is surface water (about 48%) followed by groundwater (about 45%). For agriculture, the major source is again surface water (about 71%) followed by groundwater (about 17%), whereas for energy and industry, the respective ratios are about 87% and 12%. The statistics quoted here are taken from the World Water Development Report 3 (2009).

39.6 TRENDS IN STREAMFLOW

From an analysis of 195 stream gauging stations worldwide, it has been shown (WWDR3, 2009) that 25% of stations in Africa having increasing trend while there is no increasing trend in Asia. Majority of the gauging stations (except Africa) show no trend whereas significant decreasing trends can be seen in Africa (50%) and Asia (38%). This observation appears to be contrary to the claims made by climate change proponents.

39.7 WATER SUPPLY AND SANITATION

Over one billion people on earth do not have access to safe drinking water, and it is estimated that over 2.4 billion people will have no access to basic sanitation if the present trends continue (WWDR3, 2009). The millennium development goals (MDG) were aimed at halving the number of people who do not have access to safe drinking water and basic sanitation facilities by the year 2015. In East Asia, the coverage has increased from 68% in 1990 to 86% in 2006. Much of the improvements have been limited to urban areas.

The statistics about access to safe drinking water and safely managed sanitation vary from region to region and sometimes within regions too. In general, access to both these services is better in urban areas compared with rural areas. As of 2015, 29% of the world population (2.4 billion people) did not have access to safely managed drinking water sources. As a result of the implementation of the Millennium Development Goals, the global population using at least the basic drinking water service has increased from 81% to 89% during the period 2000–2015 whereas the percentage of world population having access to basic sanitation facilities has increased from 59% to 68% during the same period. Coverage of safely managed drinking water services ranged from a low of 24% in sub-Saharan Africa to a high of 94% in Europe and North America. By 2015, 181 countries have achieved a coverage of over 75% for basic drinking water services. On the sanitation side, 39% of the world population had access to safely managed sanitation facilities by 2015. In contrast, about 159 million people of which 59% live in sub-Saharan Africa still use untreated water directly from water sources. The statistics quoted in this paragraph are based on the information given in WWDR (2019) and should be interpreted with caution.

Investment in water brings in more benefits than associated costs. Improved water supplies have resulted in improved health and reduced infant mortalities sharply. Exposure to much of the preventable diseases such as diarrhoea, malaria, cholera, dysentery, etc. can be eliminated with improved water supply and sanitation at regional and national levels. It is estimated that 1$ invested in improving water supply and sanitation will bring about a yield equivalent to between $4 and $12, depending on the type of investment. Providing safe drinking water and proper sanitation facilities is one of the most efficient ways of improving the health of a nation. Whether the world population will be able to achieve the goal of having basic sanitation facilities by 2030 remains to be seen.

Global cost–benefit studies have demonstrated that water, sanitation and hygiene (WASH) services provide good social and economic returns when compared with their costs, with a global average benefit–cost ratio of 5.5 for improved sanitation and 2.0 for improved drinking water.

One of the key sustainable development goals (SDG's) is SDG6 on safe drinking water and sanitation. It involves health, dignity, environmental sustainability and the survival of the planet and requires efforts at regional, national and international levels. On July 28, 2010, the UN General Assembly adopted a resolution recognizing 'the right to safe and clean drinking water and sanitation as a human right that is essential for the full enjoyment and all human rights'. Regions and nations, therefore, have an obligation to work towards achieving universal access to safe drinking water and sanitation for all human beings without any discrimination while prioritizing those most in need.

The SDG6 also defines accessibility to safely managed drinking water as water available at the premises when needed that is not contaminated by faecal and chemical matter. In terms of access, there is no legal standard for physical accessibility, but a maximum of 30 min round trip travel time to a location of managed water supply is considered as a norm. For basic sanitation, the facilities should not be shared and available at the premises.

On the affordability issue, while it is not a human right to expect free access to safe and clean water, disconnection of water services due to lack of means is considered as a violation of human rights. Sometimes water rights may impinge on human rights. Water rights is a legal issue and can be taken away from an individual but human rights cannot be taken away from an individual.

39.8 WATER AND ENERGY

Water and energy are closely linked. Energy is required for providing water services, and water resources are required for the production of energy. They have a symbiotic relationship. All services in the water sector, such as pumping and distribution of water (including lift irrigation), water supply, wastewater treatment and desalination require energy. The energy sector also requires water to cool thermal power plants, generate hydropower and grow biofuels. Energy production comes from fossil fuel, nuclear fuel, hydro power, geothermal, wind, tidal and solar power. Of this hydro power is the largest renewable source for power generation in the world meeting about 16% of global electricity needs. Approximately 90% of global energy production is water intensive. In addition to hydropower, other forms of energy production except geothermal and photovoltaic require water for steam production and cooling purposes. With increasing demand for energy, there is an indirect increasing demand for water.

The quantities of water required for energy production depend upon the type of energy, and sometimes the demand for water for energy production can conflict with other demands for water. Because of the interdependencies and linkages, optimal allocation of water for different uses becomes very important. Usually, people who are deprived of water services are also deprived of energy services (electricity).

The amount of energy needed to provide potable water varies from source to source. It is reported (WWDR, 2014, p. 24) that if the source is from lakes and rivers, the energy required to produce 1 m^3 of potable water is 0.37 kWh, groundwater 0.48 kWh, wastewater treatment 0.62–0.87 kWh, from seawater 2.58–8.5 kWh.

The provision of water for agriculture, which needs no treatment, requires energy for delivering the water from the source to the demand centre which depends upon the quantity of water delivered. In the case of surface waters, it is usually by gravity or by pumping. The former needs no external energy if the source is at a higher elevation than the demand centre. If on the other hand the demand centre is at a higher elevation than the source, the energy required depends upon the elevation difference between the source and the demand centre and the distance between them. In the case of groundwater, the energy requirements depend upon the depth of the aquifer as well as the hydraulic properties of the water bearing formation.

Energy requirements for water treatment depend upon the quality of the source water, the level of treatment and the treatment process. For drinking water, there are many stages and levels of treatment, some of which are energy-intensive. For example, reverse osmosis requires energy of the order of 1.5–3.5 kWh/m^3 whereas ultraviolet treatment requires much less energy (of the order of 0.01–0.04 kWh/m^3).

Desalination is the most expensive method of providing potable water. It is highly energy intensive and therefore does not appear to be a viable option for developing countries.

It is also important to note that the wastewater produced needs treatments before they can be discharged into receiving waters. The processes involved are equally energy intensive. Looking at the reverse process of the symbiotic relationship between water and energy, water is used in many industries that produce fuels such as coal, uranium, oil and gas. Water is an input for growing biofuels such as corn and sugar cane. Water is also crucial for cooling purposes in thermal power plants and is the driving force for hydroelectric and steam turbines.

Water is also a medium of conveyance for the transport of fuel through barges and sometimes via pipelines. A relatively new type of demand for water is in the area of hydraulic fracturing (or fracking) for extracting oil and/or gas from subterranean rocky formation. The process involves injection of fracking fluid (water containing sand or other proppants suspended with the aid of thickening agents) at high pressure into a borehole to create cracks in the subterranean rocky formations to allow the natural gas and other petroleum products to flow more easily. Fracking fluid typically consists of 98% sand and water and 2% chemicals (acids, surfactants, biocides and scaling inhibitors). The quantities of water injected are of the order of 8–30 million litres per well. Some of the injected water comes back as wastewater.

39.9 WATER HAZARDS

A hazard becomes a disaster when the people or the region affected are vulnerable and lack the coping capacity. Disasters can be natural or human-induced. The former type is difficult if not impossible to prevent whereas the latter type is preventable. In terms of the cost and damage induced by various types of natural disasters, 'water-related disasters' by far exceed those by any other natural disaster. In this context, water-related disasters include all types of floods, land and mudslides, storm surges, tsunamis, tidal waves, debris flow, avalanches, droughts and all types of cyclones. In addition to such geophysical disasters, water-related biological disasters such as epidemics and endemics also take a significant toll in terms of human lives. Human-induced disasters include various types of pollution, accidents and wildfires, among others. In the modern world, pollution of the water environment is a major environmental disaster in many regions, with some places reaching irreversible conditions.

Natural disasters have taken place from time immemorial. In the past, biotic populations living under natural conditions and in harmony with nature were able to live with disasters by adapting their lifestyles or by changing their habitats. With the exponential increase in human population and increasing urbanization, natural conditions no longer exist in many places. With increased population density and high value-added infrastructure, the impacts have increased manifold.

Definition of a disaster depends upon the agency or organization that collects and disseminates data. There is a wide variation in the criteria used for inclusion in databases. One of the comprehensive databases on disaster information is EMDAT, which is located in the University Catholic Louvain, Brussels, Belgium (http://www.EMDAT.net) and which is updated regularly. They define an event as a disaster if there have been more than 10 deaths or more than 100 people displaced, or if the government of the affected country has declared a state of emergency and asked for international assistance.

Over the period 1995–2015, floods accounted for 43% of all documented natural disasters, affecting 2.3 billion people, killing over 157,000 and causing US$662 billion in damage. Droughts accounted for 5% of natural disasters, affecting 1.1 billion people, killing over 22,000 and causing US$100 billion in damage over the same 20-year period. Over the course of one decade (1995–2014) the number of floods rose from an annual average of 127 in 1995 to 171 in 2004 (CRED/UNISDR, 2015).

According to an ICHARM report (Adikari and Yoshitani, 2009) based on data compiled by EMDAT, there have been 3,050 incidents of flood disasters during the period 1900–2006 causing economic damage to the extent of some US$342 billion. During the same period, there have been 2,758 incidences of windstorm disasters causing US$536 billion worth of damage. Figure 39.2 illustrates the trends for water-related disasters on a 3-year period basis. The numbers of people who lost their lives have been in excess of 6.8 million and 1.2 million respectively for flood and windstorm disasters. These two types of disasters alone accounted for over 56% of all natural disasters in that period. Of the 1,000 worst natural disasters in terms of the number of human casualties that occurred during 1900–2006, floods accounted for 345, windstorms for 252 and droughts for 273 (Figure 39.3). A summary of

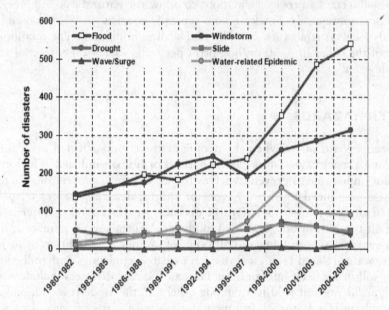

Figure 39.2 Trends in different types of disasters.

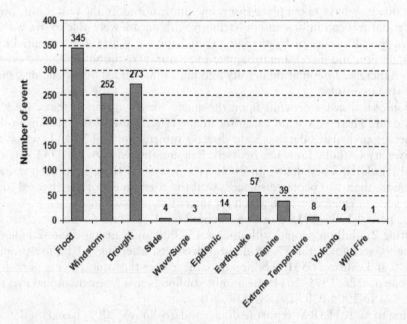

Figure 39.3 Comparison of the numbers of disasters by origin.

the historical major flood disasters in the world, including damages caused for the period 1860–2008 is given in WWDR3 (2009, Table 12.1). All these facts and figures illustrate the importance of hydro-meteorological disasters. It is also important to note that not only the numbers of disasters are increasing but also the numbers of people affected too because of migration of people into areas with better economic prospects.

Except for tsunami and tidal waves, all the hydro-meteorological disasters are caused by rain or snowfall. Drought, which is lack of sufficient rainfall, can lead to a shortage of water for agriculture, industry and domestic use that can lead to a disaster if it continuously prolongs for long periods of time. Under such conditions, the quality of water gets deteriorated resulting in undesirable microorganisms to grow that can cause water-borne diseases, which in uncontrollable situations may lead to epidemics. Thus rainfall can be considered as the triggering cause of almost all water-related disasters.

In the Asia Pacific Region, flooding caused by typhoons which bring large amounts of rainfall is a common and periodic occurrence during the summer months. Casualties and economic damages depend upon the severity of the typhoon. Countries in the region that have been seriously affected by flooding in recent years include Bangladesh, China, India, Japan, Nepal, Pakistan, Philippines and Vietnam. The most recent flooding caused by the typhoon Hagibis (meaning rapidly or strong in Tagalog, the Philippine language) during the period October 12–14, 2019 in Japan brought in 40% of the annual rainfall in just 2 days. The damages include casualties exceeding 80 people, evacuation of over 230,000 people, 146 landslides, 47 river bank breaches and damages beyond repair to 120 Shinkansen (bullet train) carriages. In the resort area of Hakone, a record rainfall of 939.5 mm have been recorded over a period of 12 h. In many parts of the main island Honshu, rainfalls between 200 and 500 mm in 2 days which amounts to 30%–40% of the annual rainfall have been recorded. Many river gauging stations have recorded discharges exceeding a 100 year return period. The estimated damage resulting from Hagibis is said to be running into 8–16 billion US Dollars.

39.10 WATER-RELATED CONFLICTS

There are some 286 international rivers and 592 transboundary aquifers shared by 153 countries (United Nations, 2018). In a non-homogeneous world, it is therefore natural to expect some conflicts about the way the water has to be shared. In addition to international conflicts, there are also internal conflicts within countries when there are disparities in the use and accessibility to water from different communities. With the per capita share of water availability decreasing with time, such conflicts could also be expected to increase in the future.

Conflicts are categorized as Trigger, Weapon and Casualty based on the use, impact or effect that water has within the conflict. Trigger refers to the situation when there is a dispute over the control of water or water systems or where economic or physical access to water or scarcity of water triggers violence. Weapon refers to the situation when water resources or water systems themselves are used as tools or weapons in a violent conflict. Casualty refers to the situation where water resources, or water systems, are intentional or incidental casualties or targets of violence.

Water conflicts can arise because of territorial disputes, competition over resources and/or political reasons. During the period 2000–2009, there had been 94 registered conflicts where water played a role (49 as a Trigger, 20 as a Weapon and 34 as a Casualty[2]).

---

[2] The different categories add up to more than the total number, because some conflicts have been listed in more than one category.

The period 2010–2018 (up to May 2018) reported 263 registered conflicts (123 with water as a Trigger, 29 as a Weapon, and 133 as a Casualty) (WWDR, 2019).

Water conflicts have existed in pre-historic times, historic times and now in the present world. The earliest conflict may have been in Mesopotamia over a dispute about depriving water to Girsu, a city in Umma. Since then, there have been conflicts of all three types in ancient Western Asia, Northern Africa, Southern Europe, Eastern Asia, North America, Latin America, until the beginning of the 20th century. In the 20th century, conflicts extended to sub-Saharan Africa and Southern Asia. More recent conflicts include the division of Indus River between India and Pakistan, Arab forces cutting water to Israel which led to the first Arab-Israel War, North Korea releasing flood waters to damage floating bridges operated by UN troops in the Pukhan Valley, several conflicts of all types among Israel, Syria and Jordan, trigger type conflict between Egypt and Sudan over Nile River water, destroying irrigation facilities in North Vietnam by US bombing, between Paraguay and Brazil over Parana River, and Cuba cutting off water supply to US bases in Guantanamo Bay. In the 21st century, a conflict arose in Bolivia as a result of water privatization.

In the recent times, three conflicts of international concern are over the Brahmaputra River which has its headwaters in Tibet, China and finally flows through India and Bangladesh, the Grand Ethiopian Renaissance Dam across Nile River which will control downstream water flow to Egypt and the Ilisu Dam across Tigris River in Turkey which will control downstream flow to Iraq and Syria. All these three conflicts are about dams been built on the upstream of rivers which will control/reduce downstream flow. Egypt is sensitive about water from the Nile as it depends almost entirely on Nile water for drinking, farming and industrial water supplies. Very recently, it is reported that Ethiopian Attorney General has filed charges related to the 6,450-MW Grand Ethiopian Renaissance Dam project (https://www.hydroreview.com/2020/01/02/ethiopian-attorney-general-files-charges-related-to-the-6450-mw-grand-ethiopian-renaissance-dam-project/?utm_medium=email&utm_campaign=hydro_weekly_newsletter&utm_source=enl&utm_content=2020-01-07).

Major river basins that have had conflicts and are likely to continue to have conflicts include River Nile Basin with riparian rights shared by Burundi, Democratic Republic of Congo, Egypt, Eritrea, Ethiopia, Kenya, Rwanda, South Sudan, Sudan, Tanzania and Uganda, Euphrates-Tigris Basin with riparian rights shared by Turkey, Syria, Iraq and Iran, Mekong River Basin with riparian rights shared by China, Myanmar, Cambodia, Laos, Thailand and Vietnam and Jordan River Basin with riparian rights shared by Israel, Lebanon, Jordan and the State of Palestine. At times, the conflicts in these regions have led to 'water wars'.

In Southeast Asia, the construction of dams in the Upper Mekong River has already raised serious concerns in the downstream riparian countries. However, the downstream countries themselves have also been planning, in recent years, the construction of a number of dams in the Lower Mekong River to meet their various water demands. Since these downstream countries are only starting to develop now, the demand for Mekong's water will continue to grow in the foreseeable future. These only add to the already existing concerns about the sustainability of the Mekong River basin and, consequently, highlight further complications in the management of its water resources.

The upper Mekong River (also known as Langcang Jiang) has a high potential for hydropower development. As of now, there are plans for the construction of 22 hydropower stations with a total installed capacity of 3,200 MW. It is the fourth largest in China in terms of hydropower potential, only after Chang Jiang (Yangtze), Yarlung Tsangpo (upper Brahmaputra) and Nu Jiang (upper Salween). A cascade of 11 dams have already been built across the mainstream of Langcang Jiang in the Yunnan Province in China. Several others, including some in Tibet, are either under construction or being planned. The combined installed capacity of these hydropower projects would run into over 31,665 MW. The construction of dams on the upper reaches of Mekong (Langcang Jiang) has both positives and negatives for the downstream riparian countries. On the positive side, such dams control flood flows during the rainy season and, thus, mitigate damages to life and properties that would otherwise occur; they also augment flows during the dry season. At the same time, however, there are also significant concerns that the construction of dams on the upper reaches will result in a significant reduction of flows to the downstream riparian countries and will adversely affect the ecological balance in such countries, including those related to sediment transport and fish migration.

To be fair and equitable, the development of the Mekong River's resources needs to be done without limiting the right to the use of the river and its resources by any riparian country. Benefits of any development activity may come in different forms, and so are the associated costs. In this context, a fair and equitable guiding principle is to have the beneficiaries bear the costs. It is also important to realize that the development of the river and its resources should take into account not only the human population that benefit from the river but also other biotic populations as well. To avoid conflicts and minimize negative outcomes, not only engineers but also many other professionals from many diverse disciplines, such as biologists, geographers, geologists, social scientists, economists, politicians, etc. have a synergistic role to play to sustain the life and services of the river and its environment. The way forward should be to follow a holistic approach in which the river is considered as belonging to all riparian countries, their people, and other biotic populations that depend on the river when development activities are planned. To this end, the Mekong River Commission, in its present form, can only be considered, at best, a means to an end, rather than an end in itself. There is, therefore, still a long way to go to achieve a holistic approach (Jayawardena and Sivakumar, 2016).

River Nile is considered as the longest river (6,695 km) in the world with its headwaters originating in Burundi. Its three major tributaries are the Blue Nile and Atbara Rivers, both originating primarily from Ethiopia and the White Nile from the upland nations. Of all her riparian countries, Egypt depends on external sources to the tune of about 97% of her water needs while the other riparian countries are able to somehow manage with resources from within their borders. In ancient times, Egypt has dominated use of the Nile waters but with agreements made in 1929 and 1959, most water rights were awarded to Egypt while lesser rights were awarded to Sudan, disregarding Ethiopia and upper Nile countries resulting in questioning the validity of the agreements. The Nile Basin is considered as a high risk area for 'water wars'.

The Jordan River which is only 251 km long, with over most of its length flowing below sea level, and its waters are central to both the Arab–Israeli conflict (including Israel–Palestine conflict), as well as the more recent Syrian Civil War. Its waters originate from the high precipitation areas in and near the Anti-Lebanon mountains in the north and flow through the Sea of Galilee and Jordan River Valley ending in the Dead Sea at an elevation of minus 400 m, in the south.

Internally, India has had several inter-state water conflicts over sharing of water. Such disputes are settled by tribunals on a case by case basis. Major disputes include the Godavari Water Disputes Tribunal, Krishna Water Disputes Tribunal–I, Narmada Water Disputes Tribunal, Ravi and Beas Water Tribunal, Cauvery Water Disputes Tribunal, Krishna Water Disputes Tribunal–II, Vansadhara Water Disputes Tribunal, and Mahadayi Water Disputes Tribunal.

The Indus Water Treaty of 1960 brokered by the World Bank and signed between India and Pakistan and recognized as one of the successful water treaties in the world provides control of the waters of the three eastern rivers, namely Sutlej, Beas and Ravi to India and the three western rivers, namely West Indus, Jhelum and Chenab to Pakistan. The technical details of the treaty have been a subject of negotiation from time to time but the treaty has so far been a success.

Brahmaputra River has caused tension between India and China for not sharing information on the status of the river upstream during the run-up to landslides in Tibet in the year 2000 which caused flooding in northeastern India and Bangladesh. Between India and Bangladesh, an ad-hoc water sharing agreement reached in 1983, whereby the two countries were allocated 39% and 36% of the water flow, respectively. The new bilateral treaty expands upon this agreement by proposing an equal allocation of the Teesta River water which flows through the northern part of Bengal and merges with Brahmaputra River after entering Bangladesh. The conflict has taken a political stance in recent years.

REFERENCES

Adikari, Y. and Yoshitani, J. (2009): Global trends in water-related disasters: An insight for policy-makers, United Nations World Water Assessment Programme side publication, p. 26.

Allen, J. A. (1993): Fortunately there are substitutes for water, otherwise our hydro-political futures would be impossible. In: *Priorities for Water Resources Allocation and Management*, Overseas Development Administration, London, pp. 13–26.

Centre for Research on Epidemiology of Disasters/United Nations Office for Disaster Risk Reduction, CRED/UNISDR (2015): The human cost of weather related disasters 1995–2015, Brussels, Geneva. www.unisdr.org/docs/climatechange/COP21_WeatherDisasterReport_2015_FINAL.pdf.

Chapagain, A. K. and Hoekstra, A. Y. (2004): *Water Footprints of Nations*, Value of Water Research Report Series; No. 16, Unesco-IHE Institute for Water Education, Delft.

Chikamori, H., Liu, H. and Daniell, T. (Editors) (2012): *Catalogue of Rivers for Southeast Asia and the Pacific - Volume VI*, UNESCO-IHP Regional Steering Committee for Southeast Asia and the Pacific, UNESCO-IHP Publication, Paris, 99 p.

Falkenmark, M. (1989): The massive water scarcity threatening Africa: Why isn't it being addressed, *Ambio*, vol. 18, no. 2, pp. 112–118.

Falkenmark, M., Lundquist, J. and Widstrand, C. (1989): Macro-scale water scarcity requires micro-scale approaches: Aspects of vulnerability in semi-arid development, *Natural Resources Forum*, vol. 13, no. 4, pp. 258–267.

Hoekstra, A. Y. and Hung, P. Q. (2002): Virtual water trade a quantification of virtual water flows between nations in relation to international crop trade, September 2002, Research Report 11, September 2002, IHE, Delft.

Ibbitt, R., Takara, K., Desa, M. N. M. and Pawitan, H. (Editors) (2002): *Catalogue of Rivers for Southeast Asia and the Pacific - Volume IV*, UNESCO-IHP Regional Steering Committee for Southeast Asia and the Pacific, UNESCO-IHP Publication, Paris, 338 p.

Jayawardena, A. W. and Sivakumar, B. (2016): Mekong river, Chapter 102, In: (Editor -in-Chief: V. P. Singh), *Handbook of Applied Hydrology*, (2nd Edition), McGaw Hill Education, pp. 102-1 to 102-9.

Jayawardena, A. W., Takeuchi, K. and Machbub, B. (Editors) (1997): *Catalogue of Rivers for Southeast Asia and the Pacific - Volume II*, UNESCO-IHP Regional Steering Committee for Southeast Asia and the Pacific, UNESCO-IHP Publication, Paris, 285 p.

Lvovich, M. I. (1979): *World Water Resources and Their Future*, English translation edited by R. L. Nace, American Geophysical Union, Washington, DC.

Mekonnen, M. M. and Hoekstra, A. Y. (2010): A global and high-resolution assessment of the green, blue and grey water footprint of wheat, *Hydrology and Earth System Sciences*, vol. 14, no. 7, pp. 1259–1276.

Mekonnen, M. M. and Hoekstra, A. Y. (2016): Four billion people facing severe water scarcity, *Science Advances*, vol. 2, no. 2, doi: 10.1126/sciadv.1500323.

Pawitan, H., Jayawardena, A. W., Takeuchi, K., and Lee, S. (Editors) (2000): *Catalogue of Rivers for Southeast Asia and the Pacific – Volume III*, UNESCO-IHP Regional Steering Committee for Southeast Asia and the Pacific, UNESCO-IHP Publication, Paris, 268 p.

Seckler, D. et al. (1998): World water demand and supply, 1990 to 2025: Scenarios and issues, International Water Management Institute (IWMI) Research Report 19, IWMI, Colombo, Sri Lanka.

Singh, V. P. (Editor) (2016): *Hanbook of Applied Hydrology*, (2nd Edition), McGraw Hill Education, New York, 1440 p.

Tachikawa, Y., James, R., Abdulla, K., and Desa, M. N. M. (Editors) (2004): *Catalogue of Rivers for Southeast Asia and the Pacific - Volume V*, UNESCO-IHP Regional Steering Committee for Southeast Asia and the Pacific, UNESCO-IHP Publication, Paris, 285 p.

Takeuchi, K., Jayawardena, A. W. and Takahasi, Y. (Editors) (1995): *Catalogue of Rivers for Southeast Asia and the Pacific - Volume I*, UNESCO-IHP Regional Steering Committee for Southeast Asia and the Pacific, UNESCO-IHP Publication, Paris, 289 p.

United Nations (2018): Sustainable development goal 6: Synthesis Report 2018 on Water and Sanitation, United Nations.

UNESCO (2009): Water in a changing world, World Water Assessment Programme, World Water Development Report 3.

UNESCO (2019): Leaving no one behind, World Water Assessment Programme, World Water Development Report.

WEBSITES FOR WATER INFORMATION

http://www.unesco.org/water/wwap/.
http://seer.cancer.gov/Publications/CSR7393/.
http://www.agr.ca/pfra/water/groundw.htm.
http://www.agric.gov.ab.ca/water/wells/index.html.
http://www.agric.gov.ab.ca/water/wells/module1.html.
http://www.angelfire.com/nh/cpkumar/hydrology.html.
http://www.chi.on.ca/swmmqa.html.
http://www.chula.ac.th/international/index_en.html.
http://www.epa.gov/safewater/ars/arsenic.html.
http://www.fluoridealert.org/f-arsenic.htm.
http://www.lboro.ac.uk/departments/cv/wedc/education/dl.htm.
http://www.nrdc.org/water/drinking/arsenic/aolinx.asp.
http://www.pacinst.org/naw.html.
https://waterfootprint.org/en/water-footprint/what-is-water-footprint/.
https://waterfootprint.org/en/resources/waterstat/.
http://www.rhodes.ac.za/institutes/iwr/.
http://www.undp.org.vn/dmu/.
http://www.worldwater.org/.

Chapter 40

Sources and demand for water

40.1 INTRODUCTION

Sources of water can be from rain, surface waters, groundwater, sea water, soil moisture, imported water, recycled water and stormwater as a non-potable source. The important requirements for a potable source of water are as follows:

- The source should be near the demand area,
- Water should be available in sufficiently large quantities,
- Water should be clean, and,
- The elevation of water level with respect to the service area should be sufficiently high to make use of gravity flow wherever possible.

40.2 RAINWATER

Rainwater is widely used in rural areas in developing countries as a source of water and is being encouraged by agencies such as the WHO and UN Centre for Appropriate Technology because of the low level of technology involved. It has a long history dating back to Roman villas. In the present days, rainwater is used as a potable source of water in countries such as Australia, Bermuda, Israel, Thailand, Kenya, Tanzania and USA (California and Hawaii). Because of its importance to the International Drinking Water Decade, an International Conference on Rainwater Cistern Systems was held in Hawaii in June 1982. Basically, there are two types of Rainwater Catchment Systems (RWCS).

40.2.1 Prepared catchments

Usually, the catchments should be upland ones to enable a clean supply. Rainwater is stored in reservoirs constructed by damming a valley. Water loss by seepage is prevented by covering with a plastic sheet or by applying a sealant at the bed. Potential yield is estimated from the runoff characteristics. Usually, the runoff is of the order of 40% of the rainfall.

The quality of rainwater from upland catchment is good, soft and free from man-made pollutants. However, sterilization is preferred if kept for a long time in view of microbial growth.

40.2.2 Rooftop catchments

Rainwater is collected by rooftops and stored in cisterns. The first flush of such collected water is discarded as it could be contaminated with dust and other particles contained in the atmosphere and on the collecting surfaces as well as bird droppings, etc. Usually, rooftop catchments are suitable in dry regions and in atoll[1] islands where there are no substantial groundwater resources. For instance, in Bermuda, which is a coral atoll, it is required by law to have facilities to collect water using rooftop catchments for use by those living in the building.

Rooftop catchment systems require little or no technology. The main problem for the designer is to design a system to provide potable water to the user at the most economical cost. In most instances, RWCS will not provide total water requirements. In such situations, the optimum combination with another source must be sought. The variables involved in the design are the rainfall, the rooftop area (roof area projected on a horizontal plane), the storage capacity and the demand. Of these, the rooftop area is pre-determined from other considerations. Rainfall is a random process and therefore the only other variable that has to be determined by the designer is the capacity of the storage tank. It is estimated based on the per capita consumption and the percentage of the per capita consumption that is expected to be provided by the RWCS.

In Bermuda, the design code specifies a minimum permissible roof area of 11 m^2 per person and a minimum storage capacity of 6.8 m^3 per person. These specifications are unrealistic in an urban environment with high population density.

Storage tanks are constructed either underground or overground. Underground tanks keep the water cool but require energy for lifting. Overground tanks require the necessary structural strength and need to be covered.

The quality of RWCS water is good, soft and acidic if collected on a clean surface and kept covered and sealed. However, it lacks the minerals the body needs from a nutritional point of view.

40.2.3 Cloud seeding

Cloud seeding is a method of weather modification. It is done by introducing a seeding agent to a cloud. It may be carried out by deploying the seeding agent from above or inside the cloud, from below the cloud and allowing it to disperse naturally by updrafts or thermals, or by ground-based generators for low-hanging cold clouds in mountainous areas. The most widely used purpose is to increase precipitation.

The common seeding agents are silver iodide, potassium iodide, dry ice (solid carbon dioxide) and hygroscopic particles such as common salt. Introduction of a seeding agent that normally has a crystalline structure similar to that of ice will induce nucleation and freezing. The results of cloud seeding are mixed. There are no statistically significant results to prove that cloud seeding will always enhance precipitation although there have been observations made in some places where precipitation has been produced as a result of cloud seeding.

Cloud seeding is practiced in many countries in Asia, Europe, USA, UAE, Russia, Canada, Australia and Africa. Attempts to pre-empt rainfall by cloud seeding were carried out in Beijing prior to the 2008 Olympics to prevent rainfall during the games. One of the better known examples of cloud seeding in Asia is the Royal Rainmaking Project in Thailand in late 1950s. It has become so popular in Thailand and in neighbouring countries to the extent that in 2005 the European Patent Office granted King Bhumibol Adulyadej the patent on Weather Modification by Royal rainmaking technology.

---

[1] Ring-shaped coral reef enclosing lagoon.

40.2.4 Extracting water from fog – Occult precipitation

Occult precipitation, also known as horizontal precipitation, is caused by the interception of the fog by the vegetation. The small droplets of water existing in the fog which normally would remain suspended in the atmosphere would precipitate after intercepting the vegetation and coalescing and becoming raindrops. Fog can be considered as any cloud that intercepts a topographic surface. The presence of an obstacle such as vegetation can enhance the interception of such small droplets that will coalesce and become raindrops that fall upon the ground. The factors that affect occult precipitation include the type, size, density and homogeneity of the vegetation as well as the exposure to winds. Occult precipitation can occur only when there is air movement, fog and vegetation simultaneously. It can occur in places where the humid air is forced to rise due to orographic features or by wind. The humid air cools adiabatically due to the lowering of pressure and condenses in tiny droplets that form clouds and fog.

Based on studies carried out in Madeira island, it has been reported that in addition to the contribution to replenish the soil water, occult precipitation also contributes to nutrient cycling and ecosystem bio-geochemistry (http://aprenderamadeira.net/en/occult-precipitation/). The fog is usually richer than rain in nitrogen which is essential for plant growth. Occult precipitation is not normally recorded by a rain gauge but is reported that it contributes 7%–28% of the total rainfall in some places (Cardenas et al., 2017).

40.3 SURFACE WATER

40.3.1 Streams and lakes by continuous draft

40.3.1.1 Streams

This method is suitable for perennial rivers and drafts in which the minimum flows are greater than the maximum demand for the season. Abstraction may be either by direct tapping from the river itself or indirect tapping from a nearby water table. In the latter case, filtration is automatically achieved. Intake structures are required for both types of tapping, but the cost of these will be small in comparison with headworks required for impounding reservoirs. The usual problems associated with this type of sources are that

- The quality is not satisfactory because all sewage and other effluents are usually discharged into the river,
- The flow is not sufficient for all the time, and,
- The flow usually carries a large amount of silt.

40.3.1.2 Lakes

Lakes are natural reservoirs, and the water in lakes is generally clearer than in rivers because of the settling of silt and other suspended matter. However, lakes are not found everywhere. Their use is limited to areas where there are freshwater lakes. Some of the world's large fresh water lakes are

- Lake Superior in the USA (surface area 84,650 km²)
- Lake Victoria in East Central Africa (surface area 69,700 km²)
- Lake Huron in the USA (surface area 61,200 km²)
- Lake Michigan in the USA (surface area 59,600 km²)
- Lake Nyasa in East Africa (surface area 37,800 km²).

Other small lakes are found in Scotland, Finland, Canada, Denmark, Germany, Switzerland, China, Cambodia, etc. Lakes rarely dry up in temperate climates. But in tropics and sub-tropics where the rate of evaporation may exceed the rate of inflow, freshwater lakes may dry up. It is also possible for such lakes to become highly saline like the Dead Sea. The science that deals with the study of lakes is called Limnology.

The quality of water in lakes is dependent on factors such as

- Inflow water, catchment and/or rivers,
- Effluent discharges,
- Location of intake,
- Degree of eutrophication (if any),
- Depth of intake, and,
- Season.

Abstraction from lakes is by pumping or by gravity flow if the elevation is high.

40.3.2 Rivers and streams dammed to form impounding reservoirs

Impounding reservoirs are used to store water during the high flow periods and release during low flow periods. They are constructed in upland areas by damming a watercourse. There are many advantages of having an impounding reservoir as a source of potable water supply. For example,

- Being usually at a high elevation, distribution to treatment works and to consumers (sometimes) can be by gravity;
- Runoff which is stored in the reservoir will improve its clarity, turbidity, etc. when allowed to remain in the reservoir for a period of time;
- Indirectly serves as a means of regulating flow downstream;
- Can be used for other activities, e.g. fish farming, recreation, etc.

40.4 GROUNDWATER

40.4.1 Natural springs

When the water table surfaces due to geological formations, a natural spring appears. Artesian springs usually have a large tributary watershed at some distance from the spring, and the water appears usually under pressure.

'Thousand Springs' in the Snake River plain in Idaho, USA, is reported to be perennial and yielding large quantities of water (660 mgd or 2.5×10^6 m³/day). Other large springs include Silver Springs in Florida, USA with a yield of about 900 mgd and Vaucluse in France with a yield of about 2,400 mgd.

40.4.2 Wells

Wells derive their water from the aquifers. Shallow wells usually derive water from unconfined aquifers (water table) whereas deep wells derive water from confined aquifers.

40.4.2.1 Shallow wells

These are easily vulnerable to pollution and are often the main source of water supply in many rural areas. Shallow wells usually dry up during the dry season.

40.4.2.2 Deep wells

Deep wells are more reliable and can provide large quantities of water. They are suitable for public supplies or for large private users. Quality of water from deep wells is generally good but hard.

40.4.3 Infiltration galleries

An infiltration gallery may be thought of as a horizontal well. It is constructed by opening a ditch from the surface. It needs to be covered from top to prevent polluted water entering. They may also be buried porous conduits. They are better than boreholes in situations where the seawater freshwater interface is near the abstraction area.

Infiltration basins are areas where floodwater or polluted water is detained in basins for infiltration into the water table to take place. Water may then be pumped by means of a well.

Infiltration galleries can be found in Iran (qanats), in Xinjiang in the grape valley, etc.

40.5 SEAWATER – DESALINATION

This is the most abundant source of water on earth (approximately 97%). Although very large in quantity, the quality of seawater is not good enough for domestic or agricultural use (except for flushing and cooling purposes). However, by an expensive process of purification, it is possible to convert seawater into freshwater.

In the hydrological cycle, natural desalination is taking place in large quantities by solar distillation. The distilled (evaporated) water falls upon the earth as rainwater, which is the primary source of all other sources of freshwater. However, it is uneconomical and unrealistic as a stable source except in regions where there is plenty of sunlight and no other alternative. Other methods of desalination require an external source of energy.

Desalination is the process of removal of dissolved solids from water. It may be used to convert seawater to freshwater or polluted freshwater to freshwater. The salinity present in seawater is normally expressed in mg/l (ppm) of dissolved solids, Cl^- ion or NaCl. Typical solids concentrations in certain types of waters are as follows:

- Brackish water – 1,000–5,000 mg/l
- Moderately saline water – inland waters with 2,000–10,000 mg/l
- Severely saline water – coastal and inland water 10,000–30,000 mg/l
- Sea water > 30,000 mg/l ($\cong$35,000)
- Potable water – max 500 mg/l (not more than 250 ppm of NaCl).

The different methods of desalting seawater can be classified as follows and are briefly described in the sections that follow:

- Distillation (Heat consuming)
  - Multiple effect
  - Multiple-stage flash - >85% of all desalting in the world

- Solar
- Vapour compression
- Vapour compression with forced circulation
- Electrodialysis (power consuming)
- Ion-exchange
- Solvent extraction
- Freeze separation (power consuming)
- Reverse osmosis (Power consuming).

The main consideration for a desalting plant is the cost. It is a variable factor, which depends upon

- Size of plant (unit price decreases with increasing size)
- Quality of seawater
- Fuel cost
- Location of plant
- Efficiency (thermal) of the plant – running cost is a substantial part of the total cost. Therefore, high efficiency is very essential, i.e. for each unit of energy input, the weight of water processed must be a maximum.
- Sharing of fuel costs, i.e. desalting is done in conjunction with another process where the rejected heat from the desalting plants can be utilized or vice versa.

40.5.1 Multiple-effect distillation

It is the oldest form of distillation. The source of heat energy is high-temperature and high-pressure steam from an external source. The basic component of the plant is a heat exchanger, which is usually of the shell and tube type. This type permits a large amount of heat transfer surface (area) within a given volume. The purpose of the heat exchanger is to cool or to heat. The hot fluid can be either in the bundle of tubes or in the shell itself. In the case of the hot fluid in the bundle of tubes, condensation takes place within the tubes while evaporation takes place outside of the tubes, i.e. inside of the shell. The inlet (high pressure and temperature) steam condenses while transferring the heat to the seawater. This heat evaporates some of the seawater at a slightly reduced pressure and temperature.

This increases the temperature of the seawater, which is fed into the second heat exchanger. It receives the vapour generated from the previous exchanger. To cause the flow of heat from the condensing steam to the water inside the tubes, it is necessary to maintain the water temperature less than that of the condensing steam. The vapour formed will be at the same temperature as the water so that the vapour in the second effect will have a slightly lower temperature and pressure than the first. This process can be repeated any number of times until the vapour temperature approaches the temperature of cooling water, i.e. when there is no further evaporation. The heat in condensation is absorbed into the cooling water as a temperature rise.

In this system, the heating effect of the external steam has been multiplied. The amount of product water formed is proportional to the amount of external steam+number of effects. An index used to specify the steam economy is the 'Gained Output Ratio (GOR)', which is the ratio of the amount of distillate to the amount of heating steam used in the first evaporator. It varies from one plant to another, but as a first approximation, it can be taken as the number of effects multiplied by 0.8.

From the point of view of steam economy, it is desirable to use as many effects as possible. But the capital cost increases with each added new unit. Often a compromise is needed.

In general, if fuel is cheap, a fewer number of effects is used whereas in places where fuel is expensive, a large number of effects is preferred (as much as 40 have been used).

40.5.2 Multi-stage flash distillation

The principle is simply that of evaporating seawater and obtaining pure water by condensation. This method differs from the multiple-effect distillation in that no evaporation of the seawater occurs during the heating process, but rather the seawater is heated to its maximum temperature. The hot water is then admitted to the first flash chamber by an orifice, which reduces its pressure slightly. The pre-heated seawater evaporates immediately (flashes) under reduced pressure and is condensed on tubes cooled by incoming seawater, thereby increasing the temperature of the seawater flowing towards the heat input section. The condensed water is collected in trays.

Original plants of this type had separate heat exchangers for each stage, but modern units combine several stages in one casing. The excess saline water passes through an orifice into the second stage at a slightly lower temperature and pressure carrying out the second stage of cooling and evaporation. This process is repeated in each compartment at successively lower pressures and temperatures as the water flows through the remaining stages of the plant. The circulation rate of the flow of saline water in a multi-stage flash distillation type plant must be very high because of the nature of the flashing process.

The amount of water circulated through the plant is about ten times the amount of water evaporated compared with multi-effect distillation where this ratio is about 2. It has been found that the best heat economy is obtained when the saline water temperature at the beginning of the flashing process is between 175°C and 200°C, the exact value depending upon other parameters. There is no direct relationship between the number of stages and the steam economy unlike in the case of multiple-effect distillation.

40.5.3 Vapour compression distillation

This process uses mechanical energy rather than heat energy for the performance of the evaporation – condensation cycle. The cycle is regenerative so that no heat rejection section is needed. Once the process is started, it does not require any more heat energy. It needs only mechanical energy to keep the compressor running. The process may be made more efficient by pre-heating the brine going in.

40.5.4 Solar distillation

Solar distillation in nature is the process by which natural rainfall is produced. In a smaller scale, this process can be duplicated by having the seawater in a closed confined space covered with a transparent roof (glass or polythene). The effect is similar to that of a greenhouse. The roof permits solar energy to enter but prevents any evaporated water from leaving the enclosure. To distil 1 gpd requires approximately 1 m^2 under average conditions. The method can also be used to treat wastewater.

40.5.5 Electrodialysis

This process reduces salinity by transferring ions from feed compartments through membranes under the influence of an electrical potential difference. Saline feed water contains dissolved salts composed of positively and negatively charged ions. They will move towards oppositely charged electrodes immersed in the solution. The positively charged ions (Na$^+$)

will move towards the cathode (negative electrical) while the negatively charged ions (Cl$^-$) will move towards the anode.

Special membranes, which are respectively permeable to cations and anions, are placed alternately, parallel to each other and separated by a small gap. The migration of ions takes place, leaving alternate compartments with high and low ionic concentrations. Compartments with low concentrations contain the product water. This process reduces the salinity by about 40%. Passage through a second stack in series will reduce it by a further 40% and so on. It is especially good for treating brackish waters with low salinity.

The electricity required will depend upon the stack geometry, the salinity of water, the resistance of the membranes and the current density. The costs are less than for distillation although the efficiencies are low. Typical energy is in the range of 5–20 amperes/cm^2 for 4,000–5,000 ppm saline water. Problems include scale formation, polarization, and, membrane fouling.

40.5.6 Ion – exchange

This method uses the selective properties of granular exchange materials which are organic resins containing H$^+$ and OH$^-$. These when placed in saline water exchange the unwanted anions and cations with H$^+$ and OH$^-$. Therefore, the water leaving the system is free from originally dissolved ions and instead will have accumulated H$^+$ and OH$^-$ equivalent to its original ionic concentration. Regeneration is necessary.

40.5.7 Solvent extraction

This method makes use of solvents, which absorb water preferentially to salts. The solvents include amines.

40.5.8 Freeze – separation

The principle of purification is similar to that of distillation. The salts are removed by freezing. When seawater is frozen, the dissolved solids remain in liquid brine. Only the pure water freezes. It is also called 'cold distillation' and is attractive from an energy standpoint because latent heat of freezing is only about one-seventh that of vaporization. However, there is no great variation of freezing point with pressure variation and therefore, there is no equivalent to the multi-effect process in distillation.

Freezing can be accomplished by using water as a refrigerant or using a liquid hydrocarbon (Butane) which will not mix with water. The latter is preferred because of design considerations and capital costs (a very high vacuum and a large vapour compressor is required for the former). The method consists of vaporizing Butane while in direct contact with saline water. The heat transfer causes cooling of the pre-cooled saline water, which forms an ice–brine slurry. The ice crystals are separated and washed to remove brine and melted to give product water. The melting of the ice converts the butane gas back into a liquid. Advantages include low energy cost and no corrosion.

40.5.9 Reverse osmosis

This method uses the property of an osmotic membrane which permits water molecules to pass more readily than dissolved ions. When such a membrane is placed between solutions of different concentrations, the water molecules will pass from the fresh to the saline until equilibrium is attained. This process is known as osmosis. The force that drives the water through the semi-permeable membrane is due to osmotic pressure. If pressure in excess of this is applied to the saline solution, then fresh water will pass from the saline to the water, which is 'reverse osmosis'. The membrane used at present is cellulose acetate. It is weak and must be supported to withstand the high pressures. The flow rates of product water with present membranes are about 500 l/day/m^2 of membrane.

40.6 IMPORTED WATER

Imported water refers to water that is transported from areas of abundance to areas of shortages, e.g. from Guangdong Province of China to Hong Kong.

40.7 SOIL MOISTURE

Soil moisture can be used as means of survival in deserts.

40.8 RECYCLED WATER

The utilization of wastewater for irrigation and fertilization of agricultural lands has a long history. With increasing demand for water, reuse of wastewater has been receiving attention in recent years. Wastewater consists of water that goes down the drains in households, toilet flushing water, industrial wastewater of different degrees of contamination from factories, and wastewater coming from agricultural and livestock farms. Recycled wastewater can be used in agriculture and industry with minimum treatment although with the present level of technology, wastewater can be made biologically and chemically acceptable for human consumption too. However, there is still a psychological stigma towards the use of recycled water as a source of potable water. Indirect potable reuse (IPR), whereby treated wastewater is blended with ground or surface sources that receives further treatment and eventually ends up as drinking water, has become increasingly common. In such cases, the wastewater after tertiary treatment is discharged to storage reservoirs (surface or groundwater) where it is kept for periods of 6 months or more to assure public fears about 'toilet to tap' concerns. When available water is diverted for human use, sensitive ecosystems where wildlife, fish and plants become vulnerable. In such cases, reclaimed water can replenish the ecosystem needs. It can also be used to recharge coastal groundwater aquifers to prevent saltwater intrusion. Table 40.1 gives an approximate distribution of reclaimed water for various uses.

According to FAO's AQUASTAT, around 3,928 km^3 of water per year is withdrawn worldwide, of which 44% (1,716 km^3 per year) is consumed and 56% (2,212 km^3 per year) is released as wastewater. The wastewater includes agricultural drainage and municipal wastewater. Water reuse is done in many rivers around the world where wastes and wastewater are discharged into waterways at several upstream locations. In developed countries, the discharge of waste and wastewater is only after they have been treated to levels acceptable for release into rivers where water is tapped at downstream locations. In some developing countries where treatment facilities for waste and wastewater are unavailable or unaffordable, the raw waste and wastewater are discharged into watercourses at different upstream locations. In such cases, the downstream water is contaminated by bacteria, nitrates, phosphates and solvents, resulting in negative health and environmental consequences. Untreated wastewater in 2015 ranged from 30% in high income countries

Table 40.1 Global water reuse after advanced tertiary treatment

Water reuse	Percentage
Agricultural irrigation	32
Landscape irrigation	20
Industry	19.3
Non-potable urban uses	8.3
Environmental enhancements	8
Recreational	6.4
Indirect potable	2.3
GW recharge	2.1
Other	1.5

Source: WWDR (2017).

to a 92% in low income countries. The target in 2030 is to reduce these to 15% and 46%, respectively (WWDR, 2017). Of particular concern is the presence of microbeads in the waste water. They are found in certain consumer products, such as facial cleansers and toothpaste. After use, these spherical particles made of polyethylene or polypropylene end up in wastewater. Once microbeads enter the wastewater system, few wastewater treatment facilities are able to remove them from the water streams. Risks to aquatic life and public health are not yet well understood, but the particles themselves may contain toxins or attract other toxins in the water. The good news is that in recent years (2016–2018) governments of the US, Canada and UK have banned the use of microbeads in cosmetics and personal care products.

In recent years, wastewater is considered as an 'untapped resource' rather than as a waste product. This paradigm shift is highlighted in WWDR (2017). Absolute necessity due to lack of alternative sources, health and environmental considerations, and cost are some of the factors driving the need for water reuse whereas public acceptability is a distracting factor. Countries like Australia, Israel, Namibia and Singapore, as well as some states in USA, including California, Virginia and New Mexico are already drinking treated wastewater.

Singapore, which has a high population density and lacks land as catchments to collect rainwater, has undertaken a scheme to reclaim water which their Public Utilities Board names as NEWater. The system involves microfiltration which removes microscopic particles, including some bacteria, reverse osmosis, a process by which undesirable contaminants are removed, followed by ultraviolet disinfection. Chemicals are added to restore the pH balance. The product is a high grade recycled water. It is used mainly in industry but is also added to reservoirs to blend with raw water. NEWater is more energy-efficient and cost-efficient than desalination. At present, there are five NEWater plants supplying up to 40% of Singapore's current water needs. By 2060, NEWater is expected to meet up to 55% of Singapore's future water demand. NEWater surpasses World Health Organization drinking water standards.

Namibia, the most arid country in southern Africa, which produces 35% of the water needs for the capital city Windhoek by reclaiming wastewater, has been drinking recycled water since 1969. With increasing population, the demand for water is likely to exceed the supplies at least in certain regions and cities and during certain times. Coupled with advances in wastewater treatment technology, reclaimed water will become a major resource in the future.

40.9 NON-POTABLE USE OF STORMWATER

40.9.1 Collection

- Catchments
- Permeable pavements (with porous asphalt, porous pavement, modular interlocking concrete blocks)
- Road surfaces
- Roof tops
- Car parks.

40.9.2 Storage

- Open ponds
- Underground storage tanks.

40.9.3 Treatment

- Screening
- Sedimentation
- Filtration (depend upon the proposed use)
- Disinfection
- Chlorination
- Ozonation (no odour or taste, but more expensive than Chlorination)
- UV radiation.

Examples of non-potable use include those in USA, UK, Australia, and in many small islands and atolls.

40.10 WATER AVAILABILITY

40.10.1 Some indicators of per capita water availability

- <1,000 m^3/year catastrophically low
- 1,100–2,000 m^3/year very low
- 2,100–5,000 m^3/year low
- 5,100–10,000 m^3/year average
- 10,000–20,000 m^3/year high
- >20,000 m^3/year very high.

Highest (in 1995) water availability was in Canada with 170,000–180,000 m^3/year whereas North Africa and Arabian Peninsula had only about 200–300 m^3/year.

China in 1990 had 2,427 m^3/year, which is slightly less than the threshold for being classified as a water-stressed country. However, by 2025, when the population is projected to exceed the 1.5 billion mark, China will have only 1,818 m^3/year. In Northern China, the situation is even worse.

Oil rich countries such as Kuwait, Qatar, Bahrain, Saudi Arabia and the UAE are among the countries with least water per capita.

Israel and Jordan are also among the water-scarce countries. The scarcity has also led to conflicts. Israel controls Palestinian use of water in the occupied 5,890 km^2 West Bank. The per capita water availability in Jordan in 1990 had been only 327 m^3/year.

40.11 WATER DEMAND

Water demand can be broadly classified into four categories – domestic demand, agricultural demand, industrial demand and recreational demand.

40.11.1 Domestic demand

This arises out of human's biological requirements for sustaining life and the amenities required for living a lifestyle of the present civilization.

40.11.1.1 Human's biological requirements

An adult's body consists of about 60% of its weight in water; approximately 26.5 l inside cells, 12 l flowing between cells and 3.5 l in the blood (total of 42 l). To maintain the metabolism, an adult needs approximately 2.5 l of water per day. This is made up as 1.3 l in drinks, 1.2 l in solid food (0.85 l of free water+0.35 l of metabolic water) since

$$Food + oxygen = CO_2 + energy + water$$

This intake can vary from place to place, depending upon the environmental conditions. For instance, in hot climates, a greater amount of intake is needed to prevent dehydration. The approximate outflow of 2.5 l of water from a human body in a temperate climate is as follows:

- 0.4 l expired from lungs
- 0.1 l in faecal matter
- 1.5 l in urine
- 0.5 l in respiration (or perspiration).

In hot climates, respiration may be as high as 11 l, i.e. 22 times. Since a reduction in the body's water of anything over 10% is fatal, the same amount of water must be taken to maintain proper metabolism.

The figure of 2.5 l is slightly misleading because it reflects only what a human directly consumes. It is important to note that the food chain which provides energy also requires large amounts of water.

In composition, most food substances contain large amounts of water. For example, Lettuce, Cucumber, Spinach, etc. contain about 95% water (in fact, cucumber contains more water than sea water – the weight of dry matter in a given weight of cucumber is less than the same in seawater); bamboo shoots, mushrooms, tomatoes, etc. contain about 90% water; potatoes contain about 80% water; bread about 33%; dried beans about 10%.

If an adult needs about 1 kg of food by dry weight per day, this may mean 1 kg of bread per day for a simple vegetarian diet. Bread is made from wheat which has a transpiration ratio of 500. (transpiration ratio is defined as the ratio of the weight of water circulating through the plant in a growing season to the weight of dry matter produced during the same period). Assuming that 2 kg of wheat is required to make 1 kg of bread, 1,000 kg (or 1 m³) of water is needed for 1 kg of bread.

Animal proteins lengthen the food chain and increase the water requirements. For example, the water requirements for a diet of 0.5 kg of beef+1 kg of vegetable matter may be worked out as follows:

It takes about 2 years to raise a beef cow, and it may yield about 350 kg of beef in 2 years, which is approximately 0.5 kg/day over the 2-year period. Therefore to have a continuous supply, it is necessary to have one beef cow per person continuously. A beef cow's food requirements are approximately 10–15 kg of Alphalpha/day and about 50 l of water per day. The transpiration ratio of Alphalpha is about 800. Therefore 10 kg will require 8,000 kg (or 8 m^3) of water. It can be seen that the water cost of 1 kg of meat is approximately 16 times that of 1 kg of vegetable matter. In summary, it can be said that 1–9 m^3 of water are needed to sustain life for a naked human being.

40.11.1.2 Human's requirements demanded by the present day living habits

In the ancient times, people lived as nomads roaming from place to place and lived near lakes and rivers. Their water requirements were only for their survival. But since the present-day civilization began, humans began to harness resources not only for survival but also for improving the quality of life. The standards of living began to change. Humans looked for more conveniences resulting in a complete change of lifestyle. Instead of going to the place of the natural occurrence of water, humans devised ways of bringing water to their places of living. The society also demanded a minimum of hygienic standards such as cleanliness of body, clothing, kitchen utensils, place of living, etc.

All these contributed to a heavy demand of domestic water. The present-day domestic water consumption rates are highly varied. They change from place to place, season to season and even day to day. They also depend upon the availability, cost, etc. For instance, in a place where water for domestic purposes is provided on a communal basis, the consumption will depend upon the distance of the consumer to the standpipe. If it is closer, the consumption will be higher. Cost is also another factor. In many countries, water is not metered. Therefore, there is no incentive to conserve or to save. On the other hand, if water is metered and if there are different tariffs for different rates of consumption, many instances of unnecessary use and wastage could be avoided.

Typical domestic consumption rates vary from 10 to 20 litres per capita per day (lpcd) in under-developed countries to about 100–250 lpcd in developed countries. Kuwait has the highest water consumption rate, over 800 lpcd, followed by USA which has a rate of about 575 lpcd. Japan has a rate of about 250, whereas Nigeria has a rate as low as 30 lpcd. Gambia is about 5 lpcd. The highest consumption frequency is for the consumption rate range of 105–115 lpcd. Of the water consumed at home, the distribution in a developed area is as follows:

Toilet	31%
Washing machine	20%
Bath	15%
Kitchen	15%
Wash basin	9%
Shower	5%
Car & garden	4%
Dishwasher	1%
Total	100%

These figures are based on a study carried out by F. Memon and D. Butler (2001).

In Kuwait, the consumption rates are very high because of two factors. The climate is very dry and therefore more water is needed for even normal living. Secondly, the water in

Kuwait comes from seawater through desalination. Since the country is oil rich, the cost of water is insignificant for their economy. In USA, particularly in California, the consumption rates are high because of higher standards of living. For instance, automatic washing machines consume large amounts of water (about 160 l/wash). Garbage grinders, lawns, swimming pools and even toilets use more water in USA than anywhere else. (Toilets in USA use 33 l/flush compared to 11 l/flush in the UK.) However, water consumption in appliances has gone down over the years as a result of more efficient machines. For example, reduction in water use in dishwashers has come down from 60 l/cycle in 1970 to about 10 l/cycle in 1999 and in washing machines from about 180 l/wash to about 50 l/cycle during the same period.

In addition to these demands at the consumer's end, allowance must be made for leakages from pipes, plumbing, taps, etc. As a general guideline for estimating demands, the following figures may be assumed:

20 lpcd	Absolute minimum
150 lpcd	Average for a city
250 lpcd	Tropical and semi-tropical countries

It can be seen that the biological need is only a fraction of the actual demand.

WHO surveys in developing countries have indicated that as consumption increases the general health of the community also improves. It is also true that the wastage of water also increases as the per capita consumption increases.

40.11.2 Agricultural demand

Agricultural demand can be considered as the second most important need for water after the biological requirements. It is necessary for survival and sustenance of life. Agricultural demand in the future is likely to exceed all other demands.

Food for humans comes directly or indirectly from plants, and water is an essential ingredient in their life. It carries nutrients to cells and wastes from cells. In the natural environment water requirement for plant life is provided by rainwater. But in places where the rainfall is low, improved crops can be made to grow with the provision of irrigation. Irrigation has been practiced from as far back as 4000 BC in countries like Egypt, Syria, Persia, India, and Sri Lanka in the eastern hemisphere and about 2000 years ago in countries such as Mexico and Peru in the western hemisphere.

Irrigation requires large amounts of water, and in most cases, it cannot be reused except through the natural processes in the hydrological cycle. Irrigation requirements vary from crop to crop. Some require flooding, e.g. rice; others require only sprinkling. Rice needs flooding to about 150–200 mm. For an area of one-hectare, flooding with 200 mm requires 2,000 m^3. Typical requirements for some crops are as follows:

Sugar beet	1,000 Tons/Tons during the growing season
Wheat	1,500 Tons/Tons during the growing season
Rice	4,000 Tons/Tons during the growing season

Irrigation requirements excepting the gardening done in individual houses do not come under 'Water Supply Engineering'. The quality of irrigation water need not be as high as for domestic water.

40.11.3 Industrial demand

Industrial demand is relatively new compared with the former two types. Perhaps it started in the 20th century when the industrial revolution began to change lifestyles. It is not a need. It is not necessary for survival but serves humans with improved living standards. In industry, water is used for the following:

- **In boilers**: For making steam for thermal power generation. High-quality water is needed.
- **For cooling**: By far the largest user in industry (~66%), e.g. power stations, air conditioning, manufacturing, etc. Quality can be low and the water may be recycled.
- **Manufacturing**: Food and drinks, chemicals, textiles, plastics. Used either as a solvent or for washing and rinsing. Quality depends upon the particular industry.
- **Fire fighting**: It is an essential service in urban living. Although it is not a significant percentage of the total industrial use, the rates of use are very high and last only for a short time.

Typical requirements for some selected industries are as follows:

Nitrogenous fertilizer	600 m³/Ton
Steel	150
Paper	250
Oil	180
Sugar	100
Artificial silk	1,000
Brick	2
Plastics	750–2,000 m³/Ton

Source: Falkenmark and Lindh (1976, p. 24).

As a result of the rapid expansion of industrial activity, industrial water consumption in industrialized countries has risen sharply during this century. In the mid 1960s

Sweden consumed	1,800 lpcd
USA consumed	3,000 lpcd
World (average)	1,200 lpcd

Because of the large amounts of water needed for certain types of industries, provision is made in most cases to have independent sources so that public supply is not affected excessively.

40.11.4 Recreational demand

This demand arises out of the need to have leisure activities such as swimming, camping, etc. In most instances, the water is used only as a medium. Usually, municipal and public water supplies are designed for domestic consumption only. Separate supplies are provided for other needs, e.g. industry, agriculture, etc. because of the different standards of water quality expected for different purposes. A substantial portion of the cost incurred in providing water is used in the purification process. Therefore it is not economical to provide water for industrial use such as cooling from a domestic supply.

REFERENCES

Cárdenas, M. F., Tobón, C. and Buytaert, W. (2017): Contribution of occult precipitation to the water balance of Páramo ecosystems in the Colombian Andes, *Hydrological Processes*, vol. 31, no. 24, pp. 4440–4449. doi: 10.1002/hyp.11374.

Falkenmark, M. and Lindh, G. (1976): *Water for a Starving World*, Westview Press, Boulder, CO, p. 24.

Memon, F. and Butler, D. (2001): Water consumption trends and domestic demand forecasts, *Watersave Network*, Second Meeting, December 4, 2001. (Web presentation).

UNESCO (2017): An untapped resource, World Water Assessment Programme, World Water Development Report.

Chapter 41

Challenges in coping with water problems

41.1 INTRODUCTION

Water is a precious commodity necessary for all forms of life. Its role and importance in societies are exemplified in the following visionary statements in history:

- 'To control China one must first control water' — Guan Zhong, politician and statesman during the Spring and Autumn period of Chinese history (c 720–645 BC)
- 'Not a drop of water that falls from the heavens shall flow into oceans without being utilized by man' – King Parakramabahu the Great (King of Sri Lanka, 1153–1186)
- 'One who solves water problems in the world deserves two Nobel Prizes, one for Peace and one for Science' – Late President of USA, John F. Kennedy (1917–1963).
- 'The deficit of freshwater is becoming increasingly severe and large-scale – whereas, unlike other resources, there is no substitute for water' – Mikhael Gorbachev, the founding president of Green Cross International and President of the former Soviet Union (March 12, 2012 at the World Water Forum).

In the ancient world, civilizations began along the banks of rivers: Yellow and Yangtze in China, Nile in Egypt, Indus in India, among others. The Chinese, during the Wu, Qi and Sui Dynasties, (starting from 486 BC) built the Grand Canal which runs from Hanzhou to Beijing – a distance of about 1,800 km. It is the oldest and the longest water conveyance system in the world – longer than the Panama and Suez canals combined. Hydraulic civilization in Sri Lanka started in the 6th century BC. Romans built aqueducts in the 7th century BC. In Persia (now Iran), qanats were built some 3000 years ago. Archeological findings point to the existence of aqueducts in India, Mexico, Madeira, among other places, in ancient times. More recently, the Nanzenji aqueduct in Kyoto was built in 1890 during Meiji era. The main purpose of all these engineering feats was to provide water for irrigation and drinking. The kings or the rulers had immense power over the subjects as they controlled the flow of water.

One of the main water problems around the world today is the lack of potable water. After oxygen water is the most vital ingredient for sustaining life. Food, which requires water for growing comes next. Over one billion people in the world do not have access to safe drinking water and about twice as many do not have access to proper sanitation. The consequences of not having access to potable water are serious particularly in developing countries which suffer from various types of water-borne diseases some of which result in premature death. In terms of quantities, the requirements for domestic and municipal supplies (approximately 8% of total freshwater resources of earth) are far less than those for growing or producing food. The two major users of water are agriculture and industry (see also Chapter 39). The main challenge for future is how to guarantee water and food security to all inhabitants of the world.

The second challenge comes as a result of human intervention of the environment over the years. Many of the water resources that were in pristine condition many years ago are now heavily polluted. They are in a state where they are totally unfit for human use, or that they can only be restored to levels suitable for human consumption at a very high cost. Associated with any development activity, there is always an environmental cost. Waterbodies such as rivers, streams, and lakes also act as waste receiving bodies in many regions. When the release of wastes into such waterbodies exceed their capacities to self-purify, they will become 'dead'. No living organism can survive in a 'dead' waterbody. How the benefits arising from development activities should be balanced against environmental costs is a major issue that needs to be considered holistically in the context of the total water environment rather than as isolated systems.

A third challenge lies in the area of water-related disasters. Water, despite being essential for all forms of life, can also at times be destructive. Floods, landslides and debris flow are all triggered by excess water. Many regions in the world are vulnerable to water-related disasters and the damages as well as the resulting casualties are on the increase. Figure 39.2 in Chapter 39 illustrates the rising damages from different types of natural disasters over the years. It is also important to note that not only the numbers of disasters are increasing but also the numbers of people affected too because of the migration of people into areas with better economic prospects.

41.2 WATER STRESS AND SCARCITY (SEE ALSO CHAPTER 39)

Water stress occurs when the available supply is unable to meet the demand during a certain period or in a certain area. Water stress can occur as a result of the physical lack of water, poor quality or being economically unaffordable. Water stress occurs in many countries, some to a lesser degree than in others. The sub-Saharan African region is perhaps the most vulnerable. Quantitatively, it is defined as the ratio of water consumption to water availability. The table below gives a measure of the water stress level in a region:

Low	<0.1
Moderate	0.1–0.2
Medium–high	0.2–0.4
High	>0.4

Water stress and water scarcity can lead to economic, social, environmental and health problems. Lack of water in sufficient quantities affects agricultural and industrial production whereas lack of water in acceptable quality affects the environment and health of the community. Poor quality water is fertile ground for growth and spreading of pathogenic organisms.

Water stress can also be caused by political conflicts in landlocked countries. When the upstream countries control the resources of transboundary rivers inequitably, the downstream countries suffer. Water has been the root cause of many international conflicts which in some cases have led to wars.

Although water is freely available in nature, it has an economic cost when it is delivered to the consumer for domestic, agricultural or industrial use. Therefore, water stress and even scarcity can occur in a place with abundant natural water if the community is unable to afford it. Such instances have led to many political and social issues when attempts to privatize the water industry are made.

41.3 CHALLENGES FOR THE FUTURE

41.3.1 Population increase (see also Chapter 39)

It is a fact that the world population is increasing and the increases are in regions or countries with slow economic development. Coupled with the aspirations of those in less developed countries to catch up with the developed countries, the demand for water will continue to increase in the foreseeable future. A major challenge for the future is to contain the population at a sustainable level where the earth's resources can be shared equitably.

41.3.2 Drinking water security

The health of a nation depends upon the level of cleanliness of the domestic water supply. It is a problem that is often ignored or sidelined by the developed countries as it is only a problem of the poor and the developing countries. Approximately one billion people, or one-seventh of the world population, do not have access to clean and safe drinking water. Coupled with the accompanying sanitation problem, which affects approximately twice that number, the situation if allowed to continue can be disastrous. Unlike a major flood or an earthquake which affect a small region with high population density, the effect of the lack of drinking water is spread over vast areas with relatively low population densities. From the media point of view, such widespread and prolonged suffering receives much less attention compared with that received for high impact type of disasters such as earthquakes and major floods.

Water scarcity can be arising from physical lack of water, poor quality or due to lack of capacity for developing and maintaining a reliable supply. Arid and desert areas suffer from physical lack of water and such areas also experience pollution problems. In such situations, traditional techniques such as rainwater harvesting and groundwater exploitation would be better appropriate technologies than conventional water supply technology. Since such areas are sparsely populated, achieving individual household self-sufficiency is more favoured and should be encouraged than traditional water distribution systems where the conveyance cost can be excessively high. It is also important to introduce low-cost water filters that can be used in individual households. It is estimated that $1 invested in improving access to safe drinking water can increase the GDP by $3–$14.

41.3.2.1 Unaccounted-for water

Unaccounted-for water (UFW) refers to the difference between 'net production' (the volume of water delivered into a network) and 'consumption' (the volume of water that can be accounted for by legitimate consumption, whether metered or not). Non-revenue water (NRW) refers to the difference between the volume of water delivered into a network and billed authorized consumption. Water loss is expressed as a percentage of net water production delivered to the distribution system. It cannot be completely eliminated and varies from city to city and from distribution system to distribution system. Physical water losses can be due to leakage from distribution systems and household connections as well as overflows from storage reservoirs. There can also be losses due to incorrect meter readings, illegal tapping and incorrect billing. Such losses are considered as commercial losses. In Asian cities, the non-revenue water ranges from a low of 7% to a high of 62% (Table 41.1).

Physical losses due to leakage in the distribution system can be avoided with proper and regular maintenance, but the most important contributor to leakage is the pressure in the distribution system. Pressure management, therefore, is the key to leakage reduction in

Table 41.1 Non revenue water in some Asia cities

City	Losses as a percentage of the net production
Manila	62
Colombo	55
Delhi	53
Jakarta	51
Kuala Lumpur	43
Dhaka	40
Hochiming	38
Karachi	30
Hong Kong	25
Chendu	18
Osaka	7

Source: Asian Development Bank (2004).

water distribution systems. Leakages usually take place at joints, and it has been suggested that joints could be replaced by pressure reduction valves. A procedure to optimally locate the pressure reduction valves for a pre-determined leakage loss is given in a paper by Covelli et al. (2016).

Illegal tapping is quite common in developing countries in cities where there are significant differences in income levels. Although issues arising out of 'water theft' can be resolved through legal means on a temporary basis, the broader issue of income disparity needs to be addressed to eliminate the problem on a permanent basis.

41.3.2.2 Privatization of water

Water is a vital resource essential for all forms of life but not always available when and where it is needed. The UN resolution of July 29, 2010 acknowledges the 'right to water' to everyone, without discrimination physically and economically to a sufficient amount that is safe to drink. According to the report 'The World's Water' updated every two years by the Pacific Institute, just under 65% of drinking water is located in just 13 countries – Brazil (14.9%), Russia (8.2%), Canada (6%), USA (5.6%), Indonesia (5.2%), China (5.1%), Colombia (3.9%), India (3.5%), Peru (3.5%), Congo (2.3%), Venezuela (2.2%), Bangladesh (2.2%) and Myanmar (1.9%).

Water is free when it falls down on the surface of the earth but cannot be free when it is collected, stored, treated and conveyed to the consumer. Collection from the source, storage, treatment and distribution of water costs money and someone has to pay for it. In many countries, domestic water supply is managed by the state sector and the costs are recovered either as a tax or as a tariff based upon the consumption. Such an approach works satisfactorily in many situations. However, there is a perceived impression that the state sector is generally inefficient and subject to corrupt practices as a result of interferences by politicians and others with vested interests. The alternative then is privatization by entrusting water services to private management.

Privatization of water can be interpreted in different ways depending upon the region, practice and culture. In terms of property rights, a person who owns a property also owns what is above his/her property and what is below. In this context, the property owner also owns the water that falls within his/her property as well as any groundwater beneath his/her property, and therefore, the owner can sell that water. In Europe, water is considered as a public property owned by everyone and the consumers can buy the right to use but not own. Privatization in this sense implies entrusting water services to private management. Another interpretation is private financing of infrastructures and services that allow private entities to finance water management in exchange for the exploitation of the resource and other benefits.

One of the arguments put forward by proponents of privatization is that it would increase the efficiency of management compared with the state sector. The arguments against privatization include increased costs, not fulfilling their obligation to the poor and those who do not consume much, lack of public input, lack of transparency and lack of competition as there are not too many players in this business. The water managers in the private sector have the sole objective of making profits. The negative arguments for privatization far outweigh the positive arguments. The big players in this business are Suez and Vivendi (recently renamed Veolia Environment) of France, RWE-AG of Germany, Bouygues SAUR, Thames Water (owned by RWE) and Bechtel-United Utilities. Veolia Environment and RWE-AG deliver water and wastewater services to almost 300 million customers in over 100 countries.

World Bank has been the principal financier of privatization through its International Finance Corporation (IFC). In many instances, a condition of World Bank loans for governments (particularly in developing countries) is to transfer the management of utilities such as water and wastewater from public sector to private sector.

Water privatization started in the 1980s when Britain sold off its entire water industry to the private sector. Historically, there have been many instances where privatization of water services has not produced the expected results. An often cited example of failure is the famous 'Cochabamba Water War' of April 2000 and uprising in La Paz/El Alto in January 2005, both in Bolivia, which were direct results of World Bank initiatives involving a Bechtel subsidiary. When the price of water tripled after privatization, thousands took to the streets until the government backed down and told the company to leave. Bechtel sued the government of Bolivia for loss of future profits (see World Bank's ICSID to Hear Case on Bolivia Water Privatization, Economic Justice News, October 2002), but later in 2006, both parties agreed to drop any financial claims against each other. Similar cases of failure include water and sewerage services to the cities of Buenos Aires by Suez, Senegal by SAUR, Lagos as well as a national program called PROMAQUA in Mexico, among others.

One of the rare success stories of water privatization is 'Manila Water'. It was a World Bank project in which IFC designed a 25-year, $2.7 billion concession in 1997, giving Manila Water part-ownership alongside other companies. The IFC claims the project is a success story because it has provided an extra 1.7 million people with clean water, reduced diarrhoea cases by 51% and offered customers significant savings – 20 times less than per-cubic-metre rates previously charged by water vendors. But others claim that Manila's water privatization has led to continual price hikes, legal challenges, investigations, failures to provide certain districts with water, and has given the companies unfair returns for their work.

An alternative to privatization is public-public partnerships (PuP). Public-public partnerships (PuPs) allow several public water utilities or non-governmental organizations to join hands. The synergy of the partnership enables pooling resources and expertise, thereby enabling the delivery of services at higher efficiencies and lower costs. The partnerships can be national or international. They provide the collaborative advantage without the profit-making focus of private operators. Public partners are more cost-effective than private water companies. They are also more responsive, and making profits is not their sole objective. Inter-municipal cooperation, inter-local agreements and bulk purchasing consortia can improve public services and reduce costs while allowing communities to retain local control.

PuP's generally have lower costs and greater focus on capacity building. The main objective is to provide water to the public in an equitable manner paying special attention to poor consumers. Public private partnerships (PPP's) will not take into consideration those who are unable to pay.

An example of the shift in focus from PPP's to PuP's under the tenth European Development Fund is in the Africa-Caribbean-Pacific (ACP) countries. PuPs are new partnerships developed between public water operators, communities, trade unions and other key groups, without profit motive and on the basis of equality. The performance of PPPs has been widely debated, but PPPs continue to be promoted with the goal of improving efficiency, revenue collection and technical performance. PuPs have received increasing attention in recent years as an alternative approach that could improve the performance of struggling utilities. At their core is a spirit of public service and solidarity.

41.3.3 Food security

Food and Agriculture Organization (FAO) and the World Water Council (FAO and WWC, 2015) have concluded that, with appropriate investment and policy interventions, food production will be sufficient to support a global population of 9–10 billion in 2050 although food and nutritional insecurity will persist in many regions. The global average food intake has increased from 2,250 kcal in 1961 to 2,800 kcal in 2000, although, in South Asia and sub-Saharan Africa, it still remains at 2,450 and 2,230 kcal, respectively (IWMI, 2006). This increase may be attributed to a number of factors. Land and water productivity has increased with average grain production from 1.4 to 2.7 Tons/ha during the last four decades (IWMI, 2006). Global trade in food products also has increased, thereby increasing the flow of virtual water. On the negative side are the facts that the population is still increasing, and that the increases are in areas where productivity is low and with inadequate human and economic capacities to upgrade their production.

It is a fact that grains have been the basic form of food for all humans. However, changing lifestyles have also changed the dietary habits of many societies that have attained some degree of affluence. Meat, milk and fish consumption have increased substantially, thereby exerting an additional water cost to food products. For example, the water cost of 1 kg of grain varies between 500 and 4,000 litres whereas the water cost of 1 kg of meat is about 10,000 litres. To produce the grains that are needed for meat and milk production, vast areas of land are irrigated, resulting in high water and environmental costs. On a global scale, agriculture uses about 70% of the world's freshwater resources, followed by industry which uses about 22%. In recent years, many countries have embarked on the production and use of biofuels in place of fossil fuels. This practice also has added a further burden on earth's freshwater resources because of the additional quantities of water needed for growing the bio-species from which the biofuels are extracted.

The challenge in this context is how to produce our food (and energy sources) at the least water cost. Large scale crop production necessarily depends on irrigation to ensure guaranteed successful harvest. Rainfed cultivation, on the other hand, is weather and climate dependent and therefore does not guarantee a successful harvest every year. The water efficiency in irrigated cultivation is always low. Flooded irrigation results in evaporation from the free water surface as well as from the bare soil. To increase the water productivity, it is necessary to reduce crop evapo-transpiration. Better and more efficient techniques such as drip irrigation, low pressure sprinklers are currently being used to increase water productivity.

Changing lifestyles of people in emerging economies will also add extra burden on available water resources as they aspire to catch-up with those in advanced economies who consume more meat products which require more water.

Food trade is another area that requires attention. It is not meaningful to attempt to grow food in water-poor regions at high costs and consequent environmental degradation. With global trade expanding, it is quite logical to grow more food in water-rich regions and make them available as food products to those in water-poor regions. However, in the present context, many countries aspire to be self-sufficient in food for strategic reasons. Such aspirations will not be needed if fair and reasonable trade agreements and treaties to share trans-boundary water resources are in place. Above all, concern for the well-being of other human beings should be the guiding principle for sharing the water resources on earth.

41.3.4 Water productivity

The productivity of water in agriculture is generally quantified using the crop water productivity (CWP) which varies with the crop type, evapo-transpiration as well as with climatic condition, soil type and several other factors. The main crops which end up as staple food in the world are rice, wheat and maize which are all cereals. Of these, rice consumes more water than other cereals, and thus the water productivity of rice is significantly lower than that of other cereals. Rice has a worldwide average growing season of 136 days from crop emergence to harvest, including mountain and lowland rice. Reported values of CWP vary from region to region and from source to source too. Based on data for the period 1998–2008, a global average value of CWP for rice is reported to be 0.98 kg/m^3 (Bastiaanssen and Steduto, 2016) whereas a high value of 2.2 kg/m^3 has been reported for rice crop in the Zhangye irrigation system in China (Dong et al., 2001 as cited by Bastiaanssen and Steduto, 2016). The lowest CWP is in sub-Saharan Africa with values ranging between 0.10 and 0.25 kg/m^3. In USA, values range from 0.9 to 1.9 kg/m^3 with higher values in the north than in the south and the highest in the north-western regions (Cai and Rosegrant, 2003). The projected values of water productivity and water use for the period 2021–2025 for rice in developing countries have been reported to be 0.53 kg/m^3 and 8,445 m^3/ha. The corresponding figures in developed countries are 0.57 kg/m^3 and 9,730 m^3/ha, respectively (Cai and Rosegrant, 2003).

Other indicators of water productivity in agriculture include the yield per unit area and the water consumption per unit area. In sub-Saharan Africa, which have the lowest values in the world, the average yield for rice is 1.4 Tons/ha, and the water consumption is close to 9,500 m^3/ha. In the developed world, the yield and water consumption for rice are 4.7 Tons/ha and 10,000 m^3/ha, respectively, whereas in the developing world the corresponding figures are 3.3 Tons/ha and 8,600 m^3/ha (Cai and Rosegrant, 2003).

In a study by Cai and Rosegrant (2003), the projected values of three indicators (basin efficiency, water withdrawals and irrigation consumptive use) for the period 2021–2025 under different scenarios (baseline, high efficiency and high efficiency with low water withdrawal) have been reported. Of these, the basin efficiencies in developing countries are 0.59 as baseline, 0.77 for high basin efficiency and 0.77 with high basin efficiency and low water withdrawal, respectively, whereas the corresponding figures in developed countries are 0.69, 0.81 and 0.81, respectively. The water withdrawals in developing countries are 3,486, 3,347 and 3,043 km^3, respectively, as baseline, high basin efficiency and high basin efficiency and low water withdrawal whereas the corresponding figures in developed countries are 1,277, 1,228 and 1,183 km^3, respectively. The irrigation consumptive use in developing countries are 1,214, 1,135 and 283 km^3, respectively, whereas in developed countries, the corresponding figures are 274, 250 and 227 km^3, respectively (Cai and Rosegrant, 2003).

The challenge, therefore, is to increase the water productivity in food production. Any saving in the water needs for food production could be used for other water needs. It is reported that a 1% increase in water productivity in food production generates a potential of water use of 24 l/capita/day (http://www.fao.org/3/y4525e/y4525e06.htm). Increasing the water productivity in the agriculture sector appears to be the best way of freeing water for other purposes. This can be achieved by minimizing the outflow from paddy fields, re-use of any outflows and drainage from fields, adopting aerobic irrigation in place of flooded irrigation, reducing evaporation from bare soil and reducing seepage and percolation. Aerobic irrigation reduces the yield but is compensated by less water requirement compared with flooded irrigation. Wet seeding is another method of increasing the water productivity. In this method, farmers soak the rice seeds for 24 h and then sow them in puddled or muddy fields. By this method which is popular in parts of Thailand, Vietnam and the Philippines, the water requirements are about 20%–25% less than for traditional methods. Intermittent irrigation, where the fields are flooded and allowed to dry and flooded again is a practice adopted in China where the per capita freshwater availability is among the lowest in Asia. Land levelling where the slopes are reduced to improve uniform field conditions, thereby requiring less water to flood is another approach.

Using pricing policies to increase water productivity requires government intervention and even so it does not seem feasible at least in the Asian context as it would be difficult to recover the true cost of providing water for agriculture. In the Indian subcontinent, it has been suggested that the charge required to affect demand significantly would be about ten times the cost of operation and maintenance of irrigation systems. It is also not feasible to charge agricultural water users on a volumetric basis because of the high cost of installing water meters as well as maintaining them. This applies to surface waters as well as groundwater which require energy for pumping.

In some regions, crop substitution has been used as a means of increasing water productivity. Farmers tend to grow crops that require less water but have higher market values such as flowers and vegetables which can also be exported. Such practices can conflict with national policies that aim at self-sufficiency in staple foods. Aiming at the long-term sustainability of water and food security should be the way forward.

41.3.5 Climate change

Global warming, which is an indicator of climate change, is currently a hot topic. The Intergovernmental Panel on Climate Change (IPCC) has concluded that there has been significant temperature rises since the 1970s which they attribute to global greenhouse gas emissions. The issue has also received endorsements and publicity from powerful circles and personalities. There is no doubt that discernible warming is taking place in some parts of the globe as evidenced by melting of ice caps and glaciers, sea level rises, temperature rises, among other changes. At the same time, there is another school of thought, though not as powerful as those advocating the global warming phenomena, who take the view that the issue is blown out of proportion, and that warming exists locally and that it is premature to conclude that the issue is a global phenomenon. A critical analysis of issues related to climate change as well as impediments to the implementation of Paris agreement can be found in separate publications by the author (Jayawardena, 2015, 2019; 2020).

Notwithstanding the arguments for and against global warming, it is a fact that the earth has gone through cycles of warming and cooling in the past. Changes have persisted over decades and sometimes over centuries. Although instrumental measurement of temperature started around 1850, various proxy methods (such as tree rings, ice cores, corals, etc.) have been used to understand paleoclimatology. Examples include the Holocene warm period (circa 1800–4000 BP), the Roman warm period (circa 200–500 AD), the medieval warm period (circa 1000–1100 AD) and the little ice age (circa 1200–1800 AD) (Rundt, 2008). It has also been shown that there is a 1500-year cycle of global warming (Avery, 2008).

Some scientists believe that the warming has peaked and that the earth is more resilient than predicted (http://news.bbc.co.uk/2/hi/science/nature/7329799.stm). This would mean that the temperature has not risen globally since the 1998 El Nino warm period. There are evidences of cyclical changes of climate in the recent histories of China, UK and Greenland where the warmth has been measured by the ability to grow plant species such as vine and the ability for animal species to survive. It is also argued that the issues of climate change and global warming, in particular, have been used as inhibitors to economic progress in less developed countries. Whether there is global warming or not, earth's resources should not be unnecessarily wasted but should be shared in an equitable manner. Adaptation, rather than prevention, should be the way forward.

41.3.6 Environmental pollution

Environmental pollution is a by-product of economic development and goes unabated in many water bodies as a result of indiscriminate dumping of domestic, agricultural and industrial wastes. Slow accumulation of pollutants over the years in many rivers (e.g., in China and India) has made them aesthetically unpleasant and biologically and chemically toxic. Restoration of such rivers to environmentally acceptable levels is costly and a fair and reasonable approach to recover costs is to follow the polluter pay principle. Many countries have enacted legislation to address this issue, but the enforcement becomes difficult as the costs are passed back to the taxpayers by the polluters as increased costs of their commercial products. Incidents of pollution caused by accidents such as the one that occurred in Songhwa River in Northeast China are also on the increase as more and more toxic industrial ingredients are conveyed too frequently and over long distances. Introduction of advanced methods of wastewater treatment such as membrane technology, recycling, reclamation of waste water, etc. help alleviate the pollution problem to some extent. In the long term, an integrated approach of water management in which all aspects of the water sector are considered and optimized within the framework of a single ecosystem appears to be the way forward.

41.3.7 Disaster reduction (see also Chapter 39)

Natural hazards are not preventable. In terms of the damages and the casualties, the main fresh water-related hazards are floods, landslides and debris flow. A hazard becomes a disaster when the region and the community are vulnerable and lack the coping capacity. Therefore any approach for mitigating the consequence of a disaster needs to focus on reducing the vulnerability and enhancing the coping capacity. Although there are many international and regional initiatives aimed at disaster reduction, their implementation is slow and lacks high priority due to political, cultural and economic issues particularly in developing countries. It should also be recognized that capital intensive engineered approaches of disaster reduction practiced in developed countries may not be applicable in developing countries. Rather practices which take into account the local culture, economic status, as well as the political environment would be more effective and implementable. It is only when the community attains a certain degree of affluence that people will begin to think about disasters and invest in disaster mitigation measures. For those living at or below the poverty line, day to day survival by itself is a disaster, and it is very difficult to get them involved proactively on implementing mitigative measures. Although investment in disaster mitigation is considered as a development issue in developed countries, there are other areas of higher priority where investments need to be channelled in developing countries. More can be achieved by promoting non-engineered approaches of coping with disasters as well as assisting in upgrading the living standards of those in less developed countries.

41.3.8 Research and development

Understanding a problem is a prerequisite to solve it. This applies to the water sector too. Technical issues in the water sector are all contained within the confines of the hydrological cycle in which meteorology plays the upstream role and oceanology the downstream role. In between, there are several processes taking place. They can be quantified using the basic concepts of fluid mechanics, hydraulics, hydrology and environmental engineering. Quantification is done using the basic laws of physics such as conservation of mass, momentum and energy. Governing equations are obtained using the Lagrangian approach which follows the motion of the same mass of fluid in space and time, or the Eulerian approach which follows the fluid passing through a given fixed position in space. The latter approach uses the control volume concept and leads to the differential form of mathematical representation whereas the former leads to the integral form. The catchment can be considered as the basic unit of spatial domain within which all such processes are considered.

Catchment is a topographically demarcated region which is influenced by the atmosphere from above, the geosphere from below and the biosphere and the hydrosphere from within. The main challenges of catchment hydrology arise as a result of the interactions of influences from these various components, influences brought about by human activities and the need to ensure that the catchment processes are sustainable. In the past, such influences may have existed but to a lesser degree with no conspicuous adverse effects. In recent times, human influences have accelerated, and the cry for a sustainable future has become louder. A better understanding of the dynamics of the catchment is the key to face such challenges.

Understanding the dynamics of the catchment can be achieved in many ways. For example, collection of better and more comprehensive data could be a starting point. Without data no theory or technique can be validated. Almost all practical tools of hydrology in the past have been empirical. Over the years, techniques of data collection have also changed – from on-site measurements to remotely sensed. Space-based remotely sensed data have become accessible to almost anyone, anywhere, but not without problems. They are grid-based, and one data point may represent several hundreds of square kilometres. They then have to be down-scaled to the catchment scale, and the methods of downscaling are not perfected yet. Another challenge is that the governing equations of the catchment processes may not be the same at different scales. This is another challenge.

Understanding can also be achieved by modelling the catchment dynamics using conceptual, data driven and/or physics-based approaches. Nowadays, there is no shortage of models. However, there is no hydrological model that has universal applicability and as a result, more and more models seem to originate at a rate faster than many hydrologists can digest. In fact, there are simple models, not-so-simple models, complex models and more complex models with each type having its own pros and cons. There has been a proliferation of models and modelling techniques in the past few decades, and as a result, it is confusing even to an experienced hydrologist. Establishing guidelines on the choice of models under different constraints is another challenge. Simple models have fewer parameters and are relatively easier to calibrate. However, such models do not represent the catchment processes adequately and, therefore, would in many instances be over-simplified. On the other hand, complex models can potentially describe the catchment processes to any degree of sophistication, but at a price. With increasing complexity of the model, the number of parameters also increases, and the principle of parsimony is often violated. The interactions of the multi-parameters give rise to the problem of 'equifinality', implying that there is no unique set of parameter values that will give a set of output results but rather a 'Pareto' set of feasible parameter space.

There are also emerging techniques of modelling such as for example artificial neural networks that emulate the brain, genetic algorithms and genetic programming that emulate genetic evolution of biological species, phase space re-construction methods that uses the theory of chaos, fuzzy logic that takes into consideration partial truths for dealing with imprecise information and their various hybrid forms. These methods can be considered as belonging to 'data mining' which attempts to uncover hidden information contained in the data. Such techniques have been mainly developed in the mathematics, statistics and control engineering domains and are gradually finding their way into Civil Engineering applications. Embarking on research in these areas would be challenging and rewarding.

41.3.9 Education

Education plays an important role in the sustainable management of the water environment. Earth's water resources belong to all living things but are controlled by humans only. For its fair and equitable use (and misuse), all human beings should have at least a basic under-standing of the water cycle, including its relationship with the environment. This is best achieved by introducing related subjects in the school curricula very early in life. Advanced and specialized knowledge should be provided at the university level for the professionals who would be managing the resource including their impacts on the environment. A more important aspect is continuing education. Learning is a life-long experience. Knowledge is never at steady state and will never be. Many engineering professional bodies promote, and in some cases require, engineers to update their knowledge by attending continuing profes-sional development (CPD) courses to retain their memberships. It is also important to have such CPD courses accredited by reputed and relevant learned societies.

41.3.10 Integrated water resources management (IWRM)

Water is fundamental for sustainable development. It affects health, sanitation, poverty alle-viation, disaster reduction and ecosystem conservation, and cuts across all eight Millennium Development Goals.

All processes in the hydrological cycle are interconnected, and IWRM is emerging as an alternative to sector-by-sector top down management style in the past. It helps to protect the environment, enhances economic and sustainable agricultural growth, promotes participa-tory governance, and improves human health.

IWRM is defined as a process that promotes the coordinated development and manage-ment of water, land and related resources to maximize the resultant economic and social welfare in an equitable manner without compromising the sustainability of vital ecosystems (Global Water Partnership (GWP), 2000, p. 22). It requires the synergy of different target groups with a broader governance framework.

IWRM is based on social equity, economic efficiency and environmental sustainabil-ity (the three 'e's). Collective decisions made should adhere to these three principles. The synergy and the collective wisdom of decision-makers from various disciplines as well as views and opinions of various stakeholders make IWRM a comprehensive package for managing and developing water resources in a way that will balance the social and eco-nomic needs and sustainability of ecosystems for future generations.

The Dublin Statement on Water and Sustainable Development was agreed at the International Conference on Water and the Environment (ICWE) during January 26–31, 1992, which was a preparatory meeting of the United Nations Conference on Environment and Development (UNCED) held later that year. The Dublin Statement, which included four principles on water, was submitted to the UNCED in Rio de Janeiro during June 3–14, 1992, also known as The Earth Summit. Hence the name DublinRio principles. GWP adapted and elaborated these principles to reflect an international understanding of the 'equitable and efficient management and sustainable use of water'. The four principles are

- Freshwater is a finite and vulnerable resource, essential to sustain life, development and the environment,
- Water development and management should be based on a participatory approach, involving users, planners and policymakers at all levels,
- Women play a central part in the provision, management and safeguarding of water, and
- Water has an economic value in all its competing uses and should be recognized as an economic good.

Important ingredients for successful implementation of IWRM include political will and commitment that can help unite all stakeholders, a clear vision of the river basin management taking into consideration the plans of individual stakeholders, information sharing among all stakeholders and decision-makers, capacity development to enable continuity and responding to changing circumstances, a clearly defined legal framework, adequate financial plans and ways of cost recovery and comprehensive monitoring.

Since the river basin is the basic spatial unit for water management, implementing IWRM at the river basin level is therefore an essential element in managing water resources more sustainably leading to long-term social, economic and environmental benefits. The term 'Integrated River Basin Management' refers to implementation of IWRM at the river basin level. A basin-level approach enables integration of downstream and upstream issues, quantity and quality, surface water and groundwater, and land use and water resources in a practical manner. Provision should be made for inter-basin water transfer if circumstances permit. IWRM is a step-by-step process and takes time. It is a holistic approach that seeks to integrate the management of the physical environment within that of the broader socio-economic and political framework.

IWRM has no beginning or end. It can be modelled as a spiral framework which is a convenient conceptualization of the iterative, evolutionary and adaptive adjusting to new needs, constraints, circumstances and societal goals. One turn of the spiral starts from recognizing and identifying, conceptualizing, coordinating and planning, implementing, monitoring and evaluating. The process is repeated in a cyclic manner. Depending on the issues involved, one turn of the spiral can extend to several years. IWRM does not aim at arriving at a steady state solution to water management. Rather it seeks an evolving solution. The spiral model allows the process to start at any point in time, allows capacity building over time, promotes cooperation and integration and provides a step-by-step process that can accommodate new and innovative ideas and solutions.

IWRM is not prescriptive since the goals and constraints vary from country to country and sometimes from region to region and even from river basin to river basin. The goals also vary from time to time. The trend however is to shift the focus from supply-side management which requires expensive infrastructure and a top-down approach to demand-side management through economic incentives and an inclusive approach. IWRM has also created an increased awareness of the importance of environmental and social considerations towards sustainable water management. It also places increasing emphasis on stakeholder collaboration and community involvement in the decision-making process, thereby reducing the risk of conflicts over water-related issues. There are of course pros and cons due to the lack of a prescriptive approach to IWRM, but the flexibility can help different countries, regions and river basins to adopt different approaches that suit their respective unique circumstances.

The recognition of ecosystem requirements, pollution, risk of declining water availability and increasing demand driven by population growth and aspirations for more water use, and managing risks associated with water-related disasters are essential considerations for effective IWRM. Achieving water security requires trade-offs to maintain an equitable balance between various sectors' conflicting needs. It is also important to cope with evolving environmental, economical and social circumstances.

41.3.11 Water governance

An important consideration in water governance is that it should be inclusive. Water governance should not be solely by governments. It should be 'beyond governments' and include the civil society, private sector, professionals, social workers, farmers and anyone who has a stake in the water sector. It is difficult if not impossible to solve local water problems from a global perspective. The basic spatial unit for water governance is the river catchment, and the problems need to be addressed within that spatial framework. Water was not even an issue discussed at the UN Conference on Environment and Development held in Rio de Janeiro in 1992. It was only after 1998 that water became an issue at a global level when the UN Commission on Sustainable Development adopted the 'Strategic Approach to Freshwater Management'. Recently, water has become an issue of extreme importance in the global community as evidenced by the UN General Assembly adopting a resolution recognizing 'the right to safe and clean drinking water and sanitation as a human right that is essential for the full enjoyment and all human rights' on July 28, 2010. Regions and nations, therefore, have an obligation to work towards achieving universal access to safe drinking water and sanitation for all human beings without any discrimination while prioritizing those most in need.

The main obstacle for effective water governance is the human component. In the modern world, social developments take place at a much slower pace compared to technological developments. Human component involves agreements or disagreements between stakeholders with different interests, politicians attempting to have a controlling interest in water affairs as a means of pursuing their political goals, and business enterprises whose sole objective is to make profits. If each party pursues its own goals only, it becomes difficult if not impossible to work towards a fair and equitable governance structure. Flexibility and compromise are essential requirements. One way of water governance is to implement the Integrated Water Resources Management (IWRM) approach.

Governance becomes extremely important at times of crisis. It is best to adapt to prevent a crisis than to respond to it once it has occurred, following the well-known saying that 'prevention is always better than cure'. This is particularly relevant when dealing with issues such as sharing the waters of transboundary rivers and/or aquifers. Governance, in such cases, requires institutions that cross traditional political boundaries.

41.3.12 Way forward

Sustainable management of earth's water resources needs the attention, commitment and dedication of all stakeholders. The quantities are dwindling, and the qualities are deteriorating. Many issues including societal and cultural matters are interconnected. The problem of water management therefore needs to be addressed in a holistic way. Responding to different challenges sometimes need addressing conflicting interests. If any of the above challenges are taken in isolation, other challenges may have to be overlooked or ignored. An approach in which a balance is sought between development and conservation, and where modern technology and traditional practices go hand in hand would be ideal, but the implementation of such an approach requires the will and commitment of all stakeholders. A guiding principle would be to share the resources of the planet earth in an equitable manner. Failure to do so will result in a situation whereby the water-rich countries can starve the water-poor countries when conflicts reach critical stages. In the not so distant future, water will take the place of fossil fuel as a political and economic tool that can be used to manipulate communities and governments.

REFERENCES

Asian Development Bank (2004): Water in Asian cities: Utilities' performance and civil society views. https://www.adb.org/sites/default/files/publication/28452/water-asian-cities.pdf.

Avery, D. T. (2008): Global warming every 1500 years: Implications for an Engineering Vision, *Leadership in Management and Engineering*, vol. 8, no. 3; *Transactions, AGU*, vol. 89, no. 19, pp. 153–159

Bastiaanssen, W. G. M. and Steduto, P. (2016): The water productivity score (WPS) at global and regional level: Methodology and first results from remote sensing measurements of wheat, rice and maize, *Science of the Total Environment*. doi: 10.1016/j.scitotenv.2016.09.032.

Cai, X. and Rosegrant, M. W. (2003): World water productivity: Current situation and future options. In: J. W. Kijne, R. Barker and D. Molden (eds.), *Water Productivity in Agriculture: Limits and Opportunities for Improvement*, Wallingford, UK: CABI; Colombo, Sri Lanka: International Water Management Institute (IWMI) pp.163-178. (Comprehensive Assessment of Water Management in Agriculture Series 1).

Covelli, C., Cimorelli, L., Cozzolino, L., Morte, R. D. and Pianese, D. (2016): Reduction in water losses in water distribution systems using pressure reduction valves, *Water Supply*, vol. 16, no. 4, pp. 1033–1045. doi: 10.2166/ws.2016.020.

Dong, B., Loeve, R., Li, Y. H., Chen, C. D., Deng, L. and Molden, D. (2001): Water productivity in the Zhangye Irrigation System: Issue of scale. In: R. Barker, et al. (eds.), *Water Saving for Rice*, pp. 97–115. *Proceedings of an International Workshop Held in Wuhan*, China, March 23–25, 2001.

FAO and WWC (2015): Towards a water and food secure future – Critical perspectives for policy-makers, FAO, Rome; WWC, Marseille, 61pp

Global Water Partnership – Technical Advisory Committee (2000): Integrated water resources management. https://www.gwp.org/globalassets/global/toolbox/publications/background-papers/04-integrated-water-resources-management-2000-english.pdf.

International Water Management Institute (IWMI) (2006): *Insights from the Comprehensive Assessment of Water Management in Agriculture*, IWMI: Colombo, Sri Lanka.

Jayawardena, A. W. (2015): Climate change: Is it the cause or the effect? *KSCE Journal of Civil Engineering*, vol. 19, no. 2, pp. 359–365.

Jayawardena, A. W. (2019): Climate change: Myths, realities and prospects of a global deal to curb climate change, *Sixth World Conference on Climate Change*, Berlin, Germany, September 02–03, 2019.

Jayawardena, A. W. (2020) : Climate change and impediments to the implementation of COP21 Paris agreement. *Journal of Earth Science and Climate Change*, vol. 11, no. 5, 10 pp.

Rundt, K. (2008): Global warming–Man-made or natural? http://www.factsandarts.com/articles/global-warming-man-made-or-natural.

Subject Index

Author Index

Printed in the United States
By Bookmasters